Instructor's Annotated Edition

Basic College Mathematics:

An Applied Approach

5th Edition

Instructor's Annotated Edition

Basic College Mathematics: An Applied Approach

Richard N. Aufmann
Palomar College

Vernon C. Barker
Palomar College

HOUGHTON MIFFLIN COMPANY Boston Toronto
Geneva, Illinois Palo Alto Princeton, New Jersey

Senior Sponsoring Editor: *Maureen O'Connor*
Senior Associate Editor: *Robert Hupp*
Editorial Production Manager: *Nancy Doherty Schmitt*
Production/Design Coordinator: *Carol Merrigan*
Senior Manufacturing Coordinator: *Marie Barnes*
Marketing Manager: *Charles Baker*

Credits

Cover design by Len Massiglia, LMA Communications.
Cover image by Herman Cherry, PUSH COMES TO SHOVE 1988–89. Oil on canvas, 66 × 60 inches. Courtesy Luise Ross Gallery, NY. Photo credit: Regina Cherry.

Chapter opening art by Patrice Rossi. Interior art by Network Graphics and Nugraphic Design Studio.

Page 20, table sources, *USA Today,* August 24, 1993 and *The Merger Yearbook,* U.S./International Ed. (New York: Securities Data Publishing Co., 1994). Page 39, table sources, *USA Today,* September 15, 1993 and Craig Carter, *Complete Baseball Record Book—1994* (The Sporting News). Page 55, art, copyright 1993, USA TODAY. Reprinted with permission. Page 124, table sources, *USA Today,* August 23, 1993 and Lee Wolverton, "Rooney, Steelers Expect Prosperity Despite Cost Spike," *Pittsburgh Business Times,* June 7, 1993. Page 126, art, middle of the page, copyright 1993, USA TODAY. Reprinted with permission. Page 130, table at the bottom of the page source, Bureau of Labor Statistics. Page 196, art, copyright 1993, USA TODAY. Reprinted with permission. Page 208, art, copyright 1993, USA TODAY. Reprinted with permission. Page 250, table source, *Money* Magazine, March 1994. Page 277, art at the top of the page source, Paine-Webber Semi-Annual Report, August 31, 1993. Page 281, art at the bottom of the page source, Bureau of Labor Statistics. Page 282, art at the bottom of the page from *Mathematics Teacher,* Vol. 86, No. 8, p. 629. Reprinted with permission from the *Mathematics Teacher,* copyright 1993 by the National Council of Teachers of Mathematics. Page 444, art copyright 1993, USA TODAY. Reprinted with permission.

Printed in the U.S.A.

ISBN Numbers:
Student Text: 0-395-70830-3
Instructor's Annotated Edition: 0-395-71214-9

123456789-B-98 97 96 95 94

Contents

6 Applications for Business and Consumers: A Calculator Approach 217

7 Statistics 269

8 U.S. Customary Units of Measurement 299

9 The Metric System of Measurement 327

10 Rational Numbers 357

11 Introduction to Algebra 399

Preface

The fifth edition of *Basic College Mathematics: An Applied Approach* provides mathematically sound and comprehensive coverage of the topics considered essential in a basic college mathematics course. The text has been designed not only to meet the needs of the traditional college student but also to serve the needs of returning students whose mathematical proficiency may have declined during years away from formal education.

In this new edition of *Basic College Mathematics: An Applied Approach,* careful attention has been given to implementing the standards suggested by NCTM. Each chapter begins with a mathematical vignette in which there may be a historical note, application, or curiosity related to mathematics. At the end of each section there are "Applying the Concepts" exercises that include writing, synthesis, critical thinking, and challenge problems. The chapter ends with "Projects in Mathematics" that can be used for cooperative learning activities.

INSTRUCTIONAL FEATURES

Interactive Approach

Basic College Mathematics: An Applied Approach uses an interactive style that provides a student with an opportunity to try a skill as it is presented. Each section is divided into objectives, and every objective contains one or more sets of matched-pair examples. The first example in each set is worked out; the second example, called "You Try It," is for the student to work. By solving this problem, the student practices concepts as they are presented in the text. There are *complete* worked-out solutions to these examples in an appendix at the end of the book. By comparing their solution to the solution in the appendix, students are able to obtain immediate feedback on and reinforcement of the concept.

Emphasis on Problem-Solving Strategies

Basic College Mathematics: An Applied Approach features a carefully developed approach to problem-solving that emphasizes developing strategies to solve problems. Students are encouraged to develop their own strategies, to draw diagrams, and to write strategies as part of their solution to a problem. In each case, model strategies are presented as guides for students to follow as they attempt the "You Try It" problem. Having students provide strategies is a natural way to incorporate writing into the math curriculum.

Emphasis on Applications

The traditional approach to teaching algebra covers only the straightforward manipulation of numbers and variables and thereby fails to teach students the practical value of algebra. By contrast, *Basic College Mathematics: An Applied Approach* contains an extensive collection of contemporary application problems. Wherever appropriate, the last objective of a section presents applications that require the student to use the skills covered in that section to solve practical problems. This carefully integrated applied approach generates student awareness of the value of algebra as a real-life tool.

Completely Integrated Learning System Organized by Objectives

Each chapter begins with a list of the learning objectives included within that chapter. Each of the objectives is then restated in the chapter to remind the student of the current topic of discussion. The same objectives that organize the text are also used as the structure for exercises, testing programs, and the Computer Tutor. For each objective in the text, there is a corresponding computer tutorial and a corresponding set of test questions.

AN INTERACTIVE APPROACH

Instructors have long realized the need for a text that requires students to use a skill as it is being taught. *Basic Mathematics: An Applied Approach* uses an interactive technique that meets this need. Every objective, including the one shown below, contains at least one pair of examples. One of the examples is worked. The second example in the pair (You Try It) is not worked so that students may "interact" with the text by solving it. To provide immediate feedback, a complete solution to this example is provided in the answer section at the end of the book. The benefit of this interactive style is that students can check that a new skill has been learned before attempting a homework assignment.

An explanatory passage begins each skill objective.

Paired examples follow the explanatory passage.

The interactive key is the You Try It. It has not been worked so that the student may practice the skill, referring to the worked example at the left, if necessary.

Reference to the Answer Section allows the student to check solutions immediately.

Section 2.7 / Division of Fractions and Mixed Numbers 97

2.7 Division of Fractions and Mixed Numbers

Objective A ***To divide fractions***

The **reciprocal** of a fraction is the fraction with the numerator and denominator interchanged. For instance, the reciprocal of $\frac{2}{3}$ is $\frac{3}{2}$. The process of interchanging the numerator and denominator of a fraction is called **inverting** the fraction.

To find the reciprocal of a whole number, first write the whole number as a fraction with a denominator of 1; then find the reciprocal of that fraction.

The reciprocal of 5 is $\frac{1}{5}$. $\left(\text{Think } 5 = \frac{5}{1}\right)$

Reciprocals are used to rewrite division problems as related multiplication problems. Look at the following two problems:

$8 \div 2 = 4$ — 8 divided by 2 is 4.

$8 \times \frac{1}{2} = 4$ — 8 times the reciprocal of 2 is 4.

"Divided by" means the same as "times the reciprocal of." Thus "÷ 2" can be replaced with "$\times \frac{1}{2}$," and the answer will be the same. Fractions are divided by making this replacement.

➡ Divide: $\frac{2}{3} \div \frac{3}{4}$

$$\frac{2}{3} \div \frac{3}{4} = \frac{2}{3} \times \frac{4}{3} = \frac{2 \cdot 4}{3 \cdot 3} = \frac{2 \cdot 2 \cdot 2}{3 \cdot 3} = \frac{8}{9}$$

Example 1 Divide: $\frac{5}{8} \div \frac{4}{9}$

Solution $\frac{5}{8} \div \frac{4}{9} = \frac{5}{8} \times \frac{9}{4} = \frac{5 \cdot 9}{8 \cdot 4}$
$= \frac{5 \cdot 3 \cdot 3}{2 \cdot 2 \cdot 2 \cdot 2 \cdot 2} = \frac{45}{32} = 1\frac{13}{32}$

You Try It 1 Divide: $\frac{3}{7} \div \frac{2}{3}$

Your solution

Example 2 Find the quotient of $\frac{3}{5}$ and $\frac{12}{25}$.

Solution $\frac{3}{5} \div \frac{12}{25} = \frac{3}{5} \times \frac{25}{12} = \frac{3 \cdot 25}{5 \cdot 12}$
$= \frac{\overset{1}{\cancel{3}} \cdot \overset{1}{\cancel{5}} \cdot 5}{\underset{1}{\cancel{5}} \cdot 2 \cdot 2 \cdot \underset{1}{\cancel{3}}} = \frac{5}{4} = 1\frac{1}{4}$

You Try It 2 Divide: $\frac{3}{4} \div \frac{9}{10}$

Your solution

Solutions on p. A16

Objective B ***To divide whole numbers, mixed numbers, and fractions***

To divide a fraction and a whole number, first write the whole number as a fraction with a denominator of 1.

➡ Divide: $\frac{3}{7} \div 5$

$$\frac{3}{7} \div \boxed{5} = \frac{3}{7} \div \boxed{\frac{5}{1}} = \frac{3}{7} \times \frac{1}{5} = \frac{3 \cdot 1}{7 \cdot 5} = \frac{3}{35}$$

• **Write 5 with a denominator of 1; then divide the fractions.**

The traditional teaching approach neglects the difficulties that students have in making the transition from arithmetic to algebra. One of the most troublesome and uncomfortable transitions for the student is from concrete arithmetic to symbolic algebra. *Basic College Mathematics: An Applied Approach* recognizes the formidable task the student faces by introducing variables in a very natural way—through applications of mathematics. A secondary benefit of this approach is that the student becomes aware of the value of algebra as a real-life tool.

The solution of an application problem in *Basic College Mathematics: An Applied Approach* is always accompanied by two parts: **Strategy** and **Solution.** The strategy is a written description of the steps that are necessary to solve the problem; the solution is the implementation of the strategy. This format provides students with a structure for problem solving. It also encourages students to write strategies for solving problems which, in turn, fosters organizing problem-solving strategies in a logical way.

A strategy which the student may use in solving an application problem is stated.

This strategy is used in the solution of the worked example.

Students are encouraged to write a strategy for the application problem they try to solve.

198 Chapter 5 / Percents

Objective B To solve application problems

To solve percent problems, remember that it is necessary to identify the percent, base, and amount. Usually the base follows the phrase "percent of."

Example 4
The Kaminski family has an income of $1500 each month and makes a car payment of $180 a month. What percent of the monthly income is the car payment?

Strategy
To find what percent of the income the car payment is, write and solve the basic percent equation using n to represent the unknown percent. The base is $1500 and the amount is $180.

Solution $n \times \$1500 = \$180$

$n = \$180 \div \$1500$

$n = 0.12 = 12\%$

The car payment is 12% of the monthly income.

You Try It 4
Tomo Nagata had an income of $20,000 and paid $3000 in income tax. What percent of the income is the income tax?

Your strategy

Your solution

Example 5
Claudia Sloan missed 25 questions out of 200 on a state nursing license exam. What percent of the questions did she answer correctly?

Strategy
To find what percent of the questions she answered correctly:

- Find the number of questions the nurse answered correctly (200 − 25).
- Write and solve a basic percent equation using n to represent the unknown percent. The base is 200 and the amount is the number of questions answered correctly.

Solution 200 − 25 = 175 questions answered correctly.

$n \times 200 = 175$

$n = 175 \div 200$

$n = 0.875 = 87.5\%$

Claudia answered 87.5% of the questions correctly.

You Try It 5
A survey of 1000 people showed that 667 people favored Minna Oliva for governor of the state. What percent of the people surveyed did not favor her?

Your strategy

Your solution

Solutions on p. A27

OBJECTIVE-SPECIFIC APPROACH

Many mathematics texts are not organized in a manner that facilitates management of learning. Typically, students are left to wander through a maze of apparently unrelated lessons, exercise sets, and tests. *Basic Mathematics: An Applied Approach* solves this problem by organizing all lessons, exercise sets, computer tutorials, and tests around a carefully constructed hierarchy of objectives. The advantage of this objective-by-objective organization is that it enables the student who is uncertain at any step in the learning process to refer easily to the original presentation and review that material.

The Objective-Specific Approach also gives the instructor greater control over the management of student progress. The Computerized Test Generator and the printed Test Bank are organized by the same objectives as the text. These references are provided with the answers to the test items, thereby allowing the instructor to quickly determine those objectives for which a student may need additional instruction.

The Computer Tutor is also organized around the objectives of the text. As a result, supplemental instruction is available for any objectives that are troublesome for a student.

A numbered objective statement names the topic of each lesson.

The Least Common Multiple and Greatest Common Factor

Objective A *To find the least common multiple (LCM)*

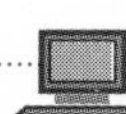

The exercise sets correspond to the objectives in the text.

2.1 Exercises

Objective A

Find the LCM.

1. 5, 8 **2.** 3, 6 **3.** 3, 8 **4.** 2, 5 **5.** 5, 6

The answers to the Chapter Test show the objective to study if the student incorrectly answers an item.

CHAPTER TEST *pages 113–114*

1. 120 (2.1A) **2.** 8 (2.1B) **3.** $\frac{11}{4}$ (2.2A) **4.** $3\frac{3}{5}$ (2.2B) **5.** $\frac{49}{5}$ (2.2B)
6. $\frac{5}{8} = \frac{45}{72}$ (2.3A) **7.** $\frac{5}{8}$ (2.3B) **8.** $1\frac{11}{12}$ (2.4A) **9.** $1\frac{61}{90}$ (2.4B) **10.** $22\frac{4}{15}$ (2.4C)

The answers to the Cumulative Review exercises also show the objective that relates to the exercise.

CUMULATIVE REVIEW *pages 115–116*

1. 290,000 (1.1D) **2.** 291,278 (1.3B) **3.** 73,154 (1.4B) **4.** 540 R2 (1.5C)
5. 1 (1.6B) **6.** $2 \cdot 2 \cdot 11$ (1.7B) **7.** 210 (2.1A) **8.** 20 (2.1B) **9.** $\frac{23}{3}$ (2.2B)

ADDITIONAL LEARNING AIDS

Projects in Mathematics

The Project in Mathematics feature occurs at the end of each chapter. These projects can be used as extra credit or cooperative learning activities. The projects cover various aspects of mathematics including the use of calculators, extended applications, additional problem-solving strategies, and other topics related to mathematics.

Chapter Summaries

At the end of each chapter there is a Chapter Summary that includes Key Words and Essential Rules that were covered in the chapter. These chapter summaries provide a single point of reference as the student prepares for a test.

Study Skills

The To the Student passage provides suggestions for using this text and approaches to creating good study habits.

Computer Tutor

The Computer Tutor is a networkable, interactive, algorithmically-driven software package. This powerful ancillary features full-color graphics, a glossary, extensive hints, animated solution steps, and a comprehensive class management system.

Glossary

New to this edition, a Glossary at the end of the book includes definitions of terms used in the text.

EXERCISES

End-of-Section Exercises

Basic College Mathematics: An Applied Approach contains more than **6000** exercises. At the end of each section there are exercise sets that are keyed to the corresponding learning objective. The exercises are carefully developed to ensure that students can apply the concepts in the section to a variety of problem situations.

Applying the Concepts Exercises

The End-of-Section Exercises are followed by Applying the Concepts Exercises. These sections contain a variety of exercise types, including:

- challenge problems
- problems that require that the student determine if a statement is always true, sometimes true, or never true
- problems that ask students to determine incorrect procedures
- problems that require a calculator to determine a solution

Writing Exercises

Within the "Applying the Concepts Exercises," there are Writing Exercises denoted by [W]. These exercises ask students to write about a topic in the section or to research and report on a related topic.

Chapter Review Exercises

Review Exercises are found at the end of each chapter. These exercises are selected to help the student integrate all of the topics presented in the chapter. The answers to all review exercises are given in the answer section at the end of the book.

Cumulative Review Exercises

Cumulative Review Exercises, which appear at the end of each chapter (beginning with Chapter 2), help students maintain skills learned in previous chapters. The answers to all Cumulative Review Exercises are given in the answer section. Along with the answer, there is a reference to the objective that pertains to each exercise.

NEW TO THIS EDITION

Topical Coverage

Geometry has been moved from Chapter 10 to Chapter 12. This change was requested by users of the previous editions so that some algebra could be used with geometry problems. The topical coverage of this chapter has been increased to include properties of parallel lines.

The material in Chapter 6, *Applications for Business and Consumers,* has been updated to reflect current interest rates and prices.

One of the main challenges for students is the ability to translate verbal phrases into mathematical expressions. One reason for this difficulty is that students are not exposed to verbal phrases until later in most texts. In *Basic College Mathematics: An Applied Approach,* we introduce verbal phrases for operations as we introduce the operation. For instance, after addition concepts have been presented, we provide exercises which say "Find the sum of . . . ." or "What is 6 more than 7?" In this way, students are constantly confronted with verbal phrases and must make a mathematical connection between the phrase and a mathematical operation.

Point of Interest notes are interspersed throughout the text. These notes are interesting sidelights of the topic being discussed. In addition, there are *Instructor Notes* which are printed only in the Instructor's Annotated Edition. These notes provide suggestions for presenting the material or related material that can be used in class.

Computer Tutor

The Computer Tutor has been completely revised. It is now an algorithmically-based tutor that includes color and animation. The algorithmic feature essentially provides an infinite number of practice problems for students to attempt. The algorithms have been carefully crafted to present a variety of problem types from easy to difficult. A complete solution to each problem is available.

There is an interactive feature of the Computer Tutor that requires students to respond to questions about the topic in the current lesson. In this way, students can assess their understanding of concepts as they are presented. There is a Glossary which can be accessed at any time so that students can look up words whose definitions that they may have forgotten.

When the student completes a lesson, a printed report is available. This optional report also gives the student's name, the objective studied, the number of problems attempted, the number of problems correct, and the percent correct. The instructor also has the option of creating a cumulative report via the Tutor's new class management system.

Applying the Concepts

The end-of-section exercises are followed by Applying the Concepts Exercises. These exercises include challenge problems and applications of the concepts presented in the section in a way that requires a student to combine several skills to solve a problem. These exercises also include Writing Exercises, denoted by [W], that ask students to write about a topic in the section or to research and report on a related topic.

Projects in Mathematics

Through the Project in Mathematics feature, some of the standards suggested by the NCTM can be implemented. These Projects offer an opportunity for students to explore several topics relating to calculators, applications of math, logic, problem solving, and related subjects.

Glossary

The Glossary contains the definitions or the descriptions of all the key words highlighted in the text.

SUPPLEMENTS FOR THE INSTRUCTOR

Instructor's Annotated Edition

The Instructor's Annotated Edition is an exact replica of the student text except that answers to all exercises are given in the text. Also, there are Instructor Notes in the margin that offer suggestions for presenting the material in that objective.

Instructor's Resource Manual with Chapter Tests

The Instructor's Resource Manual contains the printed Test Bank, which is the first of three sources of testing material. Four printed tests, two free response and two multiple choice, are provided for each chapter. In addition, there are cumulative tests after Chapters 4, 8, and 11, and a final exam. The Instructor's Resource Manual also includes suggestions for course sequencing and outlines for the answers to the Writing Exercises.

Computerized Test Generator

The Computerized Test Generator is the second source of testing material. The data base contains more than 3000 test items. The Test Generator is designed to provide an unlimited number of tests for each chapter, cumulative chapter tests, and a final exam. It is available for the IBM PC and compatible computers and the Macintosh.

Printed Test Bank

The printed Test Bank, the third component of the testing material, is a printout of all items in the Computerized Test Generator. Instructors who do not have access to a computer can use the Test Bank to select items to include on a test being prepared by hand.

Solutions Manual

The Solutions Manual contains worked-out solutions for all end-of-section exercises, chapter review exercises, and cumulative review exercises.

Videotapes

The videotape series consists of 12 30-minute lessons to accompany *Basic College Mathematics: An Applied Approach*. These lessons are closely tied to specific sections of the text. Each tape begins with footage of a real-life application, which is solved during the lesson.

SUPPLEMENTS FOR THE STUDENT

Student Solutions Manual

The Student Solutions Manual contains the complete solution to all odd-numbered exercises in the text.

Computer Tutor

The Computer Tutor is an interactive instructional computer program for student use. Each objective of the text is supported by a tutorial in the Computer Tutor. These tutorials contain an interactive lesson that covers the material in the objective. Following the lesson are randomly generated exercises for the student to attempt. A record of the student's progress is available.

The Computer Tutor can be used in several ways: (1) to cover material the student missed because of an absence; (2) to reinforce instruction on a concept that the student has not yet mastered; (3) to review material in preparation for exams. This tutorial is available for the IBM PC and compatible computers running Windows and the Macintosh.

ACKNOWLEDGMENTS

The authors would like to thank the people who reviewed this manuscript and provided many valuable suggestions:

Ronald J. Bautch, *Cook County College/Denton Center, TX*

Rebecca C. Benson-Beaver, *Valencia Community College, FL*

Martha J. Eagle, *Longview Community College, MO*

Diane Fariss, *McLennan Community College, TX*

Nancy C. Hall, *Muskingham Area Technical College, OH*

William C. Hoston

Laura L. Hoye, *Trident Technical College, SC*

Robert L. Jones, *IVY Technical College/Terre Haute, IN*

JoAnne Kennedy, *LaGuardia Community College, NY*

Gerald L. LaPage, *Bristol Community College, MA*

Lynda D. MacLeod, *Robeson Community College, NC*

James J. Meissner, *Butler County Community College, PA*

Linda J. Murphy, *Northern Essex Community College, MA*

Nancy K. Nickerson, *Northern Essex Community College, MA*

Elizabeth Ogilvie, *Horry-Georgetown Technical College, SC*

Rose L. Pugh, *Bellevue Community College, SC*

James P. Ryan, *Madera Community College, CA*

William D. Summons, *Shelby State Community College, TN*

To the Student

Many students feel that they will never understand math, while others appear to do very well with little effort. Oftentimes what makes the difference is that successful students take an active role in the learning process.

Learning mathematics requires your *active* participation. Although doing homework is one way you can actively participate, it is not the only way. First, you must attend class regularly and become an active participant in class. Secondly, you must become actively involved with the textbook.

Basic College Mathematics: An Applied Approach was written and designed with you in mind as a participant. Here are some suggestions on how to use the features of this textbook.

There are 12 chapters in this text. Each chapter is divided into sections, and each section is subdivided into learning objectives. Each learning objective is labeled with a letter from A–D.

First, read each objective statement carefully so you will understand the learning goal that is being presented. Next, read the objective material carefully, being sure to note each bold word. These words indicate important concepts that you should familiarize yourself with. Study each in-text example carefully, noting the techniques and strategies used to solve the example.

You will then come to the key learning feature of this text, the *boxed examples*. These examples have been designed to aid you in a very specific way. Notice that in each example box, the example on the left is completely worked out and the "You Try It" example on the right is not. *You* are expected to work the right-hand example (in the space provided) in order to immediately test your understanding of the material you have just studied.

You should study the worked-out example carefully by working through each step presented. This allows you to focus on each step and reinforces the technique for solving that type of problem. You can then use the worked-out example as a model for solving similar problems.

Next, try to solve the "You Try It" example using the problem-solving techniques that you have just studied. When you have completed your solution, check your work by turning to the page at the end of the book where the complete solution can be found. The page number on which the solution appears is printed at the bottom of the example box in the right-hand corner. By checking your solution, you will know immediately whether or not you fully understand the skill you just studied.

When you have completed studying an objective, do all of the exercises in the exercise set that corresponds with that objective. The exercises will be labeled with the same letter as the objective. Math is a subject that needs to be learned in small sections and practiced continually in order to be mastered. Doing all of the exercises in each exercise set will help you master the problem-solving techniques necessary for success.

Once you have completed the exercises to an objective, you should check your answers to the odd-numbered exercises with those found in the back of the book.

After completing a chapter, read the Chapter Summary. This summary highlights the important topics covered in the chapter. Following the Chapter Summary are Chapter Review Exercises, a Chapter Test, and a Cumulative Review (beginning with Chapter 2). Doing the review exercises is an important way of testing your understanding of the chapter. The answer to each review exercise is given at the back of the book. Each answer to the Chapter Test and Cumulative Review is followed by a reference that tells which objective that exercise was taken from. For example, (4.2B) means Section 4.2, Objective B. After checking your answers, restudy any objective that you missed. It may be very helpful to retry some of the exercises for that objective to reinforce your problem-solving techniques.

The Chapter Test should be used to prepare for an exam. We suggest that you try the Chapter Test a few days before your actual exam. Take the test in a quiet place and try to complete the test in the same amount of time you will be allowed for your exam. When taking the Chapter Test, practice the strategies of successful test takers: 1) scan the entire test to get a feel for the questions; 2) read the directions carefully; 3) work the problems that are easiest for you first; and, perhaps most importantly, 4) try to stay calm.

When you have completed the Chapter Test, check your answers. If you missed a question, review the material in that objective and rework some of the exercises from that objective. This will strengthen your ability to perform the skills outlined in that objective.

The Cumulative Review allows you to refresh the skills you learned in previous chapters. This is very important in mathematics. By consistently reviewing previous material, you will retain the skills already learned as you build new ones.

Remember, to be successful: attend class regularly; read the textbook carefully; actively participate in class; work with your textbook using the "You Try It" examples for immediate feedback and reinforcement of each skill; do all the homework assignments; review constantly; and work carefully.

Index of Applications

CONSUMER ECONOMICS

CONSUMPTION OF RESOURCES

ECONOMICS

FOOD–NUTRITION

EDUCATION

ENVIRONMENT AND BIOLOGICAL SCIENCES

GEOGRAPHY

GEOMETRY

HEALTH AND PHYSICAL FITNESS

SOCIAL SCIENCE

SPORTS

WAGES AND EARNED INCOME

Chapter

Whole Numbers

Objectives

Section 1.1

To identify the order relation between two numbers

To write whole numbers in words and in standard form

To write whole numbers in expanded form

To round a whole number to a given place value

Section 1.2

To add whole numbers

To solve application problems

Section 1.3

To subtract whole numbers without borrowing

To subtract whole numbers with borrowing

To solve application problems

Section 1.4

To multiply a number by a single digit

To multiply larger whole numbers

To solve application problems

Section 1.5

To divide by a single digit with no remainder in the quotient

To divide by a single digit with a remainder in the quotient

To divide by larger whole numbers

To solve application problems

Section 1.6

To simplify expressions that contain exponents

To use the Order of Operations Agreement to simplify expressions

Section 1.7

To factor numbers

To find the prime factorization of a number

Family Tree for Numbers

Our number system is called the Hindu-Arabic system because it has its ancestry in India and was refined by the Arabs. But despite the influence of these cultures on our system, there is some evidence that our system may have originated in China around 1400 B.C. That is 34 centuries ago.

The family tree below illustrates the most widely believed account of the history of our number system. In the 16th century, with Gutenberg's invention of the printing press, symbols for our numbers started to become standardized.

1.1 Introduction to Whole Numbers

Objective A To identify the order relation between two numbers

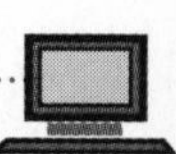

POINT OF INTEREST

The numerals that we use have not always appeared as they do today. In the second century, the symbols for 1 through 9 were

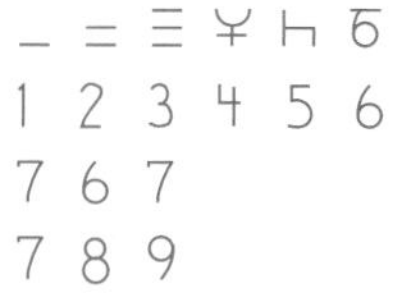

By the eighth century, the symbols 2 and 3 were used instead of horizontal lines, but five looked like ㄐ and 9 was shown as Ɛ.

The **whole numbers** are 0, 1, 2, 3, 4, 5, 6, 7, 8, 9, 10, 11, 12, 13, 14, ...

The three dots mean that the list continues on and on and that there is no largest whole number.

Just as distances are associated with the markings on the edge of a ruler, the whole numbers can be associated with points on a line. This line is called the **number line.** The arrow on the number line indicates that there is no largest whole number.

The Number Line

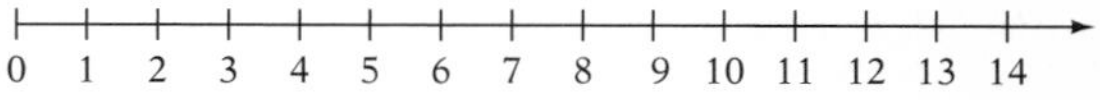

The **graph of a whole number** is shown by placing a heavy dot on the number line directly above the number. Here is the graph of 7 on the number line:

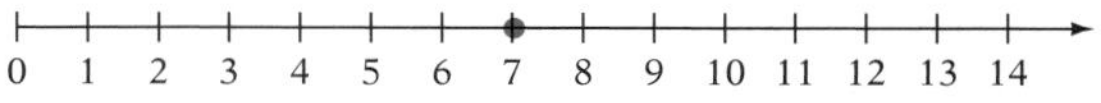

INSTRUCTOR NOTE

One of the main pedagogical features of this text is the paired You-Try-It example. This example is for the student to work, using the example on the left as a model. A *complete solution* can be found on the page referenced at the bottom of the box.

The number line can be used to show the order of whole numbers. A number that appears to the left of a given number is **less than** the given number. The symbol for "is less than" is <. A number that appears to the right of a given number is **greater than** the given number. The symbol for "is greater than" is >.

Four is less than seven.
$4 < 7$

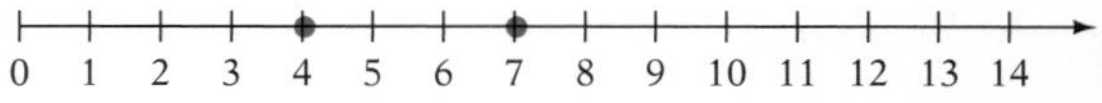

Twelve is greater than seven.
$12 > 7$

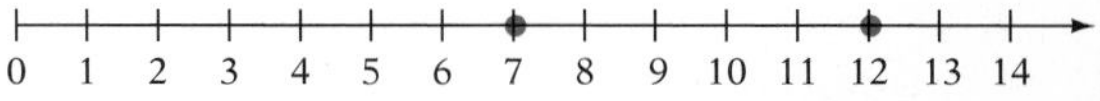

Example 1 Graph 11 on the number line.

Solution

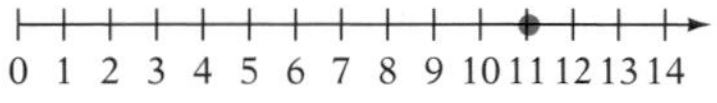

You Try It 1 Graph 9 on the number line.

Your solution

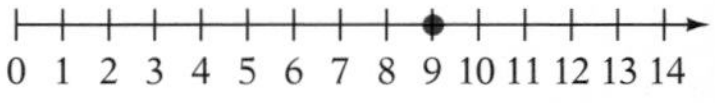

Example 2 Place the correct symbol, < or >, between the two numbers.

a. 39 24

b. 0 51

Solution **a.** $39 > 24$

b. $0 < 51$

You Try It 2 Place the correct symbol, < or >, between the two numbers.

a. 45 29

b. 27 0

Your solution **a.** $45 > 29$

b. $27 > 0$

Solutions on p. A7

Objective B To write whole numbers in words and in standard form

POINT OF INTEREST

The Babylonians had a place-value system based on 60. Its influence is still with us in angle measurement and time: 60 seconds in 1 minute, 60 minutes in 1 hour. It appears that the earliest record of a base 10 place-value system for natural numbers was developed in the eighth century.

When a whole number is written using the digits 0, 1, 2, 3, 4, 5, 6, 7, 8, and 9, it is said to be in **standard form.** The position of each digit in the number determines the digit's **place value.** The diagram below shows a **place-value chart** naming the first twelve place values. The number 37,462 is in standard form and has been entered in the chart.

In the number 37,462, the position of the digit 3 determines that its place value is ten-thousands.

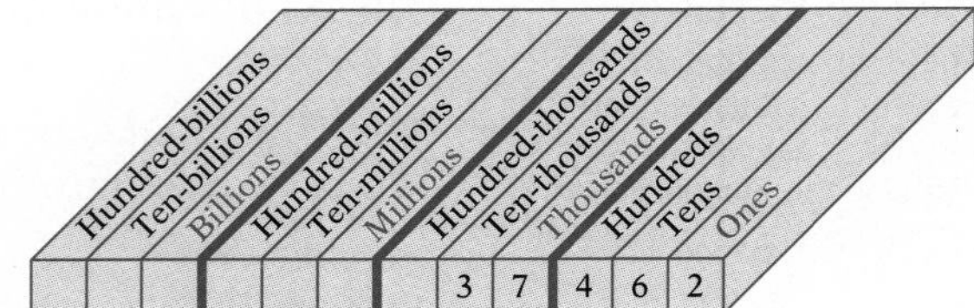

When a number is written in standard form, each group of digits separated by a comma is called a **period.** The number 3,786,451,294 has four periods. The period names are shown in red in the place-value chart above.

To write a number in words, start from the left. Name the number in each period. Then write the period name in place of the comma.

3,786,451,294 is read "three billion seven hundred eighty-six million four hundred fifty-one thousand two hundred ninety-four."

To write a whole number in standard form, write the number named in each period, and replace each period name with a comma.

Four million sixty-two thousand five hundred eighty-four is written 4,062,584. The zero is used as a place holder for the hundred-thousands' place.

Example 3 Write 25,478,083 in words.

Solution Twenty-five million four hundred seventy-eight thousand eighty-three

You Try It 3 Write 36,462,075 in words.

Your solution Thirty-six million four hundred sixty-two thousand seventy-five

Example 4 Write three hundred three thousand three in standard form.

Solution 303,003

You Try It 4 Write four hundred fifty-two thousand seven in standard form.

Your solution 452,007

Solutions on p. A7

Objective C To write whole numbers in expanded form

The whole number 26,429 can be written in **expanded form** as

20,000 + 6000 + 400 + 20 + 9

The place-value chart can be used to find the expanded form of a number.

Hundred-billions Ten-billions Billions Hundred-millions Ten-millions Millions Hundred-thousands Ten-thousands Thousands Hundreds Tens Ones
2 6 4 2 9

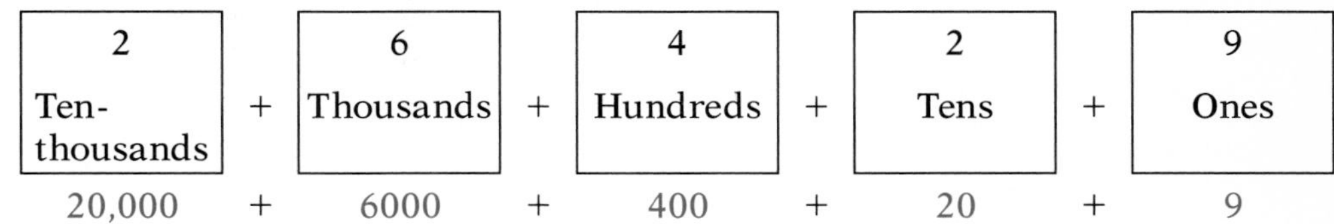

Write the number 420,806 in expanded form.

Note the effect of having zeros in the number.

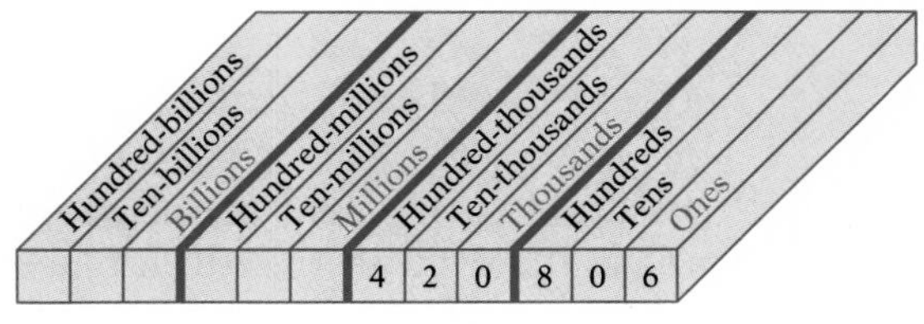

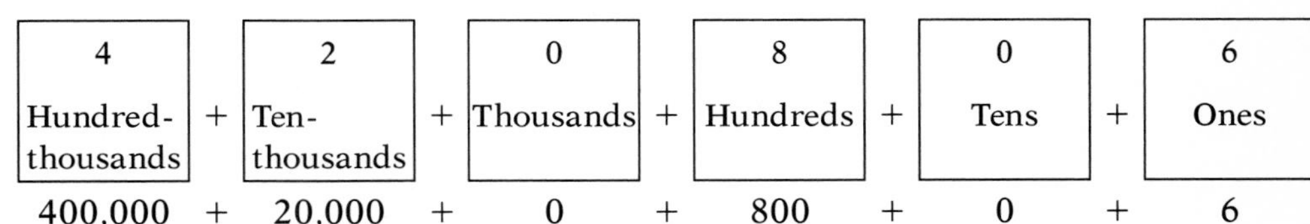

or simply 400,000 + 20,000 + 800 + 6

Example 5 Write 23,859 in expanded form.

Solution 20,000 + 3000 + 800 + 50 + 9

You Try It 5 Write 68,281 in expanded form.

Your solution 60,000 + 8000 + 200 + 80 + 1

Example 6 Write 709,542 in expanded form

Solution 700,000 + 9000 + 500 + 40 + 2

You Try It 6 Write 109,207 in expanded form.

Your solution 100,000 + 9000 + 200 + 7

Solutions on p. A7

Objective D To round a whole number to a given place value

When the distance to the moon is given as 240,000 miles, the number represents an approximation to the true distance. Giving an approximate value for an exact number is called **rounding.** A number is always rounded to a given place value.

37 is closer to 40 than it is to 30. 37 rounded to the nearest ten is 40.

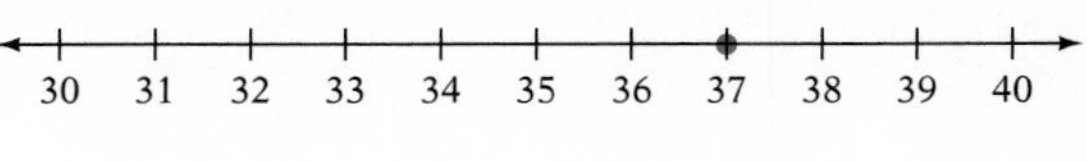

3673 rounded to the nearest ten is 3670. 3673 rounded to the nearest hundred is 3700.

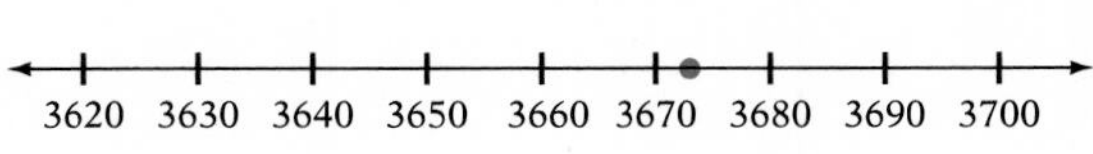

A whole number is rounded to a given place value without using the number line by looking at the first digit to the right of the given place value.

If the digit to the right of the given place value is less than 5, that digit and all digits to the right are replaced by zeros.

➡ Round 13,834 to the nearest hundred.

Given place value
13,834
$3 < 5$

13,834 rounded to the nearest hundred is 13,800.

If the digit to the right of the given place value is greater than or equal to 5, increase the digit in the given place value by 1, and replace all other digits to the right by zeros.

➡ Round 386,217 to the nearest ten-thousand.

Given place value
386,217
$6 > 5$

386,217 rounded to the nearest ten-thousand is 390,000.

Example 7 Round 525,453 to the nearest ten-thousand.

Solution

Given place value
525,453
$5 = 5$

525,453 rounded to the nearest ten-thousand is 530,000.

You Try It 7 Round 368,492 to the nearest ten-thousand.

Your solution 370,000

Example 8 Round 1972 to the nearest hundred.

Solution

Given place value
1972
$7 > 5$

1972 rounded to the nearest hundred is 2000.

You Try It 8 Round 3962 to the nearest hundred.

Your solution 4000

Solutions on p. A7

1.1 Exercises

Objective A

Graph the number on the number line.

1. 3

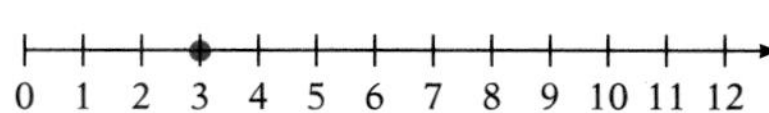

2. 5

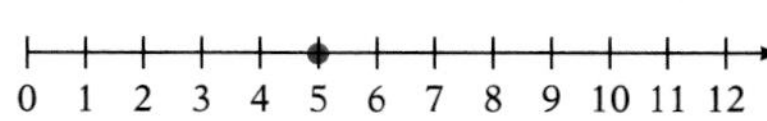

3. 9

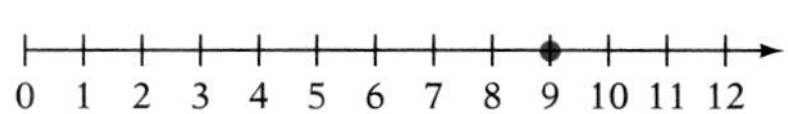

4. 0

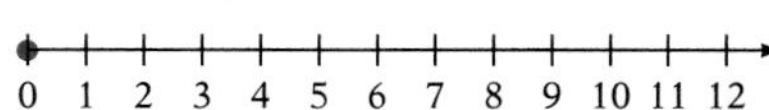

Place the correct symbol, < or >, between the two numbers.

5. 37 < 49

6. 58 > 21

7. 101 > 87

8. 16 > 5

9. 245 > 158

10. 2701 > 2071

11. 0 < 45

12. 107 > 0

13. 815 < 928

Objective B

Write the number in words.

14. 2675
Two thousand six hundred seventy-five

15. 3790
Three thousand seven hundred ninety

16. 42,928
Forty-two thousand nine hundred twenty-eight

17. 58,473
Fifty-eight thousand four hundred seventy-three

18. 356,943
Three hundred fifty-six thousand nine hundred forty-three

19. 498,512
Four hundred ninety-eight thousand five hundred twelve

20. 3,697,483
Three million six hundred ninety-seven thousand four hundred eighty-three

21. 6,842,715
Six million eight hundred forty-two thousand seven hundred fifteen

Write the number in standard form.

22. Eighty-five
85

23. Three hundred fifty-seven
357

24. Three thousand four hundred fifty-six
3456

25. Sixty-three thousand seven hundred eighty
63,780

26. Six hundred nine thousand nine hundred forty-eight
609,948

27. Seven million twenty-four thousand seven hundred nine
7,024,709

Objective C

Write the number in expanded form.

28. 5287
5000 + 200 + 80 + 7

29. 6295
6000 + 200 + 90 + 5

30. 58,943
50,000 + 8000 + 900 + 40 + 3

31. 453,921
400,000 + 50,000 + 3000 + 900 + 20 + 1

32. 200,583
200,000 + 500 + 80 + 3

33. 301,809
300,000 + 1000 + 800 + 9

34. 403,705
400,000 + 3000 + 700 + 5

35. 3,000,642
3,000,000 + 600 + 40 + 2

Objective D

Round the number to the given place value.

36. 926 Tens
930

37. 845 Tens
850

38. 1439 Hundreds
1400

39. 3973 Hundreds
4000

40. 43,607 Thousands
44,000

41. 52,715 Thousands
53,000

42. 647,989 Ten-thousands
650,000

43. 253,678 Ten-thousands
250,000

APPLYING THE CONCEPTS

Answer true or false for Exercise 44 and 45. If the answer is false, give an example to show that it is false.

44. If you are given two distinct whole numbers, then one of the numbers is always greater than the other number.
True

45. A rounded-off number is always less than its exact value.
False, 8270 rounded to the nearest hundred is 8300

46. What is the largest three-digit whole number. What is the smallest five-digit whole number?
999 10,000

47. In the roman numeral system, IV = 4 and VI = 6. Does the position of the I in this system change the value of the number it represents? Determine the value of IX and XI.
Yes. IX = 9 XI = 11

48. If 3846 is rounded off to the nearest ten and then that number is rounded to the nearest hundred, is the result the same as what you get when you round 3846 to the nearest hundred? If not, which of the two methods is correct for rounding to the nearest hundred?
No. Round 3846 to the nearest hundred

1.2 Addition of Whole Numbers

Objective A To add whole numbers

Addition is the process of finding the total of two or more numbers.

POINT OF INTEREST
The first use of the plus sign appeared in 1489 in *Mercantile Arithmetic.* It was used to indicate a surplus and not as the symbol for addition. That use did not appear until around 1515.

By counting, we see that the total of \$3 and \$4 is \$7.

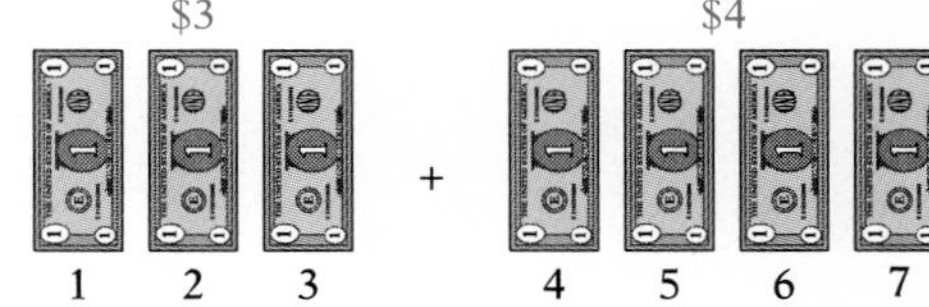

\$3	+	\$4	=	\$7
Addend		**Addend**		**Sum**

Addition can be illustrated on the number line by using arrows to represent the addends. The size, or magnitude, of a number can be represented on the number line by an arrow.

The number 3 can be represented anywhere on the number line by an arrow that is 3 units in length.

To add on the number line, place the arrows representing the addends head to tail, with the first arrow starting at zero. The sum is represented by an arrow starting at zero and stopping at the tip of the last arrow.

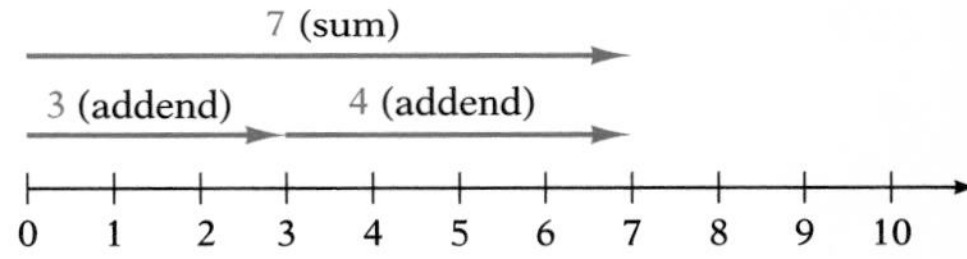

$$3 + 4 = 7$$

More than two numbers can be added on the number line.

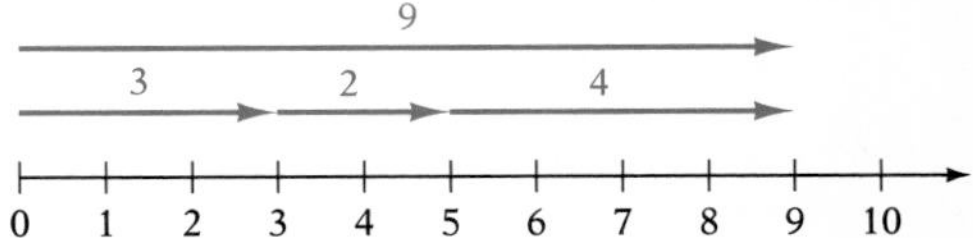

$$3 + 2 + 4 = 9$$

Some special properties of addition that are used frequently are given below.

Addition Property of Zero

Zero added to a number does not change the number.

$$4 + 0 = 4$$
$$0 + 7 = 7$$

Commutative Property of Addition

Two numbers can be added in either order; the sum will be the same.

$$4 + 8 = 8 + 4$$
$$12 = 12$$

Associative Property of Addition

Grouping the addition in any order gives the same result. The parentheses are grouping symbols and have the meaning "do the operations inside the parentheses first."

$$\underbrace{(4 + 2)}_{6} + 3 = 4 + \underbrace{(2 + 3)}_{5}$$
$$6 + 3 = 4 + 5$$
$$9 = 9$$

The number line is not useful for adding large numbers. The basic addition facts for adding one digit to one digit are listed in the Appendix on page A2 for your review. Addition of larger numbers requires the repeated use of the basic addition facts.

To add large numbers, begin by arranging the numbers vertically, keeping the digits of the same place value in the same column.

➡ Add: 321 + 6472

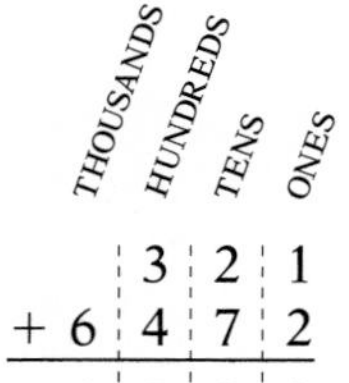

	3	2	1
+ 6	4	7	2
6	7	9	3

- **Add the digits in each column.**

There are several words or phrases in English that indicate the operation of addition. Here are some examples.

added to	3 added to 5	5 + 3
more than	7 more than 5	5 + 7
the sum of	the sum of 3 and 9	3 + 9
increased by	4 increased by 6	4 + 6
the total of	the total of 8 and 3	8 + 3
plus	5 plus 10	5 + 10

➡ What is the sum of 24 and 71?

The phrase *the sum of* means to add.

$$\begin{array}{r} 24 \\ +\ 71 \\ \hline 95 \end{array}$$

The sum of 24 and 71 is 95.

INSTRUCTOR NOTE

Carrying can be modeled with money. For instance, to add \$87 + \$45 think \$7 + \$5 is \$12, which can be exchanged for 1 ten dollar bill and 2 one dollar bills. Add the 1 ten dollar bill to the 8 tens and 4 tens. The result is 13 ten dollar bills which can be exchanged for 1 one hundred dollar bill and 3 ten dollar bills.

When the sum of the digits in a column exceeds 9, the addition will involve "carrying."

➡ Add: 487 + 369

HUNDREDS TENS ONES

$$\begin{array}{r} \scriptstyle 1\ \\ 4\ 8\ 7 \\ +\ 3\ 6\ 9 \\ \hline 6 \end{array}$$

- **Add the ones' column.**
 7 + 9 = 16 (1 ten + 6 ones).
 Write the 6 in the ones' column and carry the 1 ten to the tens' column.

$$\begin{array}{r} \scriptstyle 1\ 1\ \ \\ 4\ 8\ 7 \\ +\ 3\ 6\ 9 \\ \hline 5\ 6 \end{array}$$

- **Add the tens' column.**
 1 + 8 + 6 = 15 (1 hundred + 5 tens).
 Write the 5 in the tens' column and carry the 1 hundred to the hundreds' column.

$$\begin{array}{r} \scriptstyle 1\ 1\ \ \\ 4\ 8\ 7 \\ +\ 3\ 6\ 9 \\ \hline 8\ 5\ 6 \end{array}$$

- **Add the hundreds' column.**
 1 + 4 + 3 = 8 (8 hundreds).
 Write the 8 in the hundreds' column

Example 1 Find the total of 17, 103, and 8.

Solution

$$\begin{array}{r} 17 \\ 103 \\ +\quad 8 \\ \hline 128 \end{array}$$

You Try It 1 What is 347 increased by 12,453?

Your solution 12,800

Example 2 Add: 89 + 36 + 98

Solution

$$\begin{array}{r} {}^{2} \\ 89 \\ 36 \\ +\ 98 \\ \hline 223 \end{array}$$

You Try It 2 Add: 95 + 88 + 67

Your solution 250

Example 3 Add:

$$\begin{array}{r} 41{,}395 \\ 4{,}327 \\ 497{,}625 \\ +\ \ 32{,}991 \\ \hline \end{array}$$

Solution

$$\begin{array}{r} {}^{1\,1\,2\ \ 2\,1} \\ 41{,}395 \\ 4{,}327 \\ 497{,}625 \\ +\ \ 32{,}991 \\ \hline 576{,}338 \end{array}$$

You Try It 3 Add:

$$\begin{array}{r} 392 \\ 4{,}079 \\ 89{,}035 \\ +\ \ 4{,}992 \\ \hline \end{array}$$

Your solution 98,498

Solutions on p. A7

ESTIMATION

INSTRUCTOR NOTE
Estimating is an important skill. Students should use this skill anytime a calculator is used.

Estimation and Calculators

At some places in the text, you will be asked to use your calculator. Effective use of a calculator requires that you estimate the answer to the problem. This helps ensure that you have entered the numbers correctly and pressed the correct keys.

For example, if you use your calculator to find 22,347 + 5896 and the answer in the calculator's display is 131,757,912, you should realize that you have entered some part of the calculation incorrectly. In this case, you pressed [×] instead of [+]. By estimating the answer to a problem, you can help ensure the accuracy of your calculations. The symbol ≈ is used to denote **approximately equal.**

For example, to estimate the answer to 22,347 + 5896, round off each number to the same place value. In this case we will round to the nearest thousand. Then add.

$$\begin{array}{rcr} 22{,}347 & \approx & 22{,}000 \\ +\ \ 5{,}896 & \approx & +\ \ 6{,}000 \\ \hline & & 28{,}000 \end{array}$$

The sum 22,347 + 5896 is approximately 28,000. Knowing this, you would know that 131,757,912 is much too large and is therefore incorrect.

To estimate the sum of two numbers, first round each whole number to the same place value and then add. Compare this answer with the calculator's answer.

Objective B To solve application problems

To solve an application problem, first read the problem carefully. The **Strategy** involves identifying the quantity to be found and planning the steps that are necessary to find that quantity. The **Solution** involves performing each operation stated in the Strategy and writing the answer.

INSTRUCTOR NOTE

There are many everyday instances when a table is used to present information. We have included several tables in this text and have asked students to answer questions based on the data.

A table displaying the number of pounds of the best selling ready-to-eat cereals for the year ending in 1992 is shown at the right.

Company	*Cereal*	*Amount (lb)*
General Mills	Cheerios	201,600,000
Kellogg's	Corn Flakes	128,200,000
Kellogg's	Frosted Flakes	113,300,000
Kellogg's	Raisin Bran	81,100,000
Kellogg's	Rice Krispies	74,900,000
Kellogg's	Mini-Wheats	67,100,000
General Mills	Total	52,900,000

➡ Find the total amount of ready-to-eat cereal sold for Kellogg's five best-selling cereals for the year ending in 1992.

INSTRUCTOR NOTE

Every application problem in this text has a written *strategy*. Encourage students to write a strategy for each application problem they solve.

Strategy To find the total amount of cereal sold by Kellogg's, read the table to find the amount of each kind of cereal sold by Kellogg's. Then add the amounts.

Solution

$$\begin{array}{r} 128{,}200{,}000 \\ 113{,}300{,}000 \\ 81{,}100{,}000 \\ 74{,}900{,}000 \\ +\ 67{,}100{,}000 \\ \hline 464{,}600{,}000 \end{array}$$

For the year ended in 1992, Kellogg's sold 464,600,000 pounds of ready-to-eat cereal.

Example 4

Your paycheck shows deductions of $225 for savings, $98 for taxes, and $27 for insurance. Find the total of the three deductions.

Strategy

To find the total of the deductions, add the three amounts ($225, $98, and $27).

Solution

$$\begin{array}{r} \$225 \\ 98 \\ +\ 27 \\ \hline \$350 \end{array}$$

The total of the three deductions is $350.

You Try It 4

Anna Barrera has a monthly budget of $475 for food, $275 for car expenses, and $120 for entertainment. Find the total amount budgeted for the three items each month.

Your strategy

Your solution $870

Solution on p. A7

1.2 Exercises

Objective A

Add.

1. $\begin{array}{r} 17 \\ +\ 11 \\ \hline 28 \end{array}$	**2.** $\begin{array}{r} 25 \\ +\ 63 \\ \hline 88 \end{array}$	**3.** $\begin{array}{r} 83 \\ +\ 42 \\ \hline 125 \end{array}$	**4.** $\begin{array}{r} 63 \\ +\ 94 \\ \hline 157 \end{array}$
5. $\begin{array}{r} 77 \\ +\ 25 \\ \hline 102 \end{array}$	**6.** $\begin{array}{r} 63 \\ +\ 49 \\ \hline 112 \end{array}$	**7.** $\begin{array}{r} 56 \\ +\ 98 \\ \hline 154 \end{array}$	**8.** $\begin{array}{r} 86 \\ +\ 68 \\ \hline 154 \end{array}$
9. $\begin{array}{r} 658 \\ +\ 831 \\ \hline 1489 \end{array}$	**10.** $\begin{array}{r} 842 \\ +\ 936 \\ \hline 1778 \end{array}$	**11.** $\begin{array}{r} 735 \\ +\ 93 \\ \hline 828 \end{array}$	**12.** $\begin{array}{r} 189 \\ +\ 50 \\ \hline 239 \end{array}$
13. $\begin{array}{r} 859 \\ +\ 725 \\ \hline 1584 \end{array}$	**14.** $\begin{array}{r} 637 \\ +\ 829 \\ \hline 1466 \end{array}$	**15.** $\begin{array}{r} 470 \\ +\ 749 \\ \hline 1219 \end{array}$	**16.** $\begin{array}{r} 427 \\ +\ 690 \\ \hline 1117 \end{array}$
17. $\begin{array}{r} 36{,}925 \\ +\ 65{,}392 \\ \hline 102{,}317 \end{array}$	**18.** $\begin{array}{r} 56{,}772 \\ +\ 51{,}239 \\ \hline 108{,}011 \end{array}$	**19.** $\begin{array}{r} 50{,}873 \\ +\ 28{,}453 \\ \hline 79{,}326 \end{array}$	**20.** $\begin{array}{r} 34{,}872 \\ +\ 46{,}079 \\ \hline 80{,}951 \end{array}$
21. $\begin{array}{r} 878 \\ 737 \\ +\ 189 \\ \hline 1804 \end{array}$	**22.** $\begin{array}{r} 768 \\ 461 \\ +\ 669 \\ \hline 1898 \end{array}$	**23.** $\begin{array}{r} 319 \\ 348 \\ +\ 912 \\ \hline 1579 \end{array}$	**24.** $\begin{array}{r} 292 \\ 579 \\ +\ 315 \\ \hline 1186 \end{array}$
25. $\begin{array}{r} 9409 \\ 3253 \\ +\ 7078 \\ \hline 19{,}740 \end{array}$	**26.** $\begin{array}{r} 8188 \\ 8020 \\ +\ 7104 \\ \hline 23{,}312 \end{array}$	**27.** $\begin{array}{r} 2038 \\ 2243 \\ +\ 3139 \\ \hline 7420 \end{array}$	**28.** $\begin{array}{r} 4252 \\ 6882 \\ +\ 5235 \\ \hline 16{,}369 \end{array}$
29. $\begin{array}{r} 67{,}428 \\ 32{,}171 \\ +\ 20{,}971 \\ \hline 120{,}570 \end{array}$	**30.** $\begin{array}{r} 52{,}801 \\ 11{,}664 \\ +\ 89{,}638 \\ \hline 154{,}103 \end{array}$	**31.** $\begin{array}{r} 76{,}290 \\ 43{,}761 \\ +\ 87{,}402 \\ \hline 207{,}453 \end{array}$	**32.** $\begin{array}{r} 43{,}901 \\ 98{,}301 \\ +\ 67{,}943 \\ \hline 210{,}145 \end{array}$

Add.

33. 20,958 + 3218 + 42
24,218

34. 80,973 + 5168 + 29
86,170

35. 392 + 37 + 10,924 + 621
11,974

36. 694 + 62 + 70,129 + 217
71,102

37. 294 + 1029 + 7935 + 65
9323

38. 692 + 2107 + 3196 + 92
6087

39. 97 + 7234 + 69,532 + 276
77,139

40. 87 + 1698 + 27,317 + 727
29,829

41. What is 9874 plus 4509? 14,383

42. What is 7988 plus 5678? 13,666

43. What is 3487 increased by 5986? 9473

44. What is 99,567 added to 126,863? 226,430

45. What is 23,569 more than 9678? 33,247

46. What is 7894 more than 45,872? 53,766

47. What is 479 added to 4579? 5058

48. What is 23,902 added to 23,885? 47,787

49. Find the total of 659, 55, and 1278. 1992

50. Find the total of 4561, 56, and 2309. 6926

51. Find the sum of 34, 67,892, 329, and 8.
68,263

52. Find the sum of 32,876, 45, 1289, and 7.
34,217

Estimate by rounding to the nearest hundreds. Then use your calculator to add.

53. 1234 + 9780 + 6740
Est.: 17,700
Cal.: 17,754

54. 919 + 3642 + 8796
Est.: 13,300
Cal.: 13,357

55. 241 + 569 + 390 + 1672
Est.: 2900
Cal.: 2872

56. 107 + 984 + 1035 + 2904
Est.: 5000
Cal.: 5030

Estimate by rounding off to the nearest thousands. Then use your calculator to add.

57. 32,461 + 9,844 + 59,407
Est.: 101,000
Cal.: 101,712

58. 29,036 + 22,904 + 7,903
Est.: 60,000
Cal.: 59,843

59. 25,432 + 62,941 + 70,390
Est.: 158,000
Cal.: 158,763

60. 66,541 + 29,365 + 98,742
Est.: 195,000
Cal.: 194,648

Estimate by rounding to the nearest ten-thousand. Then use your calculator to add.

61.
```
  67,421
  82,984
  66,361
  10,792
+ 34,037
```
Est.: 260,000
Cal.: 261,595

62.
```
  21,896
   4,235
  62,544
  21,892
+  1,334
```
Est.: 100,000
Cal.: 111,901

63.
```
  281,421
    9,874
   34,394
  526,398
+  94,631
```
Est.: 940,000
Cal.: 946,718

64.
```
  542,698
   97,327
    7,235
   73,667
+ 173,201
```
Est.: 890,000
Cal.: 894,128

Estimate by rounding to the nearest million. Then use your calculator to add.

65.
```
  28,627,052
     983,073
+  3,081,496
```
Est.: 33,000,000
Cal.: 32,691,621

66.
```
   1,792,085
  29,919,301
+  3,406,882
```
Est.: 35,000,000
Cal.: 35,118,268

67.
```
  12,377,491
   3,409,723
   7,928,026
+ 10,705,682
```
Est.: 34,000,000
Cal.: 34,420,922

68.
```
  46,751,070
   6,095,832
     280,011
+  1,563,897
```
Est.: 55,000,000
Cal.: 54,690,810

Objective B *Application Problems*

69. The attendance at a Friday night rock concert was 2114, and the attendance at the Saturday night concert was 3678. Find the total attendance at the two concerts. 5792 people

70. The attendance at a Friday night San Diego Padres game was 16,542. Attendance at the Saturday night game was 20,763. Find the total attendance for the two games. 37,305 people

71. Dan Marino threw 5 passes for 42 yards in the first quarter, 7 passes for 117 yards in the second quarter, 9 passes for 66 yards in the third quarter, and 8 passes for 82 yards in the fourth quarter. Find the total number of yards gained by passing. 307 yards

72. The high school basketball team scored 17 points in the first quarter, 26 points in the second quarter, 24 points in the third quarter, and 22 points in the fourth quarter. Find the total number of points scored. 89 points

73. Ken Owens, an account executive, received commissions of $3,168 during October, $2,986 during November, and $819 during December. Find the total commission Ken received for the three-month period. $6973

74. The attendance figures for the first four games of the 1992 National League playoffs between the Atlanta Braves and the Pittsburg Pirates were 51,971, 51,975, 56,610 and 57,164. Find the total attendance for the four games. 217,720 people

75. There were 207,429 paid admissions for the first four games of the 1992 World Series. There were 52,268 and 51,763 paid admissions for the fifth and sixth games of the series.
a. Find the total paid attendance for the fifth and sixth games. 104,031 people
b. Find the total paid attendance for the entire series. 311,460 people

76. A computer software store in the Westside shopping center sold 3578 compact disks during the first 9 months of the year. The store sold 359 compact disks in October, 259 in November, and 1027 in December.
   a. How many compact disks were sold during the last three months of the year? 1645 disks
   b. What was the number of compact disks sold during the year? 5223 disks

77. A student has $2135 in a checking account to be used for the fall semester. During the summer the student makes deposits of $518, $678, and $468.
   a. Find the total amount deposited. $1664
   b. Find the new checking account balance, assuming there are no withdrawals. $3799

78. The odometer on a moving van reads 68,692. The driver plans to drive 515 miles the first day, 492 miles the second day, and 278 miles the third day.
   a. How many miles will be driven during the three days? 1285 miles
   b. What will the odometer reading be at the end of the trip? 69,977 miles

APPLYING THE CONCEPTS

79. How many two-digit numbers are there? How many three-digit numbers are there? 90, 900

80. If you roll two ordinary six-sided dice and add the two numbers that appear on top, how many different sums are possible? 11 different sums

81. If you add two *different* whole numbers, is the sum always greater than either one of the numbers? If not, give an example. No. $0 + 2 = 2$

82. If you add two whole numbers, is the sum always greater than either one of the numbers? If not, give an example. (Compare this with the previous exercise.) No. $0 + 0 = 0$

83. [W] Make up a word problem for which the answer is the sum of 34 and 28. Answers will vary

84. Call a number "lucky" if it ends in a 7. How many lucky numbers are less than 100? 10 numbers

The accompanying table shows the average amount of money all Americans have invested in selected assets and the amounts invested for Americans between the ages of 16 and 34.

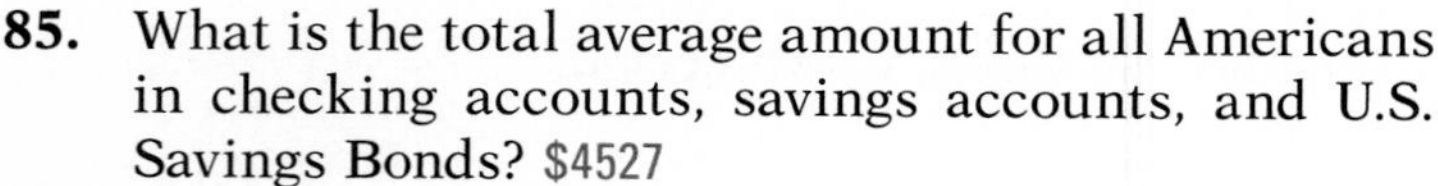

85. What is the total average amount for all Americans in checking accounts, savings accounts, and U.S. Savings Bonds? $4527

86. What is the total average amount for Americans ages 16 to 34 in checking accounts, savings accounts, and U.S. Savings Bonds? $1796

87. What is the total average amount for Americans ages 16 to 34 in all categories except home equity and retirement? $7838

88. Is the sum of the average amounts invested in home equity and retirement for all Americans greater than or less than that same sum for Americans between the ages of 16 and 34? Greater than

	All Americans	*Ages 16 to 34*
Checking Accounts	$487	$375
Savings Accounts	3,494	1,155
U.S. Savings Bonds	546	266
Money Market	10,911	4,427
Stocks/mutual Funds	4,510	1,615
Home Equity	43,070	17,184
Retirement	9,016	4,298

1.3 Subtraction of Whole Numbers

Objective A To subtract whole numbers without borrowing

Subtraction is the process of finding the difference between two numbers.

POINT OF INTEREST

The use of the minus sign dates from the same period as the plus sign, around 1515.

By counting, we see that the difference between \$8 and \$5 is \$3.

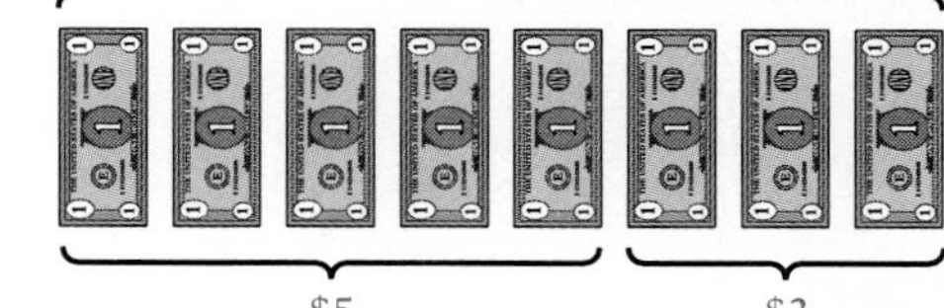

\$8	−	\$5	=	\$3
Minuend		**Subtrahend**		**Difference**

The difference 8 − 5 can be shown on the number line.

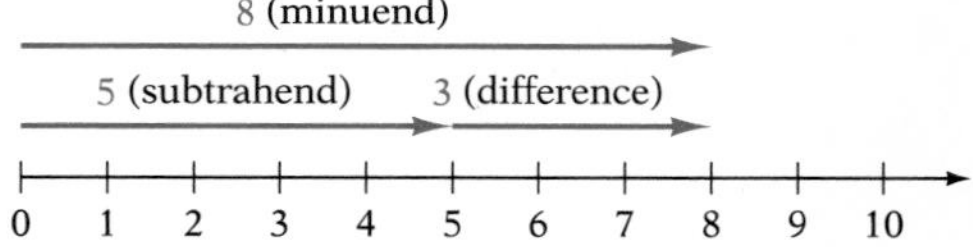

Note from the number line that addition and subtraction are related.

Subtrahend	5
+ Difference	+ 3
= Minuend	8

The fact that the sum of the subtrahend and the difference equals the minuend can be used to check subtraction.

To subtract large numbers, begin by arranging the numbers vertically, keeping the digits that have the same place value in the same column. Then subtract the digits in each column.

➡ Subtract 8955 − 2432 and check.

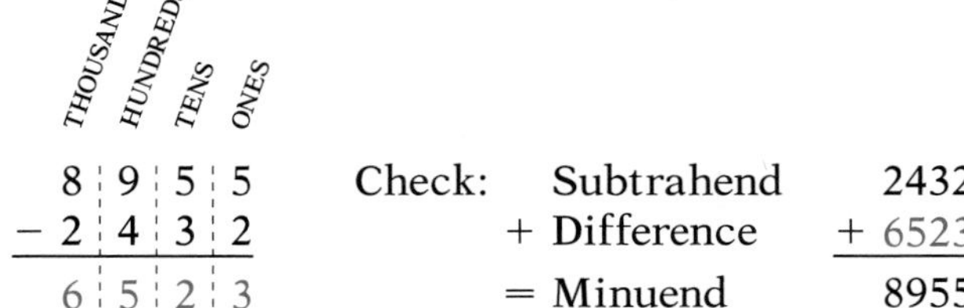

THOUSANDS	HUNDREDS	TENS	ONES
8	9	5	5
− 2	4	3	2
6	5	2	3

Check:

Subtrahend	2432
+ Difference	+ 6523
= Minuend	8955

Example 1 Subtract 6594 − 3271 and check.

Solution

```
  6594   Check:   3271
− 3271          + 3323
  3323            6594
```

You Try It 1 Subtract 8925 − 6413 and check.

Your solution 2512

Example 2 Subtract 15,762 − 7541 and check.

Solution

```
  15,762   Check:   7,541
−  7,541          + 8,221
   8,221           15,762
```

You Try It 2 Subtract 17,504 − 9302 and check.

Your solution 8202

Solutions on p. A8

Objective B To subtract whole numbers with borrowing

INSTRUCTOR NOTE

Borrowing can be related to money. For instance, if Kelly has $27 as 2 ten dollar bills and 7 one dollar bills and Chris wants to borrow $9, then Kelly can exchange a ten dollar bill for 10 one dollar bills. Kelly then has 1 ten dollar bill and 17 one dollar bills. Kelly now can give Chris 9 one dollar bills. This leaves Kelly with 1 ten and 8 one dollar bills.

In all the subtraction problems in the previous objective, for each place value the lower digit was not larger than the upper digit. When the lower digit is larger than the upper digit, subtraction will involve "borrowing."

➡ Subtract: 692 − 378

HUNDREDS	TENS	ONES
	8+1	
6	9	2
− 3	7	8

Because 8 > 2, borrowing is necessary.
9 tens = 8 tens + 1 ten.

HUNDREDS	TENS	ONES
	8+①	10
6	9	2
− 3	7	8

Borrow 1 ten from the tens' column and write 10 in the ones' column.

HUNDREDS	TENS	ONES
	8	12
6	9	2
− 3	7	8

Add the borrowed 10 to 2.

HUNDREDS	TENS	ONES
	8	12
6	9	2
− 3	7	8
3	1	4

Subtract the digits in each column.

INSTRUCTOR NOTE

The phrases that indicate subtraction are much more difficult for a student; in particular, 2 less than 7 as 7 − 2. Contrast that phrase with "the difference between" 7 and 2 which is 7 − 2.

The phrases below are used to indicate the operation of subtraction. An example is shown at the right of each phrase.

minus	8 minus 5	8 − 5
less	9 less 3	9 − 3
less than	2 less than 7	7 − 2
the difference between	the difference between 8 and 2	8 − 2
decreased by	5 decreased by 1	5 − 1

➡ Find the difference between 1234 and 485 and check.

From the phrases that indicate subtraction, the difference between 1234 and 485 is 1234 − 485.

```
      2 14         1 12 14       0 11 12 14              11
  1 2 3 4        1 2  3  4       1  2  3  4    Check:   485
−   4 8 5      −   4  8  5    −     4  8  5           + 749
        9             4  9          7  4  9            1234
```

Subtraction with a zero in the minuend involves repeated borrowing.

➡ Subtract: 3904 − 1775

```
                      9                 9
    8 10            8 10 14           8 10 14
  3 9 0 4         3 9  0  4         3 9  0  4
− 1 7 7 5       − 1 7  7  5       − 1 7  7  5
                                    2 1  2  9
```

5 > 4
There is a 0 in the tens' column. Borrow 1 hundred (= 10 tens) from the hundreds' column and write 10 in the tens' column.

Borrow 1 ten from the tens' column and add 10 to the 4 in the ones' column.

Subtract the digits in each column.

Example 3 Subtract 4392 − 678 and check.

Solution

```
 3 13  8 12
 4  3  9  2    Check:    678
 −  6  7  8           + 3714
 ----------           ------
 3  7  1  4             4392
```

You Try It 3 Subtract 3481 − 865 and check.

Your solution 2616

Example 4 Find 23,954 less than 63,221 and check.

Solution

```
  5 12  11 11 11
  6  3 , 2  2  1   Check:   23,954
− 2  3 , 9  5  4          + 39,267
----------------          --------
  3  9 , 2  6  7            63,221
```

You Try It 4 Find 54,562 decreased by 14,485 and check.

Your solution 40,077

Example 5 Subtract 46,005 − 32,167 and check.

Solution

```
   5 10
  4 6,0 0 5
− 3 2,1 6 7
```

• There are two zeros in the minuend. Borrow 1 thousand from the thousands' column and write 10 in the hundreds' column.

```
     9
   5 10 10
  4 6,0 0 5
− 3 2,1 6 7
```

• Borrow 1 hundred from the hundreds' column and write 10 in the tens' column.

```
     9  9
   5 10 10 15
  4 6,0 0 5
− 3 2,1 6 7
-----------
  1 3,8 3 8
```

• Borrow 1 ten from the tens' column and add 10 to the 5 in the ones' column.

```
Check:   32,167
       + 13,838
       --------
         46,005
```

You Try It 5 Subtract 64,003 − 54,936 and check.

Your solution 9067

Solutions on p. A8

ESTIMATION

Estimating the Difference Between Two Whole Numbers

Estimate and then use your calculator to find 323,502 − 28,912.

To estimate the difference between two numbers, round each number to the same place value. In this case we will round to the nearest ten-thousand. Then subtract. The estimated answer is 290,000.

```
  323,502 ≈    320,000
−  28,912 ≈  −  30,000
             ---------
               290,000
```

Now use your calculator to find the exact result. The exact answer is 294,590.

323502 [−] 28912 [=] 294590

Objective C To solve application problems

The table at the right shows the number and value of corporate takeovers during the years 1987 through 1993. Use the table for Example 6 and You Try It 6.

Takeover deals are bouncing back this year. Number and value of deals each year through August 20:

	Number of Deals	Value (billions)
1987	1891	$135
1988	2447	$199
1989	3509	$199
1990	3609	$122
1991	3301	$ 88
1992	3465	$ 82
1993	3809	$135

Example 6
How many more corporate takeovers occurred in 1993 than in 1987?

Strategy
To find the number, subtract the number of takeovers that occurred in 1987 (1891) from the number that occurred in 1993 (3809).

Solution

$$\begin{array}{r} 3809 \\ -\ 1891 \\ \hline 1918 \end{array}$$

1918 more takeovers occurred in 1993.

You Try It 6
Find the difference between the value of the takeovers that occurred in 1992 and those that occurred in 1993.

Your strategy

Your solution $53 billion

Example 7
You had a balance of $815 in your checking account. You then wrote checks in the amount of $112 for taxes, $57 for food, and $39 for shoes. What is your new checking account balance?

Strategy
To find your new checking account balance:

- Add to find the total of the three checks ($112 + $57 + $39).
- Subtract the total of the three checks from the old balance ($815).

Solution

$$\begin{array}{r} 112 \\ 57 \\ +\ 39 \\ \hline 208 \end{array} \text{ total of checks} \qquad \begin{array}{r} 815 \\ -\ 208 \\ \hline 607 \end{array}$$

Your new checking account balance is $607.

You Try It 7
Your total salary is $638. Deductions of $127 for taxes, $18 for insurance, and $35 for savings are taken from your pay. Find your take-home pay.

Your strategy

Your solution $458

Solutions on p. A8

1.3 Exercises

Objective A

Subtract.

1. $9 - 5$ 4 **2.** $8 - 7$ 1 **3.** $8 - 4$ 4 **4.** $7 - 3$ 4 **5.** $10 - 0$ 10 **6.** $7 - 0$ 7

7. $11 - 4$ 7 **8.** $12 - 8$ 4 **9.** $19 - 8$ 11 **10.** $15 - 6$ 9 **11.** $16 - 7$ 9 **12.** $18 - 9$ 9

13. $25 - 3$ 22 **14.** $55 - 4$ 51 **15.** $68 - 8$ 60 **16.** $77 - 3$ 74 **17.** $89 - 23$ 66 **18.** $68 - 43$ 25

19. $54 - 21$ 33 **20.** $88 - 57$ 31 **21.** $1202 - 701$ 501 **22.** $1305 - 404$ 901 **23.** $1763 - 801$ 962

24. $1497 - 706$ 791 **25.** $8974 - 3972$ 5002 **26.** $2836 - 1711$ 1125 **27.** $8976 - 7463$ 1513 **28.** $9273 - 6142$ 3131

29. 17 − 8 9 **30.** 15 − 7 8 **31.** 12 − 5 7 **32.** 93 − 71 22 **33.** 83 − 52 31

34. 77 − 36 41 **35.** 129 − 82 47 **36.** 132 − 61 71 **37.** 969 − 44 925 **38.** 1347 − 103 1244

39. 4865 − 304 4561 **40.** 1525 − 702 823 **41.** 9999 − 6794 3205 **42.** 7806 − 3405 4401 **43.** 8843 − 7621 1222

44. What is 3795 minus 1092? 2703

45. What is 9071 minus 6050? 3021

46. Find the difference between 9763 and 541. 9222

47. Find the difference between 6094 and 3072. 3022

48. What is 3701 less than 6932? 3231

49. What is 2031 less than 5071? 3040

50. Find 6509 decreased by 3102. 3407

51. Find 7994 decreased by 7782. 212

52. Find 23,907 less 12,705. 11,202

53. Find 65,986 less 5741. 60,245

Objective B

Subtract.

54. $\begin{array}{r} 71 \\ -\ 18 \\ \hline \end{array}$ 53

55. $\begin{array}{r} 93 \\ -\ 28 \\ \hline \end{array}$ 65

56. $\begin{array}{r} 47 \\ -\ 18 \\ \hline \end{array}$ 29

57. $\begin{array}{r} 44 \\ -\ 27 \\ \hline \end{array}$ 17

58. $\begin{array}{r} 71 \\ -\ 67 \\ \hline \end{array}$ 4

59. $\begin{array}{r} 37 \\ -\ 29 \\ \hline \end{array}$ 8

60. $\begin{array}{r} 50 \\ -\ 27 \\ \hline \end{array}$ 23

61. $\begin{array}{r} 70 \\ -\ 33 \\ \hline \end{array}$ 37

62. $\begin{array}{r} 993 \\ -\ 537 \\ \hline \end{array}$ 456

63. $\begin{array}{r} 681 \\ -\ 328 \\ \hline \end{array}$ 353

64. $\begin{array}{r} 250 \\ -\ 192 \\ \hline \end{array}$ 58

65. $\begin{array}{r} 840 \\ -\ 783 \\ \hline \end{array}$ 57

66. $\begin{array}{r} 768 \\ -\ 194 \\ \hline \end{array}$ 574

67. $\begin{array}{r} 679 \\ -\ 519 \\ \hline \end{array}$ 160

68. $\begin{array}{r} 770 \\ -\ 395 \\ \hline \end{array}$ 375

69. 674 − 337
337

70. 3526 − 387
3139

71. 1712 − 289
1423

72. 4350 − 729
3621

73. 1702 − 948
754

74. 1607 − 869
738

75. 5933 − 3754
2179

76. 7293 − 3748
3545

77. 8143 − 2417
5726

78. 7236 − 1978
5258

79. 5714 − 2367
3347

80. 8462 − 3575
4887

81. 9407 − 2918
6489

82. 3706 − 2957
749

83. 8605 − 7716
889

84. 8052 − 2709
5343

85. 80,305 − 9176
71,129

86. 70,702 − 4239
66,463

87. 10,004 − 9306
698

88. 80,009 − 63,419
16,590

89. 70,618 − 41,213
29,405

90. 80,053 − 27,649
52,404

91. 70,700 − 21,076
49,624

92. 80,800 − 42,023
38,777

93. $\begin{array}{r} 2600 \\ -\ 1972 \\ \hline \end{array}$ 628

94. $\begin{array}{r} 8400 \\ -\ 3762 \\ \hline \end{array}$ 4638

95. $\begin{array}{r} 9003 \\ -\ 2471 \\ \hline \end{array}$ 6532

96. $\begin{array}{r} 6004 \\ -\ 2392 \\ \hline \end{array}$ 3612

97. $\begin{array}{r} 8202 \\ -\ 3916 \\ \hline \end{array}$ 4286

98. $\begin{array}{r} 7050 \\ -\ 4137 \\ \hline \end{array}$ 2913

99. $\begin{array}{r} 7015 \\ -\ 2973 \\ \hline \end{array}$ 4042

100. $\begin{array}{r} 4207 \\ -\ 1624 \\ \hline \end{array}$ 2583

101. $\begin{array}{r} 7005 \\ -\ 1796 \\ \hline \end{array}$ 5209

102. $\begin{array}{r} 8003 \\ -\ 2735 \\ \hline \end{array}$ 5268

103. $\begin{array}{r} 20{,}005 \\ -\ 9{,}627 \\ \hline \end{array}$ 10,378

104. $\begin{array}{r} 80{,}004 \\ -\ 8{,}237 \\ \hline \end{array}$ 71,767

105. Find 10,051 less 9027. 1024

106. Find 17,031 less 5792. 11,239

107. Find the difference between 1003 and 447. 556

108. Find the difference between 9000 and 763. 8237

109. What is 223,491 minus 96,987? 126,504

110. What is 29,874 minus 21,392? 8482

111. What is 29,797 less than 68,005? 38,208

112. What is 69,379 less than 70,004? 625

113. What is 25,432 decreased by 7994? 17,438

114. What is 86,701 decreased by 9976? 76,725

Estimate by rounding to the nearest ten-thousand. Then use your calculator to subtract.

115. 80,032 − 19,605
Est.: 60,000
Cal.: 60,427

116. 90,765 − 60,928
Est.: 30,000
Cal.: 29,837

117. 32,574 − 10,961
Est.: 20,000
Cal.: 21,613

118. 96,430 − 59,762
Est.: 40,000
Cal.: 36,668

119. 567,423 − 208,444
Est.: 360,000
Cal.: 358,979

120. 300,712 − 198,714
Est.: 100,000
Cal.: 101,998

Objective C Application Problems

121. You have $304 in your checking account. If you write a check for $139, how much is left in your checking account? $165

122. Sam Akyol has $521 in a checking account. If Sam writes a check for $247, how much is left in the checking account? $274

123. The tennis coach at a high school purchased a video camera that costs $1079 and makes a down payment of $180. Find the amount that remains to be paid. $899

124. Rod Guerra, an engineer, purchased a used car that cost $5225 and made a down payment of $450. Find the amount that remains to be paid. $4775

125. The odometer of a rental car read 3,459 miles after a trip of 297 miles. What was the odometer reading at the start of the trip? 3162 miles

126. At the end of a vacation trip, the odometer of your car read 63,459 miles. If the odometer reading at the start of the trip was 62,963, what was the length of the trip? 496 miles

127. Florida had 20,188,506 acres of wetlands 200 years ago. Today Florida has 10,901,793 acres of wetlands. How many acres of wetlands has Florida lost over the last 200 years? 9,286,713 acres

128. The seating capacity of the Kingdome in Seattle is 57,748. The seating capacity of Fenway Park in Boston is 34,182. Find the difference between the seating capacity of the Kingdome and that of Fenway Park. 23,566

The graph at the right shows the number of months required to recover all interest and principal from the Social Security fund after a person retires. Use this graph for Exercises 129 and 130.

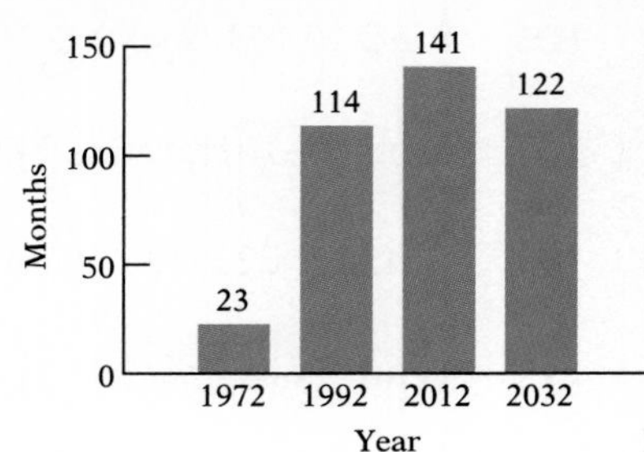

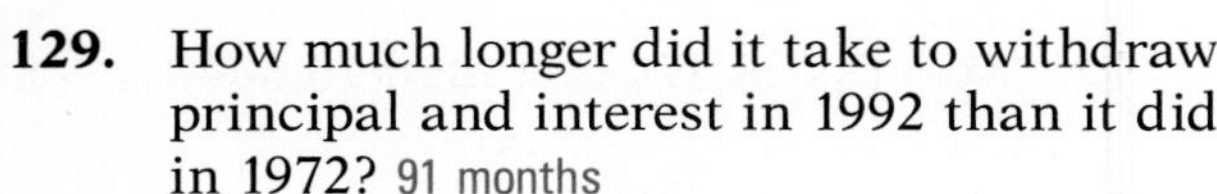

129. How much longer did it take to withdraw principal and interest in 1992 than it did in 1972? 91 months

130. How much longer will it take to withdraw principal and interest in 2012 than it did in 1992? 27 months

APPLYING THE CONCEPTS

131. Answer true or false.
a. The phrases "the difference between 9 and 5" and "5 less than 9" mean the same thing. True
b. $9 - (5 - 3) = (9 - 5) - 3$. False
c. Subtraction is an associative operation. *Hint*: See part b of this exercise. False

132. [W] Explain how you can check the answer to a subtraction problem. Answers will vary

133. [W] Make up a word problem for which the difference between 15 and 8 is the answer. Answers will vary

The table at the right shows the cost of two cars and of the various options the car buyer wants to have. Use this table for Exercises 134 and 135.

	Ford	*Chevrolet*
Basic car	$8589	$8799
Air	1025	978
Anti-lock brakes	675	533
Stereo AM/FM	425	515
Power windows	915	1025

134. Which of these two cars, equipped with all the options, costs more? Chevrolet

135. What is the difference in cost between the two basic cars? $210

The data in the table at the right show the approximate total average score on the Scholastic Aptitude Test (SAT) in the United States for a five-year period. Use this table for Exercises 136 and 137.

Year	*Score*
1989	903
1990	899
1991	896
1992	898
1993	902

136. Between which two consecutive years did the SAT scores decrease the most? What was the amount of the decrease? 1989–1990 4 points

137. Between which two years did the SAT scores increase the most? What was the amount of the increase? 1992–1993 4 points

1.4 Multiplication of Whole Numbers

Objective A *To multiply a number by a single digit*

Six boxes of toasters are ordered. Each box contains eight toasters. How many toasters are ordered?

This problem can be worked by adding 6 eights.

$8 + 8 + 8 + 8 + 8 + 8 = 48$

This problem involves repeated addition of the same number and can be worked by a shorter process called **multiplication.** Multiplication is the repeated addition of the same number.

$8 + 8 + 8 + 8 + 8 + 8 = 48$

The numbers that are multiplied are called **factors.** The answer is called the **product.**

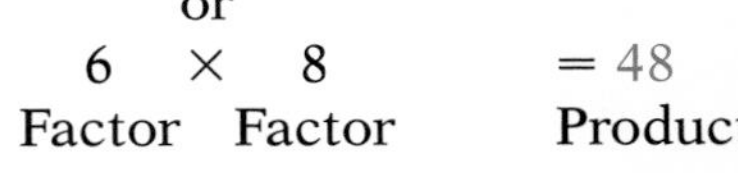

or

6	×	8	= 48
Factor		Factor	Product

The product of 6×8 can be represented on the number line. The arrow representing the whole number 8 is repeated 6 times. The result is the arrow representing 48.

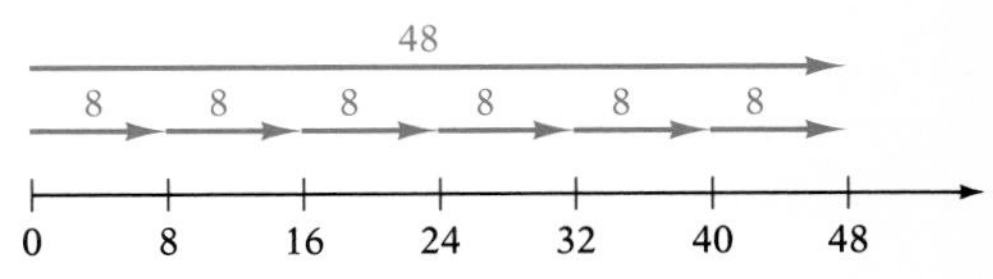

The times sign "×" is one symbol that is used to mean multiplication. Another common symbol used is a dot placed between the numbers.

$$7 \times 8 = 56 \qquad 7 \cdot 8 = 56$$

As with addition, there are some useful properties of multiplication.

Multiplication Property of Zero

The product of a number and zero is zero.

$0 \times 4 = 0$
$7 \times 0 = 0$

Multiplication Property of One

The product of a number and one is the number.

$1 \times 6 = 6$
$8 \times 1 = 8$

Commutative Property of Multiplication

Two numbers can be multiplied in either order. The product will be the same.

$4 \times 3 = 3 \times 4$
$12 = 12$

Associative Property of Multiplication

Grouping the numbers to be multiplied in any order gives the same result. Do the multiplication inside the parentheses first.

$\underbrace{(4 \times 2)}_{8} \times 3 = 4 \times \underbrace{(2 \times 3)}_{6}$
$8 \times 3 = 4 \times 6$
$24 = 24$

The basic facts for multiplying one-digit numbers are listed for your review on page A2 in the Appendix. Multiplication of larger numbers requires the repeated use of the basic multiplication facts.

➡ Multiply: 37 × 4

INSTRUCTOR NOTE
Here is an extra credit problem that students might enjoy:

Solve: STRAW × 4 = WARTS

Each letter represents the same digit. Ans: 21,978 × 4 = 87,912

$$\begin{array}{r} {}^{2}\\ 37 \\ \times\ 4 \\ \hline 8 \end{array}$$

- 4 × 7 = 28 (2 tens + 8 ones). Write the 8 in the ones' column and carry the 2 to the tens' column.

$$\begin{array}{r} {}^{2}\\ 37 \\ \times\ 4 \\ \hline 148 \end{array}$$

- The 3 in 37 is 3 tens.
  4 × 3 tens = 12 tens
  Add the carry digit. + 2 tens
  14 tens
- Write the 14.

The phrases below are used to indicate the operation of multiplication. An example is shown at the right of each phrase.

times	7 times 3	7 · 3
the product of	the product of 6 and 9	6 · 9
multiplied by	8 multiplied by 2	2 · 8

Example 1 Multiply: 735 × 9

Solution

$$\begin{array}{r} {}^{3\,4}\\ 735 \\ \times\ 9 \\ \hline 6615 \end{array}$$

You Try It 1 Multiply: 648 × 7

Your solution 4536

Solution on p. A8

Objective B To multiply larger whole numbers

Note the pattern when the following numbers are multiplied.

Multiply the nonzero part of the factors.

Now attach the same number of zeros to the product as the total number of zeros in the factors.

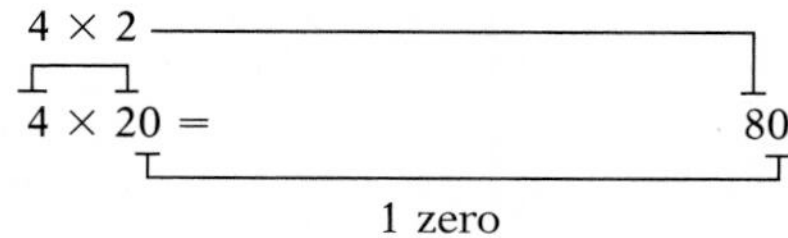

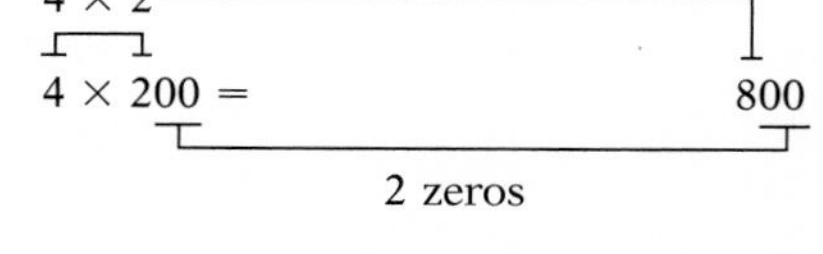

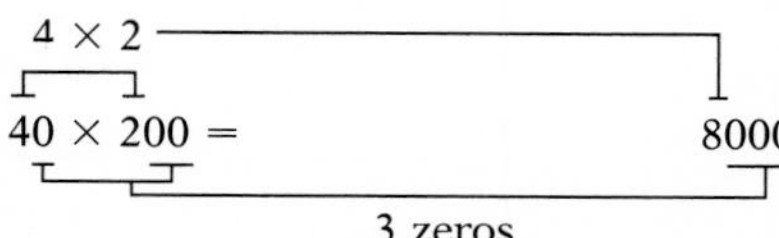

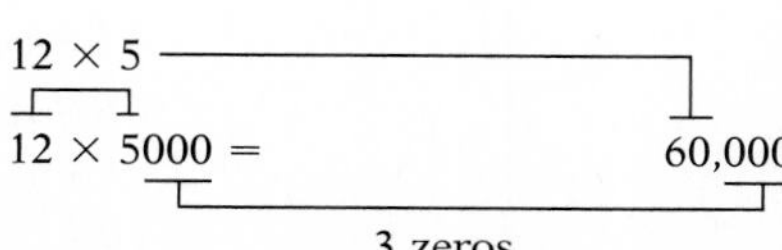

➡ Find the product of 47 and 23.

Multiply by the ones' digit.	Multiply by the tens' digit.	Add.
47 × 23 141 (= 47 × 3)	47 × 23 141 940 (= 47 × 20) Writing the 0 is optional.	47 × 23 141 94 1081

The place-value chart illustrates the placement of the products.

Note the placement of the products when we are multiplying by a factor that contains a zero.

➡ Multiply: 439 × 206

```
   439
 × 206
  2634
  000      0 × 439
 878
 90,434
```

When working the problem, we usually write only one zero. Writing this zero ensures the proper placement of the products.

```
   439
 × 206
  2634
 8780
 90,434
```

Example 2 Find 829 multiplied by 603.

Solution

```
   829
 × 603
  2487
 49740
499,887
```

You Try It 2 Multiply: 756 × 305

Your solution 230,580

Solution on p. A8

ESTIMATION

Estimating the Product of Two Whole Numbers

Estimate and then use your calculator to find 3267 × 389.

To estimate a product, round each number so that all the digits are zero except the first digit. Then multiply. The estimated answer is 1,200,000.

$$\begin{array}{rcr} 3267 & \approx & 3000 \\ \times\ 389 & \approx & \times\ \ 400 \\ & & 1{,}200{,}000 \end{array}$$

Now use your calculator to find the exact answer. The exact answer is 1,270,863.

3267 [×] 389 [=] 1270863

Objective C To solve application problems

Example 3
An auto mechanic receives a salary of \$525 each week. How much does the auto mechanic earn in 4 weeks?

Strategy
To find the mechanic's earnings for 4 weeks, multiply the weekly salary (\$525) by the number of weeks (4).

Solution

$$\begin{array}{r} \$525 \\ \times \quad 4 \\ \hline \$2100 \end{array}$$

The mechanic earns \$2100 in 4 weeks.

You Try It 3
A new-car dealer receives a shipment of 37 cars each month. Find the number of cars the dealer will receive in 12 months.

Your strategy

Your solution 444 cars

Example 4
A press operator earns \$320 for working a 40-hour week. This week the press operator also worked 7 hours of overtime at \$13 an hour. Find the press operator's total pay for the week.

Strategy
To find the press operator's income for the week:
- Find the overtime pay by multiplying the hours of overtime (7) by the overtime rate of pay (\$13).
- Add the weekly salary (\$320) to the overtime pay.

Solution

$$\begin{array}{r} \$13 \\ \times \quad 7 \\ \hline \$91 \end{array} \text{ overtime pay} \qquad \begin{array}{r} \$320 \\ + \quad 91 \\ \hline \$411 \end{array}$$

The press operator earned \$411 this week.

You Try It 4
The buyer for Ross Department Store can buy 80 men's suits for \$4800. Each sports jacket will cost the store \$23. The manager orders 80 men's suits and 25 sports jackets. What is the total cost of the order?

Your strategy

Your solution \$5375

Solutions on p. A9

1.4 Exercises

Objective A

Multiply.

1. 3 × 4 = 12	**2.** 2 × 8 = 16	**3.** 5 × 7 = 35	**4.** 6 × 4 = 24	**5.** 5 × 5 = 25
6. 7 × 7 = 49	**7.** 0 × 7 = 0	**8.** 8 × 0 = 0	**9.** 8 × 9 = 72	**10.** 7 × 6 = 42
11. 66 × 3 = 198	**12.** 70 × 4 = 280	**13.** 67 × 5 = 335	**14.** 127 × 9 = 1143	**15.** 623 × 4 = 2492
16. 802 × 5 = 4010	**17.** 607 × 9 = 5463	**18.** 300 × 5 = 1500	**19.** 600 × 7 = 4200	**20.** 906 × 8 = 7248
21. 703 × 9 = 6327	**22.** 127 × 5 = 635	**23.** 632 × 3 = 1896	**24.** 559 × 4 = 2236	**25.** 632 × 8 = 5056
26. 780 × 7 = 5460	**27.** 690 × 5 = 3450	**28.** 465 × 4 = 1860	**29.** 382 × 7 = 2674	**30.** 367 × 3 = 1101
31. 524 × 4 = 2096	**32.** 337 × 5 = 1685	**33.** 841 × 6 = 5046	**34.** 6709 × 7 = 46,963	**35.** 3608 × 5 = 18,040
36. 8568 × 7 = 59,976	**37.** 5495 × 4 = 21,980	**38.** 4780 × 4 = 19,120	**39.** 3690 × 5 = 18,450	**40.** 9895 × 2 = 19,790

41. Find the product of 5, 7, and 4. 140

42. Find the product of 6, 2, and 9. 108

43. Find the product of 5304 and 9. 47,736

44. Find the product of 458 and 8. 3664

45. What is 3208 multiplied by 7? 22,456

46. What is 5009 multiplied by 4? 20,036

47. What is 3105 times 6? 18,630

48. What is 8957 times 8? 71,656

Objective B

Multiply.

49. $\begin{array}{r} 16 \\ \times\ 21 \\ \hline \end{array}$ 336

50. $\begin{array}{r} 18 \\ \times\ 24 \\ \hline \end{array}$ 432

51. $\begin{array}{r} 35 \\ \times\ 26 \\ \hline \end{array}$ 910

52. $\begin{array}{r} 27 \\ \times\ 72 \\ \hline \end{array}$ 1944

53. $\begin{array}{r} 39 \\ \times\ 46 \\ \hline \end{array}$ 1794

54. $\begin{array}{r} 37 \\ \times\ 25 \\ \hline \end{array}$ 925

55. $\begin{array}{r} 67 \\ \times\ 23 \\ \hline \end{array}$ 1541

56. $\begin{array}{r} 95 \\ \times\ 33 \\ \hline \end{array}$ 3135

57. $\begin{array}{r} 693 \\ \times\ 91 \\ \hline \end{array}$ 63,063

58. $\begin{array}{r} 581 \\ \times\ 72 \\ \hline \end{array}$ 41,832

59. $\begin{array}{r} 419 \\ \times\ 80 \\ \hline \end{array}$ 33,520

60. $\begin{array}{r} 727 \\ \times\ 60 \\ \hline \end{array}$ 43,620

61. $\begin{array}{r} 8279 \\ \times\ 46 \\ \hline \end{array}$ 380,834

62. $\begin{array}{r} 9577 \\ \times\ 35 \\ \hline \end{array}$ 335,195

63. $\begin{array}{r} 6938 \\ \times\ 78 \\ \hline \end{array}$ 541,164

64. $\begin{array}{r} 8875 \\ \times\ 67 \\ \hline \end{array}$ 594,625

65. $\begin{array}{r} 7035 \\ \times\ 57 \\ \hline \end{array}$ 400,995

66. $\begin{array}{r} 6702 \\ \times\ 48 \\ \hline \end{array}$ 321,696

67. $\begin{array}{r} 3009 \\ \times\ 35 \\ \hline \end{array}$ 105,315

68. $\begin{array}{r} 6003 \\ \times\ 57 \\ \hline \end{array}$ 342,171

69. $\begin{array}{r} 809 \\ \times\ 530 \\ \hline \end{array}$ 428,770

70. $\begin{array}{r} 607 \\ \times\ 460 \\ \hline \end{array}$ 279,220

71. $\begin{array}{r} 800 \\ \times\ 325 \\ \hline \end{array}$ 260,000

72. $\begin{array}{r} 700 \\ \times\ 274 \\ \hline \end{array}$ 191,800

73. $\begin{array}{r} 987 \\ \times\ 349 \\ \hline \end{array}$ 344,463

74. $\begin{array}{r} 688 \\ \times\ 674 \\ \hline \end{array}$ 463,712

75. $\begin{array}{r} 312 \\ \times\ 134 \\ \hline \end{array}$ 41,808

76. $\begin{array}{r} 423 \\ \times\ 427 \\ \hline \end{array}$ 180,621

77. $\begin{array}{r} 379 \\ \times\ 500 \\ \hline \end{array}$ 189,500

78. $\begin{array}{r} 684 \\ \times\ 700 \\ \hline \end{array}$ 478,800

79. $\begin{array}{r} 985 \\ \times\ 408 \\ \hline \end{array}$ 401,880

80. $\begin{array}{r} 758 \\ \times\ 209 \\ \hline \end{array}$ 158,422

81. $\begin{array}{r} 3407 \\ \times\ 309 \\ \hline \end{array}$ 1,052,763

82. $\begin{array}{r} 5207 \\ \times\ 902 \\ \hline \end{array}$ 4,696,714

83. $\begin{array}{r} 4258 \\ \times\ 986 \\ \hline \end{array}$ 4,198,388

84. $\begin{array}{r} 6327 \\ \times\ 876 \\ \hline \end{array}$ 5,542,452

85. What is 5763 times 45? 259,335

86. What is 7349 times 7? 51,443

87. Find the product of 2, 19, and 34. 1292

88. Find the product of 6, 73, and 43. 18,834

89. What is 376 multiplied by 402? 151,152

90. What is 842 multiplied by 309? 260,178

91. Find the product of 233,489 times 3005. 701,634,445

92. Find the product of 34,985 times 9007. 315,109,895

Estimate and then use your calculator to multiply.

93. 8745 × 63
Est.: 540,000
Cal.: 550,935

94. 4732 × 93
Est.: 450,000
Cal.: 440,076

95. 39,246 × 29
Est.: 1,200,000
Cal.: 1,138,134

96. 64,409 × 67
Est.: 4,200,000
Cal.: 4,315,403

97. 2937 × 206
Est.: 600,000
Cal.: 605,022

98. 8941 × 726
Est.: 6,300,000
Cal.: 6,491,166

99. 3097 × 1025
Est.: 3,000,000
Cal.: 3,174,425

100. 6379 × 2936
Est.: 18,000,000
Cal.: 18,728,744

101. 32,508 × 591
Est.: 18,000,000
Cal.: 19,212,228

102. 62,504 × 923
Est.: 54,000,000
Cal.: 57,691,192

103. 81,405 × 902
Est.: 72,000,000
Cal.: 73,427,310

104. 66,735 × 844
Est.: 56,000,000
Cal.: 56,324,340

Objective C Application Problems

105. Rob Hupp owns a compact car that averages 43 miles on 1 gallon of gas. How many miles could the car travel on 12 gallons of gas?
516 miles

106. A plane flying from Los Angeles to Boston uses 865 gallons of jet fuel each hour. How many gallons of jet fuel were used on a 6-hour flight? 5190 gallons

107. Assume that a machine at the Coca Cola Bottling Company can fill and cap 4200 bottles of Coke in 1 hour. How many bottles of Coke can the machine fill and cap in 40 hours? 168,000 bottles

108. A computer can store 368,640 bytes of information on 1 disk. How many bytes of information can be stored on 6 disks?
2,211,840 bytes

109. A computer graphics screen has 640 rows of pixels, and there are 480 pixels per row. Find the total number of pixels on the screen.
307,200 pixels

110. Anthony Dias owns 350 shares of Public Service of Colorado that has a yearly dividend of $2 a share paid in 4 quarterly payments. Find the yearly income from the 350 shares of stock. $700

111. A food server buys a car by making payments of $235 each month for 48 months. Find the total cost of the car. $11,280

A lighting consultant to a bank suggests that the bank lobby contain 43 can lights, 15 high-intensity lights, 20 fire safety lights, and one chandelier. The table at the right gives the costs for each type of light from two companies. Use this table for Exercises 112 and 113.

	Company A	*Company B*
Can Lights	$2 each	$3 each
High-Intensity	$6 each	$4 each
Fire Safety	$12 each	$11 each
Chandelier	$998 each	$1089 each

112. Which company offers the lighting designer the lights for the lower total price? Company A

113. How much can the lighting designer save by purchasing the lights from the company that offers the lower total price? $84

The table at the right shows the hourly wages of four different job classifications at a small construction company. Use this table for Exercises 114 to 117.

Type of Work	*Wage per Hour*
Electrician	$17
Plumber	$15
Clerk	$8
Bookkeeper	$10

114. The owner of this company wants to provide the electrical installation for a new house. Based on the architectural plans for the house, it is estimated that it will require 3 electricians each working 50 hours to complete the job. What is the estimated cost for the electricians' labor? $2550

115. Carlos Vasquez, a plumbing contractor, hires 4 plumbers from this company at the hourly wage given in the table. If each plumber works 23 hours, what are the total wages paid by Carlos? $1380

116. The owner of this company estimates that remodeling a kitchen will require 1 electrician working 30 hours and 1 plumber working 33 hours. This project also requires 3 hours of clerical work and 4 hours of bookkeeping. What is the total cost for these four components of this remodeling? $1069

117. An estimate to remodel a bathroom included 1 electrician working 5 hours and 1 plumber working 18 hours. There was 1 hour estimated for clerical work and 2 hours for bookkeeping. What is the total cost of these four components of this remodeling? $383

APPLYING THE CONCEPTS

118. Determine whether each of the following statements is always true, sometimes true, or never true.
a. A whole number times zero is zero. True
b. A whole number times one is the whole number. True
c. The product of two whole numbers is greater than either one of the whole numbers. Sometimes true

119. [W] According to the National Safety Council, in a recent year a death resulting from an accident occurred at the rate of 1 every 5 min. At this rate, how many accidental deaths occurred each hour? Each day? Throughout the year? Explain how you arrived at your answers. 12 deaths each hour 288 deaths each day 105,120 deaths each year

120. [W] In Brazil, 713 acres are deforested each hour. At this rate, how many acres of deforestation occur each day? Each month? Throughout the year? Explain how you arrived at your answers. 17,112 acres each day 513,360 acres each month 6,245,880 acres each year

121. [W] Pick your favorite number between 1 and 9. Multiply the number by 3. Multiply that product by 37037. How is the product related to your favorite number? Explain why this works. (Suggestion: Multiply 3 and 37037 first.) Answers will vary

122. There are quite a few tricks based on whole numbers. Here's one about birthdays. Write down the month in which you were born. Multiply by 5. Add 7. Multiply by 20. Subtract 100. Add the day of the month on which you were born. Multiply by 4. Subtract 100. Multiply by 25. Add the year you were born. Subtract 3400. The answer is the month/day/year of your birthday. Answers will vary

1.5 Division of Whole Numbers

Objective A ***To divide by a single digit with no remainder in the quotient***

Division is used to separate objects into equal groups.

A store manager wants to display 24 new objects equally on 4 shelves. From the diagram, we see that the manager would place 6 objects on each shelf.

The manager's division problem can be written as follows:

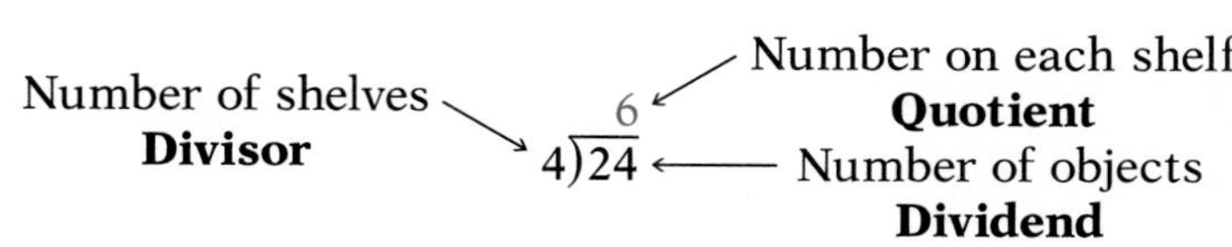

Note that the quotient multiplied by the divisor equals the dividend.

$\begin{array}{r}6\\4\overline{)24}\end{array}$	because	6 Quotient	×	4 Divisor	=	24 Dividend
$\begin{array}{r}6\\9\overline{)54}\end{array}$	because	6	×	9	=	54
$\begin{array}{r}5\\8\overline{)40}\end{array}$	because	5	×	8	=	40

Here are some important quotients and the properties of zero in division:

Important Quotients

Any whole number, except zero, divided by itself is 1. $\begin{array}{r}1\\8\overline{)8}\end{array}$ $\begin{array}{r}1\\14\overline{)14}\end{array}$ $\begin{array}{r}1\\10\overline{)10}\end{array}$

Any whole number divided by 1 is the whole number. $\begin{array}{r}9\\1\overline{)9}\end{array}$ $\begin{array}{r}27\\1\overline{)27}\end{array}$ $\begin{array}{r}10\\1\overline{)10}\end{array}$

INSTRUCTOR NOTE

One method to help students understand that division by zero is not allowed is to relate it to the problem of the store manager above. Ask how the manager can display 24 items on 4 shelves; 3 shelves; 2 shelves; 1 shelf; 0 shelves!

Properties of Zero in Division

Zero divided by any other whole number is zero. $\begin{array}{r}0\\7\overline{)0}\end{array}$ $\begin{array}{r}0\\13\overline{)0}\end{array}$ $\begin{array}{r}0\\10\overline{)0}\end{array}$

Division by zero is not allowed. $\begin{array}{r}\boxed{?}\\0\overline{)8}\end{array}$ There is no number whose product with 0 is 8.

When the dividend is a larger whole number, the digits in the quotient are found in steps.

➡ Divide $4\overline{)3192}$ and check.

$$\begin{array}{r} 7 \\ 4\overline{)3192} \\ -28 \\ \hline 39 \end{array}$$

- Think $4\overline{)31}$.
- Subtract 7 × 4.
- Bring down the 9.

$$\begin{array}{r} 79 \\ 4\overline{)3192} \\ -28 \\ \hline 39 \\ -36 \\ \hline 32 \end{array}$$

- Think $4\overline{)39}$.
- Subtract 9 × 4.
- Bring down the 2.

$$\begin{array}{r} 798 \\ 4\overline{)3192} \\ -28 \\ \hline 39 \\ -36 \\ \hline 32 \\ -32 \\ \hline 0 \end{array}$$

- Think $4\overline{)32}$.
- Subtract 8 × 4.

Check:
$$\begin{array}{r} 798 \\ \times4 \\ \hline 3192 \end{array}$$

The place-value chart can be used to show why this method works.

	HUNDREDS	TENS	ONES	
	7	9	8	
4) 3	1	9	2	
−2	8	0	0	7 hundreds × 4
	3	9	2	
	−3	6	0	9 tens × 4
		3	2	
		−3	2	8 ones × 4
			0	

There are other ways of expressing division.

54 divided by 9 equals 6.

54 ÷ 9 equals 6.

$\frac{54}{9}$ equals 6.

Example 1 Divide $7\overline{)56}$ and check.

Solution

$$\begin{array}{r} 8 \\ 7\overline{)56} \end{array}$$

Check: $8 \times 7 = 56$

You Try It 1 Divide $9\overline{)63}$ and check.

Your solution 7

Example 2 Divide $2808 \div 8$ and check.

Solution

$$\begin{array}{r} 351 \\ 8\overline{)\ 2808} \\ -24 \\ \hline 40 \\ -40 \\ \hline 08 \\ -\ 8 \\ \hline 0 \end{array}$$

Check: $351 \times 8 = 2808$

You Try It 2 Divide $4077 \div 9$ and check.

Your solution 453

Example 3 Divide $7\overline{)2856}$ and check.

Solution

$$\begin{array}{r} 408 \\ 7\overline{)\ 2856} \\ -28 \\ \hline 05 \\ -\ 0 \\ \hline 56 \\ -56 \\ \hline 0 \end{array}$$

- **Think $7\overline{)5}$.** **Place 0 in quotient.** **Subtract 0×7.** **Bring down the 6.**

Check: $408 \times 7 = 2856$

You Try It 3 Divide $9\overline{)6345}$ and check.

Your solution 705

Solutions on p. A9

Objective B To divide by a single digit with a remainder in the quotient

Occasionally it is not possible to separate objects into a whole number of equal groups.

A warehouse clerk must place 14 objects into 3 boxes. From the diagram, we see that the clerk would place 4 objects in each box and have 2 objects left over. The 2 is called the **remainder.**

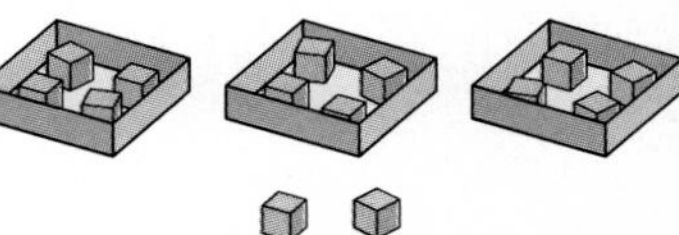

INSTRUCTOR NOTE
Some students have difficulty with the concept of remainder. Have these students try to give 15 pennies to 4 students so that each student has the same number of pennies.

The clerk's division problem could be written

	4 ↙	Quotient (Number in each box)
Divisor (Number of boxes) →	3) 14 ←	Dividend (Total number of objects)
	−12	
	2 ←	Remainder (Number left over)

The answer to a division problem with a remainder is frequently written

$$3\overline{)14}\ \ 4\ \text{r}2$$

Note that $\underset{\text{Quotient}}{4} \times \underset{\text{Divisor}}{3} + \underset{\text{Remainder}}{2} = \underset{\text{Dividend}}{14}$

Example 4 Divide $4\overline{)2522}$ and check.

Solution

```
    630 r2
4) 2522
  -24
    12
   -12
    02    • Think 4)2.
   - 0    • Place 0 in
     2      quotient.
            Subtract 0 × 4.
```

Check: (630 × 4) + 2 =
2520 + 2 = 2522

You Try It 4 Divide $6\overline{)5225}$ and check.

Your solution 870 r5

Example 5 Divide $9\overline{)27,438}$ and check.

Solution

```
    3,048 r6
9) 27,438
  -27
    0 4     • Think 9)4.
   -  0     • Subtract 0 × 9.
      43
     -36
       78
      -72
        6
```

Check: (3048 × 9) + 6 =
27,432 + 6 = 27,438

You Try It 5 Divide $7\overline{)21,409}$ and check.

Your solution 3058 r3

Solutions on pp. A9 and A10

Objective C To divide by larger whole numbers

When the divisor has more than one digit, estimate at each step by using the first digit of the divisor. If that product is too large, lower the guess by 1 and try again.

➡ Divide $34\overline{)1598}$ and check.

```
       5
34) 1598      • Think 3)15.
   -170       • Subtract 5 × 34.
```

170 is too large. Lower guess by 1 and try again.

```
       4
34) 1598
   -136       • Subtract 4 × 34.
     238
```

```
      47
34) 1598
   -136
     238      • Think 3)23.
    -238      • Subtract 7 × 34.
       0
```

Check:

```
    47
   ×34
   188
  141
  1598
```

The phrases below are used to indicate the operation of division. An example is shown at the right of each phrase.

the quotient of	the quotient of 9 and 3	9 ÷ 3
divided by	6 divided by 2	6 ÷ 2

Example 6 Find 7077 divided by 34 and check.

Solution

```
     208 r5
34) 7077
   -68
     27      • Think 34)27.
    - 0      • Place 0 in
     277       the quotient.
    -272     • Subtract 0 × 34.
       5
```

Check: $(208 \times 34) + 5 =$
$7072 + 5 = 7077$

You Try It 6 Divide 4578 ÷ 42 and check.

Your solution 109

Solution on p. A10

Example 7 Find the quotient of 21,312 and 56 and check.

Solution

```
      380 r32
56)21,312
  -16 8
    4 51
   -4 48
      32
     - 0
      32
```

- Think 5)21. 4 × 56 is too large. Try 3.

Check: (380 × 56) + 32 =
21,280 + 32 = 21,312

You Try It 7 Divide 18,359 ÷ 39 and check.

Your solution 470 r29

Example 8 Divide 427)24,782 and check.

Solution

```
       58 r16
427)24,782
   -21 35
     3 432
    -3 416
        16
```

Check: (58 × 427) + 16 =
24,766 + 16 = 24,782

You Try It 8 Divide 534)33,219 and check.

Your solution 62 r111

Example 9 Divide 386)206,149 and check.

Solution

```
        534 r25
386)206,149
   -193 0
     13 14
    -11 58
      1 569
     -1 544
         25
```

Check: (534 × 386) + 25 =
206,124 + 25 = 206,149

You Try It 9 Divide 515)216,848 and check.

Your solution 421 r33

Solutions on p. A10

ESTIMATION

Estimating the Quotient of Two Whole Numbers

Estimate and then use your calculator to find 36,936 ÷ 54.

To estimate a quotient, round each number so that all the digits are zero except the first digit. Then divide.

$36{,}936 \div 54 \approx$
$40{,}000 \div 50 = 800$

The estimated answer is 800.

Now use your calculator to find the exact answer.

36936 [÷] 54 [=] 684

The exact answer is 684.

Objective D *To solve application problems*

The *average* of the total of several events is the quotient of that total and the number of events. For instance, if the total of 6 of your test scores was 498, then

$$\text{average test score} = \frac{498}{6} = 83$$

➡ The table at the right shows the number of home dates and the total home attendance of the baseball teams in the National League. Find the average home attendance for the Chicago team. Round to the nearest whole number.

Team	Home Dates	Home Att.
Atlanta	69	3,296,967
Chicago	74	2,446,912
Cincinnati	69	2,189,722
Colorado	67	3,872,511
Florida	67	2,629,117
Houston	74	1,913,944
Los Angeles	71	2,796,973
Montreal	72	1,374,708
New York	70	1,708,606
Philadelphia	74	2,869,808
Pittsburgh	70	1,514,206
St. Louis	72	2,578,292
San Diego	73	1,277,268
San Francisco	73	2,362,524

Strategy

Determine, from the table, the total attendance and the number of home dates for the Chicago team. To find the average home attendance, divide the total attendance (2,446,912) by the number of home dates (74).

Solution

```
      33,066
74) 2,446,912
   -2 22
      226
     -222
        4 9
      -   0
        4 91
       -4 44
          472
         -444
           28
```

• When rounding to the nearest whole number, compare twice the remainder to the divisor. If twice the remainder is less than the divisor, drop the remainder. If twice the remainder is greater than or equal to the divisor, add 1 to the units digit of the quotient.

• Twice the remainder is $2 \times 28 = 56$. Because $56 < 74$, drop the remainder.

The average attendance to the nearest whole number is 33,066.

Example 10
Ngan Hui, a freight supervisor, shipped 192,600 bushels of wheat in 9 railroad cars. Find the amount of wheat shipped in each car?

Strategy
To find the amount of wheat shipped in each car, divide the number of bushels (192,600) by the number of cars (9)

Solution

$$\begin{array}{r} 21{,}400 \\ 9\overline{)\,192{,}600} \\ -18\phantom{0{,}600} \\ \hline 12\phantom{{,}600} \\ -9\phantom{{,}600} \\ \hline 36 \\ -36 \\ \hline 0 \end{array}$$

Each car carried 21,400 bushels of wheat.

You Try It 10
Suppose a Firestone retail outlet can store 270 tires on 15 shelves. How many tires can be stored on each shelf?

Strategy

Your solution 18 tires

Example 11
The car you are buying costs $11,216. A down payment of $2000 is required. The remaining balance is paid in 48 equal monthly payments. What is the monthly payment?

Strategy
To find the monthly payment:

- Find the remaining balance by subtracting the down payment ($2000) from the total cost of the car ($11,216).
- Divide the remaining balance by the number of equal monthly payments (48).

Solution

$$\begin{array}{r} 11{,}216 \\ -\ \ 2{,}000 \\ \hline 9{,}216 \end{array}$$

remaining balance

$$\begin{array}{r} 192 \\ 48\overline{)\,9216} \\ -48 \\ \hline 441 \\ -432 \\ \hline 96 \\ -96 \\ \hline 0 \end{array}$$

The monthly payment is $192.

You Try It 11
A soft-drink manufacturer produces 12,600 cans of soft drink each hour. Cans are packed 24 to a case. How many cases of soft drink are produced in 8 hours?

Your strategy

Your solution 4200 cases

Solutions on p. A11

1.5 Exercises

Objective A

Divide.

1. $\begin{array}{r}2\\4\overline{)8}\end{array}$
2. $\begin{array}{r}3\\3\overline{)9}\end{array}$
3. $\begin{array}{r}6\\6\overline{)36}\end{array}$
4. $\begin{array}{r}9\\9\overline{)81}\end{array}$
5. $\begin{array}{r}5\\5\overline{)25}\end{array}$
6. $\begin{array}{r}7\\7\overline{)49}\end{array}$
7. $\begin{array}{r}16\\5\overline{)80}\end{array}$
8. $\begin{array}{r}16\\6\overline{)96}\end{array}$
9. $\begin{array}{r}80\\6\overline{)480}\end{array}$
10. $\begin{array}{r}70\\9\overline{)630}\end{array}$
11. $\begin{array}{r}210\\4\overline{)840}\end{array}$
12. $\begin{array}{r}230\\3\overline{)690}\end{array}$
13. $\begin{array}{r}44\\7\overline{)308}\end{array}$
14. $\begin{array}{r}29\\7\overline{)203}\end{array}$
15. $\begin{array}{r}206\\7\overline{)1442}\end{array}$
16. $\begin{array}{r}703\\9\overline{)6327}\end{array}$
17. $\begin{array}{r}530\\4\overline{)2120}\end{array}$
18. $\begin{array}{r}910\\8\overline{)7280}\end{array}$
19. $\begin{array}{r}902\\9\overline{)8118}\end{array}$
20. $\begin{array}{r}908\\8\overline{)7264}\end{array}$
21. $\begin{array}{r}21{,}560\\3\overline{)64{,}680}\end{array}$
22. $\begin{array}{r}12{,}690\\4\overline{)50{,}760}\end{array}$
23. $\begin{array}{r}3580\\6\overline{)21{,}480}\end{array}$
24. $\begin{array}{r}3610\\5\overline{)18{,}050}\end{array}$
25. Find the quotient of 1446 and 3. 482
26. Find the quotient of 4123 and 7. 589
27. What is 7525 divided by 7? 1075
28. What is 32,364 divided by 4? 8091

Objective B

Divide.

29. $\begin{array}{r}2 \text{ r1}\\4\overline{)9}\end{array}$
30. $\begin{array}{r}3 \text{ r1}\\2\overline{)7}\end{array}$
31. $\begin{array}{r}5 \text{ r2}\\5\overline{)27}\end{array}$
32. $\begin{array}{r}9 \text{ r7}\\9\overline{)88}\end{array}$
33. $\begin{array}{r}13 \text{ r1}\\3\overline{)40}\end{array}$
34. $\begin{array}{r}16 \text{ r1}\\6\overline{)97}\end{array}$
35. $\begin{array}{r}10 \text{ r3}\\8\overline{)83}\end{array}$
36. $\begin{array}{r}10 \text{ r4}\\5\overline{)54}\end{array}$
37. $\begin{array}{r}90 \text{ r2}\\7\overline{)632}\end{array}$
38. $\begin{array}{r}90 \text{ r3}\\4\overline{)363}\end{array}$
39. $\begin{array}{r}230 \text{ r1}\\4\overline{)921}\end{array}$
40. $\begin{array}{r}120 \text{ r5}\\7\overline{)845}\end{array}$
41. $\begin{array}{r}204 \text{ r3}\\8\overline{)1635}\end{array}$
42. $\begin{array}{r}309 \text{ r3}\\5\overline{)1548}\end{array}$
43. $\begin{array}{r}1347 \text{ r3}\\7\overline{)9432}\end{array}$
44. $\begin{array}{r}1058 \text{ r4}\\6\overline{)6352}\end{array}$
45. $\begin{array}{r}778 \text{ r2}\\9\overline{)7004}\end{array}$
46. $\begin{array}{r}857 \text{ r2}\\7\overline{)6001}\end{array}$
47. $\begin{array}{r}391 \text{ r4}\\6\overline{)2350}\end{array}$
48. $\begin{array}{r}796 \text{ r2}\\8\overline{)6370}\end{array}$

49. $7\overline{)8124}$ 1160 r4

50. $3\overline{)5162}$ 1720 r2

51. $5\overline{)3542}$ 708 r2

52. $8\overline{)3274}$ 409 r2

53. $4\overline{)15,300}$ 3825

54. $7\overline{)43,500}$ 6214 r2

55. $8\overline{)72,354}$ 9044 r2

56. $5\overline{)43,542}$ 8708 r2

57. Find the quotient of 3107 and 8. 388 r3

58. Find the quotient of 8642 and 8. 1080 r2

59. What is 45,738 divided by 4? Round to the nearest ten. 11,430

60. What is 37,896 divided by 9? Round to the nearest hundred. 4200

61. What is 3572 divided by 7? Round to the nearest ten. 510

62. What is 78,345 divided by 4? Round to the nearest hundred. 19,600

Objective C

Divide.

63. $27\overline{)96}$ 3 r15

64. $44\overline{)82}$ 1 r38

65. $42\overline{)87}$ 2 r3

66. $67\overline{)93}$ 1 r26

67. $41\overline{)897}$ 21 r36

68. $32\overline{)693}$ 21 r21

69. $23\overline{)784}$ 34 r2

70. $25\overline{)772}$ 30 r22

71. $74\overline{)600}$ 8 r8

72. $92\overline{)500}$ 5 r40

73. $70\overline{)329}$ 4 r49

74. $50\overline{)467}$ 9 r17

75. $36\overline{)7225}$ 200 r25

76. $44\overline{)8821}$ 200 r21

77. $19\overline{)3859}$ 203 r2

78. $32\overline{)9697}$ 303 r1

79. $88\overline{)3127}$ 35 r47

80. $92\overline{)6177}$ 67 r13

81. $33\overline{)8943}$ 271

82. $27\overline{)4765}$ 176 r13

83. $22\overline{)98,654}$ 4484 r6

84. $77\overline{)83,629}$ 1086 r7

85. $64\overline{)38,912}$ 608

86. $78\overline{)31,434}$ 403

87. $206\overline{)3097}$ 15 r7

88. $504\overline{)6504}$ 12 r456

89. $654\overline{)1217}$ 1 r563

90. $546\overline{)2344}$ 4 r160

91. $169\overline{)8542}$ 50 r92

92. $456\overline{)7723}$ 16 r427

93. $223\overline{)8927}$ 40 r7

94. $467\overline{)9344}$ 20 r4

95. Find the quotient of 5432 and 21. 258 r14

96. Find the quotient of 8507 and 53. 160 r27

97. What is 37,294 divided by 72? 517 r70

98. What is 76,788 divided by 46? 1669 r14

99. Find 23,457 divided by 43. Round to the nearest hundred. 500

100. Find 341,781 divided by 43. Round to the nearest ten. 7950

Estimate and then use your calculator to divide.

101. $76\overline{)389{,}804}$ Est.: 5000 Cal.: 5129

102. $53\overline{)117{,}925}$ Est.: 2000 Cal.: 2225

103. $29\overline{)637{,}072}$ Est.: 20,000 Cal.: 21,968

104. $67\overline{)738{,}072}$ Est.: 10,000 Cal.: 11,016

105. $38\overline{)934{,}648}$ Est.: 22,500 Cal.: 24,596

106. $34\overline{)906{,}304}$ Est.: 30,000 Cal.: 26,656

107. $309\overline{)876{,}324}$ Est.: 3000 Cal.: 2836

108. $642\overline{)323{,}568}$ Est.: 500 Cal.: 504

109. $209\overline{)632{,}016}$ Est.: 3000 Cal.: 3024

110. $614\overline{)332{,}174}$ Est.: 500 Cal.: 541

111. $179\overline{)5{,}734{,}444}$ Est.: 30,000 Cal.: 32,036

112. $374\overline{)7{,}712{,}254}$ Est.: 20,000 Cal.: 20,621

Objective D *Application Problems*

113. Assume that the American Red Cross, the Boys Club, the Girls Club, and the North County Crisis Center collected $548,000 to promote and provide community services. What amount did each organization receive if the money was divided evenly? $137,000

114. An insurance agent drives a car with a 16-gallon gas tank. The agent used 128 gallons of gas in traveling 3456 miles. Find the average number of miles traveled on each gallon of gas. 27 miles per gallon

The graph at the right shows the maximum amount paid by an employee for Social Security benefits for the years shown. Use this graph for Exercises 115 to 118. In each case, assume that the employee makes the maximum contribution for the year given.

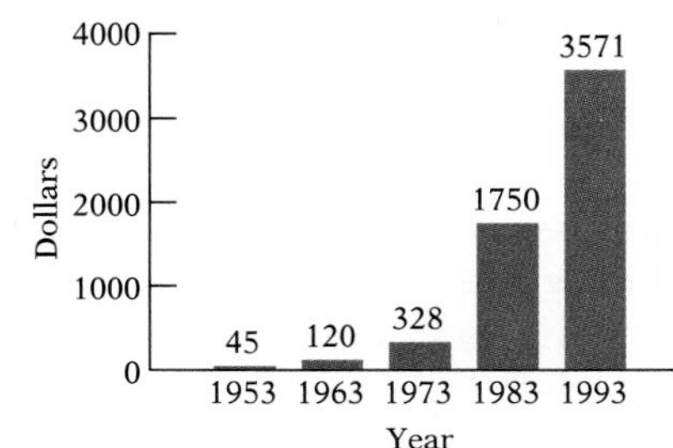

115. In 1993, what was the monthly contribution of an employee, to the nearest dollar? $298

116. What is the difference in monthly contribution, to the nearest dollar, between 1993 and 1983? $152

117. To the nearest whole number, how many times greater was the annual contribution in 1993 than the annual contribution in 1953? 79 times

118. If the maximum Social Security contribution for the years 1993 to 2003 increases by the same amount as it increased from 1983 to 1993, what will the maximum contribution be in 2003? $5392

119. A computer can store 2,211,840 bytes of information on 6 disks. How many bytes of information can be stored on 1 disk? 368,640 bytes

120. Ken Martinez, a computer analyst, received $5376 for working 168 hours on a computer consulting project. Find the hourly rate that Ken charged. $32

121. A family bought a living room set for $2856, which includes finance charges. This amount is to be repaid in 24 equal payments. Find the amount of each payment. $119

122. Marie Sarro, a design consultant, made a down payment of $1560 on a car costing $10,536.
a. What is the remaining balance to be paid? $8976
b. The balance is to be paid in 48 equal monthly payments. Find the monthly payment. $187

Each year *Forbes* magazine publishes what is called the *Forbes* Celebrity list. It is a list of the highest paid entertainers in the United States. Selected entries from the 1993 list are given at the right. Use this list for Exercises 123 to 126.

Name	*Annual Income*
Oprah Winfrey	$98,000,000
Steven Spielberg	$72,000,000
Bill Cosby	$66,000,000
Garth Brooks	$48,000,000
Madonna	$36,000,000

123. The average worker in the United States works approximately 2000 hours per year. Assuming that Oprah Winfrey worked 2000 hours in 1993, find her average hourly wage. $49,000 per hour

124. What is the difference between the average monthly salary of Steven Spielberg and that of Bill Cosby? $500,000

125. How many times more was Steven Spielberg's annual income than Madonna's annual income? 2 times

126. If Garth Brooks worked 1200 hours in 1993 and Bill Cosby worked 1500 hours that year, which of the two entertainers had the smaller hourly wage? Garth Brooks

APPLYING THE CONCEPTS

127. A palindromic number is a whole number that remains unchanged when its digits are written in reverse order. For instance, 292 is a palindromic number. Find the smallest three-digit palindromic number that is divisible by 4. 212

128. Find the smallest four-digit palindromic number (see Exercise 127) that is divisible by 8. (Helpful fact: A whole number whose last three digits are divisible by 8 is itself divisible by 8.) 2112

129. The number 10,981 is not divisible by 4. By rearranging the digits, find the largest possible number that is divisible by 4. 91,180

130. Determine whether each of the following statements is true or false.
a. Any whole number divided by zero is zero. False
b. $\frac{0}{0} = 1$ False
c. Zero divided by any whole number, except zero, is zero. True

131. Explain how the answer to a division problem can be checked.
Quotient × Divisor + Remainder = Dividend

1.6 Exponential Notation and the Order of Operations Agreement

Objective A *To simplify expressions that contain exponents*

Repeated multiplication of the same factor can be written in two ways:

$3 \cdot 3 \cdot 3 \cdot 3 \cdot 3$ or 3^5 ← **exponent**

The exponent indicates how many times the factor occurs in the multiplication. The expression 3^5 is in **exponential notation.**

It is important to be able to read numbers written in exponential notation.

$6 = 6^1$ is read "six to the first power" or just "six." Usually the exponent 1 is not written.

$6 \cdot 6 = 6^2$ is read "six squared" or "six to the second power."

$6 \cdot 6 \cdot 6 = 6^3$ is read "six cubed" or "six to the third power."

$6 \cdot 6 \cdot 6 \cdot 6 = 6^4$ is read "six to the fourth power."

$6 \cdot 6 \cdot 6 \cdot 6 \cdot 6 = 6^5$ is read "six to the fifth power."

Each place value in the place-value chart can be expressed as a power of 10.

$$\begin{aligned}
\text{Ten} &= 10 = 10 = 10^1\\
\text{Hundred} &= 100 = 10 \cdot 10 = 10^2\\
\text{Thousand} &= 1000 = 10 \cdot 10 \cdot 10 = 10^3\\
\text{Ten-Thousand} &= 10{,}000 = 10 \cdot 10 \cdot 10 \cdot 10 = 10^4\\
\text{Hundred-Thousand} &= 100{,}000 = 10 \cdot 10 \cdot 10 \cdot 10 \cdot 10 = 10^5\\
\text{Million} &= 1{,}000{,}000 = 10 \cdot 10 \cdot 10 \cdot 10 \cdot 10 \cdot 10 = 10^6
\end{aligned}$$

To simplify a numerical expression containing exponents, write each factor as many times as indicated by the exponent and carry out the indicated multiplication.

$4^3 = 4 \cdot 4 \cdot 4 = 64$

$2^2 \cdot 3^4 = (2 \cdot 2) \cdot (3 \cdot 3 \cdot 3 \cdot 3) = 4 \cdot 81 = 324$

Example 1 Write $3 \cdot 3 \cdot 3 \cdot 5 \cdot 5$ in exponential notation.

Solution $3 \cdot 3 \cdot 3 \cdot 5 \cdot 5 = 3^3 \cdot 5^2$

You Try It 1 Write $2 \cdot 2 \cdot 2 \cdot 2 \cdot 3 \cdot 3 \cdot 3$ in exponential notation.

Your solution $2^4 \cdot 3^3$

Example 2 Write $10 \cdot 10 \cdot 10 \cdot 10$ as a power of ten.

Solution $10 \cdot 10 \cdot 10 \cdot 10 = 10^4$

You Try It 2 Write $10 \cdot 10 \cdot 10 \cdot 10 \cdot 10 \cdot 10 \cdot 10$ as a power of 10.

Your solution 10^7

Example 3 Simplify $3^2 \cdot 5^3$.

Solution $3^2 \cdot 5^3 = (3 \cdot 3) \cdot (5 \cdot 5 \cdot 5) = 9 \cdot 125 = 1125$

You Try It 3 Simplify $2^3 \cdot 5^2$.

Your solution 200

Solutions on p. A11

Objective B To use the Order of Operations Agreement to simplify expressions

INSTRUCTOR NOTE
Students may want to try part of the Projects in Mathematics at the end of this chapter and determine if their calculator uses the Order of Operations Agreement.

More than one operation may occur in a numerical expression. The answer may be different, depending on the order in which the operations are performed. For example, consider $3 + 4 \times 5$.

Multiply first, then add.

$$\underbrace{3 + 20}_{23} \quad \text{from} \quad 3 + \underbrace{4 \times 5}$$

Add first, then multiply.

$$\underbrace{7 \times 5}_{35} \quad \text{from} \quad \underbrace{3 + 4} \times 5$$

An Order of Operations Agreement is used so that only one answer is possible.

The Order of Operations Agreement

Step 1 Do all the operations inside parentheses.
Step 2 Simplify any number expressions containing exponents.
Step 3 Do multiplication and division as they occur from left to right.
Step 4 Do addition and subtraction as they occur from left to right.

Simplify $3 \times (2 + 1) - 2^2 + 4 \div 2$ by using the Order of Operations Agreement.

$3 \times (2 + 1) - 2^2 + 4 \div 2$ 1. Perform operations in parentheses.
$3 \times 3 - 2^2 + 4 \div 2$ 2. Simplify expressions with exponents.
$3 \times 3 - 4 + 4 \div 2$ 3. Do multiplications and divisions as they occur from left to right.
$9 - 4 + 4 \div 2$
$9 - 4 + 2$ 4. Do additions and subtractions as they occur from left to right.
$5 + 2$
7

One or more of the above steps may not be needed to simplify an expression. In that case, proceed to the next step in the Order of Operations Agreement.

Simplify $5 + 8 \div 2$. No parentheses or exponents. Proceed to step 3 of the Agreement.

$5 + 8 \div 2$ 3. Do multiplication or division.
$5 + 4$ 4. Do addition or subtraction.
9

Example 4 Simplify:
$16 \div (8 - 4) \cdot 9 - 5^2$

Solution
$16 \div (8 - 4) \cdot 9 - 5^2$
$16 \div 4 \cdot 9 - 5^2$
$16 \div 4 \cdot 9 - 25$
$4 \cdot 9 - 25$
$36 - 25$
11

You Try It 4 Simplify:
$5 \cdot (8 - 4) \div 4 - 2$

Your solution 3

Solution on p. A11

1.6 Exercises

Objective A

Write the number in exponential notation.

1. $2 \cdot 2 \cdot 2$
   2^3
2. $7 \cdot 7 \cdot 7 \cdot 7 \cdot 7$
   7^5
3. $6 \cdot 6 \cdot 6 \cdot 7 \cdot 7 \cdot 7 \cdot 7$
   $6^3 \cdot 7^4$
4. $3 \cdot 3 \cdot 3 \cdot 5 \cdot 5 \cdot 5 \cdot 5$
   $3^3 \cdot 5^4$
5. $6 \cdot 6 \cdot 6 \cdot 11 \cdot 11 \cdot 11 \cdot 11 \cdot 11$
   $6^3 \cdot 11^5$
6. $2 \cdot 2 \cdot 2 \cdot 2 \cdot 2 \cdot 2 \cdot 3 \cdot 3 \cdot 3$
   $2^6 \cdot 3^3$
7. $3 \cdot 10 \cdot 10 \cdot 10 \cdot 10$
   $3 \cdot 10^4$
8. $7 \cdot 10 \cdot 10 \cdot 10 \cdot 10 \cdot 10 \cdot 10 \cdot 10$
   $7 \cdot 10^7$
9. $2 \cdot 2 \cdot 3 \cdot 3 \cdot 3 \cdot 5 \cdot 5 \cdot 5 \cdot 5$
   $2^2 \cdot 3^3 \cdot 5^4$
10. $7 \cdot 7 \cdot 11 \cdot 11 \cdot 11 \cdot 13 \cdot 13 \cdot 13 \cdot 13$
    $7^2 \cdot 11^3 \cdot 13^4$
11. $2 \cdot 3 \cdot 3 \cdot 7 \cdot 7 \cdot 11$
    $2 \cdot 3^2 \cdot 7^2 \cdot 11$
12. $5 \cdot 7 \cdot 7 \cdot 9 \cdot 9 \cdot 11$
    $5 \cdot 7^2 \cdot 9^2 \cdot 11$
13. $2 \cdot 2 \cdot 7 \cdot 7 \cdot 7 \cdot 7 \cdot 11 \cdot 11 \cdot 11 \cdot 11$
    $2^2 \cdot 7^4 \cdot 11^4$
14. $3 \cdot 3 \cdot 7 \cdot 7 \cdot 7 \cdot 7 \cdot 17 \cdot 17 \cdot 17 \cdot 17$
    $3^2 \cdot 7^4 \cdot 17^4$

Simplify.

15. 2^3
    8
16. 2^6
    64
17. $2^4 \cdot 5^2$
    400
18. $2^6 \cdot 3^2$
    576
19. $3^2 \cdot 10^2$
    900
20. $2^3 \cdot 10^4$
    80,000
21. $6^2 \cdot 3^3$
    972
22. $4^3 \cdot 5^2$
    1600
23. $5 \cdot 2^3 \cdot 3$
    120
24. $6 \cdot 3^2 \cdot 4$
    216
25. $2^2 \cdot 3^2 \cdot 10$
    360
26. $3^2 \cdot 5^2 \cdot 10$
    2250
27. $0^2 \cdot 4^3$
    0
28. $6^2 \cdot 0^3$
    0
29. $3^2 \cdot 10^4$
    90,000
30. $5^3 \cdot 10^3$
    125,000
31. $2^2 \cdot 3^3 \cdot 5$
    540
32. $5^2 \cdot 7^3 \cdot 2$
    17,150
33. $2 \cdot 3^4 \cdot 5^2$
    4050
34. $6 \cdot 2^6 \cdot 7^2$
    18,816
35. $5^2 \cdot 3^2 \cdot 7^2$
    11,025
36. $4^2 \cdot 9^2 \cdot 6^2$
    46,656
37. $3^4 \cdot 2^6 \cdot 5$
    25,920
38. $4^3 \cdot 6^3 \cdot 7$
    96,768
39. $4^2 \cdot 3^3 \cdot 10^4$
    4,320,000
40. $5^2 \cdot 2^3 \cdot 10^3$
    200,000
41. $6^2 \cdot 4^4 \cdot 10$
    92,160

Objective B

Simplify by using the Order of Operations Agreement.

42. $4 - 2 + 3$
5

43. $6 - 3 + 2$
5

44. $6 \div 3 + 2$
4

45. $8 \div 4 + 8$
10

46. $6 \cdot 3 + 5$
23

47. $5 \cdot 9 + 2$
47

48. $3^2 - 4$
5

49. $5^2 - 17$
8

50. $4 \cdot (5 - 3) + 2$
10

51. $3 + (4 + 2) \div 3$
5

52. $5 + (8 + 4) \div 6$
7

53. $8 - 2^2 + 4$
8

54. $16 \cdot (3 + 2) \div 10$
8

55. $12 \cdot (1 + 5) \div 12$
6

56. $10 - 2^3 + 4$
6

57. $5 \cdot 3^2 + 8$
53

58. $16 + 4 \cdot 3^2$
52

59. $12 + 4 \cdot 2^3$
44

60. $16 + (8 - 3) \cdot 2$
26

61. $7 + (9 - 5) \cdot 3$
19

62. $2^2 + 3 \cdot (6 - 2)$
16

63. $3^3 + 5 \cdot (8 - 6)$
37

64. $2^2 \cdot 3^2 + 2 \cdot 3$
42

65. $4 \cdot 6 + 3^2 \cdot 4^2$
168

66. $16 - 2 \cdot 4$
8

67. $12 + 3 \cdot 5$
27

68. $3 \cdot (6 - 2) + 4$
16

69. $5 \cdot (8 - 4) - 6$
14

70. $8 - (8 - 2) \div 3$
6

71. $12 - (12 - 4) \div 4$
10

72. $8 + 2 - 3 \cdot 2 \div 3$
8

73. $10 + 1 - 5 \cdot 2 \div 5$
9

74. $3 \cdot (4 + 2) \div 6$
3

APPLYING THE CONCEPTS

75. Memory in computers is measured in bytes. One kilobyte (1 K) is 2^{10} bytes. Write this number in standard form. 1024

76. [W] Explain the difference in the order of operations for **a.** $\frac{14 - 2}{2} \div 2 \cdot 3$ and **b.** $\frac{14 - 2}{2} \div (2 \cdot 3)$. Work the two problems. What is the difference between the larger and the smaller answer?
8. Answers will vary

77. [W] If a number is divisible by 6 and 10, is the number divisible by 30? If so, explain why. If not, give an example. Yes. Answers will vary

1.7 Prime Numbers and Factoring

Objective A To factor numbers

Whole-number factors of a number divide that number evenly (there is no remainder).

1, 2, 3, and 6 are whole-number factors of 6 because they divide 6 evenly.

$$1\overline{)6}^{\,6} \quad 2\overline{)6}^{\,3} \quad 3\overline{)6}^{\,2} \quad 6\overline{)6}^{\,1}$$

Note that both the divisor and the quotient are factors of the dividend.

To find the factors of a number, try dividing the number by 1, 2, 3, 4, 5,.... Those numbers that divide the number evenly are its factors. Continue this process until the factors start to repeat.

➡ Find all the factors of 42.

$42 \div 1 = 42$	1 and 42 are factors	
$42 \div 2 = 21$	2 and 21 are factors	
$42 \div 3 = 14$	3 and 14 are factors	
$42 \div 4$	Will not divide evenly	
$42 \div 5$	Will not divide evenly	
$42 \div 6 = 7$	6 and 7 are factors	Factors are repeating; all the
$42 \div 7 = 6$	7 and 6 are factors	factors of 42 have been found.

1, 2, 3, 6, 7, 14, 21, and 42 are factors of 42.

The following rules are helpful in finding the factors of a number.

2 is a factor of a number if the last digit of the number is 0, 2, 4, 6, or 8.

436 ends in 6; therefore, 2 is a factor of 436. ($436 \div 2 = 218$)

3 is a factor of a number if the sum of the digits of the number is divisible by 3.

The sum of the digits of 489 is $4 + 8 + 9 = 21$. 21 is divisible by 3. Therefore, 3 is a factor of 489. ($489 \div 3 = 163$)

5 is a factor of a number if the last digit of the number is 0 or 5.

520 ends in 0; therefore, 5 is a factor of 520. ($520 \div 5 = 104$)

Example 1 Find all the factors of 30.

Solution
$30 \div 1 = 30$
$30 \div 2 = 15$
$30 \div 3 = 10$
$30 \div 4$ Will not divide evenly
$30 \div 5 = 6$
$30 \div 6 = 5$

1, 2, 3, 5, 6, 10, 15 and 30 are factors of 30.

You Try It 1 Find all the factors of 40.

Your solution 1, 2, 4, 5, 8, 10, 20, 40

Solution on p. A11

Objective B To find the prime factorization of a number

POINT OF INTEREST

Prime numbers are an important part of cryptology, the study of secret codes. To ensure that codes cannot be broken, a cryptologist uses prime numbers that have hundreds of digits.

A number is a **prime number** if its only whole-number factors are 1 and itself. 7 is prime because its only factors are 1 and 7. If a number is not prime, it is called a **composite number.** Because 6 has factors of 2 and 3, 6 is a composite number. The number 1 is not considered a prime number; therefore it is not included in the following list of prime numbers less than 50.

2, 3, 5, 7, 11, 13, 17, 19, 23, 29, 31, 37, 41, 43, 47

The **prime factorization** of a number is the expression of the number as a product of its prime factors. We use a "T-diagram" to find the prime factors of 60. Begin with the smallest prime number as a trial divisor, and continue with prime numbers as trial divisors until the final quotient is 1.

	60	
2	30	$60 \div 2 = 30$
2	15	$30 \div 2 = 15$
3	5	$15 \div 3 = 5$
5	1	$5 \div 5 = 1$

The prime factorization of 60 is $2 \cdot 2 \cdot 3 \cdot 5$.

Finding the prime factorization of larger numbers can be more difficult. Try each prime number as a trial divisor. Stop when the square of the trial divisor is greater than the number being factored.

➡ Find the prime factorization of 106.

	106
2	53
53	1

- 53 cannot be divided evenly by 2, 3, 5, 7, or 11. Prime numbers greater than 11 need not be tested because 11^2 is greater than 53.

The prime factorization of 106 is $2 \cdot 53$.

Example 2 Find the prime factorization of 315.

Solution

	315
3	105
3	35
5	7
7	1

$315 = 3 \cdot 3 \cdot 5 \cdot 7$

You Try It 2 Find the prime factorization of 44.

Your solution $2 \cdot 2 \cdot 11$

Example 3 Find the prime factorization of 201.

Solution

	201
3	67
67	1

- Try only 2, 3, 5, 7, and 11 because $11^2 > 67$.

$201 = 3 \cdot 67$

You Try It 3 Find the prime factorization of 177.

Your solution $3 \cdot 59$

Solutions on p. A11

1.7 Exercises

Objective A

Find all the factors of the number.

1. 4
1, 2, 4

2. 6
1, 2, 3, 6

3. 10
1, 2, 5, 10

4. 20
1, 2, 4, 5, 10, 20

5. 5
1, 5

6. 7
1, 7

7. 12
1, 2, 3, 4, 6, 12

8. 9
1, 3, 9

9. 8
1, 2, 4, 8

10. 16
1, 2, 4, 8, 16

11. 13
1, 13

12. 17
1, 17

13. 18
1, 2, 3, 6, 9, 18

14. 24
1, 2, 3, 4, 6, 8, 12, 24

15. 25
1, 5, 25

16. 56
1, 2, 4, 7, 8, 14, 28, 56

17. 36
1, 2, 3, 4, 6, 9, 12, 18, 36

18. 45
1, 3, 5, 9, 15, 45

19. 28
1, 2, 4, 7, 14, 28

20. 32
1, 2, 4, 8, 16, 32

21. 29
1, 29

22. 33
1, 3, 11, 33

23. 22
1, 2, 11, 22

24. 26
1, 2, 13, 26

25. 44
1, 2, 4, 11, 22, 44

26. 52
1, 2, 4, 13, 26, 52

27. 49
1, 7, 49

28. 82
1, 2, 41, 82

29. 37
1, 37

30. 79
1, 79

31. 57
1, 3, 19, 57

32. 69
1, 3, 23, 69

33. 48
1, 2, 3, 4, 6, 8, 12, 16, 24, 48

34. 64
1, 2, 4, 8, 16, 32, 64

35. 87
1, 3, 29, 87

36. 95
1, 5, 19, 95

37. 46
1, 2, 23, 46

38. 54
1, 2, 3, 6, 9, 18, 27, 54

39. 50
1, 2, 5, 10, 25, 50

40. 75
1, 3, 5, 15, 25, 75

41. 66
1, 2, 3, 6, 11, 22, 33, 66

42. 77
1, 7, 11, 77

43. 80
1, 2, 4, 5, 8, 10, 16, 20, 40, 80

44. 100
1, 2, 4, 5, 10, 20, 25, 50, 100

45. 84
1, 2, 3, 4, 6, 7, 12, 14, 21, 28, 42, 84

46. 96
1, 2, 3, 4, 6, 8, 12, 16, 24, 32, 48, 96

47. 85
1, 5, 17, 85

48. 90
1, 2, 3, 5, 6, 9, 10, 15, 18, 30, 45, 90

49. 101
1, 101

50. 123
1, 3, 41, 123

Objective B

Find the prime factorization.

51. 6
2 · 3

52. 14
2 · 7

53. 17
Prime

54. 83
Prime

55. 16
2 · 2 · 2 · 2

56. 24
2 · 2 · 2 · 3

57. 12
2 · 2 · 3

58. 27
3 · 3 · 3

59. 9
3 · 3

60. 15
3 · 5

61. 36
2 · 2 · 3 · 3

62. 40
2 · 2 · 2 · 5

63. 19
Prime

64. 37
Prime

65. 50
2 · 5 · 5

66. 90
2 · 3 · 3 · 5

67. 65
5 · 13

68. 115
5 · 23

69. 80
2 · 2 · 2 · 2 · 5

70. 87
3 · 29

71. 18
2 · 3 · 3

72. 26
2 · 13

73. 28
2 · 2 · 7

74. 49
7 · 7

75. 29
Prime

76. 31
Prime

77. 42
2 · 3 · 7

78. 62
2 · 31

79. 81
3 · 3 · 3 · 3

80. 51
3 · 17

81. 22
2 · 11

82. 39
3 · 13

83. 101
Prime

84. 89
Prime

85. 70
2 · 5 · 7

86. 66
2 · 3 · 11

87. 86
2 · 43

88. 74
2 · 37

89. 95
5 · 19

90. 105
3 · 5 · 7

91. 67
Prime

92. 78
2 · 3 · 13

93. 55
5 · 11

94. 46
2 · 23

95. 122
2 · 61

96. 130
2 · 5 · 13

97. 118
2 · 59

98. 102
2 · 3 · 17

99. 150
2 · 3 · 5 · 5

100. 125
5 · 5 · 5

101. 120
2 · 2 · 2 · 3 · 5

102. 144
2 · 2 · 2 · 2 · 3 · 3

103. 160
2 · 2 · 2 · 2 · 2 · 5

104. 175
5 · 5 · 7

105. 343
7 · 7 · 7

106. 216
2 · 2 · 2 · 3 · 3 · 3

107. 400
2 · 2 · 2 · 2 · 5 · 5

108. 625
5 · 5 · 5 · 5

109. 225
3 · 3 · 5 · 5

110. 180
2 · 2 · 3 · 3 · 5

APPLYING THE CONCEPTS

111. Twin primes are two prime numbers that differ by 2. For instance, 17 and 19 are twin primes. Find three sets of twin primes, not including 17 and 19. 3 and 5, 5 and 7, 11 and 13 Other answers are possible

112. In 1742, Christian Goldbach conjectured that every even number greater than 2 could be expressed as the sum of two prime numbers. Show that this conjecture is true for 8, 24, and 72. (*Note*: Mathematicians have not yet been able to determine whether Goldbach's conjecture is true or false.)
$8 = 3 + 5$ $24 = 11 + 13$ $72 = 29 + 43$

113. [W] Explain why 2 is the only even prime number. Answers will vary

114. [W] Explain the method of finding prime numbers using the Sieve of Erathosthenes. Answers will vary

Projects in Mathematics

Order of Operations

INSTRUCTOR NOTE

These projects are a source of activities for students. They may be used for enrichment, extra credit, or group activities.

Does your calculator use the Order of Operations Agreement? To find out, try this problem:

$$2 + 4 \cdot 7$$

If your answer is 30, then the calculator uses the Order of Operations Agreement. If your answer is 42, it does not use that agreement.

Even if your calculator does not use the order of Operations Agreement, you can still correctly evaluate numerical expressions. The parentheses keys, [( ] and [ )], are used for this purpose.

Remember that $2 + 4 \cdot 7$ means $2 + (4 \cdot 7)$ because the multiplication must be completed before the addition. To evaluate this expression, enter the following:

Enter:	2	[+]	[( ]	4	[×]	7	[ )]	[=]	
Display:	2	2	[( ]	4	4	7	28	30	

When using your calculator to evaluate numerical expressions, insert parentheses around multiplications or divisions. This has the effect of forcing the calculator to do the operations in the order you want rather than in the order the calculator wants.

Exercises:

Evaluate

1. $3 \cdot (15 - 2 \cdot 3) - 36 \div 3$

2. $4 \cdot 2^2 - (12 + 24 \div 6) - 5$

3. $16 \div 4 \cdot 3 + (3 \cdot 4 - 5) + 2$

4. $15 \cdot 3 \div 9 + (2 \cdot 6 - 3) + 4$

Patterns in Mathematics

For the circle at the left, use a straight line to connect each dot on the circle with every other dot on the circle. How many different straight lines are there?

Follow the same procedure for each of the circles shown below. How many different straight lines are there in each?

Find a pattern to describe the number of dots on a circle and the corresponding number of different lines drawn. Use the pattern to determine the number of different lines that would be drawn in a circle with 7 dots and in a circle with 8 dots.

You are arranging a tennis tournament with 9 players. Following the pattern of finding the number of lines drawn on a circle, find the number of singles matches that can be played among the 9 players if each player plays each of the other players only once.

Chapter Summary

Key Words The *whole numbers* are 0, 1, 2, 3, 4, 5, 6, 7, 8, 9, 10, ...

The symbol for "*is less than*" is <.

The symbol for "*is greater than*" is >.

The position of a digit in a number determines the digit's *place value.*

Giving an approximate value for an exact number is called *rounding.*

An expression of the form 4^2 is in *exponential notation,* where 4 is the *base* and 2 is the *exponent.*

A number is *prime* if it's only whole-number factors are 1 and itself.

The *prime factorization* of a number is the expression of the number as a product of its prime factors.

Essential Rules

The Addition Property of Zero Zero added to a number does not change the number.
5 + 0 = 0 + 5 = 5

The Commutative Property of Addition Two numbers can be added in either order.
6 + 5 = 5 + 6 = 11

The Associative Property of Addition Grouping numbers to be added in any order gives the same result.
(2 + 3) + 5 = 2 + (3 + 5) = 10

The Multiplication Property of Zero The product of a number and zero is zero.
0 × 3 = 3 × 0 = 0

The Multiplication Property of 1 The product of a number and 1 is the number.
1 × 8 = 8 × 1 = 8

The Commutative Property of Multiplication Two numbers can be multiplied in any order.
5 × 2 = 2 × 5 = 10

The Associative Property of Multiplication Grouping numbers to be multiplied in any order gives the same result.
(3 × 2) × 5 = 3 × (2 × 5) = 30

The Properties of Zero in Division Zero divided by any other number is zero. Division by zero is not allowed.

Order of Operations Agreement

Step 1 Perform operations inside grouping symbols.
Step 2 Simplify expressions with exponents.
Step 3 Do multiplications and divisions as they occur from left to right.
Step 4 Do additions and subtractions as they occur from left to right.

Chapter Review

SECTION 1.1

1. Place the correct symbol, $<$ or $>$, between the two numbers. 101 87
   $101 > 87$

2. Write 276,057 in words.
   Two hundred seventy-six thousand fifty-seven

3. Write two million eleven thousand forty-four in standard form.
   2,011,044

4. Write 10,327 in expanded form.
   $10{,}000 + 300 + 20 + 7$

SECTION 1.2

5. Add:
$$\begin{array}{r} 298 \\ 461 \\ +\ 322 \\ \hline \end{array}$$
   1081

6. Find the sum of 5,894, 6301, and 298.
   12,493

7. An insurance account executive received commissions of $723, $544, $812, and $488 during a 4-week period. Find the total income from commissions for the 4 weeks.
   $2567

8. You had a balance of $516 in your checking account before making deposits of $88 and $213. Find the total amount deposited, and determine your new checking balance.
   $301, $817

SECTION 1.3

9. Subtract:
$$\begin{array}{r} 4926 \\ -\ 3177 \\ \hline \end{array}$$
   1749

10. What is 10,134 decreased by 4725?
    5409

The graph at the right shows the amount that companies from different countries spend on business trips. Use the graph for Exercises 11–13.

Big (Travel) Spenders

U.S. companies spend more than twice the amount many European companies spend for business travel and entertainment. Spending per employee:

U.S.A. $3,113
Germany $1,767
France $1,443
U.K. $1,422
Italy $1,128

11. How much more does a U.S. company spend for business travel and entertainment than does a company from Italy? $1985 per person

12. How much more does a U.S. company spend for business travel and entertainment than the combined spending of a company from the U.K. and a company from Italy? $563

13. Does a company from the United States outspend the combined spending of a company from Germany and a company from the U.K.? No

SECTION 1.4

14. Multiply: $\begin{array}{r} 843 \\ \times\ 27 \\ \hline \end{array}$

22,761

15. What is 2019 multiplied by 307?

619,833

16. You have a car payment of $123 per month. What is the total of the car payments over a 12-month period?

$1476

17. Vincent Meyers, a sales assistant, earns $240 for working a 40-hour week. Last week Vincent worked an additional 12 hours at $12 an hour. Find Vincent's total pay for last week's work.

$384

SECTION 1.5

18. Divide: $7\overline{)14{,}945}$

2135

19. Find the quotient of 109,763 and 84.

1306 r59

20. Louis Reyes, a sales executive, drove a car 351 miles on 13 gallons of gas. Find the number of miles driven per gallon of gasoline.

27 miles/gallon

21. A car is purchased for $8940, with a down payment of $1500. The balance is paid in 48 equal monthly payments. Find the monthly car payment.

$155

SECTION 1.6

22. Write $2 \cdot 2 \cdot 2 \cdot 2 \cdot 5 \cdot 5 \cdot 5$ in exponential notation.

$2^4 \cdot 5^3$

23. Write $5 \cdot 5 \cdot 7 \cdot 7 \cdot 7 \cdot 7 \cdot 7$ in exponential notation.

$5^2 \cdot 7^5$

24. Simplify: $3 \cdot 2^3 \cdot 5^2$

600

25. Simplify: $2^3 - 3 \cdot 2$

2

26. Simplify: $3^2 + 2^2 \cdot (5 - 3)$

17

27. Simplify: $8 \cdot (6 - 2) \div 4$

8

SECTION 1.7

28. Find all the factors of 18.

1, 2, 3, 6, 9, 18

29. Find all the factors of 30.

1, 2, 3, 5, 6, 10, 15, 30

30. Find the prime factorization of 42.

$2 \cdot 3 \cdot 7$

31. Find the prime factorization of 72.

$2 \cdot 2 \cdot 2 \cdot 3 \cdot 3$

Chapter Test

1. Place the correct symbol, $<$ or $>$, between the two numbers. 21 $>$ 19 [1.1A]*

2. Write 207,068 in words.
Two hundred seven thousand sixty-eight [1.1B]

3. Write one million two hundred four thousand six in standard form.
1,204,006 [1.1B]

4. Write 906,378 in expanded form.
900,000 + 6000 + 300 + 70 + 8 [1.1C]

5. Round 74,965 to the nearest hundred.
75,000 [1.1D]

6. Add: $\begin{array}{r} 25{,}492 \\ +\ 71{,}306 \\ \hline \end{array}$
96,798 [1.2A]

7. Find the sum of 89,756, 9094, and 37,065.
135,915 [1.2A]

8. A family drives 425 miles the first day, 187 miles the second day, and 243 miles the third day of their vacation. The odometer read 47,626 miles at the start of the vacation.
a. How many miles were driven during the 3 days? 855 miles
b. What is the odometer reading at the end of the 3 days? 48,481 miles [1.2B]

9. Subtract: $\begin{array}{r} 17{,}495 \\ -\ 8{,}162 \\ \hline \end{array}$
9,333 [1.3A]

10. Find the difference between 29,736 and 9814.
19,922 [1.3B]

The data in the table at the right are based on information from the California Department of Insurance in 1993. This table shows the annual insurance premiums of some companies for renters insurance. Use it for Exercises 11 and 12.

State Farm	\$139
Allstate	\$129
Farmers	\$147
Safeco	\$220
Auto Club (AAA)	\$136
CNA Insurance Co.	\$164
Transamerica	\$189

11. What is the difference between the annual premium charged by CNA Insurance Co. and that charged by State Farm? \$25 [1.3C]

12. What is the difference in premium between the company with the highest premium and the one with the lowest premium? \$91 [1.3C]

*Note: The notation [1.1A] indicates the objective that the student should review if that question was answered incorrectly. For example, [1.1A] means Chapter 1, Section 1, Objective A. This notation will be used for all Chapter Tests and Cumulative Reviews throughout the text.

13. Find the product of 90,763 and 8.
726,104 [1.4A]

14. Multiply: $\begin{array}{r} 9736 \\ \times\ 704 \\ \hline \end{array}$
6,854,144 [1.4B]

15. An investor receives $237 each month from a corporate bond fund. How much will the investor receive over a 12-month period? $2844 [1.4C]

16. Find the quotient of 5624 and 8.
703 [1.5A]

17. Divide: $7\overline{)60,972}$
8710 r2 [1.5B]

18. Divide: $97\overline{)108,764}$
1,121 r27 [1.5C]

19. A farmer harvested 48,290 pounds of lemons from one grove and 23,710 pounds of lemons from another grove. The lemons were packed in boxes with 24 pounds of lemons in each box. How many boxes were needed to pack the lemons? 3000 boxes [1.5D]

20. Write $3 \cdot 3 \cdot 3 \cdot 7 \cdot 7$ in exponential notation.
$3^3 \cdot 7^2$ [1.6A]

21. Simplify: $3^3 \cdot 4^2$
432 [1.6A]

22. Simplify: $4^2 \cdot (4 - 2) \div 8 + 5$
9 [1.6B]

23. Simplify: $16 \div 4 \times 2 - (6 - 5)^3$
7 [1.6B]

24. Find all the factors of 20.
1, 2, 4, 5, 10, 20 [1.7A]

25. Find the prime factorization of 84.
$2 \cdot 2 \cdot 3 \cdot 7$ [1.7B]

Chapter

2

Fractions

Objectives

Section 2.1

To find the least common multiple (LCM)

To find the greatest common factor (GCF)

Section 2.2

To write a fraction that represents part of a whole

To write an improper fraction as a mixed number or a whole number, and a mixed number as an improper fraction

Section 2.3

To find equivalent fractions by raising to higher terms

To write a fraction in simplest form

Section 2.4

To add fractions with the same denominator

To add fractions with unlike denominators

To add whole numbers, mixed numbers, and fractions

To solve application problems

Section 2.5

To subtract fractions with the same denominator

To subtract fractions with unlike denominators

To subtract whole numbers, mixed numbers, and fractions

To solve application problems

Section 2.6

To multiply fractions

To multiply whole numbers, mixed numbers, and fractions

To solve application problems

Section 2.7

To divide fractions

To divide whole numbers, mixed numbers, and fractions

To solve application problems

Section 2.8

To identify the order relation between two fractions

To simplify expressions containing exponents

To use the Order of Operations Agreement to simplify expressions

Egyptian Fractions

The Rhind papyrus is one of the earliest written accounts of mathematics.[1] In the papyrus, a scribe named Ahmes gives an early account of the concept of fractions. A portion of the Rhind papyrus is rendered below with its hieroglyphic transcription.

The early Egyptians primarily used unit fractions. This is a fraction in which the numerator is a 1. To write a fraction, a small oval was placed above a series of lines. The number of lines indicated the denominator. Some examples of these fractions are

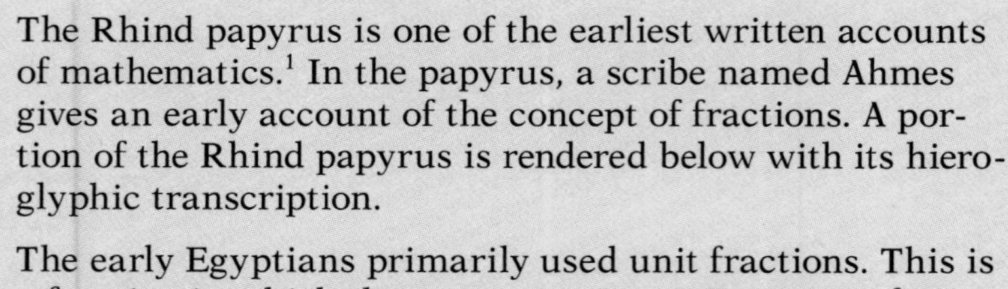

In the first example, each line represents a 1. Because there are 4 lines, the fraction is $\frac{1}{4}$.

The second example is the special symbol that was used for the fraction $\frac{1}{2}$.

[1]Papyrus comes from the stem of a plant. The stem was dried and then pounded thin. The resulting material served as a primitive type of paper.

2.1 The Least Common Multiple and Greatest Common Factor

Objective A *To find the least common multiple (LCM)*

The **multiples** of a number are the products of that number and the numbers 1, 2, 3, 4, 5, ...

$3 \times 1 = 3$
$3 \times 2 = 6$
$3 \times 3 = 9$
$3 \times 4 = 12$ The multiples of 3 are 3, 6, 9, 12, 15, ...
$3 \times 5 = 15$
⋮

A number that is a multiple of two or more numbers is a **common multiple** of those numbers.

The multiples of 4 are 4, 8, 12, 16, 20, 24, 28, 32, 36, ...
The multiples of 6 are 6, 12, 18, 24, 30, 36, 42, ...
Some common multiples of 4 and 6 are 12, 24, and 36.

The **least common multiple** (LCM) is the smallest common multiple of two or more numbers.

The least common multiple of 4 and 6 is 12.

Listing the multiples of each number is one way to find the LCM. Another way to find the LCM uses the prime factorization of each number.

To find the LCM of 450 and 600, find the prime factorization of each number and write the factorization of each number in a table. Circle the largest product in each column. The LCM is the product of the circled numbers.

	2	3	5
450 =	2	3 · 3	5 · 5
600 =	2 · 2 · 2	3	5 · 5

In the column headed by 5, the products are equal. Circle just one product.

The LCM is the product of the circled numbers.
The LCM = 2 · 2 · 2 · 3 · 3 · 5 · 5 = 1800.

Example 1 Find the LCM of 24, 36, and 50.

Solution

	2	3	5
24 =	2 · 2 · 2	3	
36 =	2 · 2	3 · 3	
50 =	2		5 · 5

The LCM = 2 · 2 · 2 · 3 · 3 · 5 · 5 = 1800.

You Try It 1 Find the LCM of 50, 84, and 135.

Your solution 18,900

Solution on p. A12

Objective B To find the greatest common factor (GCF)

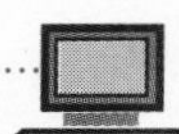

Recall that a number that divides another number evenly is a factor of that number. 64 can be evenly divided by 1, 2, 4, 8, 16, 32, and 64. 1, 2, 4, 8, 16, 32, and 64 are factors of 64.

A number that is a factor of two or more numbers is a **common factor** of those numbers.

The factors of 30 are 1, 2, 3, 5, 6, 10 15, and 30.
The factors of 105 are 1, 3, 5, 7, 15, 21, 35, and 105.
The common factors of 30 and 105 are 1, 3, 5, and 15.

INSTRUCTOR NOTE
The following model may help some students with the LCM and GCF.

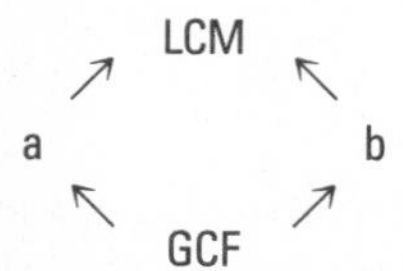

The arrows indicate "divides into."

The **greatest common factor** (GCF) is the largest common factor of two or more numbers.

The greatest common factor of 30 and 105 is 15.

Listing the factors of each number is one way of finding the GCF. Another way to find the GCF uses the prime factorization of each number.

To find the GCF of 126 and 180, find the prime factorization of each number and write the factorization of each number in a table. Circle the smallest product in each column that does not have a blank. The GCF is the product of the circled numbers.

	2	3	5	7
126 =	2	3 · 3		7
180 =	2 · 2	3 · 3	5	

In the column headed by 3, the products are equal. Circle just one product.
Columns 5 and 7 have a blank, so 5 and 7 are not common factors of 126 and 180. Do not circle any number in these columns.

The GCF is the product of the circled numbers.
The GCF = 2 · 3 · 3 = 18.

Example 2 Find the GCF of 90, 168, and 420.

Solution

	2	3	5	7
90 =	2	3 · 3	5	
168 =	2 · 2 · 2	3		7
420 =	2 · 2	3	5	7

The GCF = 2 · 3 = 6.

You Try It 2 Find the GCF of 36, 60, and 72.

Your solution 12

Example 3 Find the GCF of 7, 12, and 20.

Solution

	2	3	5	7
7 =				7
12 =	2 · 2	3		
20 =	2 · 2		5	

Since no numbers are circled, the GCF = 1.

You Try It 3 Find the GCF of 11, 24, and 30.

Your solution 1

Solutions on p. A12

2.1 Exercises

Objective A

Find the LCM.

1. 5, 8
40

2. 3, 6
6

3. 3, 8
24

4. 2, 5
10

5. 5, 6
30

6. 5, 7
35

7. 4, 6
12

8. 6, 8
24

9. 8, 12
24

10. 12, 16
48

11. 5, 12
60

12. 3, 16
48

13. 8, 14
56

14. 6, 18
18

15. 3, 9
9

16. 4, 10
20

17. 8, 32
32

18. 7, 21
21

19. 9, 36
36

20. 14, 42
42

21. 44, 60
660

22. 120, 160
480

23. 102, 184
9384

24. 123, 234
9594

25. 4, 8, 12
24

26. 5, 10, 15
30

27. 3, 5, 10
30

28. 2, 5, 8
40

29. 3, 8, 12
24

30. 5, 12, 18
180

31. 9, 36, 64
576

32. 18, 54, 63
378

33. 16, 30, 84
1680

34. 9, 12, 15
180

Objective B

Find the GCF.

35. 3, 5
1

36. 5, 7
1

37. 6, 9
3

38. 18, 24
6

39. 15, 25
5

40. 14, 49
7

41. 25, 100
25

42. 16, 80
16

43. 32, 51
1

44. 21, 44
1

45. 12, 80
4

46. 8, 36
4

47. 16, 140
4

48. 12, 76
4

49. 8, 14
2

50. 24, 30
6

51. 48, 144
48

52. 44, 96
4

53. 18, 32
2

54. 16, 30
2

55. 3, 5, 11
1

56. 6, 8, 10
2

57. 7, 14, 49
7

58. 6, 15, 36
3

59. 8, 12, 16
4

60. 10, 15, 20
5

61. 12, 18, 20
2

62. 24, 40, 72
8

63. 3, 17, 51
1

64. 17, 31, 81
1

65. 14, 42, 84
14

66. 25, 125, 625
25

67. 12, 68, 92
4

68. 28, 35, 70
7

69. 1, 49, 153
1

70. 32, 56, 72
8

71. 24, 36, 48
12

APPLYING THE CONCEPTS

72. Define the phrase *relatively prime numbers.* List three pairs of relatively prime numbers.
Numbers with no common factors except the factor 1. 4, 5 8, 9 16, 21 (variable)

73. Joe Salvo, a clerk, works three days and then has a day off. A friend works five days and then has a day off. How many days after Joe and his friend have a day off together will they have another day off together?
12 days

74. Find the LCM of the following pairs of numbers: 2 and 3, 5 and 7, and 11 and 19. Can you draw a conclusion about the LCM of two prime numbers? Suggest a way of finding the LCM of three prime numbers.
The LCM of 2 and 3 is 6, of 5 and 7 is 35, of 11 and 19 is 209. The LCM of three prime numbers is the product of the three numbers.

75. Find the GCF of the following pairs of numbers: 3 and 5, 7 and 11, and 29 and 43. Can you draw a conclusion about the GCF of two prime numbers. What is the GCF of three prime numbers?
The GCF of 3 and 5 is 1, of 7 and 11 is 1, of 29 and 43 is 1. The GCF of three prime numbers is 1.

76. Is the LCM of two numbers always divisible by the GCF of the two numbers? If so, explain why. If not, give an example.
Yes. Each number contains all the factors of the GCF.

77. Using the pattern for the first two triangles below, determine the center number of the last triangle. 4

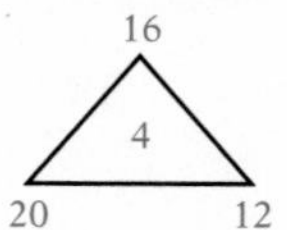

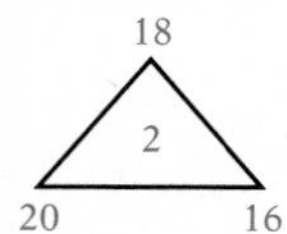

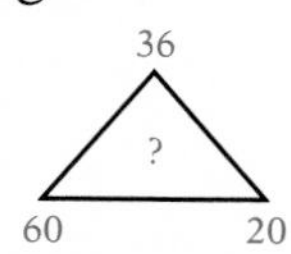

78. The ancient Mayans used two calendars, a solar calendar of 365 days and a ritual calendar of 220 days. If a solar year and the ritual year begin on the same day, how many solar years and how many ritual years will pass until this situation occurs again?
(*Mathematics Teacher*, November 1993, page 104)
44 solar years 73 ritual years

2.2 Introduction to Fractions

Objective A To write a fraction that represents part of a whole

A **fraction** can represent the number of equal parts of a whole.

The shaded portion of the circle is represented by the fraction $\frac{4}{7}$. Four-sevenths of the circle are shaded.

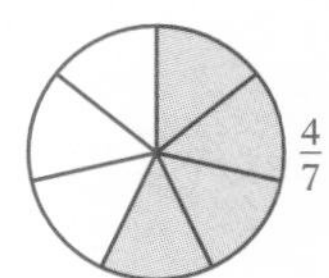

Each part of a fraction has a name.

POINT OF INTEREST
The fraction bar was first used in 1050 by al-Hassar. It is also called a vinculum.

Fraction bar → $\frac{4}{7}$ ← **Numerator** / ← **Denominator**

A **proper fraction** is a fraction less than 1. The numerator of a proper fraction is smaller than the denominator. The shaded portion of the circle can be represented by the proper fraction $\frac{3}{4}$.

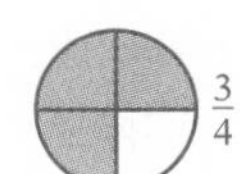

A **mixed number** is a number greater than 1 with a whole-number part and a fractional part. The shaded portion of the circles can be represented by the mixed number $2\frac{1}{4}$.

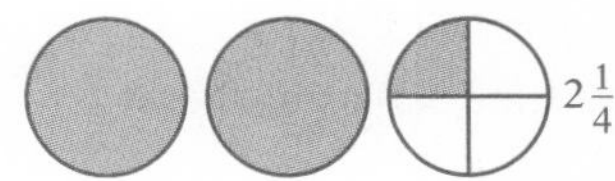

An **improper fraction** is a fraction greater than or equal to 1. The numerator of an improper fraction is greater than or equal to the denominator. The shaded portion of the circles can be represented by the improper fraction $\frac{9}{4}$. The shaded portion of the square can be represented by $\frac{4}{4}$.

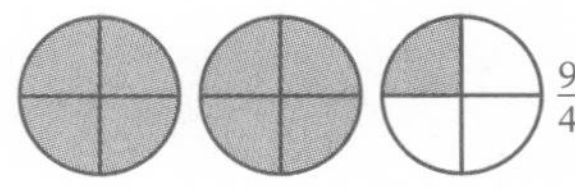

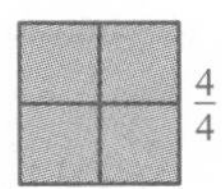

Example 1
Express the shaded portion of the circled as a mixed number.

 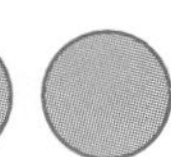 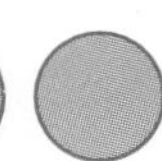 

Solution $3\frac{2}{5}$

You Try It 1
Express the shaded portion of the circles as a mixed number.

   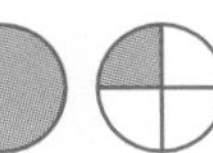

Your solution $4\frac{1}{4}$

Example 2
Express the shaded portion of the circles as an improper fraction.

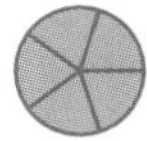 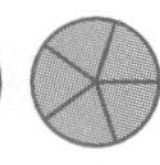 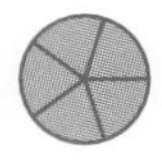 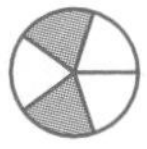

Solution $\frac{17}{5}$

You Try It 2
Express the shaded portion of the circles as an improper fraction.

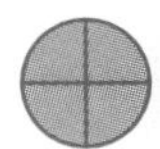 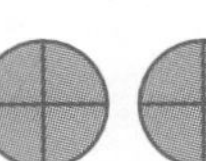 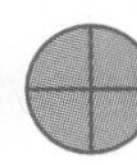 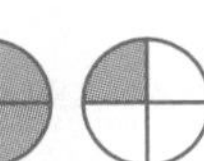

Your solution $\frac{17}{4}$

Solutions on p. A12

Objective B To write an improper fraction as a mixed number or a whole number, and a mixed number as an improper fraction

INSTRUCTOR NOTE

As a classroom exercise, ask students to give examples of mixed numbers. Some possible answers: stock market tables, carpentry, sewing, recipes.

Note from the diagram that the mixed number $2\frac{3}{5}$ and the improper fraction $\frac{13}{5}$ both represent the shaded portion of the circles.

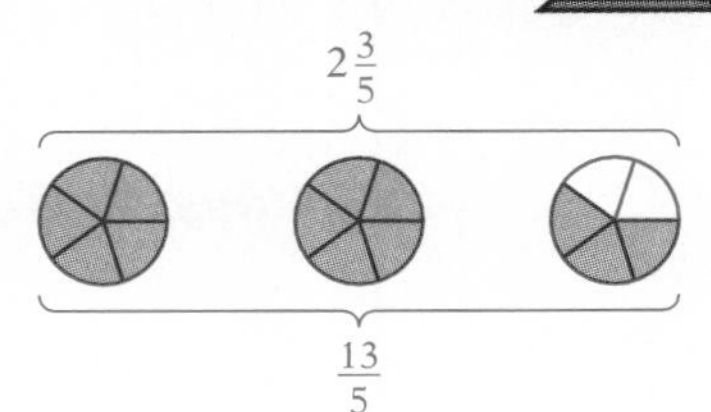

$$2\frac{3}{5} = \frac{13}{5}$$

An improper fraction can be written as a mixed number.

➡ Write $\frac{13}{5}$ as a mixed number.

Divide the numerator by the denominator.

$$\begin{array}{r} 2 \\ 5\overline{)\ 13} \\ -10 \\ \hline 3 \end{array}$$

To write the fractional part of the mixed number, write the remainder over the divisor.

$$\begin{array}{r} 2\frac{3}{5} \\ 5\overline{)\ 13} \\ -10 \\ \hline 3 \end{array}$$

Write the answer.

$$\frac{13}{5} = 2\frac{3}{5}$$

POINT OF INTEREST

Archimedes (c. 287–212 B.C.) is the person who calculated that $\pi \approx 3\frac{1}{7}$. He actually showed that $3\frac{10}{71} < \pi < 3\frac{1}{7}$. The approximation $3\frac{10}{71}$ is more accurate but more difficult to use.

To write a mixed number as an improper fraction, multiply the denominator of the fractional part by the whole-number part. The sum of this product and the numerator of the fractional part is the numerator of the improper fraction. The denominator remains the same.

➡ Write $7\frac{3}{8}$ as an improper fraction.

$$7\frac{3}{8} = \frac{(8 \times 7) + 3}{8} = \frac{56 + 3}{8} = \frac{59}{8} \qquad 7\frac{3}{8} = \frac{59}{8}$$

Example 3 Write $\frac{21}{4}$ as a mixed number.

Solution

$$\begin{array}{r} 5 \\ 4\overline{)\ 21} \\ -20 \\ \hline 1 \end{array} \qquad \frac{21}{4} = 5\frac{1}{4}$$

You Try It 3 Write $\frac{22}{5}$ as a mixed number.

Your solution $4\frac{2}{5}$

Example 4 Write $\frac{18}{6}$ as a whole number.

Solution

$$\begin{array}{r} 3 \\ 6\overline{)\ 18} \\ -18 \\ \hline 0 \end{array} \qquad \frac{18}{6} = 3$$

Note. The remainder is zero.

You Try It 4 Write $\frac{28}{7}$ as a whole number.

Your solution 4

Example 5 Write $21\frac{3}{4}$ as an improper fraction.

Solution $21\frac{3}{4} = \frac{84 + 3}{4} = \frac{87}{4}$

You Try It 5 Write $14\frac{5}{8}$ as an improper fraction.

Your solution $\frac{117}{8}$

Solutions on p. A12

2.2 Exercises

Objective A

Express the shaded portion of the circle as a fraction.

1. 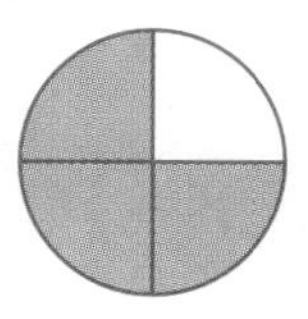 $\frac{3}{4}$

2. 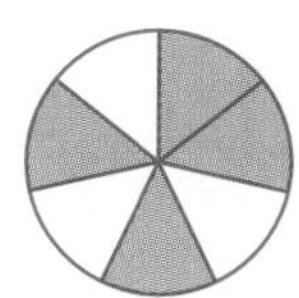 $\frac{4}{7}$

3. 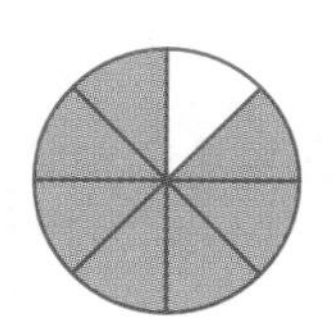 $\frac{7}{8}$

4. 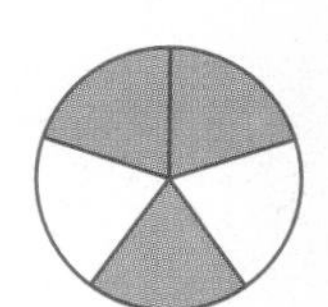 $\frac{3}{5}$

Express the shaded portion of the circles as a mixed number.

5. 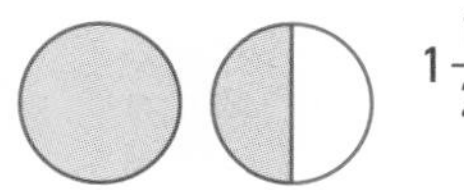 $1\frac{1}{2}$

6. 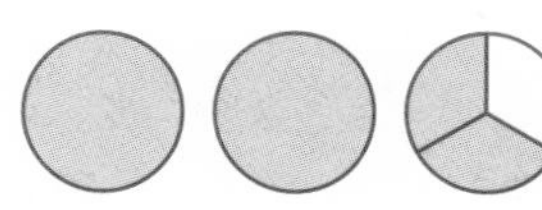 $2\frac{2}{3}$

7.  $2\frac{5}{8}$

8. 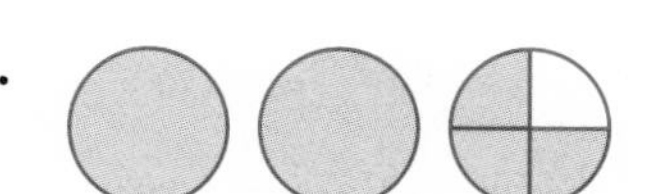 $2\frac{3}{4}$

9. 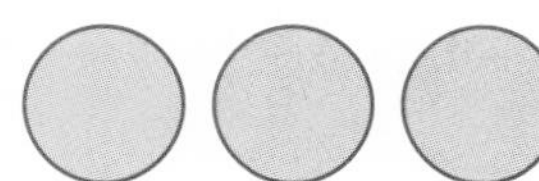 $3\frac{3}{5}$

10. 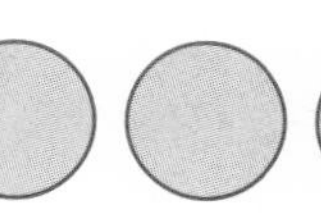 $3\frac{5}{6}$

Express the shaded portion of the circles as an improper fraction.

11. 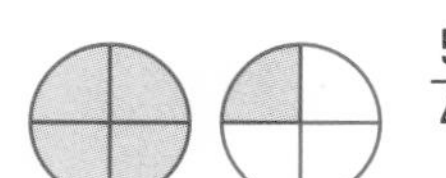 $\frac{5}{4}$

12. 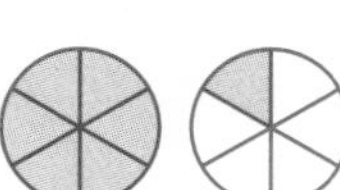 $\frac{7}{6}$

13. 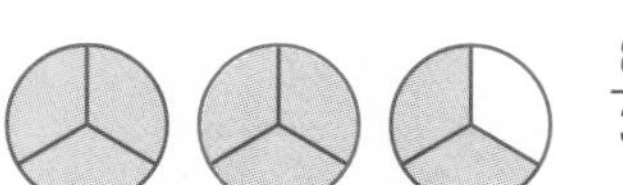 $\frac{8}{3}$

14. 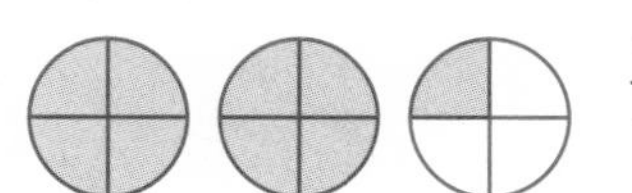 $\frac{9}{4}$

15. 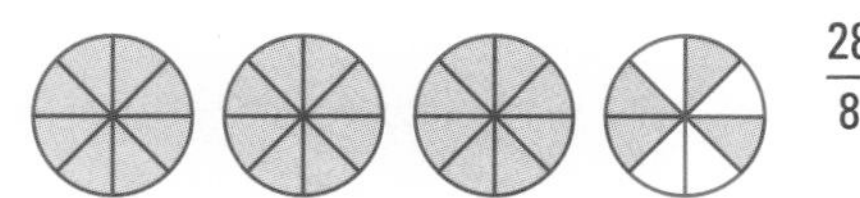 $\frac{28}{8}$

16. 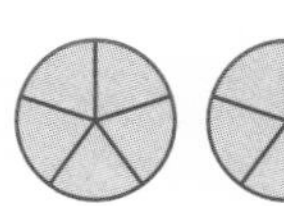 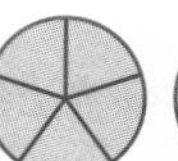 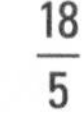 $\frac{18}{5}$

17. Shade $\frac{5}{6}$ of 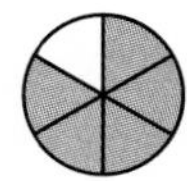

18. Shade $1\frac{2}{5}$ of 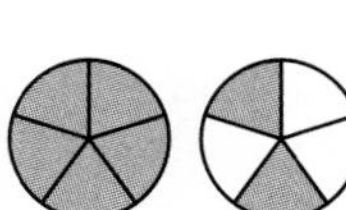

19. Shade $\frac{6}{5}$ of 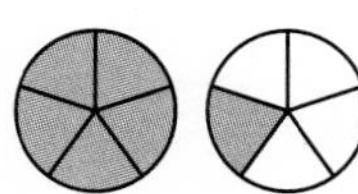

Objective B

Write the improper fraction as a mixed number or a whole number.

20. $\frac{11}{4}$ $2\frac{3}{4}$ **21.** $\frac{16}{3}$ $5\frac{1}{3}$ **22.** $\frac{20}{4}$ 5 **23.** $\frac{18}{9}$ 2 **24.** $\frac{9}{8}$ $1\frac{1}{8}$ **25.** $\frac{13}{4}$ $3\frac{1}{4}$

26. $\frac{23}{10}$ $2\frac{3}{10}$ **27.** $\frac{29}{2}$ $14\frac{1}{2}$ **28.** $\frac{48}{16}$ 3 **29.** $\frac{51}{3}$ 17 **30.** $\frac{8}{7}$ $1\frac{1}{7}$ **31.** $\frac{16}{9}$ $1\frac{7}{9}$

32. $\frac{7}{3}$ $2\frac{1}{3}$ **33.** $\frac{9}{5}$ $1\frac{4}{5}$ **34.** $\frac{16}{1}$ 16 **35.** $\frac{23}{1}$ 23 **36.** $\frac{17}{8}$ $2\frac{1}{8}$ **37.** $\frac{31}{16}$ $1\frac{15}{16}$

38. $\frac{12}{5}$ $2\frac{2}{5}$ **39.** $\frac{19}{3}$ $6\frac{1}{3}$ **40.** $\frac{9}{9}$ 1 **41.** $\frac{40}{8}$ 5 **42.** $\frac{72}{8}$ 9 **43.** $\frac{3}{3}$ 1

Write the mixed number as an improper fraction.

44. $2\frac{1}{3}$ $\frac{7}{3}$ **45.** $4\frac{2}{3}$ $\frac{14}{3}$ **46.** $6\frac{1}{2}$ $\frac{13}{2}$ **47.** $8\frac{2}{3}$ $\frac{26}{3}$ **48.** $6\frac{5}{6}$ $\frac{41}{6}$ **49.** $7\frac{3}{8}$ $\frac{59}{8}$

50. $9\frac{1}{4}$ $\frac{37}{4}$ **51.** $6\frac{1}{4}$ $\frac{25}{4}$ **52.** $10\frac{1}{2}$ $\frac{21}{2}$ **53.** $15\frac{1}{8}$ $\frac{121}{8}$ **54.** $8\frac{1}{9}$ $\frac{73}{9}$ **55.** $3\frac{5}{12}$ $\frac{41}{12}$

56. $5\frac{3}{11}$ $\frac{58}{11}$ **57.** $3\frac{7}{9}$ $\frac{34}{9}$ **58.** $2\frac{5}{8}$ $\frac{21}{8}$ **59.** $12\frac{2}{3}$ $\frac{38}{3}$ **60.** $1\frac{5}{8}$ $\frac{13}{8}$ **61.** $5\frac{3}{7}$ $\frac{38}{7}$

62. $11\frac{1}{9}$ $\frac{100}{9}$ **63.** $12\frac{3}{5}$ $\frac{63}{5}$ **64.** $3\frac{3}{8}$ $\frac{27}{8}$ **65.** $4\frac{5}{9}$ $\frac{41}{9}$ **66.** $6\frac{7}{13}$ $\frac{85}{13}$ **67.** $8\frac{5}{14}$ $\frac{117}{14}$

APPLYING THE CONCEPTS

68. What fraction of the states in the United States of America begins with the letter M? What fraction of the states begins and ends with a vowel?
$\frac{4}{25}$ $\frac{4}{25}$

69. [W] Find the business section of your local newspaper. Choose a stock and record the fluctuations in the stock price for one week. Explain the part that fractions play in reporting the price and change in price of the stock.
Answers will vary

70. [W] Explain in your own words the procedure for rewriting a mixed number as an improper fraction.
Answers will vary

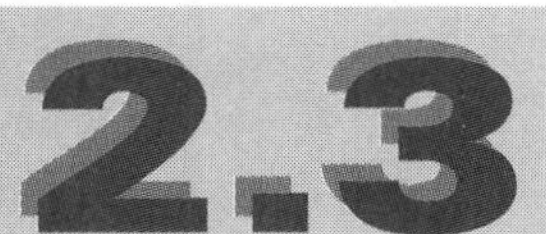

Writing Equivalent Fractions

Objective A To find equivalent fractions by raising to higher terms

INSTRUCTOR NOTE

Using a pizza can help some students understand equivalent fractions. By cutting the pizza into say 8 pieces, students are able to see

$\frac{1}{2} = \frac{4}{8}$

$\frac{1}{4} = \frac{2}{8}$

Equal fractions with different denominators are called **equivalent fractions.**

$\frac{4}{6}$ is equivalent to $\frac{2}{3}$.

Remember that the Multiplication Property of One stated that the product of a number and one is the number. This is true for fractions as well as whole numbers. This property can be used to write equivalent fractions.

$$\frac{2}{3} \times 1 = \frac{2}{3} \times \frac{1}{1} = \frac{2 \cdot 1}{3 \cdot 1} = \frac{2}{3}$$

$$\frac{2}{3} \times 1 = \frac{2}{3} \times \boxed{\frac{2}{2}} = \frac{2 \cdot 2}{3 \cdot 2} = \frac{4}{6}$$ $\frac{4}{6}$ is equivalent to $\frac{2}{3}$.

$$\frac{2}{3} \times 1 = \frac{2}{3} \times \boxed{\frac{4}{4}} = \frac{2 \cdot 4}{3 \cdot 4} = \frac{8}{12}$$ $\frac{8}{12}$ is equivalent to $\frac{2}{3}$.

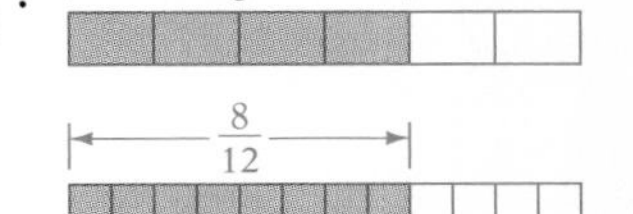

$\frac{2}{3}$ was rewritten as the equivalent fractions $\frac{4}{6}$ and $\frac{8}{12}$.

➡ Write a fraction that is equivalent to $\frac{5}{8}$ and has a denominator of 32.

$8\overline{)32}$ = 4 • Divide the larger denominator by the smaller.

$$\frac{5}{8} = \frac{5 \cdot 4}{8 \cdot 4} = \frac{20}{32}$$ • Multiply the numerator and denominator of the given fraction by the quotient (4).

$\frac{20}{32}$ is equivalent to $\frac{5}{8}$.

Example 1 Write $\frac{2}{3}$ as an equivalent fraction that has a denominator of 42.

Solution $3\overline{)42}$ = 14 $\quad \frac{2}{3} = \frac{2 \cdot 14}{3 \cdot 14} = \frac{28}{42}$

$\frac{28}{42}$ is equivalent to $\frac{2}{3}$.

You Try It 1 Write $\frac{3}{5}$ as an equivalent fraction that has a denominator of 45.

Your solution $\frac{27}{45}$

Example 2 Write 4 as a fraction that has a denominator of 12.

Solution Write 4 as $\frac{4}{1}$.

$1\overline{)12}$ = 12 $\quad 4 = \frac{4 \cdot 12}{1 \cdot 12} = \frac{48}{12}$

$\frac{48}{12}$ is equivalent to 4.

You Try It 2 Write 6 as a fraction that has a denominator of 18.

Your solution $\frac{108}{18}$

Solutions on p. A12

Objective B To write a fraction in simplest form

A fraction is in **simplest form** when there are no common factors in the numerator and the denominator.

The fractions $\frac{4}{6}$ and $\frac{2}{3}$ are equivalent fractions.

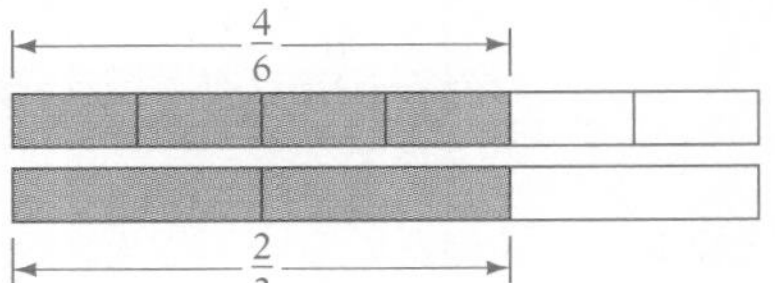

$\frac{4}{6}$ has been written in simplest form as $\frac{2}{3}$.

The Multiplication Property of One can be used to write fractions in simplest form. Write the numerator and denominator of the given fraction as a product of factors. Write factors common to both the numerator and denominator as an improper fraction equivalent to 1.

$$\frac{4}{6} = \frac{2 \cdot 2}{2 \cdot 3} = \boxed{\frac{2}{2}} \cdot \frac{2}{3} = \boxed{\frac{2}{2}} \cdot \frac{2}{3} = 1 \cdot \frac{2}{3} = \frac{2}{3}$$

INSTRUCTOR NOTE

As mentioned earlier, one of the main pedagogical features of this text is the paired examples. Using the model of the Example, students should work the You Try It. A *complete solution* can be found in the appendix so that students can not only check the answer, but also their work.

The process of eliminating common factors is displayed with slashes through the common factors as shown at the right.

$$\frac{4}{6} = \frac{\overset{1}{\cancel{2}} \cdot 2}{\underset{1}{\cancel{2}} \cdot 3} = \frac{2}{3}$$

To write a fraction in simplest form, eliminate the common factors.

$$\frac{18}{30} = \frac{\overset{1}{\cancel{2}} \cdot \overset{1}{\cancel{3}} \cdot 3}{\underset{1}{\cancel{2}} \cdot \underset{1}{\cancel{3}} \cdot 5} = \frac{3}{5}$$

An improper fraction can be changed to a mixed number.

$$\frac{22}{6} = \frac{\overset{1}{\cancel{2}} \cdot 11}{\underset{1}{\cancel{2}} \cdot 3} = \frac{11}{3} = 3\frac{2}{3}$$

Example 3 Write $\frac{15}{40}$ in simplest form.

Solution $\frac{15}{40} = \frac{3 \cdot \overset{1}{\cancel{5}}}{2 \cdot 2 \cdot 2 \cdot \underset{1}{\cancel{5}}} = \frac{3}{8}$

You Try It 3 Write $\frac{16}{24}$ in simplest form.

Your solution $\frac{2}{3}$

Example 4 Write $\frac{6}{42}$ in simplest form.

Solution $\frac{6}{42} = \frac{\overset{1}{\cancel{2}} \cdot \overset{1}{\cancel{3}}}{\underset{1}{\cancel{2}} \cdot \underset{1}{\cancel{3}} \cdot 7} = \frac{1}{7}$

You Try It 4 Write $\frac{8}{56}$ in simplest form.

Your solution $\frac{1}{7}$

Example 5 Write $\frac{8}{9}$ in simplest form.

Solution $\frac{8}{9} = \frac{2 \cdot 2 \cdot 2}{3 \cdot 3} = \frac{8}{9}$

$\frac{8}{9}$ is already in simplest form because there are no common factors in the numerator and denominator.

You Try It 5 Write $\frac{15}{32}$ in simplest form.

Your solution $\frac{15}{32}$

Example 6 Write $\frac{30}{12}$ in simplest form.

Solution $\frac{30}{12} = \frac{\overset{1}{\cancel{2}} \cdot \overset{1}{\cancel{3}} \cdot 5}{\underset{1}{\cancel{2}} \cdot 2 \cdot \underset{1}{\cancel{3}}} = \frac{5}{2} = 2\frac{1}{2}$

You Try It 6 Write $\frac{48}{36}$ in simplest form.

Your solution $1\frac{1}{3}$

Solutions on p. A13

2.3 Exercises

Objective A

Write an equivalent fraction with the given denominator.

1. $\frac{1}{2} = \frac{5}{10}$ **2.** $\frac{1}{4} = \frac{4}{16}$ **3.** $\frac{3}{16} = \frac{9}{48}$ **4.** $\frac{5}{9} = \frac{45}{81}$ **5.** $\frac{3}{8} = \frac{12}{32}$

6. $\frac{7}{11} = \frac{21}{33}$ **7.** $\frac{3}{17} = \frac{9}{51}$ **8.** $\frac{7}{10} = \frac{63}{90}$ **9.** $\frac{3}{4} = \frac{12}{16}$ **10.** $\frac{5}{8} = \frac{20}{32}$

11. $3 = \frac{27}{9}$ **12.** $5 = \frac{125}{25}$ **13.** $\frac{1}{3} = \frac{20}{60}$ **14.** $\frac{1}{16} = \frac{3}{48}$ **15.** $\frac{11}{15} = \frac{44}{60}$

16. $\frac{3}{50} = \frac{18}{300}$ **17.** $\frac{2}{3} = \frac{12}{18}$ **18.** $\frac{5}{9} = \frac{20}{36}$ **19.** $\frac{5}{7} = \frac{35}{49}$ **20.** $\frac{7}{8} = \frac{28}{32}$

21. $\frac{5}{9} = \frac{10}{18}$ **22.** $\frac{11}{12} = \frac{33}{36}$ **23.** $7 = \frac{21}{3}$ **24.** $9 = \frac{36}{4}$ **25.** $\frac{7}{9} = \frac{35}{45}$

26. $\frac{5}{6} = \frac{35}{42}$ **27.** $\frac{15}{16} = \frac{60}{64}$ **28.** $\frac{11}{18} = \frac{33}{54}$ **29.** $\frac{3}{14} = \frac{21}{98}$ **30.** $\frac{5}{6} = \frac{120}{144}$

31. $\frac{5}{8} = \frac{30}{48}$ **32.** $\frac{7}{12} = \frac{56}{96}$ **33.** $\frac{5}{14} = \frac{15}{42}$ **34.** $\frac{2}{3} = \frac{28}{42}$ **35.** $\frac{17}{24} = \frac{102}{144}$

36. $\frac{5}{13} = \frac{65}{169}$ **37.** $\frac{3}{8} = \frac{153}{408}$ **38.** $\frac{9}{16} = \frac{153}{272}$ **39.** $\frac{17}{40} = \frac{340}{800}$ **40.** $\frac{9}{25} = \frac{360}{1000}$

Objective B

Write the fraction in simplest form.

41. $\frac{4}{12}$ $\frac{1}{3}$ **42.** $\frac{8}{22}$ $\frac{4}{11}$ **43.** $\frac{22}{44}$ $\frac{1}{2}$ **44.** $\frac{2}{14}$ $\frac{1}{7}$ **45.** $\frac{2}{12}$ $\frac{1}{6}$ **46.** $\frac{6}{25}$ $\frac{6}{25}$

47. $\frac{50}{75}$ $\frac{2}{3}$

48. $\frac{40}{36}$ $1\frac{1}{9}$

49. $\frac{12}{8}$ $1\frac{1}{2}$

50. $\frac{36}{9}$ 4

51. $\frac{0}{30}$ 0

52. $\frac{10}{10}$ 1

53. $\frac{9}{22}$ $\frac{9}{22}$

54. $\frac{14}{35}$ $\frac{2}{5}$

55. $\frac{75}{25}$ 3

56. $\frac{8}{60}$ $\frac{2}{15}$

57. $\frac{16}{84}$ $\frac{4}{21}$

58. $\frac{14}{45}$ $\frac{14}{45}$

59. $\frac{20}{44}$ $\frac{5}{11}$

60. $\frac{12}{35}$ $\frac{12}{35}$

61. $\frac{8}{36}$ $\frac{2}{9}$

62. $\frac{28}{44}$ $\frac{7}{11}$

63. $\frac{12}{16}$ $\frac{3}{4}$

64. $\frac{48}{35}$ $1\frac{13}{35}$

65. $\frac{16}{12}$ $1\frac{1}{3}$

66. $\frac{24}{18}$ $1\frac{1}{3}$

67. $\frac{24}{40}$ $\frac{3}{5}$

68. $\frac{44}{60}$ $\frac{11}{15}$

69. $\frac{8}{88}$ $\frac{1}{11}$

70. $\frac{9}{90}$ $\frac{1}{10}$

71. $\frac{144}{36}$ 4

72. $\frac{140}{297}$ $\frac{140}{297}$

73. $\frac{48}{144}$ $\frac{1}{3}$

74. $\frac{32}{120}$ $\frac{4}{15}$

75. $\frac{60}{100}$ $\frac{3}{5}$

76. $\frac{33}{110}$ $\frac{3}{10}$

77. $\frac{36}{16}$ $2\frac{1}{4}$

78. $\frac{80}{45}$ $1\frac{7}{9}$

79. $\frac{32}{160}$ $\frac{1}{5}$

APPLYING THE CONCEPTS

80. Use the diagram to express the following ratios in lowest form.

a. The ratio of the dots in the triangle to the total number of dots. $\frac{1}{4}$

b. The ratio of the number of dots in the triangle to the number of dots in the rectangle. $\frac{1}{2}$

c. The ratio of the number of dots common to the triangle and rectangle to the dots in the rectangle. $\frac{1}{4}$

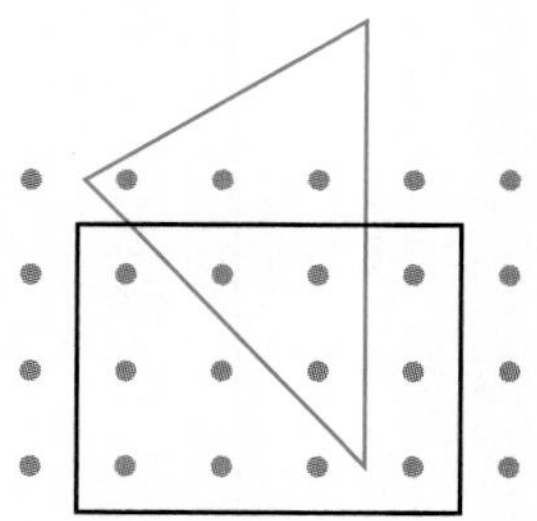

81. Show that $\frac{15}{24} = \frac{5}{8}$ by using a diagram.

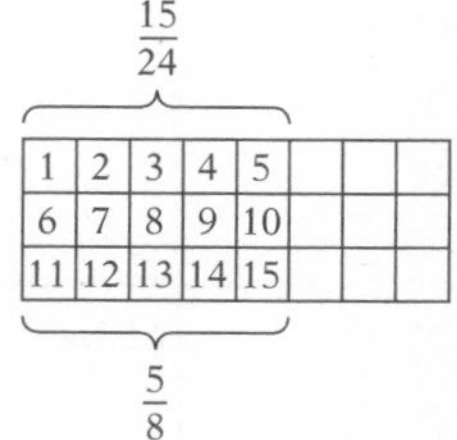

82. Explain the procedure for finding equivalent fractions.
[W] Answers will vary

83. Explain the procedure for reducing fractions. Answers will vary
[W]

2.4 Addition of Fractions and Mixed Numbers

Objective A To add fractions with the same denominator

Fractions with the same denominator are added by adding the numerators and placing the sum over the common denominator. After adding, write the sum in simplest form.

➡ Add: $\frac{2}{7} + \frac{4}{7}$

$$\begin{array}{r} \frac{2}{7} \\ + \frac{4}{7} \\ \hline \frac{6}{7} \end{array}$$

• Add the numerators and place the sum over the common denominator.

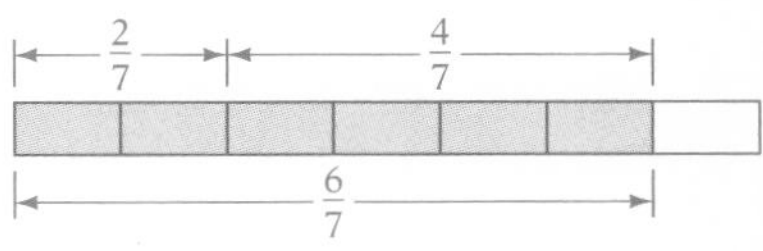

$$\frac{2}{7} + \frac{4}{7} = \frac{2+4}{7} = \frac{6}{7}$$

Example 1 Add: $\frac{5}{12} + \frac{11}{12}$

Solution

$$\begin{array}{r} \frac{5}{12} \\ + \frac{11}{12} \\ \hline \frac{16}{12} = \frac{4}{3} = 1\frac{1}{3} \end{array}$$

You Try It 1 Add: $\frac{3}{8} + \frac{7}{8}$

Your solution $1\frac{1}{4}$

Solution on p. A13

Objective B To add fractions with unlike denominators

INSTRUCTOR NOTE

Students would like to simply add numerators and denominators. They can be shown this does not yield a correct result by using money.

$\$\frac{1}{4} + \$\frac{1}{2} = \$\frac{3}{4}$, *not* $\$\frac{1}{3}$

To add fractions with unlike denominators, first rewrite the fractions as equivalent fractions with a common denominator. The common denominator is the LCM of the denominators of the fractions.

➡ Find the total of $\frac{1}{2}$ and $\frac{1}{3}$.

The common denominator is the LCM of 2 and 3. LCM = 6. The LCM of denominators is sometimes called the **least common denominator** (LCD).

Write equivalent fractions using the LCM.

$$\begin{array}{r} \frac{1}{2} = \frac{3}{6} \\ + \frac{1}{3} = \frac{2}{6} \\ \hline \end{array}$$

Add the fractions.

$$\begin{array}{r} \frac{1}{2} = \frac{3}{6} \\ + \frac{1}{3} = \frac{2}{6} \\ \hline \frac{5}{6} \end{array}$$

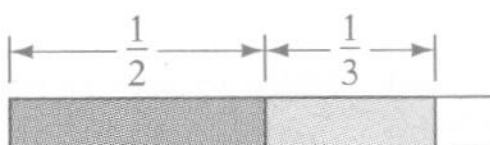

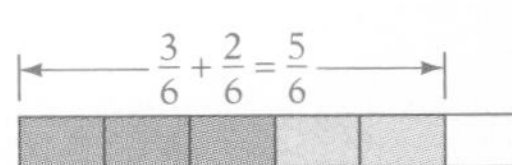

Example 2 Find $\frac{7}{12}$ more than $\frac{3}{8}$.

Solution

$$\begin{array}{r} \frac{3}{8} = \frac{9}{24} \\ + \frac{7}{12} = \frac{14}{24} \\ \hline \frac{23}{24} \end{array}$$

• The LCM of 8 and 12 is 24.

You Try It 2 Find the sum of $\frac{5}{12}$ and $\frac{9}{16}$.

Your solution $\frac{47}{48}$

Example 3 Add: $\frac{5}{8} + \frac{7}{9}$

Solution

$$\begin{array}{r} \frac{5}{8} = \frac{45}{72} \\ + \frac{7}{9} = \frac{56}{72} \\ \hline \frac{101}{72} = 1\frac{29}{72} \end{array}$$

• The LCM of 8 and 9 is 72.

You Try It 3 Add: $\frac{7}{8} + \frac{11}{15}$

Your solution $1\frac{73}{120}$

Example 4 Add: $\frac{2}{3} + \frac{3}{5} + \frac{5}{6}$

Solution

$$\begin{array}{r} \frac{2}{3} = \frac{20}{30} \\ \frac{3}{5} = \frac{18}{30} \\ + \frac{5}{6} = \frac{25}{30} \\ \hline \frac{63}{30} = 2\frac{3}{30} = 2\frac{1}{10} \end{array}$$

• The LCM of 3, 5, and 6 is 30.

You Try It 4 Add: $\frac{3}{4} + \frac{4}{5} + \frac{5}{8}$

Your solution $2\frac{7}{40}$

Solutions on p. A13

Objective C To add whole numbers, mixed numbers, and fractions

The sum of a whole number and a fraction is a mixed number.

➡ Add: $2 + \frac{2}{3}$

$$\boxed{2} + \frac{2}{3} = \boxed{\frac{6}{3}} + \frac{2}{3} = \frac{8}{3} = 2\frac{2}{3}$$

To add a whole number and a mixed number, write the fraction, then add the whole numbers.

➡ Add: $7\frac{2}{5} + 4$

Write the fraction.

$$\begin{array}{r} 7\frac{2}{5} \\ + 4 \\ \hline \frac{2}{5} \end{array}$$

Add the whole numbers.

$$\begin{array}{r} 7\frac{2}{5} \\ + 4 \\ \hline 11\frac{2}{5} \end{array}$$

To add two mixed numbers, add the fractional parts and then add the whole numbers. Remember to reduce the sum to simplest form.

➡ What is $5\frac{4}{9}$ added to $6\frac{14}{15}$?

The LCM of 9 and 15 is 48.

Add the fractional parts.

$$\begin{array}{r} 5\frac{4}{9} = 5\frac{20}{45} \\ +\ 6\frac{14}{15} = 6\frac{42}{45} \\ \hline \frac{62}{45} \end{array}$$

Add the whole numbers.

$$\begin{array}{r} 5\frac{4}{9} = 5\frac{20}{45} \\ +\ 6\frac{14}{15} = 6\frac{42}{45} \\ \hline 11\frac{62}{45} = 11 + 1\frac{17}{45} = 12\frac{17}{45} \end{array}$$

Example 5 Add: $5 + \frac{3}{8}$

Solution $5 + \frac{3}{8} = 5\frac{3}{8}$

You Try It 5 What is 7 added to $\frac{6}{11}$?

Your solution $7\frac{6}{11}$

Example 6 Find 17 increased by $3\frac{3}{8}$.

Solution

$$\begin{array}{r} 17 \\ +\ 3\frac{3}{8} \\ \hline 20\frac{3}{8} \end{array}$$

You Try It 6 Find the sum of 29 and $17\frac{5}{12}$.

Your solution $46\frac{5}{12}$

Example 7 Add: $5\frac{2}{3} + 11\frac{5}{6} + 12\frac{7}{9}$

Solution

$$\begin{array}{r} 5\frac{2}{3} = 5\frac{12}{18} \\ 11\frac{5}{6} = 11\frac{15}{18} \\ +\ 12\frac{7}{9} = 12\frac{14}{18} \\ \hline 28\frac{41}{18} = 30\frac{5}{18} \end{array}$$

• LCM = 18.

You Try It 7 Add: $7\frac{4}{5} + 6\frac{7}{10} + 13\frac{11}{15}$

Your solution $28\frac{7}{30}$

Example 8 Add: $11\frac{5}{8} + 7\frac{5}{9} + 8\frac{7}{15}$

Solution

$$\begin{array}{r} 11\frac{5}{8} = 11\frac{225}{360} \\ 7\frac{5}{9} = 7\frac{200}{360} \\ +\ 8\frac{7}{15} = 8\frac{168}{360} \\ \hline 26\frac{593}{360} = 27\frac{233}{360} \end{array}$$

• LCM = 360.

You Try It 8 Add: $9\frac{3}{8} + 17\frac{7}{12} + 10\frac{14}{15}$

Your solution $37\frac{107}{120}$

Solutions on p. A13

Objective D To solve application problems

Example 9

A rain gauge collected $2\frac{1}{3}$ inches of rain in October, $5\frac{1}{2}$ inches in November, and $3\frac{3}{8}$ inches in December. Find the total rainfall for the 3 months.

Strategy

To find the total rainfall for the 3 months, add the 3 amounts of rainfall $\left(2\frac{1}{3}, 5\frac{1}{2}, \text{ and } 3\frac{3}{8}\right)$.

Solution

$$\begin{aligned} 2\frac{1}{3} &= 2\frac{8}{24} \\ 5\frac{1}{2} &= 5\frac{12}{24} \\ +\ 3\frac{3}{8} &= 3\frac{9}{24} \\ \hline & 10\frac{29}{24} = 11\frac{5}{24} \end{aligned}$$

The total rainfall for the 3 months was $11\frac{5}{24}$ inches.

You Try It 9

On Monday, you spent $4\frac{1}{2}$ hours in class, $3\frac{3}{4}$ hours studying, and $1\frac{1}{3}$ hours driving. Find the number of hours spent on these three activities.

Your strategy

Your solution $9\frac{7}{12}$ hours

Example 10

Barbara Walsh worked 4 hours, $2\frac{1}{3}$ hours, and $5\frac{2}{3}$ hours this week on a part-time job. Barbara is paid \$6 an hour. How much did she earn this week?

Strategy

To find how much Barbara earned:

- Find the total number of hours worked.
- Multiply the total number of hours worked by the hourly wage (\$6).

Solution

$$\begin{array}{r} 4 \\ 2\frac{1}{3} \\ +\ 5\frac{2}{3} \\ \hline 11\frac{3}{3} \end{array} = 12 \text{ hours worked} \qquad \begin{array}{r} 12 \\ \times\ \$6 \\ \hline \$72 \end{array}$$

Barbara earned \$72 this week.

You Try It 10

Jeff Sapone, a carpenter, worked $1\frac{2}{3}$ hours of overtime on Monday, $3\frac{1}{3}$ hours of overtime on Tuesday, and 2 hours of overtime on Wednesday. At an overtime hourly rate of \$24, find Jeff's overtime pay for these 3 days.

Your strategy

Your solution \$168

Solutions on p. A14

2.4 Exercises

Objective A

Add.

1. $\frac{2}{7} + \frac{1}{7}$

$\frac{3}{7}$

2. $\frac{3}{11} + \frac{5}{11}$

$\frac{8}{11}$

3. $\frac{1}{2} + \frac{1}{2}$

1

4. $\frac{1}{3} + \frac{2}{3}$

1

5. $\frac{8}{11} + \frac{7}{11}$

$1\frac{4}{11}$

6. $\frac{9}{13} + \frac{7}{13}$

$1\frac{3}{13}$

7. $\frac{8}{5} + \frac{9}{5}$

$3\frac{2}{5}$

8. $\frac{5}{3} + \frac{7}{3}$

4

9. $\frac{3}{5} + \frac{8}{5} + \frac{3}{5}$

$2\frac{4}{5}$

10. $\frac{3}{8} + \frac{5}{8} + \frac{7}{8}$

$1\frac{7}{8}$

11. $\frac{3}{4} + \frac{1}{4} + \frac{5}{4}$

$2\frac{1}{4}$

12. $\frac{2}{7} + \frac{4}{7} + \frac{5}{7}$

$1\frac{4}{7}$

13. $\frac{3}{8} + \frac{7}{8} + \frac{1}{8}$

$1\frac{3}{8}$

14. $\frac{5}{12} + \frac{7}{12} + \frac{1}{12}$

$1\frac{1}{12}$

15. $\frac{4}{15} + \frac{7}{15} + \frac{11}{15}$

$1\frac{7}{15}$

16. $\frac{3}{4} + \frac{3}{4} + \frac{1}{4}$

$1\frac{3}{4}$

17. $\frac{3}{16} + \frac{5}{16} + \frac{7}{16}$

$\frac{15}{16}$

18. $\frac{5}{18} + \frac{11}{18} + \frac{17}{18}$

$1\frac{5}{6}$

19. $\frac{3}{11} + \frac{5}{11} + \frac{7}{11}$

$1\frac{4}{11}$

20. $\frac{5}{7} + \frac{4}{7} + \frac{5}{7}$

2

21. Find the sum of $\frac{4}{9}$ and $\frac{5}{9}$.

1

22. Find the sum of $\frac{5}{12}$, $\frac{1}{12}$, and $\frac{11}{12}$.

$1\frac{5}{12}$

23. Find the total of $\frac{5}{8}$, $\frac{3}{8}$, and $\frac{7}{8}$.

$1\frac{7}{8}$

24. Find the total of $\frac{4}{13}$, $\frac{7}{13}$, and $\frac{11}{13}$.

$1\frac{9}{13}$

Objective B

Add.

25. $\frac{1}{2} + \frac{2}{3}$

$1\frac{1}{6}$

26. $\frac{2}{3} + \frac{1}{4}$

$\frac{11}{12}$

27. $\frac{3}{14} + \frac{5}{7}$

$\frac{13}{14}$

28. $\frac{3}{5} + \frac{7}{10}$

$1\frac{3}{10}$

29. $\frac{5}{9} + \frac{7}{15}$

$1\frac{1}{45}$

30. $\frac{8}{15} + \frac{7}{20}$

$\frac{53}{60}$

31. $\frac{1}{6} + \frac{7}{9}$

$\frac{17}{18}$

32. $\frac{3}{8} + \frac{9}{14}$

$1\frac{1}{56}$

33. $\frac{5}{12} + \frac{5}{16}$

$\frac{35}{48}$

34. $\frac{13}{40} + \frac{12}{25}$

$\frac{161}{200}$

35. $\frac{3}{20} + \frac{7}{30}$

$\frac{23}{60}$

36. $\frac{5}{12} + \frac{7}{30}$

$\frac{13}{20}$

37. $\frac{2}{3} + \frac{6}{19}$

$\frac{56}{57}$

38. $\frac{1}{2} + \frac{3}{29}$

$\frac{35}{58}$

39. $\frac{3}{7} + \frac{4}{21}$

$\frac{13}{21}$

Add.

40. $\frac{1}{3}+\frac{5}{6}+\frac{7}{9}$
$1\frac{17}{18}$

41. $\frac{2}{3}+\frac{5}{6}+\frac{7}{12}$
$2\frac{1}{12}$

42. $\frac{5}{6}+\frac{1}{12}+\frac{5}{16}$
$1\frac{11}{48}$

43. $\frac{2}{9}+\frac{7}{15}+\frac{4}{21}$
$\frac{277}{315}$

44. $\frac{2}{3}+\frac{1}{5}+\frac{7}{12}$
$1\frac{9}{20}$

45. $\frac{3}{4}+\frac{4}{5}+\frac{7}{12}$
$2\frac{2}{15}$

46. $\frac{1}{4}+\frac{4}{5}+\frac{5}{9}$
$1\frac{109}{180}$

47. $\frac{2}{3}+\frac{3}{5}+\frac{7}{8}$
$2\frac{17}{120}$

48. $\frac{5}{16}+\frac{11}{18}+\frac{17}{24}$
$1\frac{91}{144}$

49. $\frac{3}{10}+\frac{14}{15}+\frac{9}{25}$
$1\frac{89}{150}$

50. $\frac{2}{3}+\frac{5}{8}+\frac{7}{9}$
$2\frac{5}{72}$

51. $\frac{1}{3}+\frac{2}{9}+\frac{7}{8}$
$1\frac{31}{72}$

52. Find $\frac{7}{8}$ added to $\frac{11}{12}$.
$1\frac{19}{24}$

53. Find $\frac{13}{18}$ added to $\frac{5}{12}$.
$1\frac{5}{36}$

54. What is $\frac{3}{8}$ added to $\frac{3}{5}$?
$\frac{39}{40}$

55. What is $\frac{5}{9}$ added to $\frac{7}{12}$?
$1\frac{5}{36}$

56. Find the sum of $\frac{3}{8}$, $\frac{5}{6}$, and $\frac{7}{12}$.
$1\frac{19}{24}$

57. Find the sum of $\frac{11}{12}$, $\frac{13}{24}$, and $\frac{4}{15}$.
$1\frac{29}{40}$

58. Find the total of $\frac{1}{2}$, $\frac{5}{8}$, and $\frac{7}{9}$.
$1\frac{65}{72}$

59. Find the total of $\frac{5}{14}$, $\frac{3}{7}$, and $\frac{5}{21}$.
$1\frac{1}{42}$

Objective C

Add.

60. $1\frac{1}{2} + 2\frac{1}{6}$
$3\frac{2}{3}$

61. $2\frac{2}{5} + 3\frac{3}{10}$
$5\frac{7}{10}$

62. $4\frac{1}{2} + 5\frac{7}{12}$
$10\frac{1}{12}$

63. $3\frac{3}{8} + 2\frac{5}{16}$
$5\frac{11}{16}$

64. $2\frac{7}{9} + 3\frac{5}{12}$
$6\frac{7}{36}$

65. $4\frac{7}{15} + 3\frac{11}{12}$
$8\frac{23}{60}$

66. $4 + 5\frac{2}{7}$
$9\frac{2}{7}$

67. $6\frac{8}{9} + 12$
$18\frac{8}{9}$

68. $3\frac{5}{8} + 2\frac{11}{20}$
$6\frac{7}{40}$

69. $4\frac{5}{12} + 6\frac{11}{18}$
$11\frac{1}{36}$

70. $10\frac{2}{7} + 7\frac{21}{35}$
$17\frac{31}{35}$

71. $16\frac{2}{3} + 8\frac{1}{4}$
$24\frac{11}{12}$

72. $7\frac{5}{12}+2\frac{9}{16}$
$9\frac{47}{48}$

73. $9\frac{1}{2}+3\frac{3}{11}$
$12\frac{17}{22}$

74. $6\frac{1}{3}+2\frac{3}{13}$
$8\frac{22}{39}$

75. $8\frac{21}{40}+6\frac{21}{32}$
$15\frac{29}{160}$

76. $8\frac{29}{30}+7\frac{11}{40}$
$16\frac{29}{120}$

77. $17\frac{5}{16}+3\frac{11}{24}$
$20\frac{37}{48}$

78. $17\frac{3}{8}+7\frac{7}{20}$
$24\frac{29}{40}$

79. $14\frac{7}{12}+29\frac{13}{21}$
$44\frac{17}{84}$

80. $5\frac{7}{8} + 27\frac{5}{12}$ $33\frac{7}{24}$

81. $7\frac{5}{6} + 3\frac{5}{9}$ $11\frac{7}{18}$

82. $7\frac{5}{9} + 2\frac{7}{12}$ $10\frac{5}{36}$

83. $3\frac{1}{2} + 2\frac{3}{4} + 1\frac{5}{6}$ $8\frac{1}{12}$

84. $2\frac{1}{2} + 3\frac{2}{3} + 4\frac{1}{4}$ $10\frac{5}{12}$

85. $3\frac{1}{3} + 7\frac{1}{5} + 2\frac{1}{7}$ $12\frac{71}{105}$

86. $3\frac{1}{2} + 3\frac{1}{5} + 8\frac{1}{9}$ $14\frac{73}{90}$

87. $6\frac{5}{9} + 6\frac{5}{12} + 2\frac{5}{18}$ $15\frac{1}{4}$

88. $2\frac{3}{8} + 4\frac{7}{12} + 3\frac{5}{16}$ $10\frac{13}{48}$

89. $2\frac{1}{8} + 4\frac{2}{9} + 5\frac{17}{18}$ $12\frac{7}{24}$

90. $6\frac{5}{6} + 17\frac{2}{9} + 18\frac{5}{27}$ $42\frac{13}{54}$

91. $4\frac{7}{20} + \frac{17}{80} + 25\frac{23}{60}$ $29\frac{227}{240}$

92. Find the sum of $2\frac{4}{9}$ and $5\frac{7}{12}$. $8\frac{1}{36}$

93. Find the sum of $4\frac{7}{16}$ and $7\frac{11}{12}$. $12\frac{17}{48}$

94. Find $2\frac{3}{8}$ more than $5\frac{1}{3}$. $7\frac{17}{24}$

95. Find $5\frac{5}{6}$ more than $3\frac{3}{8}$. $9\frac{5}{24}$

96. What is $4\frac{3}{4}$ added to $9\frac{1}{3}$? $14\frac{1}{12}$

97. What is $4\frac{8}{9}$ added to $9\frac{1}{6}$? $14\frac{1}{18}$

98. Find the total of $2\frac{2}{3}$, $4\frac{5}{8}$, and $2\frac{2}{9}$. $9\frac{37}{72}$

99. Find the total of $1\frac{5}{8}$, $3\frac{5}{6}$, and $7\frac{7}{24}$. $12\frac{3}{4}$

Objective D *Application Problems*

100. A family of four finds that $\frac{1}{3}$ of their income is spent on housing, $\frac{1}{8}$ is spent on transportation, and $\frac{1}{4}$ is spent on food. Find the total fractional amount of their income that is spent on these three items. $\frac{17}{24}$

101. A table 36 inches high has a top that is $1\frac{1}{8}$ inches thick. Find the total thickness of the table top after a $\frac{3}{16}$-inch veneer is applied. $1\frac{5}{16}$ inches

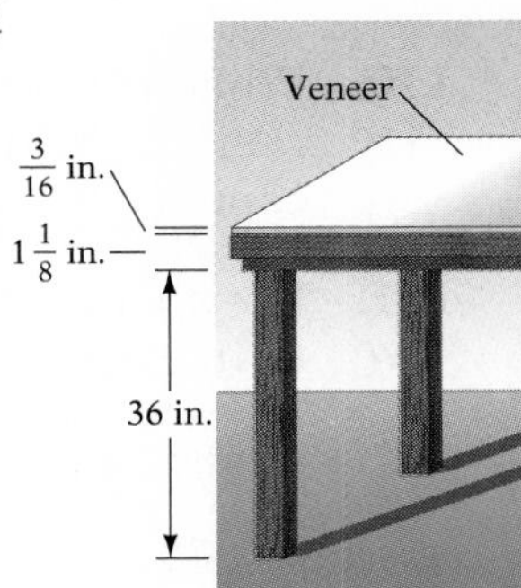

102. On March 29, the stock of AT&T was selling for $\$56\frac{3}{4}$. Then the price of the stock gained $\$7\frac{1}{8}$ per share by July 23. Find the price of the stock on July 23. $63\frac{7}{8}$

103. Find the length of the shaft.

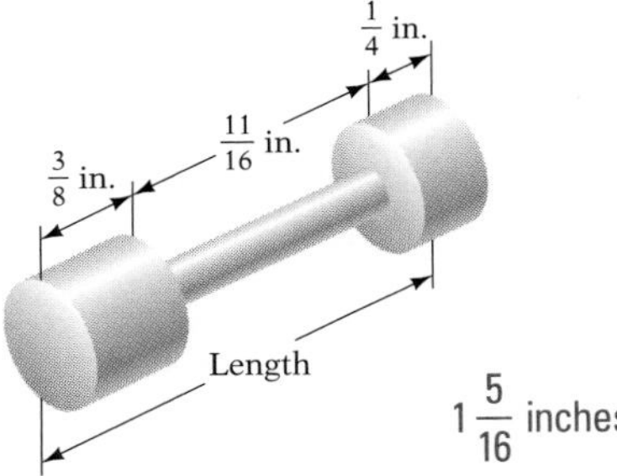

$1\frac{5}{16}$ inches

104. Find the length of the shaft.

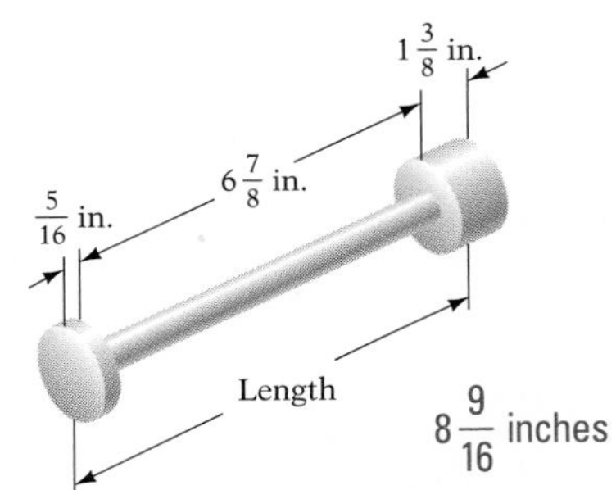

$8\frac{9}{16}$ inches

105. Assume that at the beginning of the month, American Airlines stock was selling at $\$158\frac{3}{8}$ per share. During the month, the stock gained $\$28\frac{3}{4}$ per share. Find the price of the stock at the end of the month. $\$187\frac{1}{8}$

106. You are working a part-time job that pays $7 an hour. You worked 5, $3\frac{3}{4}$, $2\frac{1}{3}$, $1\frac{1}{4}$, and $7\frac{2}{3}$ hours during the last five days.
a. Find the total number of hours worked during the last five days. 20 hours
b. Find your total wages for the five days. $140

107. Fred Thomson, a nurse, worked $2\frac{2}{3}$ hours of overtime on Monday, $1\frac{1}{4}$ hours on Wednesday, $1\frac{1}{3}$ on Friday, and $6\frac{3}{4}$ hours on Saturday.
a. Find the total number of overtime hours worked during the week. 12 hours
b. At an overtime hourly wage of $22 per hour, how much overtime pay does Fred receive? $264

108. Mt. Baldy had $5\frac{3}{4}$ inches of snow in December, $15\frac{1}{2}$ inches in January, and $9\frac{5}{8}$ inches in February. Find the total snowfall for the three months. $30\frac{7}{8}$ inches

109. The price of a utility stock was $\$26\frac{3}{16}$ at the end of one month. The monthly gains for the next three months were $\$1\frac{1}{2}$, $\$\frac{5}{16}$, and $\$2\frac{5}{8}$ per share. Find the value of the stock at the end of the three months. $\$30\frac{5}{8}$

APPLYING THE CONCEPTS

110. [W] What is a unit fraction? Find the sum of the three largest unit fractions. Is there a smallest unit fraction? If so, write it down. If not, explain why. A unit fraction is a fraction with a numerator of one. $1\frac{1}{12}$. No.

111. [W] Use a model to illustrate and explain the addition of fractions with unlike denominators. Answers will vary

112. [W] A survey was conducted to determine people's favorite color from among blue, green, red, purple, or other. The surveyor claims that $\frac{1}{3}$ of the people responded blue, $\frac{1}{6}$ responded green, $\frac{1}{8}$ responded red, $\frac{1}{12}$ responded purple, and $\frac{2}{5}$ responded some other color. Is this possible? Explain your answer. No

113. The following are the average portions of each day that a person spends for each activity: sleeping, $\frac{1}{3}$; working, $\frac{1}{3}$; personal hygiene, $\frac{1}{24}$; eating, $\frac{1}{8}$; rest and relaxation, $\frac{1}{12}$. Do these five activities account for an entire day? Explain your answer. No. These activities account for only 22 hours.

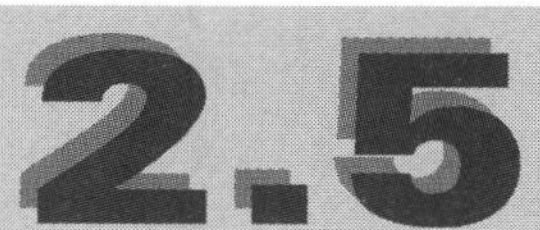

2.5 Subtraction of Fractions and Mixed Numbers

Objective A To subtract fractions with the same denominator

Fractions with the same denominator are subtracted by subtracting the numerators and placing the difference over the common denominator. After subtracting, write the fraction in simplest form.

➡ Subtract: $\frac{5}{7} - \frac{3}{7}$

$$\begin{array}{r} \frac{5}{7} \\ -\frac{3}{7} \\ \hline \frac{2}{7} \end{array}$$

• Subtract the numerators and place the difference over the common denominator

$\frac{5}{7}$

$\frac{3}{7}$ $\frac{2}{7}$

$$\frac{5}{7} - \frac{3}{7} = \frac{5-3}{7} = \frac{2}{7}$$

Example 1 Find $\frac{17}{30}$ less $\frac{11}{30}$.

Solution

$$\begin{array}{r} \frac{17}{30} \\ -\frac{11}{30} \\ \hline \frac{6}{30} = \frac{1}{5} \end{array}$$

You Try It 1 Subtract: $\frac{16}{27} - \frac{7}{27}$

Your solution $\frac{1}{3}$

Solution on p. A14

Objective B To subtract fractions with unlike denominators

INSTRUCTOR NOTE

Another money example that may reinforce the common denominator concept is: "Find 3 quarters minus 7 dimes." A common denominator would be to exchange all coins for nickels. Three quarters equals 15 nickels; 7 dimes equals 14 nickels.

$$\frac{3}{4} - \frac{7}{10} = \frac{15}{20} - \frac{14}{20} = \frac{1}{20}$$

To subtract fractions with unlike denominators, first rewrite the fractions as equivalent fractions with a common denominator. As with adding fractions, the common denominator is the LCM of the denominators of the fractions.

➡ Subtract: $\frac{5}{6} - \frac{1}{4}$

The common denominator is the LCM of 6 and 4. LCM = 12.

Build equivalent fractions using the LCM.

$$\begin{array}{r} \frac{5}{6} = \frac{10}{12} \\ -\frac{1}{4} = \frac{3}{12} \\ \hline \end{array}$$

Subtract the fractions.

$$\begin{array}{r} \frac{5}{6} = \frac{10}{12} \\ -\frac{1}{4} = \frac{3}{12} \\ \hline \frac{7}{12} \end{array}$$

$\frac{5}{6}$

$\frac{1}{4}$

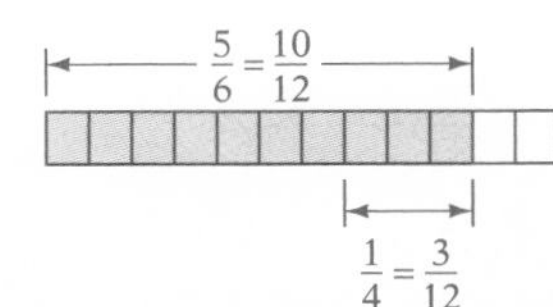

$\frac{5}{6} = \frac{10}{12}$

$\frac{1}{4} = \frac{3}{12}$

$\frac{10}{12} - \frac{3}{12} = \frac{7}{12}$

Example 2 Subtract: $\frac{11}{16} - \frac{5}{12}$

Solution

$$\begin{aligned} \frac{11}{16} &= \frac{33}{48} \\ -\frac{5}{12} &= \frac{20}{48} \\ \hline &\quad \frac{13}{48} \end{aligned}$$

• LCM = 48

You Try It 2 Subtract: $\frac{13}{18} - \frac{7}{24}$

Your solution $\frac{31}{72}$

Solution on p. A14

Objective C To subtract whole numbers, mixed numbers, and fractions

To subtract mixed numbers without borrowing, subtract the fractional parts and then subtract the whole numbers.

➡ Subtract: $5\frac{5}{6} - 2\frac{3}{4}$

Subtract the fractional parts.

$$\begin{aligned} 5\frac{5}{6} &= 5\frac{10}{12} \\ -2\frac{3}{4} &= 2\frac{9}{12} \\ \hline &\quad \frac{1}{12} \end{aligned}$$

• The LCM of 6 and 4 is 12.

Subtract the whole numbers.

$$\begin{aligned} 5\frac{5}{6} &= 5\frac{10}{12} \\ -2\frac{3}{4} &= 2\frac{9}{12} \\ \hline &\quad 3\frac{1}{12} \end{aligned}$$

Subtraction of mixed numbers sometimes involves borrowing.

➡ Subtract: $5 - 2\frac{5}{8}$

Borrow 1 from 5.

$$\begin{aligned} 5 &= \overset{4}{\cancel{5}}1 \\ -2\frac{5}{8} &= 2\frac{5}{8} \\ \hline \end{aligned}$$

Write 1 as a fraction so that the fractions have the same denominators.

$$\begin{aligned} 5 &= 4\frac{8}{8} \\ -2\frac{5}{8} &= 2\frac{5}{8} \\ \hline \end{aligned}$$

Subtract the mixed numbers.

$$\begin{aligned} 5 &= 4\frac{8}{8} \\ -2\frac{5}{8} &= 2\frac{5}{8} \\ \hline &\quad 2\frac{3}{8} \end{aligned}$$

➡ Subtract: $7\frac{1}{6} - 2\frac{5}{8}$

Write equivalent fractions using the LCM.

$$\begin{aligned} 7\frac{1}{6} &= 7\frac{4}{24} \\ -2\frac{5}{8} &= 2\frac{15}{24} \\ \hline \end{aligned}$$

Borrow 1 from 7. Add the 1 to $\frac{4}{24}$. Write $1\frac{4}{24}$ as $\frac{28}{24}$.

$$\begin{aligned} 7\frac{1}{6} &= \overset{6}{\cancel{7}}1\frac{4}{24} = 6\frac{28}{24} \\ -2\frac{5}{8} &= \ \ 2\frac{15}{24} = 2\frac{15}{24} \\ \hline \end{aligned}$$

Subtract the mixed numbers.

$$\begin{aligned} 7\frac{1}{6} &= 6\frac{28}{24} \\ -2\frac{5}{8} &= 2\frac{15}{24} \\ \hline &\quad 4\frac{13}{24} \end{aligned}$$

Example 3 Subtract: $15\frac{7}{8} - 12\frac{2}{3}$

Solution LCM = 24.

$$\begin{array}{r} 15\frac{7}{8} = 15\frac{21}{24} \\ -\ 12\frac{2}{3} = 12\frac{16}{24} \\ \hline 3\frac{5}{24} \end{array}$$

You Try It 3 Subtract: $17\frac{5}{9} - 11\frac{5}{12}$

Your solution $6\frac{5}{36}$

Example 4 Subtract: $9 - 4\frac{3}{11}$

Solution LCM = 11.

$$\begin{array}{r} 9 \quad = 8\frac{11}{11} \\ -\ 4\frac{3}{11} = 4\frac{3}{11} \\ \hline 4\frac{8}{11} \end{array}$$

You Try It 4 Subtract: $8 - 2\frac{4}{13}$

Your solution $5\frac{9}{13}$

Example 5 Find $11\frac{5}{12}$ decreased by $2\frac{11}{16}$.

Solution LCM = 48.

$$\begin{array}{r} 11\frac{5}{12} = 11\frac{20}{48} = 10\frac{68}{48} \\ -\ \ 2\frac{11}{16} = \ \ 2\frac{33}{48} = \ \ 2\frac{33}{48} \\ \hline 8\frac{35}{48} \end{array}$$

You Try It 5 Subtract: $21\frac{7}{9}$ and $7\frac{11}{12}$.

Your solution $13\frac{31}{36}$

Solutions on p. A14

Objective D To solve application problems

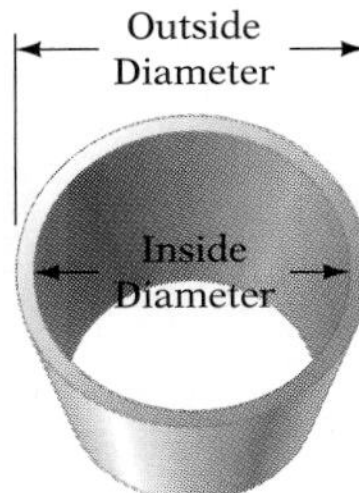

Find the inside diameter of the bushing with the outside diameter of $3\frac{3}{8}$ in. and the wall thickness is $\frac{1}{4}$ in.

$$\begin{array}{r} \frac{1}{4} \\ +\ \frac{1}{4} \\ \hline \frac{2}{4} = \frac{1}{2} \end{array}$$

- **Add $\frac{1}{4}$ to $\frac{1}{4}$ to find the total thickness of the two walls.**

$$\begin{array}{r} 3\frac{3}{8} = 3\frac{3}{8} = 2\frac{11}{8} \\ -\ \frac{1}{2} = \frac{4}{8} = \ \ \frac{4}{8} \\ \hline 2\frac{7}{8} \end{array}$$

- **Subtract the total thickness of the two walls to find the inside diameter.**

The inside diameter of the busing is $2\frac{7}{8}$ inches.

Example 6

A $2\frac{2}{3}$-inch piece is cut from a $6\frac{5}{8}$-inch board. How much of the board is left?

Strategy
To find the length remaining, subtract the length of the piece cut from the total length of the board.

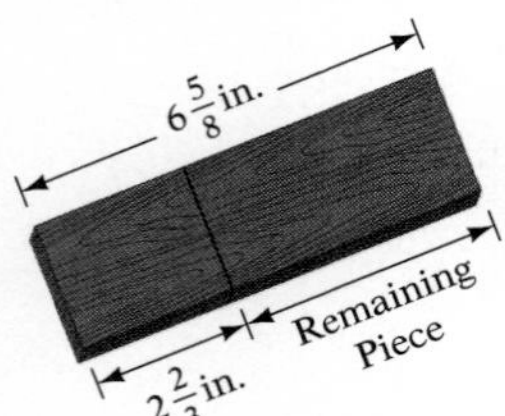

Solution

$$\begin{array}{r} 6\frac{5}{8} = 6\frac{15}{24} = 5\frac{39}{24} \\ -\ 2\frac{2}{3} = 2\frac{16}{24} = 2\frac{16}{24} \\ \hline 3\frac{23}{24} \end{array}$$

$3\frac{23}{24}$ inches of the board are left.

You Try It 6

A flight from New York to Los Angeles takes $5\frac{1}{2}$ hours. After the plane is in the air for $2\frac{3}{4}$ hours, how much flight time remains?

Your strategy

Your solution $2\frac{3}{4}$ hours

Example 7

Two painters are staining a house. In 1 day one painter stained $\frac{1}{3}$ of the house, and the other stained $\frac{1}{4}$ of the house. How much of the job remains to be done?

Strategy
To find how much of the job remains:

- Find the total amount of the house already stained $\left(\frac{1}{3} + \frac{1}{4}\right)$.
- Subtract the amount already stained from 1, which represents the complete job.

Solution

$$\begin{array}{r} \frac{1}{3} = \frac{4}{12} \\ +\ \frac{1}{4} = \frac{3}{12} \\ \hline \frac{7}{12} \end{array} \qquad \begin{array}{r} 1 = \frac{12}{12} \\ -\ \frac{7}{12} = \frac{7}{12} \\ \hline \frac{5}{12} \end{array}$$

$\frac{5}{12}$ of the house remains to be stained.

You Try It 7

A patient is put on a diet to lose 24 pounds in 3 months. The patient lost $7\frac{1}{2}$ pounds the first month and $5\frac{3}{4}$ pounds the second month. How much weight must be lost the third month to achieve the goal?

Your strategy

Your solution $10\frac{3}{4}$ pounds

Solutions on p. A15

2.5 Exercises

Objective A

Subtract.

1. $\dfrac{9}{17} - \dfrac{7}{17}$ $\dfrac{2}{17}$

2. $\dfrac{11}{15} - \dfrac{3}{15}$ $\dfrac{8}{15}$

3. $\dfrac{11}{12} - \dfrac{7}{12}$ $\dfrac{1}{3}$

4. $\dfrac{13}{15} - \dfrac{4}{15}$ $\dfrac{3}{5}$

5. $\dfrac{9}{20} - \dfrac{7}{20}$ $\dfrac{1}{10}$

6. $\dfrac{27}{40} - \dfrac{13}{40}$ $\dfrac{7}{20}$

7. $\dfrac{48}{55} - \dfrac{13}{55}$ $\dfrac{7}{11}$

8. $\dfrac{42}{65} - \dfrac{17}{65}$ $\dfrac{5}{13}$

9. $\dfrac{11}{24} - \dfrac{5}{24}$ $\dfrac{1}{4}$

10. $\dfrac{23}{30} - \dfrac{13}{30}$ $\dfrac{1}{3}$

11. $\dfrac{17}{42} - \dfrac{5}{42}$ $\dfrac{2}{7}$

12. $\dfrac{29}{48} - \dfrac{13}{48}$ $\dfrac{1}{3}$

13. What is $\frac{5}{14}$ less than $\frac{13}{14}$?
$\frac{4}{7}$

14. What is $\frac{7}{19}$ less than $\frac{17}{19}$?
$\frac{10}{19}$

15. Find the difference between $\frac{7}{8}$ and $\frac{5}{8}$.
$\frac{1}{4}$

16. Find the difference between $\frac{7}{12}$ and $\frac{5}{12}$.
$\frac{1}{6}$

17. What is $\frac{18}{23}$ minus $\frac{9}{23}$?
$\frac{9}{23}$

18. What is $\frac{7}{9}$ minus $\frac{3}{9}$?
$\frac{4}{9}$

19. Find $\frac{17}{24}$ decreased by $\frac{11}{24}$.
$\frac{1}{4}$

20. Find $\frac{19}{30}$ decreased by $\frac{11}{30}$.
$\frac{4}{15}$

Objective B

Subtract.

21. $\dfrac{2}{3} - \dfrac{1}{6}$ $\dfrac{1}{2}$

22. $\dfrac{7}{8} - \dfrac{5}{16}$ $\dfrac{9}{16}$

23. $\dfrac{5}{8} - \dfrac{2}{7}$ $\dfrac{19}{56}$

24. $\dfrac{5}{6} - \dfrac{3}{7}$ $\dfrac{17}{42}$

25. $\dfrac{5}{7} - \dfrac{3}{14}$ $\dfrac{1}{2}$

26. $\dfrac{7}{10} - \dfrac{3}{5}$ $\dfrac{1}{10}$

27. $\dfrac{5}{9} - \dfrac{7}{15}$ $\dfrac{4}{45}$

28. $\dfrac{8}{15} - \dfrac{7}{20}$ $\dfrac{11}{60}$

29. $\dfrac{7}{9} - \dfrac{1}{6}$ $\dfrac{11}{18}$

30. $\dfrac{9}{14} - \dfrac{3}{8}$ $\dfrac{15}{56}$

31. $\dfrac{5}{12} - \dfrac{5}{16}$ $\dfrac{5}{48}$

32. $\dfrac{12}{25} - \dfrac{13}{40}$ $\dfrac{31}{200}$

Subtract.

33. $\frac{46}{51} - \frac{3}{17}$ $\frac{37}{51}$

34. $\frac{9}{16} - \frac{17}{32}$ $\frac{1}{32}$

35. $\frac{21}{35} - \frac{5}{14}$ $\frac{17}{70}$

36. $\frac{19}{40} - \frac{3}{16}$ $\frac{23}{80}$

37. $\frac{29}{60} - \frac{3}{40}$ $\frac{49}{120}$

38. What is $\frac{3}{5}$ less than $\frac{11}{12}$? $\frac{19}{60}$

39. What is $\frac{5}{9}$ less than $\frac{11}{15}$? $\frac{8}{45}$

40. Find the difference between $\frac{11}{24}$ and $\frac{7}{18}$. $\frac{5}{72}$

41. Find the difference between $\frac{9}{14}$ and $\frac{5}{42}$. $\frac{11}{21}$

42. Find $\frac{11}{12}$ decreased by $\frac{11}{15}$. $\frac{11}{60}$

43. Find $\frac{17}{20}$ decreased by $\frac{7}{15}$. $\frac{23}{60}$

44. What is $\frac{13}{20}$ minus $\frac{1}{6}$? $\frac{29}{60}$

45. What is $\frac{5}{6}$ minus $\frac{7}{9}$? $\frac{1}{18}$

Objective C

Subtract.

46. $5\frac{7}{12} - 2\frac{5}{12}$ $3\frac{1}{6}$

47. $16\frac{11}{15} - 11\frac{8}{15}$ $5\frac{1}{5}$

48. $72\frac{21}{23} - 16\frac{17}{23}$ $56\frac{4}{23}$

49. $19\frac{16}{17} - 9\frac{7}{17}$ $10\frac{9}{17}$

50. $6\frac{1}{3} - 2$ $4\frac{1}{3}$

51. $5\frac{7}{8} - 1$ $4\frac{7}{8}$

52. $10 - 6\frac{1}{3}$ $3\frac{2}{3}$

53. $3 - 2\frac{5}{21}$ $\frac{16}{21}$

54. $6\frac{2}{5} - 4\frac{4}{5}$ $1\frac{3}{5}$

55. $16\frac{3}{8} - 10\frac{7}{8}$ $5\frac{1}{2}$

56. $25\frac{4}{9} - 16\frac{7}{9}$ $8\frac{2}{3}$

57. $8\frac{3}{7} - 2\frac{6}{7}$ $5\frac{4}{7}$

58. $16\frac{2}{5} - 8\frac{4}{9}$ $7\frac{43}{45}$

59. $23\frac{7}{8} - 16\frac{2}{3}$ $7\frac{5}{24}$

60. $6 - 4\frac{3}{5}$ $1\frac{2}{5}$

61. $65\frac{8}{35} - 16\frac{11}{14}$ $48\frac{31}{70}$

62. $82\frac{4}{33} - 16\frac{5}{22}$ $65\frac{59}{66}$

63. $101\frac{2}{9} - 16$ $85\frac{2}{9}$

64. $77\frac{5}{18} - 61$ $16\frac{5}{18}$

65. $17 - 7\frac{8}{13}$ $9\frac{5}{13}$

66. What is $5\frac{3}{8}$ less than $8\frac{1}{9}$?
$2\frac{53}{72}$

67. What is $7\frac{3}{5}$ less than $23\frac{3}{20}$?
$15\frac{11}{20}$

68. Find the difference between $9\frac{2}{7}$ and $3\frac{1}{4}$.
$6\frac{1}{28}$

69. Find the difference between $12\frac{3}{8}$ and $7\frac{5}{12}$.
$4\frac{23}{24}$

70. What is $5\frac{3}{5}$ minus $2\frac{5}{6}$?
$2\frac{23}{30}$

71. What is $10\frac{5}{9}$ minus $5\frac{11}{15}$?
$4\frac{37}{45}$

72. Find $6\frac{1}{3}$ decreased by $3\frac{3}{5}$.
$2\frac{11}{15}$

73. Find $8\frac{5}{8}$ decreased $6\frac{5}{6}$.
$1\frac{19}{24}$

Objective D *Application Problems*

74. Find the missing dimension.

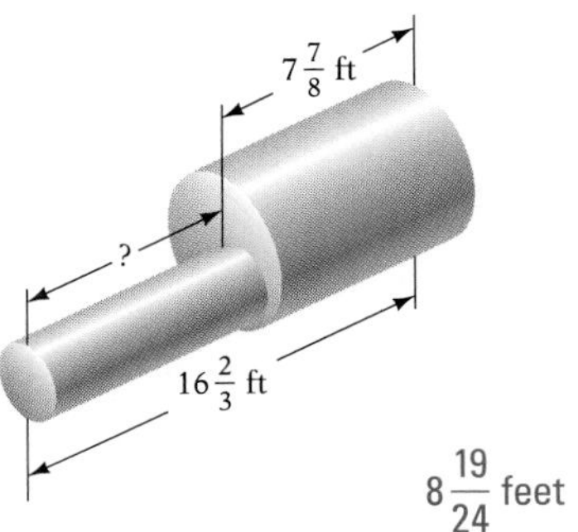

$8\frac{19}{24}$ feet

75. Find the missing dimension.

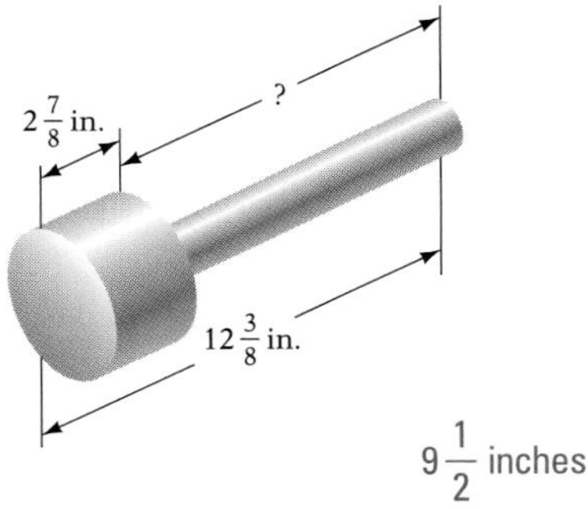

$9\frac{1}{2}$ inches

76. The horses in the Kentucky Derby run $1\frac{1}{4}$ mile. In the Belmont Stakes they run $1\frac{1}{2}$ mile, and in the Preakness Stakes they run $1\frac{3}{16}$ mile. How much farther do the horses run in the Kentucky Derby than in the Preakness Stakes? How much farther do they run in the Belmont Stakes than in the Preakness Stakes? $\frac{1}{16}$ mile $\frac{5}{16}$ mile

77. In the running high jump in the 1948 Summer Olympic Games, Alice Coachman's distance was $66\frac{1}{8}$ in. In the same event in the 1956 Summer Olympics, Mildred McDaniel jumped $69\frac{1}{4}$ in., and in the 1988 Olympic Games Louise Ritter jumped 80 in. Find the difference between Ritter's distance and Coachman's distance. Find the difference between Ritter's distance and McDaniel's distance.

$13\frac{7}{8}$ inches $10\frac{3}{4}$ inches

78. Two hikers plan a 3-day $27\frac{1}{2}$-mile backpack trip carrying a total of 80 pounds. The hikers plan to travel $7\frac{3}{8}$ miles the first day and $10\frac{1}{3}$ miles the second day.
a. How many miles do the hikers plan to travel the first two days?
b. How many miles will be left to travel on the third day?

a. $17\frac{17}{24}$ miles b. $9\frac{19}{24}$ miles

79. A 12-mile walkathon has three checkpoints. The first is $3\frac{3}{8}$ miles from the starting point. The second checkpoint is $4\frac{1}{3}$ miles from the first.

a. How many miles is it from the starting point to the second checkpoint? $7\frac{17}{24}$ miles

b. How many miles is it from the second checkpoint to the finish line? $4\frac{7}{24}$ miles

80. A patient with high blood pressure is put on a diet to lose 38 pounds in 3 months. The patient loses $12\frac{1}{2}$ pounds the first month and $16\frac{3}{8}$ pounds the second month. How much weight must be lost the third month for the goal to be achieved? $9\frac{1}{8}$ pounds

81. A wrestler is entered in the 172-pound weight class in the conference finals coming up in 3 weeks. The wrestler needs to lose $12\frac{3}{4}$ pounds. If $5\frac{1}{4}$ pounds are lost the first week and $4\frac{1}{4}$ the second week, how many pounds must be lost the third week for the desired weight to be reached? $3\frac{1}{4}$ pounds

APPLYING THE CONCEPTS

The figure at the right shows a selected portion of the New York Stock Exchange stock prices. The figure shows the closing Friday price of the stocks, the change in price for the week, and the Friday change in the price of the stocks. To find the price of "AsiaPc" at the beginning of the week, subtract $2\frac{1}{2}$ from 23. Use the figure for Exercises 82–85.

	Div.	PE	Wk. vol (100's)	Rel Stance	Fri. Close	Fri. Chg.	Wk. Chg.
AquilaGas n	...	13	2486	41	13	+1/8	+1 1/4
Aracrz n	...	...	5750	86	12	+3/8	-1/4
Acadn n	2.42	33	1600	71	22 3/8	...	-1/8
ArchDn b	.10	17	68078	42	23 3/4	+1/4	+1 3/8
ArcCh	2.50	22	673	53	43 1/2	+1/2	+1 1/4
Argent n	1.72	9	2566	83	22 1/8	-1/4	-1/8
ArgFd n q	...	...	2537	86	14 1/4	...	-3/8
AriP pfW	1.23	...	1401	42	24 3/4	...	...
Arkla	.28	22	18140	14	7 3/4	...	+1/4
Armco	...	...	23472	39	6 1/4	+1/4	+1/4
ArmWI	1.20	33	9380	98	49	+5/8	+2 1/2
ArrowE	...	16	6323	69	38	+1/4	+1/8
Artra	...	...	745	69	6 3/4	+1/4	-1/8
Arvin	.76	17	4222	41	30 3/8	+3/8	+2 3/4
ARX	...	13	2073	88	3 5/8	...	+1/8
ASA q	2.00	...	3972	86	48 5/8	-5/8	-5/8
Asarco	.40	...	7814	48	22 1/4	+3/4	-1/2
AshOil	1.00	14	7429	68	32	+3/8	-1 1/8
AshOil pf	3.12	26	757	61	58 1/2	+1/8	-1 1/2
AsiaPc e q	1.74	...	8926	97	23	+5/8	+2 1/2

82. Find the price of one share of ArcCh stock at the beginning of the week. $\$42\frac{1}{4}$

83. Find the price of one share of Arvin stock at the beginning of the week. $\$27\frac{5}{8}$

84. Find the price of one share of Arvin stock at the beginning of Friday's trading. $30

85. Find the price of one share of ArmWl at the beginning of the week. $\$46\frac{1}{2}$

86. Fill in the square to produce a true statement: $5\frac{1}{3} - \square = 2\frac{1}{2}$. $2\frac{5}{6}$

87. Fill in the square to produce a true statement: $\square - 4\frac{1}{2} = 1\frac{5}{8}$. $6\frac{1}{8}$

88. Fill-in the blank squares at the right so that the sum of the numbers along any row, column, or diagonal is the same. The resulting square is called a magic square.

$\frac{3}{8}$	$\frac{3}{4}$	$\frac{3}{4}$
1	$\frac{5}{8}$	$\frac{1}{4}$
$\frac{1}{2}$	$\frac{1}{2}$	$\frac{7}{8}$

89. If $\frac{4}{15}$ of an electrician's income is spent for housing, what fraction of the electrician's income is not spent for housing? $\frac{11}{15}$

2.6 Multiplication of Fractions and Mixed Numbers

Objective A *To multiply fractions*

The product of two fractions is the product of the numerators over the product of the denominators.

➡ Multiply: $\frac{2}{3} \times \frac{4}{5}$

$$\frac{2}{3} \times \frac{4}{5} = \frac{2 \cdot 4}{3 \cdot 5} = \frac{8}{15}$$

- Multiply the numerators.
- Multiply the denominators.

The product $\frac{2}{3} \times \frac{4}{5}$ can be read "$\frac{2}{3}$ times $\frac{4}{5}$" or "$\frac{2}{3}$ of $\frac{4}{5}$."

Reading the times sign as "of" is useful in application problems.

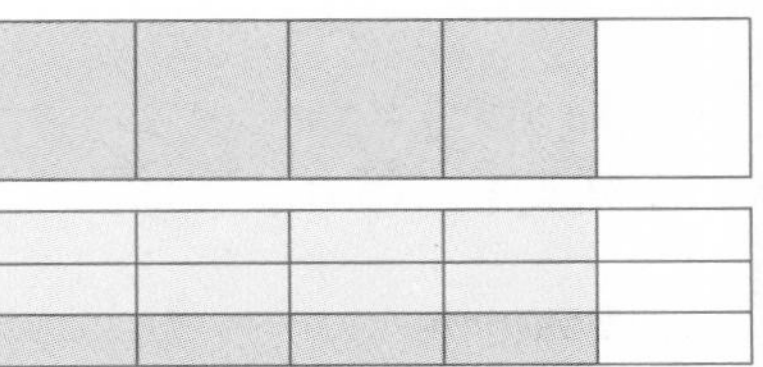

$\frac{4}{5}$ of the bar is shaded.

Shade $\frac{2}{3}$ of the $\frac{4}{5}$ already shaded.

$\frac{8}{15}$ of the bar is then shaded.

$\frac{2}{3}$ of $\frac{4}{5} = \frac{2}{3} \times \frac{4}{5} = \frac{8}{15}$

After multiplying two fractions, write the product in simplest form.

INSTRUCTOR NOTE

Some students will work this problem as follows:

$$\frac{\overset{1}{\cancel{3}}}{\underset{2}{\cancel{4}}} \times \frac{\overset{7}{\cancel{14}}}{\underset{5}{\cancel{15}}} = \frac{7}{10}$$

This method is essentially the same as writing the prime factorization and then dividing by common factors.

➡ Multiply: $\frac{3}{4} \times \frac{14}{15}$

$$\frac{3}{4} \times \frac{14}{15} = \frac{3 \cdot 14}{4 \cdot 15}$$

- Multiply numerators.
- Multiply denominators.

$$= \frac{3 \cdot 2 \cdot 7}{2 \cdot 2 \cdot 3 \cdot 5}$$

- Write the prime factorization of each number.

$$= \frac{\overset{1}{\cancel{3}} \cdot \overset{1}{\cancel{2}} \cdot 7}{\underset{1}{\cancel{2}} \cdot 2 \cdot \underset{1}{\cancel{3}} \cdot 5} = \frac{7}{10}$$

- Eliminate the common factors. Then multiply the factors of the numerator and denominator.

This example could also be worked by using the GCF.

$$\frac{3}{4} \times \frac{14}{15} = \frac{42}{60}$$

- Multiply numerators.
- Multiply denominators.

$$= \frac{6 \cdot 7}{6 \cdot 10}$$

- The GCF of 42 and 60 is 6. Factor 6 from 42 and 60.

$$= \frac{\overset{1}{\cancel{6}} \cdot 7}{\underset{1}{\cancel{6}} \cdot 10} = \frac{7}{10}$$

- Eliminate the GCF.

Example 1

Multiply: $\frac{4}{15}$ and $\frac{5}{28}$

Solution

$$\frac{4}{15} \times \frac{5}{28} = \frac{4 \cdot 5}{15 \cdot 28} = \frac{\overset{1}{\cancel{2}} \cdot \overset{1}{\cancel{2}} \cdot \overset{1}{\cancel{5}}}{3 \cdot \underset{1}{\cancel{5}} \cdot \underset{1}{\cancel{2}} \cdot \underset{1}{\cancel{2}} \cdot 7} = \frac{1}{21}$$

You Try It 1

Multiply: $\frac{4}{21}$ and $\frac{7}{44}$

Your solution $\frac{1}{33}$

Example 2

Find the product of $\frac{9}{20}$ and $\frac{33}{35}$.

Solution

$$\frac{9}{20} \times \frac{33}{35} = \frac{9 \cdot 33}{20 \cdot 35} = \frac{3 \cdot 3 \cdot 3 \cdot 11}{2 \cdot 2 \cdot 5 \cdot 5 \cdot 7} = \frac{297}{700}$$

You Try It 2

Find the product of $\frac{2}{21}$ and $\frac{10}{33}$.

Your solution $\frac{20}{693}$

Example 3

What is $\frac{14}{9}$ times $\frac{12}{7}$?

Solution

$$\frac{14}{9} \times \frac{12}{7} = \frac{14 \cdot 12}{9 \cdot 7} = \frac{2 \cdot \overset{1}{\cancel{7}} \cdot 2 \cdot 2 \cdot \overset{1}{\cancel{3}}}{3 \cdot \underset{1}{\cancel{3}} \cdot \underset{1}{\cancel{7}}} = \frac{8}{3} = 2\frac{2}{3}$$

You Try It 3

What is $\frac{16}{5}$ times $\frac{15}{24}$?

Your solution 2

Solutions on p. A15

Objective B To multiply whole numbers, mixed numbers, and fractions

To multiply a whole number by a fraction or mixed number, first write the whole number as a fraction with a denominator of 1.

➡ Multiply: $4 \times \frac{3}{7}$

$$4 \times \frac{3}{7} = \frac{4}{1} \times \frac{3}{7} = \frac{4 \cdot 3}{1 \cdot 7} = \frac{2 \cdot 2 \cdot 3}{7} = \frac{12}{7} = 1\frac{5}{7}$$

- Write 4 with a denominator of 1; then multiply the fractions.

When one or more of the factors in a product is a mixed number, write the mixed number as an improper fraction before multiplying.

➡ Multiply: $2\frac{1}{3} \times \frac{3}{14}$

$$2\frac{1}{3} \times \frac{3}{14} = \frac{7}{3} \times \frac{3}{14} = \frac{7 \cdot 3}{3 \cdot 14} = \frac{\overset{1}{\cancel{7}} \cdot \overset{1}{\cancel{3}}}{\underset{1}{\cancel{3}} \cdot 2 \cdot \underset{1}{\cancel{7}}} = \frac{1}{2}$$

- Write $2\frac{1}{3}$ as an improper fraction; then multiply the fractions.

Example 4

Multiply: $4\frac{5}{6} \times \frac{12}{13}$

Solution

$$4\frac{5}{6} \times \frac{12}{13} = \frac{29}{6} \times \frac{12}{13} = \frac{29 \cdot 12}{6 \cdot 13}$$

$$= \frac{29 \cdot \overset{1}{\cancel{2}} \cdot 2 \cdot \overset{1}{\cancel{3}}}{\underset{1}{\cancel{2}} \cdot \underset{1}{\cancel{3}} \cdot 13} = \frac{58}{13} = 4\frac{6}{13}$$

You Try It 4

Multiply: $5\frac{2}{5} \times \frac{5}{9}$

Your solution

3

Example 5

Find $5\frac{2}{3}$ times $4\frac{1}{2}$.

Solution

$$5\frac{2}{3} \times 4\frac{1}{2} = \frac{17}{3} \times \frac{9}{2} = \frac{17 \cdot 9}{3 \cdot 2}$$

$$= \frac{17 \cdot \overset{1}{\cancel{3}} \cdot 3}{\underset{1}{\cancel{3}} \cdot 2} = \frac{51}{2} = 25\frac{1}{2}$$

You Try It 5

Multiply: $3\frac{2}{5} \times 6\frac{1}{4}$

Your solution

$21\frac{1}{4}$

Example 6

Multiply: $4\frac{2}{5} \times 7$

Solution

$$4\frac{2}{5} \times 7 = \frac{22}{5} \times \frac{7}{1} = \frac{22 \cdot 7}{5 \cdot 1}$$

$$= \frac{2 \cdot 11 \cdot 7}{5} = \frac{154}{5} = 30\frac{4}{5}$$

You Try It 6

Multiply: $3\frac{2}{7} \times 6$

Your solution

$19\frac{5}{7}$

Solutions on p. A15 and A16

Objective C To solve application problems

Length (ft)	*Weight (lb/ft)*
$6\frac{1}{2}$	$\frac{3}{8}$
$8\frac{5}{8}$	$1\frac{1}{4}$
$10\frac{3}{4}$	$2\frac{1}{2}$
$12\frac{7}{12}$	$4\frac{1}{3}$

The table at the left lists the length of steel rods and the weight per foot. The weight per foot is measured in pounds for each foot of rod and is abbreviated as (lb/ft).

➡ Find the weight of the steel bar that is $10\frac{3}{4}$ feet long.

Strategy
To find the weight of a steel bar, multiply its length by the weight per foot.

Solution $10\frac{3}{4} \times 2\frac{1}{2} = \frac{43}{4} \times \frac{5}{2} = \frac{43 \cdot 5}{4 \cdot 2} = \frac{215}{8} = 26\frac{7}{8}$

The weight of the $10\frac{3}{4}$-ft rod is $26\frac{7}{8}$ lb.

Example 7

An electrician earns $150 for each day worked. What are the electrician's earnings for working $4\frac{1}{2}$ days?

Strategy

To find the electrician's total earnings, multiply the daily earnings ($150) by the number of days worked $\left(4\frac{1}{2}\right)$.

Solution $150 \times 4\frac{1}{2} = \frac{150}{1} \times \frac{9}{2}$

$$= \frac{150 \cdot 9}{1 \cdot 2}$$

$$= 675$$

The electrician's earnings are $675.

You Try It 7

Over the last 10 years, a house increased in value by $3\frac{1}{2}$ times. The price of the house 10 years ago was $30,000. What is the value of the house today?

Your strategy

Your solution $105,000

Example 8

The value of a small office building and the land on which it is built is $90,000. The value of the land is $\frac{1}{4}$ the total value. What is the value of the building (in dollars)?

Strategy

To find the value of the building:

- Find the value of the land $\left(\frac{1}{4} \times 90{,}000\right)$.
- Subtract the value of the land from the total value.

Solution $90{,}000 \times \frac{1}{4} = \frac{90{,}000}{4}$

$= 22{,}500$ value of the land

$$\begin{array}{r} 90{,}000 \\ -\ 22{,}500 \\ \hline 67{,}500 \end{array}$$

The value of the building is $67,500.

You Try It 8

A paint company bought a drying chamber and an air compressor for spray painting. The total cost of the two items was $60,000. The drying chamber's cost was $\frac{4}{5}$ of the total cost. What was the cost of the air compressor?

Your strategy

Your solution $12,000

Solutions on p. A16

2.6 Exercises

Objective A

Multiply.

1. $\frac{2}{3} \times \frac{7}{8}$ $\frac{7}{12}$

2. $\frac{1}{2} \times \frac{2}{3}$ $\frac{1}{3}$

3. $\frac{5}{16} \times \frac{7}{15}$ $\frac{7}{48}$

4. $\frac{3}{8} \times \frac{6}{7}$ $\frac{9}{28}$

5. $\frac{1}{2} \times \frac{5}{6}$ $\frac{5}{12}$

6. $\frac{1}{6} \times \frac{1}{8}$ $\frac{1}{48}$

7. $\frac{2}{5} \times \frac{5}{6}$ $\frac{1}{3}$

8. $\frac{11}{12} \times \frac{6}{7}$ $\frac{11}{14}$

9. $\frac{11}{12} \times \frac{3}{5}$ $\frac{11}{20}$

10. $\frac{2}{5} \times \frac{4}{9}$ $\frac{8}{45}$

11. $\frac{1}{6} \times \frac{6}{7}$ $\frac{1}{7}$

12. $\frac{3}{5} \times \frac{10}{11}$ $\frac{6}{11}$

13. $\frac{1}{5} \times \frac{5}{8}$ $\frac{1}{8}$

14. $\frac{6}{7} \times \frac{14}{15}$ $\frac{4}{5}$

15. $\frac{4}{9} \times \frac{15}{16}$ $\frac{5}{12}$

16. $\frac{8}{9} \times \frac{27}{4}$ 6

17. $\frac{3}{5} \times \frac{3}{10}$ $\frac{9}{50}$

18. $\frac{5}{6} \times \frac{1}{2}$ $\frac{5}{12}$

19. $\frac{3}{8} \times \frac{5}{12}$ $\frac{5}{32}$

20. $\frac{3}{10} \times \frac{6}{7}$ $\frac{9}{35}$

21. $\frac{16}{9} \times \frac{27}{8}$ 6

22. $\frac{5}{8} \times \frac{16}{15}$ $\frac{2}{3}$

23. $\frac{3}{2} \times \frac{4}{9}$ $\frac{2}{3}$

24. $\frac{5}{3} \times \frac{3}{7}$ $\frac{5}{7}$

25. $\frac{5}{16} \times \frac{12}{11}$ $\frac{15}{44}$

26. $\frac{7}{8} \times \frac{3}{14}$ $\frac{3}{16}$

27. $\frac{2}{9} \times \frac{1}{5}$ $\frac{2}{45}$

28. $\frac{1}{10} \times \frac{3}{8}$ $\frac{3}{80}$

29. $\frac{5}{12} \times \frac{6}{7}$ $\frac{5}{14}$

30. $\frac{1}{3} \times \frac{9}{2}$ $1\frac{1}{2}$

31. $\frac{15}{8} \times \frac{16}{3}$ 10

32. $\frac{5}{6} \times \frac{4}{15}$ $\frac{2}{9}$

33. $\frac{1}{2} \times \frac{2}{15}$ $\frac{1}{15}$

34. $\frac{3}{8} \times \frac{5}{16}$ $\frac{15}{128}$

35. $\frac{2}{9} \times \frac{15}{17}$ $\frac{10}{51}$

36. $\frac{5}{7} \times \frac{14}{15}$ $\frac{2}{3}$

37. $\frac{3}{8} \times \frac{15}{41}$ $\frac{45}{328}$

38. $\frac{5}{12} \times \frac{42}{65}$ $\frac{7}{26}$

39. $\frac{16}{33} \times \frac{55}{72}$ $\frac{10}{27}$

40. $\frac{8}{3} \times \frac{21}{32}$ $1\frac{3}{4}$

41. $\frac{12}{5} \times \frac{5}{3}$ 4

42. $\frac{17}{9} \times \frac{81}{17}$ 9

43. $\frac{16}{85} \times \frac{125}{84}$ $\frac{100}{357}$

44. $\frac{19}{64} \times \frac{48}{95}$ $\frac{3}{20}$

45. $\frac{18}{25} \times \frac{15}{40}$ $\frac{27}{100}$

46. Multiply $\frac{7}{12}$ and $\frac{15}{42}$.
$\frac{5}{24}$

47. Multiply $\frac{32}{9}$ and $\frac{3}{8}$.
$1\frac{1}{3}$

48. Find the product of $\frac{5}{9}$ and $\frac{3}{20}$.
$\frac{1}{12}$

49. Find the product of $\frac{7}{3}$ and $\frac{15}{14}$.
$2\frac{1}{2}$

50. What is $\frac{1}{2}$ times $\frac{8}{15}$?
$\frac{4}{15}$

51. What is $\frac{3}{8}$ times $\frac{12}{17}$?
$\frac{9}{34}$

Objective B

Multiply.

52. $4 \times \frac{3}{8}$
$1\frac{1}{2}$

53. $14 \times \frac{5}{7}$
10

54. $\frac{2}{3} \times 6$
4

55. $\frac{5}{12} \times 40$
$16\frac{2}{3}$

56. $\frac{1}{3} \times 1\frac{1}{3}$
$\frac{4}{9}$

57. $\frac{2}{5} \times 2\frac{1}{2}$
1

58. $1\frac{7}{8} \times \frac{4}{15}$
$\frac{1}{2}$

59. $2\frac{1}{5} \times \frac{5}{22}$
$\frac{1}{2}$

60. $55 \times \frac{3}{10}$
$16\frac{1}{2}$

61. $\frac{5}{14} \times 49$
$17\frac{1}{2}$

62. $4 \times 2\frac{1}{2}$
10

63. $9 \times 3\frac{1}{3}$
30

64. $2\frac{1}{7} \times 3$
$6\frac{3}{7}$

65. $5\frac{1}{4} \times 8$
42

66. $3\frac{2}{3} \times 5$
$18\frac{1}{3}$

67. $4\frac{2}{9} \times 3$
$12\frac{2}{3}$

68. $\frac{1}{2} \times 3\frac{3}{7}$
$1\frac{5}{7}$

69. $\frac{3}{8} \times 4\frac{4}{5}$
$1\frac{4}{5}$

70. $6\frac{1}{8} \times \frac{4}{7}$
$3\frac{1}{2}$

71. $5\frac{1}{3} \times \frac{5}{16}$
$1\frac{2}{3}$

72. $5\frac{1}{8} \times 5$
$25\frac{5}{8}$

73. $6\frac{1}{9} \times 2$
$12\frac{2}{9}$

74. $\frac{3}{8} \times 4\frac{1}{2}$
$1\frac{11}{16}$

75. $\frac{5}{7} \times 2\frac{1}{3}$
$1\frac{2}{3}$

76. $6 \times 2\frac{2}{3}$
16

77. $6\frac{1}{8} \times 0$
0

78. $1\frac{1}{3} \times 2\frac{1}{4}$
3

79. $2\frac{5}{8} \times \frac{3}{23}$
$\frac{63}{184}$

80. $2\frac{5}{8} \times 3\frac{2}{5}$
$8\frac{37}{40}$

81. $5\frac{3}{16} \times 5\frac{1}{3}$
$27\frac{2}{3}$

82. $3\frac{1}{7} \times 2\frac{1}{8}$
$6\frac{19}{28}$

83. $16\frac{5}{8} \times 1\frac{1}{16}$
$17\frac{85}{128}$

84. $2\frac{2}{5} \times 3\frac{1}{12}$ $7\frac{2}{5}$

85. $2\frac{2}{3} \times \frac{3}{20}$ $\frac{2}{5}$

86. $5\frac{1}{5} \times 3\frac{1}{13}$ 16

87. $3\frac{3}{4} \times 2\frac{3}{20}$ $8\frac{1}{16}$

88. $10\frac{1}{4} \times 3\frac{1}{5}$ $32\frac{4}{5}$

89. $12\frac{3}{5} \times 1\frac{3}{7}$ 18

90. $5\frac{3}{7} \times 5\frac{1}{4}$ $28\frac{1}{2}$

91. $6\frac{1}{2} \times 1\frac{3}{13}$ 8

92. Multiply $2\frac{1}{2}$ and $3\frac{3}{5}$. 9

93. Multiply $4\frac{3}{8}$ and $3\frac{3}{5}$. $15\frac{3}{4}$

94. Find the product of $2\frac{1}{8}$ and $\frac{5}{17}$. $\frac{5}{8}$

95. Find the product of $12\frac{2}{5}$ and $3\frac{7}{31}$. 40

96. What is $1\frac{3}{8}$ times $2\frac{1}{5}$? $3\frac{1}{40}$

97. What is $3\frac{1}{8}$ times $2\frac{4}{7}$? $8\frac{1}{28}$

Objective C *Application Problems*

98. A car can travel $23\frac{1}{2}$ miles on 1 gallon of gas. How far can the car travel on $\frac{1}{2}$ gallon of gas? $11\frac{3}{4}$ miles

99. Maria Rivera can walk $3\frac{1}{2}$ miles in 1 hour. At this rate, how far can Maria walk in $\frac{1}{3}$ hour? $1\frac{1}{6}$ miles

100. A Honda Civic travels 38 miles on each gallon of gasoline. How many miles can the car travel on $9\frac{1}{2}$ gallons of gasoline? 361 miles

101. A board is $5\frac{3}{4}$ feet long. One-third of the board is cut off. What is the length of the piece cut off? $1\frac{11}{12}$ feet

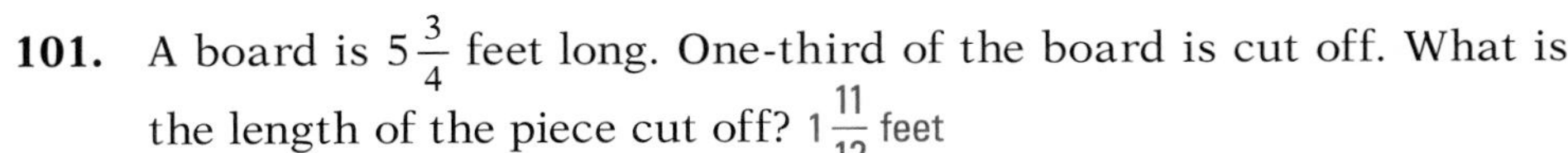

102. A cook is using a recipe that calls for $3\frac{1}{2}$ cups of flour. The cook wants to double the recipe. How much flour should the cook use? 7 cups

103. The F-1 engine in the first stage of the Saturn 5 rocket burns 214,000 gallons of propellant in 1 minute. The first stage burns $2\frac{1}{2}$ minutes before burnout. How much propellant is used before burnout? 535,000 gallons

104. A family budgets $\frac{2}{5}$ of its monthly income of \$3200 per month for housing and utilities.
a. What amount is budgeted for housing and utilities? \$1280
b. What amount remains for purposes other than housing and utilities? \$1920

105. Ann Rodman read $\frac{2}{3}$ of a book containing 432 pages.
a. How many pages did Ann read? 288 pages
b. How many pages remain to be read? 144 pages

106. The parents of the Newton Junior High School Choir members are making robes for the choir. Each robe requires $2\frac{5}{8}$ yards of material at a cost of \$8 per yard. Find the total cost of 24 choir robes. \$504

107. A college spends $\frac{5}{8}$ of its monthly income on employee salaries. During one month the college had an income of \$712,000. How much of the monthly income remained after the employee's salaries were paid? \$267,000

The table at the right shows the length of steel rods and their weight per foot. Use this table for Exercises 108–110.

Length (ft)	*Weight (lb/ft)*
$6\frac{1}{2}$	$\frac{3}{8}$
$8\frac{5}{8}$	$1\frac{1}{4}$
$10\frac{3}{4}$	$2\frac{1}{2}$
$12\frac{7}{12}$	$4\frac{1}{3}$

108. Find the weight of $6\frac{1}{2}$ feet of steel rod. $2\frac{7}{16}$ pounds

109. Find the weight of $12\frac{7}{12}$ feet of steel rod. $54\frac{19}{36}$ pounds

110. Find the total weight of $8\frac{5}{8}$ feet and $10\frac{3}{4}$ feet of steel rod. $37\frac{21}{32}$ pounds

APPLYING THE CONCEPTS

111. The product of 1 and a number is $\frac{1}{2}$. Find the number. $\frac{1}{2}$

112. Our calendar is based on the solar year, which is $365\frac{1}{4}$ days. Use this fact to explain leap years. Every 4 years, we must add one day.

113. If two positive fractions, each less than 1, are multiplied, is the product less than 1? Yes

114. Is the product of two positive fractions always greater than either one of the two numbers? If so, explain why. If not, give an example. No. $\frac{1}{4} \times \frac{1}{2} = \frac{1}{8}$ $\frac{1}{8}$ is less than either $\frac{1}{4}$ or $\frac{1}{2}$

115. Determine the square of $\frac{2}{3}$. $\frac{4}{9}$

116. A state park has reserved $\frac{4}{5}$ of its total acreage for a wildlife preserve. Three-fourths of the wildlife preserve is heavily wooded. What fraction of the state park is heavily wooded? $\frac{3}{5}$

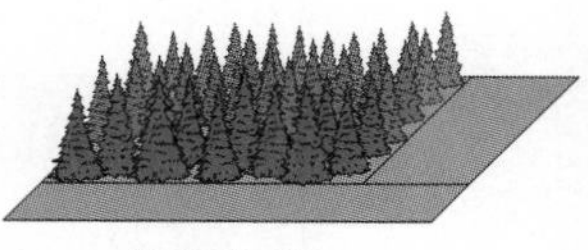
State Park

117. The manager of a mutual fund has one-half of the portfolio invested in bonds. Of the amount invested in bonds, $\frac{3}{8}$ is invested in corporate bonds. What fraction of the total portfolio is invested in corporate bonds? $\frac{3}{16}$

118. Which of the labeled points on the number line at the right could be the graph of the product of *B* and *C*? *A*

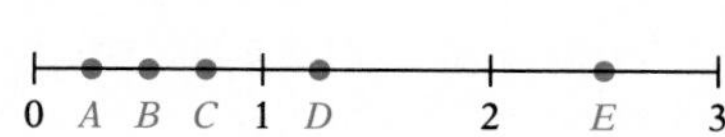

119. Fill-in the circles on the square at the right with the fractions $\frac{1}{6}, \frac{5}{18}, \frac{4}{9}, \frac{5}{9}, \frac{2}{3}, \frac{3}{4}, 1\frac{1}{9}$, $1\frac{1}{2}$, and $2\frac{1}{4}$ so that the product of any three connected circles is equal to $\frac{5}{18}$. (*Note*: There is more than one answer.)

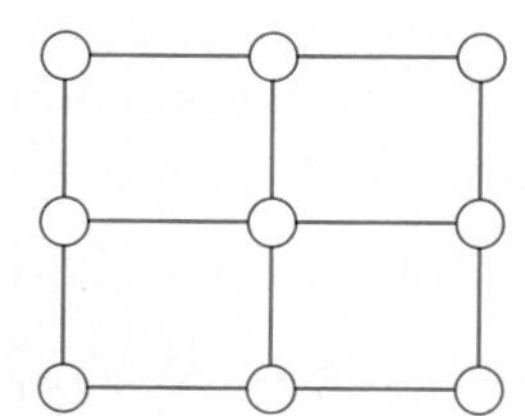

$\frac{2}{3}$	$\frac{3}{4}$	$\frac{5}{9}$
$\frac{10}{9}$	$\frac{1}{6}$	$\frac{3}{2}$
$\frac{9}{4}$	$\frac{5}{18}$	$\frac{4}{9}$

2.7 Division of Fractions and Mixed Numbers

Objective A To divide fractions

The **reciprocal** of a fraction is the fraction with the numerator and denominator interchanged. For instance, the reciprocal of $\frac{2}{3}$ is $\frac{3}{2}$. The process of interchanging the numerator and denominator of a fraction is called **inverting** the fraction.

To find the reciprocal of a whole number, first write the whole number as a fraction with a denominator of 1; then find the reciprocal of that fraction.

The reciprocal of 5 is $\frac{1}{5}$. $\left(\text{Think } 5 = \frac{5}{1}\right)$

INSTRUCTOR NOTE

Here is an extra credit or class exercise:

One-quarter is the same part of one-third as one-half is of what number?

Reciprocals are used to rewrite division problems as related multiplication problems. Look at the following two problems:

$8 \div 2 = 4$ — 8 divided by 2 is 4.

$8 \times \frac{1}{2} = 4$ — 8 times the reciprocal of 2 is 4.

"Divided by" means the same as "times the reciprocal of." Thus "÷ 2" can be replaced with "× $\frac{1}{2}$," and the answer will be the same. Fractions are divided by making this replacement.

➡ Divide: $\frac{2}{3} \div \frac{3}{4}$

$$\frac{2}{3} \div \frac{3}{4} = \frac{2}{3} \times \frac{4}{3} = \frac{2 \cdot 4}{3 \cdot 3} = \frac{2 \cdot 2 \cdot 2}{3 \cdot 3} = \frac{8}{9}$$

Example 1 Divide: $\frac{5}{8} \div \frac{4}{9}$

Solution

$$\frac{5}{8} \div \frac{4}{9} = \frac{5}{8} \times \frac{9}{4} = \frac{5 \cdot 9}{8 \cdot 4} = \frac{5 \cdot 3 \cdot 3}{2 \cdot 2 \cdot 2 \cdot 2 \cdot 2} = \frac{45}{32} = 1\frac{13}{32}$$

You Try It 1 Divide: $\frac{3}{7} \div \frac{2}{3}$

Your solution $\frac{9}{14}$

Example 2 Find the quotient of $\frac{3}{5}$ and $\frac{12}{25}$.

Solution

$$\frac{3}{5} \div \frac{12}{25} = \frac{3}{5} \times \frac{25}{12} = \frac{3 \cdot 25}{5 \cdot 12} = \frac{\overset{1}{\cancel{3}} \cdot \overset{1}{\cancel{5}} \cdot 5}{\underset{1}{\cancel{5}} \cdot 2 \cdot 2 \cdot \underset{1}{\cancel{3}}} = \frac{5}{4} = 1\frac{1}{4}$$

You Try It 2 Divide: $\frac{3}{4} \div \frac{9}{10}$

Your solution $\frac{5}{6}$

Solutions on p. A16

Objective B To divide whole numbers, mixed numbers, and fractions

To divide a fraction and a whole number, first write the whole number as a fraction with a denominator of 1.

➡ Divide: $\frac{3}{7} \div 5$

$$\frac{3}{7} \div \boxed{5} = \frac{3}{7} \div \boxed{\frac{5}{1}} = \frac{3}{7} \times \frac{1}{5} = \frac{3 \cdot 1}{7 \cdot 5} = \frac{3}{35}$$

• Write 5 with a denominator of 1; then divide the fractions.

INSTRUCTOR NOTE
Here is a classic problem that students frequently misinterpret: "What is 8 divided by one-half?"

When one of the numbers in a quotient is a mixed number, write the mixed number as an improper fraction before dividing.

➡ Divide: $4\frac{2}{3} \div \frac{8}{15}$

Write $4\frac{2}{3}$ as an improper fraction; then divide the fractions.

$$4\frac{2}{3} \div \frac{8}{15} = \frac{14}{3} \div \frac{8}{15} = \frac{14}{3} \times \frac{15}{8} = \frac{14 \cdot 15}{3 \cdot 8} = \frac{\overset{1}{\cancel{2}} \cdot 7 \cdot \overset{1}{\cancel{3}} \cdot 5}{\underset{1}{\cancel{3}} \cdot 2 \cdot \underset{1}{\cancel{2}} \cdot 2} = \frac{35}{4} = 8\frac{3}{4}$$

➡ Divide: $1\frac{13}{15} \div 4\frac{4}{5}$

Write the mixed numbers as improper fractions. Then divide the fractions.

$$1\frac{13}{15} \div 4\frac{4}{5} = \frac{28}{15} \div \frac{24}{5} = \frac{28}{15} \times \frac{5}{24} = \frac{28 \cdot 5}{15 \cdot 24} = \frac{\overset{1}{\cancel{2}} \cdot \overset{1}{\cancel{2}} \cdot 7 \cdot \overset{1}{\cancel{5}}}{3 \cdot \underset{1}{\cancel{5}} \cdot \underset{1}{\cancel{2}} \cdot \underset{1}{\cancel{2}} \cdot 2 \cdot 3} = \frac{7}{18}$$

Example 3 Divide $\frac{4}{9}$ by 5.

Solution

$$\frac{4}{9} \div 5 = \frac{4}{9} \div \frac{5}{1} = \frac{4}{9} \times \frac{1}{5}$$
$$= \frac{4 \cdot 1}{9 \cdot 5} = \frac{2 \cdot 2}{3 \cdot 3 \cdot 5} = \frac{4}{45}$$

You Try It 3 Divide $\frac{5}{7}$ by 6.

Your solution $\frac{5}{42}$

Example 4

Find the quotient of $\frac{3}{8}$ and $2\frac{1}{10}$.

Solution

$$\frac{3}{8} \div 2\frac{1}{10} = \frac{3}{8} \div \frac{21}{10} = \frac{3}{8} \times \frac{10}{21}$$
$$= \frac{3 \cdot 10}{8 \cdot 21} = \frac{\overset{1}{\cancel{3}} \cdot \overset{1}{\cancel{2}} \cdot 5}{\underset{1}{\cancel{2}} \cdot 2 \cdot 2 \cdot \underset{1}{\cancel{3}} \cdot 7} = \frac{5}{28}$$

You Try It 4

Find the quotient of $12\frac{3}{5}$ and 7.

Your solution $1\frac{4}{5}$

Example 5 Divide: $2\frac{3}{4} \div 1\frac{5}{7}$

Solution

$$2\frac{3}{4} \div 1\frac{5}{7} = \frac{11}{4} \div \frac{12}{7} = \frac{11}{4} \times \frac{7}{12} = \frac{11 \cdot 7}{4 \cdot 12}$$
$$= \frac{11 \cdot 7}{2 \cdot 2 \cdot 2 \cdot 2 \cdot 3} = \frac{77}{48} = 1\frac{29}{48}$$

You Try It 5 Divide: $3\frac{2}{3} \div 2\frac{2}{5}$

Your solution $1\frac{19}{36}$

Solutions on pp. A16 and A17

Example 6 Divide: $1\frac{13}{15} \div 4\frac{1}{5}$

Solution

$$1\frac{13}{15} \div 4\frac{1}{5} = \frac{28}{15} \div \frac{21}{5} = \frac{28}{15} \times \frac{5}{21} = \frac{28 \cdot 5}{15 \cdot 21}$$

$$= \frac{2 \cdot 2 \cdot \overset{1}{\cancel{7}} \cdot \overset{1}{\cancel{5}}}{3 \cdot \underset{1}{\cancel{5}} \cdot 3 \cdot \underset{1}{\cancel{7}}} = \frac{4}{9}$$

You Try It 6 Divide: $2\frac{5}{6} \div 8\frac{1}{2}$

Your solution $\frac{1}{3}$

Example 7 Divide: $4\frac{3}{8} \div 7$

Solution

$$4\frac{3}{8} \div 7 = \frac{35}{8} \div \frac{7}{1} = \frac{35}{8} \times \frac{1}{7}$$

$$= \frac{35 \cdot 1}{8 \cdot 7} = \frac{5 \cdot \overset{1}{\cancel{7}}}{2 \cdot 2 \cdot 2 \cdot \underset{1}{\cancel{7}}} = \frac{5}{8}$$

You Try It 7 Divide: $6\frac{2}{5} \div 4$

Your solution $1\frac{3}{5}$

Solutions on p. A17

Objective C To solve application problems

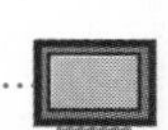

Example 8

A car used $15\frac{1}{2}$ gallons of gasoline on a 310-mile trip. How many miles can this car travel on 1 gallon of gasoline?

Strategy

To find the number of miles, divide the number of miles traveled by the number of gallons of gasoline used.

Solution

$$310 \div 15\frac{1}{2} = \frac{310}{1} \div \frac{31}{2}$$

$$= \frac{310}{1} \times \frac{2}{31}$$

$$= \frac{310 \cdot 2}{1 \cdot 31}$$

$$= 20$$

The car travels 20 miles on 1 gallon of gasoline.

You Try It 8

Leon Dern purchased a $\frac{1}{2}$-ounce gold coin for \$195. What was the price of 1 ounce?

Your strategy

Your solution \$390

Solution on p. A17

Example 9

A 12-foot board is cut into pieces $2\frac{1}{4}$ feet long for use as bookshelves. What is the length of the remaining piece after as many shelves as possible are cut?

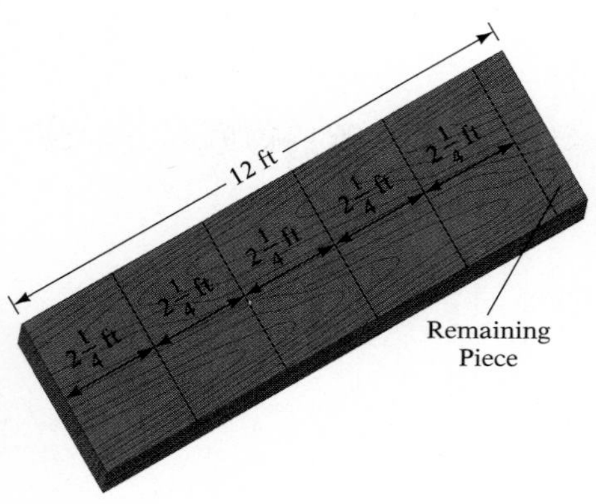

Strategy

To find the length of the remaining piece:

- Divide the total length by the length of each shelf $\left(12 \div 2\frac{1}{4}\right)$. This will give you the number of shelves cut, with a certain fraction of a shelf left over.
- Multiply the fraction left over by the length of a shelf to determine the actual length of the piece remaining.

Solution

$$12 \div 2\frac{1}{4} = \frac{12}{1} \div \frac{9}{4} = \frac{12}{1} \times \frac{4}{9}$$
$$= \frac{12 \cdot 4}{1 \cdot 9} = \frac{16}{3} = 5\frac{1}{3}$$

5 pieces $2\frac{1}{4}$ feet long

1 piece $\frac{1}{3}$ of $2\frac{1}{4}$ feet long

$$\frac{1}{3} \times 2\frac{1}{4} = \frac{1}{3} \times \frac{9}{4} = \frac{1 \cdot 9}{3 \cdot 4} = \frac{3}{4}$$

The length of the piece remaining is $\frac{3}{4}$ foot.

You Try It 9

A 16-foot board is cut into pieces $3\frac{1}{3}$ feet long for shelves for a bookcase. What is the length of the remaining piece after as many shelves as possible are cut?

Your strategy

Your solution $2\frac{2}{3}$ feet

Solution on p. A17

2.7 Exercises

Objective A

Divide.

1. $\frac{1}{3} \div \frac{2}{5}$ $\frac{5}{6}$

2. $\frac{3}{7} \div \frac{3}{2}$ $\frac{2}{7}$

3. $\frac{31}{40} \div \frac{11}{15}$ $1\frac{5}{88}$

4. $\frac{3}{7} \div \frac{3}{7}$ 1

5. $0 \div \frac{1}{2}$ 0

6. $0 \div \frac{3}{4}$ 0

7. $\frac{16}{33} \div \frac{4}{11}$ $1\frac{1}{3}$

8. $\frac{5}{24} \div \frac{15}{36}$ $\frac{1}{2}$

9. $\frac{11}{15} \div \frac{1}{12}$ $8\frac{4}{5}$

10. $\frac{2}{9} \div \frac{16}{19}$ $\frac{19}{72}$

11. $\frac{15}{16} \div \frac{16}{39}$ $2\frac{73}{256}$

12. $\frac{2}{15} \div \frac{3}{5}$ $\frac{2}{9}$

13. $\frac{8}{9} \div \frac{4}{5}$ $1\frac{1}{9}$

14. $\frac{11}{15} \div \frac{5}{22}$ $3\frac{17}{75}$

15. $\frac{12}{13} \div \frac{4}{9}$ $2\frac{1}{13}$

16. $\frac{1}{9} \div \frac{2}{3}$ $\frac{1}{6}$

17. $\frac{10}{21} \div \frac{5}{7}$ $\frac{2}{3}$

18. $\frac{2}{5} \div \frac{4}{7}$ $\frac{7}{10}$

19. $\frac{3}{8} \div \frac{5}{12}$ $\frac{9}{10}$

20. $\frac{5}{9} \div \frac{15}{32}$ $1\frac{5}{27}$

21. $\frac{1}{2} \div \frac{1}{4}$ 2

22. $\frac{1}{3} \div \frac{1}{9}$ 3

23. $\frac{1}{5} \div \frac{1}{10}$ 2

24. $\frac{4}{15} \div \frac{2}{5}$ $\frac{2}{3}$

25. $\frac{7}{15} \div \frac{14}{5}$ $\frac{1}{6}$

26. $\frac{5}{8} \div \frac{15}{2}$ $\frac{1}{12}$

27. $\frac{14}{3} \div \frac{7}{9}$ 6

28. $\frac{7}{4} \div \frac{9}{2}$ $\frac{7}{18}$

29. $\frac{5}{9} \div \frac{25}{3}$ $\frac{1}{15}$

30. $\frac{5}{16} \div \frac{3}{8}$ $\frac{5}{6}$

31. $\frac{2}{3} \div \frac{1}{3}$ 2

32. $\frac{4}{9} \div \frac{1}{9}$ 4

33. $\frac{5}{7} \div \frac{2}{7}$ $2\frac{1}{2}$

34. $\frac{5}{6} \div \frac{1}{9}$ $7\frac{1}{2}$

35. $\frac{2}{3} \div \frac{2}{9}$ 3

36. $\frac{5}{12} \div \frac{5}{6}$ $\frac{1}{2}$

37. $4 \div \frac{2}{3}$ 6

38. $\frac{2}{3} \div 4$ $\frac{1}{6}$

39. $\frac{3}{2} \div 3$ $\frac{1}{2}$

40. $3 \div \frac{3}{2}$ 2

41. Divide $\frac{7}{8}$ by $\frac{3}{4}$.
$1\frac{1}{6}$

42. Divide $\frac{7}{12}$ by $\frac{3}{4}$.
$\frac{7}{9}$

43. Find the quotient of $\frac{5}{7}$ and $\frac{3}{14}$.
$3\frac{1}{3}$

44. Find the quotient of $\frac{6}{11}$ and $\frac{9}{32}$.
$1\frac{31}{33}$

Objective B

Divide.

45. $\frac{5}{6} \div 25$
$\frac{1}{30}$

46. $22 \div \frac{3}{11}$
$80\frac{2}{3}$

47. $6 \div 3\frac{1}{3}$
$1\frac{4}{5}$

48. $5\frac{1}{2} \div 11$
$\frac{1}{2}$

49. $3\frac{1}{3} \div \frac{3}{8}$
$8\frac{8}{9}$

50. $6\frac{1}{2} \div \frac{1}{2}$
13

51. $\frac{3}{8} \div 2\frac{1}{4}$
$\frac{1}{6}$

52. $\frac{5}{12} \div 4\frac{4}{5}$
$\frac{25}{288}$

53. $1\frac{1}{2} \div 1\frac{3}{8}$
$1\frac{1}{11}$

54. $2\frac{1}{4} \div 1\frac{3}{8}$
$1\frac{7}{11}$

55. $8\frac{1}{4} \div 2\frac{3}{4}$
3

56. $3\frac{5}{9} \div 32$
$\frac{1}{9}$

57. $4\frac{1}{5} \div 21$
$\frac{1}{5}$

58. $6\frac{8}{9} \div \frac{31}{36}$
8

59. $5\frac{3}{5} \div \frac{7}{10}$
8

60. $\frac{11}{12} \div 2\frac{1}{3}$
$\frac{11}{28}$

61. $\frac{7}{8} \div 3\frac{1}{4}$
$\frac{7}{26}$

62. $\frac{5}{16} \div 5\frac{3}{8}$
$\frac{5}{86}$

63. $\frac{9}{14} \div 3\frac{1}{7}$
$\frac{9}{44}$

64. $27 \div \frac{3}{8}$
72

65. $35 \div \frac{7}{24}$
120

66. $\frac{3}{8} \div 2\frac{3}{4}$
$\frac{3}{22}$

67. $\frac{11}{18} \div 2\frac{2}{9}$
$\frac{11}{40}$

68. $\frac{21}{40} \div 3\frac{3}{10}$
$\frac{7}{44}$

69. $\frac{6}{25} \div 4\frac{1}{5}$
$\frac{2}{35}$

70. $2\frac{1}{16} \div 2\frac{1}{2}$
$\frac{33}{40}$

71. $7\frac{3}{5} \div 1\frac{7}{12}$
$4\frac{4}{5}$

72. $1\frac{2}{3} \div \frac{3}{8}$
$4\frac{4}{9}$

73. $16 \div \frac{2}{3}$
24

74. $2\frac{1}{2} \div 3\frac{5}{14}$
$\frac{35}{47}$

75. $1\frac{5}{8} \div 4$
$\frac{13}{32}$

76. $13\frac{3}{8} \div \frac{1}{4}$
$53\frac{1}{2}$

77. $16 \div 1\frac{1}{2}$
$10\frac{2}{3}$

78. $9 \div \frac{7}{8}$
$10\frac{2}{7}$

79. $15\frac{1}{3} \div 2\frac{2}{9}$
$6\frac{9}{10}$

80. $16\frac{5}{8} \div 1\frac{2}{3}$
$9\frac{39}{40}$

81. $24\frac{4}{5} \div 2\frac{3}{5}$
$9\frac{7}{13}$

82. $1\frac{1}{3} \div 5\frac{8}{9}$
$\frac{12}{53}$

83. $13\frac{2}{3} \div 0$
Undefined

84. $82\frac{3}{5} \div 19\frac{1}{10}$ $4\frac{62}{191}$

85. $45\frac{3}{5} \div 15$ $3\frac{1}{25}$

86. $102 \div 1\frac{1}{2}$ 68

87. $44\frac{5}{12} \div 32\frac{5}{6}$ $1\frac{139}{394}$

88. $0 \div 3\frac{1}{2}$ 0

89. $8\frac{2}{7} \div 1$ $8\frac{2}{7}$

90. $6\frac{9}{16} \div 1\frac{3}{32}$ 6

91. $5\frac{5}{14} \div 3\frac{4}{7}$ $1\frac{1}{2}$

92. $8\frac{8}{9} \div 2\frac{13}{18}$ $3\frac{13}{49}$

93. $10\frac{1}{5} \div 1\frac{7}{10}$ 6

94. $7\frac{3}{8} \div 1\frac{27}{32}$ 4

95. $4\frac{5}{11} \div 2\frac{1}{3}$ $1\frac{10}{11}$

96. Divide $7\frac{7}{9}$ by $5\frac{5}{6}$. $1\frac{1}{3}$

97. Divide $2\frac{3}{4}$ by $1\frac{23}{32}$. $1\frac{3}{5}$

98. Find the quotient of $8\frac{1}{4}$ and $1\frac{5}{11}$. $5\frac{43}{64}$

99. Find the quotient of $\frac{14}{17}$ and $3\frac{1}{9}$. $\frac{9}{34}$

Objective C *Application Problems*

100. Individual cereal boxes contain $\frac{3}{4}$ ounce of cereal. How many boxes can be filled with 600 ounces of cereal? 800 boxes

101. A box of bran flakes contains 20 ounces of cereal. How many $1\frac{1}{4}$-ounce portions can be served from this box? 16 servings

102. A $\frac{5}{8}$-carat diamond was purchased for $1200. What would a similar diamond weighing 1 carat cost? $1920

103. The Inverness Investor Group bought $8\frac{1}{3}$ acres of land for $200,000. What was the cost of each acre? $24,000

104. KU Energy stock is offered for $31\frac{5}{8}$ per share. How many shares can you buy for $1265? 40 shares

105. The Tuesday attendance at the dinner theater was 280. The theater was $\frac{2}{3}$ full. What is the capacity of the theater? 420

106. The Hammond Company purchased $9\frac{3}{4}$ acres for a housing project. One and one-half acres were set aside for a park.

a. How many acres are available for housing? $8\frac{1}{4}$ acres

b. How many $\frac{1}{4}$-acre parcels of land can be sold after the land for the park is set aside? 33 parcels

107. A chef purchased a roast that weighed $10\frac{3}{4}$ pounds. After the fat was trimmed and the bone removed, the roast weighed $9\frac{1}{3}$ pounds.

a. What was the total weight of the fat and bone? $1\frac{5}{12}$ pounds

b. How many $\frac{1}{3}$-pound servings can be cut from the roast? 28 servings

108. One-tenth of a shipment of $18\frac{1}{3}$ pounds of grapes was spoiled. How many $\frac{1}{2}$-pound packages of unspoiled grapes can be packaged from this shipment? 33 packages

109. A 15-foot board is cut into pieces $3\frac{1}{2}$ feet long for a bookcase. What is the length of the piece remaining after as many shelves as possible have been cut? 1 foot

110. A scale of $\frac{1}{2}$ inch to 1 foot is used to draw the plans for a house. The scale measurements for three walls are given in the table at the right. Complete the table to determine the actual wall lengths for the three walls a, b, and c.

Wall	*Scale*	*Actual Wall Length*
a	$6\frac{1}{4}$ in.	? $12\frac{1}{2}$ ft
b	9 in.	? 18 ft
c	$7\frac{7}{8}$ in.	? $15\frac{3}{4}$ ft

APPLYING THE CONCEPTS

111. On a map, two cities are $4\frac{5}{8}$ inches apart. If $\frac{3}{8}$ inch on the map represents 60 miles, what is the number of miles between the two cities? 740 miles

112. [W] Is the quotient always less than the dividend in a division problem? Explain. No. $4 \div \frac{1}{2} = 8.$ Answers will vary

113. Fill in the box to make a true statement.

a. $\frac{3}{4} \cdot \square = \frac{1}{2}$ $\frac{2}{3}$ b. $\frac{2}{3} \cdot \square = 1\frac{3}{4}$ $\frac{21}{8}$

114. A page of type in a certain textbook is $7\frac{1}{2}$ inches wide. If the page is divided into three equal columns, with $\frac{3}{8}$ inch between columns, how wide is each column? $2\frac{1}{4}$ inches

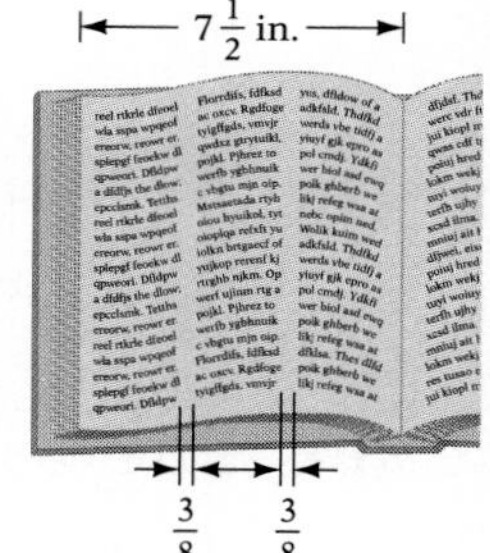

115. A whole number is both multiplied and divided by the same proper fraction. Which is greater, the product or the quotient? The quotient

116. Fractions are multiplied by multiplying the numerators and multiplying the denominators. Consider dividing fractions in a similar manner. Divide the numerators and divide the denominators.

$$\frac{4}{15} \div \frac{2}{5} = \frac{4 \div 2}{15 \div 5} = \frac{2}{3}$$

This gives the same answer as the traditional method.

$$\frac{4}{15} \div \frac{2}{5} = \frac{4}{15} \cdot \frac{5}{2} = \frac{20}{30} = \frac{2}{3}$$

Try this method for the following division problems.

a. $\frac{5}{6} \div \frac{1}{3}$ $\frac{5}{2}$ b. $\frac{4}{21} \div \frac{2}{7}$ $\frac{2}{3}$ c. $\frac{5}{9} \div \frac{5}{3}$ $\frac{1}{3}$ d. $\frac{15}{16} \div \frac{3}{8}$ $\frac{5}{2}$

117. Does the method of dividing fractions shown above always work? Give an example of when the traditional method would be a better choice. Yes. Answers will vary

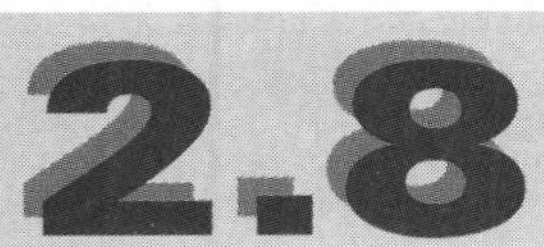

2.8 Order, Exponents, and the Order of Operations Agreement

Objective A To identify the order relation between two fractions

POINT OF INTEREST
Leonardo of Pisa, who was also called Fibonacci (c. 1175–1250), is credited with bringing the Hindu-Arabic number system to the western world and promoting its use instead of the cumbersome Roman numeral system.

He was also influential in promoting the idea of the fraction bar. His notation, however, was very different from what we have today. For instance, he wrote $\frac{3\ 5}{4\ 7}$ to mean $\frac{5}{7} + \frac{3}{7 \cdot 4}$.

Recall that whole numbers can be graphed as points on the number line. Fractions can also be graphed as points on the number line.

The graph of $\frac{3}{4}$ on the number line

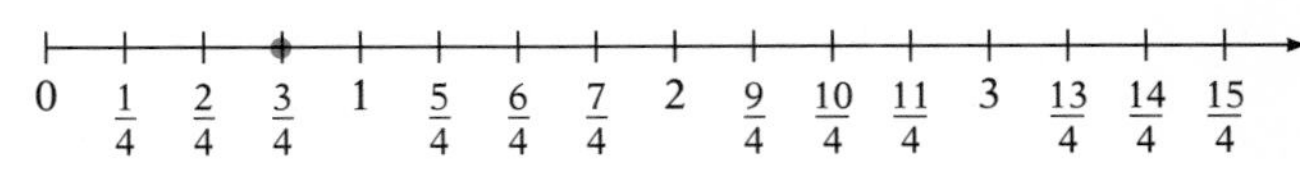

The number line can be used to determine the order relation between two fractions. A fraction that appears to the left of a given fraction is less than the given fraction. A fraction that appears to the right of a given fraction is greater than the given fraction.

$\frac{1}{8} < \frac{3}{8}$ $\quad$ $\frac{6}{8} > \frac{3}{8}$

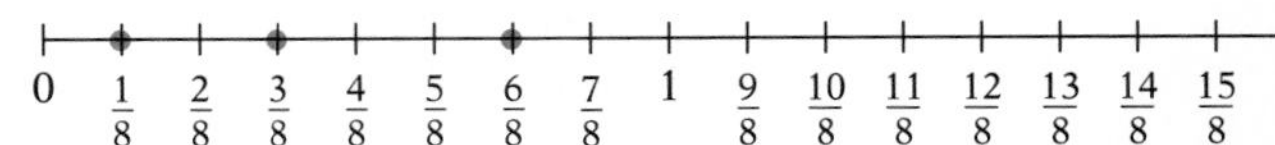

To find the order of relation between two fractions with the same denominator, compare numerators. The fraction that has the smaller numerator is the smaller fraction. When the denominators are different, begin by writing equivalent fractions with a common denominator; then compare numerators.

➡ Find the order relation between $\frac{11}{18}$ and $\frac{5}{8}$.

The LCM of 18 and 8 is 72.

$\frac{11}{18} = \frac{44}{72}$ ← Smaller numerator

$\frac{5}{8} = \frac{45}{72}$ ← Larger numerator

$\frac{11}{18} < \frac{5}{8}$ or $\frac{5}{8} > \frac{11}{18}$

Example 1 Place the correct symbol, < or >, between the numbers.

$\frac{5}{12}$ $\quad$ $\frac{7}{18}$

Solution $\frac{5}{12} = \frac{15}{36}$ $\quad$ $\frac{7}{18} = \frac{14}{36}$

$\frac{5}{12} > \frac{7}{18}$

You Try It 1 Place the correct symbol, < or >, between the numbers.

$\frac{9}{14}$ $\quad$ $\frac{13}{21}$

Your solution $\frac{9}{14} > \frac{13}{21}$

Solution on p. A18

Objective B To simplify expressions containing exponents

Repeated multiplication of the same fraction can be written in two ways:

$$\frac{1}{2} \cdot \frac{1}{2} \cdot \frac{1}{2} \cdot \frac{1}{2} \text{ or } \left(\frac{1}{2}\right)^4 \leftarrow \text{Exponent}$$

The exponent indicates how many times the fraction occurs as a factor in the multiplication. The expression $\left(\frac{1}{2}\right)^4$ is in exponential notation.

Example 2 Simplify: $\left(\frac{5}{6}\right)^3 \cdot \left(\frac{3}{5}\right)^2$

Solution $\left(\frac{5}{6}\right)^8 \cdot \left(\frac{3}{5}\right)^2 = \left(\frac{5}{6} \cdot \frac{5}{6} \cdot \frac{5}{6}\right) \cdot \left(\frac{3}{5} \cdot \frac{3}{5}\right)$

$$= \frac{\overset{1}{\cancel{5}} \cdot \overset{1}{\cancel{5}} \cdot 5 \cdot \overset{1}{\cancel{3}} \cdot \overset{1}{\cancel{3}}}{2 \cdot \underset{1}{\cancel{3}} \cdot 2 \cdot \underset{1}{\cancel{3}} \cdot 2 \cdot 3 \cdot \underset{1}{\cancel{5}} \cdot \underset{1}{\cancel{5}}} = \frac{5}{24}$$

You Try It 2 Simplify: $\left(\frac{7}{11}\right)^2 \cdot \left(\frac{2}{7}\right)$

Your solution $\frac{14}{121}$

Solution on p. A18

Objective C To use the Order of Operations Agreement to simplify expressions

The Order of Operations Agreement is used for fractions as well as whole numbers.

Step 1 Do all operations inside parentheses.
Step 2 Simplify any number expressions containing exponents.
Step 3 Do multiplications and divisions as they occur from left to right.
Step 4 Do additions and subtractions as they occur from left to right.

⇒ Simplify $\frac{14}{15} - \left(\frac{1}{2}\right)^2 \times \left(\frac{2}{3} + \frac{4}{5}\right)$ by using the Order of Operations Agreement.

$\frac{14}{15} - \left(\frac{1}{2}\right)^2 \times \left(\frac{2}{3} + \frac{4}{5}\right)$ 1. Perform operations in parentheses.

$\frac{14}{15} - \left(\frac{1}{2}\right)^2 \times \frac{22}{15}$ 2. Simplify expressions with exponents.

$\frac{14}{15} - \frac{1}{4} \times \frac{22}{15}$ 3. Do multiplications and divisions as they occur from left to right.

$\frac{14}{15} - \frac{11}{30}$ 4. Do additions and subtractions as they occur from left to right.

$\frac{17}{30}$

One or more of the above steps may not be needed to simplify an expression. In that case, proceed to the next step in the Order of Operations Agreement.

Example 3 Simplify: $\left(\frac{3}{4}\right)^2 \div \left(\frac{3}{8} - \frac{1}{12}\right)$

Solution $\left(\frac{3}{4}\right)^2 \div \left(\frac{3}{8} - \frac{1}{12}\right)$

$\left(\frac{3}{4}\right)^2 \div \left(\frac{7}{24}\right)$

$\frac{9}{16} \div \frac{7}{24} = \frac{27}{14} = 1\frac{13}{14}$

You Try It 3 Simplify: $\left(\frac{1}{13}\right)^2 \cdot \left(\frac{1}{4} + \frac{1}{6}\right) \div \frac{5}{13}$

Your solution $\frac{1}{156}$

Solution on p. A18

2.8 Exercises

Objective A

Place the correct symbol, $<$ or $>$, between the two numbers.

1. $\frac{11}{40} < \frac{19}{40}$

2. $\frac{92}{103} > \frac{19}{103}$

3. $\frac{2}{3} < \frac{5}{7}$

4. $\frac{2}{5} > \frac{3}{8}$

5. $\frac{5}{8} > \frac{7}{12}$

6. $\frac{11}{16} < \frac{17}{24}$

7. $\frac{7}{9} < \frac{11}{12}$

8. $\frac{5}{12} < \frac{7}{15}$

9. $\frac{13}{14} > \frac{19}{21}$

10. $\frac{13}{18} > \frac{7}{12}$

11. $\frac{7}{24} < \frac{11}{30}$

12. $\frac{13}{36} < \frac{19}{48}$

Objective B

Simplify.

13. $\left(\frac{3}{8}\right)^2$ $\frac{9}{64}$

14. $\left(\frac{5}{12}\right)^2$ $\frac{25}{144}$

15. $\left(\frac{2}{9}\right)^3$ $\frac{8}{729}$

16. $\left(\frac{5}{7}\right)^3$ $\frac{125}{343}$

17. $\left(\frac{1}{2}\right) \cdot \left(\frac{2}{3}\right)^2$ $\frac{2}{9}$

18. $\left(\frac{2}{3}\right) \cdot \left(\frac{1}{2}\right)^4$ $\frac{1}{24}$

19. $\left(\frac{1}{3}\right)^2 \cdot \left(\frac{3}{5}\right)^3$ $\frac{3}{125}$

20. $\left(\frac{2}{5}\right)^2 \cdot \left(\frac{5}{8}\right)^3$ $\frac{5}{128}$

21. $\left(\frac{2}{5}\right)^3 \cdot \left(\frac{5}{7}\right)^2$ $\frac{8}{245}$

22. $\left(\frac{5}{9}\right)^3 \cdot \left(\frac{18}{25}\right)^2$ $\frac{4}{45}$

23. $\left(\frac{1}{3}\right)^4 \cdot \left(\frac{9}{11}\right)^2$ $\frac{1}{121}$

24. $\left(\frac{8}{15}\right)^3 \cdot \left(\frac{5}{8}\right)^4$ $\frac{5}{216}$

25. $\left(\frac{1}{2}\right)^6 \cdot \left(\frac{32}{35}\right)^2$ $\frac{16}{1225}$

26. $\left(\frac{2}{3}\right)^4 \cdot \left(\frac{81}{100}\right)^2$ $\frac{81}{625}$

27. $\left(\frac{1}{6}\right) \cdot \left(\frac{6}{7}\right)^2 \cdot \left(\frac{2}{3}\right)$ $\frac{4}{49}$

28. $\left(\frac{2}{7}\right) \cdot \left(\frac{7}{8}\right)^2 \cdot \left(\frac{8}{9}\right)$ $\frac{7}{36}$

29. $3 \cdot \left(\frac{3}{5}\right)^3 \cdot \left(\frac{1}{3}\right)^2$ $\frac{9}{125}$

30. $4 \cdot \left(\frac{3}{4}\right)^3 \cdot \left(\frac{4}{7}\right)^2$ $\frac{27}{49}$

31. $11 \cdot \left(\frac{3}{8}\right)^3 \cdot \left(\frac{8}{11}\right)^2$ $\frac{27}{88}$

32. $5 \cdot \left(\frac{3}{5}\right)^3 \cdot \left(\frac{2}{3}\right)^4$ $\frac{16}{75}$

33. $\left(\frac{2}{7}\right)^2 \cdot \left(\frac{7}{9}\right)^2 \cdot \left(\frac{9}{11}\right)^2$ $\frac{4}{121}$

Objective C

Simplify using the Order of Operations Agreement.

34. $\frac{1}{2} - \frac{1}{3} + \frac{2}{3}$
$\frac{5}{6}$

35. $\frac{2}{5} + \frac{3}{10} - \frac{2}{3}$
$\frac{1}{30}$

36. $\frac{1}{3} \div \frac{1}{2} + \frac{3}{4}$
$1\frac{5}{12}$

37. $\frac{3}{5} \div \frac{6}{7} + \frac{4}{5}$
$1\frac{1}{2}$

38. $\frac{4}{5} + \frac{3}{7} \cdot \frac{14}{15}$
$1\frac{1}{5}$

39. $\frac{2}{3} + \frac{5}{8} \cdot \frac{16}{35}$
$\frac{20}{21}$

40. $\left(\frac{3}{4}\right)^2 - \frac{5}{12}$
$\frac{7}{48}$

41. $\left(\frac{3}{5}\right)^3 - \frac{3}{25}$
$\frac{12}{125}$

42. $\frac{5}{6} \cdot \left(\frac{2}{3} - \frac{1}{6}\right) + \frac{7}{18}$
$\frac{29}{36}$

43. $\frac{3}{4} \cdot \left(\frac{11}{12} - \frac{7}{8}\right) + \frac{5}{16}$
$\frac{11}{32}$

44. $\frac{7}{12} - \left(\frac{2}{3}\right)^2 + \frac{5}{8}$
$\frac{55}{72}$

45. $\frac{11}{16} - \left(\frac{3}{4}\right)^2 + \frac{7}{12}$
$\frac{17}{24}$

46. $\frac{3}{4} \cdot \left(\frac{4}{9}\right)^2 + \frac{1}{2}$
$\frac{35}{54}$

47. $\frac{9}{10} \cdot \left(\frac{2}{3}\right)^3 + \frac{2}{3}$
$\frac{14}{15}$

48. $\left(\frac{1}{2} + \frac{3}{4}\right) \div \frac{5}{8}$
2

49. $\left(\frac{2}{3} + \frac{5}{6}\right) \div \frac{5}{9}$
$2\frac{7}{10}$

50. $\frac{3}{8} \div \left(\frac{5}{12} + \frac{3}{8}\right)$
$\frac{9}{19}$

51. $\frac{7}{12} \div \left(\frac{2}{3} + \frac{5}{9}\right)$
$\frac{21}{44}$

52. $\left(\frac{3}{8}\right)^2 \div \left(\frac{3}{7} + \frac{3}{14}\right)$
$\frac{7}{32}$

53. $\left(\frac{5}{6}\right)^2 \div \left(\frac{5}{12} + \frac{2}{3}\right)$
$\frac{25}{39}$

54. $\frac{2}{5} \div \frac{3}{8} \cdot \frac{4}{5}$
$\frac{64}{75}$

APPLYING THE CONCEPTS

55. $\frac{2}{3} < \frac{3}{4}$. Is $\frac{2+3}{3+4}$ less than $\frac{2}{3}$, greater than $\frac{2}{3}$, or between $\frac{2}{3}$ and $\frac{3}{4}$?
Between $\frac{2}{3}$ and $\frac{3}{4}$

56. [W] A farmer died and left 17 horses to be divided among 3 children. The first child was to receive $\frac{1}{2}$ of the horses, the second child $\frac{1}{3}$ of the horses, and the third child $\frac{1}{9}$ of the horses. The executor for the family's estate realized that 17 horses could not be divided by halves, thirds, or ninths and so added a neighbor's horse to the farmer's. With 18 horses, the executor gave 9 horses to the first child, 6 horses to the second child, and 2 horses to the third child. This accounted for the 17 horses, so the executor returned the borrowed horse to the neighbor. Explain why this worked. Answers will vary

Projects in Mathematics

Music In musical notation, notes are printed on a staff, which is a set of five horizontal lines and the spaces between them. The notes of a musical composition are grouped into measures, or bars. Vertical lines separate measures on a staff. The shape of a note indicates how long it should be held. The whole note has the longest time value of any note. Each time value is divided by 2 in order to find the next smallest note value.

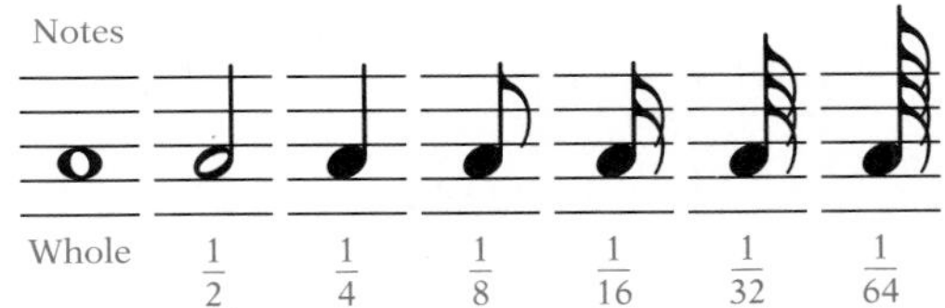

The time signature is a fraction that appears at the beginning of a piece of music. The numerator of the fraction indicates the number of beats in a measure. The denominator indicates what kind of note receives one beat. For example, music written in $\frac{2}{4}$ time has 2 beats to a measure, and a quarter note receives one beat. One measure in $\frac{2}{4}$ time may have 1 half note, 2 quarter notes, and 4 eighth notes or any other combination of notes totaling 2 beats. Other common time signatures include $\frac{4}{4}$, $\frac{3}{4}$, and $\frac{6}{8}$.

Exercises:

1. Explain the meaning of the 6 and the 8 in the time signature $\frac{6}{8}$. Give some possible combinations of notes in one measure of a piece written in $\frac{4}{4}$ time.
2. What does a dot at the right of a note indicate? What is the effect of a dot at the right of a half note? A quarter note? An eighth note?
3. Symbols called rests are used to indicate periods of silence in a piece of music. What symbols are used to indicate the different time values of rests?
4. Find some examples of musical compositions written in different time signatures. Use a few measures from each to show that the sum of the time values of the notes and rests in each measure equals the numerator of the time signature.

Construction Suppose you are involved in building your own home. Design a stairway from the first floor of the house to the second floor. Some of the questions you will need to answer follow.

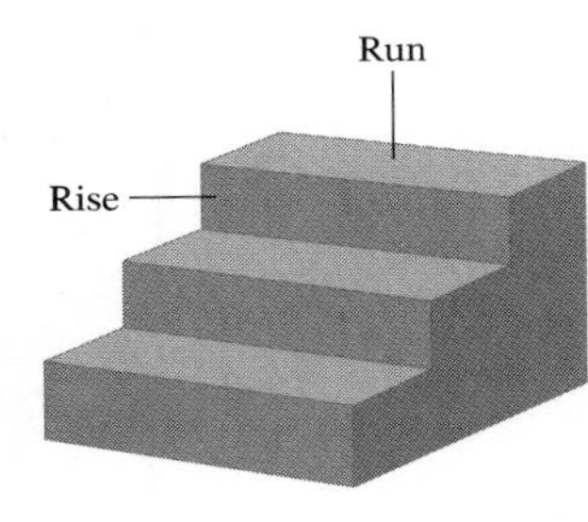

What is the distance from the floor of the first story to the floor of the second story?

Typically, what is the number of steps in a stairway?

What is a reasonable length for the run of each step?

What width wood is being used to build the staircase?

In designing the stairway, remember that each riser should be the same height and each run should be the same length. And the width of the wood used for the steps will have to be incorporated in the calculation.

Chapter Summary

Key Words The *least common multiple* (LCM) is the smallest common multiple of two or more numbers.

The *greatest common factor* (GCF) is the largest common factor of two or more numbers.

A *fraction* can represent the number of equal parts of a whole.

A *proper fraction* is a fraction less than 1.

A *mixed number* is a number greater than 1 with a whole number part and a fractional part.

An *improper fraction* is a fraction greater than or equal to 1.

Equal fractions with different denominators are called *equivalent fractions.*

A fraction is in *simplest form* when there are no common factors in the numerator and the denominator.

The *reciprocal* of a fraction is the fraction with the numerator and denominator interchanged.

Inverting is the process of finding the reciprocal of a fraction.

Essential Rules ***Addition of Fractions with Like Denominators***

To add fractions with like denominators, add the numerators and place the sum over the common denominator.

Addition of Fractions with Unlike Denominators

To add fractions with unlike denominators, first rewrite the fractions as equivalent fractions with the same denominator. Then add the numerators and place the sum over the common denominator.

Subtraction of Fractions with Like Denominators

To subtract fractions with like denominators, subtract the numerators and place the difference over the common denominator.

Subtraction of Fractions with Unlike Denominators

To subtract fractions with unlike denominators, rewrite the fractions as equivalent fractions with the same denominator. Then subtract the numerators and place the difference over the common denominator.

Multiplication of Fractions

To multiply two fractions, multiply the numerators and place the product over the product of the denominators.

Division of Fractions

To divide two fractions, multiply by the reciprocal of the divisor.

Chapter Review

SECTION 2.1

1. Find the LCM of 18 and 12.
36

2. Find the LCM of 18 and 27.
54

3. Find the GCF of 20 and 48.
4

4. Find the GCF of 15 and 25.
5

SECTION 2.2

5. Express the shaded portion of the circles as an improper fraction.

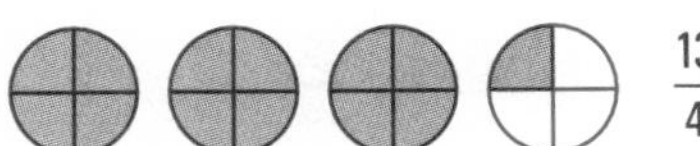

$\frac{13}{4}$

6. Express the shaded portion of the circles as a mixed number.

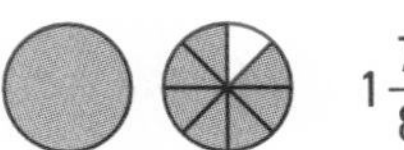

$1\frac{7}{8}$

7. Write $\frac{17}{5}$ as a mixed number.
$3\frac{2}{5}$

8. Write $2\frac{5}{7}$ as an improper fraction.
$\frac{19}{7}$

SECTION 2.3

9. Write an equivalent fraction with the given denominator.

$\frac{2}{3} = \frac{24}{36}$

10. Write an equivalent fraction with the given denominator.

$\frac{8}{11} = \frac{32}{44}$

11. Write $\frac{30}{45}$ in simplest form.
$\frac{2}{3}$

12. Write $\frac{16}{44}$ in simplest form.
$\frac{4}{11}$

SECTION 2.4

13. Add: $\frac{3}{8} + \frac{5}{8} + \frac{1}{8}$
$1\frac{1}{8}$

14. Find the total of $\frac{2}{3}$, $\frac{5}{6}$, and $\frac{2}{9}$.
$1\frac{13}{18}$

15. Add: $\frac{3}{8} + 1\frac{2}{3} + 3\frac{5}{6}$
$5\frac{7}{8}$

16. Add: $4\frac{4}{9} + 2\frac{1}{6} + 11\frac{17}{27}$
$18\frac{13}{54}$

17. During three months of the rainy season, $5\frac{7}{8}$, $6\frac{2}{3}$, and $8\frac{3}{4}$ inches of rain fell. Find the total rainfall for the three months. $21\frac{7}{24}$ inches

SECTION 2.5

18. Subtract: $\frac{11}{18} - \frac{5}{18}$
$\frac{1}{3}$

19. Find $\frac{17}{24}$ decreased by $\frac{3}{16}$.
$\frac{25}{48}$

20. Subtract: $18\frac{1}{6} - 3\frac{5}{7}$ $14\frac{19}{42}$

21. Subtract: $16 - 5\frac{7}{8}$ $10\frac{1}{8}$

22. A 15-mile race has three checkpoints. The first checkpoint is $4\frac{1}{2}$ miles from the starting point. The second checkpoint is $5\frac{3}{4}$ miles from the first checkpoint. How many miles is the second checkpoint from the finish line? $4\frac{3}{4}$ miles

SECTION 2.6

23. Multiply: $\frac{5}{12} \times \frac{4}{25}$ $\frac{1}{15}$

24. What is $\frac{11}{50}$ multiplied by $\frac{25}{44}$? $\frac{1}{8}$

25. Multiply: $2\frac{1}{3} \times 3\frac{7}{8}$ $9\frac{1}{24}$

26. Multiply: $2\frac{1}{4} \times 7\frac{1}{3}$ $16\frac{1}{2}$

27. A compact car gets 36 miles on each gallon of gasoline. How many miles can the car travel on $6\frac{3}{4}$ gallons of gasoline? 243 miles

SECTION 2.7

28. Divide: $\frac{5}{6} \div \frac{5}{12}$ 2

29. What is $\frac{15}{28}$ divided by $\frac{5}{7}$? $\frac{3}{4}$

30. Divide: $1\frac{1}{3} \div \frac{2}{3}$ 2

31. Divide: $8\frac{2}{3} \div 2\frac{3}{5}$ $3\frac{1}{3}$

32. A home building contractor bought $4\frac{2}{3}$ acres for $168,000. What was the cost of each acre? $36,000

SECTION 2.8

33. Place the correct symbol, < or >, between the two numbers.
$\frac{11}{18} < \frac{17}{24}$

34. Simplify: $\left(\frac{3}{4}\right)^3 \cdot \frac{20}{27}$ $\frac{5}{16}$

35. Simplify: $\frac{2}{7}\left(\frac{5}{8} - \frac{1}{3}\right) \div \frac{3}{5}$ $\frac{5}{36}$

36. Simplify: $\left(\frac{4}{5} - \frac{2}{3}\right)^2 \div \frac{4}{15}$ $\frac{1}{15}$

Chapter Test

1. Find the LCM of 24 and 40.
120 [2.1A]

2. Find the GCF of 24 and 80.
8 [2.1B]

3. Express the shaded portion of the circles as an improper fraction.

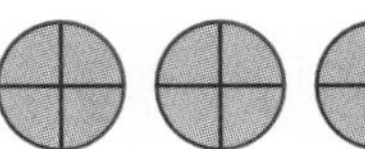

$\frac{11}{4}$ [2.2A]

4. Write $\frac{18}{5}$ as a mixed number.
$3\frac{3}{5}$ [2.2B]

5. Write $9\frac{4}{5}$ as an improper fraction.
$\frac{49}{5}$ [2.2B]

6. Write an equivalent fraction with the given denominator.
$\frac{5}{8} = \frac{}{72}$ $\frac{45}{72}$ [2.3A]

7. Write $\frac{40}{64}$ in simplest form.
$\frac{5}{8}$ [2.3B]

8. Add: $\frac{7}{12} + \frac{11}{12} + \frac{5}{12}$
$1\frac{11}{12}$ [2.4A]

9. Add: $\frac{5}{6} + \frac{7}{9} + \frac{1}{15}$
$1\frac{61}{90}$ [2.4B]

10. What is $12\frac{5}{12}$ more than $9\frac{17}{20}$?
$22\frac{4}{15}$ [2.4C]

11. The rainfall for a 3-month period was $11\frac{1}{2}$ inches, $7\frac{5}{8}$ inches, and $2\frac{1}{3}$ inches. Find the total rainfall for the 3 months. $21\frac{11}{24}$ inches [2.4D]

12. Subtract: $\frac{17}{24} - \frac{11}{24}$
$\frac{1}{4}$ [2.5A]

13. What is $\frac{9}{16}$ minus $\frac{5}{12}$?
$\frac{7}{48}$ [2.5B]

14. Subtract: $23\frac{1}{8} - 9\frac{9}{44}$
$13\frac{81}{88}$ [2.5C]

15. Chris Aguilar bought 100 shares of a utility stock at $\$24\frac{1}{2}$ per share. The stock gained $\$5\frac{5}{8}$ during the first month of ownership and lost $\$2\frac{1}{4}$ during the second month. Find the value of 1 share of the utility stock at the end of the second month. $\$27\frac{7}{8}$ [2.5D]

16. Multiply: $\frac{9}{11} \times \frac{44}{81}$

$\frac{4}{9}$ [2.6A]

17. What is $5\frac{2}{3}$ multiplied by $1\frac{7}{17}$?

8 [2.6B]

18. An electrician earns $120 for each day worked. What is the total of the electrician's earnings for working $3\frac{1}{2}$ days? $420 [2.6C]

19. Divide: $\frac{5}{9} \div \frac{7}{18}$

$1\frac{3}{7}$ [2.7A]

20. Find the quotient of $6\frac{2}{3}$ and $3\frac{1}{6}$?

$2\frac{2}{19}$ [2.7B]

21. Grant Miura bought $7\frac{1}{4}$ acres of land for a housing project. One and three-fourths acres were set aside for a park, and the remaining land was developed into $\frac{1}{2}$-acre lots. How many lots were available for sale? 11 lots [2.7C]

22. Place the correct symbol, < or >, between the two numbers.

$\frac{3}{8} < \frac{5}{12}$ [2.8A]

23. Simplify: $\left(\frac{2}{3}\right)^4 \cdot \frac{27}{32}$

$\frac{1}{6}$ [2.8B]

24. Simplify: $\left(\frac{3}{4}\right)^2 \div \left(\frac{2}{3} + \frac{5}{6}\right) - \frac{1}{12}$

$\frac{7}{24}$ [2.8C]

25. Simplify: $\left(\frac{1}{4}\right)^3 \div \left(\frac{1}{8}\right)^2 - \frac{1}{6}$

$\frac{5}{6}$ [2.8C]

Cumulative Review

1. Round 290,496 to the nearest thousand.
290,000 [1.1D]

2. Subtract: $\begin{array}{r} 390{,}047 \\ -\ 98{,}769 \\ \hline \end{array}$
291,278 [1.3B]

3. Find the product of 926 and 79.
73,154 [1.4B]

4. Divide: $57\overline{)30{,}792}$
540 r 12 [1.5C]

5. Simplify: $4 \cdot (6 - 3) \div 6 - 1$
1 [1.6B]

6. Find the prime factorization of 44.
$2 \cdot 2 \cdot 11$ [1.7B]

7. Find the LCM of 30 and 42.
210 [2.1A]

8. Find the GCF of 60 and 80.
20 [2.1B]

9. Write $7\frac{2}{3}$ as an improper fraction.
$\frac{23}{3}$ [2.2B]

10. Write $\frac{25}{4}$ as a mixed number.
$6\frac{1}{4}$ [2.2B]

11. Write an equivalent fraction with the given denominator.
$\frac{5}{16} = \frac{15}{48}$ [2.3A]

12. Write $\frac{24}{60}$ in simplest form.
$\frac{2}{5}$ [2.3B]

13. What is $\frac{9}{16}$ more than $\frac{7}{12}$?
$1\frac{7}{48}$ [2.4B]

14. Add: $\begin{array}{r} 3\frac{7}{8} \\ 7\frac{5}{12} \\ +\ 2\frac{15}{16} \\ \hline \end{array}$
$14\frac{11}{48}$ [2.4C]

15. Find $\frac{3}{8}$ less than $\frac{11}{12}$.
$\frac{13}{24}$ [2.5B]

16. Subtract: $\begin{array}{r} 5\frac{1}{6} \\ -\ 3\frac{7}{18} \\ \hline \end{array}$
$1\frac{7}{9}$ [2.5C]

17. Multiply: $\frac{3}{8} \times \frac{14}{15}$
$\frac{7}{20}$ [2.6A]

18. Multiply: $3\frac{1}{8} \times 2\frac{2}{5}$
$7\frac{1}{2}$ [2.6B]

19. Divide: $\frac{7}{16} \div \frac{5}{12}$
$1\frac{1}{20}$ [2.7A]

20. Find the quotient of $6\frac{1}{8}$ and $2\frac{1}{3}$.
$2\frac{5}{8}$ [2.7B]

21. Simplify: $\left(\frac{1}{2}\right)^3 \cdot \frac{8}{9}$
$\frac{1}{9}$ [2.8B]

22. Simplify: $\left(\frac{1}{2} + \frac{1}{3}\right) \div \left(\frac{2}{5}\right)^2$
$5\frac{5}{24}$ [2.8C]

23. Molly O'Brien had $1359 in a checking account. During the week, Molly wrote checks of $128, $54, and $315. Find the amount in the checking account at the end of the week. $862 [1.3C]

24. The tickets for a movie were $5 for an adult and $2 for a student. Find the total income from the sale of 87 adult tickets and 135 student tickets. $705 [1.4C]

25. Find the total weight of three packages that weigh $1\frac{1}{2}$ pounds, $7\frac{7}{8}$ pounds, and $2\frac{2}{3}$ pounds. $12\frac{1}{24}$ pounds [2.4D]

26. A board $2\frac{5}{8}$ feet long is cut from a board $7\frac{1}{3}$ feet long. What is the length of the remaining piece? $4\frac{17}{24}$ feet [2.5D]

27. A car travels 27 miles on each gallon of gasoline. How many miles can the car travel on $8\frac{1}{3}$ gallons of gasoline? 225 miles [2.6C]

28. Jimmy Santos purchased $10\frac{1}{3}$ acres of land to build a housing development. Jimmy donated 2 acres for a park. How many $\frac{1}{3}$-acre parcels can be sold from the remaining land? 25 parcels [2.7C]

Chapter

Decimals

Objectives

Section 3.1

To write decimals in standard form and in words

To round a decimal to a given place value

Section 3.2

To add decimals

To solve application problems

Section 3.3

To subtract decimals

To solve application problems

Section 3.4

To multiply decimals

To solve application problems

Section 3.5

To divide decimals

To solve application problems

Section 3.6

To convert fractions to decimals

To convert decimals to fractions

To identify the order relation between two decimals or between a decimal and a fraction

Decimal Fractions

How would you like to add $\frac{37{,}544}{23{,}465} + \frac{5184}{3456}$? These two fractions are very cumbersome, and it would take even a mathematician some time to get the answer.

Well, around 1550, help with such problems arrived with the publication of a book called *La Disme* (*The Tenth*), which urged the use of decimal fractions. A decimal fraction is one in which the denominator is 10, 100, 1000, 10,000, and so on.

This book suggested that all whole numbers were "units" and when written would end with the symbol ⓪. For example, the number 294⓪ would be the number two hundred ninety-four. This is very much like the way numbers are currently written (except for the ⓪).

For a fraction between 0 and 1, a unit was broken down into parts called "primes." The fraction three-tenths would be written

$\frac{3}{10}$ = 3① The ① was used to mean the end of the primes, or what are now called tenths.

Each prime was further broken down into "seconds," and each second was broken down into "thirds," and so on. The primes ended with ①, the seconds ended with ②, and the thirds ended with ③.

Examples of these numbers in the old notation and our modern fraction notation are shown below.

$\frac{37}{100}$ = 3①7② $\frac{257}{1000}$ = 2①5②7③

After completing this chapter, you might come back to the problem in the first line. Use decimals instead of fractions to find the answer. The answer is 3.1.

3.1 Introduction to Decimals

Objective A To write decimals in standard form and in words

POINT OF INTEREST

Leonardo of Pisa (Fibonacci) anticipated the decimal system. He wrote $\frac{2\ \ 7}{10\ 10}5$ to mean $5 + \frac{7}{10} + \frac{2}{10 \cdot 10}$.

The smallest human bone is found in the middle ear and measures 0.135 inch in length. The number 0.135 is in **decimal notation.**

Note the relationship between fractions and numbers written in decimal notation.

Three-tenths	Three-hundredths	Three-thousandths
$\frac{3}{10} = 0.3$	$\frac{3}{100} = 0.03$	$\frac{3}{1000} = 0.003$
1 zero 1 decimal place	2 zeros 2 decimal places	3 zeros 3 decimal places

A number written in decimal notation has three parts.

351	.	7089
Whole-number part	**Decimal point**	**Decimal part**

A number written in decimal notation is often called simply a **decimal.** The position of a digit in a decimal determines the digit's place value.

In the decimal 351.7089, the position of the digit 9 determines that its place value is ten-thousandths.

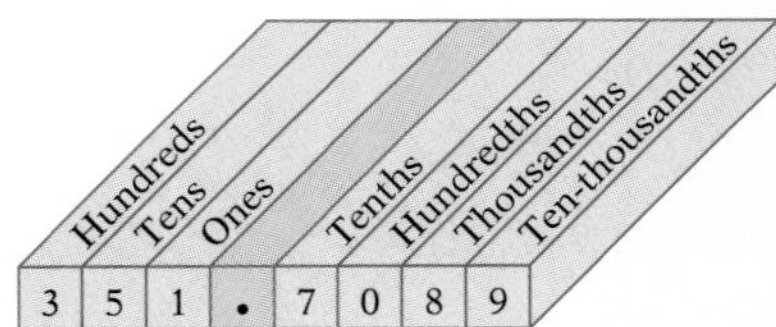

When writing a decimal in words, write the decimal part as if it were a whole number; then name the place value of the last digit.

0.6481 Six thousand four hundred eighty-one ten-thousandths

549.238 Five hundred forty-nine and two hundred thirty-eight thousandths (The decimal point is read as "and.")

To write a decimal in standard form, zeros may have to be inserted after the decimal point so that the last digit is in the given place-value position.

INSTRUCTOR NOTE

Many numbers are given as 7.3 million or 2.3 billion. As oral exercises, have students write these numbers in standard form.

Five and thirty-eight hundredths

8 is in the hundredths' place. 5.38

Nineteen and four thousandths

4 is in the thousandths' place. Insert two zeros so that 4 is in the thousandths place. 19.004

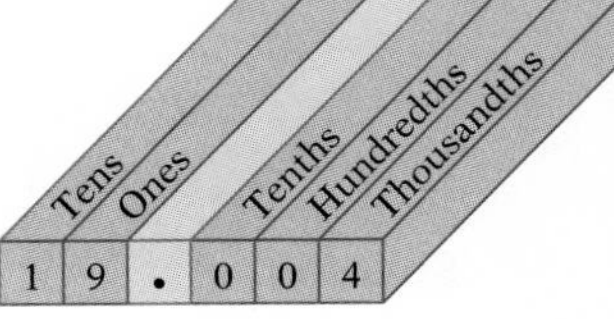

Seventy-one ten-thousandths

1 is in the ten-thousandths' place. Insert two zeros so that 1 is in the ten-thousandths place. 0.0071

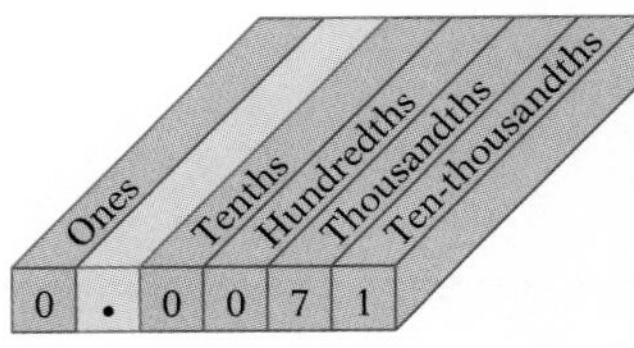

Example 1 Write 307.4027 in words.

Solution Three hundred seven and four thousand twenty-seven ten-thousandths

You Try It 1 Write 209.05838 in words.

Your solution Two hundred nine and five thousand eight hundred thirty-eight hundred-thousandths

Example 2 Write six hundred seven and seven hundred eight hundred-thousandths in standard form.

Solution 607.00708

You Try It 2 Write forty-two thousand and two hundred seven millionths in standard form.

Your solution 42,000.000207

Solutions on p. A18

Objective B To round a decimal to a given place value

INSTRUCTOR NOTE
As a calculator activity, students can determine whether their calculators round or *truncate.* Using $2 \div 3$ will serve as a good example.

Rounding decimals is similar to rounding whole numbers except that the digits to the right of the given place value are dropped instead of being replaced by zeros.

If the digit to the right of the given place value is less than 5, that digit and all digits to the right are dropped. If the digit to the right of the given place value is greater than or equal to 5, increase the given place value by 1 and drop all digits to its right.

INSTRUCTOR NOTE
Explain to students that not all rounding is done as shown here. When sales tax is computed, the decimal is always rounded up to the nearest cent. Thus, a sales tax of $.132 would be $.14.

➡ Round 26.3799 to the nearest hundredth.

26.3799 — Given place value: 7

9 > 5 Increase 7 by 1 and drop all digits to the right of 7.

26.3799 rounded to the nearest hundredth is 26.38.

Example 3 Round 0.39275 to the nearest ten-thousandth.

Solution

0.39275 — Given place value: 7

5 = 5

0.3928

You Try It 3 Round 4.349254 to the nearest hundredth.

Your solution 4.35

Example 4 Round 42.0237412 to the nearest hundred-thousandth.

Solution

42.0237412 — Given place value: 4

1 < 5

42.02374

You Try It 4 Round 3.290532 to the nearest hundred-thousandth.

Your solution 3.29053

Solutions on p. A18

3.1 Exercises

Objective A

Write each decimal in words.

1. 0.27 **2.** 0.92 **3.** 1.005 **4.** 3.067

1. Twenty-seven hundredths
2. Ninety-two hundredths
3. One and five thousandths
4. Three and sixty-seven thousandths

5. 36.4 **6.** 59.7 **7.** 0.00035 **8.** 0.00092

5. Thirty-six and four tenths
6. Fifty-nine and seven tenths
7. Thirty-five hundred-thousandths
8. Ninety-two hundred-thousandths

9. 10.007 **10.** 20.009 **11.** 52.00095 **12.** 64.00037

9. Ten and seven thousandths
10. Twenty and nine thousandths
11. Fifty-two and ninety-five hundred-thousandths
12. Sixty-four and thirty-seven hundred-thousandths

13. 0.0293 **14.** 0.0717 **15.** 6.324 **16.** 8.916

13. Two hundred ninety-three ten-thousandths
14. Seven hundred seventeen ten-thousandths
15. Six and three hundred twenty-four thousandths
16. Eight and nine hundred sixteen thousandths

17. 276.3297 **18.** 418.3115 **19.** 216.0729 **20.** 976.0317

17. Two hundred seventy-six and three thousand two hundred ninety-seven ten-thousandths
18. Four hundred eighteen and three thousand one hundred fifteen ten-thousandths
19. Two hundred sixteen and seven hundred twenty-nine ten-thousandths
20. Nine hundred seventy-six and three hundred seventeen ten-thousandths

21. 4625.0379 **22.** 2986.0925 **23.** 1.00001 **24.** 3.00003

21. Four thousand six hundred twenty-five and three hundred seventy-nine ten-thousandths
22. Two thousand nine hundred eighty-six and nine hundred twenty-five ten-thousandths
23. One and one hundred-thousandths
24. Three and three hundred-thousandths

Write each decimal in standard form.

25. Seven hundred sixty-two thousandths
0.762

26. Two hundred ninety-five thousandths
0.295

27. Sixty-two millionths
0.000062

28. Forty-one millionths
0.000041

29. Eight and three hundred four ten-thousandths
8.0304

30. Four and nine hundred seven ten-thousandths
4.0907

31. Three hundred four and seven hundredths 304.07

32. Eight hundred ninety-six and four hundred seven thousandths 896.407

Write each decimal in standard form.

33. Three hundred sixty-two and forty-eight thousandths 362.048

34. Seven hundred eighty-four and eighty-four thousandths 784.084

35. Three thousand forty-eight and two thousand two ten-thousandths 3048.2002

36. Seven thousand sixty-one and nine thousand one ten-thousandths 7061.9001

Objective B

Round each decimal to the given place value.

37. 7.359 Tenths
7.4

38. 6.405 Tenths
6.4

39. 23.009 Tenths
23.0

40. 89.19204 Tenths
89.2

41. 22.68259 Hundredths
22.68

42. 16.30963 Hundredths
16.31

43. 480.325 Hundredths
480.33

44. 670.974 Hundredths
670.97

45. 1.03925 Thousandths
1.039

46. 7.072854 Thousandths
7.073

47. 1946.3745 Thousandths
1946.375

48. 62.009435 Thousandths
62.009

49. 0.029876 Ten-thousandths
0.0299

50. 0.012346 Ten-thousandths
0.0123

51. 1.702596 Nearest whole number
2

52. 2.079239 Hundred-thousandths
2.07924

53. 0.0102903 Millionths
0.010290

54. 0.1009754 Millionths
0.100975

APPLYING THE CONCEPTS

55. [W] To what decimal place value are timed events in the Olympics recorded? Provide some specific examples of events and the winning times in each. Hundredths of a second.

56. [W] Provide an example of a situation in which a decimal is always rounded up, even if the digit to the right is less than 5. Provide an example of a situation in which a decimal is always rounded down, even if the digit to the right is 5 or greater than 5. (*Hint*: Think about situations in which money changes hands.) Answers will vary

57. Indicate which zeros of the number, if any, need not be entered on a calculator.

a. 23.500
23.5(00)

b. 0.000235
(0).000235

c. 300.0005
All zeros needed

d. 0.004050
(0).00405(0)

58. A decimal number was rounded to 6. Between what two numbers, to the nearest tenth, was the number? 5.5–6.4

59. A decimal number was rounded to 10.2. Between what two numbers to the nearest hundredth, was the number? 10.15–10.24

3.2 Addition of Decimals

Objective A To add decimals

POINT OF INTEREST

Simon Stevin (1548–1620) was the first to name decimal numbers. He wrote the number 2.345 as 2 0 3 1 4 2 5 3 . He called the whole number part the *commencement,* the tens digit was *prime,* the hundredths digit was *second,* the thousandths digit was *thirds* and so on.

To add decimals, write the numbers so that the decimal points are on a vertical line. Add as for whole numbers, and write the decimal point in the sum directly below the decimal points in the addends.

➡ Add: 0.237 + 4.9 + 27.32

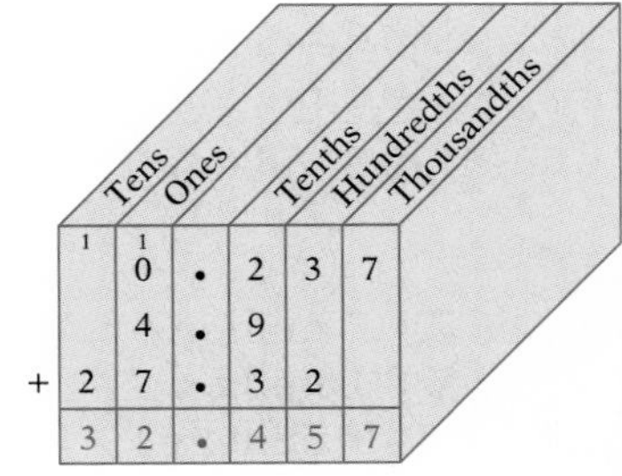

Note that by placing the decimal points on a vertical line, we make sure that digits of the same place value are added.

Example 1 Find the sum of 42.3, 162.903, and 65.0729.

Solution

$$\begin{array}{r} \scriptstyle 1\,1\,1 \\ 42.3 \\ 162.903 \\ +\ \ 65.0729 \\ \hline 270.2759 \end{array}$$

You Try It 1 Add: 4.62 + 27.9 + 0.62054

Your solution 33.14054

Example 2 Add: 0.83 + 7.942 + 15

Solution

$$\begin{array}{r} \scriptstyle 1\,1 \\ 0.83 \\ 7.942 \\ +\ 15. \\ \hline 23.772 \end{array}$$

You Try It 2 Add: 6.05 + 12 + 0.374

Your solution 18.424

Solutions on p. A18

ESTIMATION

Estimating the Sum of Two or More Decimals

Estimate and then use your calculator to find 23.037 + 16.7892.

To estimate the sum of two or more numbers, round each number to the same place value. In this case, we will round to the nearest whole number. Then add. The estimated answer is 40.

$$\begin{array}{rcr} 23.037 & \approx & 23 \\ +\ 16.7892 & \approx & +\ 17 \\ \hline & & 40 \end{array}$$

Now use your calculator to find the exact result. The exact answer is 39.8262.

23.037 [+] 16.7892 [=] 39.8262

Objective B To solve application problems

The table at the the right lists the team payrolls, in millions of dollars, of the professional football teams.

Comparing Team Payrolls

Payrolls computed to nearest $100,000

Team	1993	1992	Change
San Francisco 49ers	$41.4 mil.	$34.3 mil.	+$ 7.1
Washington Redskins	$40.3	$31.7	+$ 8.6
Buffalo Bills	$39.1	$29.9	+$ 9.2
New York Giants	$38.5	$31.5	+$ 7.0
New Orleans Saints	$36.3	$29.5	+$ 6.8
Green Bay Packers	$35.3	$28.3	+$ 7.0
Phoenix Cardinals	$35.2	$22.3	+$12.9
Kansas City Chiefs	$35.1	$24.1	+$11.0
New York Jets	$34.9	$24.8	+$10.1
Cleveland Browns	$34.4	$27.2	+$ 7.2
Miami Dolphins	$34.2	$33.9	+$ 0.3
Atlanta Falcons	$33.9	$31.6	+$ 2.3
Los Angeles Raiders	$33.3	$27.9	+$ 5.4
Indianapolis Colts	$33.2	$32.8	+$ 0.4
Chicago Bears	$31.9	$23.9	+$ 8.0
Detroit Lions	$30.7	$24.9	+$ 5.8
Denver Broncos	$30.4	$23.9	+$ 6.5
Houston Oilers	$29.6	$26.7	+$ 2.9
Los Angeles Rams	$28.4	$26.1	+$ 2.3
Tampa Bay Buccaneers	$28.2	$19.7	+$ 8.5
Minnesota Vikings	$26.7	$26.7	even
San Diego Chargers	$26.2	$25.1	+$ 1.1
Philadelphia Eagles	$26.2	$27.1	$–0.9
Pittsburgh Steelers	$26.1	$17.6	+$ 8.5
Seattle Seahawks	$25.7	$24.7	+$ 1.0
Dallas Cowboys	$24.7	$25.2	$–0.5
New England Patriots	$20.1	$21.9	$–1.8
Cincinnati Bengals	$19.1	$28.2	$–9.1

Find the combined payroll for the three highest-payroll teams and the three lowest-payroll teams in 1993.

$$\begin{array}{r} \$41.4 \text{ million} \\ \$40.3 \text{ million} \\ + \$39.1 \text{ million} \\ \hline \$120.8 \text{ million} \end{array}$$

Add the payrolls of the San Francisco 49ers, the Washington Redskins, and the Buffalo Bills. The combined payroll for the three highest payroll teams was $120.8 million.

$$\begin{array}{r} \$24.7 \text{ million} \\ \$20.1 \text{ million} \\ + \$19.1 \text{ million} \\ \hline \$63.9 \text{ million} \end{array}$$

Add the payrolls of the Dallas Cowboys, the New England Patriots, and the Cincinnati Bengals. The combined payroll for the three lowest-payroll teams was $63.9 million.

Example 3

Dan Burhoe earned a salary of $138.50 for working 5 days this week as a food server. He also received $22.92, $15.80, $19.65, $39.20, and $27.70 in tips during the 5 days. Find the total income for the week.

Strategy

To find the total income, add the tips ($22.92, $15.80, $19.65, $39.20, and $27.70) to the salary ($138.50).

Solution

$$\begin{array}{r} \$138.50 \\ 22.92 \\ 15.80 \\ 19.65 \\ 39.20 \\ + \quad 27.70 \\ \hline \$263.77 \end{array}$$

Dan's total income for the week was $263.77.

You Try It 3

Anita Khavari, an insurance executive, earns a salary of $425 every four weeks. During the past 4-week period, she received commissions of $485.60, $599.46, $326.75, and $725.42. Find her total income for the past 4-week period.

Your strategy

Your solution $2562.23

Solution on p. A19

3.2 Exercises

Objective A

Add.

1. 16.008 + 2.0385 + 132.06
150.1065

2. 17.32 + 1.0579 + 16.5
34.8779

3. 1.792 + 67 + 27.0526
95.8446

4. 8.772 + 1.09 + 26.5027
36.3647

5. 3.02 + 62.7 + 3.924
69.644

6. 9.06 + 4.976 + 59.6
73.636

7. 82.006 + 9.95 + 0.927
92.883

8. 0.826 + 8.76 + 79.005
88.591

9. 4.307 + 99.82 + 9.078
113.205

10. 0.3 + 0.07
0.37

11. 0.29 + 0.4
0.69

12. 1.007 + 2.1
3.107

13. 7.3 + 9.005
16.305

14. 4.9257 + 27.05 + 9.0063
40.9820

15. 8.72 + 99.073 + 2.9736
110.7666

16. 62.4 + 9.827 + 692.44
764.667

17. 8 + 89.43 + 7.0659
104.4959

Estimate by rounding to the nearest whole number. Then use your calculator to add.

18. 342.42 + 89.625 + 176.2
Est.: 608
Cal.: 608.245

19. 219.9 + 0.872 + 13.42
Est.: 234
Cal.: 234.192

20. 823.9 + 82.65 + 46.923
Est.: 954
Cal.: 953.473

21. 678.92 + 97.6 + 5.423
Est.: 782
Cal.: 781.943

Objective B *Application Problems*

22. A family receives an electric bill of \$70.42, a gas bill for \$54.55, and a garbage collection bill of \$23.14. Find the total bill for the three services. \$148.11

23. Find the length of the shaft.

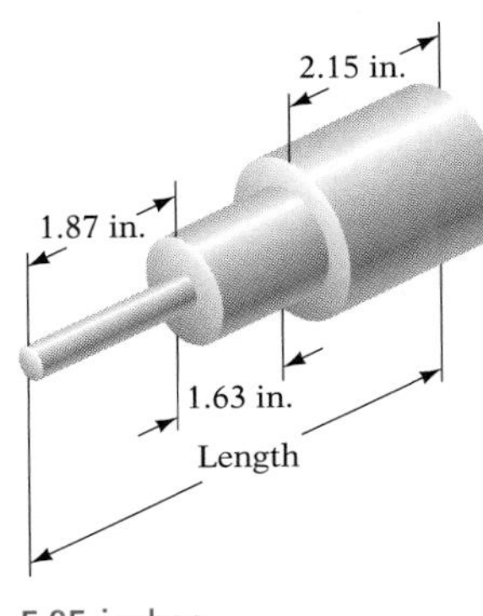

5.65 inches

24. Find the length of the shaft. 4.35 feet

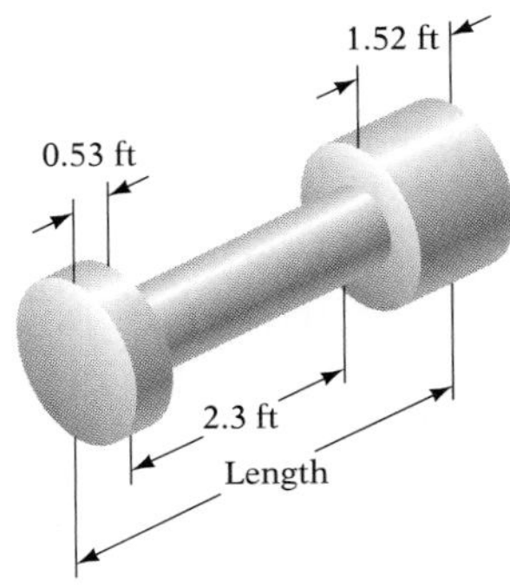

25. Commuting, Mae Chan used 12.4 gallons of gas the first week, 9.8 gallons the second week, 15.2 gallons the third week, and 10.4 gallons the fourth week. Find the total amount of gas she used during the 4 weeks. 47.8 gallons

26. The odometer on a family's car reads 24,835.9 miles. The car was driven 8.2 miles on Friday, 82.6 miles on Saturday, and 133.6 miles on Sunday.
a. How many miles was the car driven during the three days? 224.4 miles
b. Find the odometer reading at the end of the three days. 25,060.3 miles

27. You have $2143.57 in your checking account. You make deposits of $210.98, $45.32, $1236.34, and $27.99. Find the amount in your checking account after you have made the deposits if no money has been withdrawn. $3664.20

The figure at the right shows the number of viewers who watch television each night of the week. Use this figure for Exercises 28–30.

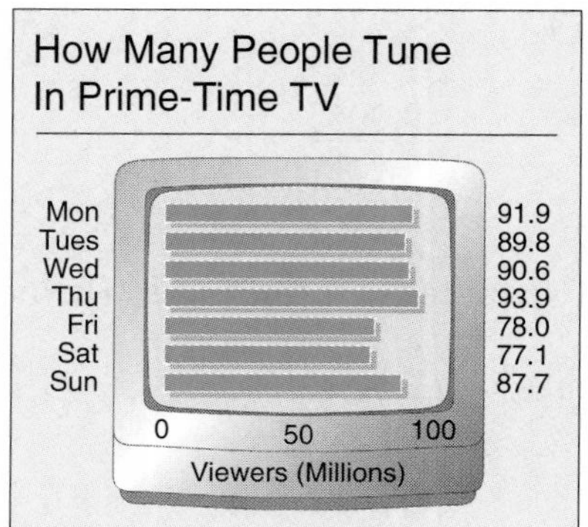

28. Find the total number of television viewers for Friday, Saturday, and Sunday nights. 242.8 million

29. Find the total number of television viewers for Monday, Tuesday, Wednesday, and Thursday nights. 366.2 million

30. Find the total number of television viewers for the week. 609 million

APPLYING THE CONCEPTS

The table at the right gives the prices for selected products in a grocery store. Use this table for Exercises 31 and 32.

Product	*Cost*
Raisin Bran	3.45
Butter	2.69
Bread	1.23
Popcorn	.89
Potatoes	1.09
Cola (6-pack)	.98
Mayonnaise	2.25
Lunch meat	3.31
Milk	2.18
Toothpaste	2.45

31. Does a customer with $10 have enough money to purchase raisin bran, bread, milk, lunch meat, and butter? No

32. Name three items that would cost more than $6 but less than $7. (There is more than one answer.) Bread, Butter, Mayonnaise } one answer

33. Can a piece of rope 4 feet long be wrapped around the box shown at the right? No

1.4 ft
1.4 ft
1.4 ft

3.3 Subtraction of Decimals

Objective A To subtract decimals

To subtract decimals, write the numbers so that the decimal points are on a vertical line. Subtract as for whole numbers, and write the decimal point in the difference directly below the decimal point in the subtrahend.

➡ Subtract 21.532 − 9.875 and check.

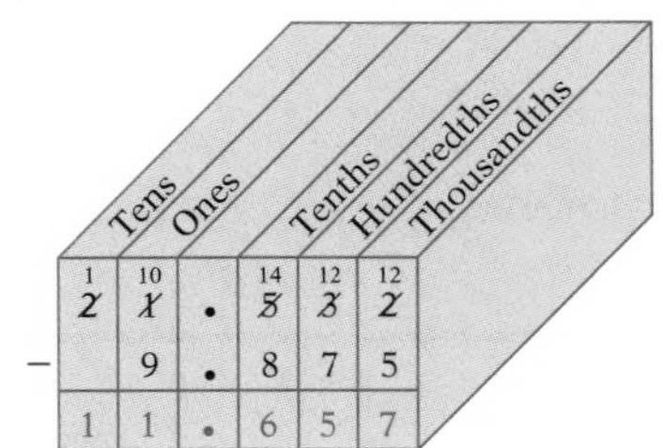

Placing the decimal points on a vertical line ensures that digits of the same place value are subtracted.

Check:	Subtrahend	9.875
	+ Difference	+ 11.657
	= Minuend	21.532

INSTRUCTOR NOTE
Inserting zeros so that each number has the same number of digits to the right of the decimal point will help some students.

➡ Subtract 4.3 − 1.7942 and check.

$$\begin{array}{r} 4.3000 \\ -\ 1.7942 \\ \hline 2.5058 \end{array}$$

If necessary, insert zeros in the minuend before subtracting.

Check: $\begin{array}{r} 1.7942 \\ +\ 2.5058 \\ \hline 4.3000 \end{array}$

Example 1 Subtract 39.047 − 7.96 and check.

Solution

$$\begin{array}{r} 39.047 \\ -\ \ 7.96\ \ \\ \hline 31.087 \end{array}$$

Check: $\begin{array}{r} 7.96 \\ +\ 31.087 \\ \hline 39.047 \end{array}$

You Try It 1 Subtract 72.039 − 8.47 and check.

Your solution 63.569

Example 2 Find 9.23 less than 29 and check.

Solution

$$\begin{array}{r} 29.00 \\ -\ \ 9.23 \\ \hline 19.77 \end{array}$$

Check: $\begin{array}{r} 9.23 \\ +\ 19.77 \\ \hline 29.00 \end{array}$

You Try It 2 Subtract 35 − 9.67 and check.

Your solution 25.33

Example 3 Subtract 1.2 − 0.8235 and check.

Solution

$$\begin{array}{r} 1.2000 \\ -\ 0.8235 \\ \hline 0.3765 \end{array}$$

Check: $\begin{array}{r} 0.8235 \\ +\ 0.3765 \\ \hline 1.2000 \end{array}$

You Try It 3 Subtract 3.7 − 1.9715 and check.

Your solution 1.7285

Solutions on p. A19

ESTIMATION

Estimating the Difference Between Two Decimals

Estimate and then use your calculator to find 820.2306 − 475.74815.

To estimate the difference between two numbers, round each number to the same place value. In this case we will round to the nearest ten. Then subtract. The estimated answer is 340.

$$\begin{array}{r} 820.2306 \\ -\ 475.74815 \\ \hline \end{array} \approx \begin{array}{r} 820 \\ -\ 480 \\ \hline 340 \end{array}$$

Now use your calculator to find the exact result. The exact answer is 344.48245.

820.2306 [−] 475.74815 [=] 344.48245

Objective B To solve application problems

Example 4
You bought a book for $15.87. How much change did you receive from a $20.00 bill?

Strategy
To find the amount of change, subtract the cost of the book ($15.87) from $20.00.

Solution

$$\begin{array}{r} \$20.00 \\ -\ 15.87 \\ \hline \$\ 4.13 \end{array}$$

You receive $4.13 in change.

You Try It 4
Your breakfast cost $3.85. How much change did you receive from a $5.00 bill?

Your strategy

Your solution $1.15

Example 5
You had a balance of $62.41 in your checking account. You then bought a cassette for $8.95, film for $3.17, and a skateboard for $39.77. After paying for these items with a check, how much do you have left in your checking account?

Strategy
To find the new balance:

- Find the total cost of the three items ($8.95 + $3.17 + $39.77).
- Subtract the total cost from the old balance ($62.41).

Solution

$$\begin{array}{r} \$\ 8.95 \\ 3.17 \\ +\ 39.77 \\ \hline \$51.89 \end{array} \text{ total cost} \qquad \begin{array}{r} \$62.41 \\ -\ 51.89 \\ \hline \$10.52 \end{array}$$

The new balance is $10.52

You Try It 5
You had a balance of $2472.69 in your checking account. You then wrote checks for $1025.60, $79.85, and $162.47. Find the new balance in your checking account.

Your strategy

Your solution $1204.77

Solutions on p. A19

3.3 Exercises

Objective A

Subtract and check.

1. 24.037 − 18.41
5.627

2. 26.029 − 19.31
6.719

3. 123.07 − 9.4273
113.6427

4. 214 − 7.143
206.857

5. 16.5 − 9.7902
6.7098

6. 13.2 − 8.6205
4.5795

7. 235.79 − 20.093
215.697

8. 463.27 − 40.095
423.175

9. 63.005 − 9.1274
53.8776

10. 23.004 − 7.2175
15.7865

11. 92 − 19.2909
72.7091

12. 41.2405 − 25.2709
15.9696

13. 7.01 − 2.325
4.685

14. 8.07 − 5.392
2.678

15. 19.0035 − 8.967
10.0365

16. 0.32 − 0.0058
0.3142

17. 0.78 − 0.0073
0.7727

18. 3.005 − 1.982
1.023

19. 6.007 − 2.734
3.273

20. 352.16 − 90.994
261.166

21. 872 − 80.753
791.247

22. 724.32 − 69
655.32

23. 625.46 − 77.509
547.951

24. 362.394 − 19.4672
342.9268

25. 421.385 − 17.5293
403.8557

26. 19 − 10.372
8.628

27. 23.4 − 0.921
22.479

Estimate by rounding to the nearest ten. Then use your calculator to subtract.

28. 620.59 − 132.79
Est.: 490
Cal.: 487.80

29. 835.07 − 244.82
Est.: 600
Cal.: 590.25

30. 67.3 − 19.793
Est.: 50
Cal.: 47.507

31. 84.1 − 48.906
Est.: 30
Cal.: 35.194

Estimate by rounding to the nearest whole number (nearest one). Then use your calculator to subtract.

32. 93.079256 − 66.09249
Est.: 27
Cal.: 26.986766

33. 3.7529 − 1.00784
Est.: 3
Cal.: 2.74506

34. 76.53902 − 45.73005
Est.: 31
Cal.: 30.80897

35. 9.07325 − 1.924
Est.: 7
Cal.: 7.14925

Objective B *Application Problems*

36. A patient has a fever of 102.3°F. Normal temperature is 98.6°F. How many degrees above normal is the patient's temperature? 3.7°

37. The manager of the Edgewater Cafe takes a reading of the cash register tape each hour. At 1:00 P.M. the tape read \$967.54; at 2:00 P.M. the tape read \$1437.15. Find the amount of sales between 1:00 P.M. and 2:00 P.M. \$469.61

38. Find the missing dimension.

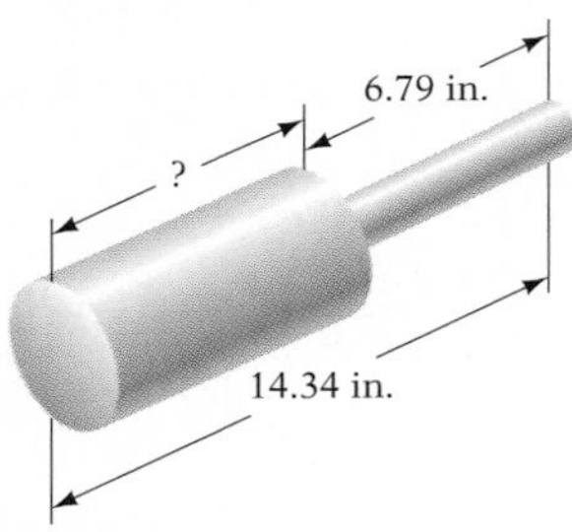

7.55 inches

39. Find the missing dimension. 2.59 feet

40. You had a balance of \$1,029.74 in your checking account. You then wrote checks for \$67.92, \$43.10, and \$496.34.
a. Find the total amount of the checks written. \$607.36
b. Find the new balance in your checking account. \$422.38

41. The price of gasoline is \$1.22 per gallon after the price rose \$.07 one month and \$.12 the second month. Find the price of gasoline before these increases in price. \$1.03

42. Rainfall for the last 3 months of the year was 1.42 inches, 5.39 inches, and 3.55 inches. The normal rainfall for the last 3 months of the year is 11.22 inches. How many inches below normal was the rainfall? 0.86 inches

The table at the right shows the amount that a wage earner would have to earn to have the buying power of a \$15,000 salary in 1977. Use the table for Exercises 43 and 44.

Keeping Up With Inflation

How much a person would have to earn each year to have the same buying power that a \$15,000 salary had in June 1977.

Year	Amount	Year	Amount
1978	\$16,729.81	1986	\$27,306.25
1979	\$18,953.85	1987	\$28,517.09
1980	\$21,326.16	1988	\$29,777.38
1981	\$23,228.94	1989	\$31,161.22
1982	\$24,118.54	1990	\$33,063.99
1983	\$25,032.85	1991	\$34,077.14
1984	\$26,021.29	1992	\$35,085.58
1985	\$27,009.74	1993[1]	\$35,683.30

1–first six months

43. How much more earnings would a wage earner need in 1992 to equal the buying power of a \$15,000 wage in 1977? \$20,085.58

44. How much more earnings would a wage earner need in 1980 to equal the buying power of a \$15,000 wage in 1977? \$6,326.16

APPLYING THE CONCEPTS

45. Find the largest amount by which the estimate of the sum of two decimals with tenths, hundredths, and thousandths places could differ from the exact sum. 0.05 0.005 0.0005

46. Grace Herrera owned 453.472 shares of a mutual fund on January 1, 1994. On December 31, 1994, she had 617.005 shares. What was the increase in the number of shares for the year? 163.533 shares

47. The average speed for the winner of the 1991 Indianapolis 500 was 176.460 mph. The average speed for the winner in 1992 was 134.477 mph. What is the difference between the average speeds for those two years? 41.983 mph

3.4 Multiplication of Decimals

Objective A To multiply decimals

POINT OF INTEREST

Benjamin Banneker (1731–1806) was the first African American to earn distinction as a mathematician and scientist. He was on the survey team that determined the boundaries of Washington, D.C. The mathematics of surveying requires extensive use of decimals.

Decimals are multiplied as if they were whole numbers; then the decimal point is placed in the product. Writing the decimals as fractions shows where to write the decimal point in the product.

$$0.\underline{3} \times 5 = \frac{3}{10} \times \frac{5}{1} = \frac{15}{10} = 1.\underline{5}$$

1 decimal place 1 decimal place

$$0.\underline{3} \times 0.\underline{5} = \frac{3}{10} \times \frac{5}{10} = \frac{15}{100} = 0.\underline{15}$$

1 decimal place 1 decimal place 2 decimal places

$$0.\underline{3} \times 0.\underline{05} = \frac{3}{10} \times \frac{5}{100} = \frac{15}{1000} = 0.\underline{015}$$

1 decimal place 2 decimal places 3 decimal places

To multiply decimals, multiply the numbers as in whole numbers. Write the decimal point in the product so that the number of decimal places in the product is the sum of the decimal places in the factors.

➡ Multiply: 21.4 × 0.36

```
   21.4      1 decimal place
 × 0.36      2 decimal places
   1284
   642
  7.704      3 decimal places
```

➡ Multiply: 0.037 × 0.08

```
    0.037    3 decimal places
 ×   0.08    2 decimal places
  0.00296    5 decimal places
```

• **Two zeros must be inserted between the 2 and the decimal point so that there are 5 decimal places in the product.**

To multiply a decimal by a power of 10 (10, 100, 1000,...), move the decimal point to the right the same number of places as there are zeros in the power of 10.

3.8925 × 10 = 38.925
1 zero 1 decimal place

3.8925 × 100 = 389.25
2 zeros 2 decimal places

3.8925 × 1000 = 3892.5
3 zeros 3 decimal places

3.8925 × 10,000 = 38,925.
4 zeros 4 decimal places

3.8925 × 100,000 = 389,250.
5 zeros 5 decimal places

Note that a zero must be inserted before the decimal point.

INSTRUCTOR NOTE

Another way to practice multiplying by powers of 10 is to relate these examples to numbers given as 3.84 million, 10.4 billion, or 2.3 trillion.

Also, multiplying or dividing (in the next section) by powers of 10 is the way one converts between various units of the metric system.

Note that if the power of 10 is written in exponential notation, the exponent indicates how many places to move the decimal point.

$3.8925 \times 10^1 = 38.925$ — 1 decimal place

$3.8925 \times 10^2 = 389.25$ — 2 decimal places

$3.8925 \times 10^3 = 3892.5$ — 3 decimal places

$3.8925 \times 10^4 = 38{,}925.$ — 4 decimal places

$3.8925 \times 10^5 = 389{,}250.$ — 5 decimal places

Example 1 Multiply: 920×3.7

Solution

$$\begin{array}{r} 920 \\ \times\ 3.7 \\ \hline 644\ 0 \\ 2760\ \ \\ \hline 3404.0 \end{array}$$

You Try It 1 Multiply: 870×4.6

Your solution 4002.0

Example 2 Find 0.00079 multiplied by 0.025.

Solution

$$\begin{array}{r} 0.00079 \\ \times \quad 0.025 \\ \hline 395 \\ 158\ \\ \hline 0.00001975 \end{array}$$

You Try It 2 Find 0.000086 multiplied by 0.057.

Your solution 0.000004902

Example 3 Find the product of 3.69 and 2.07.

Solution

$$\begin{array}{r} 3.69 \\ \times 2.07 \\ \hline 2583 \\ 7380\ \ \\ \hline 7.6383 \end{array}$$

You Try It 3 Find the product of 4.68 and 6.03.

Your solution 28.2204

Example 4 Multiply: $42.07 \times 10{,}000$

Solution $42.07 \times 10{,}000 = 420{,}700$

You Try It 4 Multiply: 6.9×1000

Your solution 6900

Example 5 Find 3.01 times 10^3.

Solution $3.01 \times 10^3 = 3010$

You Try It 5 Find 4.0273 times 10^2.

Your solution 402.73

Solutions on p. A20

ESTIMATION

Estimating the Product of Two Decimals

Estimate and then use your calculator to find 28.259 × 0.029.

To estimate a product, round each number so that there is one non-zero digit. Then multiply.

$$\begin{array}{r} 28.259 \\ \times\ 0.029 \end{array} \approx \begin{array}{r} 30 \\ \times\ 0.03 \\ \hline 0.90 \end{array}$$

The estimated answer is 0.90.

Now use your calculator to find the exact answer.

28.259 [×] 0.029 [=] 0.819511

The exact answer is 0.819511.

Objective B *To solve application problems*

The table below lists water rates and meter fees for a city. This table is used for Example 6 and You Try It 6.

Water Charges	
Commercial	\$1.39/1,000 gal
Comm Restaurant	\$1.39/1,000 gal
Industrial	\$1.39/1,000 gal
Institutional	\$1.39/1,000 gal
Res—No Sewer	
Residential—SF	
>0 <200 gal. per day	\$1.15/1,000 gal
>200 <1,500 gal. per day	\$1.39/1,000 gal
>1,500 gal. per day	\$1.54/1,000 gal

Meter Charges	
Meter	*Meter Fee*
5/8″ & 3/4″	\$13.50
1″	\$21.80
1-1/2″	\$42.50
2″	\$67.20
3″	\$133.70
4″	\$208.20
6″	\$415.10
8″	\$663.70

Example 6
Find the total bill for an industrial water user with a 6-inch meter that uses 152,000 gallons of water for July and August.

Strategy
To find the cost of water:

- Find the cost of water by multiplying the cost per 1000 gallons (1.39) by the number of 1000-gallon units.
- Add the cost of the water to the meter fee (\$415.10).

Solution

$$\text{Cost of water} = \frac{152{,}000}{1{,}000} \cdot 1.39 = 211.28$$

$$\text{Total cost} = 211.28 + 415.10 = 626.38$$

The total cost is \$626.38.

You Try It 6
Find the total bill for a commercial user that used 5000 gallons of water per day for July and August. The user has a 3-inch meter.

Your strategy

Your solution \$564.60

Solution on p. A20

Example 7

It costs \$.036 an hour to operate an electric motor. How much does it cost to operate the motor for 120 hours?

Strategy

To find the cost of running the motor for 120 hours, multiply the hourly cost (\$.036) by the number of hours the motor is run (120).

Solution

$$\begin{array}{r} \$.036 \\ \times\ 120 \\ \hline 720 \\ 36 \\ \hline \$4.320 \end{array}$$

The cost of running the motor for 120 hours is \$4.32.

You Try It 7

The cost of electricity to run a freezer for 1 hour is \$.035. This month the freezer has run for 210 hours. Find the total cost of running the freezer this month.

Your strategy

Your solution \$7.35

Example 8

Jason Ng earns a salary of \$280 for a 40-hour work week. This week he worked 12 hours of overtime at a rate of \$10.50 for each hour of overtime worked. Find his total income for the week.

Strategy

To find Jason's total income for the week:

- Find the overtime pay by multiplying the hourly overtime rate (\$10.50) by the number of hours of overtime worked (12).
- Add the overtime pay to the weekly salary (\$280).

Solution

$$\begin{array}{r} \$10.50 \\ \times\quad 12 \\ \hline 21\ 00 \\ 105\ 0 \\ \hline \$126.00 \end{array} \text{ overtime pay} \qquad \begin{array}{r} \$280.00 \\ +\ 126.00 \\ \hline \$406.00 \end{array}$$

Jason's total income for this week is \$406.00.

You Try It 8

You make a down payment of \$175 on a stereo and agree to make payments of \$37.18 a month for the next 18 months to repay the remaining balance. Find the total cost of the stereo.

Your strategy

Your solution \$844.24

Solutions on p. A21

3.4 Exercises

Objective A

Multiply.

1. 0.9×0.4 — 0.36
2. 0.7×0.9 — 0.63
3. 0.5×0.6 — 0.30
4. 0.3×0.7 — 0.21
5. 0.5×0.5 — 0.25
6. 0.7×0.7 — 0.49
7. 0.9×0.5 — 0.45
8. 0.2×0.6 — 0.12
9. 7.7×0.9 — 6.93
10. 3.4×0.4 — 1.36
11. 9.2×0.2 — 1.84
12. 2.6×0.7 — 1.82
13. 7.2×0.6 — 4.32
14. 6.8×0.4 — 2.72
15. 7.4×0.1 — 0.74
16. 3.8×0.1 — 0.38
17. 7.9×5 — 39.5
18. 9.3×7 — 65.1
19. 0.68×4 — 2.72
20. 0.83×9 — 7.47
21. 0.67×0.9 — 0.603
22. 0.84×0.3 — 0.252
23. 0.16×0.6 — 0.096
24. 0.47×0.8 — 0.376
25. 2.5×5.4 — 13.50
26. 3.9×1.9 — 7.41
27. 8.4×9.5 — 79.80
28. 7.6×5.8 — 44.08
29. 0.83×5.2 — 4.316
30. 0.24×2.7 — 0.648
31. 0.46×3.9 — 1.794
32. 0.78×6.8 — 5.304
33. 0.2×0.3 — 0.06
34. 0.3×0.3 — 0.09
35. 0.24×0.3 — 0.072
36. 0.17×0.5 — 0.085
37. $1.47 \times .09$ — 0.1323
38. 6.37×0.05 — 0.3185
39. 8.92×0.004 — 0.03568
40. 6.75×0.007 — 0.04725

Multiply.

41. $\begin{array}{r} 0.49 \\ \times\ 0.16 \\ \hline \end{array}$ 0.0784

42. $\begin{array}{r} 0.38 \\ \times\ 0.21 \\ \hline \end{array}$ 0.0798

43. $\begin{array}{r} 7.6 \\ \times\ 0.01 \\ \hline \end{array}$ 0.076

44. $\begin{array}{r} 5.1 \\ \times\ 0.01 \\ \hline \end{array}$ 0.051

45. $\begin{array}{r} 8.62 \\ \times\ \ \ 4 \\ \hline \end{array}$ 34.48

46. $\begin{array}{r} 5.83 \\ \times\ \ \ 7 \\ \hline \end{array}$ 40.81

47. $\begin{array}{r} 64.5 \\ \times\ \ \ 9 \\ \hline \end{array}$ 580.5

48. $\begin{array}{r} 37.8 \\ \times\ \ \ 8 \\ \hline \end{array}$ 302.4

49. $\begin{array}{r} 2.19 \\ \times\ 9.2 \\ \hline \end{array}$ 20.148

50. $\begin{array}{r} 1.25 \\ \times\ 5.6 \\ \hline \end{array}$ 7.000

51. $\begin{array}{r} 1.85 \\ \times\ 0.023 \\ \hline \end{array}$ 0.04255

52. $\begin{array}{r} 37.8 \\ \times\ 0.052 \\ \hline \end{array}$ 1.9656

53. $\begin{array}{r} 0.478 \\ \times\ 0.37 \\ \hline \end{array}$ 0.17686

54. $\begin{array}{r} 0.526 \\ \times\ 0.22 \\ \hline \end{array}$ 0.11572

55. $\begin{array}{r} 48.3 \\ \times\ 0.0041 \\ \hline \end{array}$ 0.19803

56. $\begin{array}{r} 67.2 \\ \times\ 0.0086 \\ \hline \end{array}$ 0.57792

57. $\begin{array}{r} 2.437 \\ \times\ 6.1 \\ \hline \end{array}$ 14.8657

58. $\begin{array}{r} 4.237 \\ \times\ 0.54 \\ \hline \end{array}$ 2.28798

59. $\begin{array}{r} 0.413 \\ \times\ 0.0016 \\ \hline \end{array}$ 0.0006608

60. $\begin{array}{r} 0.517 \\ \times\ 0.0029 \\ \hline \end{array}$ 0.0014993

61. $\begin{array}{r} 94.73 \\ \times\ 0.57 \\ \hline \end{array}$ 53.9961

62. $\begin{array}{r} 89.23 \\ \times\ 0.62 \\ \hline \end{array}$ 55.3226

63. $\begin{array}{r} 8.005 \\ \times\ 0.067 \\ \hline \end{array}$ 0.536335

64. $\begin{array}{r} 9.032 \\ \times\ 0.019 \\ \hline \end{array}$ 0.171608

65. 4.29×0.1 0.429

66. 6.78×0.1 0.678

67. 5.29×0.4 2.116

68. 6.78×0.5 3.390

69. 0.68×0.7 0.476

70. 0.56×0.9 0.504

71. 1.4×0.73 1.022

72. 6.3×0.37 2.331

73. 5.2×7.3 37.96

74. 7.4×2.9 21.46

75. 3.8×0.61 2.318

76. 7.2×0.72 5.184

77. 0.32×10 3.2

78. 6.93×10 69.3

79. 0.065×100 6.5

80. 0.039×100 3.9

81. 6.2856×1000 6285.6

Multiply.

82. 3.2954 × 1000
3295.4

83. 3.2 × 1000
3200

84. 0.006 × 10,000
60

85. 3.57 × 10,000
35,700

86. 8.52×10^1
85.2

87. 0.63×10^1
6.3

88. 82.9×10^2
8290

89. 0.039×10^2
3.9

90. 6.8×10^3
6,800

91. 4.9×10^4
49,000

92. 6.83×10^4
68,300

93. 0.067×10^2
6.7

94. 0.052×10^2
5.2

95. Find the product of 0.0035 and 3.45.
0.012075

96. Find the product of 237 and 0.34.
80.58

97. Multiply 3.005 by 0.00392.
0.0117796

98. Multiply 20.34 by 1.008.
20.50272

99. Multiply 1.348 by 0.23.
0.31004

100. Multiply 0.000358 by 3.56.
0.00127448

101. Find the product of 23.67 and 0.0035.
0.082845

102. Find the product of 0.00346 and 23.1.
0.079926

103. Find the product of 5, 0.45, and 2.3.
5.175

104. Find the product of 0.03, 23, and 9.45.
6.5205

Estimate and then use your calculator to multiply.

105. 28.5 × 3.2
Est.: 90
Cal.: 91.2

106. 86.3 × 4.4
Est.: 360
Cal.: 379.72

107. 2.38 × 0.44
Est.: 0.8
Cal.: 1.0472

108. 9.82 × 0.77
Est.: 8
Cal.: 7.5614

109. 0.866 × 4.5
Est.: 4.5
Cal.: 3.897

110. 0.239 × 8.2
Est.: 1.6
Cal.: 1.9598

111. 4.34 × 2.59
Est.: 12
Cal.: 11.2406

112. 6.87 × 9.98
Est.: 70
Cal.: 68.5626

113. 8.434 × 0.044
Est.: 0.32
Cal.: 0.371096

114. 7.037 × 0.094
Est.: 0.63
Cal.: 0.661478

115. 28.44 × 1.12
Est.: 30
Cal.: 31.8528

116. 86.57 × 7.33
Est.: 630
Cal.: 634.5581

117. 49.6854 × 39.0672
Est.: 2000
Cal.: 1941.069459

118. 2.00547 × 9.672
Est.: 20
Cal.: 19.39690584

119. 0.00456 × 0.009542
Est.: 0.00005
Cal.: 0.000043511

120. 7.00637 × .0128
Est.: 0.07
Cal.: 0.089681536

Objective B *Application Problems*

121. It costs \$.18 per mile to rent a car. Find the cost to rent a car that is driven 114 miles. \$20.52

122. An electric motor costing \$315.45 has an operating cost of \$.027 for 1 hour of operation. Find the cost to run the motor for 56 hours (Round to the nearest cent.) \$1.51

123. Four hundred empty soft drink cans weigh 18.75 pounds. A recycling center pays \$.75 per pound for the cans. Find the amount received for the 400 cans (round to the nearest cent). \$14.06

124. A recycling center pays \$.035 per pound for newspapers. Find the amount received from recycling 420 pounds of newspapers. \$14.70

125. A broker's fee for buying stock is 0.045 times the price of the stock. An investor bought 100 shares of stock at \$38.50 per share. Find the broker's fee. \$173.25

126. A broker's fee for buying stock is 0.052 times the price of the stock. You bought 100 shares of stock at \$62.75 per share. Find the broker's fee (round to the nearest cent). \$326.30

127. You bought a car for \$2000 down and made payments of \$127.50 each month for 36 months.
a. Find the amount of the payments over the 36 months. \$4590
b. Find the total cost of the car. \$6590

128. As a nurse, Rob Martinez earns a salary of \$344 for a 40-hour work week. This week he worked 15 hours of overtime at a rate of \$12.90 for each hour of overtime worked.
a. Find the amount of overtime pay. \$193.50
b. Find Rob's total income for the week. \$537.50

129. Bay Area Rental Cars charges \$12 a day and \$.12 per mile for renting a car. You rented a car for 3 days and drove 235 miles. Find the total cost of renting the car. \$64.20

130. A taxi costs \$1.50 and \$0.20 for each $\frac{1}{8}$ mile driven. Find the cost of hiring a taxi to get from the airport to the hotel—a distance of 5.5 miles. \$10.30

APPLYING THE CONCEPTS

131. Anthony Schmidt owns 236.147 shares of a mutual fund. The current value of each share is $8.67. What is the value of the shares? $2047.39

132. Show how the decimal is placed in the product of 1.3 × 2.31 by first writing each number as a fraction and then multiplying. Now change back to decimal notation. $1.3 \times 2.31 = \frac{13}{10} \times \frac{231}{100} = \frac{3003}{1000} = 3.003$

The table at the right lists three pieces of steel required for a repair project. Use this table for Exercises 133 and 134.

Grade of Steel	*Weight (Pounds per Foot)*	*Required Number of Feet*	*Cost per Pound*
1	2.2	8	$1.20
2	3.4	6.5	1.35
3	6.75	15.4	1.94

133. Find the total cost of each of the grades of steel.
Grade 1: $21.12 Grade 2: $29.84 Grade 3: $201.66

134. Find the total cost of the three pieces of steel. $252.62

A confectioner ships holiday packs of candy and nuts anywhere in the United States. Following is a price list of nuts and candy and also a table of shipping charges to zones in the United States. For any fraction of a pound, use the next higher weight. Use these tables for Exercise 135.

Code	*Description*	*Price*
112	Almonds 16 oz	4.75
116	Cashews 8 oz	2.90
117	Cashews 16 oz	5.50
130	Macadamias 7 oz	5.25
131	Macadamias 16 oz	9.95
149	Pecan halves 8 oz	6.25
155	Mixed nuts 10 oz	4.80
160	Cashew brittle 8 oz	1.95
182	Pecan roll 8 oz	3.70
199	Chocolate peanuts 8 oz	1.90

Pounds	*Zone 1*	*Zone 2*	*Zone 3*	*Zone 4*
1–3	6.55	6.85	7.25	7.75
4–6	7.10	7.40	7.80	8.30
7–9	7.50	7.80	8.20	8.70
10–12	7.90	8.20	8.60	9.10

135. Find the cost of sending the following orders to the given mail zone.

a.

Code	*Quantity*
116	2
130	1
149	3
182	4

Mail to zone 4.

b.

Code	*Quantity*
112	1
117	4
131	2
160	3
182	5

Mail to zone 3.

c.

Code	*Quantity*
117	3
131	1
155	2
160	4
182	1
199	3

Mail to zone 2.

a. $52.90 b. $79.60 c. $61.45

Chris works at B & W Garage as an auto mechanic and has just completed an engine overhaul for a customer. To determine the cost of the repair job, Chris keeps a list of times worked and parts used. A parts list and a list of the times worked are shown below. Use this table for Exercises 136–139.

Parts Used

Item	*Quantity*
Gasket set	1
Ring set	1
Valves	8
Wrist pins	8
Valve springs	16
Rod bearings	8
Main bearings	5
Valve seals	16
Timing chain	1

Time Spent

Day	*Hours*
Monday	7.0
Tuesday	7.5
Wednesday	6.5
Thursday	8.5
Friday	9.0

Price List

Item Number	*Description*	*Unit Price*
27345	Valve spring	$1.85
41257	Main bearing	3.40
54678	Valve	4.79
29753	Ring set	33.98
45837	Gasket set	48.99
23751	Timing chain	42.95
23765	Fuel pump	77.59
28632	Wrist pin	2.71
34922	Rod bearing	2.67
2871	Valve seal	0.42

136. Organize a table of data showing the parts used, the unit price for each, and the price of the quantity used. Hint: Use the following headings for the table.

Quantity	*Item Number*	*Description*	*Unit Price*	*Total*
1	45837	Gasket set	$48.99	$ 48.99
1	29753	Ring set	$33.98	$ 33.98
8	54678	Valve	$ 4.79	$ 38.32
8	28632	Wrist pin	$ 2.71	$ 21.68
16	27345	Valve spring	$ 1.85	$ 29.60
8	34922	Rod bearing	$ 2.67	$ 21.36
5	41257	Main bearing	$ 3.40	$ 17.00
16	2871	Valve seal	$ 0.42	$ 6.72
1	23751	Timing chain	$42.95	$ 42.95
				$260.60

137. Add up the numbers in the "Total" column to find the total cost of the parts. $260.60

138. If the charge for labor is $26.75 per hour, compute the cost of labor. $1029.88

139. What is the total cost for parts and labor? $1290.48

140. [W] Explain how the decimal point is placed when a number is multiplied by 10, 100, 1000, 10,000, etc. Answers will vary

141. [W] Explain how the decimal point is placed in the product of two decimals. Answers will vary

142. An emissions test for cars requires that of the total engine exhaust, less than 1 part per thousand $\left(\frac{1}{1000} = 0.001\right)$ be hydrocarbon emissions. Using this figure, determine which of the cars in the table below would fail the emissions test.

Car	*Total Engine Exhaust*	*Hydrocarbon Emission*
1	367,921	36
2	401,346	42
3	298,773	21
4	330,045	32
5	432,989	45

All will pass the emissions test.

3.5 Division of Decimals

Objective A ***To divide decimals***

To divide decimals, move the decimal point in the divisor to the right to make the divisor a whole number. Move the decimal in the dividend the same number of places to the right. Place the decimal point in the quotient directly over the decimal point in the dividend, and then divide as in whole numbers.

➡ Divide: $3.25\overline{)15.275}$

$3.25.\overline{)15.27.5}$

Move the decimal point 2 places to the right in the divisor and then in the dividend. Place the decimal point in the quotient.

$$\begin{array}{r} 4.7 \\ 325.\overline{)\ 1527.5} \\ -1300 \\ \hline 227\ 5 \\ -227\ 5 \\ \hline 0 \end{array}$$

INSTRUCTOR NOTE

The paragraph at the right gives the justification for moving the decimal point before decimal numbers are divided.

Moving the decimal point the same number of decimal places in the divisor and dividend does not change the value of the quotient, because this process is the same as multiplying the numerator and denominator of a fraction by the same number. In the example above,

$$3.25\overline{)15.275} = \frac{15.275}{3.25} = \frac{15.275 \times 100}{3.25 \times 100} = \frac{1527.5}{325} = 325\overline{)1527.5}$$

When dividing decimals, we usually round the quotient off to a specified place value, rather than writing the quotient with a remainder.

INSTRUCTOR NOTE

An alternate method for rounding quotients is

$$\begin{array}{r} 1.86 \\ 3\overline{)5.60} \\ \underline{3} \\ 2\ 6 \\ \underline{2\ 4} \\ 20 \\ \underline{18} \\ 2 \end{array}$$

Since the last remainder is more than one-half of the divisor, 1.86 is rounded to 1.87.

➡ Divide: $0.3\overline{)0.56}$
Round to the nearest hundredth.

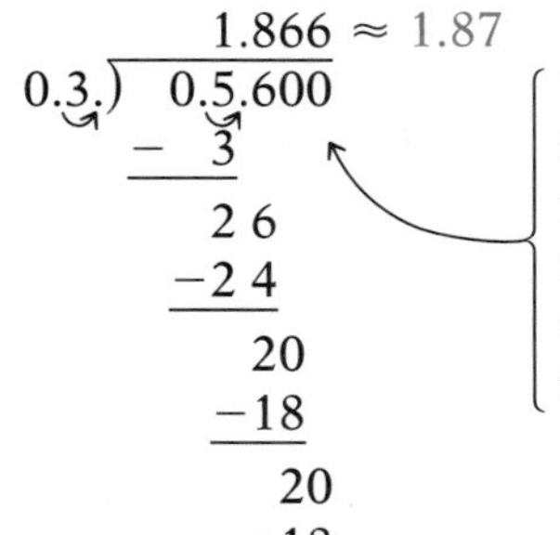

$$\begin{array}{r} 1.866 \approx 1.87 \\ 0.3.\overline{)\ 0.5.600} \\ -\ 3 \\ \hline 2\ 6 \\ -2\ 4 \\ \hline 20 \\ -18 \\ \hline 20 \\ -18 \\ \hline \end{array}$$

The division must be carried to the thousandths' place to round the quotient to the nearest hundredth. Therefore, zeros must be inserted in the dividend so that the quotient has a digit in the thousandths' place.

➡ Divide: 57.93 ÷ 3.24 Round to the nearest thousandth.

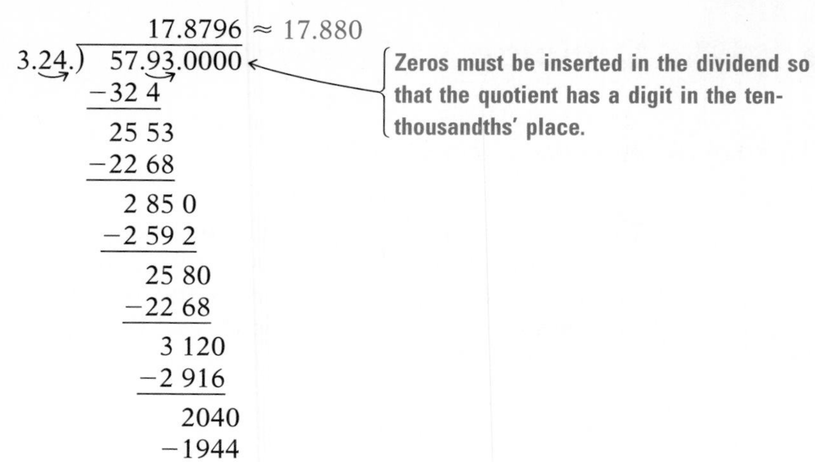

To divide a decimal by a power of 10 (10, 100, 1000, . . .), move the decimal point to the left the same number of places as there are zeros in the power of 10.

34.65 ÷ 10 = 3.465
1 zero — 1 decimal place

34.65 ÷ 100 = 0.3465
2 zeros — 2 decimal places

34.65 ÷ 1000 = 0.03465
3 zeros — 3 decimal places

Note that a zero must be inserted between the 3 and the decimal point.

34.65 ÷ 10,000 = 0.003465
4 zeros — 4 decimal places

Note that two zeros must be inserted between the 3 and the decimal point.

If the power of 10 is written in exponential notation, the exponent indicates how many places to move the decimal point.

$34.65 \div 10^1 = 3.465$ — 1 decimal place

$34.65 \div 10^2 = 0.3465$ — 2 decimal places

$34.65 \div 10^3 = 0.03465$ — 3 decimal places

$34.65 \div 10^4 = 0.003465$ — 4 decimal places

Example 1 Divide: 0.1344 ÷ 0.032

Solution

```
              4.2
0.032.)0.134.4
        -128
           6 4
          -6 4
             0
```

You Try It 1 Divide: 0.1404 ÷ 0.052

Your solution 2.7

Solution on p. A21

Example 2 Divide: $58.092 \div 82$
Round to the nearest thousandth.

Solution

$$
\begin{array}{r}
0.7084 \approx 0.708 \\
82\overline{)\ 58.0920} \\
\underline{-57\ 4} \\
69 \\
\underline{-\ \ 0} \\
692 \\
\underline{-656} \\
360 \\
\underline{-328}
\end{array}
$$

You Try It 2 Divide: $37.042 \div 76$
Round to the nearest thousandth.

Your solution 0.487

Example 3 Divide: $420.9 \div 7.06$
Round to the nearest tenth.

Solution

$$
\begin{array}{r}
59.61 \approx 59.6 \\
7.06.\overline{)\ 420.90.00} \\
\underline{-353\ 0} \\
67\ 90 \\
\underline{-63\ 54} \\
4\ 36\ 0 \\
\underline{-4\ 23\ 6} \\
12\ 40 \\
\underline{-\ 7\ 06}
\end{array}
$$

You Try It 3 Divide: $370.2 \div 5.09$
Round to the nearest tenth.

Your solution 72.7

Example 4 Divide: $402.75 \div 1000$

Solution $402.75 \div 1000 = 0.40275$

You Try It 4 Divide: $309.21 \div 10{,}000$

Your solution 0.030921

Example 5 What is 0.625 divided by 10^2?

Solution $0.625 \div 10^2 = 0.00625$

You Try It 5 What is 42.93 divided by 10^4?

Your solution 0.004293

Solutions on p. A21

ESTIMATION

Estimating the Quotient of Two Decimals

Estimate and then use your calculator to find $282.18 \div 0.485$.

To estimate a quotient, round each number so that there is one non-zero digit. Then divide.

$$282.18 \div 0.485 \approx 300 \div 0.5 = 600$$

The estimated answer is 600.

Now use your calculator to find the exact answer.

282.18 [÷] 0.485 [=] 581.814433

The exact answer is 581.814433.

Objective B To solve application problems

The table at the right shows the total amount of gasoline tax paid per gallon of gas in selected cities in 1994. Use this table for Example 6 and You Try It 6.

Gasoline Taxes in Dollars per Gallon

City	*Taxes*
Chicago	\$.5929
Tampa	\$.4380
Boston	\$.4013
Detroit	\$.3924

Example 6
Sandra Lopez's car gets 24 miles per gallon. In a year in which she drove 10,500 miles in Chicago, how much did she pay, to the nearest dollar, in gasoline taxes?

Strategy
To find the amount she paid in gasoline taxes:

- Find the total number of gallons of gas she used by dividing her total miles (10,500) by the number of miles traveled per gallon of gasoline (24).
- Multiply the total tax (\$.5929) by the total number of gallons of gasoline used.

Solution
$10{,}500 \div 24 = 437.5$
$0.5929 \times 437.5 = 259.39375$

Sandra paid approximately \$259 for gasoline taxes.

You Try It 6
In 1994, Jefferson Beckwith drove his car 13,400 miles in Boston. If his car gets 26 miles per gallon, how much did he pay, to the nearest dollar, in gasoline taxes?

Your strategy

Your solution
\$207

Example 7
In 1994, manufacturers offered consumers approximately \$183.4 billion worth of discount coupons. Of those, only \$8.3 billion were redeemed. Find the quotient of the unredeemed amount.

Strategy
To find the quotient:

- Subtract 8.3 from 183.4
- Divide the difference by 183.4

Solution
$183.4 - 8.3 = 175.1$
$175.1 \div 183.4 \approx 0.9547$

The quotient of the unredeemed amount is 0.9547.

You Try It 7
A Nielsen survey of the number of people (in millions) who watch television during the week is given in the table below.

Mon.	*Tues.*	*Wed.*	*Thu.*	*Fri.*	*Sat.*	*Sun.*
91.9	89.8	90.6	93.9	78.0	77.1	87.7

Find the average number of people watching television per day.

Your strategy

Your solution
87 million

Solutions on p. A22

3.5 Exercises

Objective A

Divide.

1. $\begin{array}{r}0.82\\3\overline{)2.46}\end{array}$

2. $\begin{array}{r}0.53\\7\overline{)3.71}\end{array}$

3. $\begin{array}{r}4.8\\0.8\overline{)3.84}\end{array}$

4. $\begin{array}{r}7.7\\0.9\overline{)6.93}\end{array}$

5. $\begin{array}{r}89\\0.7\overline{)62.3}\end{array}$

6. $\begin{array}{r}132\\0.4\overline{)52.8}\end{array}$

7. $\begin{array}{r}60\\0.4\overline{)24}\end{array}$

8. $\begin{array}{r}130\\0.5\overline{)65}\end{array}$

9. $\begin{array}{r}84.3\\0.7\overline{)59.01}\end{array}$

10. $\begin{array}{r}9.69\\0.9\overline{)8.721}\end{array}$

11. $\begin{array}{r}32.3\\0.5\overline{)16.15}\end{array}$

12. $\begin{array}{r}97\\0.8\overline{)77.6}\end{array}$

13. $\begin{array}{r}5.06\\0.7\overline{)3.542}\end{array}$

14. $\begin{array}{r}4.06\\0.6\overline{)2.436}\end{array}$

15. $\begin{array}{r}1.3\\6.3\overline{)8.19}\end{array}$

16. $\begin{array}{r}2.2\\3.2\overline{)7.04}\end{array}$

17. $\begin{array}{r}0.11\\3.6\overline{)0.396}\end{array}$

18. $\begin{array}{r}0.24\\2.7\overline{)0.648}\end{array}$

19. $\begin{array}{r}3.8\\6.9\overline{)26.22}\end{array}$

20. $\begin{array}{r}49.8\\1.7\overline{)84.66}\end{array}$

Divide. Round to the nearest tenth.

21. $55.62 \div 8.8$
6.3

22. $25.43 \div 5.4$
4.7

23. $5.427 \div 9.5$
0.6

24. $1.837 \div 1.4$
1.3

25. $18.4 \div 7.3$
2.5

26. $52.9 \div 8.1$
6.5

27. $0.183 \div 0.17$
1.1

28. $0.381 \div 0.47$
0.8

29. $6.924 \div 0.053$
130.6

Divide. Round to the nearest hundredth.

30. $4.817 \div 16$
0.30

31. $6.467 \div 8$
0.81

32. $0.0418 \div 0.53$
0.08

33. $19.08 \div 0.45$
42.40

34. $21.792 \div 0.96$
22.70

35. $38.665 \div 0.95$
40.70

36. $13.97 \div 25.4$
0.55

37. $27.738 \div 60.3$
0.46

38. $3.171 \div 45.3$
0.07

Divide. Round to the nearest thousandth.

39. 1.028 ÷ 54
0.019

40. 6.729 ÷ 27
0.249

41. 0.0437 ÷ 0.5
0.087

42. 75.469 ÷ 77.8
0.970

43. 34.31 ÷ 95.3
0.360

44. 0.2695 ÷ 2.67
0.101

45. 0.4871 ÷ 4.72
0.103

46. 0.1142 ÷ 17.2
0.007

47. 0.2307 ÷ 26.7
0.009

Divide. Round to the nearest whole number.

48. 16.5 ÷ 4
4

49. 89.76 ÷ 90
1

50. 1.94 ÷ 0.3
6

51. 1.0478 ÷ 0.413
3

52. 2.148 ÷ 0.519
4

53. 0.79 ÷ 0.778
1

54. 3.092 ÷ 0.075
41

55. 392 ÷ 6.9
57

56. 8.729 ÷ 0.075
116

Divide.

57. 4.07 ÷ 10
0.407

58. 0.039 ÷ 10
0.0039

59. 42.67 ÷ 10
4.267

60. 389.7 ÷ 100
3.897

61. 1.037 ÷ 100
0.01037

62. 237.835 ÷ 100
2.37835

63. 8.295 ÷ 1000
0.008295

64. 82,547 ÷ 1000
82.547

65. 825.37 ÷ 1000
0.82537

66. 8.35 ÷ 10
0.835

67. 0.32 ÷ 10
0.032

68. 87.65 ÷ 10
8.765

69. 23.627 ÷ 10^2
0.23627

70. 2.954 ÷ 10^2
0.02954

71. 0.0053 ÷ 10^2
0.000053

72. 289.32 ÷ 10^3
0.28932

73. 1.8932 ÷ 10^3
0.0018932

74. 0.139 ÷ 10^3
0.000139

75. Divide 44.208 by 2.4.
18.42

76. Divide 0.04664 by 0.44.
0.106

77. Find the quotient of 723.15 and 45.
16.07

78. Find the quotient of 3.3463 and 3.07.
1.09

79. Divide 13.5 by 10^3.
0.0135

80. Divide 0.045 by 10^5.
0.00000045

81. Find the quotient of 23.678 and 1000.
0.023678

82. Find the quotient of 7.005 and 10,000.
0.0007005

83. What is 0.0056 divided by 0.05?
0.112

84. What is 123.8 divided by 0.02?
6190

Estimate and then use your calculator to divide. Round your calculated answer to the nearest ten-thousandth.

85. 42.42 ÷ 3.8
Est.: 10
Cal.: 11.1632

86. 69.8 ÷ 7.2
Est.: 10
Cal.: 9.6944

87. 389 ÷ 0.44
Est.: 1000
Cal.: 884.0909

88. 642 ÷ 0.83
Est.: 750
Cal.: 773.4940

89. 6.394 ÷ 3.5
Est.: 1.5
Cal.: 1.8269

90. 8.429 ÷ 4.2
Est.: 2
Cal.: 2.0069

91. 1.235 ÷ 0.021
Est.: 50
Cal.: 58.8095

92. 7.456 ÷ 0.072
Est.: 100
Cal.: 103.5556

93. 95.443 ÷ 1.32
Est.: 100
Cal.: 72.3053

94. 423.0925 ÷ 4.0927
Est.: 100
Cal.: 103.3774

95. 1.000523 ÷ 429.07
Est.: 0.0025
Cal.: 0.0023

96. 0.03629 ÷ 0.00054
Est.: 80
Cal.: 67.2037

Objective B Application Problems

97. Ramon, a high school football player, gained 162 yards on 26 carries in a high school football game. Find the number of yards gained per carry. (Round to two decimal places.) 6.23 yards

98. Ross Lapointe earns $39,440.64 for 12 months' work as a park ranger. How much does he earn in 1 month? $3286.72

99. You pay $947.60 a year in car insurance. The insurance is paid in four equal payments. Find the amount of each payment. $236.90

100. A case of diet cola costs $6.79. If there are 24 cans in a case, find the cost per can. Round to the nearest cent. $.28

101. Anne is building bookcases that are 3.4 feet long. How many complete shelves can be cut from a 12-foot board? 3 shelves

102. You travel 295 miles on 12.5 gallons of gasoline. How many miles can you travel on 1 gallon of gasoline? 23.6 miles

103. An oil company had issued 3,541,221,500 shares of stock. The company paid $6,090,990,120 in dividends. Find the dividend for each share of stock. Round to two decimal places. $1.72

104. The total budget for the United States in 1993 was $1.47 trillion. If each person in the United States were to pay the same amount of taxes, how much would each person pay to raise that amount of money? Assume that there are 250 million people in the United States. (Round to the nearest dollar.) $5880

105. You buy a home entertainment center for $1242.58. The down payment is $400, and the balance is to be paid in 18 equal monthly payments.
a. Find the amount to be paid in monthly payments. $842.58
b. Find the amount of each monthly payment. $46.81

APPLYING THE CONCEPTS

106. [W] Explain how the decimal point is moved when dividing a number by 10, 100, 1,000, 10,000 etc. Answers will vary

107. A ball point pen priced at 50¢ was not selling. When the price was reduced to a different whole number of cents, the entire stock sold for $31.95. How many cents were charged per pen when the price was reduced? (*Hint*: There is more than one possible answer.)
Any of 1¢, 3¢, 5¢, 9¢, 15¢ or 45¢ is correct

108. [W] Explain how baseball batting averages are determined. Then find Tony Gwynn's batting average with 175 hits out of 489 at bats. (Round to three decimal places). Answers will vary. 0.358

109. [W] Explain how the decimal point is placed in the quotient when dividing by a decimal. Answers will vary

The graph at the right shows the increase in the consumption of yogurt in the U.S. for a 20-year period. Use this graph for Exercises 110 and 111.

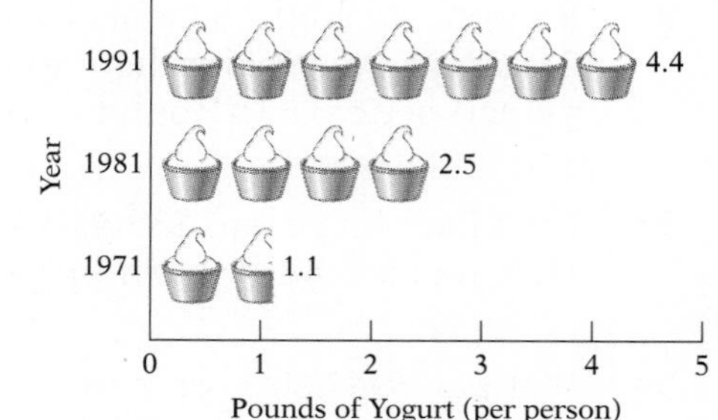

110. What was the average annual increase in yogurt consumption between 1971 and 1981?
1.4 pounds per person

111. If the average annual increase in yogurt consumption between 1981 and 1991 continued at the same rate beyond 1991, what would be the average yogurt consumption in 1995? 5.16 pounds per person

For each of the problems below, insert a +, −, ×, or ÷ into the space left for it so that the statement is true.

112. 3.45 0.5 = 6.9
÷

113. 3.46 0.24 = 0.8304
×

114. 6.009 4.68 = 1.329
−

115. 0.064 1.6 = 0.1024
×

116. 9.876 23.12 = 32.996
+

117. 3.0381 1.23 = 2.47
÷

Fill in the space to make a true statement.

118. 6.47 − = 1.253
5.217

119. 6.47 + = 9
2.53

120. 0.009 ÷ = 0.36
0.025

3.6 Comparing and Converting Fractions and Decimals

Objective A *To convert fractions to decimals*

Every fraction can be written as a decimal. To write a fraction as a decimal, divide the numerator of the fraction by the denominator. The quotient can be rounded to the desired place value.

➡ Convert $\frac{3}{7}$ to a decimal.

$7\overline{)3.00000}$ = 0.42857

$\frac{3}{7}$ rounded to the nearest hundredth is 0.43.

$\frac{3}{7}$ rounded to the nearest thousandth is 0.429.

$\frac{3}{7}$ rounded to the nearest ten-thousandth is 0.4286.

➡ Convert $3\frac{2}{9}$ to a decimal. Round to the nearest thousandth.

$3\frac{2}{9} = \frac{29}{9}$ $\quad 9\overline{)29.0000}$ = 3.2222 $\quad 3\frac{2}{9}$ rounded to the nearest thousandth is 3.222.

Example 1 Convert $\frac{3}{8}$ to a decimal. Round to the nearest hundredth.

Solution $8\overline{)3.000}$ = 0.375 ≈ 0.38

You Try It 1 Convert $\frac{9}{16}$ to a decimal. Round to the nearest tenth.

Your solution 0.6

Example 2 Convert $2\frac{3}{4}$ to a decimal. Round to the nearest tenth.

Solution $2\frac{3}{4} = \frac{11}{4}$ $\quad 4\overline{)11.00}$ = 2.75 ≈ 2.8

You Try It 2 Convert $4\frac{1}{6}$ to a decimal. Round to the nearest hundredth.

Your solution 4.17

Solutions on p. A22

Objective B *To convert decimals to fractions*

To convert a decimal to a fraction, remove the decimal point and place the decimal part over a denominator equal to the place value of the last digit in the decimal.

$0.47 = \frac{47}{100}$ (hundredths)

$7.45 = 7\frac{45}{100} = 7\frac{9}{20}$ (hundredths)

$0.275 = \frac{275}{1000} = \frac{11}{40}$ (thousandths)

$0.16\frac{2}{3} = \frac{16\frac{2}{3}}{100} = 16\frac{2}{3} \div 100 = \frac{50}{3} \times \frac{1}{100} = \frac{1}{6}$ (hundredths)

Example 3 Convert 0.82 and 4.75 to fractions.

Solution $0.82 = \frac{82}{100} = \frac{41}{50}$

$4.75 = 4\frac{75}{100} = 4\frac{3}{4}$

You Try It 3 Convert 0.56 and 5.35 to fractions.

Your solution $\frac{14}{25}, 5\frac{7}{20}$

Example 4 Convert $0.15\frac{2}{3}$ to a fraction.

Solution $0.15\frac{2}{3} = \frac{15\frac{2}{3}}{100} = 15\frac{2}{3} \div 100$

$= \frac{47}{3} \times \frac{1}{100} = \frac{47}{300}$

You Try It 4 Convert $0.12\frac{7}{8}$ to a fraction.

Your solution $\frac{103}{800}$

Solutions on p. A22

Objective C *To identify the order relation between two decimals or between a decimal and a fraction*

INSTRUCTOR NOTE
When comparing decimals, have students write each number with the same place value by adding zeros, if necessary. For example, it's easier to see that $0.375 < 0.38$ when you write it as $0.375 < 0.380$.

Decimals, like whole numbers and fractions, can be graphed as points on the number line. The number line can be used to show the order of decimals. A decimal that appears to the right of a given number is greater than the given number. A decimal that appears to the left of a given number is less than the given number.

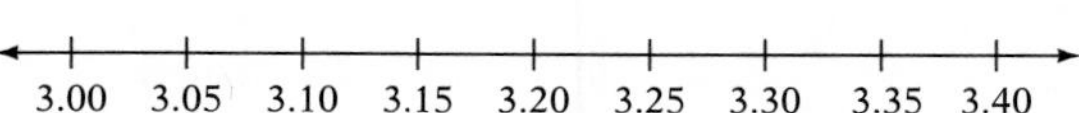

Note that 3, 3.0, and 3.00 represent the same number.

➡ Find the order relation between $\frac{3}{8}$ and 0.38.

$\frac{3}{8} = 0.375 \qquad 0.38 = 0.380$

$0.375 < 0.38$

$\frac{3}{8} < 0.38$

Example 5 Place the correct symbol, < or >, between the numbers.

$\frac{5}{16} \quad 0.32$

Solution $\frac{5}{16} \approx 0.313$

$0.313 < 0.32$

$\frac{5}{16} < 0.32$

You Try It 5 Place the correct symbol, < or >, between the numbers.

$0.63 \quad \frac{5}{8}$

Your solution $0.63 > \frac{5}{8}$

Solution on p. A22

3.6 Exercises

Objective A

Convert the fraction to a decimal. Round to the nearest thousandth.

1. $\frac{5}{8}$ 0.625

2. $\frac{7}{12}$ 0.583

3. $\frac{2}{3}$ 0.667

4. $\frac{5}{6}$ 0.833

5. $\frac{1}{6}$ 0.167

6. $\frac{7}{8}$ 0.875

7. $\frac{5}{12}$ 0.417

8. $\frac{9}{16}$ 0.563

9. $\frac{7}{4}$ 1.750

10. $\frac{5}{3}$ 1.667

11. $1\frac{1}{2}$ 1.500

12. $2\frac{1}{3}$ 2.333

13. $\frac{16}{4}$ 4.000

14. $\frac{36}{9}$ 4.000

15. $\frac{3}{1000}$ 0.003

16. $\frac{5}{10}$ 0.500

17. $7\frac{2}{25}$ 7.080

18. $16\frac{7}{9}$ 16.778

19. $37\frac{1}{2}$ 37.500

20. $87\frac{1}{2}$ 87.500

21. $\frac{3}{8}$ 0.375

22. $\frac{11}{16}$ 0.688

23. $\frac{5}{24}$ 0.208

24. $\frac{4}{25}$ 0.160

25. $3\frac{1}{3}$ 3.333

26. $8\frac{2}{5}$ 8.400

27. $5\frac{4}{9}$ 5.444

28. $3\frac{1}{12}$ 3.083

29. $\frac{5}{16}$ 0.313

30. $\frac{11}{12}$ 0.917

Objective B

Convert the decimal to a fraction.

31. 0.8 $\frac{4}{5}$

32. 0.4 $\frac{2}{5}$

33. 0.32 $\frac{8}{25}$

34. 0.48 $\frac{12}{25}$

35. 0.125 $\frac{1}{8}$

36. 0.485 $\frac{97}{200}$

37. 1.25 $1\frac{1}{4}$

38. 3.75 $3\frac{3}{4}$

39. 16.9 $16\frac{9}{10}$

40. 17.5 $17\frac{1}{2}$

41. 8.4 $8\frac{2}{5}$

42. 10.7 $10\frac{7}{10}$

43. 8.437 $8\frac{437}{1000}$

44. 9.279 $9\frac{279}{1000}$

45. 2.25 $2\frac{1}{4}$

46. 7.75 $7\frac{3}{4}$

47. $0.15\frac{1}{3}$ $\frac{23}{150}$

48. $0.17\frac{2}{3}$ $\frac{53}{300}$

49. $0.87\frac{7}{8}$ $\frac{703}{800}$

50. $0.12\frac{5}{9}$ $\frac{113}{900}$

Convert the decimal to a fraction.

51. 1.68 $1\frac{17}{25}$

52. 7.38 $7\frac{19}{50}$

53. 0.045 $\frac{9}{200}$

54. 0.085 $\frac{17}{200}$

55. 16.72 $16\frac{18}{25}$

56. 82.32 $82\frac{8}{25}$

57. 0.33 $\frac{33}{100}$

58. 0.57 $\frac{57}{100}$

59. $0.33\frac{1}{3}$ $\frac{1}{3}$

60. $0.66\frac{2}{3}$ $\frac{2}{3}$

Objective C

Place the correct symbol, $<$ or $>$, between the numbers.

61. $0.15 < 0.5$

62. $0.6 > 0.45$

63. $6.65 > 6.56$

64. $3.89 < 3.98$

65. $2.504 > 2.054$

66. $0.025 < 0.105$

67. $\frac{3}{8} > 0.365$

68. $\frac{4}{5} < 0.802$

69. $\frac{2}{3} > 0.65$

70. $0.85 < \frac{7}{8}$

71. $\frac{5}{9} > 0.55$

72. $\frac{7}{12} > 0.58$

73. $0.62 > \frac{7}{15}$

74. $\frac{11}{12} < 0.92$

75. $0.161 > \frac{1}{7}$

76. $0.623 > 0.6023$

77. $0.86 > 0.855$

78. $0.87 > 0.087$

79. $1.005 > 0.5$

80. $0.033 < 0.3$

APPLYING THE CONCEPTS

81. Which of the following is true?

a. $\frac{137}{300} = 0.456666667$ False

b. $\frac{137}{300} < 0.45666666$ False

c. $\frac{137}{300} > 0.45666666$ True

82. Is $\frac{19}{23}$ in decimal form a non-repeating decimal? Why or why not?

No. The decimal does not terminate.

83. If a number is rounded to the nearest thousandth, is it always greater than if it was rounded to the nearest hundredth? Give examples to support your answer.

No. 0.0402 rounded to hundredths is 0.04, to thousandths is 0.040

84. [W] Convert $\frac{1}{9}$, $\frac{2}{9}$, $\frac{3}{9}$, and $\frac{4}{9}$ to decimals. Describe the pattern. Use the pattern to convert $\frac{5}{9}$, $\frac{7}{9}$, and $\frac{8}{9}$ to decimals.

$\frac{1}{9} = 0.11111\ldots$ $\frac{2}{9} = 0.22222\ldots$ $\frac{3}{9} = 0.33333\ldots$ $\frac{4}{9} = 0.44444\ldots$ $\frac{5}{9} = 0.55555\ldots$ $\frac{7}{9} = 0.77777\ldots$ $\frac{8}{9} = 0.88888\ldots$

85. [W] Explain the difference between terminating, repeating, and non-repeating decimals. Give an example of each kind of decimal. Answers will vary

Project in Mathematics

Topographical Maps

Carpenters use fractions to measure the wood that is used to frame a house. For instance, a door entry may measure $42\frac{1}{2}$ inches. However, the grading contractor (the person who levels the lot on which the house is built) measures the height of the lot by using decimals. For instance, a certain place may be at a height of 554.2 feet. (The height is measured above sea level; so 554.2 means 554.2 feet above sea level.)

A surveyor provides the grading contractor with a grading plan that shows the elevation of each point of the lot. These plans are drawn so that the house can be sited in such a way that water will drain away from it. The diagram below is a **Topographical Map** for a lot. Along each closed curve, called a **contour curve,** the lot is at the same height above sea level. For instance, the curve in red means that every point on that curve is 556.5 feet above sea level.

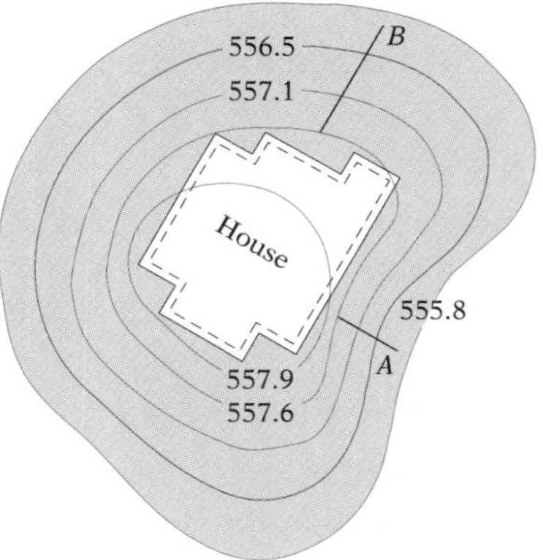

Exercises:

Using the map above, answer the following questions.

1. What is the elevation of the highest point on the lot?

2. What is the elevation of the lowest point on the lot?

3. What is the difference in elevation between the highest and lowest points on the lot?

4. Assuming this map is drawn to scale, describe the significance of how steep the slope is along line *A* compared to that along line *B*.

Chapter Summary

Key Words A number written in *decimal notation* has three parts: a whole-number part, a decimal point, and a decimal part.

The position of a digit in a *decimal* determines the digit's *place value*.

Essential Rules

To Write a Decimal in Words

To write a decimal in words, write the decimal part as if it were a whole number. Then name the place value of the last digit.

To Add Decimals

To add decimals, write the numbers so that the decimal points are on a vertical line. Add as in whole numbers, and place the decimal point in the sum directly below the decimal point in the addends.

To Subtract Decimals

To subtract decimals, place the numbers so that the decimal points are on a vertical line. Subtract as for whole numbers, and write the decimal point in the difference directly below the decimal point in the subtrahend.

To Multiply Decimals

To multiply decimals, multiply the numbers as in whole numbers. Place the decimal point in the product so that the number of decimal places in the product is the sum of the decimal places in the factors.

To Divide Decimals

To divide decimals, move the decimal point in the divisor to make it a whole number. Move the decimal point in the dividend the same number of places to the right. Place the decimal point in the quotient directly over the decimal point in the dividend. Then divide as in whole numbers.

To Write a Fraction as a Decimal

To write a fraction as a decimal, divide the numerator of the fraction by the denominator. Round the quotient to the desired number of places.

To Convert a Decimal to a Fraction

To convert a decimal to a fraction, remove the decimal point and place the decimal part over a denominator equal to the place value of the last digit in the decimal.

Chapter Review

SECTION 3.1

1. Write 22.0092 in words.
   Twenty-two and ninety-two ten thousandths

2. Write 342.37 in words.
   Three hundred forty-two and thirty-seven hundredths

3. Write thirty-four and twenty-five thousandths in standard form. 34.025

4. Write three and six thousand seven hundred fifty-three hundred-thousandths in standard form. 3.06753

5. Round 7.93704 to the nearest hundredth. 7.94

6. Round 0.05678235 to the nearest hundred-thousandths. 0.05678

SECTION 3.2

7. Add: 3.42 + 0.794 + 32.5
   36.714

8. Find the sum of 369.41, 88.3, 9.774, and 366.474. 833.958

9. You have $237.44 in your checking account. You make deposits of $56.88, $127.40, and $56.30. Find the amount in your checking account after you make the deposits. $478.02

SECTION 3.3

10. Subtract: 27.31 − 4.4465
    22.8635

11. What is 7.796 decreased by 2.9175?
    4.8785

12. You had a balance of $895.68 in your checking account. You then wrote checks of $145.72 and $88.45. Find the new balance in your checking account. $661.51

SECTION 3.4

13. Find the product of 3.08 and 2.9.
8.932

14. Multiply: $\begin{array}{r} 34.79 \\ \times\ 0.74 \\ \hline \end{array}$
25.7446

15. The state income tax on the business you own is \$560 plus 0.08 times your profit. You made a profit of \$63,000 last year. Find the amount of income tax you paid last year. \$5600

SECTION 3.5

16. Find the quotient of 3.6515 and 0.067.
54.5

17. Divide: $0.053\overline{)0.349482}$
6.594

18. A car costing \$5944.20 is bought with a down payment of \$1500 and 36 equal monthly payments. Find the amount of each monthly payment.
\$123.45

SECTION 3.6

19. Convert $\frac{7}{9}$ to a decimal. Round to the nearest thousandth. 0.778

20. Convert $2\frac{1}{3}$ to a decimal. Round to the nearest hundredth. 2.33

21. Convert 0.375 to a fraction. $\frac{3}{8}$

22. Convert $0.66\frac{2}{3}$ to a fraction. $\frac{2}{3}$

23. Place the correct symbol, < or >, between the two numbers.
$0.055 < 0.1$

24. Place the correct symbol, < or >, between the two numbers.
$\frac{5}{8} > 0.62$

Chapter Test

1. Write 45.0302 in words.
Forty-five and three hundred two ten-thousandths [3.1A]

2. Write two hundred nine and seven thousand eighty-six hundred-thousandths in standard form. 209.07086 [3.1A]

3. Round 7.0954625 to the nearest thousandth.
7.095 [3.1B]

4. Round 0.07395 to the nearest ten-thousandth. 0.0740 [3.1B]

5. What is the total of 62.3, 4.007, and 189.65?
255.957 [3.2A]

6. Add:

$$\begin{array}{r} 270.93 \\ 97. \\ 1.976 \\ +\ 88.675 \\ \hline 458.581 \end{array}$$

[3.2A]

7. You received a salary of \$363.75, a commission of \$954.82, and a bonus of \$225. Find your total income. \$1543.57 [3.2B]

8. Subtract:

$$\begin{array}{r} 13.027 \\ -\ 8.94 \\ \hline 4.087 \end{array}$$

[3.3A]

9. Find 9.23674 less than 37.003.
27.76626 [3.3A]

10. Find the missing dimension.

1.37 inches [3.3B]

11. Multiply:

$$\begin{array}{r} 1.37 \\ 0.004 \\ \hline 0.00548 \end{array}$$

[3.4A]

The table at the right shows the taxes on one gallon of gasoline in selected cities in 1994. Use this table for Questions 12 and 13.

City	*Federal Tax*	*State Tax*	*Other Taxes*
Boston	\$.1863	\$.12	\$.0005
Detroit	\$.1863	\$.15	\$.0561
Chicago	\$.1863	\$.19	\$.2166
Los Angeles	\$.1863	\$.17	\$.4747
Dallas	\$.1863	\$.20	\$.0059
New Haven	\$.1863	\$.29	\$.0528

12. Which city had the highest total tax for a gallon of gasoline? Los Angeles [3.2B]

13. Which city had the lowest total tax for a gallon of gasoline? Boston [3.2B]

14. A long-distance telephone call costs \$.85 for the first 3 minutes and \$.42 for each additional minute. Find the cost of a 12-minute long-distance telephone call. \$4.63 [3.4B]

15. Divide: $0.006\overline{)1.392}$ 232 [3.5A]

16. Find 0.0569 divided by 0.037. Round to the nearest thousandth. 1.538 [3.5A]

17. A car was bought for \$6392.60, with a down payment of \$1250. The balance was paid in 36 monthly payments. Find the amount of each monthly payment. \$142.85 [3.5B]

18. Convert $\frac{9}{13}$ to a decimal. Round to the nearest thousandth. 0.692 [3.6A]

19. Convert 0.825 to a fraction. $\frac{33}{40}$ [3.6B]

20. Place the correct symbol, $<$ or $>$, between the two numbers.
$0.66 < 0.666$ [3.6C]

Cumulative Review

1. Divide: $89\overline{)20{,}932}$
235 r17 [1.5C]

2. Simplify: $2^3 \cdot 4^2$
128 [1.6A]

3. Simplify: $2^2 - (7 - 3) \div 2 + 1$
3 [1.6B]

4. Find the LCM of 9, 12, and 24.
72 [2.1A]

5. Write $\frac{22}{5}$ as a mixed number.
$4\frac{2}{5}$ [2.2B]

6. Write $4\frac{5}{8}$ as an improper fraction.
$\frac{37}{8}$ [2.2B]

7. Build an equivalent fraction with the given denominator.
$\frac{5}{12} = \frac{\ }{60}$
$\frac{25}{60}$ [2.3A]

8. Add: $\frac{3}{8} + \frac{5}{12} + \frac{9}{16}$.
$1\frac{17}{48}$ [2.4B]

9. What is $5\frac{7}{12}$ increased by $3\frac{7}{18}$?
$8\frac{35}{36}$ [2.4C]

10. Subtract: $9\frac{5}{9} - 3\frac{11}{12}$
$5\frac{23}{36}$ [2.5C]

11. Multiply: $\frac{9}{16} \times \frac{4}{27}$
$\frac{1}{12}$ [2.6A]

12. Find the product of $2\frac{1}{8}$ and $4\frac{5}{17}$.
$9\frac{1}{8}$ [2.6B]

13. Divide: $\frac{11}{12} \div \frac{3}{4}$
$1\frac{2}{9}$ [2.7A]

14. What is $2\frac{3}{8}$ divided by $2\frac{1}{2}$?
$\frac{19}{20}$ [2.7B]

15. Simplify: $\left(\frac{2}{3}\right)^2 \cdot \left(\frac{3}{4}\right)^3$
$\frac{3}{16}$ [2.8B]

16. Simplify: $\left(\frac{2}{3}\right)^2 - \left(\frac{2}{3} - \frac{1}{2}\right) + 2$
$2\frac{5}{18}$ [2.8C]

17. Write 65.0309 in words.
Sixty-five and three hundred nine ten-thousandths [3.1A]

18. Add:
$$\begin{array}{r} 379.006 \\ 27.523 \\ 9.8707 \\ +\ 88.2994 \\ \hline \end{array}$$
504.6991 [3.2A]

19. What is 29.005 decreased by 7.9286?
21.0764 [3.3A]

20. Multiply: $\begin{array}{r} 9.074 \\ \times\ 6.09 \\ \hline \end{array}$
55.26066 [3.4A]

21. Divide: $8.09\overline{)17.42963}$. Round to the nearest thousandth. 2.154 [3.5A]

22. Convert $\frac{11}{15}$ to a decimal. Round to the nearest thousandth. 0.733 [3.6A]

23. Convert $0.16\frac{2}{3}$ to a fraction.
$\frac{1}{6}$ [3.6B]

24. Place the correct symbol, < or >, between the two numbers.
$\frac{8}{9} < 0.98$ [3.6C]

25. An airplane had 204 passengers aboard. During a stop, 97 passengers got off the plane and 127 passengers got on the plane. How many passengers were on the continuing flight? 234 passengers [1.3C]

26. An investor purchased stock at $\$32\frac{1}{8}$ per share. During the first 2 months of ownership, the stock lost $\$\frac{3}{4}$ and then gained $\$1\frac{1}{2}$. Find the value of each share of stock at the end of the 2 months.
$\$32\frac{7}{8}$ [2.5D]

27. You have a checking balance of $814.35. You then write checks for $42.98, $16.43, and $137.56. Find your checking account balance after you write the checks. $617.38 [3.3B]

28. A machine lathe takes 0.017 inch from a brass bushing that is 1.412 inches thick. Find the resulting thickness of the bushing. 1.395 inches [3.3B]

29. The state income tax on your business is $820 plus 0.08 times your profit. You made a profit of $64,860 last year. Find the amount of income tax you paid last year. $6008.80 [3.4B]

30. You bought a camera costing $210.96. The down payment is $20, and the balance is to be paid in 8 equal monthly payments. Find the monthly payment. $23.87 [3.5B]

Chapter

Ratio and Proportion

Objectives

Section 4.1

To write the ratio of two quantities in simplest form

To solve application problems

Section 4.2

To write rates

To write unit rates

To solve application problems

Section 4.3

To determine whether a proportion is true

To solve proportions

To solve application problems

Musical Scales

When a metal wire is stretched tight and then plucked, a sound is heard. Guitars, banjos, and violins are examples of instruments that use this principle to produce music. A piano is another example of this principle, but the sound is produced by the string being struck by a small hammerlike object.

After the string is plucked or struck, the string begins to vibrate. The number of times the string vibrates in 1 second is called the frequency of the vibration, or the pitch. Normally humans can hear vibrations as low as 16 cps (cycles per second) and as high as 20,000 cps. The longer the string, the lower the pitch of the sound; the shorter the string, the higher the pitch of the sound. In fact, a string half as long as another string vibrates twice as fast.

Vibrating String

Most music that is heard today is based on what is called the chromatic or twelve-tone scale. For this scale, a vibration of 261 cps is called middle C. A string half as long as the string for middle C vibrates twice as fast and produces a musical note one octave higher.

To produce the notes between the two C's, strings are placed that produce the desired pitch. Recall that as the string gets shorter, the pitch increases. A top view of a grand piano illustrates how the strings vary in length.

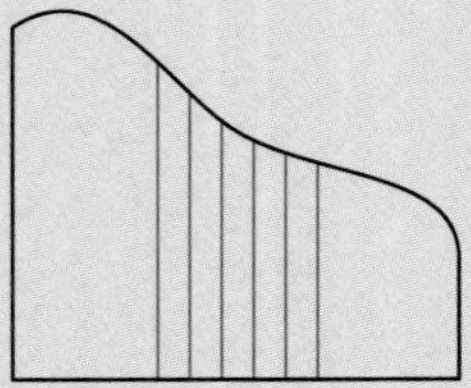

A well-tempered chromatic scale is one in which the string lengths are chosen so that the ratios of the frequencies of adjacent notes are the same.

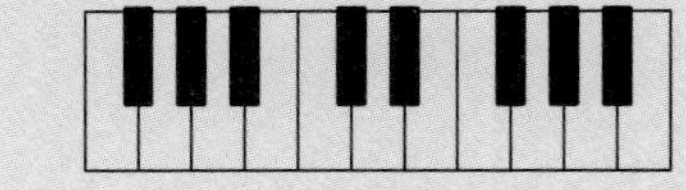

$$\frac{C}{C\#} = \frac{C\#}{D} = \frac{D}{D\#} = \frac{D\#}{E} = \frac{E}{F} = \frac{F}{F\#} = \frac{F\#}{G} = \frac{G}{G\#} = \frac{G\#}{A} = \frac{A}{A\#} = \frac{A\#}{B} = \frac{B}{C}$$

The common ratio for the chromatic scale is approximately $\frac{1}{1.0595}$.

Ratio

Objective A To write the ratio of two quantities in simplest form

INSTRUCTOR NOTE

Ratios have applications to many disciplines. Investors talk of price–earnings ratios. Accountants use the *quick* ratio, which is the ratio of current assets to current liabilities. Metallurgists use ratios to make various grades of steel.

Quantities such as 3 feet, 12 cents, and 9 cars are number quantities written with **units**.

3 feet
12 cents
9 cars

units

These are only some examples of units. Shirts, dollars, trees, miles, and gallons are further examples.

A **ratio** is the comparison of two quantities that have the *same* units. This comparison can be written three different ways:

1. As a fraction
2. As two numbers separated by a colon (:)
3. As two numbers separated by the word *to*

The ratio of the lengths of two boards, one 8 feet long and the other 10 feet long, can be written as

1. $\frac{8 \text{ feet}}{10 \text{ feet}} = \frac{8}{10} = \frac{4}{5}$
2. 8 feet:10 feet = 8:10 = 4:5
3. 8 feet to 10 feet = 8 to 10 = 4 to 5

A ratio is in **simplest form** when the two numbers do not have a common factor. Note that in a ratio, the units are not written.

This ratio means that the smaller board is $\frac{4}{5}$ the length of the longer board.

Example 1
Write the comparison \$6 to \$8 as a ratio in simplest form using a fraction, a colon, and the word *to*.

Solution $\frac{\$6}{\$8} = \frac{6}{8} = \frac{3}{4}$
\$6:\$8 = 6:8 = 3:4
\$6 to \$8 = 6 to 8 = 3 to 4

You Try It 1
Write the comparison 20 pounds to 24 pounds as a ratio in simplest form using a fraction, a colon, and the word *to*.

Your solution $\frac{5}{6}$ 5:6 5 to 6

Example 2
Write the comparison 18 quarts to 6 quarts as a ratio in simplest form using a fraction, a colon, and the word *to*.

Solution $\frac{18 \text{ quarts}}{6 \text{ quarts}} = \frac{18}{6} = \frac{3}{1}$
18 quarts:6 quarts
18:6 = 3:1
18 quarts to 6 quarts
18 to 6 = 3 to 1

You Try It 2
Write the comparison 64 miles to 8 miles as a ratio in simplest form using a fraction, a colon, and the word *to*.

Your solution $\frac{8}{1}$ 8:1 8 to 1

Solutions on p. A23

Objective B To solve application problems

Use the table below for Example 3 and You Try It 3.

Board Feet of Wood at a Lumber Store			
Pine	*Ash*	*Oak*	*Cedar*
20,000	18,000	10,000	12,000

Example 3

Find, as a fraction in simplest form, the ratio of the number of board feet of pine to board feet of oak.

Strategy

To find the ratio, write the ratio of board feet of pine (20,000) to board feet of oak (10,000) in simplest form.

Solution

$$\frac{20{,}000}{10{,}000} = \frac{2}{1}$$

The ratio is $\frac{2}{1}$.

You Try It 3

Find, as a fraction in simplest form, the ratio of the number of board feet of cedar to board feet of ash.

Your strategy

Your solution

$\frac{2}{3}$

Example 4

The cost of building a patio cover was $250 for labor and $350 for materials. What, as a fraction in simplest form, is the ratio of the cost of materials to the total cost for labor and materials?

Strategy

To find the ratio, write the ratio of the cost of materials ($350) to the total cost ($250 + $350) in simplest form.

Solution

$$\frac{\$350}{\$250 + \$350} = \frac{350}{600} = \frac{7}{12}$$

The ratio is $\frac{7}{12}$.

You Try It 4

A company spends $20,000 a month for television advertising and $15,000 a month for radio advertising. What, as a fraction in simplest form, is the ratio of the cost of radio advertising to the total cost of radio and television advertising?

Your strategy

Your solution

$\frac{3}{7}$

Solutions on p. A23

4.1 Exercises

Objective A

Write the comparison as a ratio in simplest form using a fraction, a colon (:), and the word *to*.

1. 3 pints to 15 pints
$\frac{1}{5}$ 1:5 1 to 5

2. 6 pounds to 8 pounds
$\frac{3}{4}$ 3:4 3 to 4

3. \$40 to \$20
$\frac{2}{1}$ 2:1 2 to 1

4. 10 feet to 2 feet
$\frac{5}{1}$ 5:1 5 to 1

5. 3 miles to 8 miles
$\frac{3}{8}$ 3:8 3 to 8

6. 2 hours to 3 hours
$\frac{2}{3}$ 2:3 2 to 3

7. 37 hours to 24 hours
$\frac{37}{24}$ 37:24 37 to 24

8. 29 inches to 12 inches
$\frac{29}{12}$ 29:12 29 to 12

9. 6 minutes to 6 minutes
$\frac{1}{1}$ 1:1 1 to 1

10. 8 days to 12 days
$\frac{2}{3}$ 2:3 2 to 3

11. 35 cents to 50 cents
$\frac{7}{10}$ 7:10 7 to 10

12. 28 inches to 36 inches
$\frac{7}{9}$ 7:9 7 to 9

13. 30 minutes to 60 minutes
$\frac{1}{2}$ 1:2 1 to 2

14. 25 cents to 100 cents
$\frac{1}{4}$ 1:4 1 to 4

15. 32 ounces to 16 ounces
$\frac{2}{1}$ 2:1 2 to 1

16. 12 quarts to 4 quarts
$\frac{3}{1}$ 3:1 3 to 1

17. 3 cups to 4 cups
$\frac{3}{4}$ 3:4 3 to 4

18. 6 years to 7 years
$\frac{6}{7}$ 6:7 6 to 7

19. \$5 to \$3
$\frac{5}{3}$ 5:3 5 to 3

20. 30 yards to 12 yards
$\frac{5}{2}$ 5:2 5 to 2

21. 12 quarts to 18 quarts
$\frac{2}{3}$ 2:3 2 to 3

22. \$20 to \$28
$\frac{5}{7}$ 5:7 5 to 7

23. 14 days to 7 days
$\frac{2}{1}$ 2:1 2 to 1

24. 9 feet to 3 feet
$\frac{3}{1}$ 3:1 3 to 1

Objective B Application Problems

Family Budget

Housing	*Food*	*Transportation*	*Taxes*	*Utilities*	*Miscellaneous*	*Total*
\$800	\$400	\$300	\$350	\$150	\$400	\$2400

25. Use the table to find the ratio of housing cost to total income.
$\frac{1}{3}$

26. Use the table to find the ratio of food cost to total income.
$\frac{1}{6}$

27. Use the table to find the ratio of utilities cost to food cost.
$\frac{3}{8}$

28. Use the table to find the ratio of transportation cost to housing cost.
$\frac{3}{8}$

29. According to the National Collegiate Athletic Association (NCAA), for every 1200 high school senior football players, 40 will play football as seniors in college. Write the ratio of the number of college senior football players to high school senior football players. $\frac{1}{30}$

30. The NCAA estimates that for every 2300 high school seniors that want to play basketball in college, only 40 will make a college team. Write the ratio of the number of students who make a college basketball team to the number of high school seniors who want to make a college basketball team. $\frac{2}{115}$

31. A transformer has 40 turns in the primary coil and 480 turns in the secondary coil. State the ratio of the turns in the primary coil to the number of turns in the secondary coil. $\frac{1}{12}$

32. Rita Sterling bought a computer system for $2400. Five years later she sold the computer for $900. Find the ratio of the amount she received for the computer to the cost of the computer. $\frac{3}{8}$

33. A house with an original value of $90,000 increased in value to $110,000 in 5 years.

a. Find the increase in the value of the house. $20,000

b. What is the ratio of the increase in value to the original value of the house? $\frac{2}{9}$

34. A decorator bought one box of ceramic floor tile for $21 and a box of wood tile for $33.

a. What was the total cost of the box of ceramic tile and the box of wood tile? $54

b. What is the ratio of the cost of the box of wood tile to the total cost? $\frac{11}{18}$

35. The price of gasoline jumped from $0.96 to $1.26 in 1 year. What is the ratio of the increase in price to the original price? $\frac{5}{16}$

APPLYING THE CONCEPTS

A bank uses the ratio of a borrower's total monthly debts to the borrower's total monthly income to determine the maximum monthly payment for a potential homeowner. This ratio is called the debt–income ratio. Use the homeowner's income–debt table at the right for Exercises 36 and 37.

Income	*Debts*
$3500	$900
250	170
140	160
	95

36. Compute the debt–income ratio for the potential homeowner. $\frac{265}{778}$

37. If Central Trust Bank will make a loan to a customer whose debt–income ratio is less than $\frac{1}{3}$, will the potential homeowner qualify? Explain your answer. No. $\frac{265}{778}$ is greater than $\frac{1}{3}$

38. To make a home loan, First National Bank requires a debt–income ratio that is less than $\frac{2}{5}$. Would the homeowner whose income–debt table is given at the right qualify for a loan using these standards?

Income		*Debts*	
Salary	3400	Mortgage	1,800
Interest	83	Property tax	104
Rent	650	Insurance	35
Dividends	34	Liabilities	120
		Credit card	234
		Car loan	197

No. $\frac{830}{1389}$ is greater than $\frac{2}{5}$

39. [W] Is the value of a ratio always less than 1? Explain. No. Answers will vary

4.2 Rates

Objective A To write rates

INSTRUCTOR NOTE
Emphasize the distinction between ratio, a comparison with the same units, and rate, a comparison with different units.

A **rate** is a comparison of two quantities that have *different* units. A rate is written as a fraction.

A distance runner ran 26 miles in 4 hours. The distance-to-time rate is written

$\frac{26 \text{ miles}}{4 \text{ hours}} = \frac{13 \text{ miles}}{2 \text{ hours}}$

A rate is in **simplest form** when the numbers that form the rate have no common factors. Note that the units are written as part of the rate.

Example 1 Write "6 roof supports for every 9 feet" as a rate in simplest form.

Solution $\frac{6 \text{ supports}}{9 \text{ feet}} = \frac{2 \text{ supports}}{3 \text{ feet}}$

You Try It 1 Write "15 pounds of fertilizer for 12 trees" as a rate in simplest form.

Your solution $\frac{5 \text{ pounds}}{4 \text{ trees}}$

Solution on p. A23

Objective B To write unit rates

INSTRUCTOR NOTE
Unit rates are given in many situations. The EPA evaluates cars for their miles per gallon. A more difficult unit rate for students is the one used in the airline industry: cubic feet of fresh air per person per minute. Typical rates are: economy class, 7 ft^3/min/person; first class, 50 ft^3/min/person; cockpit, 150 ft^3/min/person.

A **unit rate** is a rate in which the number in the denominator is 1.

$\frac{\$3.25}{1 \text{ pound}}$ or \$3.25/pound is read "\$3.25 per pound."

To find unit rates, divide the number in the numerator of the rate by the number in the denominator of the rate.

A car traveled 344 miles on 16 gallons of gasoline. To find the miles per gallon (unit rate), divide the numerator of the rate by the denominator of the rate.

$\frac{344 \text{ miles}}{16 \text{ gallons}}$ is the rate.

$16\overline{)344.0}$ = 21.5 21.5 miles/gallon is the unit rate.

Example 2 Write "300 feet in 8 seconds" as a unit rate.

Solution $\frac{300 \text{ feet}}{8 \text{ seconds}}$ $8\overline{)300.0}$ = 37.5

37.5 feet/second

You Try It 2 Write "260 miles in 8 hours" as a unit rate.

Your solution 32.5 miles/hour

Solution on p. A23

Objective C To solve application problems

The table at the right shows typical air fare costs for long routes.

Long Routes	*Miles*	*Fare*
New York–Los Angeles	2,475	$683
San Francisco–Dallas	1,464	$536
Denver–Pittsburgh	1,302	$525
Minneapolis–Hartford	1,050	$483

Find the cost per mile of the four routes. Which route is the most expensive, and which is the least expensive, for each mile flown?

Strategy
To find the cost per mile, divide the miles flown by the fare for each route. Compare the costs per mile to determine the most expensive and least expensive routes per mile.

Solution New York–Los Angeles $\frac{683}{2,475} \approx 0.28$

San Francisco–Dallas $\frac{536}{1,464} \approx 0.37$

Denver–Pittsburgh $\frac{525}{1,302} \approx 0.40$

Minneapolis–Hartford $\frac{483}{1,050} = 0.46$

The Minneapolis–Hartford route is the most expensive per mile, and the New York–Los Angeles route is the least expensive per mile.

Example 3
As an investor, Jung Ho purchased 100 shares of stock for $1500. One year later, Jung sold the 100 shares for $1800. What was his profit per share?

Strategy
To find Jung's profit per share:

- Find the total profit by subtracting the original cost ($1500) from the selling price ($1800).
- Find the profit per share (unit rate) by dividing the total profit by the number of shares of stock (100).

Solution

$$\begin{array}{r} \$1800 \\ -\ 1500 \\ \hline \$300 \end{array} \text{ total profit} \qquad 100\overline{)\$300}\ \ \$3$$

Jung Ho's profit per share was $3.

You Try It 3
Erik Peltier, a jeweler, purchased 5 ounces of gold for $1625. Later, he sold the 5 ounces for $1720. What was Erik's profit per ounce?

Your strategy

Your solution $19/ounce

Solution on p. A23

4.2 Exercises

Objective A

Write as a rate in simplest form.

1. 3 pounds of meat for 4 people
$\frac{3\text{ pounds}}{4\text{ people}}$

2. 30 ounces in 24 glasses
$\frac{5\text{ ounces}}{4\text{ glasses}}$

3. \$80 for 12 boards
$\frac{\$20}{3\text{ boards}}$

4. 84 cents for 6 bars of soap
$\frac{14\text{ cents}}{1\text{ bar}}$

5. 300 miles on 15 gallons
$\frac{20\text{ miles}}{1\text{ gallon}}$

6. 88 feet in 8 seconds
$\frac{11\text{ feet}}{1\text{ second}}$

7. 20 children in 8 families
$\frac{5\text{ children}}{2\text{ families}}$

8. 48 leaves on 9 plants
$\frac{16\text{ leaves}}{3\text{ plants}}$

9. 16 gallons in 2 hours
$\frac{8\text{ gallons}}{1\text{ hour}}$

10. 25 ounces in 5 minutes
$\frac{5\text{ ounces}}{1\text{ minute}}$

Objective B

Write as a unit rate.

11. 10 feet in 4 seconds
2.5 feet/second

12. 816 miles in 6 days
136 miles/day

13. \$1300 earned in 4 weeks
\$325/week

14. \$27,000 earned in 12 months
\$2250/month

15. 1100 trees planted on 10 acres
110 trees/acre

16. 3750 words on 15 pages
250 words/page

17. \$32.97 earned in 7 hours
\$4.71/hour

18. \$315.70 earned in 22 hours
\$14.35/hour

19. 628.8 miles in 12 hours
52.4 miles/hour

20. 388.8 miles in 8 hours
48.6 miles/hour

21. 344.4 miles on 12.3 gallons of gasoline
28 miles/gallon

22. 409.4 miles on 11.5 gallons of gasoline
35.6 miles/gallon

23. \$349.80 for 212 pounds
\$1.65/pound

24. \$11.05 for 3.4 pounds
\$3.25/pound

Objective C Application Problems

25. An automobile was driven 326.6 miles on 11.5 gallons of gas. Find the number of miles driven per gallon of gas. 28.4 miles per gallon

26. You drive 246.6 miles in 4.5 hours. Find the number of miles driven per hour. 54.8 miles per hour

27. The Saturn-5 rocket uses 534,000 gallons of fuel in 2.5 minutes. How much fuel does the rocket use in 1 minute? 213,600 gallons per minute

28. Shawna Monte, a ski instructor, worked 6 months at a ski resort and earned $15,900. What was her wage per month? $2650 per month

29. An investor owns 600 shares of Ford Motor Company and receives $960 in yearly dividends. Find the dividend per share. $1.60 per share

30. An investor purchased 420 shares of Charcor Corporation for $7,980. What was the cost per share? $19 per share

31. Assume that Apple Computer produced 5000 compact discs for $26,536.32. Of the discs made, 122 did not meet company standards.
a. How many compact discs did meet company standards? 4878 discs
b. What was the cost per disc for those discs that met company standards? $5.44 per disc

32. The Pierre family purchased a 250-pound side of beef for $365.75 and had it packaged. During the packaging, 75 pounds of beef were discarded as waste.
a. How many pounds of beef were packaged? 175 pounds
b. What was the cost per pound for the packaged beef? $2.09 per pound

33. The Bear Valley Fruit Stand purchased 250 boxes of strawberries for $162.50. All the strawberries were sold for $312.50. What was the profit per box of strawberries? $0.60 per box

APPLYING THE CONCEPTS

Television advertisers use the rate of $\frac{\text{cost of commercial}}{\text{one thousand viewers}}$. This rate is an indication of the cost-effectiveness of an advertisement. One graph below gives the cost of a 30-second commercial for various television shows. The second graph gives the number of people watching the show. Use these graphs for Exercises 34–37.

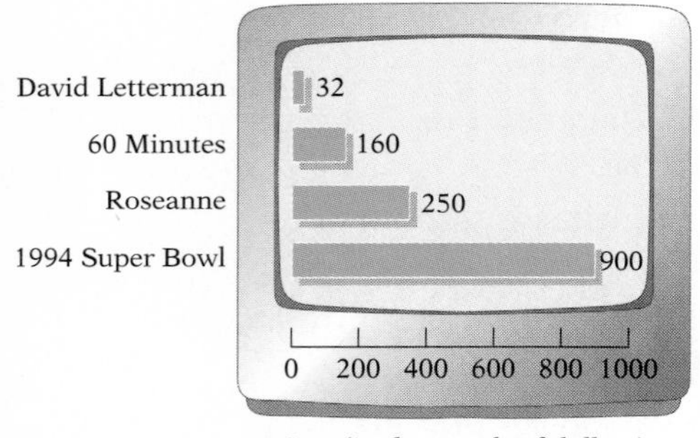

Cost (in thousands of dollars)

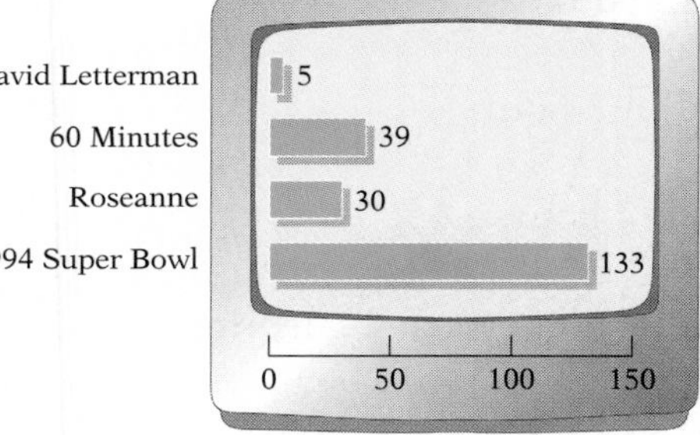

Number of Viewers (in millions)

34. What was the cost per thousand viewers for the 1994 Super Bowl? $6.77 per thousand viewers

35. What was the cost per thousand viewers for Roseanne? $8.33 per thousand viewers

36. [W] If you were an ad agency executive purchasing time for a commercial, would you try to have a low or a high cost per thousand viewers? Explain your answer. Low cost

37. Which of the four programs listed has the smallest cost per thousand viewers? 60 Minutes

38. [W] The price–earnings ratio of a company's stock is one measure used by stock market analysts to assess the financial well-being of the company. Explain the meaning of the price–earnings ratio. It is the ratio of the current selling price for a stock to the company's earnings per share of stock.

4.3 Proportions

Objective A To determine whether a proportion is true

POINT OF INTEREST
Proportions were studied by the earliest mathematicians. Clay tablets uncovered by archeologists show evidence of proportions in Egyptian and Babylonian cultures dating from 1800 B.C.

A **proportion** is an expression of the equality of two ratios or rates.

$\frac{50 \text{ miles}}{4 \text{ gallons}} = \frac{25 \text{ miles}}{2 \text{ gallons}}$ Note that the units of the numerators are the same and the units of the denominators are the same.

$\frac{3}{6} = \frac{1}{2}$

A proportion is **true** if the fractions are equal when written in lowest terms.

In any true proportion, the "cross products" are equal.

➡ Is $\frac{2}{3} = \frac{8}{12}$ a true proportion?

$\frac{2}{3} \times \frac{8}{12}$ → $3 \times 8 = 24$, $2 \times 12 = 24$

Cross products *are* equal.
$\frac{2}{3} = \frac{8}{12}$ is a true proportion.

A proportion is **not true** if the fractions are not equal when reduced to lowest terms.

If the cross products are not equal, then the proportion is not true.

➡ Is $\frac{4}{5} = \frac{8}{9}$ a true proportion?

$\frac{4}{5} \times \frac{8}{9}$ → $5 \times 8 = 40$, $4 \times 9 = 36$

Cross products *are not* equal.
$\frac{4}{5} = \frac{8}{9}$ is not a true proportion.

Example 1
Use cross products to determine whether $\frac{5}{8} = \frac{10}{16}$ is a true proportion.

Solution
$\frac{5}{8} \times \frac{10}{16}$ → $8 \times 10 = 80$, $5 \times 16 = 80$

The proportion is true.

You Try It 1
Use cross products to determine whether $\frac{6}{10} = \frac{9}{15}$ is a true proportion.

Your solution
True

Example 2
Use cross products to determine whether $\frac{62 \text{ miles}}{4 \text{ gallons}} = \frac{33 \text{ miles}}{2 \text{ gallons}}$ is a true proportion.

Solution
$\frac{62}{4} \times \frac{33}{2}$ → $4 \times 33 = 132$, $62 \times 2 = 124$

The proportion is not true.

You Try It 2
Use cross products to determine whether $\frac{\$32}{6 \text{ hours}} = \frac{\$90}{8 \text{ hours}}$ is a true proportion.

Your solution
Not true

Solutions on p. A24

Objective B To solve proportions

Sometimes one of the numbers in a proportion is unknown. In this case, it is necessary to *solve* the proportion.

INSTRUCTOR NOTE
The solution of these equations is based on the relationship between multiplication and division. You may want to show the solution by dividing each side by 9. For instance,

$$\frac{9}{6} = \frac{3}{n}$$
$$9 \times n = 6 \times 3$$
$$9 \times n = 18$$
$$\frac{9 \times n}{9} = \frac{18}{9}$$
$$n = 2$$

To **solve** a proportion, find a number to replace the unknown so that the proportion is true.

➡ Solve: $\frac{9}{6} = \frac{3}{n}$

$\frac{9}{6} = \frac{3}{n}$

$9 \times n = 6 \times 3$ • Find the cross products.

$9 \times n = 18$

$n = 18 \div 9$ • Think of $9 \times n = 18$ as $9\overline{)18}$ with quotient n.

$n = 2$

Check:

$\frac{9}{6} \quad \frac{3}{2}$ → $6 \times 3 = 18$ → $9 \times 2 = 18$

Example 3

Solve $\frac{n}{12} = \frac{25}{60}$ and check.

Solution

$n \times 60 = 12 \times 25$
$n \times 60 = 300$
$n = 300 \div 60$
$n = 5$

Check:

$\frac{5}{12} \quad \frac{25}{60}$ → $12 \times 25 = 300$ → $5 \times 60 = 300$

You Try It 3

Solve $\frac{n}{14} = \frac{3}{7}$ and check.

Your solution

$n = 6$

Example 4

Solve $\frac{4}{9} = \frac{n}{16}$. Write the answer to the nearest tenth.

Solution

$4 \times 16 = 9 \times n$
$64 = 9 \times n$
$64 \div 9 = n$
$7.1 \approx n$

Note: A rounded answer is an approximation. Therefore, the answer to a check will not be exact.

You Try It 4

Solve $\frac{5}{8} = \frac{n}{20}$. Write the answer to the nearest tenth.

Your solution

$12.5 = n$

Solutions on p. A24

Example 5

Solve $\frac{28}{52} = \frac{7}{n}$ and check.

Solution

$28 \times n = 52 \times 7$
$28 \times n = 364$
$n = 364 \div 28$
$n = 13$

Check:

$\frac{28}{52} \;\; \frac{7}{13}$ → $52 \times 7 = 364$, $28 \times 13 = 364$

You Try It 5

Solve $\frac{15}{20} = \frac{12}{n}$ and check.

Your solution

$n = 16$

Example 6

Solve $\frac{15}{n} = \frac{8}{3}$. Write the answer to the nearest hundredth.

Solution

$15 \times 3 = n \times 8$
$45 = n \times 8$
$45 \div 8 = n$
$5.63 \approx n$

You Try It 6

Solve $\frac{12}{n} = \frac{7}{4}$. Write the answer to the nearest hundredth.

Your solution

$6.86 \approx n$

Example 7

Solve $\frac{n}{9} = \frac{3}{1}$ and check.

Solution

$n \times 1 = 9 \times 3$
$n \times 1 = 27$
$n = 27 \div 1$
$n = 27$

Check:

$\frac{27}{9} \;\; \frac{3}{1}$ → $9 \times 3 = 27$, $27 \times 1 = 27$

You Try It 7

Solve $\frac{n}{12} = \frac{4}{1}$ and check.

Your solution

$n = 48$

Example 8

Solve $\frac{5}{9} = \frac{15}{n}$ and check.

Solution

$5 \times n = 9 \times 15$
$5 \times n = 135$
$n = 135 \div 5$
$n = 27$

Check:

$\frac{5}{9} \;\; \frac{15}{27}$ → $9 \times 15 = 135$, $5 \times 27 = 135$

You Try It 8

Solve $\frac{3}{8} = \frac{12}{n}$ and check.

Your solution

$n = 32$

Solutions on p. A24

Objective C To solve application problems

Example 9

A mason determines that 9 cement blocks are required for a retaining wall 2 feet long. At this rate, how many cement blocks are required for a retaining wall that is 24 feet long?

Strategy

To find the number of cement blocks for a retaining wall 24 feet long, write and solve a proportion using n to represent the number of blocks required.

Solution

$$\frac{9 \text{ cement blocks}}{2 \text{ feet}} = \frac{n \text{ cement blocks}}{24 \text{ feet}}$$

$$9 \times 24 = 2 \times n$$
$$216 = 2 \times n$$
$$216 \div 2 = n$$
$$108 = n$$

108 cement blocks are required for a 24-foot retaining wall.

You Try It 9

Twenty-four jars can be packed in 6 identical boxes. At this rate, how many jars can be packed in 15 boxes?

Your strategy

Your solution 60 jars

Example 10

The dosage of a certain medication is 2 ounces for every 50 pounds of body weight. How many ounces of this medication are required for a person who weighs 175 pounds?

Strategy

To find the number of ounces of medication for a person weighing 175 pounds, write and solve a proportion using n to represent the number of ounces of medication for a 175-pound person.

Solution

$$\frac{2 \text{ ounces}}{50 \text{ pounds}} = \frac{n \text{ ounces}}{175 \text{ pounds}}$$

$$2 \times 175 = 50 \times n$$
$$350 = 50 \times n$$
$$350 \div 50 = n$$
$$7 = n$$

7 ounces of medication are required for a 175-pound person.

You Try It 10

Three tablespoons of a liquid plant fertilizer are to be added to every 4 gallons of water. How many tablespoons of fertilizer are required for 10 gallons of water?

Your strategy

Your solution $7\frac{1}{2}$ tablespoons

Solutions on p. A25

4.3 Exercises

Objective A

Determine whether the proportion is true or not true.

1. $\frac{4}{8} = \frac{10}{20}$
True

2. $\frac{39}{48} = \frac{13}{16}$
True

3. $\frac{7}{8} = \frac{11}{12}$
Not true

4. $\frac{15}{7} = \frac{17}{8}$
Not true

5. $\frac{27}{8} = \frac{9}{4}$
Not true

6. $\frac{3}{18} = \frac{4}{19}$
Not true

7. $\frac{45}{135} = \frac{3}{9}$
True

8. $\frac{3}{4} = \frac{54}{72}$
True

9. $\frac{16}{3} = \frac{48}{9}$
True

10. $\frac{15}{5} = \frac{3}{1}$
True

11. $\frac{7}{40} = \frac{7}{8}$
Not true

12. $\frac{9}{7} = \frac{6}{5}$
Not true

13. $\frac{50 \text{ miles}}{2 \text{ gallons}} = \frac{25 \text{ miles}}{1 \text{ gallon}}$
True

14. $\frac{16 \text{ feet}}{10 \text{ seconds}} = \frac{24 \text{ feet}}{15 \text{ seconds}}$
True

15. $\frac{6 \text{ minutes}}{5 \text{ cents}} = \frac{30 \text{ minutes}}{25 \text{ cents}}$
True

16. $\frac{16 \text{ pounds}}{12 \text{ days}} = \frac{20 \text{ pounds}}{14 \text{ days}}$
Not true

17. $\frac{\$15}{4 \text{ pounds}} = \frac{\$45}{12 \text{ pounds}}$
True

18. $\frac{270 \text{ trees}}{6 \text{ acres}} = \frac{90 \text{ trees}}{2 \text{ acres}}$
True

19. $\frac{300 \text{ feet}}{4 \text{ rolls}} = \frac{450 \text{ feet}}{7 \text{ rolls}}$
Not true

20. $\frac{1 \text{ gallon}}{4 \text{ quarts}} = \frac{7 \text{ gallons}}{28 \text{ quarts}}$
True

21. $\frac{\$65}{5 \text{ days}} = \frac{\$26}{2 \text{ days}}$
True

22. $\frac{80 \text{ miles}}{2 \text{ hours}} = \frac{110 \text{ miles}}{3 \text{ hours}}$
Not true

23. $\frac{7 \text{ tiles}}{4 \text{ feet}} = \frac{42 \text{ tiles}}{20 \text{ feet}}$
Not true

24. $\frac{15 \text{ feet}}{3 \text{ yards}} = \frac{90 \text{ feet}}{18 \text{ yards}}$
True

Objective B

Solve. Round to the nearest hundredth.

25. $\frac{n}{4} = \frac{6}{8}$ 3

26. $\frac{n}{7} = \frac{9}{21}$ 3

27. $\frac{12}{18} = \frac{n}{9}$ 6

28. $\frac{7}{21} = \frac{35}{n}$ 105

29. $\frac{10}{12} = \frac{n}{24}$ 20

30. $\frac{6}{n} = \frac{24}{36}$ 9

31. $\frac{3}{n} = \frac{15}{10}$ 2

32. $\frac{n}{45} = \frac{17}{135}$ 5.67

33. $\frac{9}{4} = \frac{18}{n}$ 8

34. $\frac{7}{15} = \frac{21}{n}$ 45

35. $\frac{n}{6} = \frac{2}{3}$ 4

36. $\frac{5}{12} = \frac{n}{144}$ 60

37. $\frac{n}{5} = \frac{7}{8}$ 4.38

38. $\frac{4}{n} = \frac{9}{5}$ 2.22

39. $\frac{8}{5} = \frac{n}{6}$ 9.6

40. $\frac{n}{11} = \frac{32}{4}$ 88

41. $\frac{3}{4} = \frac{8}{n}$ 10.67

42. $\frac{5}{12} = \frac{n}{8}$ 3.33

43. $\frac{36}{20} = \frac{12}{n}$ 6.67

44. $\frac{15}{n} = \frac{65}{100}$ 23.08

45. $\frac{n}{15} = \frac{21}{12}$ 26.25

46. $\frac{40}{n} = \frac{15}{8}$ 21.33

47. $\frac{32}{n} = \frac{1}{3}$ 96

48. $\frac{5}{8} = \frac{42}{n}$ 67.2

49. $\frac{18}{11} = \frac{16}{n}$ 9.78

50. $\frac{25}{4} = \frac{n}{12}$ 75

51. $\frac{28}{8} = \frac{12}{n}$ 3.43

52. $\frac{n}{30} = \frac{65}{120}$ 16.25

53. $\frac{0.3}{5.6} = \frac{n}{25}$ 1.34

54. $\frac{1.3}{16} = \frac{n}{30}$ 2.44

55. $\frac{0.7}{9.8} = \frac{3.6}{n}$ 50.4

56. $\frac{1.9}{7} = \frac{13}{n}$ 47.89

Objective C *Application Problems*

Solve. Round to the nearest hundredth.

57. A 6-ounce package of Puffed Wheat contains 600 calories. How many calories are in a 0.5-ounce serving of the cereal? 50 calories

58. A 10-ounce container of egg noodles contains 700 milligrams of cholesterol. How many milligrams of cholesterol are in a 2-ounce serving of egg noodles? 140 milligrams

59. Ron Stokes uses 2 pounds of fertilizer for every 100 square feet of lawn for landscape maintenance. At this rate, how many pounds of fertilizer did he use on a lawn that measures 2500 square feet? 50 pounds

Solve. Round to the nearest hundredth.

60. A nursery provides a liquid plant food by adding 1 gallon of water for each 2 ounces of plant food. At this rate, how many ounces of plant food are required for 25 gallons of water? 50 ounces

61. The property tax on a $90,000 home is $1800. At this rate, what is the property tax on a home worth $125,000? $2500

62. The sales tax on a $90 purchase is $4.50. At this rate, what is the sales tax on a $300 purchase? $15

63. A brick wall 20 feet in length contains 1040 bricks. At the same rate, how many bricks would it take to build a wall 48 feet in length? 2496 bricks

64. From past experience, a farmer knows that planting 50 acres of wheat will yield 2900 bushels of wheat. At the same rate, find the number of bushels of wheat the farmer can expect from another 240 acres of wheat. 13,920 bushels

65. The scale on a map is 1.25 inch equals 10 miles. Find the distance between Carlsbad and Del Mar, which are 2 inches apart on the map. 16 miles

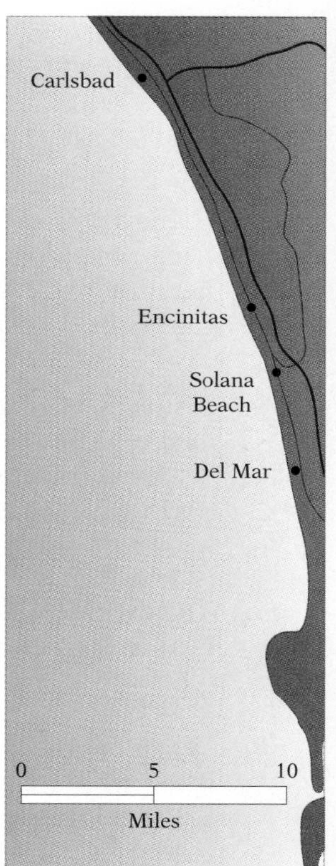

66. The scale on the plans for a new house is 1 inch equals 3 feet. Find the length and width of a room that measures 5 inches by 8 inches on the drawing. 15 feet by 24 feet

67. The dosage for a medication is $\frac{1}{3}$ ounce for every 40 pounds of body weight. At this rate, how many ounces of medication should a physician prescribe for a patient who weighs 150 pounds? 1.25 ounces

68. Randall, who is on a reducing program, has lost 3 pounds in 5 weeks. At the same rate, how long will it take him to lose 27 pounds? 45 weeks

69. A pre-election survey showed that 2 out of every 3 eligible voters would cast ballots in the county election. At this rate, how many people in a county of 240,000 eligible voters would vote in the election? 160,000 people

70. A paint manufacturer suggests using 1 gallon of paint for every 400 square feet of a wall. At this rate, how many gallons of paint would be required for a room that has 1400 square feet of wall? 3.5 gallons

71. An automobile recall was based on tests that showed 24 braking defects in 1000 cars. At this rate, how many defects would be found in 40,000 cars? 960 braking defects

72. Suppose a computer chip manufacturer knows from experience that in an average production run of 2000 circuit boards, 60 will be defective. What number of defective circuit boards can be expected from a run of 25,000 circuit boards? 750 defective boards

73. You own 240 shares of a computer stock. The company declares a stock split of 5 shares for every 3 owned. How many shares of stock will you own after the stock split? 400 shares

Solve. Round to the nearest hundredth.

74. Carlos Capasso owns 50 shares of Texas Utilities that pay dividends of $153. At this rate, what dividend would Carlos receive after buying 300 additional shares of Texas Utilities? $1071

75. The director of data processing at a college estimates that the ratio of student time to administrative time on a certain computer is 3:2. During a month in which the computer was used 200 hours for administration, how many hours was it used by students? 300 hours

APPLYING THE CONCEPTS

The table at the right shows the gross domestic product, the federal budget, and the national debt for the years 1947 and 1992. Use the table for Exercises 76 and 77.

Year	*Gross Domestic Product*	*Federal Budget*	*National Debt*
1947	223 billion	35 billion	257 billion
1992	5.87 trillion	1.38 trillion	4.00 trillion

76. If the ratio of the national debt to the federal budget had been the same in 1992 as in 1947, what would the national debt have been in 1992? What is the difference between this value and the actual debt in 1992? $10.13 trillion $6.13 trillion

77. If the ratio of the national debt to the gross domestic product had been the same in 1992 as in 1947, what would the national debt have been in 1992? What is the difference between this value and the actual debt in 1992? $6.76 trillion $2.76 trillion

78. [W] A survey of voters in a city claimed that 2 people of every 5 who voted cast a ballot in favor of city amendment A and that 3 people of every 4 who voted cast a ballot against amendment A. Is this possible? Explain your answer.

No. $\frac{2}{5} + \frac{3}{4} = \frac{23}{20}$, which is more than the number of voters

79. The ratio of weight on the moon to weight on Earth is 1:6. If a bowling ball weighs 16 pounds on Earth, what would it weigh on the moon?

$2\frac{2}{3}$ pounds

80. When engineers design a new car, they first build a model of the car. The ratio of a part on the model to the actual size is 2:5. If a door is 1.3 feet long on the model, what is the length of the door on the car?

3.25 feet

81. [W] Write a word problem that requires solving a proportion to find the answer. Answers will vary

82. [W] Choose a local pizza restaurant and a particular type of pizza. Determine the size and cost of a medium pizza and of a large pizza. (Use regular prices and no special discounts.) Is $\frac{\text{cost of medium}}{\text{size of medium}}$ approximately equal to $\frac{\text{cost of large}}{\text{size of large}}$? Explain your answer. Costs will vary

Project in Mathematics

The Golden Ratio

There are certain designs that have been repeated over and over in both art and architecture. One of these involves the **golden rectangle.**

A golden rectangle is drawn at the right. Begin with a square that measures, say, 2 inches on a side. Now measure the distance from A to B. Place this length along the bottom of the square, starting at A. The resulting rectangle is a golden rectangle.

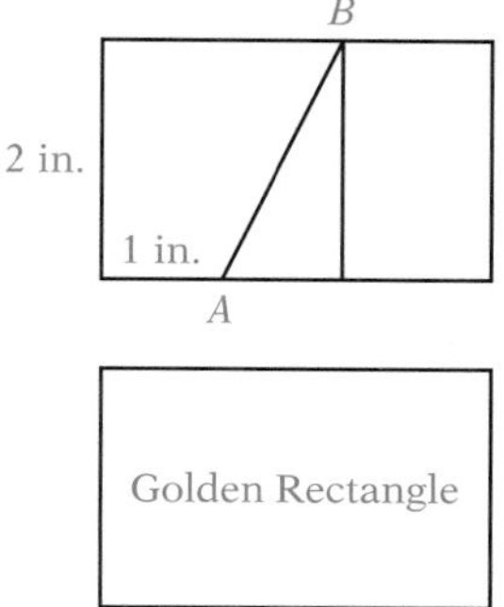

The **golden ratio** is the ratio of the length of the golden rectangle to its width. If you have drawn the rectangle following the procedure above, you will find that the golden ratio is approximately 1.6.

The golden ratio appears in many different situations. Some historians claim that some of the great pyramids of Egypt are based on the golden ratio. The drawing at the right is the Pyramid of Gizeh, which dates from approximately 2600 B.C. The ratio of the height to a side of the base is approximately 1.6.

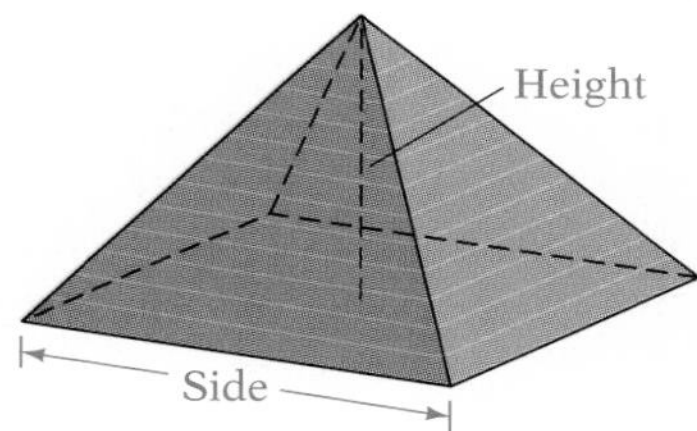

Exercises:

1. The canvas of the Mona Lisa painted by Leonardo da Vinci is a golden rectangle. However, there are other instances of the golden rectangle in the painting itself. Do some research on this painting and write a few paragraphs summarizing your findings.
2. What do 3 × 5 and 5 × 8 index cards have to do with the golden rectangle?
3. When was the United Nations building in New York built? What does the front of that building have to do with a golden rectangle?
4. When was the Parthenon in Athens, Greece, built? What does the front of that building have to do with a golden rectangle?

Chapter Summary

Key Words

Quantities such as 8 feet and 60 miles are number quantities written with *units*.

A *ratio* is the comparison of two quantities that have the same units.

A ratio is in *simplest form* when the two numbers that form the ratio have no common factors.

A *rate* is the comparison of two quantities that have different units.

A rate is in *simplest form* when the numbers that form the rate have no common factors.

A *unit rate* is a rate in which the number in the denominator is 1.

A *proportion* is an expression of the equality of two ratios or rates.

Essential Rules

To Find Unit Rates

To find unit rates, divide the number in the numerator of the rate by the number in the denominator of the rate.

To Solve a Proportion

One of the numbers in a proportion may be unknown. To solve a proportion, find a number to replace the unknown so that the proportion is true.

Ways to Express a Ratio

A ratio can be written three different ways:

1. As a fraction
2. As two numbers separated by a colon (:)
3. As two numbers separated by the word *to*

Chapter Review

SECTION 4.1

1. Write the comparison 8 feet to 28 feet as a ratio in simplest form using a fraction, a colon (:), and the word *to*.
$\frac{2}{7}$ 2:7 2 to 7

2. Write the comparison 6 inches to 15 inches as a ratio in simplest form using a fraction, a colon (:), and the word *to*.
$\frac{2}{5}$ 2:5 2 to 5

3. Write the comparison 12 days to 12 days as a ratio in simplest form using a fraction, a colon (:), and the word *to*.
$\frac{1}{1}$ 1:1 1 to 1

4. Write the comparison 32 dollars to 80 dollars as a ratio in simplest form using a fraction, a colon (:), and the word *to*.
$\frac{2}{5}$ 2:5 2 to 5

5. The high temperature during a 24-hour period was 84 degrees and the low temperature was 42 degrees. Write the ratio of the high temperature to the low temperature for the 24-hour period.
2:1

6. A house had an original value of $80,000 but increased in value to $120,000 in 2 years. Find the ratio of the increase to the original value.
$\frac{1}{2}$

7. In 5 years the price of a calculator went from $40 to $24. What is the ratio of the decrease in price to the original price?
$\frac{2}{5}$

8. A retail computer store spends $30,000 a year on TV advertising and $12,000 on newspaper advertising. Find the ratio of TV advertising to newspaper advertising.
$\frac{5}{2}$

SECTION 4.2

9. Write "100 miles in 3 hours" as a rate in simplest form.
$\frac{100 \text{ miles}}{3 \text{ hours}}$

10. Write "$15 in 4 hours" as a rate in simplest form.
$\frac{\$15}{4 \text{ hours}}$

11. Write "250 miles in 4 hours" as a unit rate.
62.5 miles/hour

12. Write "$300 earned in 40 hours" as a unit rate.
$7.50/hour

13. Write "326.4 miles on 12 gallons" as a unit rate.
27.2 miles/gallon

14. Write "$8.75 for 5 pounds" as a unit rate.
$1.75/pound

15. Pascal Hollis purchased 80 shares of stock for $3580. What is the cost per share?
$44.75 per share

16. Mahesh drove 198.8 miles in 3.5 hours. Find the average number of miles he drove per hour.
56.8 miles/hour

17. A 15-pound turkey costs $10.20. What is the cost per pound?
$.68/pound

18. The total cost of manufacturing 1000 radios was $36,600. Of the radios made, 24 did not pass inspection. Find the cost per radio of the radios that did pass inspection.
$37.50

SECTION 4.3

19. Determine whether the proportion is true or not true.
$\frac{5}{7} = \frac{25}{35}$
True

20. Determine whether the proportion is true or not true.
$\frac{2}{9} = \frac{10}{45}$
True

21. Determine whether the proportion is true or not true.
$\frac{8}{15} = \frac{32}{60}$
True

22. Determine whether the proportion is true or not true.
$\frac{3}{8} = \frac{10}{24}$
Not true

23. Solve the proportion.
$\frac{n}{8} = \frac{9}{2}$
$n = 36$

24. Solve the proportion.
$\frac{16}{n} = \frac{4}{17}$
$n = 68$

25. Solve the proportion. Round to hundredths.
$\frac{24}{11} = \frac{n}{30}$
$n = 65.45$

26. Solve the proportion. Round to hundredths.
$\frac{18}{35} = \frac{10}{n}$
$n = 19.44$

27. Monique used 1.5 pounds of fertilizer for every 200 square feet of lawn. How many pounds of fertilizer will she have to use on a lawn that measures 3000 square feet?
22.5 pounds

28. An insurance policy costs $3.87 for every $1000 of insurance. At this rate, what is the cost of $50,000 of insurance?
$193.50

29. The property tax on a $45,000 home is $900. At the same rate, what is the property tax on a home valued at $120,000?
$2400

30. A brick wall 40 feet in length contains 448 concrete blocks. At the same rate, how many blocks would it take to build a wall that is 120 feet in length?
1344 blocks

Chapter Test

1. Write the comparison 12 days to 8 days as a ratio in simplest form using a fraction, a colon (:), and the word *to*.

   $\frac{3}{2}$ 3:2 3 to 2 [4.1A]

2. Write the comparison 18 feet to 30 feet as a ratio in simplest form using a fraction, a colon (:), and the word *to*.

   $\frac{3}{5}$ 3:5 3 to 5 [4.1A]

3. Write the comparison 27 dollars to 81 dollars as a ratio in simplest form using a fraction, a colon (:), and the word *to*.

   $\frac{1}{3}$ 1:3 1 to 3 [4.1A]

4. Write the comparison 40 miles to 240 miles as a ratio in simplest form using a fraction, a colon (:), and the word *to*.

   $\frac{1}{6}$ 1:6 1 to 6 [4.1A]

5. The average summer temperature in a California desert is 112 degrees. In a city 100 miles away, the average summer temperature is 86 degrees. Write the ratio of the city temperature to the desert temperature.

   $\frac{43}{56}$ [4.1B]

6. An automobile sales company spends $25,000 each month for television advertising and $40,000 each month for radio advertising. Find, as a fraction in simplest form, the ratio of the cost of radio advertising to the total cost of advertising.

   $\frac{8}{13}$ [4.1B]

7. Write "18 supports for every 8 feet" as a rate in simplest form.

   $\frac{\text{9 supports}}{\text{4 feet}}$ [4.2A]

8. Write "$81 for 12 boards" as a rate in simplest form.

   $\frac{\$27}{\text{4 boards}}$ [4.2A]

9. Write "$22,036.80 earned in 12 months" as a unit rate.

   $1836.40/month [4.2B]

10. Write "256.2 miles on 8.4 gallons of gas" as a unit rate.

    30.5 miles/gallon [4.2B]

11. 40 feet of lumber costs $69.20. What is the per-foot cost of the lumber?
$1.73/foot [4.2C]

12. A plane travels 2421 miles in 4.5 hours. Find the plane's speed in miles per hour.
538 miles/hour [4.2C]

13. Determine whether the proportion is true or not true.
$\frac{5}{14} = \frac{25}{70}$
True [4.3A]

14. Determine whether the proportion is true or not true.
$\frac{40}{125} = \frac{5}{25}$
Not true [4.3A]

15. Solve the proportion.
$\frac{n}{18} = \frac{9}{4}$
40.5 [4.3B]

16. Solve the proportion.
$\frac{5}{12} = \frac{60}{n}$
144 [4.3B]

17. A research scientist estimates that the human body contains 88 pounds of water for every 100 pounds of body weight. At this rate, estimate the number of pounds of water in a college student who weighs 150 pounds.
132 pounds [4.3C]

18. The dosage of a medicine is $\frac{1}{4}$ ounce for every 50 pounds of body weight. How many ounces of this medication are required for a person who weighs 175 pounds? 0.875 ounce [4.3C]

19. The property tax on a house valued at $75,000 is $1500. At the same rate, find the property tax on a house valued at $140,000. $2800 [4.3C]

20. Fifty shares of a utility stock pay a dividend of $62.50. At the same rate, find the dividend paid on 500 shares of the utility stock. $625 [4.3C]

Cumulative Review

1. Subtract: $\begin{array}{r} 20{,}095 \\ -\ 10{,}937 \\ \hline \end{array}$

9158 [1.3B]

2. Write $2 \cdot 2 \cdot 2 \cdot 2 \cdot 3 \cdot 3 \cdot 3$ in exponential notation. $2^4 \cdot 3^3$ [1.6A]

3. Simplify: $4 - (5 - 2)^2 \div 3 + 2$

3 [1.6B]

4. Find the prime factorization of 160.

$2 \cdot 2 \cdot 2 \cdot 2 \cdot 2 \cdot 5$ [1.7B]

5. Find the LCM of 9, 12, and 18.

36 [2.1A]

6. Find the GCF of 28 and 42.

14 [2.1B]

7. Reduce $\frac{40}{64}$ to simplest form.

$\frac{5}{8}$ [2.3B]

8. Find $4\frac{7}{15}$ more than $3\frac{5}{6}$.

$8\frac{3}{10}$ [2.4C]

9. What is $4\frac{5}{9}$ less than $10\frac{1}{6}$?

$5\frac{11}{18}$ [2.5C]

10. Multiply: $\frac{11}{12} \times 3\frac{1}{11}$

$2\frac{5}{6}$ [2.6B]

11. Find the quotient of $3\frac{1}{3}$ and $\frac{5}{7}$.

$4\frac{2}{3}$ [2.7B]

12. Simplify: $\left(\frac{2}{5} + \frac{3}{4}\right) \div \frac{3}{2}$

$\frac{23}{30}$ [2.8C]

13. Write 4.0709 in words.

Four and seven hundred nine ten-thousandths [3.1A]

14. Round 2.09762 to the nearest hundredth.

2.10 [3.1B]

15. Divide: $8.09\overline{)16.0976}$
Round to the nearest thousandth.

1.990 [3.5A]

16. Convert $0.06\frac{2}{3}$ to a fraction.

$\frac{1}{15}$ [3.6B]

17. Write the comparison 25 miles to 200 miles as a ratio in simplest form.
$\frac{1}{8}$ [4.1A]

18. Write "87 cents for 6 bars of soap" as a rate in simplest form.
$\frac{29¢}{2\text{ bars}}$ [4.2A]

19. Write "250.5 miles on 7.5 gallons of gas" as a unit rate.
33.4 miles/gallon [4.2B]

20. Solve $\frac{40}{n} = \frac{160}{17}$. Round to the nearest hundredth.
$n = 4.25$ [4.3B]

21. A car traveled 457.6 miles in 8 hours. Find the car's speed in miles per hour.
57.2 miles/hour [4.2C]

22. Solve the proportion.
$\frac{12}{5} = \frac{n}{15}$ $n = 36$ [4.3B]

23. You had $1024 in your checking account. You then wrote checks for $192 and $88. What is your new checking account balance? $744 [1.3C]

24. Malek Khatri buys a tractor for $22,760. A down payment of $5000 is required. The balance remaining is paid in 48 equal monthly installments. What is the monthly payment? $370 [1.5D]

25. Yuko is assigned to read a book containing 175 pages. She reads $\frac{2}{5}$ of the book during Thanksgiving vacation. How many pages of the assignment remain to be read? 105 pages [2.6C]

26. A building contractor bought $2\frac{1}{3}$ acres of land for $84,000. What was the cost of each acre? $36,000 [2.7C]

27. Benjamin Eli bought a shirt for $21.79 and a tie for $8.59. He used a $50 bill to pay for the purchases. Find the amount of change. $19.62 [3.3B]

28. A college baseball player had 42 hits in 155 at-bats. Find the baseball player's batting average. (Round to three decimal places.) 0.271 [3.5B]

29. A soil conservationist estimates that a river bank is eroding at the rate of 3 inches every 6 months. At this rate, how many inches will be eroded in 50 months? 25 inches [4.3C]

30. The dosage of a medicine is $\frac{1}{2}$ ounce for every 50 pounds of body weight. How many ounces of this medication are required for a person who weighs 160 pounds? 1.6 ounces [4.3C]

Chapter 

Percents

Objectives

Section 5.1

To write a percent as a fraction or a decimal

To write a fraction or decimal as a percent

Section 5.2

To find the amount when the percent and the base are given

To solve application problems

Section 5.3

To find the percent when the base and amount are given

To solve application problems

Section 5.4

To find the base when the percent and amount are given

To solve application problems

Section 5.5

To solve percent problems using proportions

To solve application problems

Percent Symbol

The idea of using percent dates back many hundreds of years. Percents are used in business for all types of consumer and business loans, in chemistry to measure the percent concentration of an acid, in economics to measure the increases or decreases of the consumer price index (CPI), and in many other areas that affect our daily lives.

The word *percent* comes from the Latin phrase *per centum,* which means "by the hundred." The symbol that is used today for percent is %, but this was not always the symbol.

The present symbol apparently is a result of abbreviations for the word "percent."

One abbreviation was p. cent; later, p. 100 and p. $\overset{o}{c}$ were used.

From p. $\overset{o}{c}$, the abbreviation changed to $p\frac{o}{o}$ around the 17th century. This probably was a result of continual writing of p. $\overset{o}{c}$ and the eventual closing of the "c" to make an "o." By the 19th century, the "p" in front of the symbol $p\frac{o}{o}$ was no longer written. The bar that separated the o's became a slash, and the modern symbol % became widely used.

5.1 Introduction to Percents

Objective A To write a percent as a fraction or a decimal

Percent means "parts of 100." In the figure at the right, there are 100 parts. Because 13 of the 100 parts are shaded, 13% of the figure is shaded.

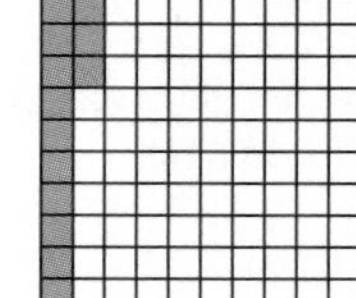

In most applied problems involving percents, it is necessary either to rewrite a percent as a fraction or a decimal or to rewrite a fraction or a decimal as a percent.

INSTRUCTOR NOTE

Example 2 and You Try It 2 are difficult for students. Here is an additional example that can be used in class.

Write $12\frac{1}{2}\%$ as a fraction.

Solution:

$$12\frac{1}{2}\% = 12\frac{1}{2} \times \frac{1}{100} = \frac{25}{2} \times \frac{1}{100} = \frac{1}{8}$$

To write a percent as a fraction, remove the percent sign and multiply by $\frac{1}{100}$.

$$13\% = 13 \times \frac{1}{100} = \frac{13}{100}$$

To rewrite a percent as a decimal, remove the percent sign and multiply by 0.01.

$$13\% = 13 \times 0.01 = 0.13$$

Move the decimal point two places to the left. Then remove the percent sign.

Example 1	Write 120% as a fraction and as a decimal.	**You Try It 1**	Write 125% as a fraction and as a decimal.
Solution	$120\% = 120 \times \frac{1}{100} = \frac{120}{100} = 1\frac{1}{5}$ $120\% = 120 \times 0.01 = 1.2$ Note that percents larger than 100 are greater than 1.	**Your solution**	$1\frac{1}{4}$, 1.25
Example 2	Write $16\frac{2}{3}\%$ as a fraction.	**You Try It 2**	Write $33\frac{1}{3}\%$ as a fraction.
Solution	$16\frac{2}{3}\% = 16\frac{2}{3} \times \frac{1}{100} = \frac{50}{3} \times \frac{1}{100} = \frac{50}{300} = \frac{1}{6}$	**Your solution**	$\frac{1}{3}$
Example 3	Write 0.5% as a decimal.	**You Try It 3**	Write 0.25% as a decimal.
Solution	$0.5\% = 0.5 \times 0.01 = 0.005$	**Your solution**	0.0025

Solutions on p. A25

Objective B To write a fraction or a decimal as a percent

INSTRUCTOR NOTE

Students will ask whether to write their answers as fractions or decimals. As a general rule, if the fraction has a terminating decimal, the answer is written in decimal form. If the answer is a repeating decimal, the answer is written as a fraction.

A fraction or a decimal can be written as a percent by multiplying by 100%.

➡ Write $\frac{3}{8}$ as a percent.

$$\frac{3}{8} = \frac{3}{8} \times 100\% = \frac{3}{8} \times \frac{100}{1}\% = \frac{300}{8}\% = 37\frac{1}{2}\% \text{ or } 37.5\%$$

➡ Write 0.37 as a percent.

$0.37 = 0.37 \times 100\% = 37\%$

Move the decimal point two places to the right. Then write the percent sign.

Example 4 Write 0.015 as a percent.

Solution $0.015 = 0.015 \times 100\%$
$= 1.5\%$

You Try It 4 Write 0.048 as a percent.

Your solution 4.8%

Example 5 Write 2.15 as a percent.

Solution $2.15 = 2.15 \times 100\% = 215\%$

You Try It 5 Write 3.67 as a percent.

Your solution 367%

Example 6 Write $0.33\frac{1}{3}$ as a percent.

Solution $0.33\frac{1}{3} = 0.33\frac{1}{3} \times 100\%$
$= 33\frac{1}{3}\%$

You Try It 6 Write $0.62\frac{1}{2}$ as a percent.

Your solution $62\frac{1}{2}\%$

Example 7 Write $\frac{2}{3}$ as a percent. Write the remainder in fractional form.

Solution $\frac{2}{3} = \frac{2}{3} \times 100\% = \frac{200}{3}\%$
$= 66\frac{2}{3}\%$

You Try It 7 Write $\frac{5}{6}$ as a percent. Write the remainder in fractional form.

Your solution $83\frac{1}{3}\%$

Example 8 Write $2\frac{2}{7}$ as a percent. Round to the nearest tenth.

Solution $2\frac{2}{7} = \frac{16}{7} = \frac{16}{7} \times 100\%$
$= \frac{1600}{7}\% \approx 228.6\%$

You Try It 8 Write $1\frac{4}{9}$ as a percent. Round to the nearest tenth.

Your solution 144.4%

Solutions on p. A25

5.1 Exercises

Objective A

Write as a fraction and as a decimal.

1. 25%
$\frac{1}{4}$, 0.25

2. 40%
$\frac{2}{5}$, 0.40

3. 130%
$1\frac{3}{10}$, 1.30

4. 150%
$1\frac{1}{2}$, 1.50

5. 100%
1, 1.00

6. 87%
$\frac{87}{100}$, 0.87

7. 73%
$\frac{73}{100}$, 0.73

8. 45%
$\frac{9}{20}$, 0.45

9. 383%
$3\frac{83}{100}$, 3.83

10. 425%
$4\frac{1}{4}$, 4.25

11. 70%
$\frac{7}{10}$, 0.70

12. 55%
$\frac{11}{20}$, 0.55

13. 88%
$\frac{22}{25}$, 0.88

14. 64%
$\frac{16}{25}$, 0.64

15. 32%
$\frac{8}{25}$, 0.32

16. 18%
$\frac{9}{50}$, 0.18

Write as a fraction.

17. $66\frac{2}{3}\%$
$\frac{2}{3}$

18. $12\frac{1}{2}\%$
$\frac{1}{8}$

19. $83\frac{1}{3}\%$
$\frac{5}{6}$

20. $3\frac{1}{8}\%$
$\frac{1}{32}$

21. $11\frac{1}{9}\%$
$\frac{1}{9}$

22. $\frac{3}{8}\%$
$\frac{3}{800}$

23. $45\frac{5}{11}\%$
$\frac{5}{11}$

24. $15\frac{3}{8}\%$
$\frac{123}{800}$

25. $4\frac{2}{7}\%$
$\frac{3}{70}$

26. $5\frac{3}{4}\%$
$\frac{23}{400}$

27. $6\frac{2}{3}\%$
$\frac{1}{15}$

28. $8\frac{2}{3}\%$
$\frac{13}{150}$

Write as a decimal.

29. 6.5%
0.065

30. 12.3%
0.123

31. 0.55%
0.0055

32. 2%
0.02

33. 8.25%
0.0825

34. 5.05%
0.0505

35. 6.75%
0.0675

36. 3.08%
0.0308

37. 0.45%
0.0045

38. 6.4%
0.064

39. 80.4%
0.804

40. 16.7%
0.167

Objective B

Write as a percent.

41. 0.16
16%

42. 0.73
73%

43. 0.05
5%

44. 0.13
13%

45. 0.01
1%

46. 0.95
95%

47. 0.70
70%

48. 1.07
107%

49. 1.24
124%

50. 2.07
207%

51. 0.004
0.4%

52. 0.37
37%

53. 0.006
0.6%

54. 1.012
101.2%

55. 3.106
310.6%

56. 0.12
12%

Write as a percent. Round to the nearest tenth.

57. $\frac{27}{50}$
54%

58. $\frac{37}{100}$
37%

59. $\frac{1}{3}$
33.3%

60. $\frac{2}{5}$
40%

61. $\frac{5}{8}$ 62.5% **62.** $\frac{1}{8}$ 12.5% **63.** $\frac{1}{6}$ 16.7% **64.** $1\frac{1}{2}$ 150%

65. $\frac{7}{40}$ 17.5% **66.** $1\frac{2}{3}$ 166.7% **67.** $1\frac{7}{9}$ 177.8% **68.** $\frac{7}{8}$ 87.5%

Write as a percent. Write the remainder in fractional form.

69. $\frac{15}{50}$ 30% **70.** $\frac{12}{25}$ 48% **71.** $\frac{7}{30}$ $23\frac{1}{3}\%$ **72.** $\frac{1}{3}$ $33\frac{1}{3}\%$

73. $2\frac{3}{8}$ $237\frac{1}{2}\%$ **74.** $1\frac{2}{3}$ $166\frac{2}{3}\%$ **75.** $2\frac{1}{6}$ $216\frac{2}{3}\%$ **76.** $\frac{7}{8}$ $87\frac{1}{2}\%$

APPLYING THE CONCEPTS

77. Determine whether the statement is true or false. If the statement is false, give an example to show that the statement is false.
a. Multiplying a number by a percent always decreases the number. False. 4(200%) = 8
b. Dividing by a percent always increases the number. False. $\frac{4}{200\%} = 2$
c. The word *percent* means "per hundred." True
d. A percent is always less than one. False. 125% = 1.25

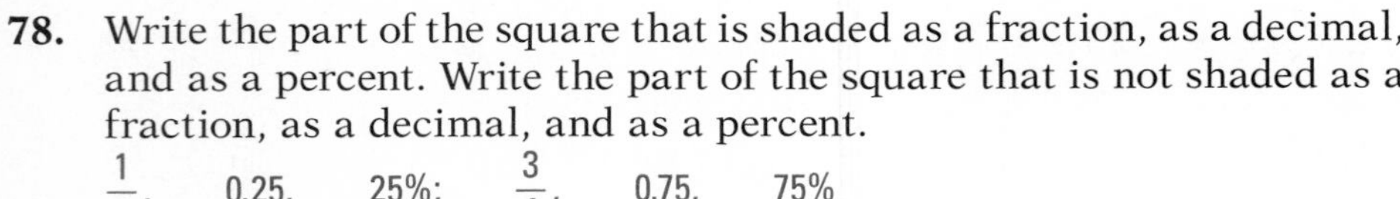

78. Write the part of the square that is shaded as a fraction, as a decimal, and as a percent. Write the part of the square that is not shaded as a fraction, as a decimal, and as a percent.
$\frac{1}{4}$, 0.25, 25%; $\frac{3}{4}$, 0.75, 75%

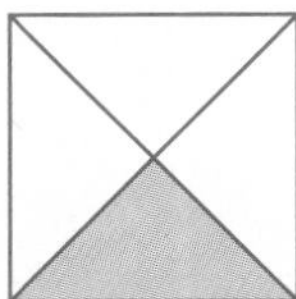

79. [W] Explain in your own words how to change a percent to a decimal and a decimal to a percent. Answers will vary

80. A sale on computers advertised $\frac{1}{3}$ off the regular price. What percent of the regular price does this represent? $33\frac{1}{3}\%$

81. A suit was priced at 50% off the regular price. What fraction of the regular price does this represent? $\frac{1}{2}$

82. If $\frac{2}{5}$ of the population voted in an election, what percent of the population did not vote? 60%

83. Is $\frac{1}{2}\%$ the same as 0.5 If not, what is the difference between 0.5 and the decimal equivalent of $\frac{1}{2}\%$? No. 0.495

84. Is 9.4% the same as $9\frac{2}{5}\%$? If not, what is the difference between the decimal equivalent of 9.4% and $9\frac{2}{5}\%$? Yes

85. How can you recognize a fraction that represents a number that is less than 1%? The fraction is less than $\frac{1}{100}$.

5.2 Percent Equations: Part I

Objective A To find the amount when the percent and base are given

A real estate broker receives a payment that is 4% of an \$85,000 sale. To find the amount the broker receives requires answering the question "4% of \$85,000 is what?"

This sentence can be written using mathematical symbols and then solved for the unknown number.

4%	of	\$85,000	is	what?
↓	↓	↓	↓	↓
percent 4%	×	base \$85,000	=	amount n
0.04	×	\$85,000	=	n
		\$3400	=	n

of is written as × (times)
is is written as = (equals)
what is written as n (the unknown number)

Note that the percent is written as a decimal.

The broker receives a payment of \$3400.

INSTRUCTOR NOTE
Effective use of the percent equation is one of the most important skills a student can acquire. This section and the next two sections are devoted to solving this equation. The last section in the chapter, Section 5.5, allows you the option of teaching this method by using proportions.

The solution was found by solving the basic percent equation for amount.

The Basic Percent Equation

Percent × base = amount

In most cases, the percent is written as a decimal before the basic percent equation is solved. However, some percents are more easily written as a fraction than as a decimal. For example,

$33\frac{1}{3}\% = \frac{1}{3}$ $66\frac{2}{3}\% = \frac{2}{3}$ $16\frac{2}{3}\% = \frac{1}{6}$ $83\frac{1}{3}\% = \frac{5}{6}$

Example 1 Find 5.7% of 160.

Solution $n = 0.057 \times 160$
$n = 9.12$

Note that the words "what is" are missing from the problem but are implied by the word "Find."

You Try It 1 Find 6.3% of 150.

Your solution $n = 9.45$

Example 2 What is $33\frac{1}{3}\%$ of 90?

Solution $n = \frac{1}{3} \times 90$
$n = 30$

You Try It 2 What is $16\frac{2}{3}\%$ of 66?

Your solution $n = 11$

Solutions on p. A26

Objective B To solve application problems

Solving percent problems requires identifying the three elements of the basic percent equation. Recall that these three parts are percent, base, and amount. Usually the base follows the phrase "percent of."

The table at the right shows the numbers of visitors to the Rocky Mountains in 1985 from selected countries. By 1992, travel to the Rocky Mountains from Japan had increased by 53%, from Germany by 187%, from the U.K. by 148%, and from Australia by 119%.

	1985
Japan	160,094
Germany	134,920
U.K.	129,986
Australia	56,295

➡ Find the number of tourists who came to the Rocky Mountains from Germany in 1992.

Strategy
Write and solve the basic percent equation to find the increase in the number of tourists. The percent is 187%; the base is 134,920.

Add the increase to the number of tourists in 1985.

Solution Percent · base = amount

$187\% \times 134{,}920 = n$

$1.87 \times 134{,}920 = n$

$252{,}300 \approx n$

$252{,}300 + 134{,}920 = 387{,}220$

There were approximately 387,220 tourists from Germany who visited the Rocky Mountains in 1992.

Example 3
A quality control inspector found that 1.2% of 2500 telephones inspected were defective. How many telephones inspected were not defective?

Strategy
To find the number of nondefective phones:

- Find the number of defective phones. Write and solve a basic percent equation using n to represent the number of defective phones (amount). The percent is 1.2% and the base is 2500.
- Subtract the number of defective phones from the number of phones inspected (2500).

Solution $1.2\% \times 2500 = n$

$0.012 \times 2500 = n$

$30 = n$ defective phones

$2500 - 30 = 2470$

2470 telephones were not defective.

You Try It 3
An electrician's hourly wage was \$13.50 before an 8% raise. What is the new hourly wage?

Your strategy

Your solution \$14.58

Solution on p. A26

5.2 Exercises

Objective A

Solve.

1. 8% of 100 is what? 8
2. 16% of 50 is what? 8
3. 27% of 40 is what? 10.8
4. 52% of 95 is what? 49.4
5. 0.05% of 150 is what? 0.075
6. 0.075% of 625 is what? 0.46875
7. 125% of 64 is what? 80
8. 210% of 12 is what? 25.20
9. Find 10.7% of 485. 51.895
10. Find 12.8% of 625. 80
11. What is 0.25% of 3000? 7.5
12. What is 0.06% of 250? 0.15
13. 80% of 16.25 is what? 13
14. 26% of 19.5 is what? 5.07
15. What is $1\frac{1}{2}$% of 250? 3.75
16. What is $5\frac{3}{4}$% of 65? 3.7375
17. $16\frac{2}{3}$% of 120 is what? 20
18. $83\frac{1}{3}$% of 246 is what? 205
19. What is $33\frac{1}{3}$% of 630? 210
20. What is $66\frac{2}{3}$% of 891? 594
21. Which is larger: 5% of 95, or 75% of 6? 5% of 95
22. Which is larger: 112% of 5, or 0.45% of 800? 112% of 5
23. Which is larger: 82% of 16, or 20% of 65? 82% of 16
24. Which is smaller: 15% of 80, or 95% of 15? 15% of 80
25. Which is larger: 22% of 120, or 84% of 32? 84% of 32
26. Which is smaller: 2% of 1500, or 72% of 40? 72% of 40
27. Find 31.294% of 82,460. 25,805.0324
28. Find 123.94% of 275,976. 342,044.65

Objective B Application Problems

29. Natasha Gomez receives a salary of $2240 per month. Of this amount 18% is deducted for income tax. Find the amount deducted for income tax. $403.20

30. In a city election for mayor, the successful candidate received 53% of the 385,000 votes cast. How many of the votes did the successful candidate receive? 204,050 votes

31. Hidden Valley Antique Emporium expects to receive $16\frac{2}{3}\%$ of the shop's sales as profit. What is the profit in a month when the total sales are $24,000? $4000

32. A student survey in a community college found that $33\frac{1}{3}\%$ of its 7500 beginning students did not complete their first year of college. How many of the beginning students did not complete the first year of college? 2500 students

33. An RV sales center offers a 7% rebate on the Roadtrek model. What rebate would a buyer receive on a Roadtrek that costs $24,000? $1680

34. A farmer is given an income tax credit of 10% of the cost of farm machinery. What tax credit would the farmer receive on farm equipment that cost $85,000? $8500

35. A sales tax of 6% of the cost of a car was added to the purchase price of $9500.
    a. How much was the sales tax? $570
    b. What is the total cost of the car including sales tax? $10,070

36. During the packaging process for oranges, spoiled oranges are discarded by an inspector. In 1 day an inspector found that 4.8% of the 20,000 pounds of oranges inspected were spoiled.
    a. How many pounds of oranges were spoiled? 960 pounds
    b. How many pounds of oranges were not spoiled? 19,040 pounds

37. Funtimes Amusement Park has 550 employees and must hire an additional 22% for the vacation season. What is the total number of employees needed for the vacation season? 671 employees

APPLYING THE CONCEPTS

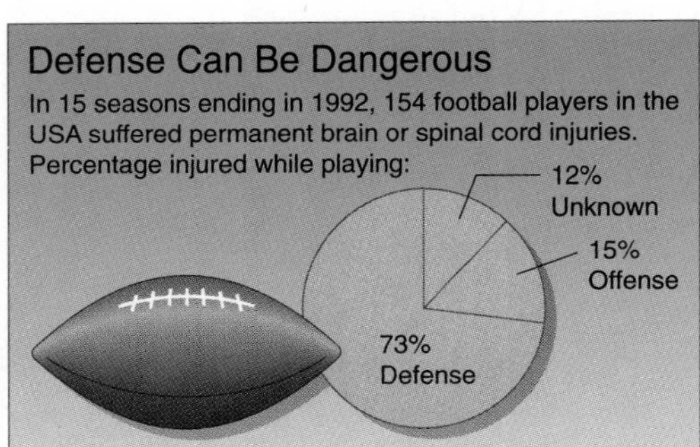

The graph at the right shows where permanent injuries to 154 football players occurred. Use this graph for Exercises 38 and 39.

38. How many permanent injuries were obtained by playing defense? 112

39. How many permanent injuries were obtained by playing offense? 23

40. According to the U.S. Census Bureau, 23.4% of California's population of 29,760,000 people had a college degree or higher. According to this information, how many Californians had a college degree or higher? 6,963,840 people

41. According to statistics on late-night television shows, David Letterman and Ted Koppel each have 4.7 million viewers on a given night. A breakdown by age shows that 41% of Letterman's viewers are between 18 and 34 years old, whereas 25% of Koppel's viewers are in the same age group. What is the difference between the number of Letterman viewers in this age group and the number of Koppel viewers in this age group? 752,000 viewers

42. A chef purchased a new knife for $37 and a stock pot for $55. If the sales tax is 7.75% of the total, how much change did the chef receive from a $100 bill? $0.87

43. A car company increased the average price of its new models by 2.6%. If the average price of a car was $7856, what was the amount of the increase? $204.26

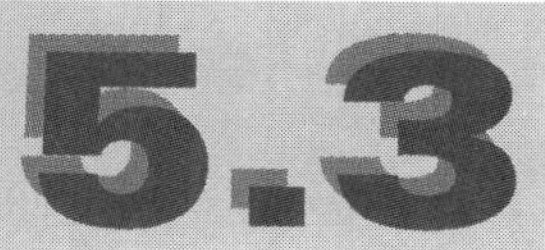

Percent Equations: Part II

Objective A *To find the percent when the base and amount are given*

INSTRUCTOR NOTE

Although not a foolproof method, the base in the basic percent equation will generally follow the phrase "percent of" in application problems. Having students find this phrase will help them identify all the components of the percent equation.

A recent promotional game at a grocery store listed the probability of winning a prize as "1 chance in 2." A percent can be used to describe the chance of winning. This requires answering the question "What percent of 2 is 1?"

The chance of winning can be found by solving the basic percent equation for *percent.*

What percent of 2 is 1

$$\boxed{\begin{matrix}\text{percent}\\ n\end{matrix}} \times \boxed{\begin{matrix}\text{base}\\ 2\end{matrix}} = \boxed{\begin{matrix}\text{amount}\\ 1\end{matrix}}$$

$$\begin{aligned} n \times 2 &= 1 \\ n &= 1 \div 2 \\ n &= 0.5 \\ n &= 50\% \end{aligned}$$

• **The solution must be written as a percent to answer the question.**

There is a 50% chance of winning a prize.

POINT OF INTEREST

Inflation increased by approximately 275% between 1971 and 1994. The price of a first class stamp has increased by 300%, not much more than inflation.

Example 1 What percent of 40 is 30?

Solution

$$\begin{aligned} n \times 40 &= 30 \\ n &= 30 \div 40 \\ n &= 0.75 \\ n &= 75\% \end{aligned}$$

You Try It 1 What percent of 32 is 16?

Your solution $n = 50\%$

Example 2 What percent of 12 is 27?

Solution

$$\begin{aligned} n \times 12 &= 27 \\ n &= 27 \div 12 \\ n &= 2.25 \\ n &= 225\% \end{aligned}$$

You Try It 2 What percent of 15 is 48?

Your solution $n = 320\%$

Example 3 25 is what percent of 75?

Solution

$$\begin{aligned} 25 &= n \times 75 \\ 25 \div 75 &= n \\ 0.33\frac{1}{3} &= n \\ 33\frac{1}{3}\% &= n \end{aligned}$$

You Try It 3 30 is what percent of 45?

Your solution $66\frac{2}{3}\% = n$

Solutions on p. A26

Objective B To solve application problems

To solve percent problems, remember that it is necessary to identify the percent, base, and amount. Usually the base follows the phrase "percent of."

Example 4
The Kaminski family has an income of \$1500 each month and makes a car payment of \$180 a month. What percent of the monthly income is the car payment?

Strategy
To find what percent of the income the car payment is, write and solve the basic percent equation using n to represent the unknown percent. The base is \$1500 and the amount is \$180.

Solution

$$n \times \$1500 = \$180$$
$$n = \$180 \div \$1500$$
$$n = 0.12 = 12\%$$

The car payment is 12% of the monthly income.

You Try It 4
Tomo Nagata had an income of \$20,000 and paid \$3000 in income tax. What percent of the income is the income tax?

Your strategy

Your solution 15%

Example 5
Claudia Sloan missed 25 questions out of 200 on a state nursing license exam. What percent of the questions did she answer correctly?

Strategy
To find what percent of the questions she answered correctly:

- Find the number of questions the nurse answered correctly (200 − 25).
- Write and solve a basic percent equation using n to represent the unknown percent. The base is 200 and the amount is the number of questions answered correctly.

Solution 200 − 25 = 175 questions answered correctly.

$$n \times 200 = 175$$
$$n = 175 \div 200$$
$$n = 0.875 = 87.5\%$$

Claudia answered 87.5% of the questions correctly.

You Try It 5
A survey of 1000 people showed that 667 people favored Minna Oliva for governor of the state. What percent of the people surveyed did not favor her?

Your strategy

Your solution 33.3%

Solutions on p. A27

5.3 Exercises

Objective A

Solve.

1. What percent of 75 is 24?
32%

2. What percent of 80 is 20?
25%

3. 15 is what percent of 90?
$16\frac{2}{3}\%$

4. 24 is what percent of 60?
40%

5. What percent of 12 is 24?
200%

6. What percent of 6 is 9?
150%

7. What percent of 16 is 6?
37.5%

8. What percent of 24 is 18?
75%

9. 18 is what percent of 100?
18%

10. 54 is what percent of 100?
54%

11. 5 is what percent of 2000?
0.25%

12. 8 is what percent of 2500?
0.32%

13. What percent of 6 is 1.2?
20%

14. What percent of 2.4 is 0.6?
25%

15. 16.4 is what percent of 4.1?
400%

16. 5.3 is what percent of 50?
10.6%

17. 1 is what percent of 40?
2.5%

18. 0.3 is what percent of 20?
1.5%

19. What percent of 48 is 18?
37.5%

20. What percent of 11 is 88?
800%

21. What percent of 2800 is 7?
0.25%

22. What percent of 400 is 12?
3%

23. 4.2 is what percent of 175?
2.4%

24. 41.79 is what percent of 99.5?
42%

25. What percent of 86.5 is 8.304?
9.6%

26. What percent of 1282.5 is 2.565?
0.2%

Objective B Application Problems

27. A company spends \$4500 of its \$90,000 budget for advertising. What percent of the budget is spent for advertising? 5%

28. Last month a thrift shop had an income of \$2812.50. The thrift store pays \$900 a month for rent. What percent of last month's income was spent for rent? 32%

29. Kodak stock is trading at $\$62\frac{1}{2}$ and pays a dividend of \$2.00. What percent of the stock price is the dividend? 3.2%

30. A home was valued at $180,000 in 1990. The value of the home in 1993 was $162,000. What percent of the 1990 value is the 1993 value? 90%

31. A television survey of 4000 families found that 2500 liked a new TV show. What percent of the families surveyed liked the new TV show? 62.5%

32. A house in Phoenix was bought for $124,000. Three years later the house was sold for $155,000.
   a. Find the increase in price. $31,000
   b. What percent of the original price is the increase? 25%

33. A 3.5-pound beef roast contains 0.7 pound of fat.
   a. What percent of the beef roast is fat? 20%
   b. At the same rate, how many pounds of fat would there be in a side of beef weighing 550 pounds? 110 pounds

34. To receive a license to sell insurance, an insurance account executive must answer correctly 70% of the 250 questions on a test. Nicholas Mosley answered 177 questions correctly. Did he pass the test? Yes

35. A test of the breaking strength of concrete slabs for freeway construction found that 3 of the 200 tested did not meet safety requirements. What percent of the slabs did meet safety requirements? 98.5%

APPLYING THE CONCEPTS

36. Make a table showing the decimal and percent equivalents of $\frac{1}{3}$, $\frac{5}{6}$, $\frac{3}{8}$, and $\frac{5}{8}$.

$\frac{1}{3}$, $0.333\overline{3}$, 33.3%
$\frac{5}{6}$, $0.833\overline{3}$, 83.3%
$\frac{3}{8}$, 0.375, 37.5%
$\frac{5}{8}$, 0.625, 62.5%

37. [W] Write a short essay on the use of percents in our everyday life. Answers will vary

38. Borland International offers its customers a $50 rebate on the purchase of a new computer program costing $189.95 before the rebate. If you purchase this program, what percent of your final cost is, to the nearest tenth of a percent, the rebate? 35.7%

The graph at the right shows how the federal government spent $1.4 trillion in 1993. Use this graph for Exercises 39 and 40.

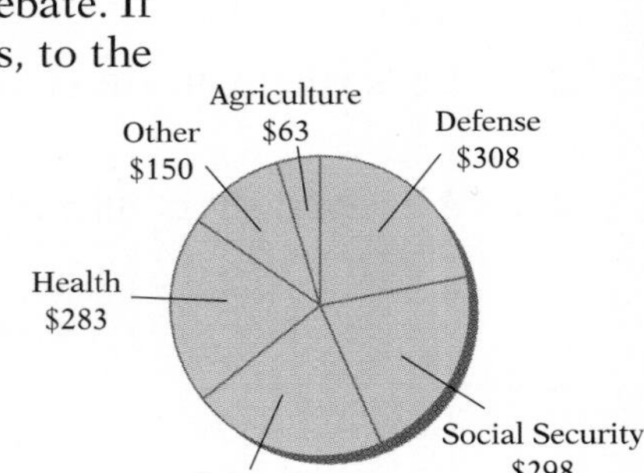

Federal Budget (in billions)

39. What percent of the total amount was spent for interest on the public debt? Round to the nearest tenth of a percent. 20.9%

40. What percent of the total amount was spent for Social Security? Round to the nearest tenth of a percent. 21.3%

41. Public utility companies will provide consumers with an analysis of their energy bills. For one customer, it was determined that $76 of a total bill of $134 was spent for home heating. What percent of the total bill was home heating? 56.7%

42. The original cost of the Statue of Liberty was approximately $24,000. The cost to refurbish the statue today is approximately $780,000. What percent of the original cost is the cost to refurbish the Statue of Liberty? 3250%

5.4 Percent Equations: Part III

Objective A To find the base when the percent and amount are given

POINT OF INTEREST
The sales of CD-ROM players is projected to increase to 22 million by 1997. That is a 900% increase over the 2.2 million units that were sold in 1992.

Each year an investor receives a payment that equals 12% of the value of an investment. This year that payment amounted to \$480. To find the value of the investment this year, the investor must answer the question "12% of what value is \$480?"

The value of the investment can be found by solving the basic percent equation for the base.

12%	of	what	is	\$480
↓	↓	↓	↓	↓
percent 12%	×	base n	=	amount \$480

$$0.12 \times n = \$480$$
$$n = \$480 \div 0.12$$
$$n = \$4000$$

This year the investment is worth \$4000.

Example 1 18% of what is 900?

Solution $0.18 \times n = 900$
$n = 900 \div 0.18$
$n = 5000$

You Try It 1 86% of what is 215?

Your solution $n = 250$

Example 2 30 is 1.5% of what?

Solution $30 = 0.015 \times n$
$30 \div 0.015 = n$
$2000 = n$

You Try It 2 15 is 2.5% of what?

Your solution $n = 600$

Example 3 $33\frac{1}{3}\%$ of what is 7?

Solution $\frac{1}{3} \times n = 7$ • Note the percent is written as a fraction.
$n = 7 \div \frac{1}{3}$
$n = 21$

You Try It 3 $16\frac{2}{3}\%$ of what is 5?

Your solution $n = 30$

Solutions on p. A27

Objective B To solve application problems

To solve percent problems, it is necessary to identify the percent, base, and amount. Usually the base follows the phrase "percent of."

Example 4
A business office bought a used copy machine for $450, which was 75% of the original cost. What was the original cost of the copier?

Strategy
To find the original cost of the copier, write and solve the basic percent equation using n to represent the original cost (base). The percent is 75% and the amount is $450.

Solution

$$75\% \times n = \$450$$
$$0.75 \times n = \$450$$
$$n = \$450 \div 0.75$$
$$n = \$600$$

The original cost of the copier was $600.

You Try It 4
A used car has a value of $3876, which is 51% of the car's original value. What was the car's original value?

Your strategy

Your solution $7600

Example 5
A carpenter's wage this year is $19.80 per hour, which is 110% of last year's wage. What was the increase in the hourly wage over last year?

Strategy
To find the increase in the hourly wage over last year:

- Find last year's wage. Write and solve the basic percent equation using n to represent last year's wage (base). The percent is 110% and the amount is $19.80.
- Subtract last year's wage from this year's wage ($19.80).

Solution

$$110\% \times n = \$19.80$$
$$1.10 \times n = \$19.80$$
$$n = \$19.80 \div 1.10$$
$$n = \$18.00. \quad \bullet \text{ last year's wage}$$

$$\$19.80 - \$18.00 = \$1.80$$

The increase in the hourly wage was $1.80.

You Try It 5
Chang's Sporting Goods has a tennis racket on sale for $44.80, which is 80% of the original price. What is the difference between the original price and the sale price?

Your strategy

Your solution $11.20

Solutions on p. A28

5.4 Exercises

Objective A

Solve.

1. 12% of what is 9?
75

2. 38% of what is 171?
450

3. 8 is 16% of what?
50

4. 54 is 90% of what?
60

5. 10 is 10% of what?
100

6. 37 is 37% of what?
100

7. 30% of what is 25.5?
85

8. 25% of what is 21.5?
86

9. 2.5% of what is 30?
1200

10. 10.4% of what is 52?
500

11. 125% of what is 24?
19.2

12. 180% of what is 21.6?
12

13. 18 is 240% of what?
7.5

14. 24 is 320% of what?
7.5

15. 4.8 is 15% of what?
32

16. 87.5 is 50% of what?
175

17. 25.6 is 12.8% of what?
200

18. 45.014 is 63.4% of what?
71

19. 0.7% of what is 0.56?
80

20. 0.25% of what is 1?
400

21. 30% of what is 2.7?
9

22. 78% of what is 3.9?
5

23. 84 is $16\frac{2}{3}$% of what?
504

24. 120 is $33\frac{1}{3}$% of what?
360

25. $66\frac{2}{3}$% of what is 72?
108

26. $83\frac{1}{3}$% of what is 13.5?
16.2

27. 6.59% of what is 469.35?
7122.1547

28. 182.3% of what is 46,253?
25,371.914

Objective B *Application Problems*

29. A mechanic estimates that the brakes of an RV still have 6000 miles of wear. This amounts to 12% of the estimated safe-life use of the brakes. What is the estimated life of the brakes? 50,000 miles

30. A used mobile home was purchased for $18,000. This amount was 64% of the cost of a new mobile home. What was the cost of a new mobile home? $28,125

31. Redwood City has a population of 42,000. This is 75% of what the population was 5 years ago. What was the city's population 5 years ago? 56,000 people

32. Solitude Ski Resort had a snowfall of 198 inches during 1993. This is 110% of the snowfall in 1992. What was the snowfall in 1992? 180 inches

33. Circuit City advertised a scientific calculator for $55.80. This amount was 120% of the cost at the May Company. What is the price of the calculator at the May Company? $46.50

34. A salesperson received a commission of $820 for selling a car. This was 5% of the selling price of the car. What was the selling price of the car? $16,400

35. During a quality control test, Micronics found that 24 computer boards were defective. This amount was 0.8% of the computer boards tested.
a. How many computer boards were tested? 3000 boards
b. How many computer boards tested were not defective? 2976 boards

36. Of the calls a directory assistance operator received, 441 were requests for telephone numbers listed in the current directory. This accounted for 98% of the calls for assistance that the operator received.
a. How many calls did the operator receive? 450 calls
b. How many telephone numbers requested were not listed in the current directory? 9 calls

APPLYING THE CONCEPTS

The table at the right contains nutrition information about a breakfast cereal. Solve Exercises 37 and 38 with information taken from this table.

NUTRITION INFORMATION

SERVING SIZE: 1.4 OZ WHEAT FLAKES WITH 0.4 OZ. RAISINS: 39.4 g. ABOUT 1/2 CUP

SERVINGS PER PACKAGE: 14

PERCENTAGE OF U.S. RECOMMENDED DAILY ALLOWANCES (U.S. RDA)

	CEREAL & RAISINS	WITH 1/2 CUP VITAMINS A & D SKIM MILK
PROTEIN	4	15
VITAMIN A	15	20
VITAMIN C	**	2
THIAMIN	25	30
RIBOFLAVIN	25	35
NIACIN	25	35
CALCIUM	**	15
IRON	100	100
VITAMIN D	10	25
VITAMIN B_6	25	25
FOLIC ACID	25	25
VITAMIN B_{12}	25	30
PHOSPHOROUS	10	15
MAGNESIUM	10	20
ZINC	25	30
COPPER	2	4

* 2% MILK SUPPLIES AN ADDITIONAL 20 CALORIES. 2 g FAT, AND 10 mg CHOLESTEROL.

** CONTAINS LESS THAN 2% OF THE U.S. RDA OF THIS NUTRIENT

37. The recommended daily amount of protein for an 80-pound child is 44 grams. Find the amount of protein in one serving of this cereal with skim milk. 6.6 grams

38. The recommended daily amount of zinc for an adult male is 18 mg. Find the amount of zinc in one serving of cereal with skim milk. 5.4 mg

39. Increase a number by 10%. Now decrease the result by 10%. Is the re-
[W] sult the original number? Explain. No

40. When the Resolution Trust Corporation (RTC) takes over a failed bank, it may be necessary to sell some of the assets of the bank at less than the amount owed. As an example, a buyer may be able to purchase a building for "25 cents on the dollar." What percent of the owed price is the buyer paying? 25%

Percent Problems: Proportion Method

Objective A To solve percent problems using proportions

INSTRUCTOR NOTE

This section explains the proportion method of solving the basic percent equation. If you choose not to use this method, you can nonetheless use the exercises as a review of problems that involve percent. This will help students review the topic and recognize the different types of percent problems.

Problems that can be solved using the basic percent equation can also be solved using proportions.

The proportion method is based on writing two ratios. One ratio is the percent ratio, written as $\frac{\text{percent}}{100}$. The second ratio is the amount-to-base ratio, written as $\frac{\text{amount}}{\text{base}}$. These two ratios form the proportion

$$\frac{\textbf{percent}}{\textbf{100}} = \frac{\textbf{amount}}{\textbf{base}}$$

To use the proportion method, first identify the percent, the amount, and the base (the base usually follows the phrase "percent of").

What is 23% of 45?

$$\frac{23}{100} = \frac{n}{45}$$
$$23 \times 45 = 100 \times n$$
$$1035 = 100 \times n$$
$$1035 \div 100 = n$$
$$10.35 = n$$

What percent of 25 is 4?

$$\frac{n}{100} = \frac{4}{25}$$
$$n \times 25 = 100 \times 4$$
$$n \times 25 = 400$$
$$n = 400 \div 25$$
$$n = 16\%$$

12 is 60% of what number?

$$\frac{60}{100} = \frac{12}{n}$$
$$60 \times n = 100 \times 12$$
$$60 \times n = 1200$$
$$n = 1200 \div 60$$
$$n = 20$$

Example 1 15% of what is 7? Round to the nearest hundredth.

Solution

$$\frac{15}{100} = \frac{7}{n}$$
$$15 \times n = 100 \times 7$$
$$15 \times n = 700$$
$$n = 700 \div 15$$
$$n \approx 46.67$$

You Try It 1 26% of what is 22? Round to the nearest hundredth.

Your solution 84.62

Example 2 30% of 63 is what?

Solution

$$\frac{30}{100} = \frac{n}{63}$$
$$30 \times 63 = n \times 100$$
$$1890 = n \times 100$$
$$1890 \div 100 = n$$
$$18.90 = n$$

You Try It 2 16% of 132 is what?

Your solution 21.12

Solutions on p. A28

Objective B To solve application problems

Example 3
An antiques dealer found that 86% of the 250 items that were sold during one month sold for under $1000. How many items sold for under $1000?

Strategy
To find the number of items that sold for under $1000, write and solve a proportion using n to represent the number of items sold (amount) for less than $1000. The percent is 86% and the base is 250.

Solution

$$\frac{86}{100} = \frac{n}{250}$$
$$86 \times 250 = 100 \times n$$
$$21{,}500 = 100 \times n$$
$$21{,}500 \div 100 = n$$
$$215 = n$$

215 items sold for under $1000.

You Try It 3
Last year it snowed 64% of the 150 days of the ski season at a resort. How many days did it snow?

Your strategy

Your solution 96 days

Example 4
In a test of the strength of nylon rope, 5 pieces of the 25 pieces tested did not meet the test standards. What percent of the nylon ropes tested did meet the standards?

Strategy
To find the percent of ropes tested that met the standards:

- Find the number of ropes that met the test standards (25 − 5).
- Write and solve a proportion using n to represent the percent of ropes that met the test standards. The base is 25. The amount is the number of ropes that met the standards.

Solution 25 − 5 = 20 ropes met test standards

$$\frac{n}{100} = \frac{20}{25}$$
$$n \times 25 = 100 \times 20$$
$$n \times 25 = 2000$$
$$n = 2000 \div 25$$
$$n = 80$$

80% of the ropes tested did meet the test standards.

You Try It 4
Five ballpoint pens in a box of 200 were found to be defective. What percent of the pens were not defective?

Your strategy

Your solution 97.5%

Solutions on p. A29

5.5 Exercises

Objective A

Solve.

1. 26% of 250 is what?
65

2. What is 18% of 150?
27

3. 37 is what percent of 148?
25%

4. What percent of 150 is 33?
22%

5. 68% of what is 51?
75

6. 126 is 84% of what?
150

7. What percent of 344 is 43?
12.5%

8. 750 is what percent of 50?
1500%

9. 82 is 20.5% of what?
400

10. 2.4% of what is 21?
875

11. What is 6.5% of 300?
19.5

12. 96% of 75 is what?
72

13. 7.4 is what percent of 50?
14.8%

14. What percent of 1500 is 693?
46.2%

15. 50.5% of 124 is what?
62.62

16. What is 87.4% of 255?
222.87

17. 120% of what is 6?
5

18. 14 is 175% of what?
8

19. What is 250% of 18?
45

20. 325% of 4.4 is what?
14.3

21. 33 is 220% of what?
15

22. 160% of what is 40?
25

Objective B *Application Problems*

23. A charity organization spent $2940 for administrative expenses. This amount is 12% of the money it collected. What is the total amount of money that the club collected? $24,500

24. A manufacturer of an anti-inflammatory drug claims that the drug will be effective for 6 hours. An independent testing service determined that the drug was effective only 80% of the length of time claimed by the manufacturer. Find the length of time the drug will be effective as determined by the testing service. 4.8 hours

25. A calculator can be purchased for $28.50. This amount is 40% of the cost of the calculator 8 years ago. What was the cost of the calculator 8 years ago? $71.25

26. The Rincon Fire Department received 24 false alarms out of a total of 200 alarms received. What percent of the alarms received were false alarms? 12%

27. Enrico made errors on 5 words on a typing test. The total length of the test was 250 words. What percent of the words did he type correctly? 98%

28. It cost a mechanic $174 to repair a car. Of this cost, 55% was for labor and 45% was for parts.
a. What was the labor cost? $95.70
b. What was the total cost for parts? $78.30

29. In 1994, the National Football League increased the amount the players in the Super Bowl would receive. The winner's share was $38,000 per player, which was a $2000 increase from the previous year. What percent of the winner's share the previous year was the increase? Round to the nearest tenth of a percent. 5.6%

30. In 1994, the National Football League increased the amount the players in the Super Bowl would receive. The loser's share was $23,500 per player, which was a $5500 increase from the previous year. What percent of the previous year's loser's share was the increase? Round to the nearest tenth of a percent. 30.6%

APPLYING THE CONCEPTS

The figure at the right shows the projections of the reduced corn harvest caused by the 1993 midwestern floods. Use the figure to answer Exercises 31 and 32.

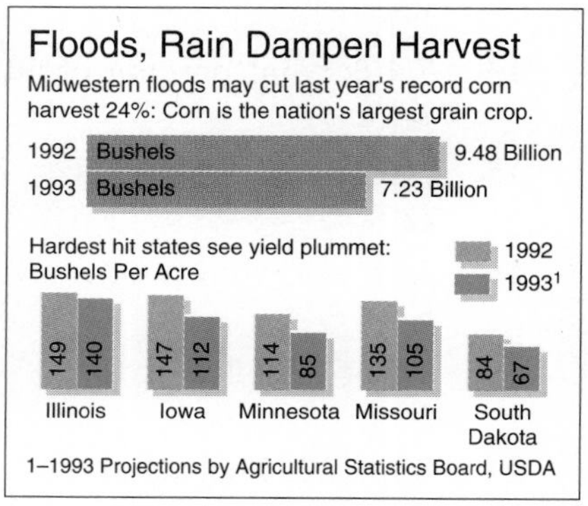

Floods, Rain Dampen Harvest
Midwestern floods may cut last year's record corn harvest 24%: Corn is the nation's largest grain crop.
1992 Bushels 9.48 Billion
1993 Bushels 7.23 Billion
Hardest hit states see yield plummet: Bushels Per Acre
1992
1993[1]
Illinois 149 140
Iowa 147 112
Minnesota 114 85
Missouri 135 105
South Dakota 84 67
1–1993 Projections by Agricultural Statistics Board, USDA

31. Find the percent decrease in the total harvest of the 5 states. Round to the nearest tenth of a percent. 19.1%

32. Which of the states shown had the largest decrease? Which had the smallest decrease? Iowa Illinois

33. A $10,000, 8% bond has a quoted price of $10,500. Look up in a business math book the meaning of *current yield*. Then find the current yield on the $10,000 bond. 7.6%

The table at the right shows the projected population increase for selected countries by 2010. Use this table for Exercises 34 and 35.

Country	*Current Population*	*Projected Increase*
Netherlands	15,200,000	1,400,000
France	56,900,000	1,700,000
Kenya	26,200,000	18,600,000
United States	260,000,000	35,500,000
Mexico	87,700,000	32,000,000

34. What percent of the Netherlands current population is the projected increase? Round to the nearest tenth of a percent. 9.2%

35. What percent of the current population of the United States is the projected increase? Round to the nearest tenth of a percent. 13.7%

36. The land area of North America is approximately 9,400,000 square miles. This represents about 16% of the total land area of the world. What is the approximate land area of the world? 58,750,000 square miles

37. A certain solution of iodine is used to study the function of the thyroid gland. Approximately every 8 days, the potency of the solution is reduced by 50% of the amount present at that time. By what percent has the potency decreased after 24 days? 87.5%

38. A certain solution of iron is used to study the formation of blood cells. Approximately every 15 hours, the potency of the solution is reduced by 50% of the amount present at that time. By what percent has the potency decreased after 60 hours? 93.75%

Project in Mathematics

Consumer Price Index

The Consumer Price Index (CPI) is a percent that is written without the percent sign. For instance, a CPI of 134.6 means 134.6%. This number means that an item that cost $100 between 1982 and 1984 (the base years) would cost $134.60 today. Determining the cost is an application of the basic percent equation.

$$\text{Base} \times \text{percent} = \text{amount}$$
$$\text{Cost in base years} \times \text{CPI} = \text{cost today}$$
$$100 \times 1.346 = 134.60$$

• 134.6% = 1.346

The table below gives the CPI for various products at the end of 1993.

Product	*CPI*
Food	144.3
Fuel oil	94.6
Medical care	195.7
Rent	116.5
Tobacco	223.1
Clothes	139.4

Exercises:

Use this table for the following exercises.

1. Of the items listed, are there any items that cost less in 1993 than they did during the base years? If so, which ones? Explain how you arrived at your answer.
2. Of the items listed, are there any items that in 1993 cost more than twice as much as they cost during the base years? If so, which ones? Explain how you arrived at your answer.
3. If it cost a family of four $122 per week for food during the base years, how much did that family spend in 1993?
4. If a raincoat cost $86 in 1993, what would a comparable raincoat have cost during the base years?

Chapter Summary

Key Words *Percent* means "parts of 100."

Essential Rules

To Write a Percent as a Fraction

To write a percent as a fraction, remove the percent sign and multiply by $\frac{1}{100}$.

To Write a Percent as a Decimal

To write a percent as a decimal, remove the percent sign and multiply by 0.01.

To Write a Decimal as a Percent

To write a decimal as a percent, multiply by 100%.

To Write a Fraction as a Percent

To write a fraction as a percent, multiply by 100%.

Basic Percent Equation Percent × base = amount

Proportion Method to Solve Percents $\frac{\text{percent}}{100} = \frac{\text{amount}}{\text{base}}$

Chapter Review

SECTION 5.1

1. Write 12% as a fraction.
$\frac{3}{25}$

2. Write $16\frac{2}{3}\%$ as a fraction.
$\frac{1}{6}$

3. Write 42% as a decimal.
0.42

4. Write 7.6% as a decimal.
0.076

5. Write 0.38 as a percent.
38%

6. Write $1\frac{1}{2}$ as a percent.
150%

SECTION 5.2

7. What is 30% of 200?
60

8. What is 7.5% of 72?
5.4

9. Find 22% of 88.
19.36

10. Find 125% of 62.
77.5

11. A company used 7.5% of its $60,000 expense budget for TV advertising. How much of the company's expense budget was spent for TV advertising?
$4500

12. Joshua purchased a video camera for $980 and paid a sales tax of 6.25% of the cost. What was the total cost of the video camera? $1041.25

SECTION 5.3

13. What percent of 20 is 30?
150%

14. 16 is what percent of 80?
20%

15. What percent of 30 is 2.2? Round to the nearest tenth of a percent.
7.3%

16. What percent of 15 is 92? Round to the nearest tenth of a percent.
613.3%

17. The stock of Advanced Board Games is trading at \$65 and paying a dividend of \$2.99. What percent of the stock price is the dividend? 4.6%

18. A house in southern California had a value of \$125,000. Two years later the house was valued at \$205,000. Find the percent increase in value during the two years. 64%

SECTION 5.4

19. 20% of what is 15?
75

20. 78% of what is 8.5? Round to the nearest tenth. 10.9

21. $66\frac{2}{3}\%$ of what is 105?
157.5

22. $16\frac{2}{3}\%$ of what is 84?
504

23. A city's population this year was 157,500, which is 225% of what the city's population was 10 years ago. What was the city's population 10 years ago? 70,000 people

24. A baseball player has a batting average of 0.315. This average is an increase of 125% over the previous year's batting average. Find the previous year's batting average. 0.252

SECTION 5.5

25. What is 62% of 320?
198.4

26. What percent of 25 is 40?
160%

27. A computer system can be purchased for \$1800. This price is 60% of what the computer cost 4 years ago. What was the cost of the computer 4 years ago? \$3000

28. Trent missed 9 out of 60 questions on a history exam. What percent of the questions did he answer correctly? 85%

Chapter Test

1. Write 97.3% as a decimal.

0.973 [5.1A]

2. Write $16\frac{2}{3}\%$ as a fraction.

$\frac{1}{6}$ [5.1A]

3. Write 0.3 as a percent.
30% [5.1B]

4. Write 1.63 as a percent.
163% [5.1B]

5. Write $\frac{3}{2}$ as a percent.

150% [5.1B]

6. Write $\frac{2}{3}$ as a percent.

$66\frac{2}{3}\%$ [5.1B]

7. What is 77% of 65?
50.05 [5.2A]

8. 47.2% of 130 is what?
61.36 [5.2A]

9. Which is larger:
7% of 120, or 76% of 13?
76% of 13 [5.2A]

10. Which is smaller:
13% of 200, or 212% of 12?
212% of 12 [5.2A]

11. A fast-food company uses 6% of its $75,000 budget for advertising. What amount of the budget is spent for advertising?
$4500 [5.2B]

12. During the packaging process for vegetables, spoiled vegetables are discarded by an inspector. In one day an inspector found that 6.4% of the 1250 pounds of vegetables were spoiled. How many pounds of vegetables were not spoiled? 1170 pounds [5.2B]

The table at the right contains nutrition information about a breakfast cereal. Solve Exercises 13 and 14 with information taken from this table.

NUTRITION INFORMATION

SERVING SIZE: 1.4 OZ WHEAT FLAKES WITH 0.4 OZ. RAISINS: 39.4 g. ABOUT 1/2 CUP
SERVINGS PER PACKAGE: 14

	CEREAL & RAISINS	WITH 1/2 CUP VITAMINS A & D SKIM MILK
CALORIES	120	180
PROTEIN, g	3	7
CARBOHYDRATE, g	28	34
FAT, TOTAL, g	1	1*
UNSATURATED, g 1		
SATURATED, g 0		
CHOLESTEROL, mg	0	0*
SODIUM, mg	125	190
POTASSIUM, mg	240	440

* 2% MILK SUPPLIES AN ADDITIONAL 20 CALORIES. 2 g FAT, AND 10 mg CHOLESTEROL.
** CONTAINS LESS THAN 2% OF THE U.S. RDA OF THIS NUTRIENT

13. The recommended amount of potassium for an adult per day is 3000 milligrams (mg). What percent, to the nearest tenth of a percent, of the daily recommended amount of potassium is provided by one serving of cereal with skim milk? 14.7%

14. The daily recommended number of calories for a 190-pound man is 2200 calories. What percent, to the nearest tenth of a percent, of the daily recommended number of calories is provided by one serving of cereal with 2% milk? 9.1%

15. The Urban Center Department Store has 125 permanent employees and must hire an additional 20 temporary employees for the holiday season. What percent of the permanent employees is the number hired as temporary employees for the holiday season? 16% [5.3B]

16. Conchita missed 7 out of 80 questions on a math exam. What percent of the questions did she answer correctly? (Round to the nearest tenth of a percent.) 91.3% [5.3B]

17. 12 is 15% of what? 80 [5.4A]

18. 42.5 is 150% of what? Round to the nearest tenth. 28.3 [5.4A]

19. A manufacturer of transistors found 384 defective transistors during a quality control study. This amount was 1.2% of the transistors tested. Find the number of transistors tested. 32,000 transistors [5.4B]

20. A new house was bought for $95,000; 5 years later the house sold for $152,000. The increase was what percent of the original price? 60% [5.4B]

21. 123 is 86% of what number? Round to the nearest tenth. 143.0 [5.5A]

22. What percent of 12 is 120? 1000% [5.5A]

23. A secretary receives a wage of $9.52 per hour. This amount is 112% of last year's salary. What is the dollar increase in the hourly wage over last year? $1.02 [5.5B]

24. A city has a population of 71,500; 10 years ago the population was 32,500. The population now is what percent of what the population was 10 years ago? 220% [5.5B]

25. The annual license fee on a car is 1.4% of the value of the car. If the license fee during a year was $91.00, what is the value of the car? $6500 [5.5B]

Cumulative Review

1. Simplify $18 \div (7 - 4)^2 + 2$.
4 [1.6B]

2. Find the LCM of 16, 24, and 30.
240 [2.1A]

3. Find the sum of $2\frac{1}{3}$, $3\frac{1}{2}$, and $4\frac{5}{8}$.
$10\frac{11}{24}$ [2.4C]

4. Subtract: $27\frac{5}{12} - 14\frac{9}{16}$
$12\frac{41}{48}$ [2.5C]

5. Multiply: $7\frac{1}{3} \times 1\frac{5}{7}$
$12\frac{4}{7}$ [2.6B]

6. What is $\frac{14}{27}$ divided by $1\frac{7}{9}$?
$\frac{7}{24}$ [2.7B]

7. Simplify: $\left(\frac{3}{4}\right)^3 \cdot \left(\frac{8}{9}\right)^2$
$\frac{1}{3}$ [2.8B]

8. Simplify: $\left(\frac{2}{3}\right)^2 - \left(\frac{3}{8} - \frac{1}{3}\right) \div \frac{1}{2}$
$\frac{13}{36}$ [2.8C]

9. Round 3.07973 to the nearest hundredth.
3.08 [3.1B]

10. Subtract: $\begin{array}{r} 3.0902 \\ -\ 1.9706 \\ \hline \end{array}$
1.1196 [3.3A]

11. Divide: $0.032\overline{)1.097}$
Round to the nearest ten-thousandth.
34.2813 [3.5A]

12. Convert $3\frac{5}{8}$ to a decimal.
3.625 [3.6A]

13. Convert 1.75 to a fraction.
$1\frac{3}{4}$ [3.6B]

14. Place the correct symbol, < or >, between the two numbers.
$\frac{3}{8} < 0.87$ [3.6C]

15. Solve the proportion $\frac{3}{8} = \frac{20}{n}$. Round to the nearest tenth.
$n = 53.3$ [4.3B]

16. Write "$76.80 earned in 8 hours" as a unit rate.
$9.60/hour [4.2B]

17. Write $18\frac{1}{3}\%$ as a fraction.
$\frac{11}{60}$ [5.1A]

18. Write $\frac{5}{6}$ as a percent.
$83\frac{1}{3}\%$ [5.1B]

19. 16.3% of 120 is what? Round to the nearest hundredth.
19.56 [5.2A]

20. 24 is what percent of 18?
$133\frac{1}{3}\%$ [5.3A]

21. 12.4 is 125% of what?
9.92 [5.4A]

22. What percent of 35 is 120? Round to the nearest tenth.
342.9% [5.5A]

23. Sergio has an income of $740 per week. One-fifth of the income is deducted for income tax payments. Find his take-home pay. $592 [2.6C]

24. Eunice bought a car for $4321, with a down payment of $1000. The balance was paid in 36 equal monthly payments. Find the monthly payment.
$92.25 [3.5B]

25. The gasoline tax is $0.19 a gallon. Find the number of gallons of gasoline used during a month in which $79.80 was paid in gasoline taxes.
420 gallons [3.5B]

26. The real estate tax on a $72,000 home is $1440. At the same rate, find the real estate tax on a home valued at $150,000. $3000 [4.3C]

27. Ken purchased a stereo set for $490 and pays $29.40 in sales tax. What percent of the purchase price was the sales tax? 6% [5.3B]

28. A survey of 300 people showed that 165 people favored a certain candidate for mayor. What percent of the people surveyed did not favor this candidate? 45% [5.3B]

29. The value of a home in the northern part of the United States was $62,000 in 1985. The same home in 1990 had a value of $155,000. What percent of the 1985 value is the 1990 value? 250% [5.5B]

30. The Environmental Protection Agency found that 990 out of 5500 children tested had levels of lead in their blood exceeding federal guidelines. What percent of the children tested had levels of lead in the blood that exceeded federal standards? 18% [5.5B]

Chapter

Applications for Business and Consumers: A Calculator Approach

Objectives

Section 6.1
To find unit cost
To find the most economical purchase
To find total cost

Section 6.2
To find percent increase
To apply percent increase to business—markup
To find percent decrease
To apply percent decrease to business—discount

Section 6.3
To calculate simple interest
To calculate compound interest

Section 6.4
To calculate the initial expenses of buying a home
To calculate ongoing expenses of owning a home

Section 6.5
To calculate the initial expenses of buying a car
To calculate ongoing expenses of owning a car

Section 6.6
To calculate commissions, total hourly wages, and salaries

Section 6.7
To calculate checkbook balances
To balance a checkbook

A Penny a Day

A fictitious job offer in a newspaper claimed that it would pay a salary of 1¢ on the first of the month and 2¢ on the second of the month. Next month, the salary would be 4¢ on the first and 8¢ on the second of the month. The next month and each succeeding month for 12 months, the procedure of doubling the previous payment would be continued the same way. Do you think you would want the job under these conditions?

This problem is an example of compounding interest, which is similar to the type of interest that is earned on bank or savings deposits. The big difference is that banks and savings and loans do not compound the interest as quickly as in the given example.

The table below shows the amount of salary you would earn for each month and then gives the total annual salary.

Month	Salary
January	\$0.01 + \$0.02 = \$0.03
February	\$0.04 + \$0.08 = \$0.12
March	\$0.16 + \$0.32 = \$0.48
April	\$0.64 + \$1.28 = \$1.92
May	\$2.56 + \$5.12 = \$7.68
June	\$10.24 + \$20.48 = \$30.72
July	\$40.96 + \$81.92 = \$122.88
August	\$163.84 + \$327.68 = \$491.52
September	\$655.36 + \$1310.72 = \$1966.08
October	\$2621.44 + \$5242.88 = \$7864.32
November	\$10,485.76 + \$20,971.52 = \$31,457.28
December	\$41,943.04 + \$83,886.08 = \$125,829.12
	Total Annual Salary = \$167,772.15

Not a bad annual salary!

6.1 Applications to Purchasing

Objective A To find unit cost

INSTRUCTOR NOTE
This chapter deals with various application problems that require the student to use the arithmetic skills of the previous chapters. Although we use a calculator to solve these applications, calculators are not required.

INSTRUCTOR NOTE
One way to help students calculate unit cost is to tell them to divide by the number that represents the quantity following the word *per*. For instance, to find cost *per* gallon, divide cost by the number of gallons.

Frequently stores advertise items for purchase as, say, 3 shirts for \$36 or 5 pounds of potatoes for 90¢.

The **unit cost** is the cost of 1 shirt or 1 pound of potatoes. To find the unit cost, divide the total cost by the number of units.

3 shirts for \$36

36 ÷ 3 = 12

\$12 is the cost of 1 shirt.

Unit cost: \$12 per shirt

5 pounds of potatoes for 90¢

90 ÷ 5 = 18

18¢ is the cost of 1 pound of potatoes.

Unit cost: 18¢ per pound

Example 1
Find the unit cost. Round to the nearest tenth of a cent.
a) 10 gallons of gasoline for \$15.49
b) 3 yards of material for \$10.

Strategy
To find the unit cost, divide the total cost by the number of units.

Solution
a) 15.49 ÷ 10 = 1.549
\$1.549 per gallon
b) 10 ÷ 3 ≈ 3.3333
\$3.333 per yard

You Try It 1
Find the unit cost. Round to the nearest tenth of a cent.
a) 5 quarts of oil for \$6.25
b) 4 ears of corn for 85¢

Your strategy

Your solution
a) \$1.25 per quart
b) 21.3¢ per ear

Solution on p. A29

Objective B To find the most economical purchase

POINT OF INTEREST
The unit cost to send a payload into orbit on a Titan IV rocket is approximately \$15,000 per pound.

Comparison shoppers often find the most economical buy by comparing unit costs.

One store sells 6 cans of cola for \$2.04, and another store sells 24 cans of the same brand for \$7.92. To find the better buy, compare the unit costs.

2.04 ÷ 6 = 0.34
Unit cost: \$0.34 per can

7.92 ÷ 24 = 0.33
Unit cost: \$0.33 per can

Because \$0.33 < \$0.34, the better buy is 24 cans for \$7.92.

Example 2
Find the more economical purchase:
5 pounds of nails for $3.25, or 4 pounds of nails for $2.58.

Strategy
To find the more economical purchase, compare the unit costs.

Solution
3.25 ÷ 5 = 0.65
2.58 ÷ 4 = 0.645
$\$.645 < \$.65$

The more economical purchase is 4 pounds for $2.58.

You Try It 2
Find the more economical purchase:
6 cans of fruit for $2.52, or 4 cans of fruit for $1.66.

Your strategy

Your solution
4 cans for $1.66

Solution on p. A29

Objective C To find total cost

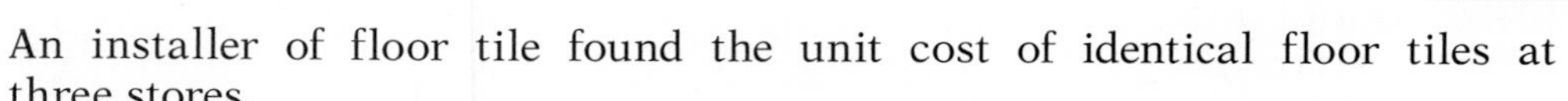

An installer of floor tile found the unit cost of identical floor tiles at three stores.

Store 1	*Store 2*	*Store 3*
$1.22 per tile	$1.18 per tile	$1.28 per tile

By comparing the unit costs, the installer determined that store 2 would provide the most economical purchase.

The installer also uses the unit cost to find the total cost of purchasing 300 floor tiles at store 2. The **total cost** is found by multiplying the unit cost by the number of units purchased.

Unit cost	×	number of units	=	total cost
1.18	×	300	=	354

The total cost is $354.

Example 3
Clear redwood lumber costs $2.43 per foot. How much would 25 feet of clear redwood cost?

Strategy
To find the total cost, multiply the unit cost ($2.43) by the number of units (25).

Solution

Unit cost	×	number of units	=	total cost
2.43	×	25	=	60.75

The total cost is $60.75.

You Try It 3
Pine saplings costs $4.96 each. How much would 7 pine saplings cost?

Your strategy

Your solution
$34.72

Solution on p. A29

6.1 Exercises

Objective A *Application Problems*

Find the unit cost. Round to the nearest tenth of a cent.

1. Syrup, 16 ounces for $0.89
   $0.056 per ounce

2. Wood, 8 feet for $11.36
   $1.42 per foot

3. Lunch meat, 12 ounces for $1.79
   $0.149 per ounce

4. Light bulbs, 6 bulbs for $2
   $0.333 per bulb

5. Wood screws, 15 screws for $0.50
   $0.033 per screw

6. Bay leaves, $\frac{1}{2}$ ounce for $0.96
   $1.92 per ounce

7. Antenna wire, 10 feet for $1.95
   $0.195 per foot

8. Sponges, 4 for $5.00
   $1.25 per sponge

9. Catsup, 24 ounces for $0.89
   $0.037 per ounce

10. Strawberry jam, 18 ounces for $1.23
    $0.068 per ounce

11. Metal clamps, 2 for $5.99
    $2.995 per clamp

12. Coffee, 3 pounds for $7.65
    $2.55 per pound

Objective B *Application Problems*

Find the more economical purchase.

13. Cheese, 12 ounces for $1.79, or 16 ounces for $2.40
    12 ounces for $1.79

14. Powder, 4 ounces for $0.53, or 9 ounces for $0.95
    9 ounces for $0.95

15. Shampoo, 8 ounces for $1.89, or 12 ounces for $2.64
    12 ounces for $2.64

16. Mayonnaise, 32 ounces for $2.37, or 16 ounces for $1.30
    32 ounces for $2.37

17. Fruit drink, 46 ounces for $0.69, or 50 ounces for $0.78
    46 ounces for $0.69

18. Detergent, 42 ounces for $2.57, or 20 ounces for $1.29
    42 ounces for $2.57

Find the more economical purchase.

19. Oil: 2 quarts for $4.79, or 5 quarts for $12.15
2 quarts for $4.79

20. Potato chips: 8 ounces for $0.99, or 10 ounces for $1.23
10 ounces for $1.23

21. Vitamins: 200 tablets for $9.98, or 150 tablets for $7.65
200 tablets for $9.98

22. Wax: 16 ounces for $2.68, or 27 ounces for $4.39
27 ounces for $4.39

23. Cereal: 18 ounces for $1.51, or 16 ounces for $1.32
16 ounces for $1.32

24. Tires: 2 for $129.98, or 4 for $258.98
4 tires for $258.98

Objective C *Application Problems*

Solve.

25. Steak costs $2.79 per pound. Find the total cost of 3 pounds.
$8.37

26. Red brick costs $0.56 per brick. Find the total cost of 50 bricks.
$28

27. Flowering plants cost $0.49 each. Find the total cost of 6 plants.
$2.94

28. Chicken costs $0.89 per pound. Find the total cost of 3.6 pounds. Round to the nearest cent.
$3.20

29. Tea costs $0.54 per ounce. Find the total cost of 6.5 ounces.
$3.51

30. Cheese costs $2.89 per pound. Find the total cost of 0.65 pounds. Round to the nearest cent.
$1.88

31. Tomatoes cost $0.22 per pound. Find the total cost of 1.8 pounds. Round to the nearest cent.
$0.40

32. Ham costs $1.49 per pound. Find the total cost of 3.4 pounds. Round to the nearest cent.
$5.07

33. Candy costs $2.68 per pound. Find the total cost of $\frac{3}{4}$ pound.
$2.01

34. Photocopying costs $0.05 per page. Find the total cost for photocopying 250 pages.
$12.50

APPLYING THE CONCEPTS

35. [W] Explain in your own words the meaning of unit pricing.
Answers will vary

36. [W] What is the UPC (Universal Product Code) and how is it used?
Answers will vary

6.2 Percent Increase and Percent Decrease

Objective A To find percent increase

INSTRUCTOR NOTE
For students to be successful calculating percent increase and percent decrease, they must remember that the base (in the basic percent equation) is always the quantity *before* the increase or decrease. For the problem at the right, the base is 50,000—the quantity before the increase.

POINT OF INTEREST
According to the U.S. Census Bureau, the number of persons aged 65 and over in the United States will increase by 67% from 1992 to 2020.

Percent increase is used to show how much a quantity has increased over its original value. The statements "car prices will show a 3.5% increase over last year's prices" and "employees were given an 11% pay increase" are illustrations of the use of percent increase.

➡ A city's population increased from 50,000 to 51,500 in 1 year. Find the percent increase of the city's population.

New value	−	original value	=	amount of increase
51,500	−	50,000	=	1500

Amount of Increase

Original Value

New Value

Now solve the basic percent equation for percent.

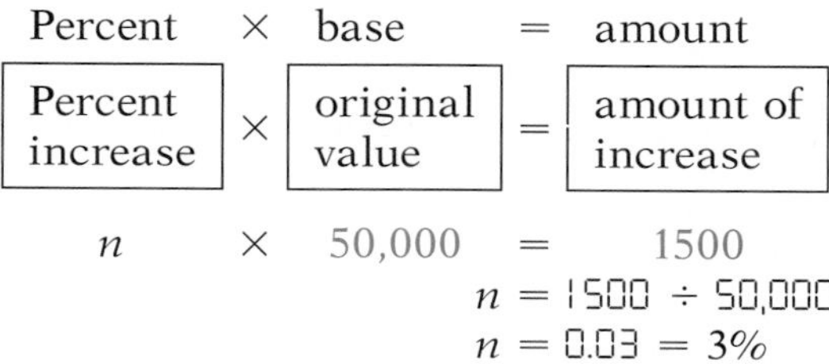

Percent	×	base	=	amount
Percent increase	×	original value	=	amount of increase
n	×	50,000	=	1500

$n = 1500 \div 50{,}000$

$n = 0.03 = 3\%$

The population increased by 3% over the previous year.

Example 1
The average price of gasoline rose from \$0.92 to \$1.15 in 4 months. What was the percent increase in the price of gasoline?

Strategy
To find the percent increase:
- Find the amount of increase.
- Solve the basic percent equation for *percent*.

Solution

New value	−	original value	=	amount of increase
1.15	−	0.92	=	0.23

Percent × base = amount

$n \times 0.92 = 0.23$

$n = 0.23 \div 0.92$

$n = 0.25 = 25\%$

The percent increase was 25%.

You Try It 1
A moving and storage company increased its number of employees from 2000 to 2250. What was the percent increase in the number of employees?

Your strategy

Your solution 12.5%

Solution on p. A30

Example 2
A seamstress was making a wage of $4.80 an hour before a 10% increase in pay. What is the new hourly wage?

Strategy
To find the new hourly wage:
- Solve the basic percent equation for *amount*.
- Add the amount of increase to the original wage.

Solution
Percent × base = amount
0.10 × 4.80 = n
0.48 = n

4.80 + 0.48 = 5.28

The new hourly wage is $5.28.

You Try It 2
A baker was making a wage of $6.50 an hour before a 14% increase in pay. What is the new hourly wage?

Your strategy

Your solution
$7.41

Solution on p. A30

Objective B To apply percent increase to business—markup

INSTRUCTOR NOTE
It will help students to know that markup is an application of percent increase. In business situations, markup can be based on *cost* or *selling price*. We have chosen cost, which is the most common practice. This means that the base in the basic percent equation is cost.
Having students focus on definitions of words in this objective will help them solve these problems.

Many expenses are involved in operating a business. Employee salaries, rent of office space, and utilities are examples of operating expenses. To pay these expenses and earn a profit, a business must sell a product at a higher price than it paid for the product.

Cost is the price that a business pays for a product, and **selling price** is the price at which a business sells a product to a customer. The difference between selling price and cost is called **markup**.

Selling price − cost = markup

or

Cost + markup = selling price

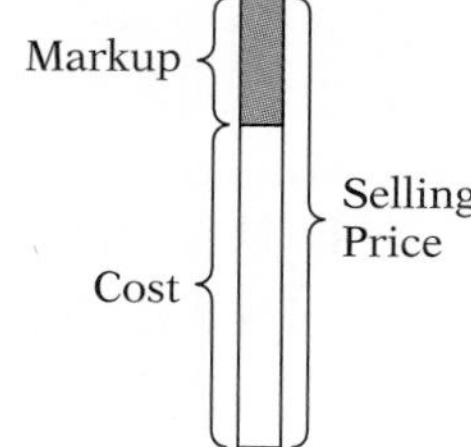

Markup is frequently expressed as a percent of a product's cost. This percent is called the **markup rate.**

Markup rate × cost = markup

➡ A bicycle store owner purchases a bicycle for $105 and sells it for $147. What markup rate does the owner use?

Selling price − cost = markup
147 − 105 = 42

• **First find the markup.**

$$\text{Percent} \times \text{base} = \text{amount}$$
$$\boxed{\text{Markup rate}} \times \boxed{\text{cost}} = \boxed{\text{markup}}$$
$$n \times \$105 = \$42$$
$$n = 42 \div 105 = 0.4$$

- **Then solve the basic percent equation for *percent*.**

The markup rate is 40%.

Example 3
The manager of a sporting goods store determines that a markup rate of 36% is necessary to make a profit. What is the markup on a pair of skis that costs the store $125?

Strategy
To find the markup, solve the basic percent equation for *amount*.

Solution

$$\text{Percent} \times \text{base} = \text{amount}$$
$$\boxed{\text{Markup rate}} \times \boxed{\text{cost}} = \boxed{\text{markup}}$$
$$0.36 \times 125 = n$$
$$45 = n$$

The markup is $45.

You Try It 3
A bookstore manager determines that a markup rate of 20% is necessary to make a profit. What is the markup on a book that costs the bookstore $8?

Your strategy

Your solution $1.60

Example 4
A plant nursery bought a citrus tree for $4.50 and used a markup rate of 46%. What is the selling price?

Strategy
To find the selling price:

- Find the markup by solving the basic percent equation for *amount*.
- Add the markup to the cost.

Solution

$$\text{Percent} \times \text{base} = \text{amount}$$
$$\boxed{\text{Markup rate}} \times \boxed{\text{cost}} = \boxed{\text{markup}}$$
$$0.46 \times 4.50 = n$$
$$2.07 = n$$

$$\boxed{\text{Cost}} + \boxed{\text{markup}} = \boxed{\text{selling price}}$$
$$4.50 + 2.07 = 6.57$$

The selling price is $6.57.

You Try It 4
A clothing store bought a suit for $72 and used a markup rate of 55%. What is the selling price?

Your strategy

Your solution $111.60

Solutions on p. A30

Objective C To find percent decrease

INSTRUCTOR NOTE
The base for percent decrease problems is the quantity before the decrease. It is shown as the original value for the problem at the right.

Percent decrease is frequently used to show how much a quantity has decreased from its original value. The statements "the unemployment rate decreased by 0.4% over last month" and "there has been a 12% decrease in the number of industrial accidents" are illustrations of the use of the percent decrease.

➡ A family's electric bill decreased from $60 per month to $52.80 per month. Find the percent decrease in the family's electric bill.

Original value	−	new value	=	amount of decrease
60	−	52.80	=	7.20

Amount of Decrease
New Value
Original Value

POINT OF INTEREST
A *bear market* is one in which the total value of all stocks decreases. During 1972 and 1973, the stock market decreased by approximately 55%. This means that an investment of $100,000 would have decreased to $45,000 during that period.

Now solve the basic percent equation for percent.

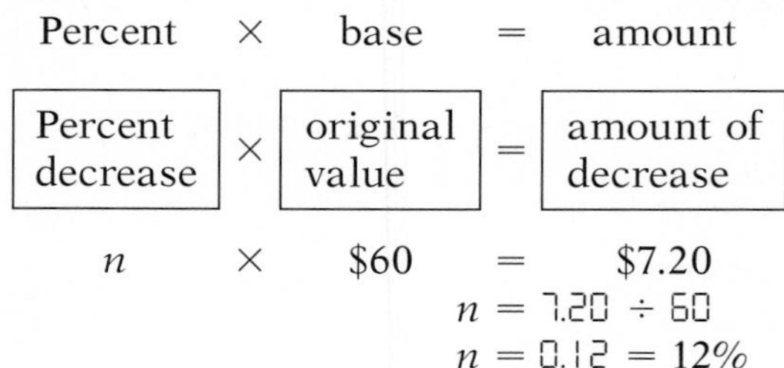

Percent × base = amount

Percent decrease	×	original value	=	amount of decrease
n	×	$60	=	$7.20

$n = 7.20 \div 60$

$n = 0.12 = 12\%$

The electric bill decreased by 12% per month.

Example 5
Because of unusually high temperatures, the number of people visiting a desert resort dropped from 650 to 525. Find the percent decrease in attendance. Round to the nearest tenth of a percent.

Strategy
To find the percent decrease:
- Find the amount of decrease.
- Solve the basic percent equation for *percent*.

Solution

Original value	−	new value	=	amount of decrease
650	−	525	=	125

Percent × base = amount

$n \times 650 = 125$

$n = 125 \div 650$

$n \approx 0.192 = 19.2\%$

The percent decrease was 19.2%.

You Try It 5
Car sales at an automobile dealership dropped from 150 in June to 120 in July. Find the percent decrease in car sales.

Your strategy

Your solution 20%

Solution on p. A30

Example 6
The total sales for December for a stationery store were $16,000. For January, total sales showed an 8% decrease from December's sales. What were the total sales for January?

Strategy
To find the total sales for January:
- Find the amount of decrease by solving the basic percent equation for *amount*.
- Subtract the amount of decrease from the December sales.

Solution
Percent × base = amount

$0.08 \times 16{,}000 = n$

$1280 = n$

The decrease in sales was $1280.

$16{,}000 - 1280 = 14{,}720$

The total sales for January were $14,720.

You Try It 6
Fog decreased the normal 5-mile visibility at an airport by 40%. What was the visibility in the fog?

Your strategy

Your solution
3 miles

Solution on p. A30

Objective D To apply percent decrease to business—discount

INSTRUCTOR NOTE
Remind students that discount is an application of percent decrease. The base in the basic percent equation is the regular price.

To promote sales, a store may reduce the regular price of some of its products temporarily. The reduced price is called the **sale price.** The difference between the regular price and the sale price is called the **discount**.

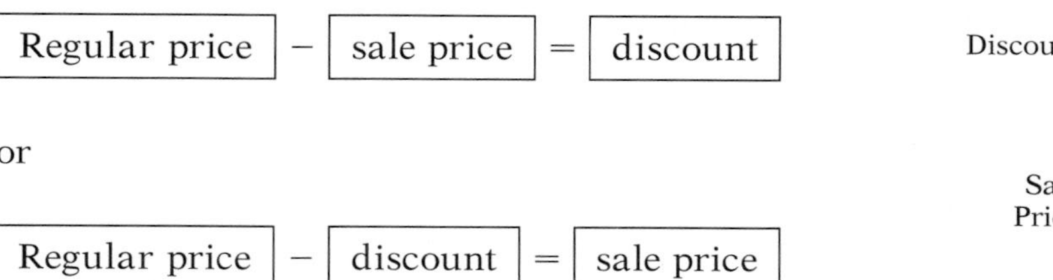

Discount is frequently stated as a percent of a product's regular price. This percent is called the **discount rate.**

Discount rate × regular price = discount

Example 7
An appliance store has a washing machine that regularly sells for $350 on sale for $297.50. What is the discount rate?

Strategy
To find the discount rate:

- Find the discount.
- Solve the basic percent equation for *percent*.

Solution

Regular price	–	sale price	=	discount
350	–	297.50	=	52.50

Percent × base = amount

Discount rate	×	regular price	=	discount

$n \times \$350 = \$52.50$

$n = 52.50 \div 350$

$n = 0.15 = 15\%$

The discount rate is 15%.

You Try It 7
A lawn mower that regularly sells for $125 is on sale for $25 off the regular price. What is the discount rate?

Your strategy

Your solution 20%

Example 8
A plumbing store is selling at 15% off the regular price sinks that regularly sell for $125. What is the sale price?

Strategy
To find the sale price:

- Find the discount by solving the basic percent equation for *amount*.
- Subtract to find the sale price.

Solution
Percent × base = amount

Discount rate	×	regular price	=	discount
0.15	×	125	=	n

$18.75 = n$

Regular price	–	discount	=	sale price
125	–	18.75	=	106.25

The sale price is $106.25.

You Try It 8
A florist is selling at 20% off the regular price potted plants that regularly sell for $10.25. What is the sale price?

Your strategy

Your solution $8.20

Solutions on p. A31

6.2 Exercises

Objective A *Application Problems*

1. Elissa had a 0.275 batting average in her junior year in high school. The next year she had a batting average of 0.297. What percent increase has occurred in her batting average? 8%

2. Rodney had a 3.80 rushing average for the Baker University football team. The next year he averaged 4.75 yards per carry. What percent increase does this amount represent? 25%

3. Josif increased his typing speed of 60 words per minute by 20 words per minute. By what percent did he increase his typing speed? $33\frac{1}{3}\%$

4. You have increased your reading speed by 70 words per minute from 280 words per minute. What is your percent increase in reading speed? 25%

5. A teacher's union negotiated a new contract calling for a 9% increase in pay over this year's pay.
   a. Find the increase in pay for a teacher who had a salary of $32,500. $2925
   b. Find the new salary for a teacher who had a salary of $32,500. $35,425

6. A new labor contract for Westside Lumber called for an 8.5% increase in pay for all employees.
   a. What is the increase in pay for an employee who makes $324 per week? $27.54
   b. What is the weekly wage of this employee after the wage increase? $351.54

7. AFLAG, whose stock is selling for $30\frac{3}{8}$ per share, plans to raise its $1.28 dividend by 25%. Find the dividend per share of AFLAG stock after the increase. $1.60

8. An amusement park increased its summer staff from 500 to 625 employees. What percent increase does this represent? 25%

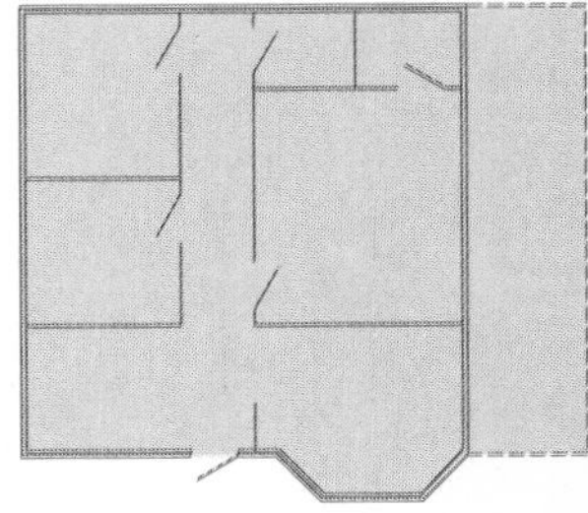

9. A family increased its 1440-square-foot home $16\frac{2}{3}\%$ by adding a recreation room. How much larger, in square feet, is the home now? 240 square feet

10. A hunter's lodge increased its weekly hours of operation from 60 hours to 70 hours. What percent increase does this amount represent? $16\frac{2}{3}\%$

Objective B *Application Problems*

11. An air conditioner cost the plumbing and air conditioning contractor $285. Find the markup on this air conditioner if the markup rate is 25%. $71.25

12. The owner of Kerr electronics uses a markup rate of 42%. What is the markup on a cassette player that costs Kerr $85? $35.70

13. The ski and tennis store uses a markup rate of 30%. What is the markup on a tennis racket that costs $42? $12.60

14. The manager of a department store determines that a markup of 38% is necessary for a profit to be made. What is the markup on a novelty item that costs $45? $17.10

15. Computer Inc. uses a markup of $975 on a computer system that costs $3250. What markup rate does this amount represent? 30%

16. Saizon Pen & Office Supply uses a markup of $12 on a calculator that costs $20. What markup rate does this amount represent? 60%

17. Giant Photo Service uses a markup rate of 48% on its Model ZA cameras, which cost the shop $162.
a. What is the markup? $77.76
b. What is the selling price? $239.76

18. The Circle R golf pro shop uses a markup rate of 45% on a set of golf clubs that costs the shop $210.
a. What is the markup? $94.50
b. What is the selling price? $304.50

19. Hauser Furniture uses a markup rate of 35% on a piece of lawn furniture that costs the store $62.
a. What is the markup? $21.70
b. What is the selling price? $83.70

20. Harvest Time Produce Inc. uses a 55% markup rate and pays $0.60 for a box of strawberries.
a. What is the markup? $0.33
b. What is the selling price of a box of strawberries? $0.93

21. Resner Builders' Hardware uses a markup rate of 42% for a table saw that costs $160. What is the selling price of the table saw? $227.20

22. Brad Burt's Magic Shop uses a markup rate of 48%. What is the selling price of an item that costs $50? $74

Objective C *Application Problems*

23. A new bridge reduced the normal 45-minute travel time between two cities by 18 minutes. What percent decrease does this represent? 40%

24. In an election, 25,400 voters went to the polls to elect a mayor. Four years later, 1778 fewer voters participated in the election for mayor. What was the percent decrease in the number of voters who went to the polls? 7%

25. By installing energy-saving equipment, the Pala Rey Youth Camp reduced its normal $800-per-month utility bill by $320. What percent decrease does this amount represent? 40%

26. During the last 40 years, the consumption of eggs in the United States has dropped from 400 eggs per person per year to 260 eggs per person per year. What percent decrease does this amount represent? 35%

27. It is estimated that the value of a new car is reduced 30% after 1 year of ownership. Using this estimate, find how much value a $11,200 new car loses after 1 year. $3360

28. Assume that a Kmart store employs 1200 people during the holiday. At the end of the holiday season, Kmart reduces the number of employees by 45%. What is the decrease in the number of employees? 540 employees

29. Because of a decrease in demand for super-8 video cameras, Kit's Cameras reduced the orders for these models from 20 per month to 8 per month.
   a. What is the amount of decrease? 12 cameras
   b. What percent decrease does this amount represent? 60%

30. A new computer system reduced the time for printing the payroll from 52 minutes to 39 minutes.
   a. What is the amount of decrease? 13 minutes
   b. What percent decrease does this amount represent? 25%

31. Juanita's average expense for gasoline was \$76. After joining a car pool, she was able to reduce the expense by 20%.
   a. What was the amount of decrease? \$15.20
   b. What is the average monthly gasoline bill now? \$60.80

32. An oil company paid a dividend of \$1.60 per share. After reorganization, the company reduced the dividend by 37.5%.
   a. What was the amount of decrease? \$0.60
   b. What is the new dividend? \$1.00

33. Because of an improved traffic pattern at a sports stadium, the average amount of time a fan waits to park decreased from 3.5 minutes to 2.8 minutes. What percent decrease does this amount represent? 20%

Objective D *Application Problems*

34. The Austin College Bookstore is giving a discount of \$8 on calculators that normally sell for \$24. What is the discount rate? $33\frac{1}{3}\%$

35. A discount clothing store is selling a \$72 sport jacket for \$24 off the regular price. What is the discount rate? $33\frac{1}{3}\%$

36. A disk player that regularly sells for \$340 is selling for 20% off the regular price. What is the discount? \$68

37. Dacor Appliances is selling its \$450 washing machine for 15% off the regular price. What is the discount? \$67.50

38. An electric grill that regularly sells for \$140 is selling for \$42 off the regular price. What is the discount rate? 30%

39. You bought a computer system for \$350 off the regular price of \$1400. What is the discount rate? 25%

40. Quick Service Gas Station has its regularly priced \$45 tune-up on sale for 16% off the regular price.
   a. What is the discount? \$7.20
   b. What is the sale price? \$37.80

41. Turkey that regularly sells for \$.85 per pound is on sale for 20% off the regular price.
   a. What is the discount? \$0.17 per pound
   b. What is the sale price? \$0.68 per pound

42. An outdoor supply store has regularly priced $160 sleeping bags on sale for $120.
a. What is the discount? $40
b. What is the discount rate? 25%

43. Standard Brands paint that regularly sells for $16 per gallon is on sale for $12 per gallon.
a. What is the discount? $4
b. What is the discount rate? 25%

APPLYING THE CONCEPTS

The figure at the right shows the closing price and the change for the day for the most active stocks on the NASDAQ stock exchange for October 13, 1993. Use this figure for Exercises 44–47. Round to the nearest tenth of a percent.

NASDAQ	Vol.	Last	Chg
Intel s	13,277,800	65	-1/2
Clsco s	4,916,400	50 7/8	+4 7/8
Wellfit s	3,910,800	53 1/4	+6 1/8
WstwOn	3,058,000	5 3/8	+7/8
SynOpt s	2,981,500	24 1/2	+7/8
Centocor	2,943,700	14 1/4	+3
GtARc	2,773,200	1 3/8	+1/4
Medgrp	2,736,800	1 17/32	+15/32
AppleC	2,735,200	24	+1/4
TelCmA	2,671,000	28 3/8	-1/4

44. Find the one-day percent increase for Cisco. 10.6%

45. Find the one-day percent increase for Centocor. 26.7%

46. Find the one-day percent decrease for Intel. 0.8%

47. One share of Wellfit increased by $\$6\frac{1}{8}$ and one share of WstwOn increased by $\frac{7}{8}$. Which of these two companies had the greater percent increase in price? WstwOn

48. A welder earning $12 per hour is given a 10% raise. To find the new wage, we can multiply $12 by 0.10 and add the product to $12. Can the new wage be found by multiplying $12 by 1.10? Try both methods and compare your answers. Yes

49. Grocers, florists, bakers, and other businesses must consider spoilage when deciding the markup of a product. For instance, a loaf of bread will become stale after a few days, and the baker will not be able to sell it. A florist purchased 200 roses at a cost of $.86 per rose. The florist wants a markup rate of 50% of the total cost of all the roses and expects 7% of the roses to wilt and therefore not be saleable. Find the selling price per rose by answering each of the following questions.
a. What is the florist's total cost for the 200 roses? $172
b. Find the total selling price without spoilage. $258
c. Find the number of roses the florist expects to sell. *Hint:* The number of roses the florist expects to sell is

% of saleable roses × number of roses purchased 186

d. To find the selling price per rose, divide the total selling price per rose by the number of roses the florist expects to sell. Round to the nearest cent. $1.39

50. [W] A promotional sale at a department store offers 25% off the sale price. The sale price is 25% off the regular price. Is this the same as a sale that offers 50% off the regular price? If not, which sale gives the better price? Explain your answer. No. 50% off the regular price

51. [W] In your own words, explain how to find a percent increase or a percent decrease. Answers will vary

Interest

Objective A To calculate simple interest

INSTRUCTOR NOTE

Emphasize that the simple interest formula requires that the interest rate and the time have comparable units. If the *annual* interest rate is given, then time must be in *years*. If a *monthly* interest rate is given (as on most credit cards), then time must be in *months*.

When money is deposited in a bank account, the bank pays the depositor for the privilege of using that money. The amount paid to the depositor is called **interest.** When money is borrowed from a bank, the borrower pays for the privilege of using that money. The amount paid to the bank is also called interest.

The original amount deposited or borrowed is called the **principal.** The amount of interest is a percent of the principal. The percent used to determine the amount of interest is the **interest rate.** Interest rates are given for specific periods of time, usually months or years.

Interest computed on the original principal is called **simple interest.** To calculate simple interest, multiply the principal by the interest rate per period by the number of time periods.

➡ Calculate the simple interest on \$1500 deposited for 2 years at an annual interest rate of 7.5%.

Principal × annual interest rate × time (in years) = interest

1500 × 0.075 × 2 = 225

The interest earned is \$225.

Example 1

Kamal borrowed \$500 from a savings and loan association for 6 months at an annual interest rate of 7%. What is the simple interest due on the loan?

Strategy

To find the simple interest, multiply:

Principal × annual interest rate × time (in years)

Solution 500 × 0.07 × 0.5 = 17.5

The interest due is \$17.50.

You Try It 1

A company borrowed \$15,000 from a bank for 18 months at an annual interest rate of 8%. What is the simple interest due on the loan?

Your strategy

Your solution \$1800

Example 2

A credit card company charges a customer 1.5% per month on the unpaid balance of charges on the credit card. What is the interest due in a month when the customer has an unpaid balance of \$54?

Strategy

To find the interest due, multiply the principal by the monthly interest rate by the time (in months).

Solution 54 × 0.015 × 1 = 0.81

The interest charge is \$0.81.

You Try It 2

A bank offers short-term loans at a simple interest rate of 1.2% per month. What is the interest due on a short-term loan of \$400 for 2 months?

Your strategy

Your solution \$9.60

Solutions on p. A31

Objective B To calculate compound interest

INSTRUCTOR NOTE
If students have a scientific calculator you might show them the compound interest formula:

$$P = A\left(1 + \frac{i}{m}\right)^{mt}$$

where P is the new principal, A is the amount invested, i is the annual interest rate written as a decimal, m is the number of compounding periods per year, and t is the number of years.

The calculator sequence for Example 3 is:

650 × (1 + .08 ÷ 2)^(2 × 5) =

Usually the interest paid on money deposited or borrowed is compound interest. **Compound interest** is computed not only on the original principal but also on interest already earned. Compound interest is usually compounded annually (once a year), semiannually (twice a year), quarterly (four times a year), or daily.

$100 is invested for 3 years at an annual interest rate of 9% compounded annually. The interest earned over the 3 years is calculated by first finding the interest earned each year.

1st year	Interest earned:	0.09 × \$100.00 = \$9.00
	New principal:	\$100.00 + \$9.00 = \$109.00
2nd year	Interest earned:	0.09 × \$109.00 = \$9.81
	New principal:	\$109.00 + \$9.81 = \$118.81
3rd year	Interest earned:	0.09 × \$118.81 ≈ \$10.69
	New principal:	\$118.81 + \$10.69 = \$129.50

To find the interest earned, subtract the original principal from the new principal.

New principal	−	original principal	=	interest earned
129.50	−	100	=	29.5

The interest earned is $29.50.

Note that the compound interest earned is $29.50. The simple interest earned on the investment would have been only $100 × 0.09 × 3 = $27.

Calculating compound interest can be very tedious, so there are tables that can be used to simplify these calculations. A portion of a compound interest table is given on pages A4 and A5.

Example 3
An investment of $650 pays 8% annual interest compounded semiannually. What is the interest earned in 5 years?

Strategy To find the interest earned:

- Find the new principal by multiplying the original principal by the factor (1.48024) found in the compound interest table.
- Subtract the original principal from the new principal.

Solution
650 × 1.48024 ≈ 962.16

The new principal is $962.16.

962.16 − 650 = 312.16

The interest earned is $312.16.

You Try It 3
An investment of $1000 pays 6% annual interest compounded quarterly. What is the interest earned in 20 years?

Your strategy

Your solution $2290.66

Solution on p. A31

6.3 Exercises

Objective A *Application Problems*

1. To finance the purchase of 15 new cars, the Tropical Car Rental Agency borrowed $100,000 for 9 months at an annual interest rate of 9%. What is the simple interest due on the loan? $6750
2. A home builder obtained a pre-construction loan of $50,000 for 8 months at an annual interest rate of 9.5%. What is the simple interest due on the loan? (Round to the nearest cent.) $3166.67
3. The Mission Valley Credit Union charges its customers an interest rate of 2% per month on money that is transferred into an account that is overdrawn. Find the interest owed to the credit union for 1 month when $800 is transferred into an overdrawn account. $16
4. Assume that Visa charges Francesca 1.6% per month on her unpaid balance. Find the interest owed to Visa when her unpaid balance for the month is $1250. $20
5. Action Machining Company purchased a robot-controlled lathe for $225,000 and financed the full amount at 8% simple annual interest for 4 years.
   a. Find the interest on the loan. $72,000
   b. Find the monthly payment.
   $\left(\text{Monthly payment} = \dfrac{\text{loan amount} + \text{interest}}{\text{number of months}}\right)$ $6187.50
6. For the purchase of an entertainment center, an $1800 loan is obtained for 2 years at a simple interest rate of 9.4%.
   a. Find the interest due on the loan. $338.40
   b. Find the monthly payment.
   $\left(\text{Monthly payment} = \dfrac{\text{loan amount} + \text{interest}}{\text{number of months}}\right)$ $89.10
7. To attract new customers, Heller Ford is offering car loans at a simple interest rate of 4.5%.
   a. Find the interest charged to a customer who finances a car loan of $12,000 for 2 years. $1080
   b. Find the monthly payment. $545
8. Cimarron Homes Inc. purchased a small plane for $57,000 and financed the full amount for 5 years at a simple annual interest rate of 9%.
   a. Find the interest due on the loan. $25,650
   b. Find the monthly payment. $1377.50
9. Dennis Pappas decided to build onto an existing structure instead of buying a new home. He borrowed $42,000 for $3\frac{1}{2}$ years at a simple interest rate of 9.5%. Find the monthly payment. $1332.50

Objective B *Application Problems*

Solve. Use the table on pages A4 and A5. Round to the nearest cent.

10. North Island Federal Credit Union pays 4% annual interest, compounded daily, on time savings deposits. Find the value of $750 deposited in this account after 1 year. $780.60

Solve. Use the table on pages A4 and A5. Round to the nearest cent.

11. What is the value after 5 years of $1000 invested at 7% annual interest compounded quarterly? $1414.78

12. An investment club invested $50,000 in a certificate of deposit that pays 5% annual interest compounded quarterly. Find the value of this investment after 10 years. $82,181

13. Tanya invested $2500 in a tax-sheltered annuity that pays 8% annual interest compounded daily. Find the value of her investment after 20 years. $12,380.43

14. Sal Trovato invested $3000 in a corporate retirement account that pays 6% annual interest compounded semiannually. Find the value of his investment after 15 years. $7281.78

15. To replace equipment, a farmer invested $20,000 in an account that pays 7% annual interest compounded semiannually. What is the value of the investment after 5 years? $28,212

16. Green River Lodge invests $75,000 in a trust account that pays 8% interest compounded quarterly.
a. What will the value of the investment be in 5 years? $111,446.25
b. How much interest will be earned in the 5 years? $36,446.25

17. To save for retirement, a couple deposited $3000 in an account that pays 7% annual interest compounded daily.
a. What will the value of the investment be in 10 years? $6040.86
b. How much interest will be earned in the 10 years? $3040.86

18. To save for a child's college education, the Petersens deposited $2500 into an account that pays 6% annual interest compounded daily. Find the amount of interest earned in this account over a 20-year period. $5799.48

APPLYING THE CONCEPTS

19. [W] Explain the fundamental difference between simple interest and compound interest. Answers will vary

20. [W] Visit a brokerage business and obtain information on annuities. Explain the advantages of an annuity as a retirement option. Answers will vary

21. Suppose you have a savings account that earns interest at the rate of 6% per year compounded monthly. On January 1, you open this account with a deposit of $100.
a. On February 1, you deposit an additional $100 into the account. What is the value of the account after the deposit? $200.50
b. On March 1, you deposit an additional $100 into the account. What is the value of the account after the deposit. *Note:* This type of savings plan, wherein equal amounts ($100) are saved at equal time intervals (every month), is called an annuity. $301.50

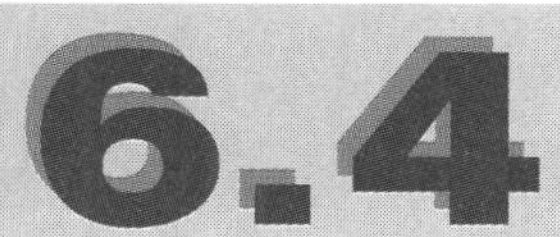

6.4 Real Estate Expenses

Objective A To calculate the initial expenses of buying a home

One of the largest investments most people ever make is the purchase of a home. The major initial expense in the purchase is the down payment. The amount of the down payment is normally a percent of the purchase price. This percent varies among banks, but it usually ranges from 5% to 25%.

The **mortgage** is the amount that is borrowed to buy real estate. The mortgage amount is the difference between the purchase price and the down payment.

➡ A home is purchased for \$85,000, and a down payment of \$12,750 is made. Find the mortgage.

Purchase price	−	down payment	=	mortgage
85,000	−	12,750	=	72,250

The mortgage is \$72,250.

Another large initial expense in buying a home is the loan origination fee, which is a fee the bank charges for processing the mortgage papers. The loan origination fee is usually a percent of the mortgage and is expressed in **points,** which is the term banks use to mean percent. For example, "5 points" means "5 percent."

Points	×	mortgage	=	loan origination fee

Example 1
A house is purchased for \$87,000, and a down payment, which is 20% of the purchase price, is made. Find the mortgage.

Strategy
To find the mortgage:

- Find the down payment by solving the basic percent equation for *amount.*
- Subtract the down payment from the purchase price.

Solution
Percent × base = amount

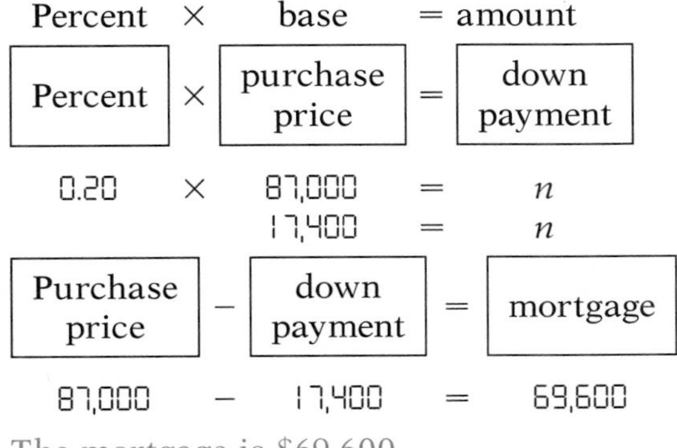

Percent	×	purchase price	=	down payment
0.20	×	87,000	=	n
		17,400	=	n

Purchase price	−	down payment	=	mortgage
87,000	−	17,400	=	69,600

The mortgage is \$69,600.

You Try It 1
An office building is purchased for \$216,000, and a down payment, which is 25% of the purchase price, is made. Find the mortgage.

Your strategy

Your solution
\$162,000

Solution on p. A32

Example 2
A home is purchased with a mortgage of $65,000. The buyer pays a loan origination fee of $3\frac{1}{2}$ points. How much is the loan origination fee?

Strategy
To find the loan origination fee, solve the basic percent equation for *amount*.

Solution

Percent × base = amount

| Points | × | mortgage | = | fee |

0.035 × 65,000 = n

2275 = n

The loan origination fee is $2275.

You Try It 2
The mortgage on a real estate investment is $80,000. The buyer paid a loan origination fee of $4\frac{1}{2}$ points. How much was the loan origination fee?

Your strategy

Your solution $3600

Solution on p. A32

Objective B To calculate ongoing expenses of owning a home

INSTRUCTOR NOTE
As an optional exercise for students with a scientific calculator, you can give them the following keystrokes to calculate a monthly payment:

$Pmt = B \times (i \div 12) \div (1 - (1 + i \div 12) \wedge n+/-) =$

where B is the amount borrowed, i is the annual interest rate as a decimal, and n is the number of months of the loan.

Besides the initial expenses of buying a home, there are continuing monthly expenses involved in owning a home. The monthly mortgage payment, utilities, insurance, and taxes are some of these ongoing expenses. Of these expenses, the largest one is normally the monthly mortgage payment.

For a fixed-rate mortgage, the monthly mortgage payment remains the same throughout the life of the loan. The calculation of the monthly mortgage payment is based on the amount of the loan, the interest rate on the loan, and the number of years required to pay back the loan. Calculating the monthly mortgage payment is fairly difficult, so tables such as the one on page A6 are used to simplify these calculations.

➡ Find the monthly mortgage payment on a 30-year $60,000 mortgage at an interest rate of 9%. Use the monthly payment table on page A6.

60,000 × 0.0080462 = 482.77
↓
from the table

The monthly mortgage payment is $482.77.

The monthly mortgage payment includes the payment of both principal and interest on the mortgage. The interest charged during any one month is charged on the unpaid balance of the loan. Therefore, during the early years of the mortgage, when the unpaid balance is high, most of the monthly mortgage payment is interest charged on the loan. During the last few years of a mortgage, when the unpaid balance is low, most of the monthly mortgage payment goes toward paying off the loan.

POINT OF INTEREST

A person receiving a loan of $125,000 for 30 years at a fixed mortgage interest rate of 7% has a monthly payment of $649.18. If the loan is kept for 30 years, the total amount of interest paid is $108,704.80.

➡ Find the interest paid on a mortgage during a month when the monthly mortgage payment is $186.26 and $58.08 of that amount goes toward paying off the principal.

Monthly mortgage payment	−	principal	=	interest
186.26	−	58.08	=	128.18

The interest paid on the mortgage is $128.18.

Property tax is another ongoing expense of owning a house. Property tax is normally an annual expense that may be paid on a monthly basis. The monthly property tax, which is determined by dividing the annual property tax by 12, is usually added to the monthly mortgage payment.

➡ A home owner must pay $534 in property tax annually. Find the property tax that must be added each month to the home owner's monthly mortgage payment.

534 ÷ 12 = 44.5

Each month, $44.50 must be added to the monthly mortgage payment for property tax.

Example 3

Serge purchased some land for $120,000 and made a down payment of $25,000. The savings and loan association charges an annual interest rate of 8% on Serge's 25-year mortgage. Find the monthly mortgage payment.

Strategy

To find the monthly mortgage payment:

- Subtract the down payment from the purchase price to find the mortgage.
- Multiply the mortgage by the factor found in the monthly payment table on page A6.

Solution

Purchase price	−	down payment	=	mort-gage
120,000	−	25,000	=	95,000

The mortgage is $95,000.

95,000 × 0.0077182 = 733.23

from the table

The monthly mortgage payment is $733.23.

You Try It 3

A new condominium project is selling townhouses for $75,000. A down payment of $15,000 is required, and a 20-year mortgage at an annual interest rate of 9% is available. Find the monthly mortgage payment.

Your strategy

Your solution $539.84

Solution on p. A32

Example 4
A home has a mortgage of $134,000 for 25 years at an annual interest rate of 7%. During a month when $375.88 of the monthly mortgage payment is principal, how much of the payment is interest?

Strategy
To find the interest:

- Multiply the mortgage by the factor found in the monthly payment table on page A6 to find the monthly mortgage payment.
- Subtract the principal from the monthly mortgage payment.

Solution
134,000 × 0.0070678 ≈ 947.09

from the table — monthly mortgage payment

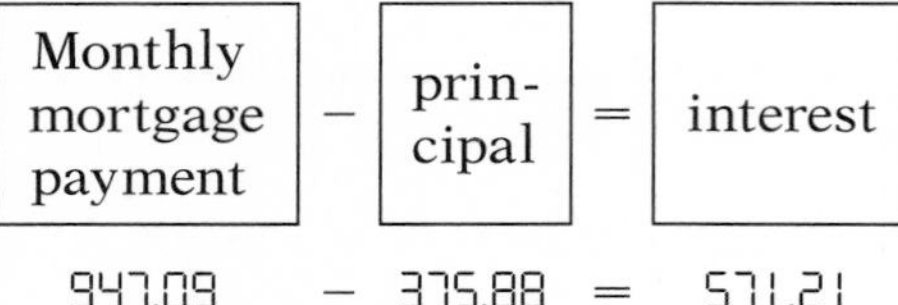

947.09 − 375.88 = 571.21

$571.21 is interest on the mortgage.

You Try It 4
An office building has a mortgage of $125,000 for 25 years at an annual interest rate of 9%. During a month when $492.65 of the monthly mortgage payment is principal, how much of the payment is interest?

Your strategy

Your solution $556.35

Example 5
The monthly mortgage payment for a home is $598.75. The annual property tax is $900. Find the total monthly payment for the mortgage and property tax.

Strategy
To find the monthly payment:

- Divide the annual property tax by 12 to find the monthly property tax.
- Add the monthly property tax to the monthly mortgage payment.

Solution
900 ÷ 12 = 75 monthly property tax
598.75 + 75 = 673.75

The total monthly payment is $673.75.

You Try It 5
The monthly mortgage payment for a home is $415.20. The annual property tax is $744. Find the total monthly payment for the mortgage and property tax.

Your strategy

Your solution $477.20

Solutions on p. A32

6.4 Exercises

Objective A *Application Problems*

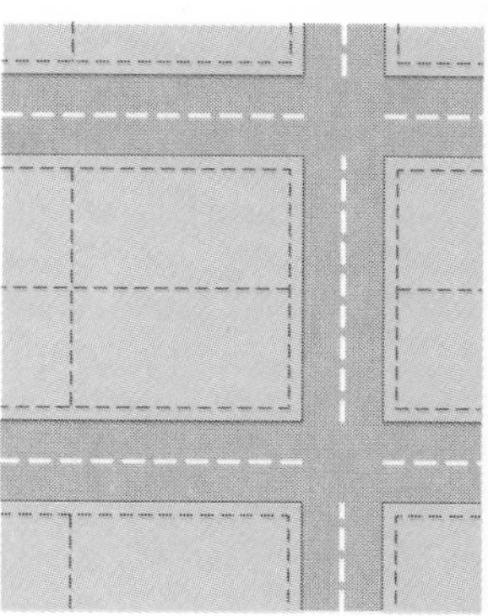

1. A condominium at Mt. Baldy Ski Resort was purchased for $97,000, and a down payment of $14,550 was made. Find the mortgage. $82,450

2. An insurance business was purchased for $173,000 and a down payment of $34,600 was made. Find the mortgage. $138,400

3. A building lot was purchased for $25,000. The lender requires a down payment of 30% of the purchase price. Find the down payment. $7500

4. Ian Goldman purchased a new home for $88,500. The lender requires a down payment of 20% of the purchase price. Find the down payment. $17,700

5. Brian Stedman made a down payment of 25% of the $850,000 purchase price of an apartment building. How much was the down payment? $212,500

6. A clothing store was purchased for $125,000, and a down payment that was 25% of the purchase price was made. How much was the down payment? $31,250

7. A loan of $150,000 is obtained to purchase a home. The loan origination fee is $2\frac{1}{2}$ points. Find the amount of the loan origination fee. $3750

8. Security Savings & Loan requires a borrower to pay $3\frac{1}{2}$ points for a loan. Find the amount of the loan origination fee for a loan of $90,000. $3150

9. Baja Construction Inc. is selling homes for $95,000. A down payment of 5% is required.
   a. Find the down payment. $4750
   b. Find the mortgage. $90,250

10. A cattle rancher purchased some land for $240,000. The bank requires a down payment of 15% of the purchase price.
    a. Find the down payment. $36,000
    b. Find the mortgage. $204,000

11. Vivian Tom purchased a home for $96,000. Find the mortgage if the down payment is 10% of the purchase price. $86,400

12. A mortgage lender requires a down payment of 5% of the $56,000 purchase price of a condominium. How much is the mortgage? $53,200

Objective B *Application Problems*

Solve. Use the monthly payment table on page A6. Round to the nearest cent.

13. An investor obtained a loan of $150,000 to buy a car wash business. The monthly mortgage payment was based on 25 years at 8%. Find the monthly mortgage payment. $1157.73

14. A beautician obtained a 20-year mortgage of $90,000 to expand the business. The credit union charges an annual interest rate of 9%. Find the monthly mortgage payment. $809.76

Solve. Use the monthly payment table of page A6. Round to the nearest cent.

15. A couple interested in buying a home determines that they can afford a monthly mortgage payment of $800. Can they afford to buy a home with a 30-year $110,000 mortgage at 8% interest? No

16. A lawyer is considering purchasing a new office building with a 20-year $400,000 mortgage at 9% interest. The lawyer can afford a monthly mortgage payment of $4000. Can the lawyer afford the monthly mortgage payment on the new office building? Yes

17. The county tax assessor has determined that the annual property tax on a $125,000 house is $1348.20. Find the monthly property tax. $112.35

18. The annual property tax on a $155,000 home is $1992. Find the monthly property tax. $166

19. Abacus Imports Inc. has a warehouse with a 25-year mortgage of $200,000 at an annual interest rate of 9%.
a. Find the monthly mortgage payment. $1678.40
b. During a month when $941.72 of the monthly mortgage payment is principal, how much of the payment is interest? $736.68

20. A vacation home has a mortgage of $135,000 for 30 years at an annual interest rate of 7%.
a. Find the monthly mortgage payment. $898.16
b. During a month when $392.47 of the monthly mortgage payment is principal, how much of the payment is interest? $505.69

21. The annual mortgage payment on a duplex is $10,844.40. The owner must pay an annual property tax of $948. Find the total monthly payment for the mortgage and property tax. $982.70

22. The monthly mortgage payment on a home is $716.40, and the homeowner pays an annual property tax of $792. Find the total monthly payment for the mortgage and property tax. $782.40

23. Maria Hernandez purchased a home for $210,000 and made a down payment of $15,000. The balance was financed for 30 years at an annual interest rate of 8%. Find the monthly mortgage payment. $1430.83

24. A customer of a savings and loan purchased a $185,000 home and made a down payment of $20,000. The savings and loan charges its customers an annual interest rate of 7% for 30 years for a home mortgage. Find the monthly mortgage payment. $1097.75

APPLYING THE CONCEPTS

25. A couple buying a home for $100,000 have a choice of loans. One loan is an 8% loan for 20 years and the other loan is at 9% for 30 years. Find the amount of interest that can be saved by choosing the 20-year loan. $88,917.60

26. [W] Find out what an adjustable-rate mortgage is. What is the difference between this type of loan and a fixed-rate mortgage. List some of the advantages and disadvantages of each. Answers will vary

6.5 Car Expenses

Objective A ***To calculate the initial expenses of buying a car***

The initial expenses in the purchase of a car usually include the down payment, the license fees, and the sales tax. The down payment may be very small or as much as 25% or 30% of the purchase price of the car, depending on the lending institution. License fees and sales tax are regulated by each state, so these expenses vary from state to state.

Example 1
A car is purchased for $8500, and the lender requires a down payment of 15% of the purchase price. Find the amount financed.

Strategy
To find the amount financed:

- Find the down payment by solving the basic percent equation for *amount*.
- Subtract the down payment from the purchase price.

Solution

Percent × base = amount

Percent	×	purchase price	=	down payment

$0.15 \times 8500 = n$

$1275 = n$

$8500 - 1275 = 7225$

The amount financed is $7225.

You Try It 1
A down payment of 20% of the $9200 purchase price of a new car is made. Find the amount financed.

Your strategy

Your solution $7360

Example 2
A sales clerk purchases a car for $6500 and pays a sales tax that is 5% of the purchase price. How much is the sales tax?

Strategy
To find the sales tax, solve the basic percent equation for *amount*.

Solution

Percent × base = amount

Percent	×	purchase price	=	sales tax

$0.05 \times 6500 = n$

$325 = n$

The sales tax is $325.

You Try It 2
A car is purchased for $7350. The car license fee is 1.5% of the purchase price. How much is the license fee?

Your strategy

Your solution $110.25

Solutions on p. A33

Objective B To calculate ongoing expenses of owning a car

INSTRUCTOR NOTE
Point out to students that the same formula that is used to calculate a mortgage payment is used to calculate a car payment.

Besides the initial expenses of buying a car, there are continuing expenses involved in owning a car. These ongoing expenses include car insurance, gas and oil, general maintenance, and the monthly car payment. The monthly car payment is calculated in the same manner as monthly mortgage payments on a home loan. A monthly payment table, such as the one on page A6, is used to simplify calculation of monthly car payments.

Example 3
At a cost of \$0.27 per mile, how much does it cost to operate a car during a year the car is driven 15,000 miles?

Strategy
To find the cost, multiply the cost per mile by the number of miles driven.

Solution
15,000 × 0.27 = 4050 The cost is \$4050.

You Try It 3
At a cost of \$0.22 per mile, how much does it cost to operate a car during a year the car is driven 23,000 miles?

Your strategy

Your solution \$5060

Example 4
During 1 month the total gasoline bill was \$84 and the car was driven 1200 miles. What was the cost per mile for gasoline?

Strategy
To find the cost per mile for gasoline, divide the cost for gasoline by the number of miles driven.

Solution 84 ÷ 1200 = 0.07
The cost per mile was \$0.07.

You Try It 4
In a year when the total car insurance bill was \$360 and the car was driven 15,000 miles, what was the cost per mile for car insurance?

Your strategy

Your solution \$0.024

Example 5
A car is purchased for \$8500 with a down payment of \$1700. The balance is financed for 3 years at an annual interest rate of 9%. Find the monthly car payment.

Strategy
To find the monthly payment:

- Subtract the down payment from the purchase price to find the amount financed.
- Multiply the amount financed by the factor found in the monthly payment table on page A6.

Solution
8500 − 1700 = 6800

The amount financed is \$6800.

6800 × 0.0317997 ≈ 216.24

The monthly payment is \$216.24.

You Try It 5
A truck is purchased for \$15,900 with a down payment of \$3975. The balance is financed for 4 years at an annual interest rate of 8%. Find the monthly payment.

Your strategy

Your solution \$291.12

Solutions on p. A33

6.5 Exercises

Objective A Application Problems

1. Amanda has saved $780 to make a down payment on a car. The car dealer requires a down payment of 12% of the purchase price. Does she have enough money to make the down payment on a car that costs $7100? No

2. A jeep was purchased for $9288. A down payment of 15% of the purchase price was required. How much was the down payment? $1393.20

3. A drapery installer bought a minivan to carry drapery samples. The purchase price of the van was $16,500, and a 4.5% sales tax was paid. How much was the sales tax? $742.50

4. A delivery truck for the Dixieline Lumber Company was purchased for $18,500. A sales tax of 4% of the purchase price was paid. Find the sales tax. $740

5. A license fee of 2% of the purchase price of a truck is to be paid on a pickup truck costing $12,500. How much is the license fee for the truck? $250

6. Your state charges a license fee of 1.5% on the purchase price of a car. How much is the license fee for a car that costs $6998? $104.97

7. An electrician bought a $12,000 flatbed truck. A state license fee of $175 and a sales tax of 3.5% of the purchase price are required.
   a. Find the sales tax. $420
   b. Find the total cost of the sales tax and the license fee. $595

8. A physical therapist bought a car for $9375 and made a down payment of $1875. The sales tax is 5% of the purchase price.
   a. Find the sales tax. $468.75
   b. Find the total cost of the sales tax and the down payment. $2343.75

9. Martin bought a motorcycle for $2200 and made a down payment that is 25% of the purchase price.
   a. Find the down payment. $550
   b. Find the amount financed. $1650

10. A carpenter bought a utility van for $14,900 and made a down payment that is 15% of the purchase price.
    a. Find the down payment. $2235
    b. Find the amount financed. $12,665

11. An author bought a sports car for $35,000 and made a down payment of 20% of the purchase price. Find the amount financed. $28,000

12. Tania purchased a new car for $13,500 and made a down payment of 25% of the cost. Find the amount financed. $10,125

Objective B Application Problems

Solve. Use the monthly payment table on page A6. Round to the nearest cent.

13. A rancher financed $14,000 for the purchase of a truck through a credit union at 9% interest for 4 years. Find the monthly truck payment. $348.39

14. A car loan of $8000 is financed for 3 years at an annual interest rate of 10%. Find the monthly car payment. $258.14

15. An estimate of the cost of owning a compact car is $0.32 per mile. Using this estimate, how much does it cost to operate a car during a year in which the car is driven 16,000 miles. $5120

16. An estimate of the cost of care and maintenance of automobile tires is $0.015 per mile. Using this estimate, how much would it cost for care and maintenance during a year in which the car is driven 14,000 miles? $210

17. A family spent $1600 on gas, oil, and car insurance during a period in which the car was driven 14,000 miles. Find the cost per mile for gas, oil, and car insurance. $0.11

18. Last year you spent $1050 for gasoline for your car. The car was driven 15,000 miles. What was your cost per mile for gasoline? $0.07

19. Elena's monthly car payment is $143.50. During a month in which $68.75 of the monthly payment is principal, how much of the payment is interest? $74.75

20. The cost for a pizza delivery truck for the year included $1870 in truck payments, $1200 for gasoline, and $675 for insurance. Find the total cost for truck payments, gasoline, and insurance for the year. $3745

21. The city of Colton purchased a fire truck for $82,000 and made a down payment of $5400. The balance is financed for 5 years at an annual rate of 9%.
    a. Find the amount financed. $76,600
    b. Find the monthly truck payment. $1590.09

22. A used car is purchased for $4995, and a down payment of $995 is made. The balance is financed for 3 years at an interest rate of 8%.
    a. Find the amount financed. $4000
    b. Find the monthly car payment. $125.35

23. An artist purchased a new car costing $27,500 and made a down payment of $5500. The balance is financed for 3 years at an annual interest rate of 10%. Find the monthly car payment. $709.88

24. A half-ton truck with a camper is purchased for $19,500, and a down payment of $2500 is made. The balance is financed for 4 years at an annual interest rate of 9%. Find the monthly payment. $423.05

APPLYING THE CONCEPTS

25. One bank offers a 4-year car loan at an annual interest rate of 10% plus a loan application fee of $45. A second bank offers 4-year car loans at an annual interest rate of 11% but charges no loan application fee. If you need to borrow $5800 to purchase a car, which of the two bank loans has the lesser loan costs? Assume you keep the car for 4 years. 10% with application fee

26. How much interest is repaid on a 5-year car loan of $9000 if the interest rate is 9%? $2209.54

6.6 Wages

Objective A ***To calculate commissions, total hourly wages, and salaries***

Commissions, hourly wage, and salary are three ways to receive payment for doing work.

Commissions are usually paid to salespersons and are calculated as a percent of total sales.

➡ As a real estate broker, Emma Smith receives a commission of 4.5% of the selling price of a house. Find the commission she earned for selling a home for $75,000.

To find the commission Emma earned, solve the basic percent equation for *amount*.

Percent	×	base	=	amount
Commission rate	×	total sales	=	commission
0.045	×	75,000	=	3375

The commission is $3375.

An employee who receives an **hourly wage** is paid a certain amount for each hour worked.

➡ A plumber receives an hourly wage of $13.25. Find the plumber's total wages for working 37 hours.

To find the plumber's total wages, multiply the hourly wage by the number of hours worked.

Hourly wage	×	number of hours worked	=	total wages
13.25	×	37	=	490.25

The plumber's total wages for working 37 hours are $490.25.

An employee who is paid a **salary** receives payment based on a weekly, biweekly (every other week), monthly, or annual time schedule. Unlike the employee who receives an hourly wage, the salaried worker does not receive additional pay for working more than the regularly scheduled work day.

➡ Ravi Basar is a computer operator who receives a weekly salary of $395. Find his salary for 1 month (4 weeks).

To find Ravi's salary for 1 month, multiply the salary per pay period by the number of pay periods.

Salary per pay period	×	number of pay periods	=	total salary
395	×	4	=	1580

Ravi's total salary for 1 month is $1580.

Example 1

A pharmacist's hourly wage is \$28. On Saturday, the pharmacist earns time and a half ($1\frac{1}{2}$ times the regular hourly wage). How much does the pharmacist earn for working 6 hours on Saturday?

Strategy

To find the pharmacist's earnings:

- Find the hourly wage for working on Saturday by multiplying the hourly wage by $1\frac{1}{2}$.
- Multiply the product by the number of hours worked.

Solution

28 × 1.5 = 42 42 × 6 = 252

The pharmacist earns \$252.

You Try It 1

A construction worker's hourly wage is \$8.50. The worker earns double time (2 times the regular hourly wage) for working overtime. How much does the worker earn for working 8 hours of overtime?

Your strategy

Your solution \$136

Example 2

An efficiency expert received a contract for \$3000. The consultant spent 75 hours on the project. Find the consultant's hourly wage.

Strategy

To find the hourly wage, divide the total earnings by the number of hours worked.

Solution

3000 ÷ 75 = 40 The hourly wage was \$40.

You Try It 2

A contractor for a bridge project receives an annual salary of \$28,224. What is the contractor's salary per month?

Your strategy

Your solution \$2352

Example 3

Dani Greene receives \$18,500 per year plus a $5\frac{1}{2}\%$ commission on sales over \$100,000. During one year, Dani sold \$150,000 worth of computers. Find Dani's total earnings for the year.

Strategy

To find the total earnings:

- Find the commission earned by multiplying the commission rate by sales over \$100,000.
- Add the commission to the annual pay.

Solution

150,000 − 100,000 = 50,000

50,000 × 0.055 = 2750 commission

18,500 + 2750 = 21,250

Dani earned \$21,250.

You Try It 3

An insurance agent receives \$12,000 per year plus a $9\frac{1}{2}\%$ commission on sales over \$50,000. During one year, the agent's sales totaled \$175,000. Find the agent's total earnings for the year.

Your strategy

Your solution \$23,875

Solutions on p. A34

6.6 Exercises

Objective A *Application Problems*

1. Lewis works in a clothing store and earns $5.50 per hour. How much does he earn in a 40-hour week? $220

2. Sasha pays a gardener an hourly wage of $5.25. How much does she pay the gardener for working 25 hours? $131.25

3. A real estate agent receives a 3% commission for selling a house. Find the commission that the agent earned for selling a house for $98,500. $2955

4. Ron Caruso works as an insurance agent and receives a commission of 40% of the first year's premium. Find Ron's commission for selling a life insurance policy with a first-year premium of $1050. $420

5. A stock broker receives a commission of 1.5% of the price of stock that is bought or sold. Find the commission on 100 shares of stock that were bought for $5600. $84

6. The owner of the Carousel Art Gallery receives a commission of 20% of paintings that are sold on consignment. Find the commission on a painting that sold for $22,500. $4500

7. As an Italian Language teacher, Keisha Brown receives an annual salary of $38,928. How much does Keisha receive each month? $3244

8. An apprentice plumber receives an annual salary of $19,680. How much does the plumber receive per month? $1640

9. An electrician's hourly wage is $15.80. For working overtime, the electrician earns double time. What is the electrician's hourly wage for working overtime? $31.60

10. Carlos receives a commission of 12% of his weekly sales as a sales representative for a medical supply company. Find the commission he earned during a week in which sales were $4500. $540

11. A golf pro receives a commission of 25% for selling a golf set. Find what commission the pro earned for selling a golf set costing $450. $112.50

12. Steven receives $1.75 per square yard to install carpet. How much does he receive for installing 160 square yards of carpet? $280

13. A typist charged $1.75 per page for typing technical material. How much does the typist earn for typing a 225-page book? $393.75

14. A nuclear chemist received $15,000 for consulting fees while working on a nuclear power plant. The chemist worked 120 hours on the project. Find the consultant's hourly wage. $125 per hour

15. Maxine received $3400 for working on a project as a computer consultant for 40 hours. Find her hourly wage. $85 per hour

16. Gil Stratton's hourly wage is $10.78. For working overtime, he receives double time.
a. What is Gil's hourly wage for working overtime? $21.56
b. How much does he earn for working 16 hours of overtime? $344.96

17. Mark is a lathe operator and receives an hourly wage of $8.50. When working on Saturday, he receives time and a half.
a. What is Mark's hourly wage on Saturday? $12.75
b. How much does he earn for working 8 hours on Saturday? $102

18. A stock clerk at a supermarket earns $6.20 an hour. For working the night shift, the clerk's wage increases by 15%.
a. What is the increase in hourly pay for working the night shift? $0.93
b. What is the clerk's hourly wage for working the night shift? $7.13

19. A nurse earns $8.50 an hour. For working the night shift, the nurse receives a 10% increase in pay.
a. What is the increase in hourly pay for working the night shift? $0.85
b. What is the hourly pay for working the night shift? $9.35

20. Tony's hourly wage as a service station attendant is $6.40. For working the night shift, his wage is increased 25%. What is Tony's hourly wage for working the night shift? $8.00

21. Nicole Tobin, a door-to-door salesperson, receives a salary of $150 per week plus a commission of 15% on all sales over $1500. Find her earnings during a week in which sales totaled $3000. $375

APPLYING THE CONCEPTS

The table at the right shows the *Money* magazine ranking of the ten most desirable careers of 1994. Complete Exercises 22–25 with information obtained from this table.

1994 Rank	*1992 Rank*	*Occupation*	*Median Annual Earnings*	*11-Year Job Growth*
1	31	Computer systems analyst	$42,700	110%
2	3	Physician	148,000	35
3	50	Physical therapist	37,200	88
4	13	Electrical engineer	59,100	24
5	9	Civil engineer	55,800	24
6	7	Pharmacist	47,500	29
7	29	Psychologist	53,000	48
8	2	Geologist	50,800	22
9	15	High school teacher	32,500	37
10	5	School principal	57,300	23

Source: *Money*, March 1994.

22. Find the monthly earnings for a civil engineer. $4650

23. Find the difference in monthly salary between a physician and an electrical engineer. $7408.33

24. What percent of a high school teacher's salary does a school principal receive? 176.3%

25. In 1994, there were 455,000 computer systems analysts in the United States. Find the projected increase in the number of computer analyst jobs between now and 2005 (11 years from now). 500,500

6.7 Bank Statements

Objective A To calculate checkbook balances

POINT OF INTEREST

There are a number of computer programs that serve as "electronic" checkbooks. With these programs, you can pay your bills by using a computer to write the check and then transmit the check over telephone lines using a modem. Paying your federal income tax is one of the few tasks that cannot be accomplished through these electronic checkbooks.

A checking account can be opened at most banks or savings and loan associations by depositing an amount of money in the bank. A checkbook contains checks and deposit slips and a checkbook register in which to record checks written and amounts deposited in the checking account. Each time a check is written, the amount of the check is subtracted from the amount in the account. When a deposit is made, the amount deposited is added to the amount in the account.

A portion of a checkbook register is shown below. The account holder had a balance of $587.93 before writing two checks, one for $286.87 and the other for $102.38, and making one deposit of $345.00.

RECORD ALL CHARGES OR CREDITS THAT AFFECT YOUR ACCOUNT

NUMBER	DATE	DESCRIPTION OF TRANSACTION	PAYMENT/DEBIT (-)		√ T	FEE (IF ANY) (-)	DEPOSIT/CREDIT (+)		BALANCE	
									$ 587	93
108	8/4	Plumber	$ 286	87		$	$		301	06
109	8/10	Car Payment	102	38					198	68
	8/14	Deposit					345	00	543	68

To find the current checking account balance, subtract the amount of each check from the previous balance. Then add the amount of the deposit.

The current checking account balance is $543.68.

Example 1 A mail carrier had a checking account balance of $485.93 before writing two checks, one for $18.98 and another for $35.72, and making a deposit of $250. Find the current checking account balance.

Strategy To find the current balance:

- Subtract the amount of each check from the old balance.
- Add the amount of the deposit.

Solution

485.93	
− 18.98	first check
466.95	
− 35.72	second check
431.23	
+ 250.00	deposit
681.23	

The current checking account balance is $681.23.

You Try It 1 A cement mason had a checking account balance of $302.46 before writing a check for $20.59 and making two deposits, one in the amount of $176.86 and another in the amount of $94.73. Find the current checking account balance.

Your strategy

Your solution $553.46

Solution on p. A35

Objective B To balance a checkbook

Each month a bank statement is sent to the account holder. The bank statement shows the checks that the bank has paid, the deposits received, and the current bank balance.

A bank statement and checkbook register are shown on the next page.

Balancing the checkbook, or determining if the checking account balance is accurate, requires a number of steps.

1. In the checkbook register, put a check (√) by each check paid by the bank and each deposit recorded by the bank.

RECORD ALL CHARGES OR CREDITS THAT AFFECT YOUR ACCOUNT

NUMBER	DATE	DESCRIPTION OF TRANSACTION	PAYMENT/DEBIT (-)		√ T	FEE (IF ANY) (-)	DEPOSIT/CREDIT (+)		BALANCE	
									$ 840	27
263	5/20	Dentist	$ 25	00	√	$	$		815	27
264	5/22	Meat Market	33	61	√				781	66
265	5/22	Gas Company	67	14					714	52
	5/29	Deposit			√		192	00	906	52
266	5/29	Pharmacy	18	95	√				887	57
267	5/30	Telephone	43	85					843	72
268	6/2	Groceries	43	19	√				800	53
	6/3	Deposit			√		215	00	1015	53
269	6/7	Insurance	103	00	√				912	53
	6/10	Deposit					225	00	1137	53
270	6/15	Clothing Store	16	63	√				1120	90
271	6/18	Newspaper	7	00					1113	90

CHECKING ACCOUNT Monthly Statement — Account Number: 924-297-8

Date	Transaction	Amount	Balance
5/20	OPENING BALANCE		840.27
5/21	CHECK	25.00	815.27
5/23	CHECK	33.61	781.66
5/29	DEPOSIT	192.00	973.66
6/1	CHECK	18.95	954.71
6/1	INTEREST	4.47	959.18
6/3	CHECK	43.19	915.99
6/3	DEPOSIT	215.00	1130.99
6/9	CHECK	103.00	1027.99
6/16	CHECK	16.63	1011.36
6/20	SERVICE CHARGE	3.00	1008.36
6/20	CLOSING BALANCE		1008.36

2. Add to the current checkbook balance all checks that have been written but have not yet been paid by the bank and any interest paid on the account.

Current checkbook balance:	1113.90
Checks: 265	67.14
267	43.85
271	7.00
Interest:	+ 4.47
	1236.36

3. Subtract any service charges and any deposits not yet recorded by the bank. This is the checkbook balance.

Service charge:	− 3.00
	1233.36
Deposit:	− 225.00
Checkbook balance:	1008.36

4. Compare the balance with the bank balance listed on the bank statement. If the two numbers are equal, the bank statement and checkbook balance.

Checking bank balance from bank statement		Checkbook balance
$1008.36	=	$1008.36

The bank statement and checkbook balance.

RECORD ALL CHARGES OR CREDITS THAT AFFECT YOUR ACCOUNT										
NUMBER	DATE	DESCRIPTION OF TRANSACTION	PAYMENT/DEBIT (-)		√ T	FEE (IF ANY) (-)	DEPOSIT/CREDIT (+)		BALANCE $ 1620	42
413	3/2	Car Payment	$ 132	15	√	$	$		1488	27
414	3/2	Utility	67	14	√				1421	13
415	3/5	Restaurant - Dinner for 4	78	14					1342	99
	3/8	Deposit			√		1842	66	3185	65
416	3/10	House Payment	672	14	√				2513	51
417	3/14	Insurance	177	10					2336	41

CHECKING ACCOUNT Monthly Statement — Account Number: 924-297-8

Date	Transaction	Amount	Balance
3/1	OPENING BALANCE		1620.42
3/4	CHECK	132.15	1488.27
3/5	CHECK	67.14	1421.13
3/8	DEPOSIT	1842.66	3263.79
3/10	INTEREST	6.77	3270.56
3/12	CHECK	672.14	2598.42
3/25	SERVICE CHARGE	2.00	2596.42
3/30	CLOSING BALANCE		2596.42

Balance the bank statement shown above.

1. In the checkbook register, put a check (√) by each check paid by the bank and each deposit recorded by the bank.

2. Add to the current checkbook balance all checks that have been written but have not yet been paid by the bank and any interest paid on the account.

   Current checkbook balance: 2336.41
   Checks: 415 78.14
   417 177.10
   Interest: + 6.77
   2598.42

3. Subtract any service charges and any deposits not yet recorded by the bank. This is the checkbook balance.

   Service charge: − 2.00
   2596.42

   Checkbook balance: 2596.42

4. Compare the balance with the bank balance listed on the bank statement. If the two numbers are equal, the bank statement and checkbook balance.

Closing bank balance from bank statement		Checkbook balance:
$2596.42	=	$2596.42

   The bank statement and checkbook balance.

RECORD ALL CHARGES OR CREDITS THAT AFFECT YOUR ACCOUNT										
NUMBER	DATE	DESCRIPTION OF TRANSACTION	PAYMENT/DEBIT (-)		√ T	FEE (IF ANY) (-)	DEPOSIT/CREDIT (+)		BALANCE	
									$ 412	64
345	1/14	Phone Bill	$ 34	75	√	$	$		377	89
346	1/19	Magazine	8	98	√				368	91
347	1/23	Theatre Tickets	45	00					323	91
	1/31	Deposit					947	00	1270	91
348	2/5	Cash	250	00	√				1020	91
349	2/12	Rent	440	00					580	91

CHECKING ACCOUNT Monthly Statement — Account Number: 924-297-8

Date	Transaction	Amount	Balance
1/10	OPENING BALANCE		412.64
1/18	CHECK	34.75	377.89
1/23	CHECK	8.98	368.91
1/31	DEPOSIT	947.00	1315.91
2/1	INTEREST	4.52	1320.43
2/10	CHECK	250.00	1070.43
2/10	CLOSING BALANCE		1070.43

Example 2

Balance the bank statement shown above.

Solution

Current checkbook balance:	580.91
Checks: 347	45.00
349	440.00
Interest:	+ 4.52
	1070.43
Service charge:	− 0.00
	1070.43
Deposit:	− 0.00
Checkbook balance:	1070.43

Closing bank balance from bank statement: $1070.43

Checkbook balance: $1070.43

The bank statement and checkbook balance.

RECORD ALL CHARGES OR CREDITS THAT AFFECT YOUR ACCOUNT

NUMBER	DATE	DESCRIPTION OF TRANSACTION	PAYMENT/DEBIT (-)		√ T	FEE (IF ANY) (-)	DEPOSIT/CREDIT (+)		BALANCE	
									$ 603	17
	2/15	Deposit	$			$	$ 523	84	1127	01
234	2/20	Mortgage	473	21	√				653	80
235	2/27	Cash	200	00	√				453	80
	3/1	Deposit					523	84	977	64
236	3/12	Insurance	275	50	√				702	14
237	3/12	Telephone	48	73					653	41

CHECKING ACCOUNT Monthly Statement — Account Number: 314-271-4

Date	Transaction	Amount	Balance
2/14	OPENING BALANCE		603.17
2/15	DEPOSIT	523.84	1127.01
2/21	CHECK	473.21	653.80
2/28	CHECK	200.00	453.80
3/1	INTEREST	2.11	455.91
3/14	CHECK	275.50	180.41
3/14	CLOSING BALANCE		180.41

You Try It 2

Balance the bank statement shown above.

Your solution

The bank statement and checkbook balance.

Solution on p. A35

6.7 Exercises

Objective A Application Problems

1. You had a checking account balance of $342.51 before making a deposit of $143.81. What is your new checking account balance? $486.32

2. Carmen had a checking account balance of $493.26 before writing a check for $48.39. What is the current checking account balance? $444.87

3. A real estate firm had a balance of $2431.76 in its rental property checking account. What is the balance in this account after a check for $1209.29 has been written? $1222.47

4. The business checking account for R and R Tires showed a balance of $1536.97. What is the balance in this account after a deposit of $439.21 has been made? $1976.18

5. A nutritionist had a checking account balance of $1204.63 before writing one check for $119.27 and another check for $260.09. Find the current checkbook balance. $825.27

6. Sam had a checking account balance of $3046.93 before writing a check for $1027.33 and making a deposit of $150.00. Find the current checkbook balance. $2169.60

7. The business checking account for Rachael's Dry Cleaning had a balance of $3476.85 before a deposit of $1048.53 was made. The store manager then wrote checks, one for $848.37 and another for $676.19. Find the current checkbook balance. $3000.82

8. Joel had a checking account balance of $427.38 before a deposit of $127.29 was made. Joel then wrote two checks, one for $43.52 and one for $249.78. Find the current checkbook balance. $261.37

9. A carpenter had a checkbook balance of $104.96 before making a deposit of $350 and writing a check for $71.29. Is there enough money in the account to purchase a refrigerator for $375? Yes

10. A taxi driver had a checkbook balance of $149.85 before making a deposit of $245 and writing a check for $387.68. Is there enough money in the account for the bank to pay the check? Yes

11. A sporting goods store has the opportunity to buy downhill skis and cross-country skis at a manufacturer's close-out sale. The downhill skis will cost $3500, and the cross-country skis will cost $2050. There is currently $5625.42 in the sporting goods store's checking account. Is there enough money in the account to make both purchases by check? Yes

12. A lathe operator's current checkbook balance is $643.42. The operator wants to purchase a utility trailer for $225 and a used piano for $450. Is there enough money in the account to make the two purchases? No

13. Balance the checkbook.

RECORD ALL CHARGES OR CREDITS THAT AFFECT YOUR ACCOUNT

NUMBER	DATE	DESCRIPTION OF TRANSACTION	PAYMENT/DEBIT (-)		√ T	FEE (IF ANY) (-)	DEPOSIT/CREDIT (+)		BALANCE	
									$ 466	79
223	3/2	Groceries	$ 67	32	√	$	$		399	47
	3/5	Deposit					560	70	960	17
224	3/5	Rent	460	00	√				500	17
225	3/7	Gas & Electric	42	35	√				457	82
226	3/7	Cash	100	00	√				357	82
227	3/7	Insurance	118	44					239	38
228	3/7	Credit Card	119	32					120	06
229	3/12	Dentist	42	00	√				78	06
230	3/13	Drug Store	17	03	√				61	03
	3/19	Deposit					560	70	621	73
231	3/22	Car Payment	141	35	√				480	38
232	3/25	Cash	100	00	√				380	38
233	3/25	Oil Company	66	40					313	98
234	3/28	Plumber	55	73	√				258	25
235	3/29	Department Store	88	39					169	86

CHECKING ACCOUNT Monthly Statement — Account Number: 122-345-1

Date	Transaction	Amount	Balance
3/1	OPENING BALANCE		466.79
3/5	DEPOSIT	560.70	1027.49
3/7	CHECK	67.32	960.17
3/8	CHECK	460.00	500.17
3/8	CHECK	100.00	400.17
3/9	CHECK	42.35	357.82
3/12	CHECK	118.44	239.38
3/14	CHECK	42.00	197.38
3/18	CHECK	17.03	180.35
3/19	DEPOSIT	560.70	741.05
3/25	CHECK	141.35	599.70
3/27	CHECK	100.00	499.70
3/29	CHECK	55.73	443.97
3/30	INTEREST	13.22	457.19
4/1	CLOSING BALANCE		457.19

The bank statement and checkbook balance.

14. Balance the checkbook.

RECORD ALL CHARGES OR CREDITS THAT AFFECT YOUR ACCOUNT

NUMBER	DATE	DESCRIPTION OF TRANSACTION	PAYMENT/DEBIT (-)		√ T	FEE (IF ANY) (-)	DEPOSIT/CREDIT (+)		BALANCE	
									$ 219	43
	5/1	Deposit	$			$	$ 219	14	438	57
515	5/2	Electric Bill	22	35					416	22
516	5/2	Groceries	55	14					361	08
517	5/4	Insurance	122	17					238	91
518	5/5	Theatre Tickets	24	50					214	41
	5/8	Deposit					219	14	433	55
519	5/10	Telephone	17	39					416	16
520	5/12	Newspaper	12	50					403	66
	5/15	Interest					7	82	411	48
	5/15	Deposit					219	14	630	62
521	5/20	Hotel	172	90					457	72
522	5/21	Credit Card	113	44					344	28
523	5/22	Eye Exam	42	00					302	28
524	5/24	Groceries	77	14					225	14
525	5/24	Deposit					219	14	444	28
526	5/25	Oil Company	44	16					400	12
527	5/30	Car Payment	88	62					311	50
528	5/30	Doctor	37	42					274	08

CHECKING ACCOUNT Monthly Statement — Account Number: 122-345-1

Date	Transaction	Amount	Balance
5/1	OPENING BALANCE		219.43
5/1	DEPOSIT	219.14	438.57
5/3	CHECK	55.14	383.43
5/4	CHECK	22.35	361.08
5/6	CHECK	24.50	336.58
5/8	CHECK	122.17	214.41
5/8	DEPOSIT	219.14	433.55
5/15	INTEREST	7.82	441.37
5/15	CHECK	17.39	423.98
5/15	DEPOSIT	219.14	643.12
5/23	CHECK	42.00	601.12
5/23	CHECK	172.90	428.22
5/24	CHECK	77.14	351.08
5/24	DEPOSIT	219.14	570.22
5/30	CHECK	88.62	481.60
6/1	CLOSING BALANCE		481.60

The bank statement and checkbook balance.

Objective B *Application Problems*

15. Balance the checkbook.

RECORD ALL CHARGES OR CREDITS THAT AFFECT YOUR ACCOUNT

NUMBER	DATE	DESCRIPTION OF TRANSACTION	PAYMENT/DEBIT (-)		√ T	FEE (IF ANY) (-)	DEPOSIT/CREDIT (+)		BALANCE $	
									1035	18
218	7/2	Mortgage	$ 284	60		$	$		750	58
219	7/4	Telephone	23	36					727	22
220	7/7	Cash	200	00					527	22
	7/12	Deposit					792	60	1319	82
221	7/15	Insurance	192	30					1127	52
222	7/18	Investment	100	00					1027	52
223	7/20	Credit Card	214	83					812	69
	7/26	Deposit					792	60	1605	29
224	7/27	Department Store	113	37					1491	92

CHECKING ACCOUNT Monthly Statement Account Number: 122-345-1

Date	Transaction	Amount	Balance
7/1	OPENING BALANCE		1035.18
7/1	INTEREST	5.15	1040.33
7/4	CHECK	284.60	755.73
7/6	CHECK	23.36	732.37
7/12	DEPOSIT	792.60	1524.97
7/20	CHECK	192.30	1332.67
7/24	CHECK	100.00	1232.67
7/26	DEPOSIT	792.60	2025.27
7/28	CHECK	200.00	1825.27
7/30	CLOSING BALANCE		1825.27

The bank statement and checkbook balance.

APPLYING THE CONCEPTS

16. When a check is written, the amount is subtracted from the balance.

17. When a deposit is made, the amount is added to the balance.

18. In checking the bank statement, add to the checkbook balance all checks that have been written but not processed.

19. In checking the bank balance, subtract any service charge and any deposits not yet recorded.

20. Define the words credit and debit as they apply to checkbooks.
[W] A credit is a deposit. A debit is a payment or withdrawal.

Project in Mathematics

Credit Card Finance Charges All credit card companies charge a *fee* (finance charge) when a credit card balance is not paid within a certain number of days (usually 25 days) of the *billing date* (the date the credit card bill is sent). There may also be an annual fee. The table below shows the charges for five banks in February 1994.

Bank	*Annual Interest Rate*	*Annual Fee*
Arkansas Federal Little Rock, Arkansas	7.75%	\$35
Federal Savings Bank Rogers, Arkansas	7.92%	\$33
Central Carolina Bank Columbus, Georgia	8.50%	\$29
People's Bank Bridgeport, Connecticut	11.50%	\$25
USAA Federal Savings San Antonio, Texas	12.50%	\$0

The amount of the monthly finance charge is based on the unpaid balance. For instance, using Arkansas Federal, if your unpaid balance for one month is \$250, then the monthly finance charge is calculated using the simple interest formula.

$$\begin{aligned}\text{Interest} &= \text{principal} \times \text{annual interest rate} \times \text{time}\\ &= 250 \times 0.0775 \times \frac{1}{12} \qquad \bullet\ 1\text{ month} = \frac{1}{12}\text{ year.}\\ &\approx 1.61458\end{aligned}$$

The monthly finance charge is \$1.61.

Exercises:

1. Suppose you have an average monthly unpaid balance of \$356.00 for 12 months. With which of the banks will your *annual* finance charges be least? USAA Federal Savings
2. Suppose you have an average monthly unpaid balance of \$650.00 for 12 months. With which of the banks will your *annual* finance charges be least? USAA Federal Savings
3. Using your calculator, experiment with some average monthly unpaid balances and determine what average monthly unpaid balance would result in essentially the same finance charges whether you used Arkansas Federal or USAA Federal Savings. approximately \$737

Chapter Summary

Key Words The *unit cost* is the cost of one item.

Percent increase is used to show how much a quantity has increased over its original value.

Cost is the price that a business pays for a product.

Selling price is the price at which a business sells a product to a customer.

Markup is the difference between selling price and cost.

Markup rate is frequently expressed as a percent of a product's cost.

Percent decrease is used to show how much a quantity has decreased from its original value.

Sale price is the price that has been reduced from the regular price.

Discount is the difference between the regular price and the sale price.

Discount rate is frequently stated as a percent of a product's regular price.

Interest is the amount of money paid for the privilege of using someone else's money.

Principal is the amount of money originally deposited or borrowed.

The percent used to determine the amount of interest is the *interest rate*.

Interest computed on the original amount is called *simple interest*.

Compound interest is computed not only on the original principal but also on interest already earned.

The *mortgage* is the amount that is borrowed to buy real estate.

The loan origination fee is usually a percent of the mortgage and is expressed in *points*.

Commissions are usually paid to the salespersons and are calculated as a percent of total sales.

An employee who receives an *hourly wage* is paid a certain amount for each hour worked.

An employee who is paid a *salary* receives payment based on a weekly, biweekly, monthly, or annual time schedule.

Essential Rules

To Find Unit Cost To find the unit cost, divide the total cost by the number of units.

To Find Total Cost To find the total cost, multiply the unit cost by the number of units.

Basic Markup Equations:

$$\text{Selling price} = \text{cost} + \text{markup}$$
$$S = C + M$$

$$\text{Markup} = \text{markup rate} \times \text{cost}$$
$$M = r \times C$$

Basic Discount Equations:

$$\text{Sale price} = \text{regular price} - \text{discount}$$
$$S = R - D$$

$$\text{Discount} = \text{discount rate} \times \text{regular price}$$
$$D = r \times R$$

Annual Simple Interest Equation:

$$\text{Principal} \times \begin{matrix}\text{annual}\\ \text{interest rate}\end{matrix} \times \begin{matrix}\text{time}\\ \text{in years}\end{matrix} = \text{interest}$$
$$P \times r \times t = I$$

Chapter Review

SECTION 6.1

1. A 20-ounce box of cereal costs $2.90. Find the unit cost.
14.5¢/ounce

2. Twenty-four ounces of a mouthwash cost $3.49. A 60-ounce container of the same kind of mouthwash costs $8.40. Which is the better buy?
60 ounces for $8.40

SECTION 6.2

3. An oil stock was bought for $\$42\frac{3}{8}$ per share; 6 months later the stock was selling for $\$55\frac{1}{4}$ per share. Find the percent increase in the price of the stock for the 6 months. Round to the nearest tenth of a percent. 30.4%

4. A professional baseball player received a salary of $1 million last year. This year the player signed a contract paying $12 million over 4 years. Find the yearly percent increase in the player's salary. 200%

5. Techno-Center uses a markup rate of 35% on all computer systems. Find the selling price of a computer system that costs the store $1540. $2079

6. Last year an oil company had earnings of $4.12 per share. This year the earnings are $4.73 per share. What is the percent increase in earnings per share? Round to the nearest percent. 15%

7. A sporting goods store uses a markup rate of 40%. What is the markup on a ski suit that costs the store $180? $72

8. A suit that regularly costs $235 is on sale for 40% off the regular price. Find the sale price. $141

SECTION 6.3

9. A contractor borrowed $100,000 from a credit union for 9 months at an annual interest rate of 9%. What is the simple interest due on the loan? $6750

10. Pros' Sporting Goods borrowed $30,000 at an annual interest rate of 8% for 6 months. Find the simple interest due on the loan. $1200

11. A computer programmer invested $25,000 in a retirement account that pays 6% interest, compounded daily. What is the value of the investment in 10 years? Use the table on page A4. Round to the nearest cent. $45,550.75

12. A fast-food restaurant invested $50,000 in an account that pays 7% annual interest compounded quarterly. What is the value of the investment in 1 year? Use the table on page A4. $53,593

SECTION 6.4

13. Paula Mason purchased a home for $125,000. The lender requires a down payment of 15%. Find the amount of the down payment. $18,750

14. A credit union requires a borrower to pay $2\frac{1}{2}$ points for a loan. Find the origination fee for a loan of $75,000. $1875

15. The monthly mortgage payment for a condominium is $523.67. The owner must pay an annual property tax of $658.32. Find the total monthly payment for the mortgage and property tax. $578.53

16. The Sweeneys bought a home for $156,000. The family made a 10% down payment and financed the remainder with a 30-year loan with an annual interest rate of 7%. Find the monthly mortgage payment. Use the monthly payment table on page A6. Round to the nearest cent. $934.08

SECTION 6.5

17. A plumber bought a truck for $13,500. A state license of $315 and a sales tax of 6.25% of the purchase price are required. Find the total cost of the sales tax and the license fee. $1158.75

18. An account executive had car expenses of $1025.58 for insurance, $605.82 for gas, $37.92 for oil, and $188.27 for maintenance during a year in which 15,320 miles were driven. Find the cost per mile for these four items. Round to the nearest tenth of a cent. 12.1¢/mi

19. Mien pays a monthly car payment of $122.78. During a month in which $25.45 is principal, how much of the payment is interest? $97.33

20. A pickup truck with a slide-in camper is purchased for $14,450. A down payment of 8% is made, and the remaining cost is financed for 4 years at an annual interest rate of 9%. Find the monthly payment. Use the monthly schedule on page A6. Round to the nearest cent. $330.82

SECTION 6.6

21. The manager of the retail store at a ski resort receives a commission of 3% on all sales at the alpine shop. Find the total commission received during a month in which the shop had $108,000 in sales. $3240

22. Richard Valdez receives $12.60 per hour for working 40 hours a week and time and a half for working over 40 hours. Find his total income during a week in which he worked 48 hours. $655.20

SECTION 6.7

23. Luke had a checking account balance of $1568.45 before writing checks for $123.76, $756.45, and $88.77. He then deposited a check for $344.21. Find Luke's current checkbook balance. $943.68

24. The business checking account of a donut shop showed a balance of $9567.44 before checks of $1023.55, $345.44, and $23.67 were written and checks of $555.89 and $135.91 were deposited. Find the current checkbook balance. $8866.58

Chapter Test

1. Twenty feet of lumber cost $138.40. What is the cost per foot? $6.92 [6.1A]

2. Find the more economical purchase: 5 pounds of tomatoes for $1.65, or 8 pounds for $2.72. 5 pounds for $1.65 [6.1B]

3. Red snapper costs $4.15 per pound. Find the cost of $3\frac{1}{2}$ pounds. (Round to the nearest cent.) $14.53 [6.1C]

4. An exercise bicycle increased in price from $415 to $498. Find the percent increase in the cost of the exercise bicycle. 20% [6.2A]

5. Fifteen years ago a painting was priced at $6000. Today the same painting has a value of $15,000. Find the percent increase in the price of the painting during the 15 years. 150% [6.2A]

6. A department store uses a 40% markup rate. Find the selling price of a compact disc player that the store purchased for $215. $301 [6.2B]

7. A bookstore bought a paperback book for $5 and used a markup rate of 25%. Find the selling price of the book. $6.25 [6.2B]

8. The price of gold dropped from $850 per ounce to $360 per ounce. What percent decrease does this amount represent? (Round to the nearest tenth of a percent.) 57.6% [6.2C]

9. The price of a video camera dropped from $1120 to $896. What percent decrease does this price drop represent? 20% [6.2C]

10. A corner hutch with a regular price of $299 is on sale for 30% off the regular price. Find the sale price. $209.30 [6.2D]

11. A box of stationery that regularly sells for $4.50 is on sale for $2.70. Find the discount rate. 40% [6.2D]

12. A construction company borrowed $75,000 for 4 months at an annual interest rate of 8%. Find the simple interest due on the loan. $2000 [6.3A]

13. Jorge, who is self-employed, placed $30,000 in an account that pays 6% annual interest compounded quarterly. How much interest was earned in 10 years? Use the table on page A4. $24,420.60 [6.3B]

14. A savings and loan institution is giving mortgage loans that have a loan origination fee of $2\frac{1}{2}$ points. Find the loan origination fee on a home purchased with a loan of $134,000. $3350 [6.4A]

15. A new housing development offers homes with a mortgage of $222,000 for 25 years at an annual interest rate of 8%. Find the monthly mortgage payment. Use the table on page A6. $1713.44 [6.4B]

16. A $17,500 minivan was purchased with an 18% down payment. Find the amount financed. $14,350 [6.5A]

17. A rancher bought a pickup for $11,500 with a down payment of 15% of the cost. The balance is financed for 3 years at an annual interest rate of 7%. Find the monthly car payment. Use the table on page A6. $301.82 [6.5B]

18. Shaney receives an hourly wage of $13.40 an hour as an emergency room nurse. When called in at night, she receives time and a half. How much does Shaney earn in a week when working 30 hours at normal rates and 15 hours during the night? $703.50 [6.6A]

19. The business checking account for a pottery store had a balance of $7349.44 before checks for $1349.67 and $344.12 were written. The store manager then made a deposit of $956.60. Find the current checkbook balance. $6612.25 [6.7A]

20. Balance the checkbook shown.

RECORD ALL CHARGES OR CREDITS THAT AFFECT YOUR ACCOUNT

NUMBER	DATE	DESCRIPTION OF TRANSACTION	PAYMENT/DEBIT (-)		√ T	FEE (IF ANY) (-)	DEPOSIT/CREDIT (+)		BALANCE $	
									422	13
	8/1	House Payment	$ 213	72		$	$		208	41
	8/4	Deposit					552	60	761	01
	8/5	Plane Tickets	162	40					598	61
	8/6	Groceries	66	44					532	17
	8/10	Car Payment	122	37					409	80
	8/15	Deposit					552	60	962	40
	8/16	Credit Card	213	45					748	95
	8/18	Doctor	92	14					656	81
	8/22	Utilities	72	30					584	51
	8/28	T. V. Repair	78	20					506	31

CHECKING ACCOUNT Monthly Statement — Account Number: 122-345-1

Date	Transaction	Amount	Balance
8/1	OPENING BALANCE		422.13
8/3	CHECK	213.72	208.41
8/4	DEPOSIT	552.60	761.01
8/8	CHECK	66.44	694.57
8/8	CHECK	162.40	532.17
8/15	DEPOSIT	552.60	1084.77
8/23	CHECK	72.30	1012.47
8/24	CHECK	92.14	920.33
9/1	CLOSING BALANCE		920.33

The bank statement and checkbook balance. [6.7B]

Cumulative Review

1. Simplify $12 - (10 - 8)^2 \div 2 + 3$.
13 [1.6B]

2. Add: $3\frac{1}{3} + 4\frac{1}{8} + 1\frac{1}{12}$
$8\frac{13}{24}$ [2.4C]

3. Find the difference between $12\frac{3}{16}$ and $9\frac{5}{12}$.
$2\frac{37}{48}$ [2.5C]

4. Find the product of $5\frac{5}{8}$ and $1\frac{9}{15}$.
9 [2.6B]

5. Divide: $3\frac{1}{2} \div 1\frac{3}{4}$
2 [2.7B]

6. Simplify $\left(\frac{3}{4}\right)^2 \div \left(\frac{3}{8} - \frac{1}{4}\right) + \frac{1}{2}$.
5 [2.8C]

7. Divide: $0.059\overline{)3.0792}$.
Round to the nearest tenth.
52.2 [3.5A]

8. Convert $\frac{17}{12}$ to a decimal. Round to the nearest thousandth.
1.417 [3.6A]

9. Write "$410 in 8 hours" as a unit rate.
$51.25/hour [4.2B]

10. Solve the proportion $\frac{5}{n} = \frac{16}{35}$.
Round to the nearest hundredth.
$n = 10.94$ [4.3B]

11. Write $\frac{5}{8}$ as a percent.
62.5% [5.1B]

12. Find 6.5% of 420.
27.3 [5.2A]

13. Write 18.2% as a decimal.
0.182 [5.1A]

14. What percent of 20 is 8.4?
42% [5.3A]

15. 30 is 12% of what?
250 [5.4A]

16. 65 is 42% of what? Round to the nearest hundredth.
154.76 [5.5A]

17. A series of late summer storms produced rainfall of $3\frac{3}{4}$, $8\frac{1}{2}$, and $1\frac{2}{3}$ inches during a 3-week period. Find the total rainfall during the 3 weeks. $13\frac{11}{12}$ inches [2.4D]

18. The Homer family pays $\frac{1}{5}$ of its total monthly income for taxes. The family has a total monthly income of $2850. Find the amount of the monthly income that the Homers pay in taxes. $570 [2.6C]

19. In 5 years, the cost of a scientific calculator went from $75 to $30. What is the ratio of the decrease in price to the original price? $\frac{3}{5}$ [4.1B]

20. A compact car drove 417.5 miles on 12.5 gallons of gasoline. Find the number of miles driven per gallon of gasoline. 33.4 miles per gallon [4.2C]

21. A 14-pound turkey costs $12.96. Find the unit cost. (Round to the nearest cent.) $0.93 per pound [4.2C]

22. Eighty shares of a stock paid a dividend of $112. At the same rate, find the dividend on 200 shares of the stock. $280 [4.3C]

23. A video camera that regularly sells for $900 is on sale for 20% off the regular price. What is the sale price? $720 [5.2B]

24. A department store bought a portable disc player for $85 and used a markup rate of 40%. Find the selling price of the disc player. $119 [6.2B]

25. Sook Kim, an elementary school teacher, received an increase in salary from $2800 per month to $3024 per month. Find the percent increase in her salary. 8% [6.2A]

26. A contractor borrowed $120,000 for 6 months at an annual interest rate of 10%. How much simple interest is due on the loan? $6000 [6.3A]

27. A red sports car was purchased for $14,000, and a down payment of $2000 was made. The balance is financed for 3 years at an annual interest rate of 9%. Find the monthly payment. Use the table on page A6. (Round to the nearest cent.) $381.60 [6.5B]

28. A family had a checking account balance of $1846.78. A check of $568.30 was deposited into the account, and checks of $123.98 and $47.33 were written. Find the new checking account balance. $2243.77 [6.7A]

29. Anna Gonzalez spent $840 on gasoline and oil, $520 on insurance, $185 on tires, and $432 on repairs. Find the cost per mile to drive the car 10,000 miles during the year. (Round to the nearest cent.) $0.20 [6.5B]

30. A house has a mortgage of $72,000 for 20 years at an annual interest rate of 11%. Find the monthly mortgage payment. Use the table on page A6. (Round to the nearest cent.) $743.18 [6.4B]

Chapter 7

Statistics

Objectives

Section 7.1
To read a pictograph
To read a circle graph

Section 7.2
To read a bar graph
To read a broken-line graph

Section 7.3
To read a histogram
To read a frequency polygon

Section 7.4
To find the mean of a set of numbers
To find the median of a set of numbers

Frequencies of Letters

ZH WKH SHRSOH

The above phrase is a cryptogram. For this phrase to be read, the cryptogram has to be decoded.

Cryptology is the study of encrypting and decrypting messages. Encrypting means to write the message in code; decrypting means breaking a secret code. One of the methods the cryptologist uses in breaking a code is statistics.

Statistics is a study of the organization and analysis of data. A cryptologist uses statistics by analyzing ordinary text, such as that in a novel or a newspaper, and determining how frequently different letters of the alphabet occur. For example, in English, the letter "e" is the most frequently occurring letter. A table of the approximate frequencies of each letter is given below:

A—7.3%	J—0.2%	S—6.3%
B—0.9%	K—0.3%	T—9.3%
C—3.0%	L—3.6%	U—2.7%
D—4.3%	M—2.5%	V—1.3%
E—13.0%	N—7.8%	W—1.6%
F—2.7%	O—7.4%	X—0.6%
G—1.7%	P—2.7%	Y—1.8%
H—3.4%	Q—0.3%	Z—0.1%
I—7.5%	R—7.3%	

Knowing these frequencies, the cryptologist reasons that the most frequently occurring letters in a coded message correspond to the most frequently occurring letters in an ordinary message. Thus, to decode the phrase above, a cryptologist might guess that the letter H in the coded message corresponds to the letter E in an ordinary message. This guess may not be correct, but it is a good first choice.

See if you can decode the above phrase.

Answer: WE THE PEOPLE. The phrase was coded by taking the letter three spaces beyond the original letter. For example, A gets coded as D, B gets coded as E, C gets coded as F, and so on. This method of coding a message is called the Caesar Cipher, after Julius Caesar, who used this method.

7.1 Pictographs and Circle Graphs

Objective A To read a pictograph

INSTRUCTOR NOTE

A good source of pictographs is *USA Today.* As an extra credit assignment, students can take a pictograph and write a paragraph that interprets the data represented.

Statistics is the branch of mathematics concerned with **data,** or numerical information. **Graphs** are displays that provide a pictorial representation of data. The advantage of graphs is that they present information in a way that is easily read. The disadvantage of graphs is that they can be misleading. (See the Project in Mathematics at the end of this chapter.)

A **pictograph** uses symbols to represent information. The pictograph in Figure 1 represents the number of applications a law school received for each of the years shown. Each symbol represents 100 applications.

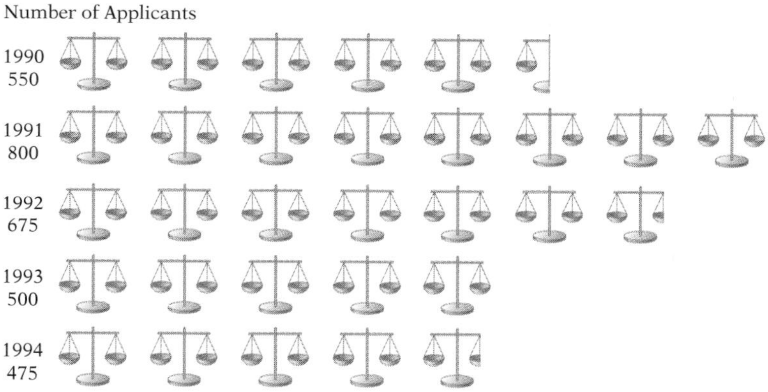

Fig. 1 Number of applications to law school.

From the pictograph, the number of applicants to the school decreased by 125 from 1991 to 1992.

The pictograph in Figure 2 represents the number of different calculators a manufacturer sold during a 1-month period. Each calculator symbol represents 5000 calculators. The total number of calculators sold during the month was 120,000.

Use the basic percent equation to determine what percent of the calculators sold were business calculators. The base B is 120,000, and the amount A is 45,000.

$$p \times B = A$$
$$p \times 120{,}000 = 45{,}000$$
$$p = \frac{45{,}000}{120{,}000}$$
$$p = 0.375$$

37.5% of the calculators sold were business calculators.

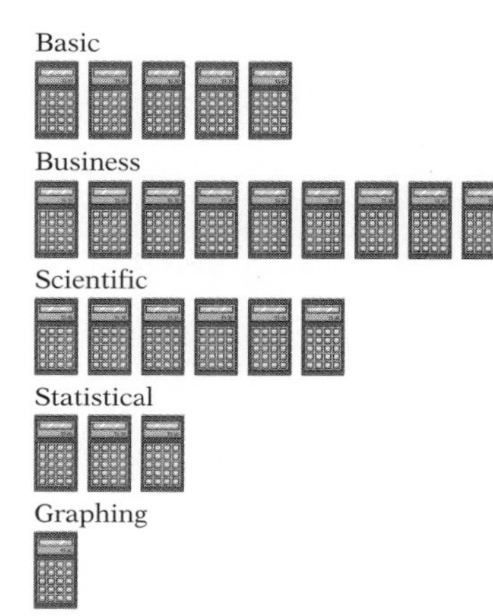

Fig. 2 Number of calculators sold.

The pictograph in Figure 3 shows the housing starts in a midwestern town during a 4-month period. Each picture represents 4 housing starts.

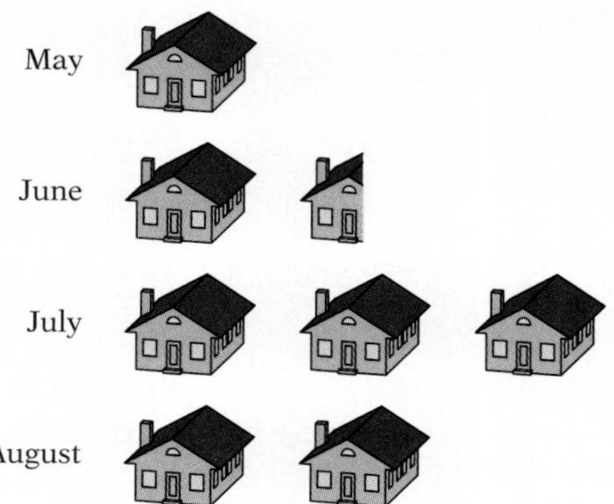

Fig. 3 Number of housing starts.

The ratio of the number of housing starts in May to the number of housing starts in August is

$$\frac{4 \text{ starts}}{8 \text{ starts}} = \frac{1}{2}$$

Example 1

Use Figure 3 to find the total number of housing starts for the 4 months.

Strategy

To find the total number of housing starts:

- Read the pictograph to determine the number of housing starts each month.
- Add the four numbers.

Solution

Starts for May: 4
Starts for June: 6
Starts for July: 12
Starts for August: 8

Total starts—30

The total number of housing starts is 30.

You Try It 1

Use Figure 3 to find what percent of the total number of housing starts the number of July housing starts represents.

Your strategy

Your solution 40%

Solution on p. A35

Objective B To read a circle graph

INSTRUCTOR NOTE

As the various types of graphs are presented, students may ask why there are so many types. Explain that each type of graph is used to convey certain types of information. Circle graphs are used when portions of a whole are important.

A **circle graph** represents data by the size of the sectors. The circle graph in Figure 4 represents the amount of money a mutual fund company has invested in different municipal bonds. The complete circle represents the total amount invested in all bonds, \$175 million. Each sector of the circle represents the amount of money invested in a different quality of bonds.

To find the percent of the total money invested in AAA-rated bonds, solve the basic percent equation for the percent (n). The base is 175 million, and the amount is 49 million.

$$\text{Percent} \times \text{base} = \text{amount}$$
$$n \times 175 = 49$$
$$n = \frac{49}{175}$$
$$n = 0.28$$

28% of the total amount invested is invested in AAA-rated bonds.

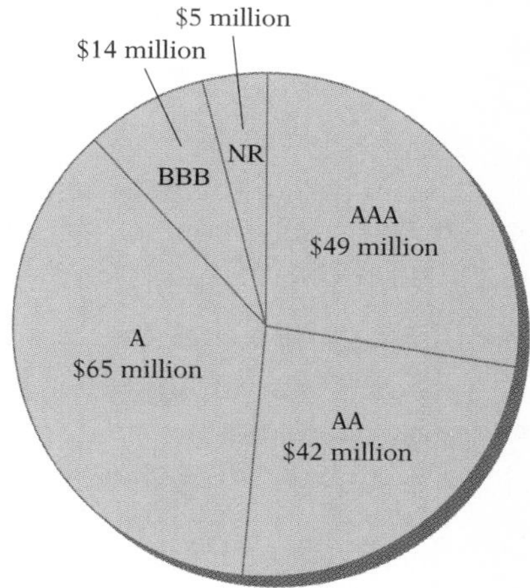

Fig. 4 Distribution of investments in municipal bonds.

The circle graph can also be used to find the ratio of the amount invested in one type of bond to the amount invested in a different type of bond. The ratio of the amount invested in AA-rated bonds to the amount invested in BBB-rated bonds is $\frac{42}{14} = \frac{3}{1}$. This indicates that 3 times as much money is invested in AA-rated bonds as in BBB-rated bonds.

➡ The circle graph in Figure 5 shows the various price categories of the approximately 9,500,000 cars sold in the United States during 1994. The complete circle graph represents 100% of all cars sold that year. Each sector of the graph represents one price range. The prices have been rounded to the nearest hundred dollars.

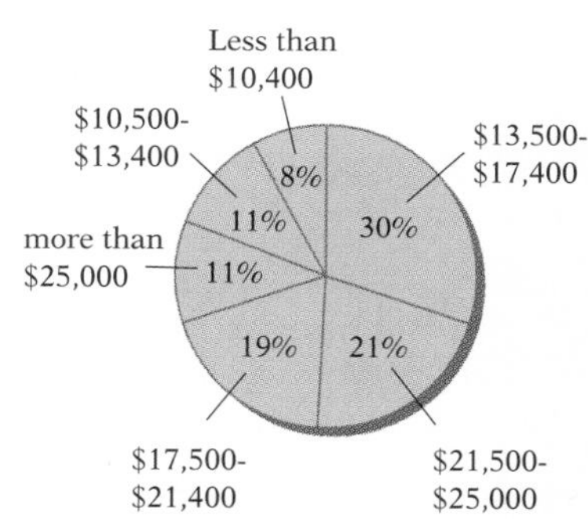

Fig. 5 Various prices of cars sold in 1994.

How many cars were sold for a price that is between \$17,500 and \$21,400?

To find the number of cars sold in the given price range, use the Basic Percent Equation.

Percent × base = amount

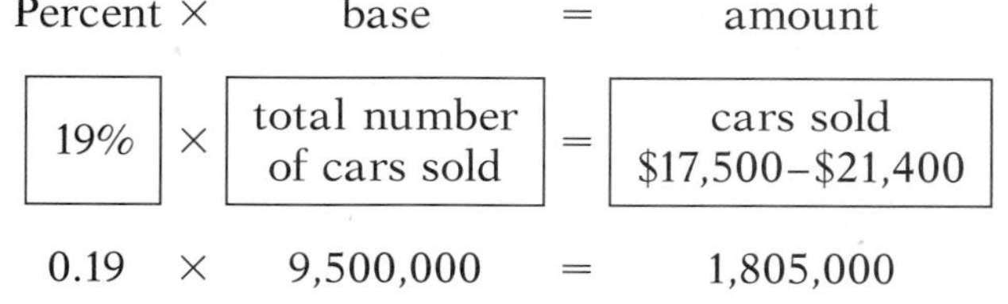

19%	×	total number of cars sold	=	cars sold \$17,500–\$21,400
0.19	×	9,500,000	=	1,805,000

There were 1,805,000 cars sold between \$17,500 and \$21,400.

The circle graph in Figure 6 shows the annual expenses of owning and operating a car.

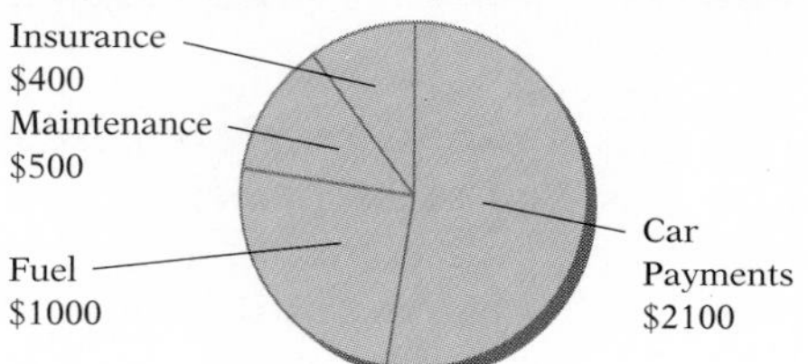

Fig. 6 Annual car expenses totaling $4000.

Example 2
Use Figure 6 to find the ratio of the annual fuel expense to the total annual cost of operating a car.

Strategy To find the ratio:

- Locate the annual fuel expense in the circle graph.
- Write the ratio of the annual fuel expense to the total annual cost of operating a car in simplest form.

Solution Annual fuel expense: $1000

$$\frac{\$1000}{\$4000} = \frac{1}{4}$$

The ratio is $\frac{1}{4}$.

You Try It 2
Use Figure 6 to find the ratio of the annual cost of insurance to the annual cost of maintenance.

Your strategy

Your solution
$\frac{4}{5}$

The circle graph in Figure 7 shows the distribution of an employee's gross monthly income.

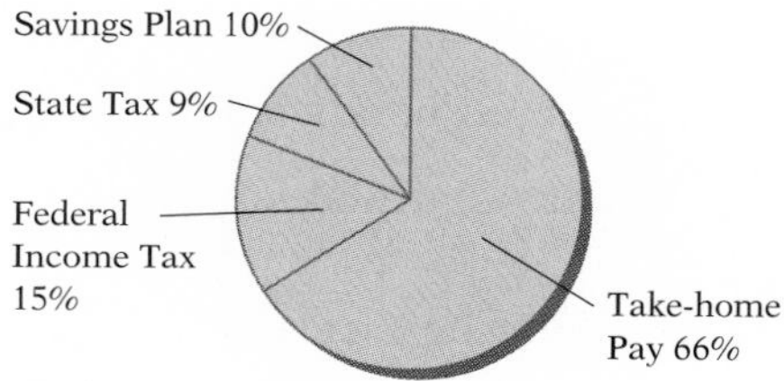

Fig. 7 Distribution of gross monthly income of $2000.

Example 3
Use Figure 7 to find the employee's take-home pay.

Strategy To find the take-home pay:

- Locate the percent of the distribution that is take-home pay in the circle graph.
- Solve the basic percent equation for *amount*.

Solution Take-home pay: 66%

Percent × base = amount
0.66 × $2000 = $1320

The employee's take-home pay is $1320.

You Try It 3
Use Figure 7 to find the federal income tax the employee paid.

Your strategy

Your solution
$300

Solutions on p. A36

7.1 Exercises

Objective A

The pictograph in Figure 8 shows the amount of gasoline a service station sold during a 4-week period. Each barrel in the pictograph represents 1000 gallons of gasoline.

Week 1

Week 2

Week 3

Week 4

Fig. 8 Barrels of gasoline sold.

1. Find the total number of gallons of gasoline sold during the month. 19,000 gallons

2. Find the ratio of the amount of gasoline sold in week 2 to the amount sold in week 4. $\frac{7}{11}$

3. Find the percent of the amount of gasoline sold during the month that was sold in week 1. (Round to the nearest tenth of a percent.) 21.1%

Of each dollar that a school receives (see Figure 9), 55¢ comes from the state government, 30¢ from local sources, and 15¢ from the federal government.

State Government

Local Sources

Federal Government

Fig. 9 Sources of school's budget.

4. Find the ratio of the amount of money that comes from local sources to the amount that comes from the state. $\frac{6}{11}$

5. Find the percent of the budget that comes from the federal government. 15%

6. If the total budget is $15,000,000, find the amount of the budget that comes from local sources. $4,500,000

The pictograph in Figure 10 shows the number of fish caught in Lost Lake during 4 years. Each picture represents 200 fish.

Fig. 10 Number of fish caught.

7. Find the total number of fish caught during the 4 years. 4000 fish

8. Find the ratio of the number of fish caught in 1991 to the number caught in 1992. $\frac{8}{7}$

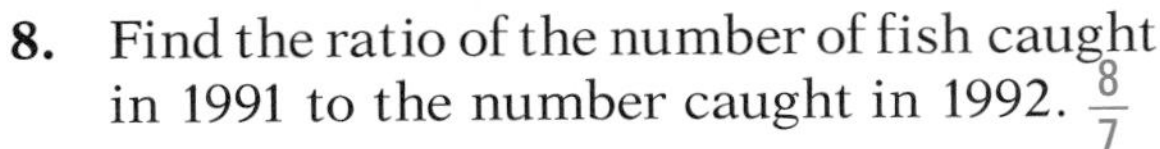

9. The number of fish caught in 1994 represents what percent of the total number of fish caught? Round to the nearest tenth of a percent. 27.5%

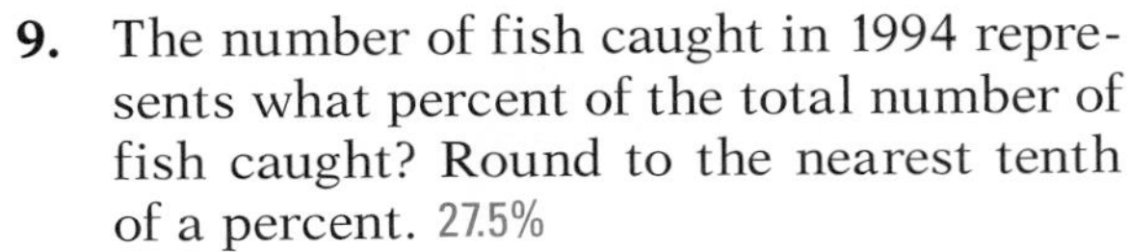

Objective B

An Accounting major recorded the number of units required in each discipline to graduate with a degree in Accounting. The results are shown in the circle graph in Figure 11.

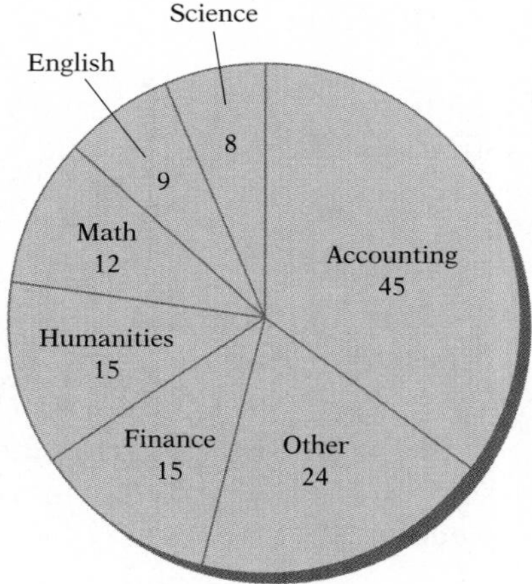

Fig. 11 Number of units required to graduate with an Accounting degree.

10. How many units are required to graduate with a degree in Accounting? 128

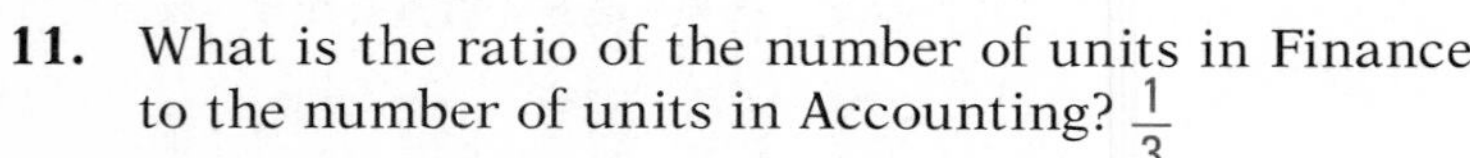

11. What is the ratio of the number of units in Finance to the number of units in Accounting? $\frac{1}{3}$

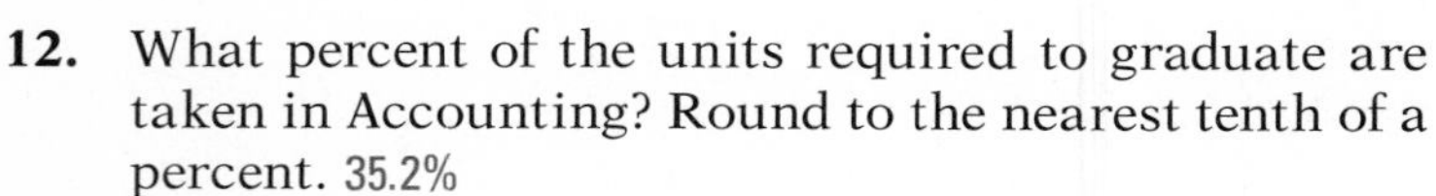

12. What percent of the units required to graduate are taken in Accounting? Round to the nearest tenth of a percent. 35.2%

13. What percent of the units required to graduate are taken in Mathematics? Round to the nearest tenth of a percent. 9.4%

The circle graph in Figure 12 shows the population of seven regions.

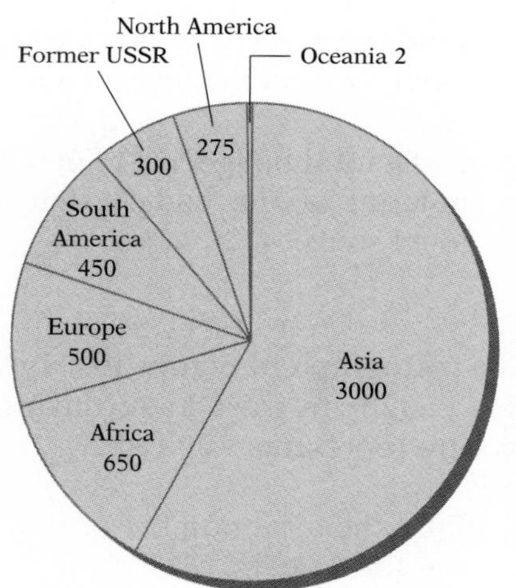

Fig. 12 Population in millions of people.

14. Find the total population of the seven regions. 5177 million people

15. What is the ratio of the population of Asia to the population of Africa? $\frac{60}{13}$

16. What is the ratio of the population of North America to the population of Asia?

$\frac{11}{120}$

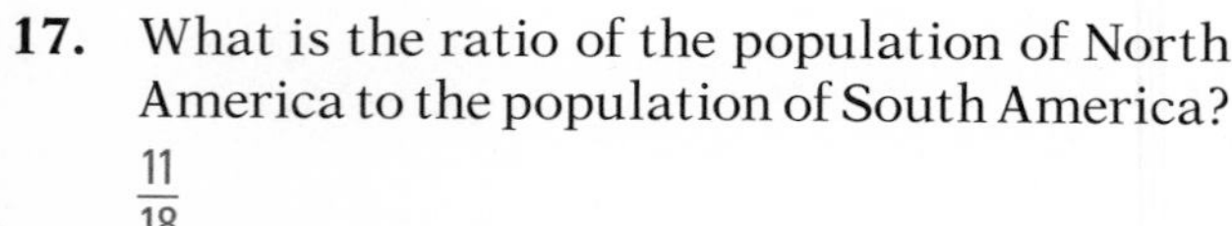

17. What is the ratio of the population of North America to the population of South America?

$\frac{11}{18}$

The circle graph in Figure 13 shows the portfolio breakdown in the Paine-Webber Income Fund.

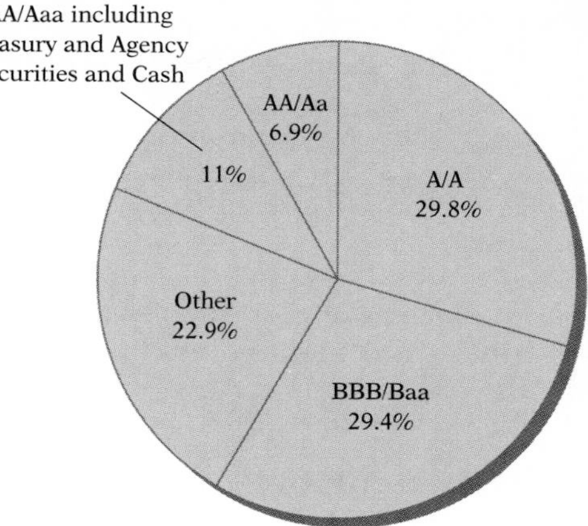

Fig. 13 Rating categories of bonds held by the Paine-Webber Income Fund. Their total value is $197,120,539.

18. Find the value of the bonds rated BBB/Baa. Round to the nearest thousand dollars. $57,953,000

19. Find the total value of the bonds rated A/A or higher. Round to the nearest thousand dollars. $94,026,000

20. What fractional part of the value of the bonds is rated AAA/Aaa? $\frac{11}{100}$

21. Find the value of the bonds rated A/A or AA/Aa. Round to the nearest thousand dollars. $72,343,000

The circle graph in Figure 14 shows the expenditures of $13,827,239 for a small midwestern college of 535 students.

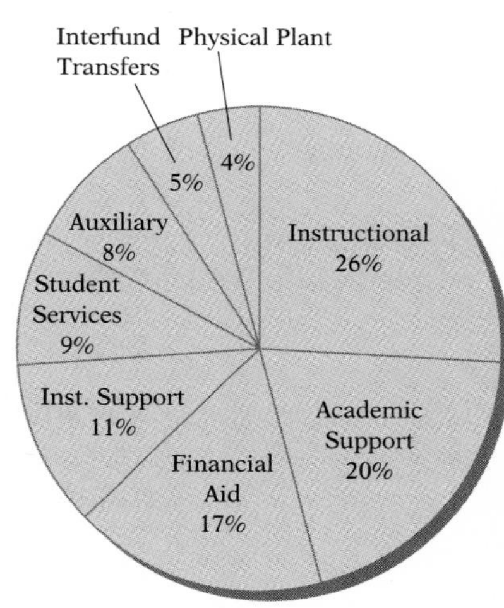

Fig. 14 A college's expenditures of $13,827,239.

22. Find the amount spent for instructional purposes. Round to the nearest dollar. $3,595,082

23. Find the amount spent for student services. Round to the nearest dollar. $1,244,452

24. Find the total amount spent for academic support and financial aid. Round to the nearest dollar. $5,116,078

25. Find the total amount spent for institutional support and the physical plant. Round to the nearest dollar. $2,074,086

The circle graph in Figure 15 shows the land area of each of the seven continents in square miles.

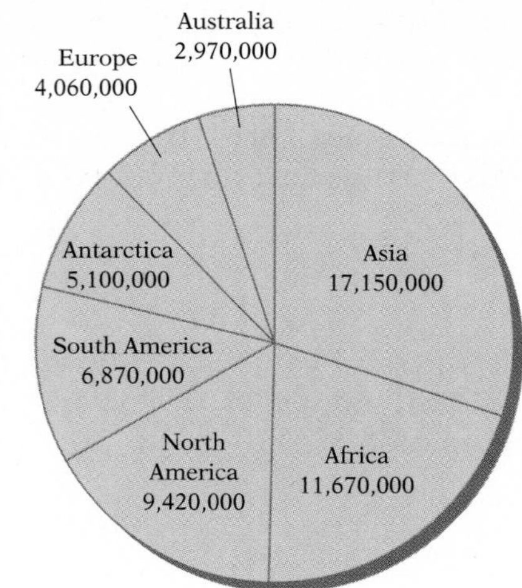

Fig. 15 Land area of the seven continents in square miles.

26. Find the total land area of the seven continents. 57,240,000 square miles

27. Find the ratio of the land area of North America to the land area of South America. $\frac{314}{229}$

28. Find the ratio of the land area of Asia to the total land area of the seven continents. $\frac{1715}{5724}$

29. Find the ratio of the land area of Australia to the total land area of the seven continents. $\frac{11}{212}$

The circle graph in Figure 16 shows the average percent distribution of a family's income of $28,400.

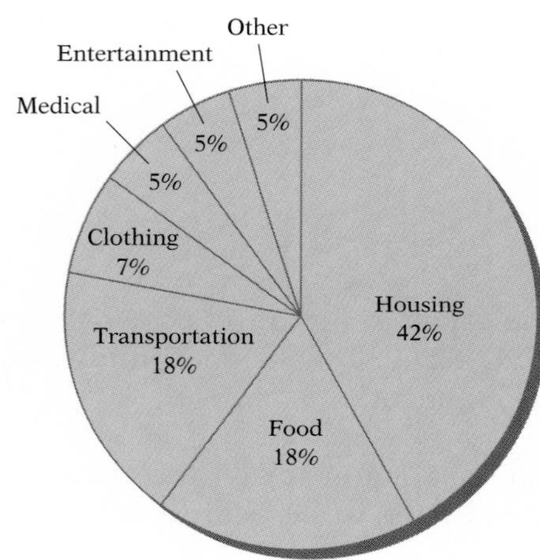

Fig. 16 Percent distribution of a family's income.

30. What sector represents the family's greatest expense? Housing

31. How much of the family's income was spent on food? $5112

32. How much of the family's income was spent on medical purposes? $1420

33. How much of the family's income was spent on clothing and entertainment? $3408

APPLYING THE CONCEPTS

34. [W] What are the advantages of presenting data in the form of a pictograph? What are the disadvantages? Answers will vary

35. [W] Make out a budget for your expenses. Record these expenses in a circle graph. Answers will vary

36. [W] Pick out a typical day, and record in a circle graph the amount of time you spent on different activities. Answers will vary

7.2 Bar Graphs and Broken-Line Graphs

Objective A To read a bar graph

INSTRUCTOR NOTE
It is important for students to note the jagged vertical axis such as is in Figure 17. One of the ways to distort the impact of data displayed graphically is to choose the vertical axis to show data in its best (or worst) light. The Project in Mathematics at the end of this chapter discusses this problem.

A **bar graph** represents data by the height of the bars. The bar graph in Figure 17 shows data from the federal government on the annual number of sales of homes each month from November 1993 to October 1994. For each month, the height of the bar indicates the number of new homes sold. The jagged line near the bottom of the graph indicates that the vertical scale is missing the numbers between 0 and 55.

The number of new homes sales in June was 64,000. (Note that the vertical scale is in 1,000.) Because the bar for February is the lowest, home sales were lowest in February.

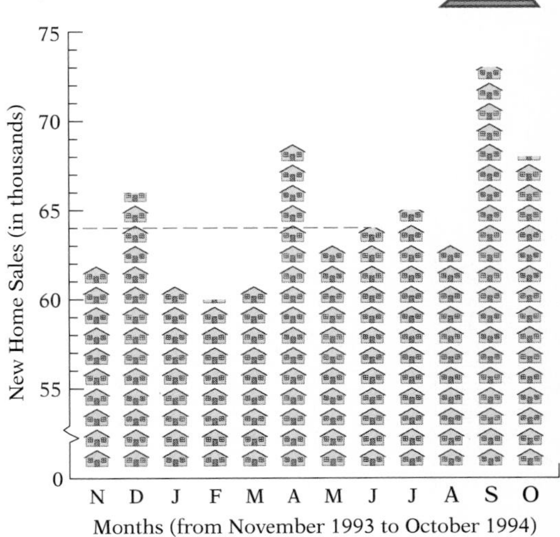

Fig. 17 Annual sales of new homes.

A double-bar graph is used to display data for purposes of comparison. The double-bar graph in Figure 18 shows the annual sales of new cars in January 1993 and January 1994 for four automotive companies.

In January 1994, General Motors (GM) sold 350,000 cars.

Each company shown on the graph had greater sales in January 1994 than in January 1993.

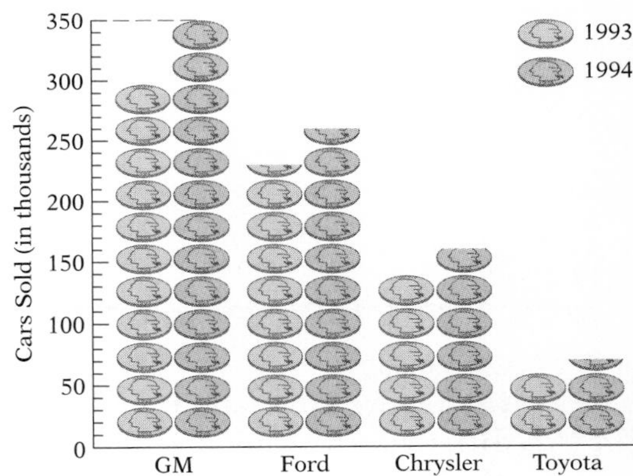

Fig. 18 Comparison of January sales.

Example 1
Use Figure 18 to find the difference in January sales between 1994 and 1993 for Ford.

Strategy To find the difference:

- Read the double-bar graph to find the sales for Ford in January of both years.
- Subtract to find the difference.

Solution
1994 Ford sales: 260,000
1993 Ford sales: 230,000
$260{,}000 - 230{,}000 = 30{,}000$

The difference in sales was 30,000 cars.

You Try It 1
Use Figure 18 to determine which company had the largest increase in sales from January 1993 to January 1994.

Your strategy

Your solution
GM

Solution on p. A36

Objective B To read a broken-line graph

A **broken-line graph** represents data by the position of the lines and shows trends and comparisons. The broken-line graph in Figure 19 shows the sales of a new album by a rock group for a five-month period. The height of each dot indicates the number of albums sold each month.

In March, 80,000 albums were sold.

In April, 65,000 albums were sold.

For the five months graphed, the fewest albums were sold in May.

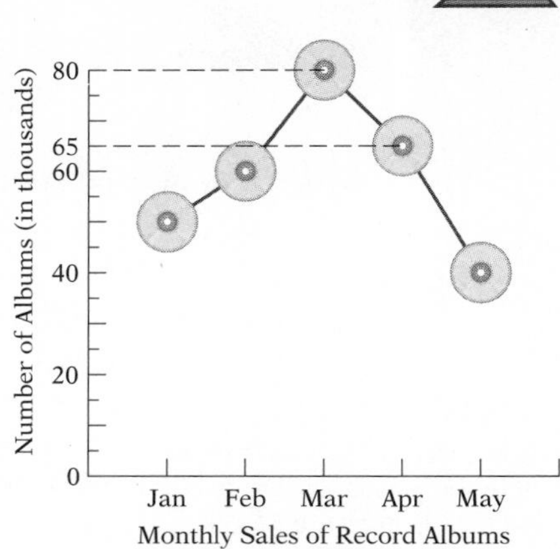

Fig. 19 Number of albums sold.

Two broken-line graphs are often shown in the same figure for comparison. Figure 20 shows the net incomes of two software companies, Math Associates and Compusoft, before their merger.

Several things can be determined from the graph:

The net income for Math Associates in 1993 was $8 million dollars.

The net income for Compusoft declined from 1990 to 1991.

The net income for Math Associates increased for each of the years shown.

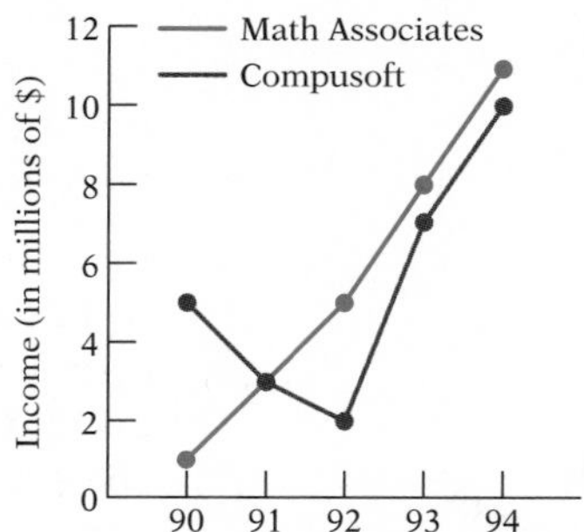

Fig. 20 Net incomes of Math Associates and Compusoft.

Example 2
Use Figure 20 to approximate the difference between the net income of Math Associates and Compusoft in 1992.

Strategy To find the difference:

- Read the line graph to determine the net income of Math Associates and Compusoft in 1992.
- Subtract to find the difference between the net incomes.

Solution
Net income for Math Associates: $5 million.
Net income for Compusoft: $2 million.

$5 - 2 = 3$

The difference between net incomes in 1992 was $3 million.

You Try It 2
Use Figure 20 to determine between which two years the net income of Math Associates increased the most.

Your strategy

Your solution
1992 and 1993

Solution on p. A36

7.2 Exercises

Objective A

The bar graph in Figure 21 shows the number of cars a dealership sold during the first 6 months of the year.

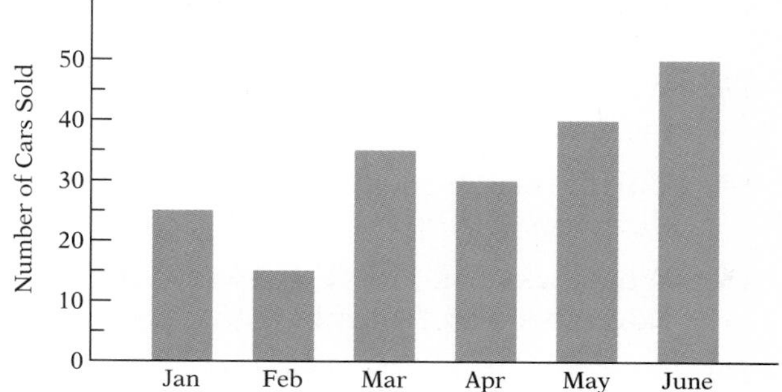

Fig. 21 Car sales.

1. How many cars were sold in May? 40 cars

2. How many cars were sold in February and March? 50 cars

3. Find the ratio of the number of cars sold in February to the number of cars sold in June. $\frac{3}{10}$

4. In which month was the greatest number of cars sold? June

The double-bar graph in Figure 22 shows some crime statistics per 100,000 population in an eastern city.

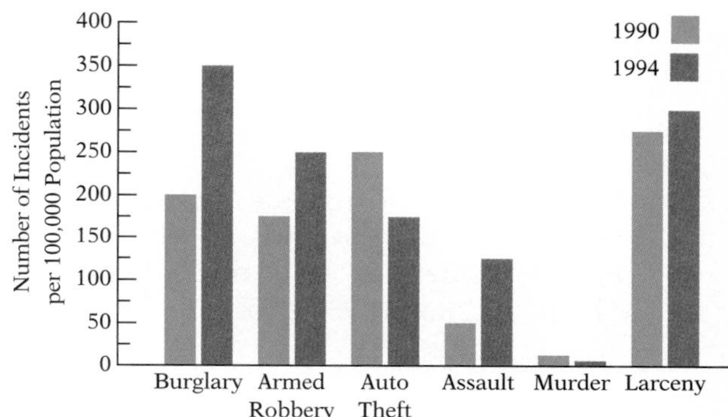

Fig. 22 Crime statistics comparisons between 1990 and 1994.

5. How many burglaries were committed per 100,000 population in 1990? 200 burglaries

6. What crimes decreased during the period 1990 to 1994? Auto theft and murder

7. Find the ratio of the number of auto thefts that occurred in 1990 to the number of auto thefts that occurred in 1994. $\frac{10}{7}$

8. Find the percent increase in armed robbery during the years 1990 to 1994. Round to the nearest tenth of a percent. 42.9%

The double-bar graph in Figure 23 shows the rates at which inflation in various medical costs has decreased from 1992 to 1993.

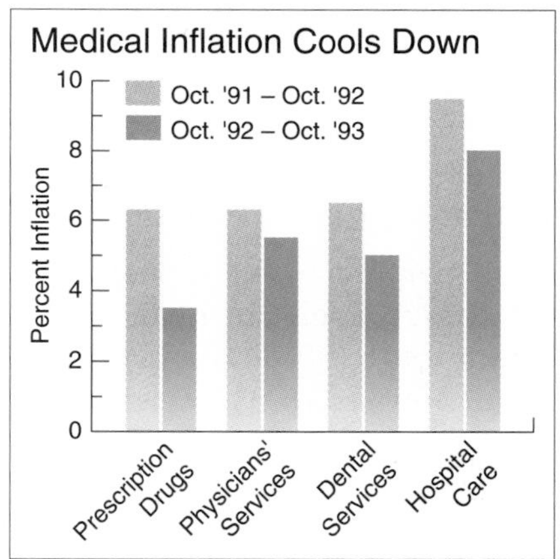

Fig. 23 Rates of inflation for various health-care categories

9. Estimate the decrease in the rate of inflation for prescription drugs from Oct. '91–Oct. '92 to Oct. '92–Oct. '93. 44%

10. Find the percent decrease in the rate of inflation for dental services from Oct. '91–Oct. '92 to Oct. '92–Oct. '93. 28%

11. Find the percent decrease in the rate of inflation for physicians services from Oct. '91–Oct. '92 to Oct. '92–Oct. '93. 14%

12. For what medical service was the rate of inflation the greatest? Hospital care

Objective B

The broken-line graph in Figure 24 shows the snowfall at High Top Ski Resort during ski season.

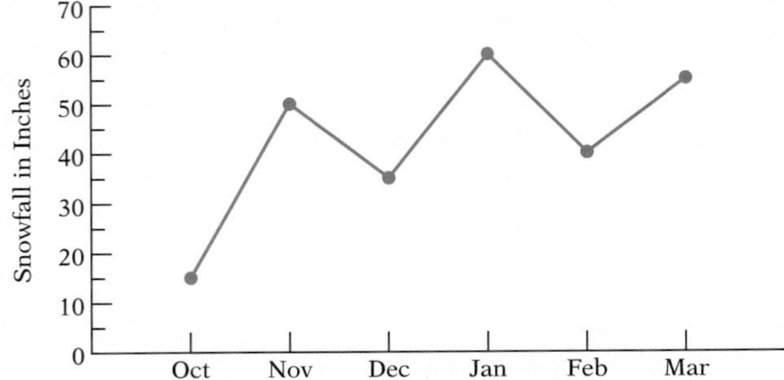

Fig. 24 Snowfall at High Top during ski season.

13. What was the amount of snowfall during January? 60 inches

14. During which month was the snowfall the lowest? October

15. What was the total snowfall during November and December? 85 inches

16. Find the ratio of the amount of snowfall in December to the amount of snowfall during January. $\frac{7}{12}$

The double-broken-line graph in Figure 25 shows the number of business calls and the number of residential calls made each hour from 9 A.M. to 5 P.M. during an 8-hour business day in a small city.

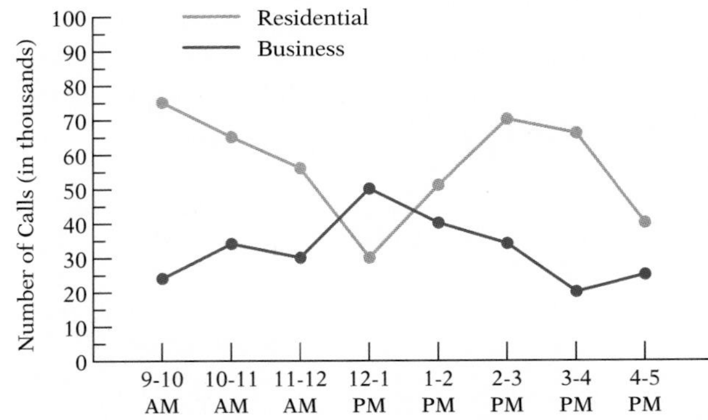

Fig. 25 Number of calls made each hour in one day.

17. What is the difference between the number of business calls and the number of residential calls made between 11 A.M. and noon? 25,000 calls

18. How many business calls were made between 9 A.M. and noon? 90,000 calls

19. How many residential calls were made between 9 A.M. and noon? 195,000 calls

APPLYING THE CONCEPTS

The double-broken-line graph in Figure 26 shows U.S. energy consumption patterns for the years 1850–1990.

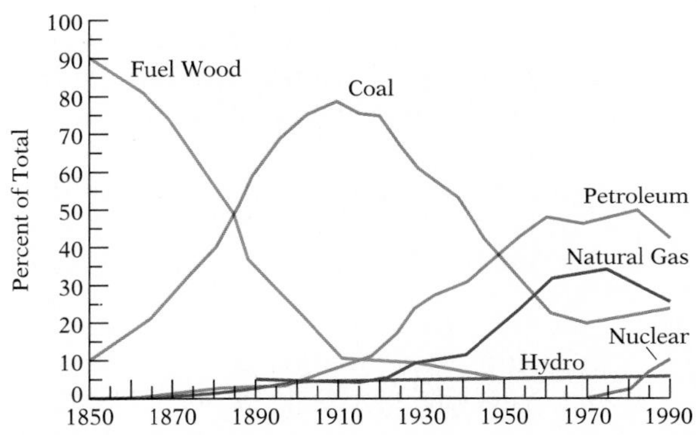

Fig. 26 United States energy consumption patterns, 1850–1990.

20. What two fuels provided the energy in the United States in 1850? Coal and wood

21. Estimate the percent of each fuel that the United States used in 1990. Does your estimate total 100%? Should your estimate total 100%? See answer below

22. [W] Write a paragraph analyzing the patterns in the energy consumption graph. What do the patterns indicate about the future consumption of petroleum and natural gas? Answers will vary

23. [W] Discuss what might be our energy consumption and needs in the year 2020 if the trends continue? Answers will vary

21. Coal: 19%, Wood: 3%, Natural Gas: 20%, Petroleum: 50%, Hydroelectric: 5%, Nuclear: 4%; your estimate should be approximately 100%.

7.3 Histograms and Frequency Polygons

Objective A To read a histogram

INSTRUCTOR NOTE
Histograms differ from bar graphs in that the vertical axis of a histogram always represents the frequency of some occurrence, which is given on the horizontal axis.

A research group measured the fuel usage of 92 cars. The results are recorded in the histogram in Figure 27. A **histogram** is a special type of bar graph. The width of each bar corresponds to a range of numbers called a **class interval.** The height of each bar corresponds to the number of occurrences of data in each class interval and is called the **class frequency.**

Class Intervals (Miles per Gallon)	*Class Frequencies (Number of Cars)*
18–20	12
20–22	19
22–24	24
24–26	17
26–28	15
28–30	5

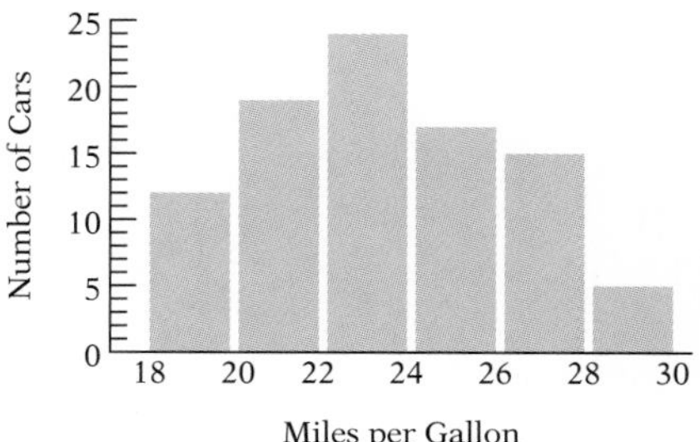

Fig. 27

Twenty-four cars get between 22 and 24 miles per gallon.

A precision tool company has 85 employees. Their hourly wages are recorded in the histogram in Figure 28.

The ratio of the number of employees whose hourly wage is between \$10 and \$12 to the total number of employees is $\frac{17 \text{ employees}}{85 \text{ employees}} = \frac{1}{5}$.

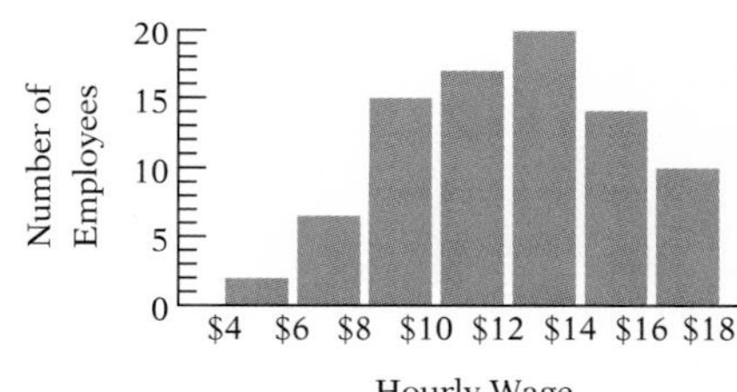

Fig. 28

Example 1
Use Figure 28 to find the number of employees whose hourly wage is between \$14 and \$18.

Strategy
To find the number of employees:
- Read the histogram to find the number of employees whose hourly wage is between \$14 and \$16 and the number whose wage is between \$16 and \$18.
- Add the two numbers.

Solution Number whose wage is between \$14 and \$16: 14; between \$16 and \$18: 10.

$14 + 10 = 24$

24 employees have an hourly wage between \$14 and \$18.

You Try It 1
Use Figure 28 to find the number of employees whose hourly wage is between \$8 and \$12.

Your strategy

Your solution 32 employees

Solution on p. A37

Objective B To read a frequency polygon

INSTRUCTOR NOTE
A frequency polygon represents information in essentially the same way as a histogram.

The speeds of 70 cars on a highway were measured by radar. The results are recorded in the frequency polygon in Figure 29. A **frequency polygon** is a graph that displays information in a manner similarly to a histogram. A dot is placed above the center of each class interval at a height corresponding to that class's frequency; the dots are then connected to form a broken-line graph. The center of a class interval is called the **class midpoint.**

Class Interval (Miles per Hour)	*Class Midpoint*	*Class Frequency*
30–40	35	7
40–50	45	13
50–60	55	25
60–70	65	21
70–80	75	4

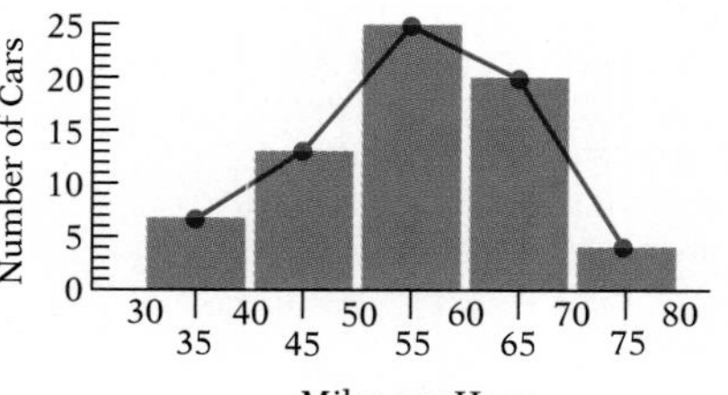

Fig. 29

Twenty-five cars were traveling between 50 and 60 miles per hour.

Sixty people took an exam for a real estate license. The scores on the exam are recorded in the frequency polygon in Figure 30.

The ratio of the number of people scoring between 70 and 80 to the total number of people tested is $\frac{24 \text{ people}}{60 \text{ people}} = \frac{2}{5}$.

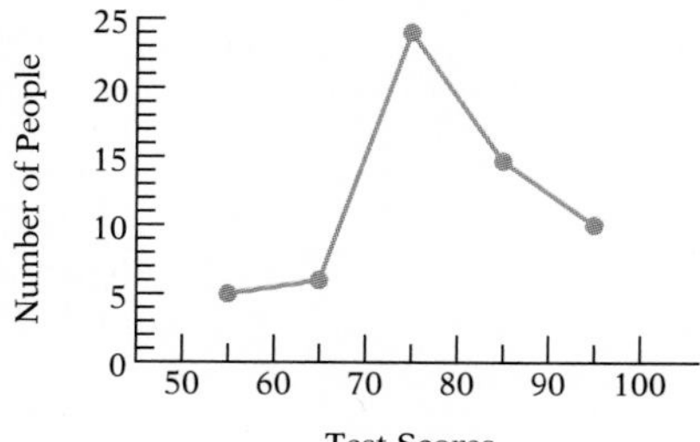

Fig. 30

Example 2
Use Figure 30 to find the number of people who scored between 80 and 100 on the real estate exam.

Strategy
To find the number of people who scored between 80 and 100 on the exam:

- Read the frequency polygon to find the number of people who scored between 80 and 90 and the number who scored between 90 and 100.
- Add the two numbers.

Solution
The number who scored between 80 and 90: 15; between 90 and 100: 10.

$15 + 10 = 25$

25 people scored between 80 and 100.

You Try It 2
Use Figure 30 to find the number of people who scored between 50 and 70 on the real estate exam.

Your strategy

Your solution
11 people

Solution on p. A37

7.3 Exercises

Objective A

The test scores of 34 students are recorded in the histogram in Figure 31.

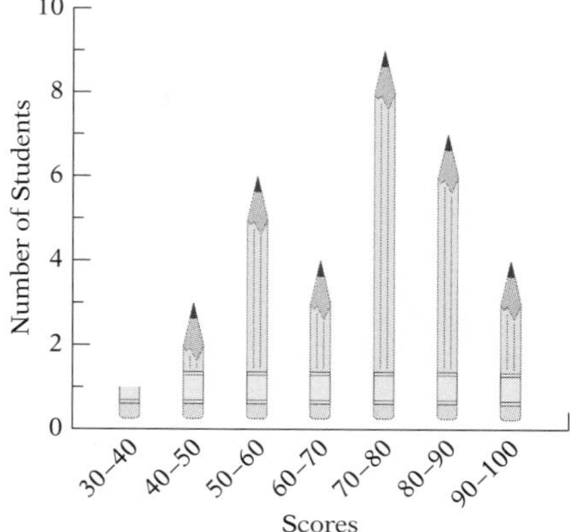

Fig. 31

1. How many students scored between 60 and 80? 13 students
2. Find the ratio of the number of students who scored between 50 and 60 to the total number of students. $\frac{3}{17}$
3. Find the number of students who scored above 80. 11 students
4. Find the percent of the students who scored below 60. Round to the nearest tenth of a percent. 29.4%

A department store keeps records of the amounts its customers spend. The histogram in Figure 32 records the dollar amounts that 153 customers spent.

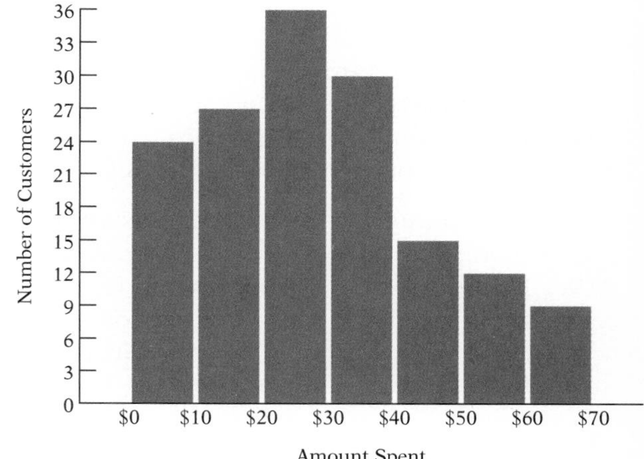

Fig. 32

5. How many customers made purchases between $20 and $30? 36 customers
6. What is the ratio of the number of customers whose purchases were between $30 and $40 to the total number of customers? $\frac{10}{51}$
7. How many customers made purchases of more than $30? 66 customers
8. What percent of the total number of customers spent more than $50? Round to the nearest tenth of a percent. 13.7%

The histogram in Figure 33 shows the number of cars sold in different price ranges.

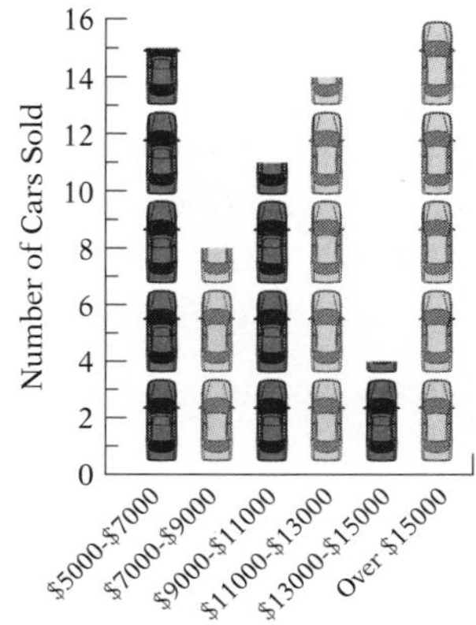

Fig. 33

9. Find the total number of cars that sold for more than $13,000. 20 cars
10. Find the number of cars sold whose price was between $5000 and $11,000. 34 cars
11. Find the ratio of the number of cars sold that were priced between $7000 and $9000 and the number sold that were priced between $11,000 and $13,000. $\frac{4}{7}$
12. What percent of the cars sold were priced over $15,000? Round to the nearest tenth of a percent. 23.5%

Objective B

A total of 34 runners ran the 100-yard dash. The results are recorded in Figure 34.

13. How many runners ran the 100-yard dash in less than 11 seconds? 4 runners

14. Find the ratio of the number of runners who ran the race in between 10 and 11 seconds to the number who ran it in between 12 and 13 seconds. $\frac{4}{11}$

15. How many runners ran the race in between 11 and 12 seconds? 19 runners

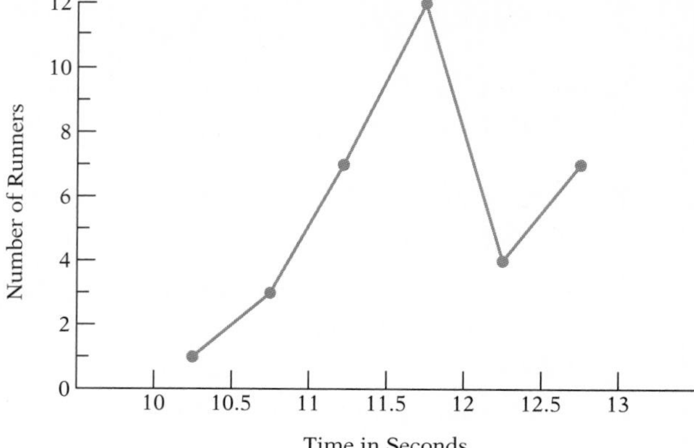

Fig. 34

A survey of 50 families was made to determine what percent of their income was spent on vacations. The results are recorded in Figure 35.

16. How many families spent between 4% and 8% of their income on vacations? 30 families

17. Find the number of families who spent more than 8% of their income on vacations. 5 families

18. How many families spent less than 4% of their income on vacations? 15 families

19. Find the ratio of the number of families who spent less than 4% of their income on vacations to the number of families who spent more than 8%. $\frac{3}{1}$

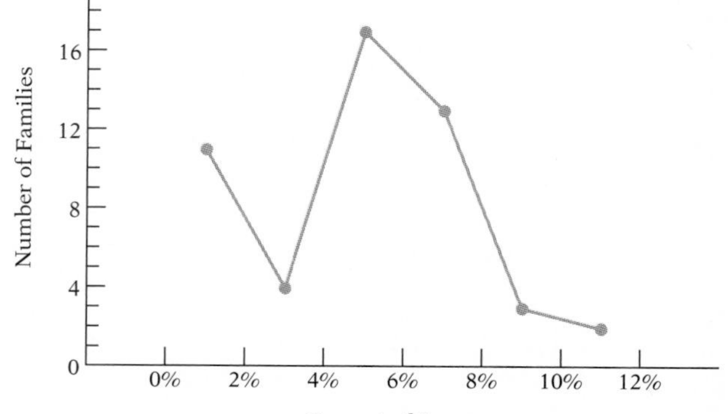

Fig. 35

Each member of a university's entering first-year class of 600 students was given a mathematics placement exam. The scores are recorded in Figure 36.

20. How many students scored between 600 and 700? 100 students

21. What is the ratio of the number of students scoring between 500 and 600 to the number of students tested? $\frac{11}{24}$

22. How many students scored below 600? 475 students

23. How many students scored above 400? 525 students

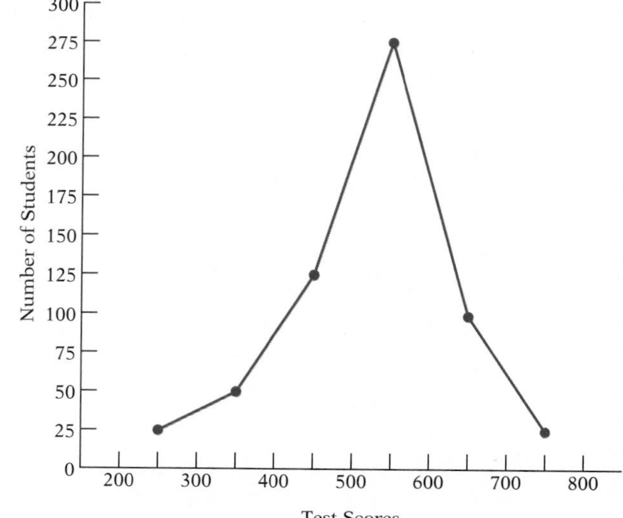

Fig. 36

APPLYING THE CONCEPTS

24. [W] Write a paragraph explaining how a histogram can be used to suggest false assumptions. Use a histogram in your explanation. Answers will vary

25. [W] In your own words, describe a frequency table. Answers will vary

7.4 Means and Medians

Objective A To find the mean of a set of numbers

A student's test scores for five tests are listed below.

Test 1	*Test 2*	*Test 3*	*Test 4*	*Test 5*
86	95	94	87	93

POINT OF INTEREST
In 1994, the approximate average annual income for a worker in the United States was $22,000.

The **average** or **mean** score is the sum of all the test scores divided by the number of tests:

$$\frac{\text{Sum of scores}}{\text{Number of tests}} = \frac{86 + 95 + 94 + 87 + 93}{5} = \frac{455}{5} = 91$$

Note that a student who scored 91 on each of the five tests would have the same grade average as the student whose test scores are given in the example above.

Example 1
For a price comparison of six supermarkets, identical items were purchased in each store. The results are listed below.

Store	*Cost*	*Store*	*Cost*
1	$40.74	4	$39.05
2	$39.45	5	$38.86
3	$38.57	6	$39.25

Find the mean cost of the items purchased.

Strategy To find the mean cost of the items purchased:

- Find the sum of the costs.
- Divide the sum of the costs by the number of stores (6).

Solution

$$\begin{array}{r} \$40.74 \\ 39.45 \\ 38.57 \\ 39.05 \\ 38.86 \\ +\ 39.25 \\ \hline \$235.92 \end{array} \text{ sum of costs} \qquad 6\overline{)\$235.92} = \$39.32$$

The mean cost of the items purchased was $39.32.

You Try It 1
The amounts that Rosie Espinosa, a sales representative, spent each month for gasoline are listed below.

Month	*Cost*	*Month*	*Cost*
Jan	$118	Apr	$141
Feb	$130	May	$134
Mar	$109	June	$136

Find the monthly mean cost for gasoline.

Your strategy

Your solution $128

Solution on p. A37

Objective B To find the median of a set of numbers

The ages of the presidents of nine corporations are 54, 38, 62, 45, 56, 60, 59, 39, 64.

POINT OF INTEREST

In 1994, the approximate median annual income for a worker in the United States was $17,000.

INSTRUCTOR NOTE

A class project would be to discuss the significance of the statistics given in the Point of Interest on this page and on the previous page. As a motivation point, the median difference between the annual salary of a worker with a college degree and that of one with a high school degree is approximately $1000 per month.

The **median** age of the presidents is the middle age when the ages are arranged from smallest to largest.

To find the median, arrange the numbers from the smallest to largest. Locate the middle number.

38, 39, 45, 54, 56, 59, 60, 62, 64

4 numbers | middle number | 4 numbers

56 is the median.

The median separates a list of numbers so that there are the same number of values below the median as there are above the median.

➡ Dominique received test scores of 89, 90, 58, 92, 84, and 96.

To find the median, arrange the numbers from smallest to largest. When there are two middle numbers, the median is the average of those two numbers.

58, 84, 89, 90, 92, 96

2 numbers | middle numbers | 2 numbers

$$\frac{89 + 90}{2} = \frac{179}{2} = 89.5$$

89.5 is the median.

Example 2

In a neighborhood survey it was found that seven houses sold during the past year had been sold for $90,000, $87,000, $86,000, $92,000, $88,000, $91,500, and $96,500. Find the median price of a house in this neighborhood.

Strategy

To find the median price, arrange the prices in order from smallest to largest. The median is the middle number.

Solution

$86,000
$87,000 } 3 numbers
$88,000
$90,000 ← middle number
$91,500
$92,000 } 3 numbers
$96,500

The median price of a house in this neighborhood is $90,000.

You Try It 2

A library's records show the number of books borrowed on each day last week. Find the median number of books borrowed.

Mon	*Tue*	*Wed*	*Thu*	*Fri*	*Sat*
450	375	420	500	480	490

Your strategy

Your solution

465 books

Solution on p. A37

7.4 Exercises

Objective A

1. You received grades of 75, 88, 94, 76, and 82 on 5 mathematics exams. Find your mean grade. 83

2. The closing prices of a utility stock for 5 days were \$36.25, \$35.75, \$36.50, \$36, and \$37.20. Find the mean closing price of the stock. \$36.34

3. The starting 5 players on a basketball team had the following heights: 70 inches, 85 inches, 80 inches, 74 inches, and 78 inches. Find the mean height of the basketball team in inches. Round to the nearest tenth of an inch. 77.4 inches

4. The prices of a pound of T-bone steak at 6 different grocery stores were \$2.58, \$2.62, \$2.49, \$2.75, \$2.66, and \$2.68. Find the mean price of T-bone steak at the 6 grocery stores. \$2.63

5. The low temperatures in degrees Fahrenheit for 5 days at a ski resort were 28°, 21°, 18°, 9°, and 11°. Find the mean low temperature. Round to the nearest tenth of a degree. 17.4°

6. The 8 everyday players on a baseball team hit 8, 15, 42, 18, 1, 24, 6, and 13 home runs during the season. Find the mean number of home runs the 8 players hit. (Round to the nearest thousandth.) 15.875 home runs

7. Sharon's pay checks for 4 months are given in the table at the right. Find the mean monthly pay check. \$1240

Jan	*Feb*	*Mar*	*Apr*
\$1200	\$1350	\$1190	\$1220

8. The taxi driver's records given at the right show the number of gallons of gasoline purchased each day on the job last week. Find the mean number of gallons of gasoline purchased. 10.1 gallons

Wed	*Thu*	*Fri*	*Sat*	*Sun*
9.4	9.3	11.8	10.3	9.7

9. A test to determine the braking distance of a car yielded the following results: 220 feet, 208 feet, 216 feet, 219 feet, 227 feet. Find the mean braking distance of the car. 218 feet

10. During the past year, 6 houses in a small town sold for the following prices: \$54,000, \$89,450, \$116,295, \$247,600, \$82,500, and \$176,300. Find the mean price of the 6 houses sold. \$127,690.83

11. The annual rainfall for each of 6 years in the rain forest on Mt. Waialeale in Kauai is recorded in the table at the right. Find the mean annual rainfall. 460 inches

Year	*Number of Inches*
1	512
2	492
3	416
4	472
5	450
6	418

12. The average time devoted each day to household TV usage for the years 1986 to 1991 is shown at the right. Find the yearly mean time, to the nearest minute, that the average household watched TV for 1986 to 1991.
6 hours, 58.6 minutes

Year	*Yearly Average*
1986–1987	7 hr 5 min
1987–1988	6 hr 55 min
1988–1989	7 hr 2 min
1989–1990	6 hr 55 min
1990–1991	6 hr 56 min

13. The daily maximum temperatures in degrees Fahrenheit for 7 days in the summer were 88°, 94°, 99°, 96°, 102°, 95°, and 87°. Find the median temperature for the week. 95°

Objective B

14. The prices of identical calculators at each of 5 department stores were \$38.75, \$44.50, \$42.95, \$41.00, and \$52.00. Find the median price of the calculator. \$42.95

15. The hourly wages for 7 job classifications at a company are \$6.42, \$9.24, \$8.98, \$6.38, \$7.24, \$6.26, and \$7.16. Find the median hourly wage. \$7.16

16. The ages of the 7 most recently hired employees at a company are 26, 45, 22, 25, 24, 30, and 34. Find the median age. 26 years

17. The number of hours the custodian of an office building worked each day last week are shown at the right. Find the median number of hours worked. $7\frac{1}{2}$

Mon	*Tue*	*Wed*	*Thu*	*Fri*	*Sat*	*Sun*
$6\frac{3}{4}$	$7\frac{1}{2}$	$8\frac{1}{4}$	$7\frac{3}{4}$	$8\frac{1}{2}$	0	0

18. The number of tickets 8 police officers gave out during a certain day are 17, 5, 9, 10, 23, 5, 13, and 20. Find the median number of tickets given out. 11.5 tickets

19. The numbers of requests for a conference room at a hotel during a 6-day period were 46, 18, 29, 48, 38, and 24. Find the median number of requests. 33.5 requests

20. During a study of the amount of coffee a vending machine dispensed, the following amounts of coffee were obtained: 3.88 ounces, 4.07 ounces, 3.89 ounces, 4.15 ounces, 4.01 ounces, 3.92 ounces, 4.11 ounces, and 3.85 ounces. Find the median amount of coffee obtained from the vending machine. 3.965 ounces

21. The monthly utility bills for 8 homes are \$86.48, \$92.81, \$48.92, \$74.16, \$112.53, \$61.92, \$86.48, and \$97.92. Find the median utility bill. \$86.48

22. The populations of the ten largest cities in the world are shown below. Find the median population. 14,300,000

Buenos Aires	12,200,000	Osaka	14,000,000
Calcutta	12,900,000	Rio de Janeiro	12,800,000
Bombay	13,500,000	Sao Paulo	21,500,000
Mexico City	24,000,000	Seoul	19,000,000
New York	14,600,000	Tokyo	28,400,000

APPLYING THE CONCEPTS

23. Tell whether each of the following statements is true or false.
   a. The mean is the total of a set of numbers multiplied by the number of addends. False
   b. The median of a set of numbers is the middle number when the numbers are organized from least to greatest. True

24. [W] Rita received scores of 82, 78, 91, and 80 on 4 tests. Elyssa received scores of 87, 83, 96, and 85 (exactly 5 points more on each test). Are the means of the two students the same? If not, what is the relationship between the means of the two students? No. The mean of the second student's score is 5 more than the mean of the first student's.

25. [W] Jason received scores of 89, 85, 94, 11, and 91 on 5 tests. Which measurement, mean or median, gives the best representation of his performance? Explain. Median

Project in Mathematics

Deceptive Graphs

A graphical representation of data can sometimes be misleading. Consider the graphs shown below. An investment firm's financial advisor claims that an investment with the firm will grow as shown in the graph on the left, whereas an investment with a competitor will grow as shown in the graph on the right. Apparently, you would accumulate more money by choosing the investment on the left.

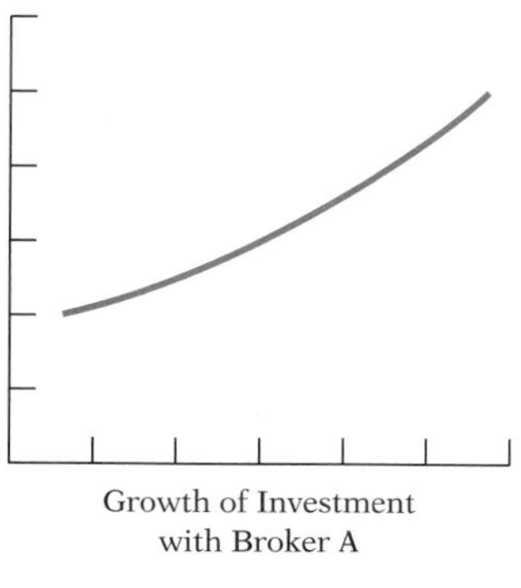
Growth of Investment with Broker A

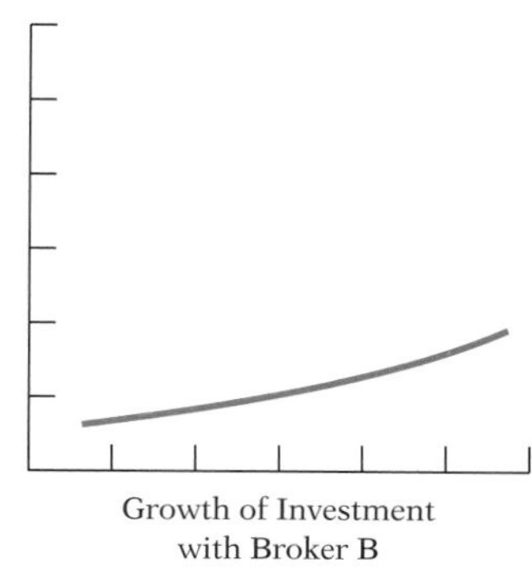
Growth of Investment with Broker B

However, these graphs have a serious flaw. There are no labels on the horizontal and vertical axes. Therefore, it is impossible to tell which investment increased more or over what time interval. When labels are not placed on the axes of a graph, the data that graph represents are meaningless. It is one way advertisers use a visual impact to distort the true meaning of data.

The graphs below are the same as those drawn above except that scales have been drawn along each axis. Now it is possible to tell how each investment has performed. Note that each one turned in exactly the same performance.

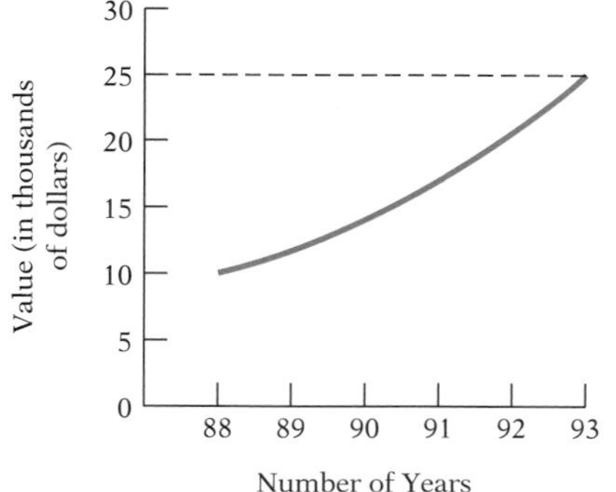

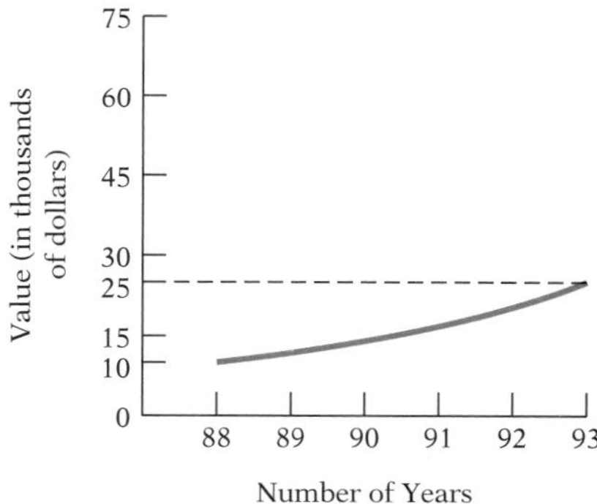

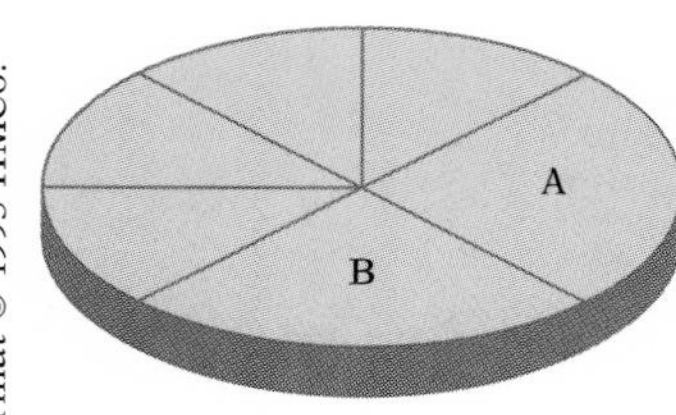

Drawing a circle graph as an oval is another way of distorting data. This is especially true if a three-dimensional representation is given. From the appearance of the circle graph at the left, one would think that region A is larger than region B. However, that isn't true. Measure the angle of each sector to see this for yourself.

Exercise:

As you read newspapers and magazines, find examples of graphs that may distort the actual data. Discuss how these graphs should be drawn to be more accurate.

Chapter Summary

Key Words *Statistics* is the branch of mathematics concerned with *data*, or numerical information.

A *graph* provides a pictorial representation of data. A *pictograph* uses symbols to represent information.

The *circle graph* represents data by the size of the sectors.

The *bar graph* represents data by the height of the bars.

The *broken-line graph* represents data by the position of the lines and shows trends and comparisons.

A *histogram* is a special kind of bar graph.

In a histogram, the width of each bar corresponds to a range of numbers called a *class interval*.

In a histogram, the height of each bar corresponds to the number of occurrences of data in each class interval and is called the *class frequency*.

A *frequency polygon* is a graph that displays information in a manner similar to a histogram. A dot is placed above the center of each class interval at a height corresponding to that class's frequency.

The *average* or *mean* score is the sum of all the test scores divided by the number of tests.

The *median* separates a list of numbers so that there are the same number of values below the median as there are above the median.

Essential Rules ***To Find the Average or Mean***

To find the average or mean of a set of numbers, divide the sum of the numbers by the number of addends.

$$\text{Mean} = \frac{\text{sum of numbers}}{\text{number of addends}}$$

To Find the Median

To find the median, arrange the numbers from smallest to largest and locate the middle number. When there are two middle numbers, the median is the average of those two numbers.

Chapter Review

SECTION 7.1

The pictograph in Figure 37 shows the number of students earning A, B, C, D, and F grades in a geology class. Each figure represents two students.

1. Find the total number of students receiving grades. 44

2. Find the ratio of the number of students receiving B grades to the number of students receiving D grades. $\frac{7}{3}$

3. Find the percent of the total number of students receiving C grades. Round to the nearest tenth of a percent. 29.5%

A
B
C
D
F

Fig. 37 Number of students receiving grades.

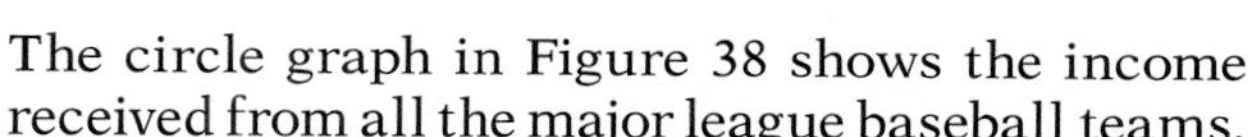

The circle graph in Figure 38 shows the income received from all the major league baseball teams.

4. Find the baseball teams' total income. $965,000,000

5. What is the ratio of the income received from tickets sold at the gate to the income received from local broadcasting? $\frac{20}{9}$

6. Find what percent of the total income is received from national broadcasting. Round to the nearest tenth of a percent. 22.8%

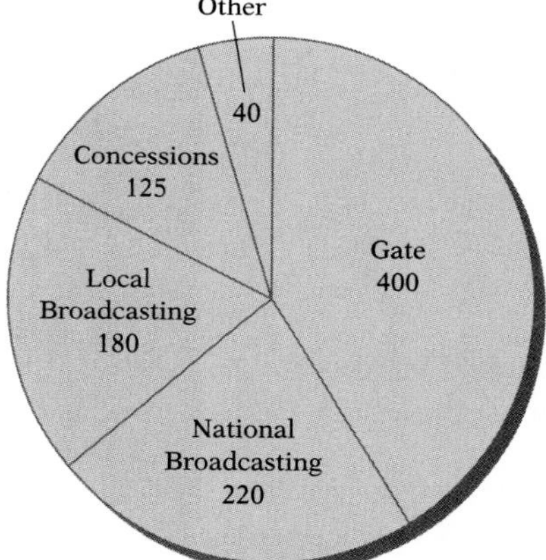

Fig. 38 Income for major league baseball (in millions of dollars).

SECTION 7.2

The double-bar graph in Figure 39 shows the maximum and minimum temperatures during 5 days.

7. Find the difference in the maximum and minimum temperatures for Tuesday. 30°

8. Find the ratio of the maximum to the minimum temperature for Friday. $\frac{7}{4}$

9. Which day had the lowest temperature? What was this temperature? Wednesday, 15°

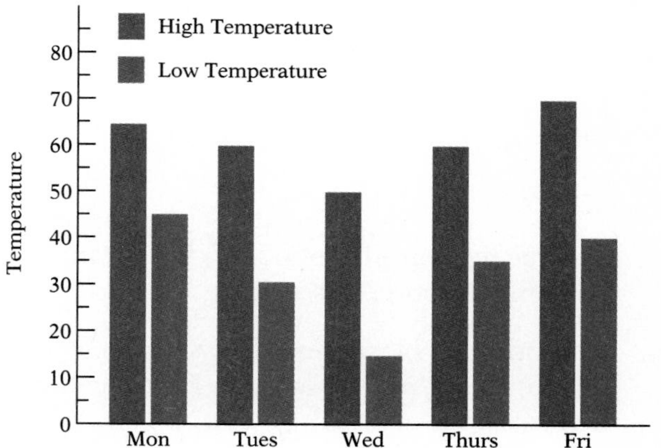

Fig. 39

The double-broken-line graph in Figure 40 shows the profit for a corporation during the four quarters of 1993 and 1994.

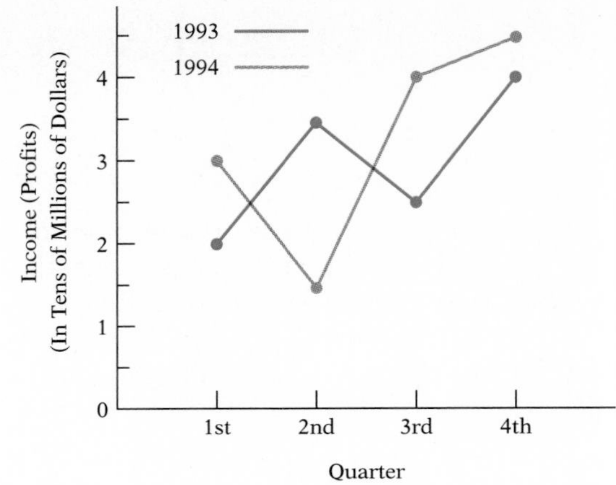

Fig. 40

10. Find the difference in profit for the second quarter of 1993 and the second quarter in 1994. $20 million

11. Find the total profit made in the third and fourth quarters of 1994. $85 million

12. Find the ratio of the profit made in the third quarter of 1993 to the profit made in the third quarter of 1994. $\frac{5}{8}$

SECTION 7.3

The histogram in Figure 41 shows the heights of 41 trees in a nursery.

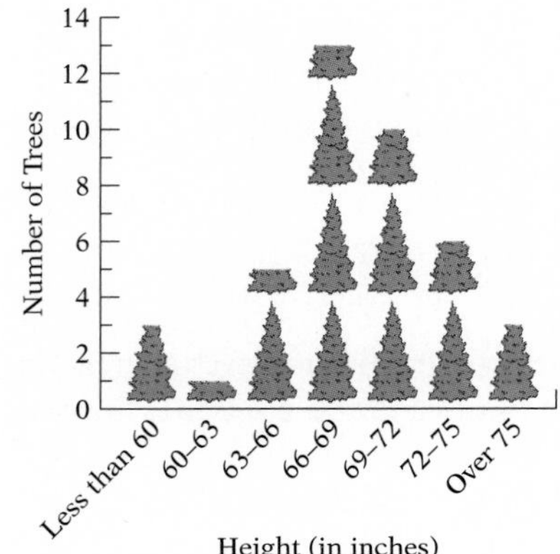

Fig. 41

13. How many of the trees were over 72 inches tall? 9 trees

14. Find the ratio of the number of trees under 66 inches to the number of trees that were between 69 to 72 inches tall. $\frac{9}{10}$

15. Find the percent of the trees that had a height between 66 to 69 inches. 31.7%

The frequency polygon in Figure 42 shows the range of the scores of a team in the national basketball association.

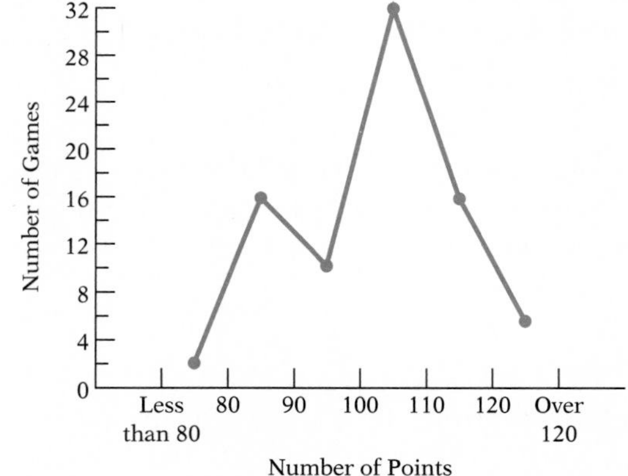

Fig. 42

16. Find the number of games in which fewer than 100 points were scored. 28 games

17. Find the ratio of the number of games in which 90 to 100 points were scored to the number in which 110 to 120 points were scored. $\frac{5}{8}$

18. In what percent of the total number of games were over 120 points scored? (Round to the nearest tenth of a percent.) 7.3%

SECTION 7.4

19. You received grades of 67, 87, 98, 83, and 77 on 5 history exams. What is your mean score? 82.4

20. The number of students in 8 classes of an elementary school are 23, 29, 25, 31, 27, 26, 19, and 28. Find the median number of students in the 8 classes. 26.5 students

Chapter Test

The pictograph in Figure 43 shows the number of students receiving grades in a geology class. Each picture in the pictograph represents two students.

1. Find the total number of students in the geology class. 30 [7.1A]

2. Find the ratio of the number of students receiving an A grade to the number of students receiving a D grade. $\frac{1}{2}$ [7.1A]

3. Find the percent of the total number of students who received a B grade. 30% [7.1A]

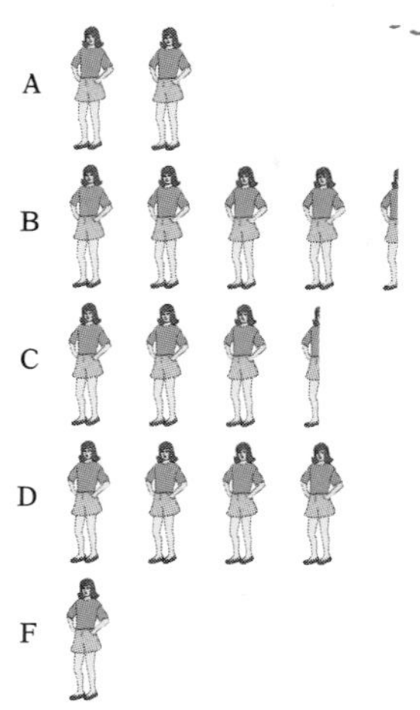

Fig. 43 Number of students receiving grades.

The circle graph in Figure 44 shows sources of income for a community college that has a total budget of $32,000,000.

4. Find the ratio of the amount of federal funds to the amount in the total budget. $\frac{1}{16}$ [7.1B]

5. Find the percent of the total budget that comes from state funds. 53.75% [7.1B]

6. What is the ratio of the amount of state funds to the amount of federal funds? $\frac{43}{5}$ [7.1B]

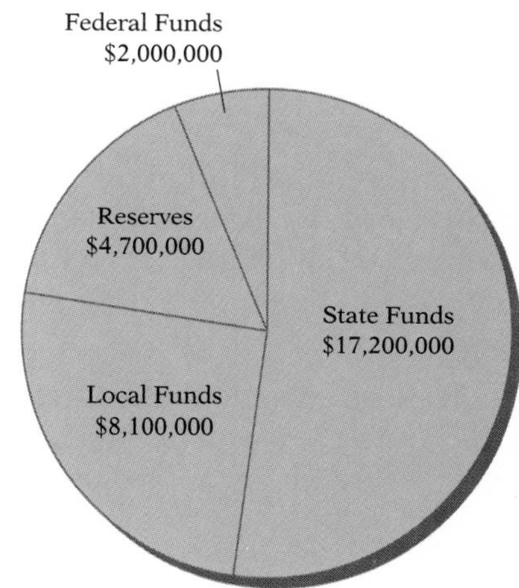

Fig. 44 Sources of income for college budget of $32,000,000.

The double-bar graph in Figure 45 shows the number of economy cars an automobile manufacturer sold during the last 6 months in 1993 and 1994.

7. Find the number of cars sold in November 1993. 27,000 cars [7.2A]

8. Find the difference between the number of cars sold in October 1993 and the number of cars sold in October 1994. 5000 cars [7.2A]

9. What is the difference between the first 3 months' sales for 1993 and 1994? 19,000 cars [7.2A]

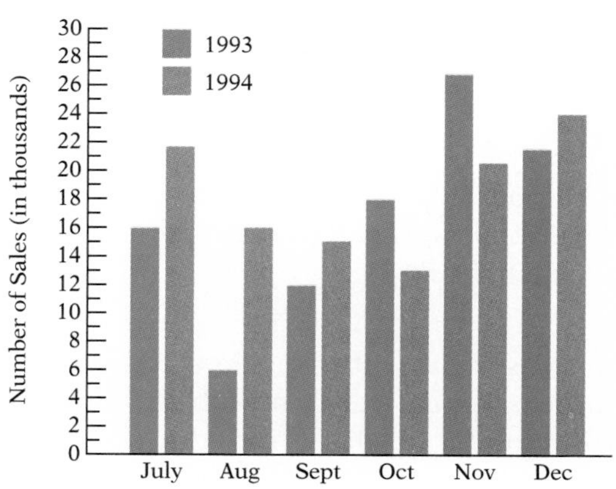

Fig. 45

The double-broken-line graph in Figure 46 shows the quarterly income for Fancy Frames for the years 1993 and 1994.

10. What is Fancy Frames' third-quarter income for 1994? $5,000,000 [7.2B]

11. Find the ratio of the second-quarter income for 1994 to the second-quarter income for 1993. $\frac{3}{4}$ [7.2B]

12. Find the difference between Fancy Frames' fourth-quarter incomes for 1993 and 1994. $1,000,000 [7.2B]

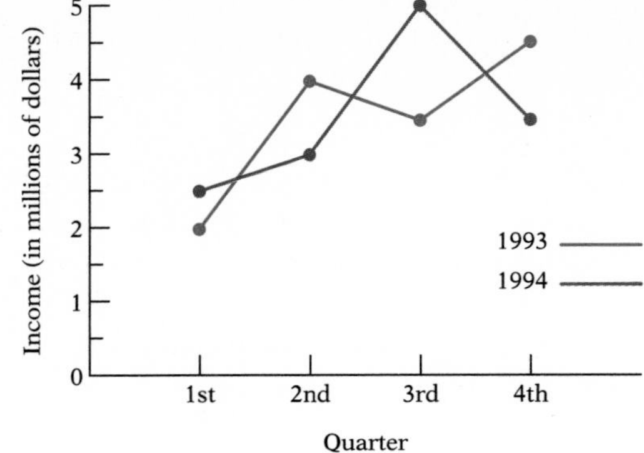

Fig. 46

The histogram in Figure 47 shows the salaries for 51 employees in a small computer company.

13. Find the number of employees who receive a salary over $25,000. 21 employees [7.3A]

14. Find the ratio of the number of employees whose salary is above $30,000 to the total number of employees. $\frac{10}{51}$ [7.3A]

15. Find the percent of the total number of employees whose salaries are between $20,000 and $30,000. Round to the nearest percent. 53% [7.3A]

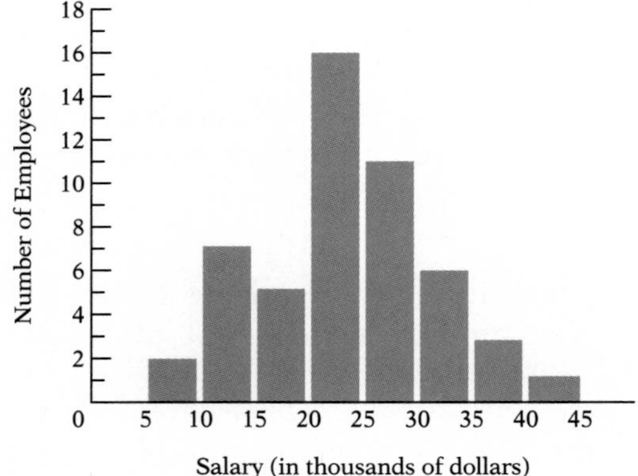

Fig. 47

A television rating service surveyed 100 families to find the number of hours they watched television. The results are recorded in a frequency polygon as shown in Figure 48.

16. How many families watched between 15 and 25 hours of television a week? 55 families [7.3B]

17. Find the ratio of the number of families that watched between 5 and 10 hours of television each week to the number of families that watched between 20 and 25 hours of television. $\frac{3}{5}$ [7.3B]

18. Find the percent of the number of families surveyed that watched over 15 hours of television each week. 60% [7.3B]

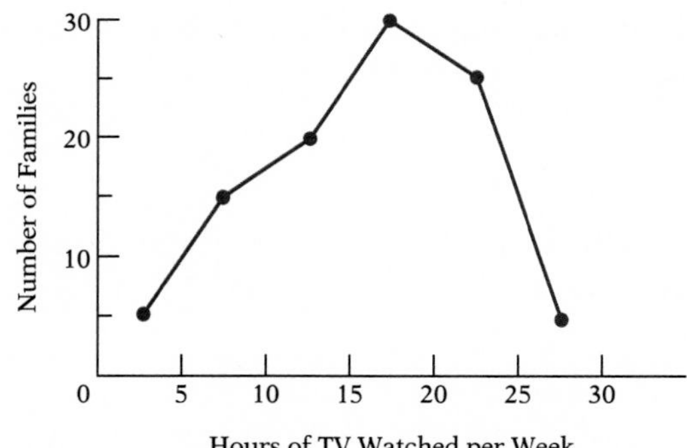

Fig. 48

19. As a sales representative, Pete Kline traveled 125, 212, 188, 231, and 189 miles, respectively, on each of 5 days during the past week. Find the mean number of miles driven during the 5 days. 189 miles [7.4A]

20. One applicant's scores on 8 exams for obtaining an insurance license were 68, 76, 98, 88, 73, 56, 78, and 83. Find the median score. 77 [7.4B]

Cumulative Review

1. Simplify: $2^2 \cdot 3^3 \cdot 5$
540 [1.6A]

2. Simplify: $3^2 \cdot (5 - 2) \div 3 + 5$
14 [1.6B]

3. Find the LCM of 24 and 40.
120 [2.1A]

4. Write $\frac{60}{144}$ in simplest form.
$\frac{5}{12}$ [2.3B]

5. Find the total of $4\frac{1}{2}$, $2\frac{3}{8}$, and $5\frac{1}{5}$.
$12\frac{3}{40}$ [2.4C]

6. Subtract: $12\frac{5}{8} - 7\frac{11}{12}$
$4\frac{17}{24}$ [2.5C]

7. Multiply: $\frac{5}{8} \times 3\frac{1}{5}$
2 [2.6B]

8. Find the quotient of $3\frac{1}{5}$ and $4\frac{1}{4}$.
$\frac{64}{85}$ [2.7B]

9. Simplify: $\frac{5}{8} \div \left(\frac{3}{4} - \frac{2}{3}\right) + \frac{3}{4}$
$8\frac{1}{4}$ [2.8C]

10. Write two hundred nine and three hundred five thousandths in standard form.
209.305 [3.1A]

11. Find the product of 4.092 and 0.69.
2.82348 [3.4A]

12. Convert $16\frac{2}{3}$ to a decimal. Round to the nearest hundredth. 16.67 [3.6A]

13. Write "330 miles on 12.5 gallons of gas" as a unit rate.
26.4 miles/gallon [4.2B]

14. Solve the proportion. $\frac{n}{5} = \frac{16}{25}$
$n = 3.2$ [4.3B]

15. Write $\frac{4}{5}$ as a percent.
80% [5.1B]

16. 8 is 10% of what?
80 [5.4A]

17. What is 38% of 43?
16.34 [5.2A]

18. What percent of 75 is 30?
40% [5.3A]

19. Tanim Kamal, a salesperson at a department store, receives $100 per week plus 2% commission on sales. Find the income for a week in which Tanim had $27,500 in sales. $650 [6.6A]

20. A life insurance policy costs $4.15 for every $1000 of insurance. At this rate, what is the cost for $50,000 of life insurance? $207.50 [4.3C]

21. A contractor borrowed $125,000 for 6 months at an annual simple interest rate of 11%. Find the interest due on the loan. $6875 [6.3A]

22. A compact disc player with a cost of $180 is sold for $279. Find the markup rate. 55% [6.2B]

23. The circle graph in Figure 49 shows how a family's monthly income of $3000 is budgeted. How much is budgeted for food? $570 [7.1B]

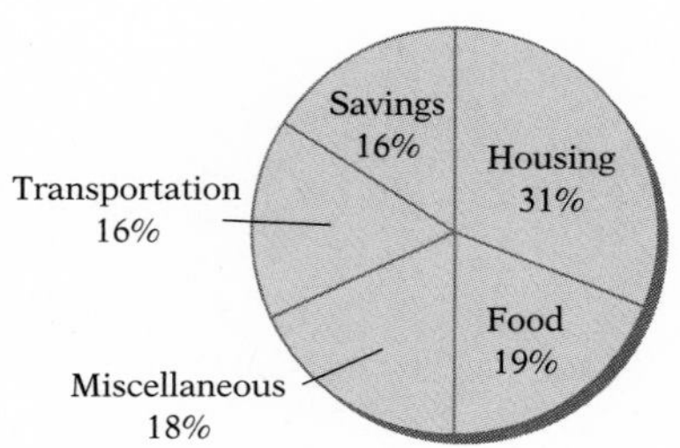

Fig. 49 Budget for a monthly income of $3000.

24. The double-broken-line graph in Figure 50 shows two students' scores on 5 math tests of 30 problems each. Find the difference between the numbers of problems the two students answered correctly on Test 1. 12 problems [7.2B]

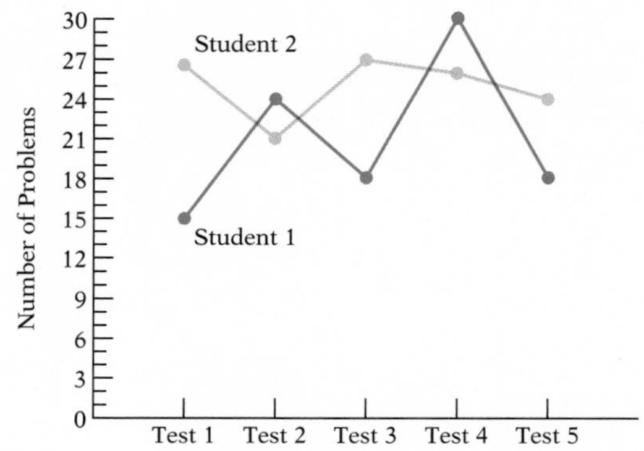

Fig. 50

25. The average daily high temperatures for a week in Newtown were 56°, 72°, 80°, 75°, 68°, 62°, and 74°. Find the mean high temperature for the week. Round to the nearest tenth of a degree. 69.6° [7.4A]

26. The salaries for six teachers in a small school are $17,000, $21,000, $28,500, $22,300, $20,200, and $25,000. Find the median salary. $21,650 [7.4B]

Chapter

8

U.S. Customary Units of Measurement

Objectives

Section 8.1

To convert measurements of length in the U.S. Customary System

To perform arithmetic operations with measurements of length

To solve application problems

Section 8.2

To convert measurements of weight in the U.S. Customary System

To perform arithmetic operations with measurements of weight

To solve application problems

Section 8.3

To convert measurements of capacity in the U.S. Customary System

To perform arithmetic operations with measurements of capacity

To solve application problems

Section 8.4

To use units of energy in the U.S. Customary System

To use units of power in the U.S. Customary System

Light Years and Astronomical Units

The measurement of a short distance between two objects normally can be made by using some kind of ruler or tape measure. For example, the distance across a room can be found with a tape measure. This distance is most likely to be given in feet.

On the other hand, a tape measure would be quite useless for finding the distance between Chicago and New York City. Even measuring that distance in feet would not be practical. Instead, an odometer is used, and the results are recorded in miles. It is approximately 1700 miles between New York City and Chicago.

But even miles become impractical for measuring the distance between two objects that are separated by a large distance. For example, the distance between the earth and Alpha Centauri, a relatively nearby star, is approximately 25 trillion miles, which is 25,000,000,000,000 miles.

Astronomers have units of measurement that are useful for measuring the vast distances in space. A unit that is used frequently is the light year. One light year is the distance light could travel in 1 year. Because light travels at 186,000 miles per second, in 1 year it could travel about 5,880,000,000,000 miles, which is close to 6 trillion miles.

Using this measure, the distance between the earth and Alpha Centauri is approximately 4.2 light years, a more convenient number for measuring that distance. On the other hand, the distance between Chicago and New York is 0.0000000029 light years—not a very convenient measure.

Another measure that astronomers use is the astronomical unit (AU). 1 AU is approximately 92,900,000 miles and is the approximate distance of the earth from the sun. Using this measure, Pluto, the farthest planet from the sun, is 39.4 AU from the sun.

8.1 Length

Objective A To convert measurements of length in the U.S. Customary System

POINT OF INTEREST

The ancient Greeks devised the foot measurement which they usually divided into 16 fingers. It was the Romans who subdivided the foot into 12 units called inches. The word *inch* is derived from the Latin word *uncia* meaning a twelfth part.

The Romans also used a unit called "pace" which equaled two steps. One thousand paces equaled one mile. The word for mile is derived from the Latin word *mille* which means 1000.

A **measurement** includes a number and a unit.

number	unit
3	feet
7	miles
12	yards

Standard units of measurement have been established to simplify trade and commerce.

The unit of length, or distance, that is called the **yard** was originally defined as the length of a specified bronze bar located in London.

The standard U.S. Customary System units of length are **inch, foot, yard,** and **mile.** Equivalences between units of length in the U.S. Customary System are

12 inches (in.) = 1 foot (ft)
3 ft = 1 yard (yd)
36 in. = 1 yard (yd)
5280 ft = 1 mile (mi)

These equivalences can be used to form conversion rates to change one unit of measurement to another. For example, because 3 ft = 1 yd, the conversion rates $\frac{3\text{ ft}}{1\text{ yd}}$ and $\frac{1\text{ yd}}{3\text{ ft}}$ are each equivalent to 1.

⇒ Convert 27 ft to yards.

$$27\text{ ft} = 27\text{ ft} \times \boxed{\frac{1\text{ yd}}{3\text{ ft}}}$$
$$= 27\,\cancel{\text{ft}} \times \frac{1\text{ yd}}{3\,\cancel{\text{ft}}}$$
$$= \frac{27\text{ yd}}{3}$$
$$= 9\text{ yd}$$

⇒ Convert 5 yd to feet.

$$5\text{ yd} = 5\text{ yd} \times \boxed{\frac{3\text{ ft}}{1\text{ yd}}}$$
$$= 5\,\cancel{\text{yd}} \times \frac{3\text{ ft}}{1\,\cancel{\text{yd}}}$$
$$= \frac{15\text{ ft}}{1}$$
$$= 15\text{ ft}$$

Note that in the conversion rate chosen, the unit in the numerator is the same as the unit desired in the answer. The unit in the denominator is the same as the unit in the given measurement.

Example 1 Convert 40 in. to feet.

Solution $40 \text{ in.} = 40 \text{ in.} \times \frac{1 \text{ ft}}{12 \text{ in.}} = 3\frac{1}{3} \text{ ft}$

You Try It 1 Convert 60 in. to feet.

Your solution 5 ft

Example 2 Convert 36 ft to yards.

Solution $36 \text{ ft} = 36 \text{ ft} \times \frac{1 \text{ yd}}{3 \text{ ft}} = 12 \text{ yd}$

You Try It 2 Convert 14 ft to yards.

Your solution $4\frac{2}{3}$ yd

Example 3 Convert $3\frac{1}{4}$ yd to feet.

Solution $3\frac{1}{4} \text{ yd} = 3\frac{1}{4} \text{ yd} \times \frac{3 \text{ ft}}{1 \text{ yd}} = 9\frac{3}{4} \text{ ft}$

You Try It 3 Convert 9800 ft to miles.

Your solution $1\frac{113}{132}$ mi

Solutions on p. A38

Objective B To perform arithmetic operations with measurements of length

When performing arithmetic operations with measurements of length, write the answer in simplest form. For example, 1 ft 14 in. should be written as 2 ft 2 in.

➡ Convert: 50 in. = ____ ft ____ in.

$$\begin{array}{r} 4 \text{ ft } 2 \text{ in.} \\ 12\overline{)\,50} \\ \underline{-48} \\ 2 \end{array}$$

• **Because 12 in. = 1 ft, divide 50 in. by 12. The whole-number part of the quotient is the number of feet. The remainder is the number of inches.**

50 in. = 4 ft 2 in.

Example 4 Convert: 17 in. = ____ ft ____ in.

Solution

$$\begin{array}{r} 1 \text{ ft } 5 \text{ in} \\ 12\overline{)\,17} \\ \underline{-12} \\ 5 \end{array}$$

17 in. = 1 ft 5 in.

You Try It 4 Convert: 42 in. = ____ ft ____ in.

Your solution 3 ft 6 in.

Example 5 Convert: 31 ft = ____ yd ____ ft

Solution

$$\begin{array}{r} 10 \text{ yd } 1 \text{ ft} \\ 3\overline{)\,31} \\ \underline{-30} \\ 1 \end{array}$$

31 ft = 10 yd 1 ft

You Try It 5 Convert: 14 ft = ____ yd ____ ft

Your solution 4 yd 2 ft

Solutions on p. A38

Example 6 Find the sum of 4 ft 4 in. and 1 ft 11 in.

Solution

$$\begin{array}{r} 4 \text{ ft} \quad 4 \text{ in.} \\ + \ 1 \text{ ft} \ 11 \text{ in.} \\ \hline 5 \text{ ft} \ 15 \text{ in.} \end{array} = 6 \text{ ft } 3 \text{ in.}$$

You Try It 6 Find the sum of 3 ft 5 in. and 4 ft 9 in.

Your solution 8 ft 2 in.

Example 7 Subtract: 9 ft 6 in. − 3 ft 8 in.

Solution

$$\begin{array}{r} \overset{8 \text{ ft.}}{\not{9} \text{ ft}} \quad \overset{18 \text{ in.}}{\not{6} \text{ in.}} \\ - \ 3 \text{ ft} \quad 8 \text{ in.} \\ \hline 5 \text{ ft} \ 10 \text{ in.} \end{array}$$

- **Borrow 1 ft (12 in.) from 9 ft and add to 6 in.**

You Try It 7 Subtract: 4 ft 2 in. − 1 ft 8 in.

Your solution 2 ft 6 in.

Example 8 Multiply: 3 yd 2 ft × 4

Solution

$$\begin{array}{r} 3 \text{ yd} \ 2 \text{ ft} \\ \times \qquad 4 \\ \hline 12 \text{ yd} \ 8 \text{ ft} \end{array} = 14 \text{ yd } 2 \text{ ft}$$

You Try It 8 Multiply: 4 yd 1 ft × 8

Your solution 34 yd 2 ft

Example 9 Find the quotient of 4 ft 3 in. and 3.

Solution

$$\begin{array}{rrr} & 1 \text{ ft} & 5 \text{ in.} \\ 3 \big) & 4 \text{ ft} & 3 \text{ in.} \\ & -3 \text{ ft} & \\ \hline & 1 \text{ ft} = & 12 \text{ in.} \\ & & 15 \text{ in.} \\ & & -15 \text{ in.} \\ \hline & & 0 \end{array}$$

You Try It 9 Find the quotient of 7 yd 1 ft and 2.

Your solution 3 yd 2 ft

Example 10 Multiply: $2\frac{3}{4}$ ft × 3

Solution

$$2\frac{3}{4} \text{ ft} \times 3 = \frac{11}{4} \text{ ft} \times 3$$
$$= \frac{33}{4} \text{ ft}$$
$$= 8\frac{1}{4} \text{ ft}$$

You Try It 10 Subtract: $6\frac{1}{4}$ ft − $3\frac{2}{3}$ ft

Your solution $2\frac{7}{12}$ ft

Solutions on p. A38

Objective C To solve application problems

Example 11

One shrub extends 15 ft along a freeway. How many shrubs are to be planted along 6450 ft on the freeway?

Strategy

To find the number of shrubs to be planted along the freeway, divide the total distance (6450 ft) by the length of one shrub (15 ft).

Solution

$$\frac{6450 \text{ ft}}{15 \text{ ft}} = 430$$

430 shrubs are to be planted along the freeway.

You Try It 11

The floor of a storage room is being tiled. Eight tiles, each a 9-inch square, fit across the width of the floor. Find the width in feet of the storage room.

Your strategy

Your solution

6 ft

Example 12

A plumber used 3 ft 9 in., 2 ft 6 in., and 11 in. of copper tubing to install a sink. Find the total length of copper tubing used.

Strategy

To find the total length of copper tubing used, add the three lengths of copper tubing (3 ft 9 in., 2 ft 6 in., and 11 in.).

Solution

$$\begin{array}{rr} & 3 \text{ ft} \quad 9 \text{ in.} \\ & 2 \text{ ft} \quad 6 \text{ in.} \\ + & 11 \text{ in.} \\ \hline & 5 \text{ ft } 26 \text{ in.} \end{array} = 7 \text{ ft } 2 \text{ in.}$$

The plumber used 7 ft 2 in. of copper tubing.

You Try It 12

A board 9 ft 8 in. is cut into four pieces of equal length. How long is each piece?

Your strategy

Your solution

2 ft 5 in.

Solutions on p. A38

8.1 Exercises

Objective A

Convert.

1. 6 ft = 72 in.
2. 9 ft = 108 in.
3. 30 in. = $2\frac{1}{2}$ ft
4. 64 in. = $5\frac{1}{3}$ ft
5. 13 yd = 39 ft
6. $4\frac{1}{2}$ yd = $13\frac{1}{2}$ ft
7. 16 ft = $5\frac{1}{3}$ yd
8. $4\frac{1}{2}$ ft = $1\frac{1}{2}$ yd
9. $2\frac{1}{3}$ yd = 84 in.
10. 5 yd = 180 in.
11. 120 in. = $3\frac{1}{3}$ yd
12. 66 in. = $1\frac{5}{6}$ yd
13. 2 mi = 10,560 ft
14. $1\frac{1}{2}$ mi = 7920 ft
15. $7\frac{1}{2}$ in. = $\frac{5}{8}$ ft

Objective B

Perform the arithmetic operation.

16. 100 in. = 8 ft 4 in.
17. 6400 ft = 1 mi 1120 ft
18. 15 in. = 1 ft 3 in.
19. 6 ft 7 in. + 3 ft 4 in. = 9 ft 11 in.
20. 9 ft 11 in. + 3 ft 6 in. = 13 ft 5 in.
21. 5 ft 3 in. − 2 ft 6 in. = 2 ft 9 in.
22. 9 yd 1 ft − 3 yd 2 ft = 5 yd 2 ft
23. 2 ft 5 in. × 6 = 14 ft 6 in.
24. $3\frac{2}{3}$ ft × 4 = $14\frac{2}{3}$ ft
25. 2)5 ft 4 in. = 2 ft 8 in.
26. $12\frac{1}{2}$ in. ÷ 3 = $4\frac{1}{6}$ in.
27. $4\frac{2}{3}$ ft + $6\frac{1}{2}$ ft = $11\frac{1}{6}$ ft
28. 3 yd 2 ft + 6 yd 2 ft = 10 yd 1 ft
29. 1 mi 4200 ft + 2 mi 3600 ft = 4 mi 2520 ft
30. 5 yd 1 ft − 2 yd 2 ft = 2 yd 2 ft

Objective C *Application Problems*

31. A kitchen counter is to be covered with tile that is 4 inches square. How many tiles can be placed along one row of a counter top that is 4 ft 8 in. long? 14 tiles

32. Thirty-two yards of material were used for making pleated draperies. How many feet of material were used? 96 ft

33. Find the missing dimension. $1\frac{5}{6}$ ft

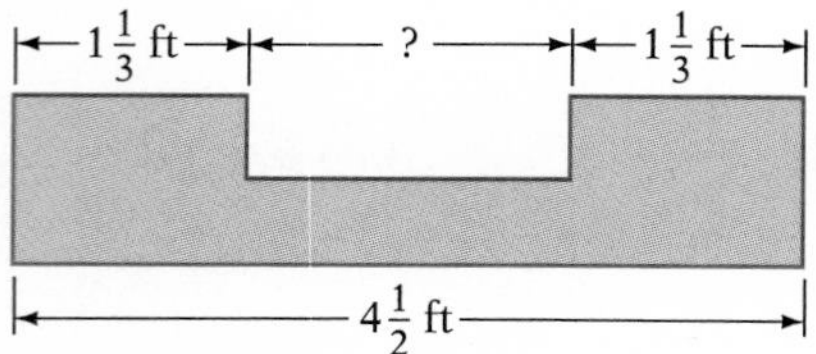

34. Find the total length of the shaft. 3 ft 1 in.

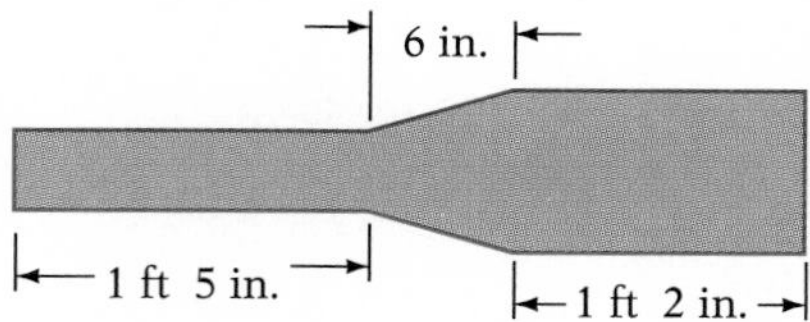

35. What length of material is needed to drill two holes 3 in. in diameter and leave $\frac{1}{2}$ inch between the holes and on either side as shown in the diagram? $7\frac{1}{2}$ in.

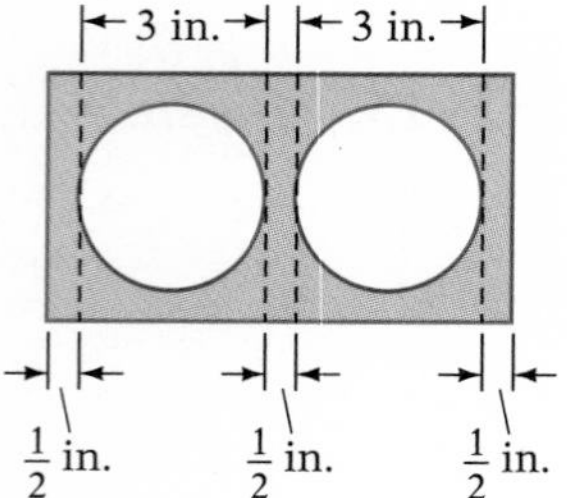

36. Find the missing dimension in the figure.

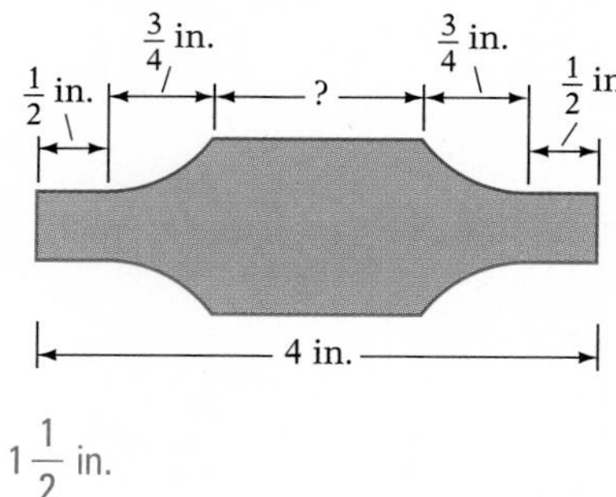

$1\frac{1}{2}$ in.

37. A board $6\frac{2}{3}$ ft is cut into four equal pieces. How long is each piece? $1\frac{2}{3}$ ft

38. How long must a board be if four pieces, each 3 ft 4 in. long, are to be cut from it? 13 ft 4 in.

39. A picture is 1 ft 9 in. high and 1 ft 6 in. wide. Find the length of framing needed to frame the picture. 6 ft 6 in.

40. You bought 32 feet of baseboard to install in the kitchen of your house. How many inches of baseboard did you purchase? 384 in.

41. Forty-five bricks, each 9 in. long, are laid end to end to make the base for a wall. Find the length of the wall in feet. $33\frac{3}{4}$ ft

42. A roof is constructed with nine rafters, each rafter 8 ft 4 in. long. Find the total number of feet of material needed to build the rafters. 75 ft

APPLYING THE CONCEPTS

43. McDonald's has sold more than 99 billion hamburgers. Assume that the average thickness of a hamburger with bun is $1\frac{1}{2}$ in. If all these hamburgers were stacked up, how far, in feet, would the stack reach into space? 12,375,000,000 ft

44. [W] How good are you at estimating lengths or distances? Estimate the length of a pencil, the width of your room, the length of a block, and the distance to the grocery store. Then measure the lengths and compare the results with your estimates. Answers will vary

8.2 Weight

Objective A To convert measurements of weight in the U.S. Customary System

POINT OF INTEREST

The Romans used two different systems of weights. In both systems, the smallest unit was the uncia, abbreviated to "oz," from which the term *ounce* is derived. In one system, there were 16 ounces to one pound. In the second system, a pound, which was called the libra, equaled 12 unciae. The abbreviation "lb" for pound comes from the word *libra*.

The avoirdupois system of measurement and the troy system of measurement have their heritage in the two Roman systems.

Weight is a measure of how strongly the earth is pulling on an object. The unit of weight called the **pound** is defined as the weight of a standard solid kept at the Bureau of Standards in Washington, D.C. The U.S. Customary System units of weight are **ounce, pound,** and **ton.**

Equivalences between units of weight in the U.S. Customary System are

$$16 \text{ ounces (oz)} = 1 \text{ pound (lb)}$$
$$2000 \text{ lb} = 1 \text{ ton}$$

These equivalences can be used to form conversion rates to change one unit of measurement to another. For example, because 16 oz = 1 lb, the conversion rates $\frac{16 \text{ oz}}{1 \text{ lb}}$ and $\frac{1 \text{ lb}}{16 \text{ oz}}$ are each equivalent to 1.

➡ Convert 62 oz to pounds.

$$62 \text{ oz} = 62 \text{ oz} \times \boxed{\frac{1 \text{ lb}}{16 \text{ oz}}}$$
$$= \frac{62 \cancel{\text{oz}}}{1} \times \frac{1 \text{ lb}}{16 \cancel{\text{oz}}}$$
$$= \frac{62 \text{ lb}}{16}$$
$$= 3\frac{7}{8} \text{ lb}$$

• The conversion rate must contain lb (the unit desired in the answer) in the numerator and oz (the original unit) in the denominator.

Example 1 Convert $3\frac{1}{2}$ tons to pounds.

Solution $3\frac{1}{2} \text{ tons} = 3\frac{1}{2} \cancel{\text{tons}} \times \frac{2000 \text{ lb}}{1 \cancel{\text{ton}}}$

$= 7000 \text{ lb}$

You Try It 1 Convert 3 lb to ounces.

Your solution 48 oz

Example 2 Convert 42 oz to pounds.

Solution $42 \text{ oz} = 42 \cancel{\text{oz}} \times \frac{1 \text{ lb}}{16 \cancel{\text{oz}}}$

$= \frac{42 \text{ lb}}{16} = 2\frac{5}{8} \text{ lb}$

You Try It 2 Convert 4200 lb to tons.

Your solution $2\frac{1}{10}$ tons

Solutions on p. A39

Objective B To perform arithmetic operations with measurements of weight

When performing arithmetic operations with measurements of weight, write the answer in simplest form. For example, 1 lb 22 oz should be written 2 lb 6 oz.

Example 3 Find the difference between 14 lb 5 oz and 8 lb 14 oz.

Solution

$$\begin{array}{rr} \overset{13\text{ lb}}{\cancel{14\text{ lb}}} & \overset{21\text{ oz}}{\cancel{5\text{ oz}}} \\ -\ 8\text{ lb} & 14\text{ oz} \\ \hline 5\text{ lb} & 7\text{ oz} \end{array}$$

• Borrow 1 lb (16 oz) from 14 lb and add to 5 oz.

You Try It 3 Find the difference between 7 lb 1 oz and 3 lb 4 oz.

Your solution 3 lb 13 oz

Example 4 Divide: 7 lb 14 oz ÷ 3

Solution

$$\begin{array}{rrr} & 2\text{ lb} & 10\text{ oz} \\ 3) & 7\text{ lb} & 14\text{ oz} \\ & -6\text{ lb} & \\ & 1\text{ lb} = & 16\text{ oz} \\ & & 30\text{ oz} \\ & & -30\text{ oz} \\ & & 0 \end{array}$$

You Try It 4 Multiply: 3 lb 6 oz × 4

Your solution 13 lb 8 oz

Solutions on p. A39

Objective C To solve application problems

Example 5

Four teachers spent their summer vacation panning for gold. How much money did each teacher receive if they found 1 lb 9 oz of gold and the price of gold was $525.80 per ounce?

Strategy To find the amount of money each teacher received:

- Convert 1 lb 9 oz to ounces.
- Multiply the number of ounces by the price per ounce ($525.80) to find the total income.
- Divide the total income by the number of teachers (4).

Solution 1 lb 9 oz = 25 oz

$$\begin{array}{r} 525.80 \\ \times\quad 25 \\ \hline \end{array} \qquad \begin{array}{r} 3{,}286.25 \\ 4\overline{)13{,}145.00} \end{array}$$

$13,145.00 total income

Each teacher received $3286.25.

You Try It 5

Find the weight in pounds of 12 bars of soap. Each bar weighs 9 oz.

Your strategy

Your solution

$6\frac{3}{4}$ lb

Solution on p. A39

8.2 Exercises

Objective A

Convert.

1. 64 oz = 4 lb

2. 36 oz = $2\frac{1}{4}$ lb

3. 4 lb = 64 oz

4. 7 lb = 112 oz

5. 3200 lb = $1\frac{3}{5}$ tons

6. 7000 lb = $3\frac{1}{2}$ tons

7. 6 tons = 12,000 lb

8. $1\frac{1}{4}$ tons = 2500 lb

9. 66 oz = $4\frac{1}{8}$ lb

10. 90 oz = $5\frac{5}{8}$ lb

11. $1\frac{1}{2}$ lb = 24 oz

12. $2\frac{5}{8}$ lb = 42 oz

13. $1\frac{3}{10}$ tons = 2600 lb

14. $\frac{4}{5}$ tons = 1600 lb

15. 500 lb = $\frac{1}{4}$ tons

16. 5000 lb = $2\frac{1}{2}$ tons

17. 180 oz = $11\frac{1}{4}$ lb

18. 12 oz = $\frac{3}{4}$ lb

Objective B

Perform the arithmetic operation.

19. 9000 lb = 4 tons 1000 lb

20. 85 oz = 5 lb 5 oz

21. 40 oz = 2 lb 8 oz

22. 4 lb 7 oz + 3 lb 12 oz
8 lb 3 oz

23. 1 ton 800 lb + 3 tons 1600 lb
5 tons 400 lb

24. 7 lb 5 oz − 3 lb 8 oz
3 lb 13 oz

25. 3 tons 500 lb − 1 ton 800 lb
1 ton 1700 lb

26. 3 lb 6 oz × 4
13 lb 8 oz

27. $5\frac{1}{2}$ lb × 6
33 lb

28. 2)3 lb 8 oz
1 lb 12 oz

29. $4\frac{2}{3}$ lb × 3
14 lb

30. 7 lb 7 oz + 6 lb 9 oz
14 lb

31. $6\frac{1}{2}$ oz + $2\frac{1}{2}$ oz
9 oz

32. $6\frac{3}{8}$ lb − $2\frac{5}{6}$ lb
$3\frac{13}{24}$ lb

33. 5 lb 12 oz ÷ 4
1 lb 7 oz

Objective C *Application Problems*

34. A machinist has 25 iron rods to mill. Each rod weighs 20 oz. Find the total weight of the rods in pounds. $31\frac{1}{4}$ lb

35. A fireplace brick weighs $2\frac{1}{2}$ lb. What is the weight of a load of 800 bricks? 2000 lb

36. A college bookstore received 1200 textbooks, each book weighing 9 oz. Find the total weight of the 1200 textbooks in pounds. 675 lb

37. A 4×4-in. tile weighs 7 oz. Find the weight in pounds of a package of 144 tiles. 63 lb

38. A farmer ordered 20 tons of feed for 100 cattle. After 15 days, the farmer has 5 tons of feed left. How many pounds of food has each cow eaten per day? 20 lb/day

39. A case of soft drink contains 24 cans, each weighing 6 oz. Find the weight, in pounds, of the case of soft drink. 9 lb

40. A baby weighed 7 lb 8 oz at birth. At 6 months of age, the baby weighed 15 lb 13 oz. Find the baby's increase in weight during the 3 months. 8 lb 5 oz

41. Shampoo weighing 5 lb 4 oz is divided equally and poured into four containers. How much shampoo is in each container? 1 lb 5 oz

42. A steel rod weighing 16 lb 11 oz is cut into three pieces. Find the weight of each piece of steel rod. 5 lb 9 oz

43. Find the cost of a ham roast weighing 5 lb 10 oz if the price per pound is \$2.40. \$13.50

44. A candy store buys candy weighing 12 lb for \$14.40. The candy is repackaged and sold in 6-oz packages for \$1.15 each. Find the markup on the 12 lb of candy. \$22.40

45. A manuscript weighing 2 lb 3 oz is mailed at the postage rate of \$0.25 per ounce. Find the cost of mailing the manuscript. \$8.75

APPLYING THE CONCEPTS

46. [W] Write a paragraph describing the growing need for precision in our measurements as civilization progressed. Include a discussion of the need for precision in the space industry. Answers will vary

47. Estimate the weight of a nickel, a textbook, a friend, and a car. Then find the actual weights and compare them with your estimates. Answers will vary

Capacity

Objective A ***To convert measurements of capacity in the U.S. Customary System***

POINT OF INTEREST

The word *quart* has its root in the Medieval Latin word *quartus* which means one-fourth. Thus, a quart is $\frac{1}{4}$ of a gallon. The same Latin word is responsible for other English words such as *quarter, quadrilateral,* and *quartet.*

Liquid substances are measured in units of **capacity.** The standard U.S. Customary units of capacity are the **fluid ounce, cup, pint, quart,** and **gallon.** Equivalences between units of capacity in the U.S. Customary System are

8 fluid ounces (fl oz) = 1 cup (c)
2 c = 1 pint (pint)
2 pt = 1 quart (qt)
4 qt = 1 gallon (gal)

These equivalences can be used to form conversion rates to change one unit of measurement to another. For example, because 8 fl oz = 1 c, the conversion rates $\frac{8 \text{ fl oz}}{1 \text{ c}}$ and $\frac{1 \text{ c}}{8 \text{ fl oz}}$ are each equivalent to 1.

➡ Convert 36 fl oz to cups.

$$36 \text{ fl oz} = 36 \text{ fl oz} \times \boxed{\frac{1 \text{ c}}{8 \text{ fl oz}}}$$
$$= \frac{36 \cancel{\text{ fl oz}}}{1} \times \frac{1 \text{ c}}{8 \cancel{\text{ fl oz}}}$$
$$= \frac{36 \text{ c}}{8}$$
$$= 4\frac{1}{2} \text{ c}$$

• The conversion rate must contain c in the numerator and fl oz in the denominator.

➡ Convert 3 qt to cups.

$$3 \text{ qt} = 3 \text{ qt} \times \boxed{\frac{2 \text{ pt}}{1 \text{ qt}}} \times \boxed{\frac{2 \text{ c}}{1 \text{ pt}}}$$
$$= \frac{3 \cancel{\text{ qt}}}{1} \times \frac{2 \cancel{\text{ pt}}}{1 \cancel{\text{ qt}}} \times \frac{2 \text{ c}}{1 \cancel{\text{ pt}}}$$
$$= \frac{12 \text{ c}}{1}$$
$$= 12 \text{ c}$$

• The direct equivalence is not given above. Use two conversion rates. First convert quarts to pints and then convert pints to cups. The unit in the denominator of the second conversion rate and the unit in the numerator of the first conversion rate must be the same in order to cancel.

Example 1 Convert 42 c to quarts.

Solution
$$42 \text{ c} = 42 \cancel{\text{ c}} \times \frac{1 \cancel{\text{ pt}}}{2 \cancel{\text{ c}}} \times \frac{1 \text{ qt}}{2 \cancel{\text{ pt}}}$$
$$= \frac{42 \text{ qt}}{4} = 10\frac{1}{2} \text{ qt}$$

You Try It 1 Convert 18 pt to gallons.

Your solution $2\frac{1}{4}$ gal

Solution on p. A39

Objective B *To perform arithmetic operations with measurements of capacity*

When performing arithmetic operations with measurements of capacity, write the answer in simplest form. For example, 1 c 12 fl oz should be written as 2 c 4 fl oz.

Example 2 What is 4 gal 1 qt decreased by 2 gal 3 qt?

Solution

$$\begin{array}{r} \overset{3\text{ gal}}{\cancel{4\text{ gal}}}\ \overset{5\text{ qt}}{\cancel{1\text{ qt}}} \\ -\ 2\text{ gal}\ 3\text{ qt} \\ \hline 1\text{ gal}\ 2\text{ qt} \end{array}$$

• Borrow 1 gal (4 qt) from 4 gal and add to 1 qt.

You Try It 2 Find the quotient of 4 gal 2 qt and 3.

Your solution 1 gal 2 qt

Solution on p. A39

Objective C *To solve application problems*

Example 3

A can of apple juice contains 25 fl oz. Find the number of quarts of apple juice in a case of 24 cans.

Strategy

To find the number of quarts of apple juice in one case:

- Multiply the number of cans (24) by the number of fluid ounces per can (25) to find the total number of fluid ounces in the case.
- Convert the number of fluid ounces in the case to quarts.

Solution

$$\begin{array}{l} \ \ 25\text{ fl oz} \\ \times\ 24 \\ \hline 600\text{ fl oz in the case} \end{array}$$

$$600\text{ fl oz} = \frac{600\ \cancel{\text{fl oz}}}{1} \cdot \frac{1\ \cancel{\text{c}}}{8\ \cancel{\text{fl oz}}} \cdot \frac{1\ \cancel{\text{pt}}}{2\ \cancel{\text{c}}} \cdot \frac{1\text{ qt}}{2\ \cancel{\text{pt}}}$$

$$= \frac{600\text{ qt}}{32} = 18\frac{3}{4}\text{ qt}$$

One case of apple juice contains $18\frac{3}{4}$ qt.

You Try It 3

Five students are going backpacking in the desert. Each student requires 1 qt of water per day. How many gallons of water should they take for a 3-day trip?

Your strategy

Your solution

$3\frac{3}{4}$ gal

Solution on p. A40

8.3 Exercises

Objective A

Convert.

1. 60 fl oz = $7\frac{1}{2}$ c
2. 48 fl oz = 6 c
3. 3 c = 24 fl oz
4. $2\frac{1}{2}$ c = 20 fl oz
5. 8 c = 4 pt
6. 5 c = $2\frac{1}{2}$ pt
7. $3\frac{1}{2}$ pt = 7 c
8. 12 pt = 6 qt
9. 22 qt = $5\frac{1}{2}$ gal
10. 10 qt = $2\frac{1}{2}$ gal
11. $2\frac{1}{4}$ gal = 9 qt
12. 7 gal = 28 qt
13. $7\frac{1}{2}$ pt = $3\frac{3}{4}$ qt
14. $3\frac{1}{2}$ qt = 7 pt
15. 20 fl oz = $1\frac{1}{4}$ pt
16. $1\frac{1}{2}$ pt = 24 fl oz
17. 17 c = $4\frac{1}{4}$ qt
18. $1\frac{1}{2}$ qt = 6 c

Objective B

Perform the arithmetic operation.

19. 14 qt = 3 gal 2 qt
20. 9 pt = 4 qt 1 pt
21. 5 pt = 2 qt 1 pt
22. 3 gal 2 qt + 4 gal 3 qt = 8 gal 1 qt
23. 4 qt 1 pt + 2 qt 1 pt = 7 qt
24. 3 gal 1 qt − 1 gal 2 qt = 1 gal 3 qt
25. 3 c 3 fl oz − 2 c 5 fl oz = 6 fl oz
26. 2 qt 1 pt × 5 = 12 qt 1 pt
27. $3\frac{1}{2}$ pt × 5 = $17\frac{1}{2}$ pt
28. 5)6 gal 1 qt = 1 gal 1 qt
29. $3\frac{1}{2}$ gal ÷ 4 = $\frac{7}{8}$ gal
30. 5 c 3 fl oz + 3 c 6 fl oz = 9 c 1 fl oz

Perform the arithmetic operation.

31. 3 gal 3 qt
+ 1 gal 2 qt
5 gal 1 qt

32. 4 c 6 fl oz
− 2 c 7 fl oz
1 c 7 fl oz

33. 3 gal
− 1 gal 2 qt
1 gal 2 qt

34. $1\frac{1}{2}$ pt + $2\frac{2}{3}$ pt
$4\frac{1}{6}$ pt

35. $4\frac{1}{2}$ gal − $1\frac{3}{4}$ gal
$2\frac{3}{4}$ gal

36. 2)3 gal 2 qt
1 gal 3 qt

Objective C *Application Problems*

37. It is estimated that 60 adults will attend a church social. Assume that each adult will drink 2 c of coffee. How many gallons of coffee should be prepared? $7\frac{1}{2}$ gal

38. The Bayside Playhouse serves punch during intermission. Assume that 200 people will each drink 1 c of punch. How many gallons of punch should be ordered? 12.5 gal

39. A solution needed for a class of 30 chemistry students required 72 oz of water, 16 oz of one solution, and 48 oz of another solution. Find the number of quarts of the final solution. $4\frac{1}{4}$ qt

40. A cafeteria sold 124 cartons of milk in 1 day. Each carton contained 1 c of milk. How many quarts of milk were sold that day? 31 qt

41. A farmer changed the oil in the tractor seven times during the year. Each oil change required 5 qt of oil. How many gallons of oil did the farmer use in the seven oil changes? $8\frac{3}{4}$ gal

42. There are 24 cans in a case of tomato juice. Each can contains 10 oz of tomato juice. Find the number of 1-cup servings in the case of tomato juice. 30 servings

43. One brand of orange juice costs $1.20 for 1 qt. Another brand of orange juice costs 96¢ for 24 oz. Which is the more economical purchase? 1 qt for $1.20

44. Mandy carried 6 qt of water for 3 days of desert camping. Water weighs 8 1/3 lb per gallon. Find the weight of water that she carried. $12\frac{1}{2}$ lb

45. A department store bought hand lotion in 5-qt containers and then repackaged the hand lotion in 8-fl oz bottles. The hand lotion costs $41.50, and each 8-fl oz bottle was sold for $4.25. How much profit was made on each 5-qt package of hand lotion? $43.50

46. Orlando, a garage mechanic, bought oil in 50-gal containers for changing the oil in customers' cars. He paid $80 for the 50 gal of oil and charged customers $1.05 per quart. Find the profit Orlando made on one 50-gal container of oil. $130

APPLYING THE CONCEPTS

47. Define the following units: grain, dram, furlong, and rod. Give an example where each would be used.
grain = 0.002285 oz dram = 0.0625 oz furlong = $\frac{1}{8}$ mi rod = 16.5 ft

48. [W] Assume that you wanted to invent a new measuring system. Discuss some of the features that would have to be incorporated into the system. Answers will vary

Energy and Power

Objective A To use units of energy in the U.S. Customary System

Energy can be defined as the ability to do work. Energy is stored in coal, in gasoline, in water behind a dam, and in one's own body.

One **foot-pound** (ft · lb) of energy from your body is required to lift 1 pound a distance of 1 foot.

To lift 50 lb a distance of 5 ft requires
$50 \times 5 = 250$ ft · lb of energy

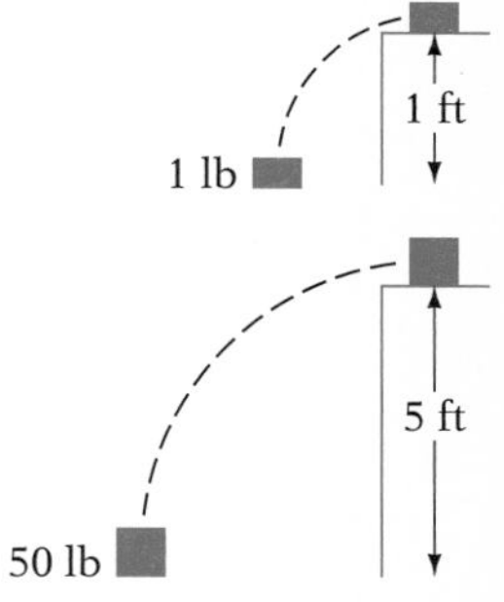

Consumer items that use energy, such as furnaces, stoves, and air conditioners, are rated in **British Thermal Units** (BTU).

A furnace with a rating of 35,000 BTU per hour releases 35,000 BTU of energy in one hour (1 h).

$$1 \text{ BTU} = 778 \text{ ft} \cdot \text{lb}$$

Therefore, the following conversion rate, equivalent to 1, can be written:

$$\frac{778 \text{ ft} \cdot \text{lb}}{1 \text{ BTU}} = 1$$

Example 1
Convert 250 BTU to foot-pounds.

Solution

$$250 \text{ BTU} = 250 \cancel{\text{BTU}} \times \frac{778 \text{ ft} \cdot \text{lb}}{1 \cancel{\text{BTU}}}$$
$$= 194{,}500 \text{ ft} \cdot \text{lb}$$

You Try It 1
Convert 4.5 BTU to foot-pounds.

Your solution
3501 ft · lb

Example 2
Find the energy required for a 125-lb person to climb a mile-high mountain.

Solution
In climbing the mountain, the person is lifting 125 lb a distance of 5280 ft.

$$\text{Energy} = 125 \text{ lb} \times 5280 \text{ ft}$$
$$= 660{,}000 \text{ ft} \cdot \text{lb}$$

You Try It 2
Find the energy required for a motor to lift 800 lb through a distance of 16 ft.

Your solution
12,800 ft · lb

Solutions on p. A40

Example 3
A furnace is rated at 80,000 BTU per hour. How many foot-pounds of energy are released in 1 h?

Solution

$$80{,}000 \text{ BTU} = 80{,}000 \cancel{\text{BTU}} \times \frac{778 \text{ ft} \cdot \text{lb}}{1 \cancel{\text{BTU}}}$$
$$= 62{,}240{,}000 \text{ ft} \cdot \text{lb}$$

You Try It 3
A furnace is rated at 56,000 BTU per hour. How many foot-pounds of energy are released in 1 h?

Your solution
43,568,000 ft · lb

Solution on p. A40

Objective B To use units of power in the U.S. Customary System

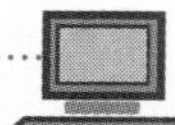

Power is the rate at which work is done, or the rate at which energy is released.

Power is measured in **foot-pounds per second** $\left(\frac{\text{ft} \cdot \text{lb}}{\text{s}}\right)$. In each of the following examples, the amount of energy released is the same, but the time taken to release the energy is different; thus the power is different.

100 lb is lifted 10 ft in 10 s.

$$\text{Power} = \frac{10 \text{ ft} \times 100 \text{ lb}}{10 \text{ s}} = 100 \frac{\text{ft} \cdot \text{lb}}{\text{s}}$$

100 lb is lifted 10 ft in 5 s.

$$\text{Power} = \frac{10 \text{ ft} \times 100 \text{ lb}}{5 \text{ s}} = 200 \frac{\text{ft} \cdot \text{lb}}{\text{s}}$$

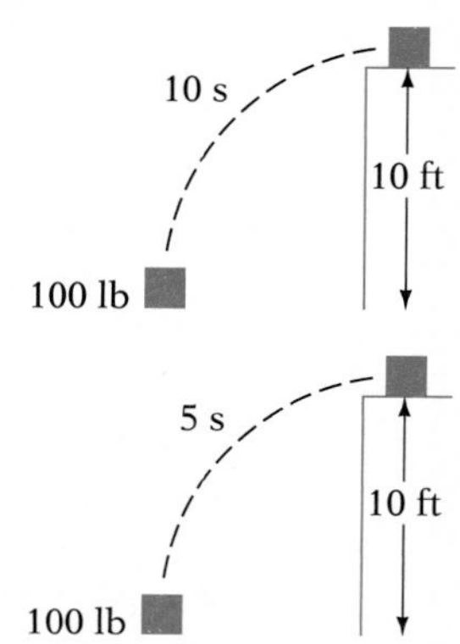

The U.S. Customary unit of power is the **horsepower.** A horse doing average work can pull 550 lb a distance of 1 ft in 1 s and can continue this work all day.

$$1 \text{ horsepower (hp)} = 500 \frac{\text{ft} \cdot \text{lb}}{\text{s}}$$

Example 4
Find the power needed to raise 300 lb a distance of 30 ft in 15 s.

Solution $\text{Power} = \frac{30 \text{ ft} \times 300 \text{ lb}}{15 \text{ s}}$

$$= 600 \frac{\text{ft} \cdot \text{lb}}{\text{s}}$$

You Try It 4
Find the power needed to raise 1200 lb a distance of 90 ft in 24 s.

Your solution $4500 \frac{\text{ft} \cdot \text{lb}}{\text{s}}$

Example 5
A motor has a power of 2750 $\frac{\text{ft} \cdot \text{lb}}{\text{s}}$. Find the horsepower of the motor.

Solution $\frac{2750}{550} = 5 \text{ hp}$

You Try It 5
A motor has a power of 3300 $\frac{\text{ft} \cdot \text{lb}}{\text{s}}$. Find the horsepower of the motor.

Your solution 6 hp

Solutions on p. A40

8.4 Exercises

Objective A

1. Convert 25 BTU to foot-pounds.
19,450 ft · lb

2. Convert 6000 BTU to foot-pounds.
4,668,000 ft · lb

3. Convert 25,000 BTU to foot-pounds.
19,450,000 ft · lb

4. Convert 40,000 BTU to foot-pounds.
31,120,000 ft · lb

5. Find the energy required to lift 150 lb a distance of 10 ft.
1500 ft · lb

6. Find the energy required to lift 300 lb a distance of 16 ft.
4800 ft · lb

7. Find the energy required to lift a 3300-lb car a distance of 9 ft.
29,700 ft · lb

8. Find the energy required to lift a 3680-lb elevator a distance of 325 ft.
1,196,000 ft · lb

9. Three tons are lifted 5 ft. Find the energy required in foot-pounds.
30,000 ft · lb

10. Seven tons are lifted 12 ft. Find the energy required in foot-pounds.
168,000 ft · lb

11. A construction worker carries 3-lb blocks up a 10-ft flight of stairs. How many foot-pounds of energy are required to carry 850 blocks up the stairs? 25,500 ft · lb

12. A crane lifts an 1800-lb steel beam to the roof of a building 36 ft high. Find the amount of energy the crane requires in lifting the beam.
64,800 ft · lb

13. A furnace is rated at 45,000 BTU per hour. How many foot-pounds of energy are released by the furnace in 1 h? 35,010,000 ft · lb

14. A furnace is rated at 22,500 BTU per hour. How many foot-pounds of energy does the furnace release in 1 h? 17,505,000 ft · lb

15. Find the amount of energy in foot-pounds given off when 1 lb of coal is burned. 1 lb of coal gives off 12,000 BTU of energy when burned.
9,336,000 ft · lb

16. Find the amount of energy in foot-pounds given off when 1 lb of gasoline is burned. 1 lb of gasoline gives off 21,000 BTU of energy when burned. 16,338,000 ft · lb

Objective B

17. Convert 1100 $\frac{\text{ft} \cdot \text{lb}}{\text{s}}$ to horsepower.
2 hp

18. Convert 6050 $\frac{\text{ft} \cdot \text{lb}}{\text{s}}$ to horsepower.
11 hp

19. Convert 4400 $\frac{\text{ft} \cdot \text{lb}}{\text{s}}$ to horsepower.
8 hp

20. Convert 1650 $\frac{\text{ft} \cdot \text{lb}}{\text{s}}$ to horsepower.
3 hp

21. Convert 5 hp to foot-pounds per second.
2750 $\frac{\text{ft} \cdot \text{lb}}{\text{s}}$

22. Convert 3 hp to foot-pounds per second.
1650 $\frac{\text{ft} \cdot \text{lb}}{\text{s}}$

23. Convert 7 hp to foot-pounds per second.
3850 $\frac{\text{ft} \cdot \text{lb}}{\text{s}}$

24. Convert 2 hp to foot-pounds per second.
1100 $\frac{\text{ft} \cdot \text{lb}}{\text{s}}$

25. Find the power in foot-pounds per second needed to raise 125 lb a distance of 12 ft in 3 s. 500 $\frac{\text{ft} \cdot \text{lb}}{\text{s}}$

26. Find the power in foot-pounds per second needed to raise 500 lb a distance of 60 ft in 8 s. 3750 $\frac{\text{ft} \cdot \text{lb}}{\text{s}}$

27. Find the power in foot-pounds per second needed to raise 3000 lb a distance of 40 ft in 25 s. 4800 $\frac{\text{ft} \cdot \text{lb}}{\text{s}}$

28. Find the power in foot-pounds per second needed to raise 12,000 lb a distance of 40 ft in 60 s. 8000 $\frac{\text{ft} \cdot \text{lb}}{\text{s}}$

29. Find the power in foot-pounds per second of an engine that can raise 180 lb to a height of 40 ft in 5 s. 1440 $\frac{\text{ft} \cdot \text{lb}}{\text{s}}$

30. Find the power in foot-pounds per second of an engine that can raise 1200 lb to a height of 18 ft in 30 s. 720 $\frac{\text{ft} \cdot \text{lb}}{\text{s}}$

31. A motor has a power of 1650 $\frac{\text{ft} \cdot \text{lb}}{\text{s}}$. Find the horsepower of the motor.
3 hp

32. A motor has a power of 16,500 $\frac{\text{ft} \cdot \text{lb}}{\text{s}}$. Find the horsepower of the motor. 30 hp

33. A motor has a power of 6600 $\frac{\text{ft} \cdot \text{lb}}{\text{s}}$. Find the horsepower of the motor. 12 hp

APPLYING THE CONCEPTS

34. [W] Pick out some source of energy. Write an article about this form of energy. Include the source, pollution, and future prospects associated with this form of energy. Answers will vary

Project in Mathematics

Averages If two towns are 150 miles apart and you drive between the two towns in 3 hours, then your

$$\textit{Average speed} = \frac{\text{total distance}}{\text{total time}} = \frac{150}{3} = 50 \text{ mi/h}$$

It is highly unlikely that your speed was *exactly* 50 mi/h the entire time of the trip. Sometimes you will have traveled faster than 50 mi/h, other times slower than 50 mi/h. Dividing the total distance you traveled by the total time it took to go that distance is an example of calculating an average.

There are many other averages that may be calculated. For instance, the Environmental Protection Agency calculates an estimated mpg (miles per gallon) for new cars. Miles per gallon is an average calculated from the formula

$$\frac{\text{gallons of gasoline}}{\text{miles traveled}}$$

For instance, the mpg for a car that travels 308 miles on 11 gallons of gas is $\frac{308}{11} = 28$ mpg.

A pilot would not use miles per gallon as a measure of fuel efficiency. Rather, pilots use gallons per hour. A plane that travels 5 hours and uses 400 gallons of fuel has an average that is calculated as

$$\frac{\text{Gallons of fuel}}{\text{Hours flown}} = \frac{400}{5} = 80 \text{ gal/h}$$

Exercises:

Using the examples above, calculate the following averages.

1. Determine the average speed of a car that travels 355 miles in 6 hours. Round to the nearest tenth.
2. Determine the mpg of a car that can travel 405 miles on 12 gallons of gasoline. Round to the nearest tenth.
3. If a plane flies 2000 miles in 5 hours and used 1000 gallons of fuel, determine the average number of gallons per hour used by the plane.

Another type of average is grade-point average (GPA). It is calculated by multiplying the units for each class by the grade points for that class, adding the results, and dividing by the total number of units taken. Here is an example using the grading scale A = 4, B = 3, C = 2, D = 1, and F = 0.

Class	*Units*	*Grade*
Math	4	B (= 3)
English	3	A (= 4)
French	5	C (= 2)
Biology	3	B (= 3)

$$\text{GPA} = \frac{4 \cdot 3 + 3 \cdot 4 + 5 \cdot 2 + 3 \cdot 3}{4 + 3 + 5 + 3} = \frac{43}{15} \approx 2.87$$

4. A grading scale that provides for plus or minus grades uses A = 4, A− = 3.7, B+ = 3.3, B = 3, B− = 2.7, C+ = 2.3, C = 2, C− = 1.7, D+ = 1.3, D = 1, D− = 0.7, and F = 0. Calculate the GPA of the student whose grades are given below.

Class	*Units*	*Grade*
Math	5	B+
English	3	C+
Spanish	5	A−
Physical Science	3	B−

Chapter Summary

Key Words A *measurement* includes a number and a unit.

The U.S. Customary units of length are *inch, foot, yard,* and *mile.*

Weight is a measure of how strongly the earth is pulling on an object.

The U.S. Customary units of weight are *ounce, pound,* and *ton.*

Liquid substances are measured in units of *capacity.*

The U.S. Customary units of capacity are *fluid ounce, cup, pint, quart,* and *gallon.*

Energy is the ability to do work.

One *foot-pound* of energy is the energy necessary to lift 1 pound a distance of 1 foot.

Furnaces, stoves, and air conditioners are rated in energy units called *British Thermal Units* (BTU).

Power is the rate at which work is done.

The U.S. Customary unit of power is the *horsepower* (hp).

Essential Rule A conversion unit is used to convert one unit of measurement to another. For example, the conversion unit 12 in./1 ft is used to convert feet to inches.

Chapter Review

SECTION 8.1

1. Convert 4 ft to inches.
48 in.

2. Convert 14 ft to yards.
$4\frac{2}{3}$ yd

3. Add: $\begin{array}{r} 3 \text{ ft } 9 \text{ in.} \\ + 5 \text{ ft } 6 \text{ in.} \\ \hline \end{array}$
9 ft 3 in.

4. Subtract: $\begin{array}{r} 5 \text{ yd } 1 \text{ ft} \\ - 3 \text{ yd } 2 \text{ ft} \\ \hline \end{array}$
1 yd 2 ft

5. Find the product of 2 ft 8 in. and 5.
13 ft 4 in.

6. What is 7 ft 6 in. divided by 3?
2 ft 6 in.

7. A board 6 ft 11 in. is cut from a board 10 ft 5 in. long. Find the length of the remaining piece of board. 3 ft 6 in.

SECTION 8.2

8. Convert $3\frac{3}{8}$ lb to ounces.
54 oz

9. Convert 2400 lb to tons.
$1\frac{1}{5}$ tons

10. Add: $\begin{array}{r} 5 \text{ lb } 11 \text{ oz} \\ + 3 \text{ lb } \ \ 8 \text{ oz} \\ \hline \end{array}$
9 lb 3 oz

11. Subtract: $\begin{array}{r} 3 \text{ tons } \ 500 \text{ lb} \\ - 1 \text{ ton } \ 1500 \text{ lb} \\ \hline \end{array}$
1 ton 1000 lb

12. Multiply: $\begin{array}{r} 5 \text{ lb } 6 \text{ oz} \\ \times \quad 8 \\ \hline \end{array}$
43 lb

13. Find the quotient of 7 lb 5 oz and 3.
2 lb 7 oz

14. A book weighing 2 lb 3 oz is mailed at the postage rate of \$0.18 per ounce. Find the cost of mailing the book. \$6.30

SECTION 8.3

15. Convert 12 c to quarts.
3 qt

16. Convert $2\frac{1}{2}$ pt to fluid ounces.
40 fl oz

17. A can of pineapple juice contains 18 fl oz. Find the number of quarts in a case of 24 cans. $13\frac{1}{2}$ qt

18. A cafeteria sold 256 cartons of milk in one school day. Each carton contains 1 c of milk. How many gallons of milk were sold that day? 16 gal

SECTION 8.4

19. Convert 50 BTU to foot-pounds. (1 BTU = 778 ft · lb)
38,900 ft · lb

20. Find the energy needed to lift 200 lb a distance of 8 ft.
1600 ft · lb

21. A furnace is rated at 35,000 BTU per hour. How many foot-pounds of energy does the furnace release in 1 h? (1 BTU = 778 ft · lb) 27,230,000 ft · lb

22. Convert $3850 \frac{\text{ft} \cdot \text{lb}}{\text{s}}$ to horsepower. $\left(1 \text{ hp} = 550 \frac{\text{ft} \cdot \text{lb}}{\text{s}}\right)$ 7 hp

23. Convert 2.5 hp to foot-pounds per second. $1375 \frac{\text{ft} \cdot \text{lb}}{\text{s}}$

24. Find the power in foot-pounds per second of an engine that can raise 800 lb to a height of 15 ft in 25 s. $480 \frac{\text{ft} \cdot \text{lb}}{\text{s}}$

Chapter Test

1. Convert $2\frac{1}{2}$ ft to inches.
30 in. [8.1A]

2. Subtract: 4 ft 2 in. − 1 ft 9 in.
2 ft 5 in. [8.1B]

3. A board $6\frac{2}{3}$ ft long is cut into five equal pieces. How long is each piece?
$1\frac{1}{3}$ ft [8.1C]

4. Seventy-two bricks, each 8 in. long, are laid end to end to make the base for a wall. Find the length of the wall in feet. 48 ft [8.1C]

5. Convert $2\frac{7}{8}$ lb to ounces.
46 oz [8.2A]

6. Convert: 40 oz = 2 lb 8 oz [8.2A]

7. Find the sum of 9 lb 6 oz and 7 lb 11 oz.
17 lb 1 oz [8.2B]

8. Divide: 6 lb 12 oz ÷ 4
1 lb 11 oz [8.2B]

9. A college bookstore received 1000 workbooks, each workbook weighing 12 oz. Find the total weight of the 1000 workbooks in pounds. 750 lb [8.2C]

10. An elementary school class gathered 800 aluminum cans for recycling. Four aluminum cans weigh 3 oz. Find the amount the class received if the rate of pay was $0.75 per pound for the aluminum cans. Round to the nearest cent. $28.13 [8.2C]

11. Convert 13 qt to gallons.
$3\frac{1}{4}$ gal [8.3A]

12. Convert $3\frac{1}{2}$ gal to pints.
28 pt [8.3A]

13. What is $1\frac{3}{4}$ gal times 7?
$12\frac{1}{4}$ gal [8.3B]

14. Add: 5 gal 2 qt + 2 gal 3 qt
8 gal 1 qt [8.3B]

15. A can of grapefruit juice contains 20 oz. Find the number of cups of grapefruit juice in a case of 24 cans. 60 c [8.3C]

16. Nick, a mechanic, bought oil in 40-gal containers for changing the oil in customers' cars. He paid $90 for the 40 gal of oil and charged customers $1.35 per quart. Find the profit Nick made on one 40-gal container of oil.
$126 [8.3C]

17. Find the energy required to lift 250 lb a distance of 15 ft. 3750 ft · lb [8.4A]

18. A furnace is rated at 40,000 BTU per hour. How many foot-pounds of energy are released by the furnace in 1 h? (1 BTU = 778 ft · lb)
31,120,000 ft · lb [8.4A]

19. Find the power needed to lift 200 lb a distance of 20 ft in 25 s.
$160 \frac{\text{ft} \cdot \text{lb}}{\text{s}}$ [8.4B]

20. A motor has a power of $2200 \frac{\text{ft} \cdot \text{lb}}{\text{s}}$. Find the motor's horsepower.
$\left(1 \text{ hp} = 550 \frac{\text{ft} \cdot \text{lb}}{\text{s}}\right)$ 4 hp [8.4B]

Cumulative Review

1. Find the LCM of 9, 12, and 15.
180 [2.1A]

2. Write $\frac{43}{8}$ as a mixed number.
$5\frac{3}{8}$ [2.2B]

3. Subtract: $5\frac{7}{8} - 2\frac{7}{12}$
$3\frac{7}{24}$ [2.5C]

4. What is $5\frac{1}{3}$ divided by $2\frac{2}{3}$?
2 [2.7B]

5. Simplify $\frac{5}{8} \div \left(\frac{3}{8} - \frac{1}{4}\right) - \frac{10}{16}$.
$4\frac{3}{8}$ [2.8C]

6. Round 2.0972 to the nearest hundredth.
2.10 [3.1B]

7. Multiply: $\begin{array}{r} 0.0792 \\ \times \quad 0.49 \\ \hline \end{array}$
0.038808 [3.4A]

8. Solve the proportion. $\frac{n}{12} = \frac{44}{60}$
$n = 8.8$ [4.3B]

9. Find $2\frac{1}{2}\%$ of 50.
1.25 [5.2A]

10. 18 is 42% of what? Round to the nearest hundredth.
42.86 [5.4A]

11. A 7.2-lb roast costs $15.48. Find the unit cost.
$2.15/lb [6.1A]

12. Add: $3\frac{2}{5}$ in. + $5\frac{1}{3}$ in.
$8\frac{11}{15}$ in. [8.1B]

13. Convert: 24 oz = ____ lb ____ oz
1 lb 8 oz [8.2B]

14. Multiply: 3 lb 8 oz × 9
31 lb 8 oz [8.2B]

15. Subtract: $4\frac{1}{3}$ qt − $1\frac{5}{6}$ qt
$2\frac{1}{2}$ qt [8.3B]

16. Find 2 lb 10 oz less than 4 lb 6 oz.
1 lb 12 oz [8.2B]

17. An investor receives a dividend of $56 from 40 shares of stock. At the same rate, find the dividend that would be received from 200 shares of stock. $280 [4.3C]

18. Anna had a balance of $578.56 in her checkbook. She wrote checks of $216.98 and $34.12 and made a deposit of $315.33. What is her new checking balance? $642.79 [6.7A]

19. An account executive receives a salary of $800 per month plus a commission of 2% on all sales over $25,000. Find the total monthly income of an account executive who has monthly sales of $140,000. $3100 [6.6A]

20. A health inspector found that 3% of a shipment of carrots were spoiled and could not be sold. Find the amount of carrots from a shipment of 2500 pounds that could be sold. 2425 lb [5.2B]

21. The scores on the final exam of a trigonometry class are recorded in the histogram in the figure. What percent of the class received a score between 80% and 90%? Round to the nearest percent. 18% [7.3A]

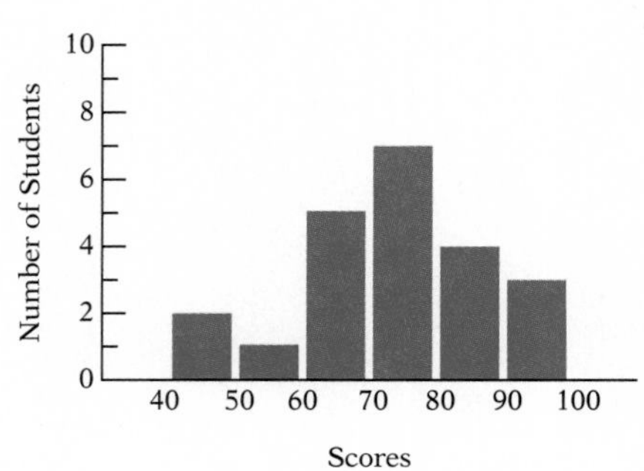

22. Hayes Department Store uses a markup rate of 40% on all merchandise. What is the selling price of a compact disc player that cost the store $220? $308 [6.2B]

23. A construction firm received a loan of $200,000 for 8 months at a simple interest rate of 11%. Find the interest paid on the loan. $14,666.67 [6.3A]

24. Six college students spent several weeks panning for gold during their summer vacation. The students obtained 1 lb 3 oz of gold. After selling the gold for $400 per ounce, how much money did each student receive? Round to the nearest dollar. $1267 [8.2C]

25. Four books were mailed at the rate of $0.15 per ounce. The books weighed 1 lb 3 oz, 13 oz, 1 lb 8 oz, and 1 lb. Find the cost of mailing the books. $10.80 [8.2C]

26. One brand of yogurt costs $0.61 for 8 oz; 36 oz of another brand can be bought for $2.70. Which purchase is the better buy? 36 oz for $2.70 [6.1B]

27. A contractor can buy a 4000-gal tank of gasoline for $3600. The pump price of the gasoline is $1.17 per gallon. How much does the contractor save by buying the 4000 gal of gasoline? $1080 [8.3C]

28. Find the energy required to lift 400 lb a distance of 8 ft. 3200 ft · lb [8.4A]

29. Find the power in foot-pounds per second needed to raise 600 lb a distance of 8 ft in 12 s. $400 \frac{\text{ft} \cdot \text{lb}}{\text{s}}$ [8.4B]

Chapter

The Metric System of Measurement

Objectives

Section 9.1

To convert units of length in the metric system of measurement

To perform arithmetic operations with measurements of length

To solve application problems

Section 9.2

To convert units of mass in the metric system of measurement

To perform arithmetic operations with measurements of mass

To solve application problems

Section 9.3

To convert units of capacity in the metric system of measurement

To perform arithmetic operations with measurements of capacity

To solve application problems

Section 9.4

To use units of energy in the metric system of measurement

Section 9.5

To convert U.S. Customary units to metric units

To convert metric units to U.S. Customary units

Is a Pound a Pound?

Which is heavier, a pound of feathers or a pound of gold? It would seem that a pound is a pound, whether it is gold or feathers. However, this is not the case.

Metals, such as gold and silver, are measured by using the Troy Weight System, whereas the weight of feathers, meat, people, and other nonmetal quantities are measured using the Avoirdupois Weight System. Each system is part of the U.S. Customary System of Measurement.

One grain is a small weight in the Customary System and is approximately 0.02 ounces. A pound of feathers weighs 7000 grains, but a pound of gold weighs only 5760 grains. Therefore, a pound of feathers weighs more than a pound of gold.

To complicate matters, 1 avoirdupois pound contains 16 ounces, but 1 troy pound contains 12 ounces, which means that there are 437.5 grains in 1 avoirdupois ounce but 480 grains in a troy ounce. Thus an ounce of feathers weighs less than an ounce of gold.

To summarize, a pound of feathers weighs more than a pound of gold, but an ounce of feathers weighs less than an ounce of gold. This kind of confusion is one reason why the metric system of measurement was developed.

9.1 Length

Objective A To convert units of length in the metric system of measurement

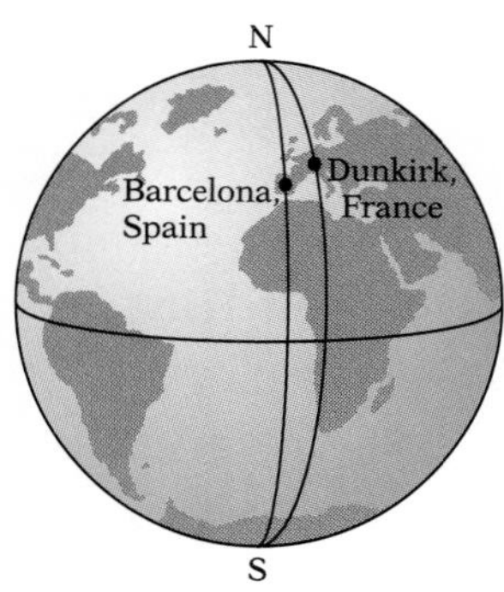

POINT OF INTEREST

One of the results of the French Revolution was a law enacted by the Constituent Assembly in 1790 that required the standardization of measurements. As a result, a commission of scientists was established to develop this new system of measurement. All units in this system were to be based on the "meter," a word that has its origins in the Greek word *metron* meaning "a measure."

On May 20, 1875, delegates from seventeen countries (including the United States, the only English-speaking country to sign) signed the Treaty of the Meter. The signing of this treaty came almost 100 years after the original attempt to define a uniform system of measurement.

In 1789, an attempt was made to standardize units of measurement internationally in order to simplify trade and commerce between nations. A commission in France developed a system of measurement known as the **metric system.** The basic unit of length in the metric system is the **meter.** One meter is approximately the distance from a doorknob to the floor. Originally the meter was defined as $\frac{1}{10,000,000}$ of the distance from the equator to the north pole. The meter is now defined as 1,650,763.73 wavelengths of a particular orange-red wavelength given off by atoms of krypton.

All units of length in the metric system are derived from the meter. Prefixes to the basic unit denote the length of each unit. For example, the prefix "centi-" means one-hundredth, so 1 centimeter is 1 one-hundredth of a meter.

kilo- = 1000	1 kilometer (km) = 1000 meters (m)
hecto- = 100	1 hectometer (hm) = 100 m
deca- = 10	1 decameter (dam) = 10 m
	1 meter (m) = 1 m
deci- = 0.1	1 decimeter (dm) = 0.1 m
centi- = 0.01	1 centimeter (cm) = 0.01 m
milli- = 0.001	1 millimeter (mm) = 0.001 m

Conversion between units of length in the metric system involves moving the decimal point to the right or to the left. Listing the units in order from largest to smallest will indicate how many places to move the decimal point and in which direction.

To convert 4200 cm to m, write the units in order from largest to smallest.

km hm dam m dm cm mm
2 positions

Converting cm to m requires moving 2 positions to the left.

4200 cm = 42.00 m
2 places

Move the decimal point the same number of places and in the same direction.

A metric measurement involving two units is customarily written in terms of one unit. Convert the smaller unit to the larger unit and then add.

To convert 8 km 32 m to km, first convert 32 m to km.

km hm dam m dm cm mm
32 m = 0.032 km

Then add the result to 8 km. 8 km 32 m = 8 km + 0.032 km = 8.032 km

Example 1 Convert 0.38 m to millimeters.

Solution 0.38 m = 380 mm

You Try It 1 Convert 3.07 m to centimeters.

Your solution 307 cm

Solution on p. A40

Example 2
Convert 4 m 62 cm to centimeters.

Solution
4 m = 400 cm
4 m 62 cm = 400 cm + 62 cm = 462 cm

You Try It 2
Convert 3 m 7 cm to meters.

Your solution
3.07 m

Solution on p. A40

Objective B To perform arithmetic operations with measurements of length

Arithmetic operations can be performed with measurements of length in the metric system. A measurement involving two units should be written in terms of a single unit before adding, subtracting, multiplying, or dividing.

➡ Add: 6 m 42 cm + 7 m 98 cm

6 m 42 cm = 6.42 m		6 m 42 cm = 642 cm
+ 7 m 98 cm = 7.98 m	or,	+ 7 m 98 cm = 798 cm
14.40 m		1440 cm

Note that the measurements can be changed to either meters or centimeters before adding. In this textbook, unless otherwise stated, the units should be changed to the larger unit before the arithmetic operation is performed.

Example 3 Find the quotient of 42 km 765 m and 14. Round to the nearest thousandth.

Solution 42 km 765 m = 42.765 km

$$14\overline{)42.765\text{ km}} = 3.0546\text{ km} \approx 3.055\text{ km}$$

You Try It 3 What is 3 m decreased by 42 cm?

Your solution 2.58 m

Solution on p. A41

Objective C To solve application problems

Example 4
A piece measuring 1 m 42 cm is cut from a board 4 m 20 cm long. Find the length of the remaining piece.

Strategy
To find the length of the remaining piece, subtract the length of the piece cut (1 m 42 cm) from the original length (4 m 20 cm).

Solution
4 m 20 cm = 4.20 m
− 1 m 42 cm = 1.42 m
2.78 m

The length of the piece remaining is 2.78 m.

You Try It 4
A bookcase 1 m 75 cm long has four shelves. Find the cost of the shelves when the price of lumber is $11.75 per meter.

Your strategy

Your solution
$82.25

Solution on p. A41

9.1 Exercises

Objective A

Convert.

1. 42 cm = 420 mm
2. 62 cm = 620 mm
3. 81 mm = 8.1 cm
4. 68.2 mm = 6.82 cm
5. 6804 m = 6.804 km
6. 3750 m = 3.750 km
7. 2.109 km = 2109 m
8. 32.5 km = 32,500 m
9. 432 cm = 4.32 m
10. 61.7 cm = 0.617 m
11. 0.88 m = 88 cm
12. 3.21 m = 321 cm
13. 6 m 42 cm = 6.42 m = 642 cm
14. 62 km 482 m = 62.482 km = 62,482 m
15. 42 cm 6 mm = 42.6 cm = 426 mm
16. 62 m 7 cm = 62.07 m = 6207 cm

Objective B

Perform the arithmetic operation.

17. 42 cm + 8 m
    8.42 m
18. 69 km 432 m + 7 km 921 m
    77.353 km
19. 98 m 67 cm − 19 m 82 cm
    78.85 m
20. 2 km − 435 m
    1.565 km
21. 7 cm 4 mm × 4
    29.6 cm
22. 16 km 691 m × 12
    200.292 km
23. 3 km 726 m ÷ 9
    0.414 km
24. 42 m 60 cm ÷ 12
    3.55 m
25. 8 km 312 m + 9 km 814 m
    18.126 km
26. 729 km 467 m + 837 km 942 m
    1567.409 km
27. 954.62 km × 944
    901,161.28 km
28. 165 km 429 cm ÷ 27
    6.11127 km

Objective C *Application Problems*

29. A carpenter needs 15 ceiling joists, each joist 4 m 60 cm long. Find the total length of ceiling joists needed in meters. 69 m

30. Neelu can walk 1 km 600 m in $\frac{1}{2}$ hour. How far can she walk in 2 hours? 6.4 km

31. Find the missing dimension in centimeters. 7.8 cm

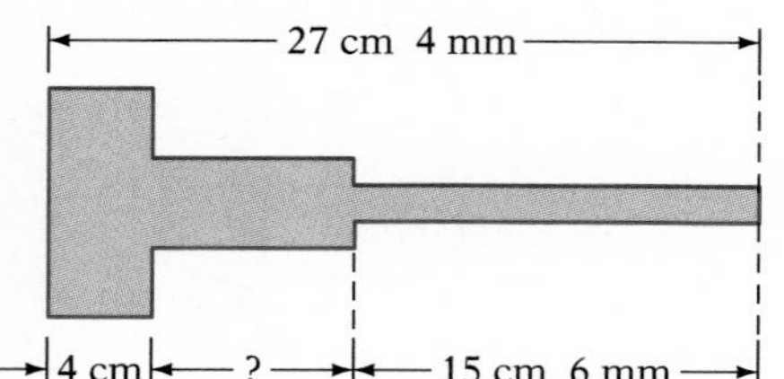

32. Find the total length in centimeters. 181 cm

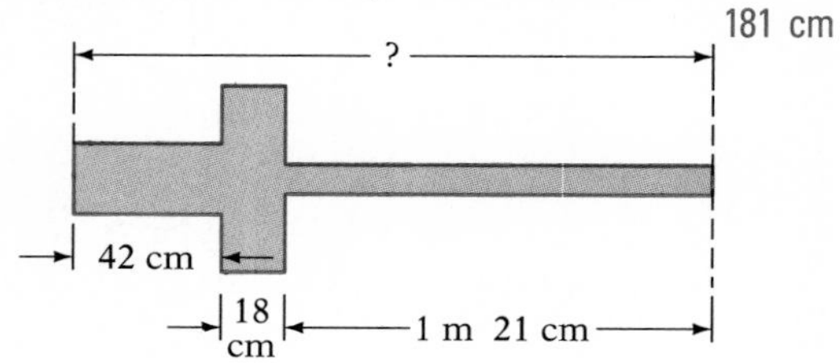

33. A bicycle race had two checkpoints. One checkpoint was 12 km 400 m from the starting point. The second checkpoint was 9 km 300 m from the first checkpoint. The second checkpoint was 8 km 800 m from the finish line. How long was the race? 30.5 km

34. How many shelves, 1 m 40 cm long, can be cut from a board that is 4 m 20 cm in length? Find the length of board remaining after the shelves are cut. 3 shelves; no length remaining

35. Twenty rivets are used to fasten two steel plates together. The plates are 3 m 40 cm long, and the rivets are equally spaced, with a rivet at each end. Find the distance between the rivets. (Round to the nearest tenth of a centimeter.) 17.9 cm

36. A frame for a picture is 1 m 10 cm long and 80 cm wide. Find the total length of framing needed to frame the picture. 3.8 m

37. Four pieces of fencing are cut from a 50-m roll to build a dog run. The dog run is 3 m 40 cm wide and 13 m 80 cm long. How much of the fencing is left on the roll after the dog run is completed? 15.6 m

38. During the week, Carmine ran 12 km 500 m, 15 km 800 m, 12 km 500 m, 13 km 200 m, and 18 km 400 m. Find the average distance run on each of these 5 days. 14.48 km

39. Light travels 3×10^8 m in 1 s. How far does the light travel in 1 year? (Astronomers refer to this distance as 1 light year.) 9.4608×10^{15} m

40. The distance around the earth is 41,000 km. Light travels 3×10^8 m in 1 s. How many times will light travel around the earth in 1 s? 7.32 times

APPLYING THE CONCEPTS

41. What units in the metric system would you use to measure the distance from Los Angeles to New York? your waist? the width of a hair? your height? the distance to the grocery store?
Kilometers Centimeters Millimeters Centimeters Kilometers

42. [W] Explain the meaning of "parsec" and how this unit is used in measuring astronomical distances. 3.258 light years

43. [W] Write a short history of the metric system. Answers will vary

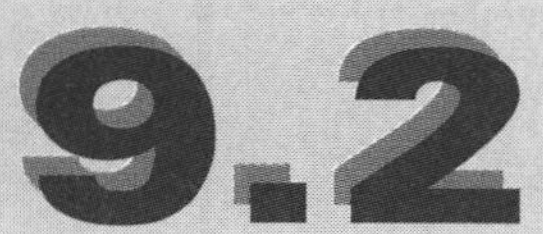

Mass

Objective A ***To convert units of mass in the metric system of measurement***

INSTRUCTOR NOTE
The difference between weight and mass is difficult for students. As another illustration, mention that the mass of an astronaut is the same on the earth or on the moon; the weight of the astronaut, however, is different.

Mass and weight are closely related. Weight is a measure of how strongly the earth is pulling on an object. Therefore, an object's weight is less in space than on the earth's surface. However, the amount of material in the object, its **mass,** remains the same. On the surface of the earth, mass and weight can be used interchangeably.

The basic unit of mass in the metric system is the **gram.** If a box that is 1 cm long on a side is filled with water, then the mass of that water is 1 gram.

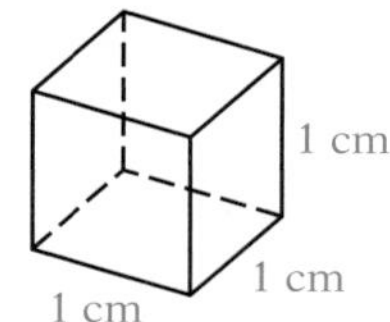

1 gram = the mass of water in the box

The gram is a very small unit of mass. A paperclip weighs about 1 gram. The kilogram (1000 grams) is a more useful unit of mass in consumer applications. This textbook weighs about 1 kilogram.

The units of mass in the metric system have the same prefixes as the units of length:

1 kilogram (kg) = 1000 grams (g)
1 hectogram (hg) = 100 g
1 decagram (dag) = 10 g
1 gram (g) = 1 g
1 decigram (dg) = 0.1 g
1 centigram (cg) = 0.01 g
1 milligram (mg) = 0.001 g

Conversion between units of mass in the metric system involves moving the decimal point to the right or to the left. Listing the units in order from largest to smallest will indicate how many places to move the decimal point and in which direction.

To convert 324 g to kg, first write the units in order from largest to smallest.

kg hg dag g dg cg mg — 3 positions

Converting g to kg requires moving 3 positions to the left.

324 g = 0.324 kg — 3 places

Move the decimal point the same number of places and in the same direction.

Example 1 Convert 4.23 g to milligrams.

Solution 4.23 g = 4230 mg

You Try It 1 Convert 42.3 mg to grams.

Your solution 0.0423 g

Solution on p. A41

Example 2
Convert 2 kg 564 g to kilograms.

Solution
564 g = 0.564 kg
2 kg 564 g = 2 kg + 0.564 kg = 2.564 kg

You Try It 2
Convert 3 g 54 mg to milligrams.

Your solution
3054 mg

Solution on p. A41

Objective B To perform arithmetic operations with measurements of mass

Arithmetic operations can be performed with measurements of mass. A measurement involving two units should be written in terms of a single unit before adding, subtracting, multiplying, or dividing.

➡ Add: 3 g 42 mg + 5 g 690 mg

3 g 42 mg = 3.042 g		3 g 42 mg = 3042 mg
+ 5 g 690 mg = 5.690 g	or	+ 5 g 690 mg = 5690 mg
8.732 g		8732 mg

Note that the units can be changed to either grams or milligrams before adding. In this textbook, unless otherwise stated, the units should be changed to the larger unit before the arithmetic operation is performed.

Example 3
Add: 3 kg + 62 g

Solution
3 kg + 62 g = 3 kg + 0.062 kg = 3.062 kg

You Try It 3
Multiply: 4 g 620 mg × 8

Your solution
36.960 g

Solution on p. A41

Objective C To solve application problems

Example 4
Find the cost of a roast weighing 3 kg 320 g if the price per kilogram is $4.17. Round to the nearest cent.

Strategy To find the cost of the roast:
- Convert 3 kg 320 g to kilograms.
- Multiply the weight by the cost per kilogram ($4.17).

Solution 3 kg 320 g = 3.320 kg

$$\begin{array}{r} 3.320 \\ \times\ 4.17 \\ \hline 13.84440 \end{array}$$

The cost of the roast is $13.84.

You Try It 4
Three hundred grams of fertilizer are used to fertilize each tree in an apple orchard. How many kilograms of fertilizer are required to fertilize 400 trees?

Your strategy

Your solution 120 kg

Solution on p. A41

9.2 Exercises

Objective A

Convert.

1. 420 g = 0.420 kg
2. 7421 g = 7.421 kg
3. 127 mg = 0.127 g
4. 43 mg = 0.043 g
5. 4.2 kg = 4200 g
6. 0.027 kg = 27 g
7. 0.45 g = 450 mg
8. 325 g = 325,000 mg
9. 1856 g = 1.856 kg
10. 1.37 kg = 1370 g
11. 4057 mg = 4.057 g
12. 0.0456 g = 45.6 mg
13. 3 kg 922 g = 3.922 kg = 3922 g
14. 1 kg 47 g = 1.047 kg = 1047 g
15. 7 g 891 mg = 7.891 g = 7891 mg
16. 209 g 42 mg = 209.042 g = 209,042 mg

Objective B

Perform the arithmetic operation. Round to the nearest thousandth.

17. 4 g − 692 mg
    3.308 g
18. 4 kg 67 g − 1 kg 296 g
    2.771 kg
19. 4 kg + 692 g
    4.692 kg
20. 362 g 419 mg + 192 g 97 mg
    554.516 g
21. 46 g × 16
    736 g
22. 3 kg 496 g × 9
    31.464 kg
23. 7432 kg 89 g − 5196 kg 72 g
    2236.017 kg
24. 832 g 498 mg × 967
    805,025.566 g
25. 325 kg 72 g ÷ 167
    1.947 kg

Objective C *Application Problems*

26. A baby weighs 3 kg 800 g at birth. At the end of 6 months, the baby weighs 8 kg 650 g. Find the increase in weight in kilograms during the 6 months. 4.85 kg

27. A doctor advises a patient weighing 108 kg 400 g to lose 20 kg of weight. How much more does the patient need to lose after losing 13 kg 800 g? 6.2 kg

28. A concrete block weighs 2 kg 350 g. Find the weight of a load of 1200 concrete blocks. 2820 kg

29. A 15×15-cm tile weighs 220 g. Find the weight of a box of 72 tiles in kilograms. 15.84 kg

30. Five scouts are taking 104 kg 600 g of supplies on a 5-day backpacking trip. How much weight in kilograms will each scout carry if the gear is divided equally? 20.92 kg

31. Doctor Futrelle recommends that her patient take 3.5 g of vitamin C a week. Find the daily dosage of the vitamin in milligrams. 500 mg

32. Find the cost of a ham weighing 4 kg 700 g if the price per kilogram is \$4.20. \$19.74

33. Eighty grams of grass seed are used for every 100 m^2 of lawn. Find the amount of seed needed to cover 2000 m^2. 1600 g

34. An overseas flight charges \$5.40 for each kilogram or part of a kilogram over 50 kg of luggage weight. How much extra must be paid for three pieces of luggage weighing 20 kg 400 g, 18 kg 300 g, and 15 kg? \$19.98

35. A variety goods store buys nuts in 10-kg containers and repackages the nuts for resale. The store packages the nuts in 200-g bags and sells them for \$1.27 per bag. Find the profit on a 10-kg container of nuts costing \$30. \$33.50

36. A train car is loaded with 15 automobiles weighing 1405 kg each. Find the total weight of the automobiles. 21,075 kg

37. During 1 year there were 2076 million kg of cotton fiber produced, 237 million kg of wood fibers produced, and 2678 million kg of synthetic fibers produced. What percent of the total fiber production is the cotton fiber production? (Round to the nearest hundredth of a percent.) 41.59%

APPLYING THE CONCEPTS

38. What units in the metric system would you use to measure the amount of protein in your diet? the medication in an aspirin? your weight? a car's weight? Grams Milligrams Kilograms Kilograms

39. Discuss the advantages and disadvantages of the U.S. Customary and metric systems of measurement. Answers will vary

9.3 Capacity

Objective A To convert units of capacity in the metric system of measurement

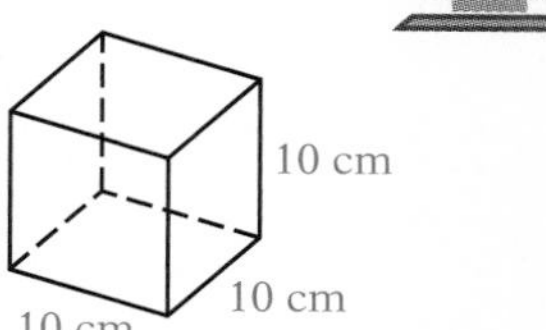

The basic unit of capacity in the metric system is the liter. One **liter** is defined as the capacity of a box that is 10 cm long on each side.

The units of capacity in the metric system have the same prefixes as the units of length:

1 kiloliter (kl) = 1000 L
1 hectoliter (hl) = 100 L
1 decaliter (dal) = 10 L
1 liter (L) = 1 L
1 deciliter (dl) = 0.1 L
1 centiliter (cl) = 0.01 L
1 milliliter (ml) = 0.001 L

The milliliter is equal to 1 **cubic centimeter** (cm^3).

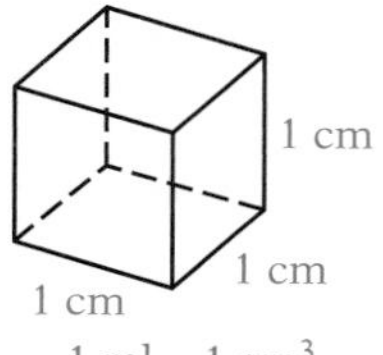

Conversion between units of capacity in the metric system involves moving the decimal point to the right or to the left. Listing the units in order from largest to smallest will indicate how many places to move the decimal point and in which direction.

To convert 824 ml to liters, first write the units in order from largest to smallest.

kl hl dal L dl cl ml
3 positions

Converting ml to L requires moving 3 positions to the left.

824 ml = 0.824 L
3 places

Move the decimal point the same number of places and in the same direction.

Example 1 Convert 4 L 32 ml to liters.

Solution 32 ml = 0.032 L
4 L 32 ml = 4 L + 0.032 L
= 4.032 L

You Try It 1 Convert 2 kl 167 L to liters.

Your solution 2167 L

Example 2 Convert 1.23 L to cubic centimeters.

Solution 1.23 L = 1230 ml = 1230 cm^3

You Try It 2 Convert 325 cm^3 to liters.

Your solution 0.325 L

Solutions on p. A42

Objective B To perform arithmetic operations with measurements of capacity

Arithmetic operations can be performed with measurements of capacity. A measurement involving two units should be written in terms of a single unit before adding, subtracting, multiplying, or dividing.

➡ Find the difference between 8 L 27 ml and 5 L 635 ml.

$$\begin{array}{r} 8\text{ L }\ 27\text{ ml} = 8.027\text{ L} \\ -\ 5\text{ L }635\text{ ml} = 5.635\text{ L} \\ \hline 2.392\text{ L} \end{array} \quad \text{or} \quad \begin{array}{r} 8\text{ L }\ 27\text{ ml} = 8027\text{ ml} \\ -\ 5\text{ L }635\text{ ml} = 5635\text{ ml} \\ \hline 2392\text{ ml} \end{array}$$

Note that the units can be changed to either liters or milliliters before subtracting. In this textbook, unless otherwise stated, the units should be changed to the larger unit before the arithmetic operation is performed.

Example 3 Multiply: 4L 147 ml × 9

Solution

$$\begin{array}{r} 4\text{ L }147\text{ ml} = 4.147\text{ L} \\ \times \qquad 9 = \qquad 9 \\ \hline 37.323\text{ L} \end{array}$$

You Try It 3 Divide: 22 kl 992 L ÷ 12

Your solution 1.916 kl

Solution on p. A42

Objective C To solve application problems

Example 4

A laboratory assistant is in charge of ordering acid for three chemistry classes of 30 students each. Each student requires 80 ml of acid. How many liters of acid should be ordered? (The assistant must order by the whole liter.)

Strategy

To find the number of liters of acid to be ordered:

- Find the amount of acid needed by multiplying the number of classes by the number of students per class by the amount of acid required by each student.
- Convert to liters.

Solution $3 \times 30 \times 80 = 7200$ ml

7200 ml = 7.2 L

The laboratory assistant should order 8 L of acid.

You Try It 4

For $42.50 a druggist buys 5 L of hand lotion and repackages it in 125-ml containers. Each container is sold for $2.79. Find the profit on the 5 L of hand lotion.

Your strategy

Your solution $69.10

Solution on p. A42

9.3 Exercises

Objective A

Convert.

1. 4200 ml = 4.2 L
2. 7.5 ml = 0.0075 L
3. 3.42 L = 3420 ml
4. 0.037 L = 37 ml
5. 423 ml = 423 cm^3
6. 0.32 ml = 0.32 cm^3
7. 642 cm^3 = 642 ml
8. 0.083 cm^3 = 0.083 ml
9. 42 cm^3 = 0.042 L
10. 3075 cm^3 = 3.075 L
11. 0.435 L = 435 cm^3
12. 2.57 L = 2570 cm^3
13. 1 L 267 ml = 1267 cm^3
14. 4 L 105 ml = 4105 cm^3
15. 3 L 23 ml = 3023 cm^3
16. 4.62 kl = 4620 L
17. 0.035 kl = 35 L
18. 1423 L = 1.423 kl
19. 3 L 42 ml = 3.042 L
 = 3042 ml
20. 1 L 127 ml = 1.127 L
 = 1127 ml
21. 3 kl 4 L = 3.004 kl
 = 3004 L
22. 6 kl 32 L = 6.032 kl
 = 6032 L

Objective B

Perform the arithmetic operation.

23. 8 L 163 ml + 4 L 275 ml
 12.438 L
24. 3 L 123 ml + 5 L 892 ml
 9.015 L
25. 3 L 162 ml × 12
 37.944 L
26. 4 L 792 ml ÷ 4
 1.198 L
27. 10 L 72 ml − 3 L 818 ml
 6.254 L
28. 437 ml + 3 L
 3.437 L
29. 198 L 625 ml + 932 L 747 ml
 1131.372 L
30. 62 L 47 ml × 3928
 243,720.616 L
31. 423 L 638 ml ÷ 764
 0.5545 L

Objective C Application Problems

32. A can of tomato juice contains 1260 ml. How many 180-ml servings are in one can of tomato juice? 7 servings

33. A Little League auxiliary serves punch to the ballplayers after the game. How many liters of punch are needed for 30 servings? Each serving contains 220 ml of punch. 6.6 L

34. A youth club uses 800 ml of chlorine each day for its swimming pool. How many liters of chlorine are used in a month of 30 days? 24 L

35. A flu vaccine is being given for the coming winter season. A medical corporation buys 12 L of flu vaccine. How many patients can be immunized if each patient receives 3 cm^3 of the vaccine? 4000 patients

36. There are 24 bottles in a case of shampoo. Each bottle of shampoo contains 320 ml. Find the number of liters in one case of shampoo. 7.68 L

37. A chemistry experiment requires 12 ml of an acid solution. How many liters of acid are used when 90 students perform the experiment? 1.08 L

38. A case of 12 1-L cans of apple juice costs $11.40. A case of 24 cans, each can containing 300 ml, of apple juice costs $6.90. Which case of apple juice is the better buy? 12 1-L cans for $11.40

39. Nineteen percent of air is oxygen. Find the amount of oxygen in 50 L of air. 9.5 L

40. Six liters of liquid soap were bought for $11.40 per liter. The soap was repackaged in 150-ml bottles and sold for $3.29 per bottle. Find the profit on the 6 L of liquid soap. $63.20

41. A pharmacy bought cough syrup in 5-L containers and then repackaged the cough syrup in 250-ml bottles. Thirteen bottles of the cough syrup have been sold. How many bottles of the cough syrup are still in stock? 7 bottles

42. A wholesale distributor purchased 32 kl of cooking oil for $34,880. The wholesaler repackaged the cooking oil in 1.25-L bottles and sold each bottle for $2.17. Find the profit on the 32 kl of cooking oil. $20,672

43. A service station operator bought 85 kl of gasoline for $19,250. The gasoline was sold for $0.329 per liter. Find the profit on the 85 kl of gasoline. $8715

APPLYING THE CONCEPTS

44. Define the metric measurement prefixes pico, mega, nano, giga, and micro. Use each prefix in an example of a measurement.
Picometer = 0.000000000001 m Megameter = 1,000,000 m Nanometer = 0.000000001 m
Gigameter = 1,000,000,000 m Micrometer = 0.000001 m

45. [W] Write an essay discussing problems in trade and manufacturing that arise between the United States and Europe because they use different systems of measurement. Answers will vary

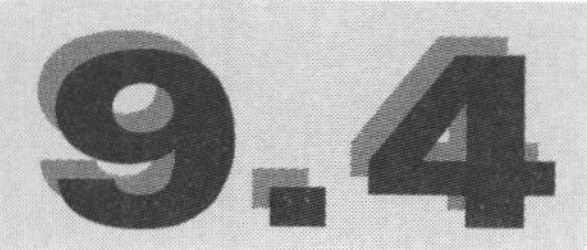

Energy

Objective A To use units of energy in the metric system of measurement

Two commonly used units of energy in the metric system are the calorie and the watt-hour.

POINT OF INTEREST
One calorie is the energy required to lift 1 kg a distance of 427 m.

The **calorie** (cal) is used for measuring heat energy. One calorie is the amount of energy that will raise the temperature of 1 kg of water 1 degree Celsius.

➡ Swimming uses 480 cal per hour. How many calories would be used by swimming $\frac{1}{2}$ hour each day for 30 days?

$\frac{1}{2} \times 30 = 15$ h
- To find the number of calories used, first find the number of hours spent swimming.

$15 \times 480 = 7200$ cal
- Multiply the number of hours spent swimming by the calories used per hour.

7200 cal would be used by swimming $\frac{1}{2}$ hour each day for 30 days.

The **watt-hour** is used for measuring electrical energy. One watthour is the amount of energy required to lift 1 kg a distance of 370 m. A light bulb rated at 100 watts (W) will emit 100 watt-hours of energy each hour.

1000 watt-hours (Wh) = 1 kilowatt-hour (kWh)

➡ A 150-W light bulb is on for 8 h. At 8¢ per kWh, find the cost of the energy used.

$150 \times 8 = 1200$ Wh
- To find the cost, first find the number of watt-hours used.

1200 Wh = 1.2 kWh
- Convert to kilowatt-hours.

$1.2 \times \$0.08 = \$0.096$
- Multiply the number of kilowatt-hours used by the cost per kilowatt-hour.

The cost of the energy used is $0.096.

Example 1
Walking uses 180 cal per hour. How many calories will you burn off by walking $\frac{3}{4}$ h each day for 1 week?

Strategy To find the number of calories:
- Multiply to find the number of hours spent walking per week.
- Multiply the product by the calories burned off per hour.

You Try It 1
Housework requires 240 cal per hour. How many calories are used in 5 days by doing $1\frac{1}{2}$ h of active housework per day?

Your strategy

Solution on p. A42

Solution

$\frac{3}{4} \times 7 = \frac{21}{4} = 5\frac{1}{4}$ h

$5\frac{1}{4} \times 180 = 945$ cal

You will burn off 945 cal.

Your solution

1800 cal

Example 2

An iron is rated at 1200 W. If the iron is used for 1.5 h, how much energy is used in kilowatt-hours?

Strategy

To find the energy used:

- Multiply to find the number of watthours used.
- Convert to kilowatt-hours.

Solution

1200 W × 1.5 h = 1800 Wh
1800 Wh = 1.8 kWh
1.8 kWh of energy are used.

You Try It 2

Find the number of kilowatt-hours of energy used when a 150-W light bulb burns for 200 h.

Your strategy

Your solution

30 kWh

Example 3

A TV set rated at 1800 W is operated an average of 3.5 h per day. At 7.2¢ per kilowatt-hour, find the cost of operating the set for 1 week. Round to the nearest cent.

Strategy

To find the cost:

- Multiply to find the total number of hours the set is used.
- Multiply the product by the number of watts to find the watt-hours.
- Convert to kilowatt-hours.
- Multiply the number of kilowatt-hours by the cost per kilowatt-hour.

Solution

3.5 × 7 = 24.5 h
24.5 × 1800 = 44,100 Wh
44,100 Wh = 44.1 kWh
44.1 × $0.072 = $3.1752

The cost is approximately $3.18.

You Try It 3

A microwave oven rated at 500 W is used an average of 20 min per day. At 8.7¢ per kilowatt-hour, find the cost of operating the oven 30 days.

Your strategy

Your solution

43.5¢

Solutions on pp. A42 and A43

9.4 Exercises

Objective A *Application Problems*

1. How many calories can be eliminated from your diet for 30 days if you omit one slice of bread per day? One slice of bread contains 110 cal. 3300 cal

2. Bruce omits one egg containing 75 cal from his usual breakfast. If he continues this practice for 90 days, how many calories will he omit from his diet? 6750 cal

3. Moderately active people need 20 cal per pound of weight to maintain their weight. How many calories would a 150-lb moderately active person need per day to maintain weight? 3000 cal

4. A person whose daily activities would be characterized as light activity needs 15 cal per pound of body weight to maintain weight. How many calories would a 135-lb person with light activity need per day to maintain body weight? 2025 cal

5. Angelina consumes 350 cal for breakfast and 650 cal for lunch. Find the percent of the total daily intake of calories left for the evening meal if she limits her daily intake to 1800 cal. Round to the nearest tenth of a percent. 44.4%

6. For a healthy diet, it is recommended that 55% of the daily intake of calories come from carbohydrates. Find the daily intake of calories from carbohydrates if you want to limit your calorie intake to 1600 cal. 880 cal

7. Playing tennis singles requires 450 cal per hour. How many calories do you burn up in 30 days playing 45 min per day? 10,125 cal

8. Cycling at 8 mi per hour requires 320 cal per hour. If Christine rides a bicycle for $1\frac{1}{2}$ h per day for 5 days a week, how many calories does she burn up in 4 weeks? 9600 cal

9. After playing golf for 3 h, Ruben had a banana split containing 550 cal. Playing golf uses 320 cal per hour. How many calories did he gain or lose from these two activities? Loss of 410 cal

10. Playing the violin requires 185 cal per hour. How many calories are consumed playing the violin for 45 min per day for 30 days? 4162.5 cal

11. Hiking requires approximately 315 cal per hour. How many hours would Avi have to hike to burn off the calories in a 375-cal sandwich, a 150-cal soda, and a 280-cal ice cream cone? Round to the nearest tenth. 2.6 h

12. Riding a bicycle requires 265 cal per hour. How many hours would Shawna have to ride a bicycle to burn off the calories in a 320-cal milkshake, a 310-cal cheeseburger, and a 150-cal apple? Round to the nearest tenth. 2.9 h

13. An oven uses 500 W of energy. How many watt-hours of energy are used to cook a 5-kg roast for $2\frac{1}{2}$ h? 1250 Wh

14. Find the cost of 560 kWh of electricity at $0.092 per kilowatt-hour. $51.52

15. A 21-in. color TV set is rated at 90 W. The TV is used an average of $3\frac{1}{2}$ h each day for a week. How many kilowatt-hours of energy are used during the week? 2.205 kWh

16. A 120-W bulb is kept burning 24 h a day. How many kilowatt-hours of electrical energy are used in 1 day? 2.88 kWh

17. A 120-W stereo set is on an average of 2 h a day. Find the cost of listening to the stereo for 2 weeks if the cost is 9.4¢ per kilowatt-hour. (Round to the nearest cent.) $0.32

18. How much does it cost to run a 2200-W air conditioner for 8 h at 9¢ per kilowatt-hour? Round to the nearest cent. $1.58

19. A 1200-W hair dryer is used an average of 5 h per week. How many kilowatt-hours of energy are used each week? 6 kWh

20. A space heater is used for 3 h. The heater uses 1400 W per hour. Electricity costs 11.1¢ per kilowatt-hour. Find the cost of using the electric heater. Round to the nearest cent. $0.47

21. A house is insulated to save energy. The house used 265 kWh of electrical energy per month before insulation and saves 45 kWh of energy per month after insulation. What percent decrease does this amount represent? Round to the nearest tenth. 17.0%

22. A household uses an average of 16.3 kWh of electrical energy each day. Electrical energy costs 10.2¢ per kilowatt-hour. Find the cost of using electrical energy for 31 days for the household. Round to the nearest cent. $51.54

23. A welder uses 6.5 kW of energy each hour. Find the cost of using the welder for 6 h a day for 30 days. The cost is 9.4¢ per kilowatt-hour. $109.98

APPLYING THE CONCEPTS

24. [W] A maintenance intake of calories allows a person to neither gain nor lose weight. By consulting a book on nutrition, make a table of the weight of a person and her or his maintenance intake of calories. Now add another column to this table indicating the calorie intake when an individual wants to lose 1 pound per week. The table should now show the appropriate caloric intake for people of different weights who want to lose 1 pound per week.

25. [W] Write an article on how to improve energy efficiency of a home.

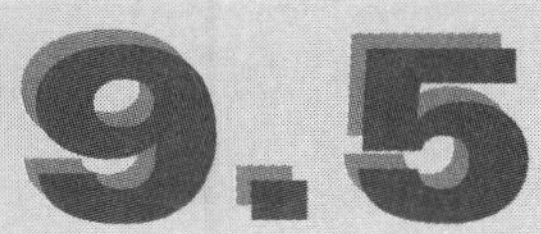

Conversion Between the U.S. Customary and the Metric Systems of Measurement

Objective A To convert U.S. Customary units to metric units

INSTRUCTOR NOTE

The major focus of this section should be approximate equivalencies. Ask questions such as "Place your hands 20 cm apart"; "Estimate the weight of a desk in kilograms"; or "I commute ____ km to school each day."

More than 90% of the world's population use the metric system of measurement. Therefore, converting U.S. Customary units to metric units is essential in trade and commerce—for example, in importing foreign goods and exporting domestic goods. Approximate equivalences between the two systems follow.

Units of Length	*Units of Weight*	*Units of Capacity*
1 m ≈ 3.28 ft	28.35 g ≈ 1 oz	1 L ≈ 1.06 qt
1 cm ≈ 0.39 in.	454 g ≈ 1 lb	
1.61 km ≈ 1 mi	0.454 kg ≈ 1 lb	
0.91 m ≈ 1 yd	1 kg ≈ 2.2 lb	
0.305 m ≈ 1 ft		
2.54 cm ≈ 1 in.		
1 m ≈ 1.09 yd		

These equivalences can be used to form conversion rates to change one unit of measurement to another. For example, because 1.61 km ≈ 1 mi, the conversion rates $\frac{1.61\text{ km}}{1\text{ mi}}$ and $\frac{1\text{ mi}}{1.6\text{ km}}$ are each approximately equal to 1.

➡ Convert 55 mi to kilometers.

$$55\text{ mi} \approx 55\text{ mi} \times \boxed{\frac{1.61\text{ km}}{1\text{ mi}}}$$

• The conversion rate must contain km in the numerator and mi in the denominator.

$$= \frac{55\text{ }\cancel{\text{mi}}}{1} \times \frac{1.61\text{ km}}{1\text{ }\cancel{\text{mi}}}$$

$$= \frac{88.55\text{ km}}{1}$$

$$55\text{ mi} \approx 88.55\text{ km}$$

Example 1

Convert 4 gal to liters. Round to the nearest tenth.

Solution

Because the conversion factor for gallons to liters is not given, convert gallons to quarts and then convert quarts to liters.

$$4\text{ gal} \approx 4\text{ }\cancel{\text{gal}} \times \frac{4\text{ }\cancel{\text{qt}}}{1\text{ }\cancel{\text{gal}}} \times \frac{1\text{ L}}{1.06\text{ }\cancel{\text{qt}}} = \frac{16\text{ L}}{1.06}$$

$$4\text{ gal} \approx 15.1\text{ L}$$

You Try It 1

Convert 10 c to liters. Round to the nearest hundredth.

Your solution

2.36 L

Solution on p. A43

Example 2
Convert 45 mi/h to kilometers per hour.

Solution

$$\frac{45\text{ mi}}{\text{h}} \approx \frac{45\cancel{\text{ mi}}}{\text{h}} \times \frac{1.61\text{ km}}{1\cancel{\text{ mi}}}$$

$$\approx 72.45\text{ km/h}$$

You Try It 2
Convert 60 ft/s to meters per second. Round to the nearest hundredth.

Your solution
18.29 m/s

Example 3
The price of gasoline is $1.52/gal. Find the cost per liter to the nearest cent.

Solution

$$\frac{\$1.52}{\text{gal}} \approx \frac{\$1.52}{\cancel{\text{gal}}} \times \frac{1\cancel{\text{ gal}}}{4\cancel{\text{ qt}}} \times \frac{1.06\cancel{\text{ qt}}}{\text{L}}$$

$$\approx \$.40/\text{L}$$

You Try It 3
The price of milk is $1.86/gal. Find the cost per liter to the nearest cent.

Your solution
$0.49/L

Solutions on p. A43

Objective B To convert metric units to U.S. Customary units

Metric units are already being used in the United States today. Gasoline may be sold by the liter, 35-mm film is available, cereal is sold by the gram, and 100-mm-long cigarettes are common. The same conversion rates used in Objective A are used for converting metric units to U.S. customary units.

Example 4
Convert 200 m to feet.

Solution

$$200\text{ m} \approx 200\cancel{\text{ m}} \times \frac{3.28\text{ ft}}{1\cancel{\text{ m}}}$$

$$200\text{ m} \approx 656\text{ ft}$$

You Try It 4
Convert 45 cm to inches. Round to the nearest hundredth.

Your solution 17.55 in.

Example 5
Convert 90 km/h to miles per hour. Round to the nearest hundredth.

Solution

$$\frac{90\text{ km}}{\text{h}} \approx \frac{90\cancel{\text{ km}}}{\text{h}} \times \frac{1\text{ mi}}{1.61\cancel{\text{ km}}}$$

$$\approx 55.90\text{ mi/h}$$

You Try It 5
Express 75 km/h in miles per hour. Round to the nearest hundredth.

Your solution 46.58 mi/h

Example 6
The price of gasoline is $0.372/L. Find the cost per gallon to the nearest cent.

Solution

$$\frac{\$0.372}{1\text{ L}} \approx \frac{\$0.372}{1\cancel{\text{ L}}} \times \frac{1\cancel{\text{ L}}}{1.06\cancel{\text{ qt}}} \times \frac{4\cancel{\text{ qt}}}{1\text{ gal}}$$

$$\approx \$1.40/\text{gal}$$

You Try It 6
The price of ice cream is $1.50/L. Find the cost per gallon to the nearest cent.

Your solution $5.66/gal

Solutions on p. A43

9.5 Exercises

Objective A

Convert. Round to the nearest hundredth.

1. Convert the 100-yd dash to meters.
91 m

2. Find the weight in kilograms of a 135-lb person.
61.36 kg

3. Find the height in meters of a person 6 ft 4 in. tall.
1.93 m

4. Find the number of liters in 1 gal of punch.
3.77 L

5. How many kilograms does a 12-lb ham weigh?
5.45 kg

6. Find the number of liters in 14.3 gal of gasoline.
53.96 liters

7. Find the number of milliliters in 1 c.
235.85 ml

8. The winning long jump at a track meet was 29 ft 2 in. Convert this distance to meters.
8.89 m

9. Express 65 mi/h in kilometers per hour.
104.65 km/h

10. Express 30 mi/h in kilometers per hour.
48.3 km/h

11. Bacon costs $1.69/lb. Find the cost per kilogram.
$3.72/kg

12. Peaches cost $0.69/lb. Find the cost per kilogram.
$1.52/kg

13. The cost of gasoline is $1.47/gal. Find the cost per liter.
$0.39/L

14. Paint costs $9.80/gal. Find the cost per liter.
$2.60/L

15. Gary is planning a 5-day backpacking trip and decides to hike an average of 5 h each day. Hiking requires an extra 320 cal per hour. How many pounds will he lose during the trip if an extra 900 cal is consumed each day? (3500 cal is equivalent to 1 lb.) 1 lb

16. Swimming requires 550 cal per hour. How many pounds could be lost by swimming $1\frac{1}{2}$ h each day for 5 days if no extra calories were consumed? (3500 cal is equivalent to 1 lb.) 1.18 lb

17. The distance around the earth is 24,887 mi. Convert this distance to kilometers. 40,068.07 km

18. The distance from the earth to the sun is 93,000,000 mi. Convert this distance to kilometers. 149,730,000 km

Objective B

Convert. Round to the nearest hundredth.

19. Convert the 100-m dash to feet.
328 ft

20. Find the weight in pounds of a 98-kg person.
215.6 lb

21. Find the number of gallons in 6 L of antifreeze.
1.59 gal

22. Leo's height is 1.65 m. Find his height in inches.
64.96 in.

23. Find the distance of the 1500-m race in feet.
4920 ft

24. Find the weight of 327 g of cereal in ounces.
11.53 oz

25. How many gallons does a 48-L tank hold?
12.72 gal

26. Find the width of 35-mm film in inches.
1.38 in.

27. A bottle of syrup contains 876 ml. Find the number of pints in 876 ml.
1.86 pt

28. A backpack tent weighs 2.1 kg. Find the weight in pounds.
4.62 lb

29. Express 80 km/h in miles per hour.
49.69 mi/h

30. Express 30 m/s in feet per second.
98.36 ft/s

31. Gasoline costs 38.5¢/L. Find the cost per gallon.
$1.45/gal

32. A turkey roast costs $3.15/kg. Find the cost per pound.
$1.43/lb

APPLYING THE CONCEPTS

33. What is a metric ton?
1000 kg

34. For the following U.S. Customary units, make an estimate of the metric equivalent. Now perform the conversion and see how close you come to the actual measurement.

60 mi/h ≈ 96.6 km/h	120 lb ≈ 54.5 kg	6 ft ≈ 1.8 m
1 mi ≈ 1.6 km	1 gal ≈ 4L	1 quarter-mile ≈ 400 m

35. Determine whether the statement is true or false.
a. A liter is more than a gallon. False
b. A meter is less than a yard. False
c. 30 mi/h is less than 60 km/h. True
d. A kilogram is greater than a pound. True
e. An ounce is less than a gram. False

36. [W] Should the United States keep the U.S. Customary system or convert to the metric system? Justify your position. Answers will vary

Project in Mathematics

Deductive Thinking

A detective studies clues to a crime in an attempt to discover the criminal. One of the skills the detective uses is *deductive reasoning*. This method of reasoning moves from one known fact to another until a conclusion based on the facts is reached.

Deductive reasoning skills are very important in Mathematics and other disciplines. The only way to sharpen those skills is to practice them on various problems. Here are two problems that require deductive reasoning. The answers are given at the bottom of the page, but try to solve each problem before you look at the answer.

Exercises:

1. At a small restaurant on the fourth planet from the star, Betelguse, a Surd was joined by three Anatarians. The people from Anatares are either Rads, who always lie, or Tads, who always tell the truth. The Surd asked the first Anatarian if she was a Rad or a Tad. She answered, but just then a starcad flew by in very bad need of a new muffler. Consequently, the Surd did not hear her answer. "She said she is a Tad," said the second Anatarian. "She is a Rad," said the third Anatarian. How many of the Anatarians are Rads and how many are Tads?
2. This problem was created by Raymond Smullyan, one of the foremost writers of problems in deductive reasoning.

   There are two doors, each with a sign as shown below.

   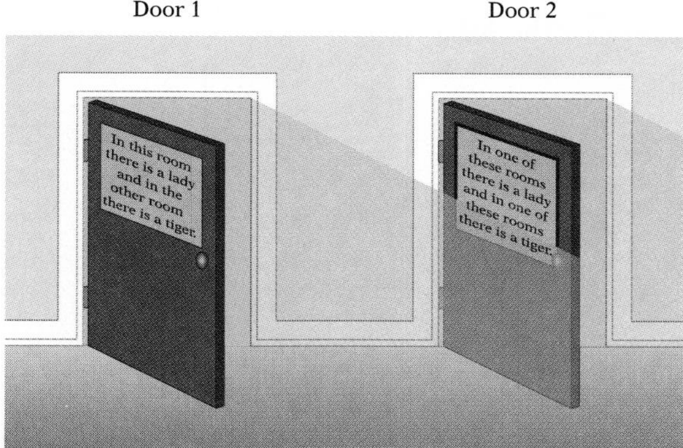

   The sign on Door 1 reads: In this room there is a lady and in the other room there is a tiger. The sign on Door 2 reads: In one of these rooms there is a lady and in one of these rooms there is a tiger.

   Behind which door is the tiger?

Answers: 1. There are 2 Rads and 1 Tad; 2. The tiger is behind door 1.

Chapter Summary

Key Words The *metric system of measurement* is a system of measurement based on the decimal system.

The basic unit of measurement of length in the metric system is the *meter*.

The basic unit of mass in the metric system is the *gram*.

The basic unit of capacity in the metric system is the *liter*.

The basic unit of energy in the metric system is the *calorie*.

The *watt-hour* is also used in the metric system for measuring electrical energy.

Essential Rules Prefixes to the basic unit denote the magnitude of each unit in the metric system

kilo-	1000
hecto-	100
deca-	10
deci-	0.1
centi-	0.01
milli-	0.001

Conversion between units in the metric system involves moving the decimal point:

1. When converting from a larger unit to a smaller unit, move the decimal point to the **right.**
2. When converting from a smaller unit to a larger unit, move the decimal point to the **left**.

Chapter Review

SECTION 9.1

1. Convert 0.37 cm to millimeters.
3.7 mm

2. Convert 1.25 km to meters.
1250 m

3. Add: 5 m 67 cm
+ 3 m 88 cm
9.55 m

4. What is 56 cm 3 mm decreased by 35 cm 8 mm? 20.5 cm

5. Multiply: 4 m 55 cm
× 8
36.4 m

6. Divide: $15\overline{)52 \text{ km } 500 \text{ m}}$
3.5 km

7. Three pieces of wire fence are cut from a 50-m roll. The three pieces measure 2 m 40 cm, 5 m 60 cm, and 4 m 80 cm. How much wire fence is left on the roll after the three pieces are cut? 37.2 m

SECTION 9.2

8. Convert 0.450 g to milligrams.
450 mg

9. Convert 4.050 kg to grams.
4050 g

10. Add: 4 g 677 mg
+ 9 g 566 mg
14.243 g

11. Subtract: 45 kg 45 g
− 32 kg 585 g
12.46 kg

12. Find the product of 3 kg 450 g and 11.
37.95 kg

13. Divide: $18\overline{)45 \text{ g } 340 \text{ mg}}$
Round to the nearest thousandth.
2.519 g

14. Find the total cost of a 7-kg 300-g turkey costing $2.79 per kilogram. Round to the nearest cent. $20.37

SECTION 9.3

15. Convert 0.0056 L to milliliters.
5.6 ml

16. Convert 1.2 L to cubic centimeters.
1200 cm^3

17. Add:

$$\begin{array}{r} 3\text{ L }\ 45\text{ ml} \\ +\ 7\text{ L } 568\text{ ml} \\ \hline \end{array}$$

10.613 L

18. Subtract:

$$\begin{array}{r} 15\text{ L } 569\text{ ml} \\ -\ 8\text{ L } 972\text{ ml} \\ \hline \end{array}$$

6.597 L

19. Multiply:

$$\begin{array}{r} 5\text{ L } 122\text{ ml} \\ \times\ \ 20 \\ \hline \end{array}$$

102.44 L

20. Find the quotient of 45 L 250 ml and 25.
1.81 L

21. One-hundred twenty-five guests are expected to attend a reception. Assuming that each person drinks 400 ml of coffee, how many liters of coffee should be prepared? 50 L

SECTION 9.4

22. A large egg contains approximately 90 cal. How many calories can be eliminated from your diet in a 30-day month by eliminating one large egg per day from your usual breakfast? 2700 cal

23. A TV uses 240 W of energy. The set is on an average of 5 h a day in a 30-day month. At a cost of 9.5¢ per kilowatt-hour, how much does it cost to run the set 30 days? $3.42

SECTION 9.5

24. Convert the 1000-m run to yards. Round to the nearest tenth. 1098.9 yd

25. A backpack tent weighs 1.90 kg. Find the weight in pounds. Round to the nearest hundredth. 4.19 lb

26. Ham costs $3.40 per pound. Find the cost per kilogram. $7.48/kg

27. Cycling burns up approximately 400 cal per hour. How many hours of cycling are necessary to lose 1 lb? (3500 cal is equivalent to 1 lb.) 8.75 h

Chapter Test

1. Convert 42.6 mm to centimeters.
4.26 cm [9.1A]

2. Convert 5 km 38 m to meters.
5038 m [9.1A]

3. Subtract:
7 m 63 cm − 2 m 98 cm
4.65 m [9.1B]

4. What is 4 m 29 cm increased by 17 m 87 cm?
22.16 m [9.1B]

5. A carpenter needs 30 rafters, each rafter 3 m 80 cm long. Find the total length of rafters needed in meters. 114 m [9.1C]

6. Twenty-five rivets are used to fasten two steel plates together. The plates are 4 m 20 cm long, and the rivets are equally spaced, with a rivet at each end. Find the distance between the rivets. 17.5 cm [9.1C]

7. Convert 3.29 kg to grams.
3290 g [9.2A]

8. Convert 3 g 89 mg to grams.
3.089 g [9.2A]

9. What is 3 kg 480 g multiplied by 7?
24.36 kg [9.2B]

10. Subtract:
17 g 164 mg − 9 g 867 mg
7.297 g [9.2B]

11. A 20 × 20-cm tile weighs 250 g. Find the weight of a box of 144 tiles in kilograms. 36 kg [9.2C]

12. Two hundred grams of fertilizer are used for each tree in an apple orchard containing 1200 trees. At $2.25 per kilogram of fertilizer, how much does it cost to fertilize the apple orchard? $540 [9.2C]

13. Convert 3.25 L to milliliters.
3250 ml [9.3A]

14. Convert 1.6 L to cubic centimeters.
1600 cm^3 [9.3A]

15. Find the quotient of 3 L 750 ml and 5.
0.75 L [9.3B]

16. Subtract:
7 L 180 ml − 3 L 249 ml
3.931 L [9.3B]

17. The community health clinic is giving flu shots for the coming flu season. Each flu shot contains 2 cm^3 of vaccine. How many liters of vaccine are needed to inoculate 2600 people? 5.2 L [9.3C]

18. A TV set rated at 1600 W is operated an average of 4 h per day. Electrical energy costs 8.5¢ per kilowatt-hour. How much does it cost to operate the TV set 30 days? $16.32 [9.4A]

19. The record ski jump for men is 636 ft. Convert this value to meters. Round to the nearest hundredth. (1 m ≈ 3.28 ft) 193.90 m [9.5A]

20. The record ski jump for women is 110 m. Convert this value to feet. Round to the nearest hundredth. (1 m ≈ 3.28 ft) 360.80 ft [9.5B]

Cumulative Review

1. Simplify:
$12 - 8 \div (6 - 4)^2 \cdot 3$
6 [1.6B]

2. Find the total of $5\frac{3}{4}$, $1\frac{5}{6}$, and $4\frac{7}{9}$.
$12\frac{13}{36}$ [2.4C]

3. Subtract: $4\frac{2}{9} - 3\frac{5}{12}$
$\frac{29}{36}$ [2.5C]

4. Divide: $5\frac{3}{8} \div 1\frac{3}{4}$
$3\frac{1}{14}$ [2.7B]

5. Simplify: $\left(\frac{2}{3}\right)^4 \cdot \left(\frac{9}{4}\right)^2$
1 [2.8B]

6. Subtract: $12.0072 - 9.937$
2.0702 [3.3A]

7. Solve the proportion $\frac{5}{8} = \frac{n}{50}$. Round to the nearest tenth.
$n = 31.3$ [4.3B]

8. Write $1\frac{3}{4}$ as a percent.
175% [5.1B]

9. 6.09 is 4.2% of what number?
145 [5.4A]

10. Convert 18 pt to gallons.
2.25 gal [8.3A]

11. Convert: 18 m 75 cm = ____ m
18.75 m [9.1A]

12. Subtract: 4 km 420 m − 1 km 892 m
2.528 km [9.1B]

13. Convert: 5 kg 50 g = ____ kg
5.05 kg [9.2A]

14. Add: 3 g + 672 mg
3.672 g [9.2B]

15. What is 12 kg 450 g divided by 15?
0.83 kg [9.2B]

16. Subtract: 6 L − 452 ml
5.548 [9.3B]

17. The Guerrero family has a monthly income of $2244 per month. The family spends one-fourth of its monthly income on rent. How much money is left after the rent is paid? $1683 [2.6C]

18. The state income tax on a business is \$620 plus 0.08 times the profit the business makes. The business made a profit of \$82,340.00 last year. Find the amount of state income tax the business paid. \$7207.20 [3.4B]

19. The property tax on a \$45,000 home is \$900. At the same rate, what is the property tax on a home worth \$75,000? \$1500 [4.3C]

20. A car dealer offers new-car buyers a 12% rebate on some models. What rebate would a new-car buyer receive on a car that cost \$13,500? \$1620 [5.2B]

21. Rob Akullian received a dividend of \$533 on an investment of \$8200. What percent of the investment is the dividend? 6.5% [5.3B]

22. You received grades of 78, 92, 45, 80, and 85 on five English exams. Find your average grade. 76 [7.4A]

23. Karla Perella, a ski instructor, receives a salary of \$22,500. Her salary will increase by 12% next year. What will be her salary next year? \$25,200 [5.2B]

24. A sporting goods store has regularly priced \$80 fishing rods on sale for \$62.40. What is the discount rate? 22% [6.2D]

25. Forty-eight blocks, each block 9 in. long, are laid end to end to make the base for a wall. Find the length of the wall in feet. 36 ft [8.1C]

26. A jar of apple juice contains 24 oz. Find the number of quarts of apple juice in a case of 16 jars. 12 qt [8.3C]

27. A garage mechanic bought oil in 40-gal containers. The mechanic bought the oil for \$2.88 per gallon and sold the oil for \$1.09 per quart. Find the profit on one 40-gal container of oil. \$59.20 [8.3C]

28. A school swimming pool uses 1 L 200 ml of chlorine each school day. How much chlorine is used for 20 days during the month? 24 L [9.3C]

29. A 1200-W hair dryer is used an average of 30 min a day. At a cost of 10.5¢ per kilowatt-hour, how much does it cost to operate the hair dryer for 30 days? \$1.89 [9.4A]

30. Convert 60 mi/h to kilometers per hour. Round to the nearest tenth. (1.61 km = 1 mi) 96.6 km/h [9.5A]

Chapter

10

Rational Numbers

Objectives

Section 10.1

To identify the order relation between two integers

To evaluate expressions that contain the absolute value symbol

Section 10.2

To add integers

To subtract integers

To solve application problems

Section 10.3

To multiply integers

To divide integers

To solve application problems

Section 10.4

To add or subtract rational numbers

To multiply or divide rational numbers

Section 10.5

To use the Order of Operations Agreement to simplify expressions

History of Negative Numbers

When a temperature is "below zero," that temperature is represented by placing the symbol $-$ before the number. Thus -4 degrees means that the temperature is 4 degrees below zero. The number -4 is read "negative four" and is an example of a negative number.

The earliest evidence for the use of negative numbers dates from China, about 250 B.C. The Chinese used two sets of calculating rods, one red for positive numbers and one black for negative numbers. The idea seems to have remained with the Chinese and did not spread quickly to the Middle East or to Europe.

Between the years 300 B.C. and A.D. 1500, there were a few further attempts to deal with negative numbers. Then, in 1500, negative numbers began appearing more and more frequently. In most cases, however, the numbers were considered "fictitious" numbers or "false" numbers. The idea of using a negative number to refer to a quantity below zero was still not an accepted idea.

By the 18th century, negative numbers were discussed in most mathematics textbooks. However, operations such as multiplying two negative numbers were not included in some books.

As the 18th century came to a close, some 2000 years after the first evidence of negative numbers from the Chinese, negative numbers were finally accepted. This concept, like many other mathematical ideas, took a long time to develop. Do you suppose some mathematical idea is floating around today that will take 2000 years before being accepted?

10.1 Introduction to Integers

Objective A To identify the order relation between two integers

POINT OF INTEREST

Chinese manuscripts dating from about 250 B.C. contain the first recorded use of negative numbers. However, it was not until late in the 14th century that most mathematicians generally accepted these numbers.

Thus far in the text, we have encountered only zero and the numbers greater than zero. The numbers greater than zero are called **positive numbers.** However, the phrases "12 degrees below zero," "$25 in debt," and "15 feet below sea level" refer to numbers less than zero. These numbers are called **negative numbers.**

The integers are $\ldots -4, -3, -2, -1, 0, 1, 2, 3, 4, \ldots$.

Each integer can be shown on a number line. The integers to the left of zero on the number line are called **negative integers** and are represented by a negative sign (−) placed in front of the number. The integers to the right of zero are called **positive integers.** The positive integers are also called natural numbers. Zero is neither a positive nor a negative integer.

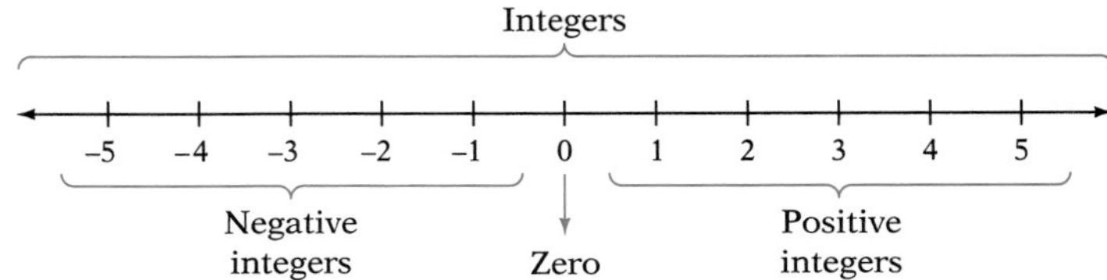

A number line can be used to visualize the order relation of two integers. A number that appears to the left of a given number is less than (<) the given number. A number that appears to the right of a given number is greater than (>) the given number.

2 is greater than negative 4.

$2 > -4$

Negative 5 is less than negative 3.

$-5 < -3$

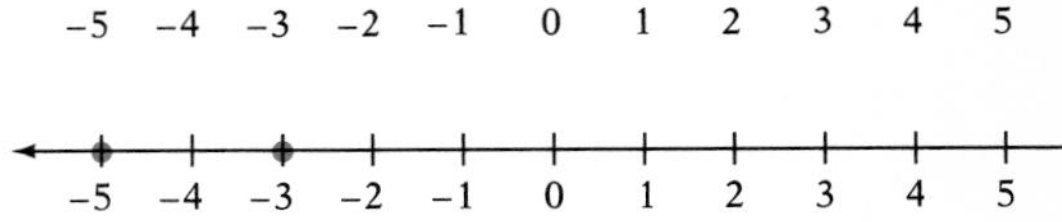

Example 1
The temperature at the North Pole was recorded as 87 degrees below zero. Represent this temperature as an integer.

Solution −87 degrees

You Try It 1
The surface of the Salton Sea is 232 ft below sea level. Represent this depth by a signed number.

Your solution −232 ft

Example 2
Graph −2 on the number line.

Solution

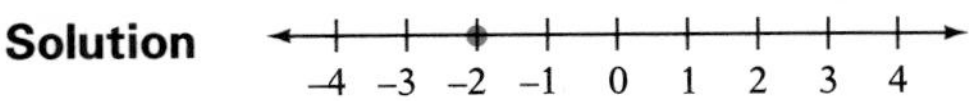

You Try It 2
Graph −4 on the number line.

Your solution

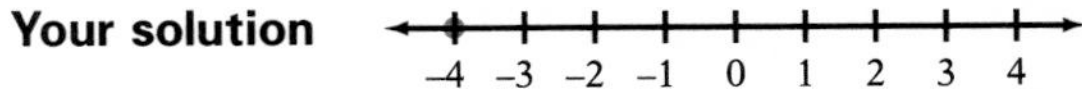

Example 3
Place the correct symbol, < or >, between the two numbers.

a. $-5 \quad -7$ **b.** $\frac{1}{2} \quad -2$

Solution **a.** $-5 > -7$ **b.** $\frac{1}{2} > -2$

You Try It 3
Place the correct symbol, < or >, between the two numbers.

a. $-12 \quad -8$ **b.** $-5 \quad 0$

Your solution a. $-12 < -8$ b. $-5 < 0$

Solutions on p. A44

Objective B To evaluate expressions that contain the absolute-value symbol

INSTRUCTOR NOTE

The [+/−] key on a calculator can be used to illustrate the idea of opposite. Entering 4 [+/−] gives the opposite of 4. Pressing the [+/−] key again gives 4, the opposite of −4.

Two numbers that are the same distance from zero on the number line but on opposite sides of zero are called **opposites.**

-4 is the opposite of 4

and

4 is the opposite of -4.

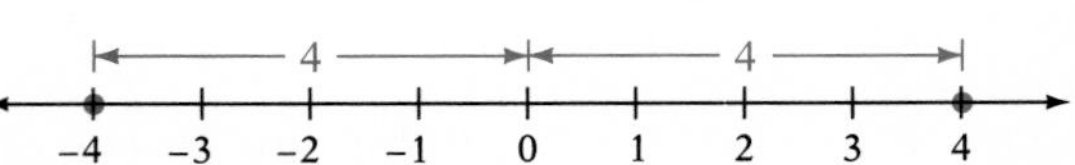

Note that a negative sign can be read as "the opposite of."

$-(4) = -4$ The opposite of positive 4 is negative 4.

$-(-4) = 4$ The opposite of negative 4 is positive 4.

INSTRUCTOR NOTE

The important point for a student to understand about absolute value is magnitude. If a student runs 5 miles west or 5 miles east, the distance is the same. Only the direction is different.

The **absolute value** of a number is the distance between zero and the number on the number line. Therefore, the absolute value of a number is a positive number or zero. The symbol for absolute value is "| |".

The distance from 0 to 4 is 4. Thus, $|4| = 4$ (the absolute value of 4 is 4).

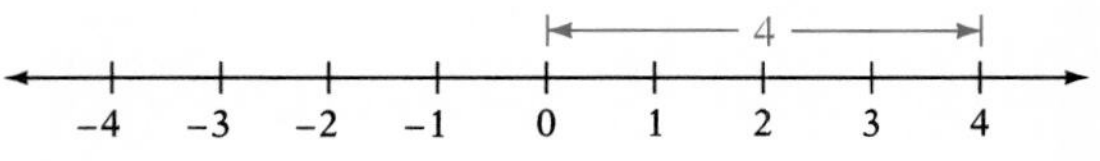

The distance from 0 to −4 is 4. Thus $|-4| = 4$ (the absolute value of -4 is 4).

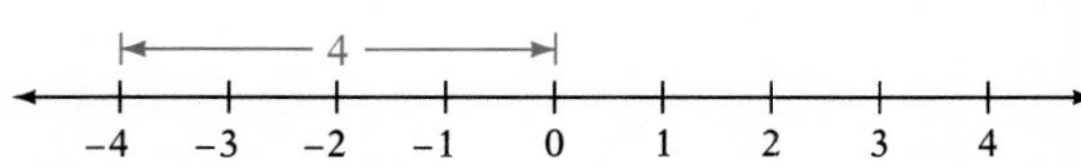

The absolute value of a positive number is the number itself. The absolute value of a negative number is the opposite of the negative number. The absolute value of zero is zero.

Example 4 Find the absolute value of 2 and −3.

Solution $|2| = 2$
$|-3| = 3$

You Try It 4 Find the absolute value of −7 and 21.

Your solution 7 21

Example 5 Evaluate $|-34|$ and $|0|$.

Solution $|-34| = 34$
$|0| = 0$

You Try It 5 Evaluate $|2|$ and $|-9|$.

Your solution 2 9

Example 6 Evaluate $-|-4|$.

Solution $-|-4| = -4$
The minus sign *in front of* the absolute-value sign is not affected by the absolute-value sign.

You Try It 6 Evaluate $-|-12|$.

Your solution −12

Solutions on p. A44

10.1 Exercises

Objective A

Place the correct symbol, $<$ or $>$, between the two numbers.

1. $3 < 5$	**2.** $7 > 4$	**3.** $-2 > -5$	**4.** $-6 < -1$
5. $-16 < 1$	**6.** $-2 < 13$	**7.** $3 > -7$	**8.** $5 > -6$
9. $-11 < -8$	**10.** $-4 > -10$	**11.** $-1 > -6$	**12.** $-9 < -4$
13. $0 > -3$	**14.** $8 > 0$	**15.** $6 > -8$	**16.** $8 > -6$
17. $-14 < 16$	**18.** $-12 < 1$	**19.** $35 > 28$	**20.** $42 > 19$
21. $-42 < 27$	**22.** $-36 < 49$	**23.** $21 > -34$	**24.** $53 > -46$
25. $-27 > -39$	**26.** $-51 < -20$	**27.** $-87 < 63$	**28.** $-75 < 92$
29. $68 > -79$	**30.** $95 > -71$	**31.** $-62 > -84$	**32.** $-91 < -70$
33. $94 > 83$	**34.** $76 < 81$	**35.** $59 > -67$	**36.** $48 > -66$
37. $-93 < -55$	**38.** $-64 > -86$	**39.** $-88 < 57$	**40.** $-58 < 82$

Objective B

Find the opposite number.

41. 4 −4	**42.** 16 −16	**43.** −2 2	**44.** −3 3
45. 22 −22	**46.** 45 −45	**47.** −31 31	**48.** −88 88

Evaluate.

49. $\|2\|$ 2	**50.** $\|-2\|$ 2	**51.** $\|-6\|$ 6	**52.** $\|6\|$ 6	**53.** $\|8\|$ 8
54. $\|5\|$ 5	**55.** $\|-9\|$ 9	**56.** $\|-1\|$ 1	**57.** $-\|-1\|$ −1	**58.** $-\|-8\|$ −8
59. $-\|-5\|$ −5	**60.** $-\|0\|$ 0	**61.** $\|16\|$ 16	**62.** $\|19\|$ 19	**63.** $\|-12\|$ 12
64. $\|-22\|$ 22	**65.** $-\|29\|$ −29	**66.** $-\|20\|$ −20	**67.** $-\|-14\|$ −14	**68.** $-\|-18\|$ −18
69. $\|-15\|$ 15	**70.** $\|-23\|$ 23	**71.** $-\|33\|$ −33	**72.** $-\|27\|$ −27	**73.** $-\|-36\|$ −36
74. $-\|-41\|$ −41	**75.** $\|32\|$ 32	**76.** $\|25\|$ 25	**77.** $\|-38\|$ 38	**78.** $\|-30\|$ 30
79. $-\|37\|$ −37	**80.** $-\|34\|$ −34	**81.** $-\|-42\|$ −42	**82.** $-\|-45\|$ −45	**83.** $\|44\|$ 44
84. $\|36\|$ 36	**85.** $\|-74\|$ 74	**86.** $\|-61\|$ 61	**87.** $-\|88\|$ −88	**88.** $-\|52\|$ −52

APPLYING THE CONCEPTS

89. Name two numbers that are 5 units from 3 on the number line. Name two numbers that are 3 units from −1 on the number line.
−2, 8   −4, 2

90. Find a number that is halfway between −7 and −5. −6

91. [W] Abdul, Becky, Carl, and Diana were being questioned by their teacher. One of the students had left an apple on the teacher's desk but the teacher did not know which one. Abdul said it was either Becky or Diana. Diana said it was neither Becky nor Carl. If both those statements are false, who left the apple on the teacher's desk? Explain how you arrived at your solution. Abdul

92. [W] In your own words, describe (a) the opposite of a number, (b) the absolute value of a number, and (c) the difference between the words "negative" and "minus." Answers will vary

10.2 Addition and Subtraction of Integers

Objective A *To add integers*

This section begins with translating the sum of two integers into words.

$8 + 5$	positive 8 plus positive 5
$(-8) + 5$	negative 8 plus positive 5
$8 + (-5)$	positive 8 plus negative 5
$(-8) + (-5)$	negative 8 plus negative 5

As shown in the examples above, we often use parentheses when writing a negative integer.

INSTRUCTOR NOTE

There are several ways to model the addition of integers. The model at the right uses arrows. Another model uses money. For instance, if you are \$8 in debt (−8) and you receive \$5, then you are only \$3 in debt (−3). This can also be related to credit card debt. If a student owes \$100 (−100) and charges \$25 more (−25), the student then owes \$125 (−125).

Giving examples like these may help students see that the rules are not arbitrary but are designed to model everyday experience. Another model of addition that is more manipulative in nature is given on the next page.

Besides an integer being graphed as a dot on a number line, an integer can be represented anywhere along a number line by an arrow. A positive number is represented by an arrow pointing to the right. A negative number is represented by an arrow pointing to the left. The absolute value of the number is represented by the length of the arrow. The integers 5 and −4 are shown on the number line in the figure below.

The sum of two integers can be shown on a number line. To add two integers, use arrows to represent the addends with the first arrow starting at zero. The sum is the number directly below the tip of the arrow that represents the second addend.

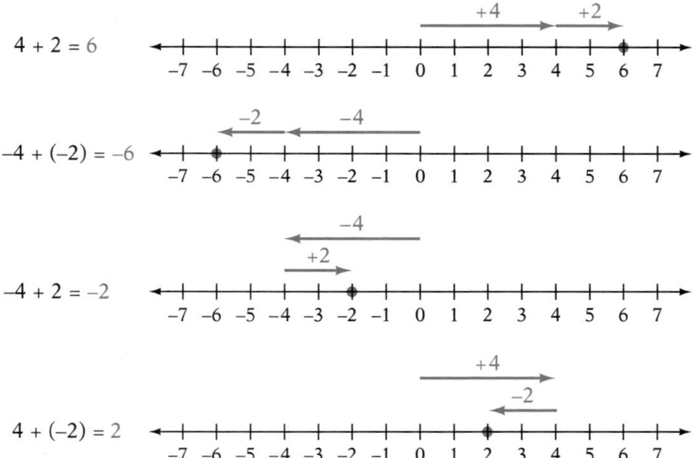

The sums of the integers shown above can be placed in two groups:

Group 1: Addends have the same sign.

$4 + 2$	*positive* 4 plus *positive* 2
$(-4) + (-2)$	*negative* 4 plus *negative* 2

Group 2: Addends have different signs.

$(-4) + 2$	*negative* 4 plus *positive* 2
$4 + (-2)$	*positive* 4 plus *negative* 2

The rule used for adding two integers depends on whether the signs of the addends are the same or different.

INSTRUCTOR NOTE

Another way to explain addition of integers involves using chips, blue chips for positive and red chips for negative. One positive chip added to one negative chip gives zero. To add -8 and 5, place 8 red chips and 5 blue chips in a region. Pair as many red and blue chips as possible and remove the pairs from the region. The remaining chips give the answer, in this case, 3 red chips, or -3.
To model $(-8) + (-5)$, place 8 red chips in the region and then 5 more red chips. There are no pairs of red and blue chips so there are 13 red chips. Therefore the answer is -13.

POINT OF INTEREST

Although mathematical symbols are fairly standard in every country, that has not always been true. Italian mathematicians in the 15th century used a "p" to indicate plus. The "p" was from the Italian word *piu.* In Germany, the familiar + sign was first used in a book called *Mercantile Arithmetic* to indicate a surplus.

Rule for Adding Two Integers

Same signs: To add integers with the same signs, add the absolute values of the numbers. Then attach the sign of the addends.

Different signs: To add integers with different signs, subtract the number with the smaller absolute value from the number with the larger absolute value. Then attach the sign of the addend with the larger absolute value.

➡ Add: $(-4) + (-9)$

$|-4| = 4, |-9| = 9$
$4 + 9 = 13$

- Because the signs of the addends are the same, add the absolute values of the numbers.

$(-4) + (-9) = -13$

- Then attach the sign of the addends.

➡ Add: $6 + (-13)$

$|6| = 6, |-13| = 13$
$13 - 6 = 7$

- Because the signs of the addends are different, subtract the number with the smaller absolute value from the number with the larger absolute value.

$6 + (-13) = -7$

- Then attach the sign of the number with the larger absolute value. Because $|-13| > |6|$, attach the negative sign.

➡ Add: $162 + (-247)$

$162 + (-247) = -85$

- Because the signs are different, find the difference between the absolute values of the numbers and attach the sign of the number with the greater absolute value.

➡ Find the sum of -14 and -47.

$-14 + (-47) = -61$

- Because the signs are the same, add the absolute values of the numbers and attach the sign of the addends.

When adding more than two integers, start from the left and add the first two numbers. Then add the sum to the third number. Continue this process until all the numbers have been added.

➡ Add: $(-4) + (-6) + (-8) + 9$

$(-4) + (-6) + (-8) + 9 = (-10) + (-8) + 9$
$= (-18) + 9$
$= -9$

- Add the first two numbers.
- Add the sum to the next number.
- Continue adding until all numbers have been added.

Example 1 What is -162 added to 98?

Solution $-162 + 98 = -64$

You Try It 1 Add: $-154 + (-37)$

Your solution -191

Solution on p. A44

Example 2
Add: $-2 + (-7) + 4 + (-6)$

Solution
$$-2 + (-7) + 4 + (-6) = -9 + 4 + (-6) = -5 + (-6) = -11$$

You Try It 2
Add: $-5 + (-2) + 9 + (-3)$

Your solution
-1

Solution on p. A44

Objective B To subtract integers

INSTRUCTOR NOTE
Although it is more complicated than the addition model, a subtraction model using the blue and red chips can also be used. To model $5 - (-3)$, place 5 blue chips in a region. Subtracting requires removing 3 red chips, but since there are no red chips in the region, add 3 pairs of a red and a blue chip (essentially adding three 0s). Now the 3 red chips can be removed. The result is 8 blue chips.

Before the rules for subtracting two integers are explained, look at the translation into words of an expression that is the difference of two integers:

$9 - 3$	positive 9 minus positive 3
$(-9) - 3$	negative 9 minus positive 3
$9 - (-3)$	positive 9 minus negative 3
$(-9) - (-3)$	negative 9 minus negative 3

Note that the sign $-$ is used two different ways. One way is as a negative sign, as in (-9), *negative* 9. The second way is to indicate the operation of subtraction, as in $9 - 3$, 9 *minus* 3.

Look at the next four subtraction expressions and decide whether the second number in each expression is a positive number or a negative number:

1. $(-10) - 8$ **2.** $(-10) - (-8)$ **3.** $10 - (-8)$ **4.** $(-10) - 8$

In expressions 1 and 4, the second number is a positive 8. In expressions 2 and 3, the second number is a negative 8.

Rule for Subtracting Two Integers

To subtract two integers, add the opposite of the second integer to the first integer.

This rule states that to subtract two integers, we rewrite the subtraction expression as the sum of the first number and the opposite of the second number.

Here are some examples:

First number	−	second number	=	first number	+	the opposite of the second number	
8	−	15	=	8	+	(−15)	= −7
8	−	(−15)	=	8	+	15	= 23
(−8)	−	15	=	(−8)	+	(−15)	= −23
(−8)	−	(−15)	=	(−8)	+	15	= 7

➡ Subtract: $(-15) - 75$

$$(-15) - 75 = (-15) + (-75) = -90$$

• **To subtract, add the opposite of the second number to the first number.**

INSTRUCTOR NOTE
The difference between a minus sign and a negative sign is confusing. It may help to give a few more examples like the ones to the right. Another way to illustrate the difference between minus and difference is to use a calculator. Subtract $5 - (-3)$ by pressing 5 [−] [−] 3 [=]. The correct method is 5 [−] 3 [+/−] [=].

➡ Subtract: $27 - (-32)$

$$27 - (-32) = 27 + 32$$
$$= 59$$

• **To subtract, add the opposite of the second number to the first number.**

When subtraction occurs several times in an expression, rewrite each subtraction as addition of the opposite and then add.

➡ Subtract: $-13 - 5 - (-8)$

$$-13 - 5 - (-8) = -13 + (-5) + 8$$
$$= -18 + 8$$
$$= -10$$

• **Rewrite each subtraction as the addition of the opposite and then add.**

Example 3
Find 8 less than −12.

Solution
$-12 - 8 = -12 + (-8)$
$= -20$

You Try It 3
Find −8 less 14.

Your solution −22

Example 4
Subtract: $8 - 6 - (-20)$

Solution
$8 - 6 - (-20)$
$= 8 + (-6) + 20$
$= 2 + 20$
$= 22$

You Try It 4
Subtract: $3 - (-4) - 15$

Your solution −8

Example 5
Subtract:
$-8 - 30 - (-12) - 7$

Solution
$-8 - 30 - (-12) - 7$
$= -8 + (-30) + 12 + (-7)$
$= -38 + 12 + (-7)$
$= -26 + (-7)$
$= -33$

You Try It 5
Subtract:
$4 - (-3) - 12 - (-7) - 20$

Your solution −18

Solutions on p. A44

Objective C To solve application problems

Example 6
Find the temperature after an increase of 9°C from −6°C.

Strategy
To find the temperature, add the increase (9) to the previous temperature (−6).

Solution
$-6 + 9 = 3°C$

You Try It 6
Find the temperature after an increase of 12°C from −10°C.

Your strategy

Your solution
2°C

Solution on p. A44

10.2 Exercises

Objective A

Add.

1. $3 + (-5)$
-2

2. $-4 + 2$
-2

3. $8 + 12$
20

4. $16 + 23$
39

5. $-3 + (-8)$
-11

6. $-12 + (-1)$
-13

7. $-4 + (-5)$
-9

8. $-12 + (-12)$
-24

9. $6 + (-9)$
-3

10. $4 + (-9)$
-5

11. $-6 + 7$
1

12. $-12 + 6$
-6

13. $2 + (-3) + (-4)$
-5

14. $7 + (-2) + (-8)$
-3

15. $-3 + (-12) + (-15)$
-30

16. $9 + (-6) + (-16)$
-13

17. $-17 + (-3) + 29$
9

18. $13 + 62 + (-38)$
37

19. $-3 + (-8) + 12$
1

20. $-27 + (-42) + (-18)$
-87

21. $13 + (-22) + 4 + (-5)$
-10

22. $-14 + (-3) + 7 + (-6)$
-16

23. $-22 + 10 + 2 + (-18)$
-28

24. $-6 + (-8) + 13 + (-4)$
-5

25. $-16 + (-17) + (-18) + 10$
-41

26. $-25 + (-31) + 24 + 19$
-13

27. $-126 + (-247) + (-358) + 339$
-392

28. $-651 + (-239) + 524 + 487$
121

29. What is -8 more than -12?
-20

30. What is -5 more than 3?
-2

31. What is -7 added to -16?
-23

32. What is 7 added to -25?
-18

33. What is -4 plus 2?
-2

34. What is -22 plus -17?
-39

35. Find the sum of -2, 8, and -12.
-6

36. Find the sum of 4, -4, and -6.
-6

37. What is the total of 2, -3, 8, and -13?
-6

38. What is the total of -6, -8, 13, and -2?
-3

Objective B

Subtract.

39. $16 - 8$
8

40. $12 - 3$
9

41. $7 - 14$
-7

42. $6 - 9$
-3

43. $-7 - 2$
-9

44. $-9 - 4$
-13

45. $7 - (-2)$
9

46. $3 - (-4)$
7

47. $-6 - (-3)$
-3

48. $-4 - (-2)$
-2

49. $6 - (-12)$
18

50. $-12 - 16$
-28

51. $-4 - 3 - 2$
-9

52. $4 - 5 - 12$
-13

53. $12 - (-7) - 8$
11

54. $-12 - (-3) - (-15)$
6

55. $4 - 12 - (-8)$
0

56. $13 - 7 - 15$
-9

57. $-6 - (-8) - (-9)$
11

58. $7 - 8 - (-1)$
0

59. $-30 - (-65) - 29 - 4$
2

60. $42 - (-82) - 65 - 7$
52

61. $-16 - 47 - 63 - 12$
-138

62. $42 - (-30) - 65 - (-11)$
18

63. $-47 - (-67) - 13 - 15$
-8

64. $-18 - 49 - (-84) - 27$
-10

65. $167 - 432 - (-287) - 359$
-337

66. $-521 - (-350) - 164 - (-299)$
-36

67. Subtract -8 from -4.
4

68. Subtract -12 from 3.
15

69. Find 4 decreased by 2.
2

70. Find 4 decreased by 8.
-4

71. What is the difference between -8 and 4?
-12

72. What is the difference between 8 and -3?
11

73. What is -4 decreased by 8?
-12

74. What is -13 decreased by 9?
-22

75. Find -2 less than 1.
3

76. Find -3 less than -5.
-2

Objective C Application Problems

77. Find the temperature after a rise of 7°C from −8°C. −1°C

78. Find the temperature after a rise of 5°C from −19°C. −14°C

79. During a card game of Hearts, Nick had a score of 11 points before his opponent "shot the moon," subtracting a score of 26 from his total. What was Nick's score after his opponent shot the moon? −15 points

80. In a card game of Hearts, Monique had a score of −19 before she shot the moon, entitling her to add 26 points to her score. What was Monique's score after she shot the moon? 7 points

81. The average temperature on the sunlit side of the moon is approximately 215°F. On the dark side, it is approximately −250°F. Find the difference between the temperature on the sunlit side of the moon and on the dark side of the moon. 465°F

82. The average temperature throughout the earth's stratosphere is −70°F. The temperature on the earth's surface is 57°F. Find the difference between these average temperatures. 127°F

83. The high temperature for the day was 9°C. The low temperature was −8°C. Find the difference between the high and low temperatures for the day. 17°C

84. The low temperature for the day was −12°F. The high temperature was 21°F. Find the difference between the high and low temperatures for the day. 33°F

The elevation, or height, of places on earth is measured in relation to sea level, or the average level of the ocean's surface. The table below shows height above sea level as a positive number and depth below sea level as a negative number.

Place	*Elevation (in feet)*
Mt. Everest	29,028
Mt. Aconcagua	23,035
Mt. McKinley	20,320
Mt. Kilimanjaro	19,340
Salinas Grandes	−131
Death Valley	−282
Qattara Depression	−436
Dead Sea	−1286

85. Use the table to find the difference in elevation between Mt. McKinley and Death Valley. 20,602 ft

86. Use the table to find the difference in elevation between Mt. Kilimanjaro and the Qattara Depression (the highest and lowest points in Africa). 19,776 ft

87. Use the table to find the difference in elevation between Mt. Everest and the Dead Sea (the highest and lowest points in Asia). 30,314 ft

88. Use the table to find the difference in elevation between Mt. Aconcagua and Salinas Grandes (the highest and lowest points in South America). 23,166 ft

89. Use the table to find the difference in elevation between Mt. Everest and Mt. Kilimanjaro. 9688 ft

90. Use the table to find the difference in elevation between the Dead Sea and Death Valley. 1004 ft

APPLYING THE CONCEPTS

91. Consider the numbers 4, −7, −5, 13, and −9. Find the largest difference that can be obtained by subtracting one number in the list from a different number in the list. What is the smallest difference? 22, 2

92. Consider the list of numbers −2, 6, −12, −8, and 3. Find the largest difference that can be obtained by subtracting one number in the list from a different number in the list. What is the smallest difference? 18, 3

93. [W] Explain the Additive Inverse Property. Is the difference between a number and its additive inverse zero?

94. [W] Make up a word problem for which the answer is the difference between 15 and −3.

95. Fill in the blank squares at the right with integers so that the sum of the integers along any row, column, or diagonal is zero.

−3	2	1
4	0	−4
−1	−2	3

The graph at the right shows the previous record and new record low temperatures that selected cities attained in 1994. Use the graph for Exercises 96 and 97.

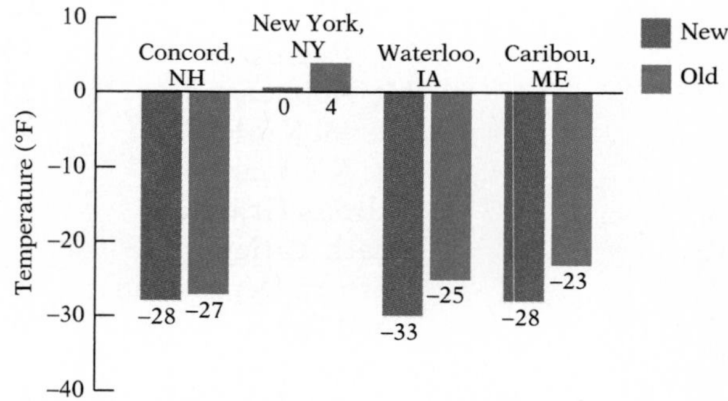

96. What is the difference between the new low temperature and the old low temperature for Waterloo, IA? −8°F

97. Which city experienced the smallest decrease in low temperature? Concord, NH

The graph at the right shows the average income per new car that car dealerships realized in 1989–1993. Use this graph for Exercises 98 and 99.

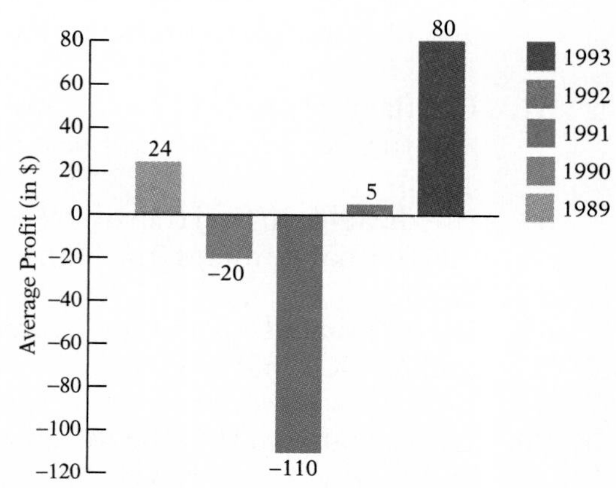

98. What was the difference between the average profit in 1989 and in 1990? $44

99. What was the difference between the average profit in 1993 and in 1991? $190

10.3 Multiplication and Division of Integers

Objective A To multiply integers

Multiplication is the repeated addition of the same number.

Several different symbols are used to indicate multiplication:

$3 \times 2 = 6$ $\qquad 3 \cdot 2 = 6$ $\qquad (3)(2) = 6$

When 5 is multiplied by a sequence of decreasing integers, each product decreases by 5.

$$5 \times 3 = 15$$
$$5 \times 2 = 10$$
$$5 \times 1 = 5$$
$$5 \times 0 = 0$$

INSTRUCTOR NOTE

Another way to suggest that positive times negative equals negative is to use repeated addition. For instance $(-5)(3)$ is $(-5) + (-5) + (-5) = (-15)$.

The pattern developed can be continued so that 5 is multiplied by a sequence of negative numbers. The resulting products must be negative in order to maintain the pattern of decreasing by 5.

$$5 \times (-1) = -5$$
$$5 \times (-2) = -10$$
$$5 \times (-3) = -15$$
$$5 \times (-4) = -20$$

This example illustrates that the product of a positive number and a negative number is negative.

When -5 is multiplied by a sequence of decreasing integers, each product increases by 5.

$$-5 \times 3 = -15$$
$$-5 \times 2 = -10$$
$$-5 \times 1 = -5$$
$$-5 \times 0 = 0$$

INSTRUCTOR NOTE

The idea that negative times negative equals positive seems arbitrary to students. Try asking them the meaning of the sentence, "It is not impossible to run a 4-minute mile."

The pattern developed can be continued so that -5 is multiplied by a sequence of negative numbers. The resulting products must be positive in order to maintain the pattern of increasing by 5.

$$-5 \times (-1) = 5$$
$$-5 \times (-2) = 10$$
$$-5 \times (-3) = 15$$
$$-5 \times (-4) = 20$$

This example illustrates that the product of two negative numbers is positive.

The pattern for multiplication shown above is summarized in the following rules for multiplying integers.

POINT OF INTEREST

Operations with negative numbers were not accepted until the late 13th century. One of the first attempts to prove that the product of two negative numbers is positive was done in the book *Ars Magna* by Girolamo Cardan in 1545.

Same Signs

To multiply numbers with the same sign, multiply the absolute values of the factors. The product is positive.

$$4 \cdot 8 = 32$$
$$(-4)(-8) = 32$$

Different Signs

To multiply numbers with different signs, multiply the absolute values of the factors. The product is negative.

$$-4 \cdot 8 = -32$$
$$(4)(-8) = -32$$

➡ Multiply: $2(-3)(-5)(-7)$

$2(-3)(-5)(-7) = -6 \cdot (-5)(-7)$
$= 30 \cdot (-7)$
$= -210$

- **To multiply more than two numbers, multiply the first two numbers.**
- **Then multiply the product by the third number.**
- **Continue until all the numbers have been multiplied.**

Example 1
Multiply: $(-2) \cdot 3 \cdot 6$

Solution
$(-2) \cdot 3 \cdot 6 = -6 \cdot 6$
$= -36$

You Try It 1
Multiply: $(-3) \cdot 4 \cdot (-5)$

Your solution 60

Example 2
Find the product of -42 and 62.

Solution
$-42 \cdot 62 = -2604$

You Try It 2
Find -38 multiplied by 51.

Your solution -1938

Example 3
Multiply: $-2 \cdot (-10) \cdot 7 \cdot 12$

Solution
$-2 \cdot (-10) \cdot 7 \cdot 12 = 20 \cdot 7 \cdot 12$
$= 140 \cdot 12$
$= 1680$

You Try It 3
Multiply: $-6 \cdot 8 \cdot (-11) \cdot 3$

Your solution 1584

Example 4
Multiply: $-5(-4)(6)(-3)$

Solution
$-5(-4)(6)(-3) = 20(6)(-3)$
$= 120(-3)$
$= -360$

You Try It 4
Multiply: $-7(-8)(9)(-2)$

Your solution -1008

Solutions on p. A44

Objective B To divide integers

For every division problem there is a related multiplication problem.

Division: $\frac{8}{2} = 4$ Related multiplication: $4 \cdot 2 = 8$

This fact can be used to illustrate the rules for dividing signed numbers.

Same Signs

The quotient of two numbers with the same sign is positive.

$\frac{12}{3} = 4$ because $4 \cdot 3 = 12$

$\frac{-12}{-3} = 4$ because $4(-3) = -12$

Different Signs

The quotient of two numbers with different signs is negative.

$\frac{12}{-3} = -4$ because $-4(-3) = 12$

$\frac{-12}{3} = -4$ because $-4 \cdot 3 = -12$

Note that $\frac{12}{-3}$, $\frac{-12}{3}$, and $-\frac{12}{3}$ are all equal to -4.

If a and b are two integers, then $\frac{a}{-b} = \frac{-a}{b} = -\frac{a}{b}$.

INSTRUCTOR NOTE

This is a good time to remind students that division by zero is not allowed.

Properties of Zero and One in Division

Zero divided by any number other than zero is zero.

Division by zero is not defined.

Any number other than zero divided by itself is 1.

$\frac{0}{a} = 0$ because $0 \cdot a = 0$

$\frac{4}{0} = ?$ $? \times 0 = 4$ There is no number whose product with zero is 4.

$\frac{a}{a} = 1$ because $1 \cdot a = a$

Example 5	Divide: $(-120) \div (-8)$	**You Try It 5**	Divide: $(-135) \div (-9)$
Solution	$(-120) \div (-8) = 15$	**Your solution**	15
Example 6	Find the quotient of -81 and 3.	**You Try It 6**	What is -72 divided by 4?
Solution	$-81 \div 3 = -27$	**Your solution**	-18
Example 7	Divide: $95 \div (-5)$	**You Try It 7**	Divide: $84 \div (-6)$
Solution	$95 \div (-5) = -19$	**Your solution**	-14

Solutions on p. A45

Objective C To solve application problems

Example 8

The combined scores of the top five golfers in a tournament equaled −10 (10 under par). What was the average score of the five golfers?

Strategy

To find the average score, divide the combined scores (−10) by the number of golfers (5).

Solution

$-10 \div 5 = -2$

The average score was −2.

You Try It 8

The melting point of mercury is −38°C. The melting point of argon is five times the melting point of mercury. Find the melting point of argon.

Your strategy

Your solution

−190°C

Example 9

The daily high temperatures during one week were recorded as follows: −9°, 3°, 0°, −8°, 2°, 1°, 4°. Find the average daily high temperature for the week.

Strategy

To find the average daily high temperature:

- Add the seven temperature readings.
- Divide by 7.

Solution

$-9 + 3 + 0 + (-8) + 2 + 1 + 4 = -7$

$-7 \div 7 = -1$

The average daily high temperature was −1°.

You Try It 9

The daily low temperatures during one week were recorded as follows: −6°, −7°, 1°, 0°, −5°, −10°, −1°. Find the average daily low temperature for the week.

Your strategy

Your solution

−4°

Solutions on p. A45

10.3 Exercises

Objective A

Multiply.

1. 14×3
42

2. 62×9
558

3. $-4 \cdot 6$
-24

4. $-7 \cdot 3$
-21

5. $-2 \cdot (-3)$
6

6. $-5 \cdot (-1)$
5

7. $(9)(2)$
18

8. $(3)(8)$
24

9. $5(-4)$
-20

10. $4(-7)$
-28

11. $-8(2)$
-16

12. $-9(3)$
-27

13. $(-5)(-5)$
25

14. $(-3)(-6)$
18

15. $(-7)(0)$
0

16. -32×4
-128

17. -24×3
-72

18. $19 \cdot (-7)$
-133

19. $6(-17)$
-102

20. $-8(-26)$
208

21. $-4(-35)$
140

22. $-5 \cdot (23)$
-115

23. $-6 \cdot (38)$
-228

24. $9(-27)$
-243

25. $8(-40)$
-320

26. $-7(-34)$
238

27. $-4(39)$
-156

28. $4 \cdot (-8) \cdot 3$
-96

29. $5 \times 7 \times (-2)$
-70

30. $8 \cdot (-6) \cdot (-1)$
48

Multiply.

31. $(-9)(-9)(2)$
162

32. $-8(-7)(-4)$
-224

33. $-5(8)(-3)$
120

34. $(-6)(5)(7)$
-210

35. $-1(4)(-9)$
36

36. $6(-3)(-2)$
36

37. $4(-4) \cdot 6(-2)$
192

38. $-5 \cdot 9(-7) \cdot 3$
945

39. $-9(4) \cdot 3(1)$
-108

40. $8(8)(-5)(-4)$
1280

41. $(-6) \cdot 7 \cdot (-10)(-5)$
-2100

42. $-9(-6)(11)(-2)$
-1188

43. What is -5 multiplied by -4?
20

44. What is 6 multiplied by -5?
-30

45. What is -8 times 6?
-48

46. What is -8 times -7?
56

47. Find the product of -4, 7, and -5.
140

48. Find the product of -2, -4, and -7.
-56

Objective B

Divide.

49. $12 \div (-6)$
-2

50. $18 \div (-3)$
-6

51. $(-72) \div (-9)$
8

52. $(-64) \div (-8)$
8

53. $0 \div (-6)$
0

54. $-49 \div 7$
-7

55. $45 \div (-5)$
-9

56. $-24 \div 4$
-6

57. $-36 \div 4$
-9

58. $-56 \div 7$
-8

59. $-81 \div (-9)$
9

60. $-40 \div (-5)$
8

61. $72 \div (-3)$
-24

62. $44 \div (-4)$
-11

63. $-60 \div 5$
-12

64. $-66 \div 6$
-11

65. $-93 \div (-3)$
31

66. $-98 \div (-7)$
14

67. $(-85) \div (-5)$
17

68. $(-60) \div (-4)$
15

69. $120 \div 8$
15

Divide.

70. $144 \div 9$
16

71. $78 \div (-6)$
-13

72. $84 \div (-7)$
-12

73. $-72 \div 4$
-18

74. $-80 \div 5$
-16

75. $-114 \div (-6)$
19

76. $-91 \div (-7)$
13

77. $-104 \div (-8)$
13

78. $-126 \div (-9)$
14

79. $57 \div (-3)$
-19

80. $162 \div (-9)$
-18

81. $-136 \div (-8)$
17

82. $-128 \div 4$
-32

83. $-130 \div (-5)$
26

84. $(-280) \div 8$
-35

85. $(-92) \div (-4)$
23

86. $-196 \div (-7)$
28

87. $-150 \div (-6)$
25

88. $(-261) \div 9$
-29

89. $204 \div (-6)$
-34

90. $165 \div (-5)$
-33

91. $-132 \div (-12)$
11

92. $-156 \div (-13)$
12

93. $-182 \div 14$
-13

94. $-144 \div 12$
-12

95. $143 \div 11$
13

96. $168 \div 14$
12

97. $-180 \div (-15)$
12

98. $-169 \div (-13)$
13

99. $154 \div (-11)$
-14

100. Find the quotient of -132 and -11.
12

101. Find the quotient of 182 and -13.
-14

102. What is -60 divided by -15?
4

103. What is 144 divided by -24?
-6

104. Find the quotient of -135 and 15.
-9

105. Find the quotient of -88 and 22.
-4

Objective C *Application Problems*

106. The daily low temperatures during one week were recorded as follows: 5°, −4°, 9°, 0°, −11°, −13°, −7°. Find the average daily low temperature for the week. −3°

107. The daily high temperatures during one week were recorded as follows: −7°, −10°, 4°, 6°, −2°, −8°, −4°. Find the average daily high temperature for the week. −3°

108. The boiling point of radon is −62°C. The boiling point of argon is three times the boiling point of radon. Find the boiling point of argon. −186°C

109. The boiling point of chlorine is −35°C. The boiling point of neon is seven times the boiling point of chlorine. Find the boiling point of neon. −245°C

110. The combined scores of the top four golfers in a tournament equaled −12 (12 under par). What was the average score of the four golfers? −3

111. The combined scores of the top four golfers in a tournament equaled −18 (18 under par). What was the average score of the four golfers? −4.5

112. The wind chill factor when the temperature is −15°F and the wind is blowing at 20 mph is five times the wind chill factor when the temperature is 25°F and the wind is blowing at 35 mph. If the wind chill factor at 25° with a 35-mph wind is −12°F, what is the wind chill factor at −15° with a 20-mph wind? −60°F

APPLYING THE CONCEPTS

113. Determine whether the statement is always true, sometimes true, or never true.

a. The product of a nonzero number and its opposite is a negative number. True

b. The product of an even number of negative numbers is a negative number. Never true

114. Find the largest possible product of two negative integers whose sum is −10. 25

115. Find the smallest possible sum of two negative numbers whose product is 16. −17

116. Use repeated addition to show that the product of two integers with different signs is a negative number.

117. [W] In your own words, describe the rules for multiplying and dividing integers. Answers will vary

118. To discourage guessing on a multiple-choice exam, Professor Perez graded the test by giving 5 points for a correct answer, −2 points for an answer left blank, and −5 points for an incorrect answer. How many points did a student score who answered 20 questions correctly, 5 questions incorrectly, and left 2 questions blank? 71

119. To discourage guessing on a multiple choice exam, Professor Lam graded the test by giving 4 points for a correct answer, −1 point for an answer left blank and −3 points for an incorrect answer. How many points did a student score who answered 38 questions correctly, answered 4 questions incorrectly, and left 8 questions blank? 132

10.4 Operations with Rational Numbers

Objective A To add or subtract rational numbers

In this section, operations with *rational numbers* will be discussed. The rational numbers include all the positive and negative fractions.

INSTRUCTOR NOTE

Once rational numbers have been defined, ask students to classify numbers as a natural number, integer, or rational. Possible examples are -3, 7, 0, $\frac{2}{7}$, and $-\frac{10}{3}$.

> **Rational Numbers**
>
> The **rational numbers** are the numbers that can be written as the ratio of two integers where the denominator is not zero.

Examples of rational numbers are $\frac{-4}{7}$ and $\frac{9}{2}$. Because $4 = \frac{4}{1}$ (the ratio of two integers), 4 is a rational number. The rational numbers include all the integers.

To add or subtract rational numbers in fractional form, first find the least common multiple (LCM) of the denominators.

➡ Simplify: $-\frac{7}{8} + \frac{5}{6}$

Find the LCM of the denominators. $8 = 2 \cdot 2 \cdot 2$; $6 = 2 \cdot 3$
LCM $= 2 \cdot 2 \cdot 2 \cdot 3 = 24$

Rewrite each fraction, using the LCM of the denominators as the common denominator. Add the numerators.

$$-\frac{7}{8} + \frac{5}{6} = -\frac{21}{24} + \frac{20}{24} = \frac{-21 + 20}{24} = \frac{-1}{24} = -\frac{1}{24}$$

➡ Simplify: $-\frac{7}{9} - \frac{5}{12}$

$9 = 3 \cdot 3$; $12 = 2 \cdot 2 \cdot 3$; LCM $= 2 \cdot 2 \cdot 3 \cdot 3 = 36$

$$-\frac{7}{9} - \frac{5}{12} = -\frac{28}{36} - \frac{15}{36} = \frac{(-28)}{36} + \frac{(-15)}{36} = \frac{(-28) + (-15)}{36} = \frac{-43}{36} = -\frac{43}{36}$$
$$= -1\frac{7}{36}$$

Because numbers written as terminating decimals can be expressed as the ratio of two integers, terminating decimals are also rational numbers. For example,

$$0.75 = \frac{75}{100} = \frac{3}{4} \quad \text{and} \quad 6.3 = 6\frac{3}{10} = \frac{63}{10}$$

To add or subtract decimals, write the numbers so that the decimal points are in a vertical line. Then use the rules for adding integers. Write the decimal point in the answer directly below the decimal points in the problem.

➡ Simplify: $47.034 + (-56.91)$

$$\begin{array}{r} 56.91 \\ -\ 47.034 \\ \hline 9.876 \end{array}$$

- Because the signs are different, find the difference between the absolute values of the numbers.

$47.034 - 56.91 = -9.876$

- Attach the sign of the number with the greater absolute value.

➡ Simplify: $-39.09 - 102.98$

$-39.09 - 102.98 = -39.09 + (-102.98)$

• To subtract two numbers, add the opposite of the second number.

$$\begin{array}{r} 39.09 \\ +\ 102.98 \\ \hline 142.07 \end{array}$$

• Because the signs are the same, find the sum of the absolute values of the numbers.

$-39.09 - 102.98 = -142.07$

• Attach the common sign.

Example 1

Simplify: $\frac{5}{16} - \frac{7}{40}$

Solution

The LCM of 16 and 40 is 80.

$$\frac{5}{16} - \frac{7}{40} = \frac{25}{80} - \frac{14}{80} = \frac{25}{80} + \frac{-14}{80} = \frac{25 + (-14)}{80} = \frac{11}{80}$$

You Try It 1

Simplify: $\frac{5}{9} - \frac{11}{12}$

Your solution $-\frac{13}{36}$

Example 2

Simplify: $-\frac{3}{4} + \frac{1}{6} - \frac{5}{8}$

Solution

The LCM of 4, 6, and 8 is 24.

$$-\frac{3}{4} + \frac{1}{6} - \frac{5}{8} = -\frac{18}{24} + \frac{4}{24} - \frac{15}{24} = \frac{-18}{24} + \frac{4}{24} + \frac{-15}{24} = \frac{-18 + 4 + (-15)}{24} = \frac{-29}{24} = -1\frac{5}{24}$$

You Try It 2

Simplify: $-\frac{7}{8} - \frac{5}{6} + \frac{2}{3}$

Your solution $-1\frac{1}{24}$

Example 3

Simplify: $42.987 - 98.61$

Solution

$$\begin{array}{r} 98.61 \\ -\ 42.987 \\ \hline 55.623 \end{array}$$

$42.987 - 98.61 = -55.623$

You Try It 3

Simplify: $16.127 - 67.91$

Your solution -51.783

Example 4

Simplify:

$1.02 + (-3.6) + 9.24$

Solution

$1.02 + (-3.6) + 9.24 = -2.58 + 9.24 = 6.66$

You Try It 4

Simplify:

$2.7 + (-9.44) + 6.2$

Your solution -0.54

Solutions on p. A45

Objective B To multiply and divide rational numbers

The product of two rational numbers written as fractions is the product of the numerators over the product of the denominators. Use the sign rules for multiplying integers.

➡ Simplify: $-\frac{3}{8} \times \frac{12}{17}$

$$-\frac{3}{8} \times \frac{12}{17} = -\left(\frac{3 \cdot 12}{8 \cdot 17}\right) = -\frac{36}{136} = -\frac{9}{34}$$

• The signs are different, the product is negative.

To divide rational numbers written as fractions, invert the divisor and then multiply.

➡ Simplify: $-\frac{3}{10} \div \left(-\frac{18}{25}\right)$

$$-\frac{3}{10} \div \left(-\frac{18}{25}\right) = \frac{3}{10} \times \frac{25}{18} = \frac{3 \cdot 25}{10 \cdot 18} = \frac{75}{180} = \frac{5}{12}$$

• The signs are the same, the quotient is positive.

To multiply or divide rational numbers written in decimal form, use the sign rules for integers.

➡ Simplify: $(-6.89) \times (-0.00035)$

6.89	2 decimal places	• The signs are the same. Multiply the absolute values.
× 0.00035	5 decimal places	
3445		
2067		
0.0024115	7 decimal places	

$(-6.89) \times (-0.00035) = 0.0024115$ The product is positive.

➡ Simplify: $1.32 \div (-0.27)$. Round to the nearest tenth.

```
       4.88   ≈   4.9
.27.)1.32.00
    -1 08
       24 0
      -21 6
        2 40
       -2 16
          24
```

• Divide the absolute values. Move the decimal point two places in the divisor and then in the dividend. Place the decimal point in the quotient.

The signs are different. The quotient is negative.

$1.32 \div (-0.27) \approx -4.9$

Example 5 Simplify: $-\frac{7}{12} \times \frac{9}{14}$

Solution The product is negative.

$$-\frac{7}{12} \times \frac{9}{14} = -\left(\frac{7 \cdot 9}{12 \cdot 14}\right) = -\frac{63}{168} = -\frac{3}{8}$$

You Try It 5 Simplify: $\left(-\frac{2}{3}\right)\left(-\frac{9}{10}\right)$

Your solution $\frac{3}{5}$

Example 6 Simplify: $-\frac{3}{8} \div \left(-\frac{5}{12}\right)$

Solution The quotient is positive.

$$-\frac{3}{8} \div \left(-\frac{5}{12}\right) = \frac{3}{8} \times \frac{12}{5} = \frac{3 \cdot 12}{8 \cdot 5} = \frac{36}{40} = \frac{9}{10}$$

You Try It 6 Simplify: $-\frac{5}{8} \div \frac{5}{40}$

Your solution -5

Example 7 Simplify: -4.29×8.2

Solution The product is negative.

```
   4.29
  × 8.2
  -----
    858
  3432
  -----
 35.178
```

$-4.29 \times 8.2 = -35.178$

You Try It 7 Simplify: -5.44×3.8

Your solution -20.672

Example 8 Simplify: $-3.2 \times (-0.4) \times 6.9$

Solution $-3.2 \times (-0.4) \times 6.9$
$= 1.28 \times 6.9$
$= 8.832$

You Try It 8 Simplify: $3.44 \times (-1.7) \times 0.6$

Your solution -3.5088

Example 9 Simplify: $-0.0792 \div (-0.42)$ Round to the nearest hundredth.

Solution

```
            0.188 ≈ 0.19
0.42.)0.07.920
        -4 2
         3 72
        -3 36
           360
          -336
            24
```

$-0.0792 \div (-0.42) \approx 0.19$

You Try It 9 Simplify: $-0.394 \div 1.7$ Round to the nearest hundredth.

Your solution -0.23

Solutions on p. A46

10.4 Exercises

Objective A

Simplify.

1. $\frac{2}{3} + \frac{5}{12}$

$\frac{13}{12} = 1\frac{1}{12}$

2. $\frac{1}{2} + \frac{3}{8}$

$\frac{7}{8}$

3. $\frac{5}{8} - \frac{5}{6}$

$-\frac{5}{24}$

4. $\frac{1}{9} - \frac{5}{27}$

$-\frac{2}{27}$

5. $-\frac{5}{12} - \frac{3}{8}$

$-\frac{19}{24}$

6. $-\frac{5}{6} - \frac{5}{9}$

$-\frac{25}{18} = -1\frac{7}{18}$

7. $-\frac{6}{13} + \frac{17}{26}$

$\frac{5}{26}$

8. $-\frac{7}{12} + \frac{5}{8}$

$\frac{1}{24}$

9. $-\frac{5}{8} - \left(-\frac{11}{12}\right)$

$\frac{7}{24}$

10. $-\frac{3}{4} - \frac{5}{8}$

$-\frac{11}{8} = -1\frac{3}{8}$

11. $-\frac{7}{12} - \left(-\frac{7}{8}\right)$

$\frac{7}{24}$

12. $\frac{5}{12} - \frac{11}{15}$

$-\frac{19}{60}$

13. $\frac{2}{5} - \frac{14}{15}$

$-\frac{8}{15}$

14. $-\frac{2}{3} - \frac{5}{8}$

$-\frac{31}{24} = -1\frac{7}{24}$

15. $-\frac{3}{4} - \left(-\frac{2}{3}\right)$

$-\frac{1}{12}$

16. $\frac{3}{4} - \frac{3}{7}$

$\frac{9}{28}$

17. $-\frac{5}{2} - \left(-\frac{13}{4}\right)$

$\frac{3}{4}$

18. $-\frac{7}{3} - \left(-\frac{3}{2}\right)$

$-\frac{5}{6}$

19. $\frac{1}{3} + \frac{5}{6} - \frac{2}{9}$

$\frac{17}{18}$

20. $\frac{1}{2} - \frac{2}{3} + \frac{1}{6}$

0

21. $-\frac{3}{8} - \frac{5}{12} - \frac{3}{16}$

$-\frac{47}{48}$

22. $-\frac{5}{16} + \frac{3}{4} - \frac{7}{8}$

$-\frac{7}{16}$

23. $\frac{1}{2} - \frac{3}{8} - \left(-\frac{1}{4}\right)$

$\frac{3}{8}$

24. $\frac{3}{4} - \left(-\frac{7}{12}\right) - \frac{7}{8}$

$\frac{11}{24}$

25. $\frac{1}{3} - \frac{1}{4} - \frac{1}{5}$

$-\frac{7}{60}$

26. $\frac{2}{3} - \frac{1}{2} + \frac{5}{6}$

1

27. $\frac{5}{16} + \frac{1}{8} - \frac{1}{2}$

$-\frac{1}{16}$

Simplify.

28. $\frac{5}{8} - \left(-\frac{5}{12}\right) + \frac{1}{3}$
$\frac{11}{8} = 1\frac{3}{8}$

29. $\frac{1}{8} - \frac{11}{12} + \frac{1}{2}$
$-\frac{7}{24}$

30. $-\frac{7}{9} + \frac{14}{15} + \frac{8}{21}$
$\frac{169}{315}$

31. $\frac{1}{2} + \left(-\frac{3}{8}\right) + \frac{5}{12}$
$\frac{13}{24}$

32. $-\frac{3}{8} + \frac{3}{4} - \left(-\frac{3}{16}\right)$
$\frac{9}{16}$

33. $\frac{1}{3} - \frac{3}{5} - \frac{5}{6}$
$-\frac{11}{10} = -1\frac{1}{10}$

34. $3.4 + (-6.8)$
-3.4

35. $-4.9 + 3.27$
-1.63

36. $-8.32 + (-0.57)$
-8.89

37. $-3.5 + 7$
3.5

38. $-4.8 + (-3.2)$
-8.0

39. $6.2 + (-4.29)$
1.91

40. $-4.6 + 3.92$
-0.68

41. $7.2 + (-8.42)$
-1.22

42. $3.09 + 6.025$
9.115

43. $-45.71 + (-135.8)$
-181.51

44. $124.09 + (-67.5)$
56.59

45. $-35.274 + 12.47$
-22.804

46. $4.2 + (-6.8) + 5.3$
2.7

47. $6.7 + 3.2 + (-10.5)$
-0.6

48. $-4.5 + 3.2 + (-19.4)$
-20.7

49. $2.09 - 6.72 - 5.4$
-10.03

50. $16.4 + 3.09 - 7.93$
11.56

51. $-18.39 + 4.9 - 23.7$
-37.19

52. $19 - (-3.72) - 82.75$
-60.03

53. $-3.07 - (-2.97) - 17.4$
-17.5

54. $-3.09 - 4.6 - 27.3$
-34.99

55. $-3.89 + (-2.9) + 4.723 + 0.2$
-1.867

56. $4.207 + 6.91 + (-3.825) + (-10.04)$
-2.748

57. $-3.005 + (-3.925) + (-6.002)$
-12.932

58. $-4.02 + 6.809 - (-3.57) - (-0.419)$
6.778

59. $0.0153 + (-1.0294) + (-1.0726)$
-2.0867

60. $0.27 + (-3.5) - (-0.27) + (-5.44)$
-8.4

Objective B

Simplify.

61. $\frac{1}{2} \times \left(-\frac{3}{4}\right)$
$-\frac{3}{8}$

62. $-\frac{2}{9} \times \left(-\frac{3}{14}\right)$
$\frac{1}{21}$

63. $\left(-\frac{3}{8}\right)\left(-\frac{4}{15}\right)$
$\frac{1}{10}$

64. $\left(-\frac{3}{4}\right)\left(-\frac{8}{27}\right)$
$\frac{2}{9}$

65. $-\frac{1}{2} \times \frac{8}{9}$
$-\frac{4}{9}$

66. $\frac{5}{12} \times \left(-\frac{8}{15}\right)$
$-\frac{2}{9}$

67. $\frac{5}{8} \times \left(-\frac{7}{12}\right) \times \frac{16}{25}$
$-\frac{7}{30}$

68. $\left(\frac{1}{2}\right)\left(-\frac{3}{4}\right)\left(-\frac{5}{8}\right)$
$\frac{15}{64}$

69. $\left(\frac{5}{12}\right)\left(-\frac{8}{15}\right)\left(-\frac{1}{3}\right)$
$\frac{2}{27}$

70. $\left(-\frac{5}{12}\right)\left(\frac{42}{65}\right)$
$-\frac{7}{26}$

71. $\left(\frac{3}{8}\right)\left(-\frac{15}{41}\right)$
$-\frac{45}{328}$

72. $\left(\frac{2}{9}\right)\left(-\frac{15}{17}\right)$
$-\frac{10}{51}$

73. $\left(-\frac{15}{8}\right)\left(-\frac{16}{3}\right)$
10

74. $\left(-\frac{5}{7}\right)\left(-\frac{14}{15}\right)$
$\frac{2}{3}$

75. $\left(\frac{3}{8}\right)\left(\frac{15}{41}\right)$
$\frac{45}{328}$

76. $\frac{1}{3} \div \left(-\frac{1}{2}\right)$
$-\frac{2}{3}$

77. $-\frac{3}{8} \div \frac{7}{8}$
$-\frac{3}{7}$

78. $\left(-\frac{3}{4}\right) \div \left(-\frac{7}{40}\right)$
$\frac{30}{7} = 4\frac{2}{7}$

79. $\frac{3}{8} \div \frac{1}{4}$
$\frac{3}{2} = 1\frac{1}{2}$

80. $\frac{5}{6} \div \left(-\frac{3}{4}\right)$
$-\frac{10}{9} = -1\frac{1}{9}$

81. $-\frac{5}{12} \div \frac{15}{32}$
$-\frac{8}{9}$

82. $-\frac{7}{8} \div \frac{4}{21}$
$-\frac{147}{32} = -4\frac{19}{32}$

83. $\frac{7}{10} \div \frac{2}{5}$
$\frac{7}{4} = 1\frac{3}{4}$

84. $-\frac{15}{64} \div \left(-\frac{3}{40}\right)$
$\frac{25}{8} = 3\frac{1}{8}$

85. $-\frac{5}{16} \div \left(-\frac{3}{8}\right)$
$\frac{5}{6}$

86. $\frac{5}{9} \div \left(-\frac{15}{32}\right)$
$-\frac{32}{27} = -1\frac{5}{27}$

87. $\left(-\frac{8}{19}\right) \div \frac{7}{38}$
$-\frac{16}{7} = -2\frac{2}{7}$

88. $\left(-\frac{2}{3}\right) \div 4$
$-\frac{1}{6}$

89. $(-6) \div \frac{4}{9}$
$-\frac{27}{2} = -13\frac{1}{2}$

90. $\left(-\frac{3}{8}\right) \div \left(-\frac{5}{12}\right)$
$\frac{9}{10}$

91. 4.2×0.9
3.78

92. $-6.7 \times (-4.2)$
28.14

93. $-8.9 \times (-3.5)$
31.15

94. -1.6×4.9
-7.84

95. 7.8×9.6
74.88

96. -14.3×7.9
-112.97

97. $(-0.78)(-0.15)$
0.117

98. $(0.49)(-3.9)$
-1.911

99. $(-1.21)(-0.03)$
0.0363

Simplify.

100. $(-8.919) \div (-0.9)$
9.91

101. $-77.6 \div (-0.8)$
97

102. $59.01 \div (-0.7)$
-84.3

103. $(-7.04) \div (-3.2)$
2.2

104. $(-84.66) \div 1.7$
-49.8

105. $-3.312 \div (0.8)$
-4.14

106. $-0.1314 \div (-0.018)$
7.3

107. $1.003 \div (-0.59)$
-1.7

108. $26.22 \div (-6.9)$
-3.8

Divide. Round to the nearest hundredth.

109. $(-19.08) \div 0.45$
-42.40

110. $21.792 \div (-0.96)$
-22.70

111. $(-38.665) \div (-9.5)$
4.07

112. $(-3.171) \div (-45.3)$
0.07

113. $27.738 \div (-60.3)$
-0.46

114. $(-13.97) \div (-25.4)$
0.55

APPLYING THE CONCEPTS

115. Determine whether the statement is always true, sometimes true, or never true.
a. The product of an odd number of negative numbers is a negative number. Always true
b. The square of a negative number is a positive number. Always true

116. Determine whether the statement is true or false.
a. Every integer is a rational number. True
b. Every whole number is an integer. True
c. Every integer is a positive number. False
d. Every rational number is an integer. False

117. The whole numbers are said to be *closed* with respect to addition because when two whole numbers are added, the result is a whole number. The whole numbers are not closed with respect to subtraction because, for example, 4 and 7 are whole numbers but $4 - 7 = -3$, and -3 is not a whole number. Complete the table below by entering a Y if the operation is closed for those numbers and an N if it is not closed. When we discuss whether multiplication and division are closed, zero is not included because division by zero is not defined.

	Add	*Subtract*	*Multiply*	*Divide*
Whole numbers	Y	N	Y	N
Integers	Y	Y	Y	N
Rational numbers	Y	Y	Y	Y

118. Find a rational number between $-\frac{3}{4}$ and $-\frac{2}{3}$. $-\frac{17}{24}$ (Other answers are possible)

119. [W] Given any two different rational numbers, is it always possible to find a rational number between them? If so, explain how. If not, give an example of two different rational numbers for which there is no rational number between them. Answers will vary

10.5 The Order of Operations Agreement

Objective A ***To use the Order of Operations Agreement to simplify expressions***

INSTRUCTOR NOTE
The exercises in this objective ask students to recall the Order of Operations Agreement and to practice a combination of operations with rational numbers.

The Order of Operations Agreement has been used throughout this textbook. In simplifying expressions with rational numbers, the same Order of Operations Agreement is used. This agreement is restated here.

The Order of Operations Agreement

Step 1 Do all operations inside parentheses.

Step 2 Simplify any expressions containing exponents.

Step 3 Do multiplication and division as they occur from left to right.

Step 4 Rewrite subtraction as the addition of the opposite. Then do additions as they occur from left to right.

Exponents may be confusing in expressions with signed numbers.

$(-3)^2 = (-3) \times (-3) = 9$

$-3^2 = -(3)^2 = -(3 \times 3) = -9$

Note that −3 is squared only when the negative sign is *inside* the parentheses.

POINT OF INTEREST
The first woman mathematician for which documented evidence exists is Hypatia (370–415). She lived in Alexandria, Egypt, and lectured at the Museum, the forerunner of our modern university. She made important contributions in Mathematics, Astronomy, and Philosophy.

➡ Simplify: $(-3)^2 - 2 \times (8 - 3) + (-5)$

$(-3)^2 - 2 \times \underbrace{(8 - 3)} + (-5)$ — 1. Perform operations inside parentheses.

$\underbrace{(-3)^2} - 2 \times 5 + (-5)$ — 2. Simplify expressions with exponents.

$9 - \underbrace{2 \times 5} + (-5)$ — 3. Do multiplications and divisions as they occur from left to right.

$9 - 10 + (-5)$

$\underbrace{9 + (-10)} + (-5)$ — 4. Rewrite subtraction as the addition of the opposite. Then add from left to right.

$\underbrace{(-1) + (-5)}$

-6

➡ Simplify: $\left(\frac{1}{4} - \frac{1}{2}\right)^2 \div \frac{3}{8}$

$\underbrace{\left(\frac{1}{4} - \frac{1}{2}\right)}^2 \div \frac{3}{8}$ — 1. Perform operations inside parentheses.

$\underbrace{\left(-\frac{1}{4}\right)^2} \div \frac{3}{8}$ — 2. Simplify expressions with exponents.

$\frac{1}{16} \div \frac{3}{8}$ — 3. Do multiplication and division as they occur from left to right.

$\underbrace{\frac{1}{16} \times \frac{8}{3}}$

$\frac{1}{6}$

Example 1 Simplify: $8 - 4 \div (-2)$

Solution
$8 - 4 \div (-2)$
$8 - (-2)$
$8 + 2$
10

You Try It 1 Simplify: $9 - 9 \div (-3)$

Your solution 12

Example 2 Simplify: $12 \div (-2)^2 + 5$

Solution
$12 \div (-2)^2 + 5$
$12 \div 4 + 5$
$3 + 5$
8

You Try It 2 Simplify: $8 \div 4 \cdot 4 - (-2)^2$

Your solution 4

Example 3 Simplify: $12 - (-10) \div (8 - 3)$

Solution
$12 - (-10) \div (8 - 3)$
$12 - (-10) \div 5$
$12 - (-2)$
$12 + 2$
14

You Try It 3 Simplify: $8 - (-15) \div (2 - 7)$

Your solution 5

Example 4 Simplify:
$(-3)^2 \times (5 - 7)^2 - (-9) \div 3$

Solution
$(-3)^2 \times (5 - 7)^2 - (-9) \div 3$
$(-3)^2 \times (-2)^2 - (-9) \div 3$
$9 \times 4 - (-9) \div 3$
$36 - (-9) \div 3$
$36 - (-3)$
$36 + 3$
39

You Try It 4 Simplify:
$(-2)^2 \times (3 - 7)^2 - (-16) \div (-4)$

Your solution 60

Example 5 Simplify: $3 \div \left(\frac{1}{2} - \frac{1}{4}\right) - 3$

Solution
$3 \div \left(\frac{1}{2} - \frac{1}{4}\right) - 3$
$3 \div \frac{1}{4} - 3$
$\frac{3}{1} \times \frac{4}{1} - 3$
$12 - 3$
$12 + (-3)$
9

You Try It 5 Simplify: $7 \div \left(\frac{1}{7} - \frac{3}{14}\right) - 9$

Your solution -107

Solutions on p. A46

10.5 Exercises

Objective A

Simplify.

1. $8 \div 4 + 2$
   4

2. $3 - 12 \div 2$
   −3

3. $4 + (-7) + 3$
   0

4. $-16 \div 2 + 8$
   0

5. $4^2 - 4$
   12

6. $6 - 2^2$
   2

7. $2 \times (3 - 5) - 2$
   −6

8. $2 - (8 - 10) \div 2$
   3

9. $4 - (-3)^2$
   −5

10. $(-2)^2 - 6$
    −2

11. $4 - (-3) - 5$
    2

12. $6 + (-8) - (-3)$
    1

13. $4 - (-2)^2 + (-3)$
    −3

14. $-3 + (-6)^2 - 1$
    32

15. $3^2 - 4 \times 2$
    1

16. $9 \div 3 - (-3)^2$
    −6

17. $3 \times (6 - 2) \div 6$
    2

18. $4 \times (2 - 7) \div 5$
    −4

19. $2^2 - (-3)^2 + 2$
    −3

20. $3 \times (8 - 5) + 4$
    13

21. $6 - 2 \times (1 - 5)$
    14

22. $4 \times 2 \times (3 - 6)$
    −24

23. $(-2)^2 - (-3)^2 + 1$
    −4

24. $4^2 - 3^2 - 4$
    3

25. $6 - (-3) \times (-3)^2$
    33

26. $4 - (-5) \times (-2)^2$
    24

27. $4 \times 2 - 3 \times 7$
    −13

28. $16 \div 2 - 9 \div 3$
    5

29. $(-2)^2 - 5 \times 3 - 1$
    −12

30. $4 - 2 \times 7 - 3^2$
    −19

31. $7 \times 6 - 5 \times 6 + 3 \times 2 - 2 + 1$
    17

32. $3 \times 2^2 + 5 \times (3 + 2) - 17$
    20

33. $-4 \times 3 \times (-2) + 12 \times (3 - 4) + (-12)$
    0

34. $3 \times 4^2 - 16 - 4 + 3 - (1 - 2)^2$
    30

Simplify.

35. $-12 \times (6 - 8) + 1^2 \times 3^2 \times 2 - 6 \times 2$
30

36. $-3 \times (-2)^2 \times 4 \div 8 - (-12)$
6

37. $10 \times 9 - (8 + 7) \div 5 + 6 - 7 + 8$
94

38. $-27 - (-3)^2 - 2 - 7 + 6 \times 3$
-27

39. $3^2 \times (4 - 7) \div 9 + 6 - 3 - 4 \times 2$
-8

40. $16 - 4 \times 8 + 4^2 - (-18) \div (-9)$
-2

41. $(-3)^2 \times (5 - 7)^2 - (-9) \div 3$
39

42. $-2 \times 4^2 - 3 \times (2 - 8) - 3$
-17

43. $(1.2)^2 - 4.1 \times 0.3$
0.21

44. $2.4 \times (-3) - 2.5$
-9.7

45. $1.6 - (-1.6)^2$
-0.96

46. $4.1 \times 8 \div (-4.1)$
-8

47. $(4.1 - 3.9) - 0.7^2$
-0.29

48. $1.8 \times (-2.3) - 2$
-6.14

49. $(-0.4)^2 \times 1.5 - 2$
-1.76

50. $(6.2 - 1.3) \times (-3)$
-14.70

51. $4.2 - (-3.9) - 6$
2.1

52. $-\frac{1}{2} + \frac{3}{8} \div \left(-\frac{3}{4}\right)$
-1

53. $\left(\frac{3}{4}\right)^2 - \frac{3}{8}$
$\frac{3}{16}$

54. $\left(\frac{1}{2}\right)^2 - \left(-\frac{1}{2}\right)^2$
0

55. $\frac{5}{16} - \frac{3}{8} + \frac{1}{2}$
$\frac{7}{16}$

56. $\frac{2}{7} \div \frac{5}{7} - \frac{3}{14}$
$\frac{13}{70}$

57. $\frac{1}{2} \times \frac{1}{4} \times \frac{1}{2} - \frac{3}{8}$
$-\frac{5}{16}$

58. $\frac{2}{3} \times \frac{5}{8} \div \frac{2}{7}$
$1\frac{11}{24}$

59. $\frac{1}{2} - \left(\frac{3}{4} - \frac{3}{8}\right) \div \frac{1}{3}$
$-\frac{5}{8}$

60. $\frac{3}{8} \div \left(-\frac{1}{2}\right)^2 + 2$
$3\frac{1}{2}$

APPLYING THE CONCEPTS

61. Evaluate: **a.** $1^3 + 2^3 + 3^3 + 4^3$ **b.** $(-1)^3 + (-2)^3 + (-3)^3 + (-4)^3$ **c.** $1^3 + 2^3 + 3^3 + 4^3 + 5^3$ Based on your answers to parts a, b, and c, evaluate **d.** $(-1)^3 + (-2)^3 + (-3)^3 + (-4)^3 + (-5)^3$.
a. 100 **b.** -100 **c.** 225 **d.** -225

62. Evaluate $2^{(3^2)}$ and $(2^3)^2$. Are the answers the same? If not, which is larger? No; $2^{(3^2)}$ is larger

Project in Mathematics

Time Zones

In 1884, a system of standard time was adopted by the International Meridian Conference. The prime meridian is a semicircle passing through Greenwich, England, and labeled 0° as shown in the diagram at the right.

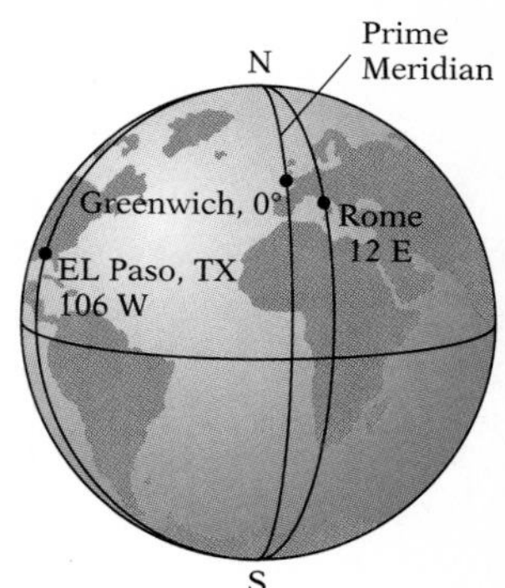

The other meridians are 15° apart, and each 15° width determines a time zone. The time zones to the east of the prime meridian are negative and the zones to the west are positive, as shown below.

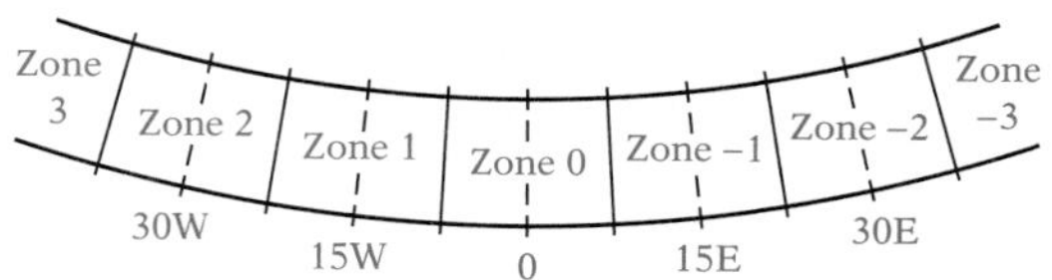

The table below gives the time zones for selected cities.

City	*Time Zone*	*City*	*Time Zone*
Athens	−2	London	0
Beijing	−8	Moscow	−3
Houston	6	Oslo	−1
Honolulu	11	Rio De Janeiro	3

To find the time for a city that is not in your time zone, subtract the time zone of the other city from the time zone of your city. Add that number to the current time in your time zone. For example, say it is 2:00 P.M. in Athens. What time is it in Houston?

From the table above, the time zone in Athens is −2 and the time zone of Houston is 6.

Subtract 6 from −2. $-2 - 6 = -8$

Add that number to the current time in your time zone. 2:00 P.M. + (−8) = 6:00 A.M.

It is 6:00 A.M. in Houston.

Exercises:

1. Find the time in Beijing when the time in Oslo is 12:00 P.M.
2. Find the time in Moscow when the time in Honolulu is 8:00 A.M.
3. Find the time in Rio De Janeiro when the time in Beijing is 3:00 P.M.
4. An office in Athens is open from 9:00 A.M. to 5:00 P.M. What are the times in Houston that a call to Athens can be made during Athens office hours.
5. An office in Honolulu is open from 8:00 to 4:00 P.M. What are the times in Moscow that a call to Honolulu can be made during Honolulu office hours?

Chapter Summary

Key Words *Positive numbers* are numbers greater than zero.

Negative numbers are numbers less than zero.

The *integers* are ... −4, −3, −2, −1, 0, 1, 2, 3, 4, ...

Negative integers are integers to the left of zero on the number line.

Positive integers are integers to the right of zero on the number line.

The *absolute value* of a number is its distance from zero on a number line.

A *rational number* is a number that can be written as the ratio of two numbers.

Essential Rules

Addition of Integers with the Same Sign

To add numbers with the same sign, add the absolute values of the numbers. Then attach the sign of the addends.

Addition of Integers with Different Signs

To add numbers with different signs, find the difference between the absolute values of the numbers. Then attach the sign of the number with the greater absolute value.

Subtraction of Integers

To subtract one integer from another, add the opposite of the second integer to the first integer.

Multiplication of Integers with the Same Sign

To multiply numbers with the same sign, multiply the absolute values of the numbers. The product is positive.

Multiplication of Integers with Different Signs

To multiply numbers with different signs, multiply the absolute values of the numbers. The product is negative.

Division of Integers with the Same Sign

The quotient of two numbers with the same sign is positive.

Division of Integers with Different Signs

The quotient of two numbers with different signs is negative.

Order of Operations Agreement

Step 1 Perform operations inside parentheses.

Step 2 Simplify exponential expressions.

Step 3 Do multiplication and division as they occur from left to right.

Step 4 Rewrite subtraction as the addition of the opposite. Then do additions as they occur from left to right.

Chapter Review

SECTION 10.1

1. Place the correct symbol, $<$ or $>$, between the two numbers.
$-2 > -40$

2. Place the correct symbol, $<$ or $>$, between the two numbers.
$0 > -3$

3. Find the opposite of -4.
4

4. Find the opposite of 22.
-22

5. Evaluate $|-5|$.
5

6. Evaluate $-|-6|$.
-6

SECTION 10.2

7. Add: $-22 + 14 + (-18)$
-26

8. Subtract: $-8 - (-2) - (-10) - 3$
1

SECTION 10.3

9. Find 2 times -13.
-26

10. Divide: $-18 \div (-3)$
6

11. Find the temperature after a rise of 18° from $-22°$. $-4°$

SECTION 10.4

12. Simplify: $\frac{5}{8} - \frac{5}{6}$

$-\frac{5}{24}$

13. Simplify: $\frac{5}{12} + \left(-\frac{2}{3}\right)$

$-\frac{1}{4}$

14. Simplify: $-\frac{3}{8} + \frac{5}{12} + \frac{2}{3}$

$\frac{17}{24}$

15. Simplify: $-\frac{5}{12} + \frac{7}{9} - \frac{1}{3}$

$\frac{1}{36}$

16. Simplify: $-0.33 + 1.98 - 1.44$
0.21

17. Simplify: $-33.4 + 9.8 - (-16.2)$
-7.4

18. Simplify: $\frac{1}{3} \times \left(-\frac{3}{4}\right)$

$-\frac{1}{4}$

19. Simplify: $\frac{6}{34} \times \frac{17}{40}$

$\frac{3}{40}$

20. Simplify: $\left(-\frac{2}{3}\right)\left(\frac{6}{11}\right)\left(-\frac{22}{25}\right)$

$\frac{8}{25}$

21. Simplify: $\left(-\frac{3}{8}\right) \div \left(-\frac{4}{5}\right)$

$\frac{15}{32}$

22. Simplify: $-\frac{7}{12} \div \left(-\frac{14}{39}\right)$

$\frac{13}{8} = 1\frac{5}{8}$

23. Simplify: $1.2 \times (-0.035)$

-0.042

24. Simplify: -0.08×16

-1.28

25. Simplify: $-1.464 \div 18.3$

-0.08

SECTION 10.5

26. Simplify: $12 - 6 \div 3$

10

27. Simplify: $16 \div 8 \times 4$

8

28. Simplify: $16 + 4 \div 2$

18

29. Simplify: $3^2 - 9 + 2$

2

30. Simplify: $16 \div 4(8 - 2)$

24

31. Simplify: $-0.4 \times 5 - (-3.33)$

1.33

32. Simplify: $\left(\frac{2}{3}\right)^2 - \frac{5}{6}$

$-\frac{7}{18}$

33. Simplify: $-\frac{1}{2} + \frac{3}{8} \div \frac{9}{20}$

$\frac{1}{3}$

Chapter Test

1. Place the correct symbol, < or >, between the two numbers.
$-8 > -10$ [10.1A]

2. Place the correct symbol, < or >, between the two numbers.
$0 > -4$ [10.1A]

3. Evaluate $-|-2|$.
-2 [10.1B]

4. Find the sum of -2, 3, and -8.
-7 [10.2A]

5. Add: $16 + (-10) + (-20)$
-14 [10.2A]

6. Subtract: $-5 - (-8)$
3 [10.2B]

7. Subtract: $16 - 4 - (-5) - 7$
10 [10.2B]

8. Find the product of -4 and 12.
-48 [10.3A]

9. Multiply: $-5 \times (-6) \times 3$
90 [10.3A]

10. Divide: $-72 \div 8$
-9 [10.3B]

11. Find the temperature after a rise of 11°C from -4°C. 7°C [10.2C]

12. The daily low temperature readings for a 3-day period were as follows: $-7°$, $9°$, $-8°$. Find the average low temperature for the 3-day period. $-2°$ [10.3C]

13. Add: $-\frac{2}{5} + \frac{7}{15}$

$\frac{1}{15}$ [10.4A]

14. Find the sum of $-\frac{1}{2}$, $\frac{1}{3}$, and $\frac{1}{4}$.

$\frac{1}{12}$ [10.4A]

15. Subtract: $-\frac{3}{8} + \frac{2}{3}$

$\frac{7}{24}$ [10.4A]

16. Subtract: $-\frac{2}{5} - \left(-\frac{7}{10}\right)$

$\frac{3}{10}$ [10.4A]

17. Multiply: $\frac{3}{8} \times \left(-\frac{5}{6}\right) \times \left(-\frac{4}{15}\right)$

$\frac{1}{12}$ [10.4B]

18. Find the quotient of $-\frac{2}{3}$ and $\frac{5}{6}$.

$-\frac{4}{5}$ [10.4B]

19. Add: $1.22 + (-3.1)$

-1.88 [10.4A]

20. What is -1.004 decreased by 3.01?

-4.014 [10.4A]

21. Subtract: $2.113 - (-1.1)$

3.213 [10.4A]

22. Find the product of 0.032 and -1.9.

-0.0608 [10.4B]

23. Divide: $-15.64 \div (-4.6)$

3.4 [10.4B]

24. Simplify: $4 \times (4 - 7) \div (-2) - 4 \times 8$

-26 [10.5A]

25. Simplify: $(-2)^2 - (-3)^2 \div (1 - 4)^2 \times 2 - 6$

-4 [10.5A]

Cumulative Review

1. Simplify:
$16 - 4 \cdot (3 - 2)^2 \cdot 4$
0 [1.6B]

2. Find the difference between $8\frac{1}{2}$ and $3\frac{4}{7}$.
$4\frac{13}{14}$ [2.5C]

3. Divide: $3\frac{7}{8} \div 1\frac{1}{2}$
$2\frac{7}{12}$ [2.7B]

4. Simplify: $\frac{3}{8} \div \left(\frac{3}{8} - \frac{1}{4}\right) \div \frac{7}{3}$
$1\frac{2}{7}$ [2.8C]

5. Subtract: $2.907 - 1.09761$
1.80939 [3.3A]

6. Solve the proportion $\frac{7}{12} = \frac{n}{32}$.
Round to the nearest hundredth.
$n = 18.67$ [4.3B]

7. 22 is 160% of what number?
13.75 [5.4A]

8. Convert:
7 qt = ____ gal ____ qt
1 gal 3 qt [8.3A]

9. Convert: 6 L 692 ml = ____ L
6.692 L [9.3A]

10. Convert 4.2 ft to meters. Round to the nearest hundredth. (1 m = 3.28 ft)
1.28 m [9.5A]

11. Find 32% of 180.
57.6 [5.2A]

12. Convert $3\frac{2}{5}$ to a percent.
340% [5.1B]

13. Add: $-8 + 5$
-3 [10.2A]

14. Add: $3\frac{1}{4} + \left(-6\frac{5}{8}\right)$
$-\frac{27}{8} = -3\frac{3}{8}$ [10.4A]

15. Subtract: $-6\frac{1}{8} - 4\frac{5}{12}$
$-\frac{253}{24} = -10\frac{13}{24}$ [10.4A]

16. Subtract:
$-12 - (-7) - 3(-8)$
19 [10.5A]

17. What is -3.2 times -1.09?
3.488 [10.4B]

18. Multiply: $-6 \times 7 \times \left(-\frac{3}{4}\right)$
$31\frac{1}{2}$ [10.4B]

19. Find the quotient of 42 and −6.
−7 [10.3B]

20. Divide: $-2\frac{1}{7} \div \left(-3\frac{3}{5}\right)$
$\frac{25}{42}$ [10.4B]

21. Simplify: $3 \times (3 - 7) \div 6 - 2$
−4 [10.5A]

22. Simplify:
$4 - (-2)^2 \div (1 - 2)^2 \times 3 + 4$
−4 [10.5A]

23. A board $5\frac{2}{3}$ ft long is cut from a board 8 ft long. What is the length of the board remaining? $2\frac{1}{3}$ ft [2.5D]

24. Nimisha had a balance of $763.56 in her checkbook before writing checks of $135.88 and $47.81 and making a deposit of $223.44. Find her new checkbook balance. $803.31 [6.7A]

25. A suit that regularly sells for $165 is on sale for $120. Find the percent decrease in price. Round to the nearest tenth of a percent. 27.3% [6.2C]

26. A reception is planned for 80 guests. How many gallons of coffee should be prepared to provide 2 cups of coffee for each guest? 10 gal [8.3C]

27. A stock selling for $$82\frac{5}{8}$ per share pays a dividend of $1.50 per share. The dividend is increased by 12%. Find the dividend per share after the increase. $1.68 [6.2A]

28. The hourly wages for five job classifications at a company are $9.40, $7.32, $13.25, $8.73, and $11.10. Find the median hourly pay.
$9.40 [7.4B]

29. A pre-election survey showed that 5 out of every 8 registered voters would cast ballots in a city election. At this rate, how many people would vote in a city of 960,000 registered voters? 600,000 [4.3C]

30. The daily high temperature readings for a 4-day period were recorded as follows: −19°, −7°, 1°, and 9°. Find the average high temperature for the 4-day period. −4° [10.3C]

Chapter 11

Introduction to Algebra

Objectives

Section 11.1

To evaluate variable expressions

To simplify variable expressions containing no parentheses

To simplify variable expressions containing parentheses

Section 11.2

To determine whether a given value is a solution of an equation

To solve an equation of the form $x + a = b$

To solve an equation of the form $ax = b$

To solve application problems

Section 11.3

To solve an equation of the form $ax + b = c$

To solve application problems

Section 11.4

To solve an equation of the form $ax + b = cx + d$

To solve an equation containing parentheses

Section 11.5

To translate a verbal expression into a mathematical expression given the variable

To translate a verbal expression into a mathematical expression by assigning the variable

Section 11.6

To translate a sentence into an equation and solve

To solve application problems

History of the Equals Sign

A portion of a page of the first book that used an equals sign, =, is shown below. This book was written in 1557 by Robert Recorde and was titled *The Whetstone of Witte*.

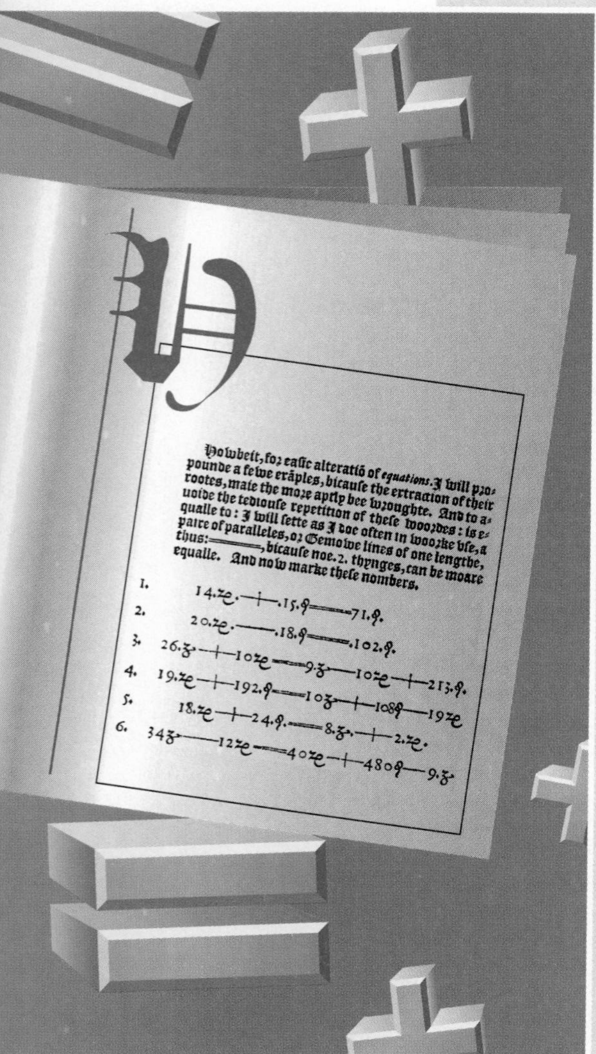

Howbeit, for easie alteratiō of *equations*. I will propounde a fewe exāples, bicause the extraction of their rootes, maie the more aptly bee wroughte. And to auoide the tediouse repetition of these woordes : is equalle to : I will sette as I doe often in woorke use, a paire of paralleles, or Gemowe lines of one lengthe, thus: =======, bicause noe. 2. thynges, can be moare equalle. And now marke these nombers.

1. 14.ze. + .15.ϕ ===== 71.ϕ.
2. 20.ze. — .18.ϕ ===== .102.ϕ.
3. 26.ʒ. + 10ze ===== 9.ʒ — 10ze + 213.ϕ.
4. 19.ze + 192.ϕ ===== 10ʒ + 108ϕ — 19ze
5. 18.ze + 24.ϕ. ===== 8.ʒ. + 2.ze.
6. 34ʒ — 12ze ===== 40ze + 480ϕ — 9.ʒ

Notice, near the end of the paragraph in the illustration, the words "bicause noe 2 thynges can be moare equalle." Recorde decided that two things could not be more equal than two parallel lines of the same length. Therefore it made sense to use this symbol to show equality.

This page also illustrates the use of the plus sign, +, and the minus sign, −. These symbols had been widely used for only about 100 years when this book was written.

11.1 Variable Expressions

Objective A To evaluate variable expressions

POINT OF INTEREST
There are historical records indicating that Mathematics has been studied for at least 4000 years. However, it has been only in the most recent 400 years that variables have been used. Prior to that, Mathematics was written in words.

Often we discuss a quantity without knowing its exact value—for example, next year's inflation rate, the price of gasoline next summer, or the interest rate on a new-car loan next fall. In mathematics, a letter of the alphabet is used to stand for a quantity that is unknown or that can change or *vary*. The letter is called a **variable.** An expression that contains one or more variables is called a **variable expression.**

A company's business manager has determined that the company will make a \$2 profit on each radio it sells. The manager wants to describe the company's total profit from the sale of radios. Because the number of radios that the company will sell is unknown, the manager lets the variable n stand for that number. Then the variable expression $2 \cdot n$, or simply $2n$, describes the company's profit for selling n radios.

The company's profit for selling n radios is $\$2 \cdot n = \$2n$.

If the company sells 12 radios, its profit is $\$2 \cdot 12 = \$24$.

If the company sells 75 radios, its profit is $\$2 \cdot 75 = \$150$.

Replacing the variable or variables in a variable expression and then simplifying the resulting numerical expression is called **evaluating the variable expression.**

INSTRUCTOR NOTE
Have students insert parentheses around the number as they replace the variable with a number. This is especially useful when negative numbers are used.

As a class example, show students the evaluation of $-x^2$ when $x = -3$ and when $x = 3$.

➡ Evaluate $3x^2 + xy - z$ when $x = -2$, $y = 3$, and $z = -4$.

$$3x^2 + xy - z$$
$$3(-2)^2 + (-2)(3) - (-4) = 3 \cdot 4 + (-2)(3) - (-4)$$
$$= 12 + (-6) - (-4)$$
$$= 12 + (-6) + 4$$
$$= 6 + 4$$
$$= 10$$

- Replace each variable in the expression with the number it stands for.
- Use the Order of Operations Agreement to simplify the resulting numerical expression.

The value of the variable expression $3x^2 + xy - z$ when $x = -2$, $y = 3$, and $z = -4$ is 10.

Example 1

Evaluate $3x - 4y$ when $x = -2$ and $y = 3$.

Solution

$3x - 4y$

$3(-2) - 4(3) = -6 - 12$

$= -6 + (-12) = -18$

You Try It 1

Evaluate $6a - 5b$ when $a = -3$ and $b = 4$.

Your solution -38

Example 2

Evaluate $-x^2 - 6 \div y$ when $x = -3$ and $y = 2$.

Solution

$-x^2 - 6 \div y$

$-(-3)^2 - 6 \div 2 = -9 - 6 \div 2$

$= -9 - 3$

$= -9 + (-3) = -12$

You Try It 2

Evaluate $-3s^2 - 12 \div t$ when $s = -2$ and $t = 4$.

Your solution -15

Example 3

Evaluate $-\frac{1}{2}y^2 - \frac{3}{4}z$ when $y = 2$ and $z = -4$.

Solution

$-\frac{1}{2}y^2 - \frac{3}{4}z$

$-\frac{1}{2}(2)^2 - \frac{3}{4}(-4) = -\frac{1}{2} \cdot 4 - \frac{3}{4}(-4)$

$= -2 - (-3)$

$= -2 + (3) = 1$

You Try It 3

Evaluate $-\frac{2}{3}m + \frac{3}{4}n^3$ when $m = 6$ and $n = 2$.

Your solution 2

Example 4

Evaluate $-2ab + b^2 + a^2$ when $a = -\frac{3}{5}$ and $b = \frac{2}{5}$.

Solution

$-2ab + b^2 + a^2$

$-2\left(-\frac{3}{5}\right)\left(\frac{2}{5}\right) + \left(\frac{2}{5}\right)^2 + \left(-\frac{3}{5}\right)^2$

$= -2\left(-\frac{3}{5}\right)\left(\frac{2}{5}\right) + \left(\frac{4}{25}\right) + \left(\frac{9}{25}\right)$

$= \frac{12}{25} + \frac{4}{25} + \frac{9}{25} = \frac{25}{25} = 1$

You Try It 4

Evaluate $-3yz - z^2 + y^2$ when $y = -\frac{2}{3}$ and $z = \frac{1}{3}$.

Your solution 1

Solutions on p. A47

Objective B To simplify variable expressions containing no parentheses

The **terms** of a variable expression are the addends of the expression. The variable expression at the right has four terms.

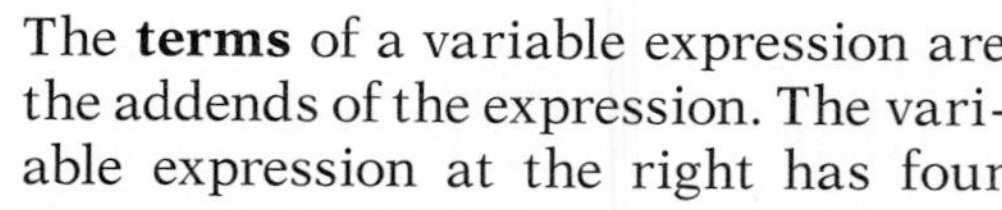

4 terms

$7x^2 + (-6xy) + x + (-8)$

variable terms — constant term

Three of the terms are **variable terms:** $7x^2$, $(-6xy)$, and x.

One of the terms is a **constant term:** (-8). A constant term has no variables.

Each variable term is composed of a **numerical coefficient** and a **variable part** (the variable or variables and their exponents). When the numerical coefficient is 1, the 1 is usually not written. $(1x = x)$

numerical coefficient

$7x^2 + (-6xy) + 1x + (-8)$

variable part

INSTRUCTOR NOTE

There are many instances of combining like terms. 2 feet plus 3 feet is 5 feet; 2 dollars plus 3 dollars is 5 dollars. When presented in this way, the idea of combining like terms seems more natural to students. Also, giving an example like 2 feet plus 3 dollars will suggest that they cannot combine unlike quantities.

Like terms of a variable expression are the terms with the same variable part. (Because $y^2 = y \cdot y$, y^2 and y are not like terms.)

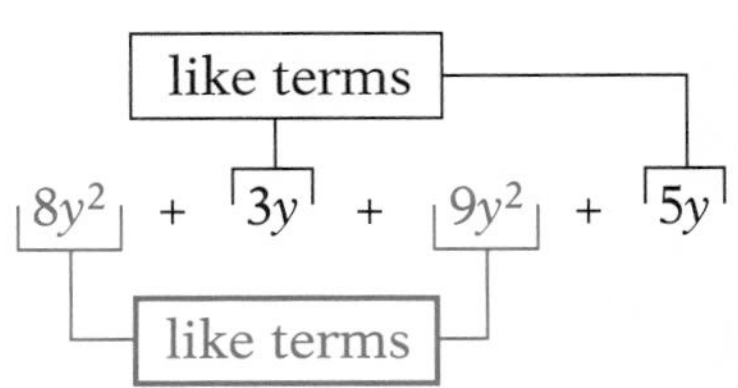

In variable expressions that contain constant terms, the constant terms are like terms.

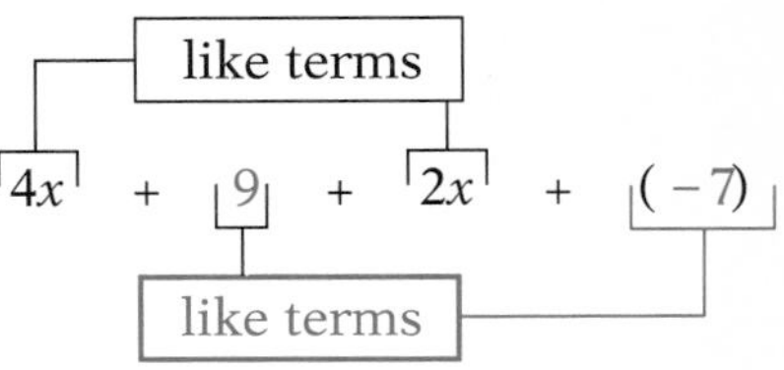

The Commutative and Associative Properties of Addition are used to simplify variable expressions. These properties can be stated in general form using variables.

Commutative Property of Addition

If a and b are two numbers, then $a + b = b + a$.

Associative Property of Addition

If a, b, and c are three numbers, then $a + (b + c) = (a + b) + c$.

To **simplify a variable expression,** *combine* like terms by adding their numerical coefficients. For example, to simplify $2y + 3y$, think

$$2y + 3y = (y + y) + (y + y + y) = 5y$$

➡ Simplify: $8z - 5 + 2z$

$$8z - 5 + 2z = 8z + 2z - 5$$
$$= 10z - 5$$

• Use the Commutative and Associative Properties of Addition to group like terms. Combine the like terms $8z + 2z = 10z$.

INSTRUCTOR NOTE
Although students will normally say that $4x + 3$ cannot be simplified, these same students will try to simplify $3 + 4x$ as $7x$.

➡ Simplify: $12a - 4b - 8a + 2b$

$12a - 4b - 8a + 2b = 12a - 8a - 4b + 2b$
$= 4a - 2b$

- Use the Commutative and Associative Properties of Addition to group like terms. Combine like terms.

➡ Simplify: $6z^2 + 3 - z^2 - 7$

$6z^2 + 3 - z^2 - 7 = 6z^2 - z^2 + 3 - 7$
$= 5z^2 - 4$

- Use the Commutative and Associative Properties to group like terms. Combine like terms.

Example 5
Simplify:
$6xy - 8x + 5x - 9xy$

Solution
$6xy - 8x + 5x - 9xy$
$= 6xy + (-8)x + 5x + (-9)xy$
$= 6xy + (-9)xy + (-8)x + 5x$
$= -3xy + (-3)x$
$= -3xy - 3x$

You Try It 5
Simplify:
$5a^2 - 6b^2 + 7a^2 - 9b^2$

Your solution $12a^2 - 15b^2$

Example 6
Simplify:
$-4z^2 + 8 + 5z^2 - 3$

Solution
$-4z^2 + 8 + 5z^2 - 3$
$= -4z^2 + 8 + 5z^2 + (-3)$
$= -4z^2 + 5z^2 + 8 + (-3)$
$= z^2 + 5$

You Try It 6
Simplify:
$-6x + 7 + 9x - 10$

Your solution $3x - 3$

Example 7
Simplify:
$\frac{1}{4}m^2 - \frac{1}{2}n^2 + \frac{1}{2}m^2$

Solution
$\frac{1}{4}m^2 - \frac{1}{2}n^2 + \frac{1}{2}m^2$
$= \frac{1}{4}m^2 + \frac{1}{2}m^2 - \frac{1}{2}n^2$
$= \frac{3}{4}m^2 - \frac{1}{2}n^2$

You Try It 7
Simplify:
$\frac{3}{8}w + \frac{1}{2} - \frac{1}{4}w - \frac{2}{3}$

Your solution $\frac{1}{8}w - \frac{1}{6}$

Solutions on p. A47

Objective C *To simplify variable expressions containing parentheses*

INSTRUCTOR NOTE
Some of the *Applying the Concept* exercises on page 410 involve using rectangles and squares to model algebraic expressions. These models can help some students better understand adding like terms and the Distributive Property.

The Commutative and Associative Properties of Multiplication and the Distributive Property are used to simplify variable expressions that contain parentheses. These properties can be stated in general form using variables.

> **Commutative Property of Multiplication**
> If a and b are two numbers, then $a \cdot b = b \cdot a$.

> **Associative Property of Multiplication**
> If a, b, and c are three numbers, then $a \cdot (b \cdot c) = (a \cdot b) \cdot c$.

The Associative and Commutative Properties of Multiplication are used to simplify variable expressions such as the following.

➡ Simplify: $-5(4x)$

$-5(4x) = (-5 \cdot 4)x$ • Use the Associative Property of Multiplication.
$= -20x$

➡ Simplify: $(6y) \cdot 5$

$(6y) \cdot 5 = 5 \cdot (6y)$ • Use the Commutative Property of Multiplication.
$= (5 \cdot 6)y = 30y$ • Use the Associative Property of Multiplication.

The Distributive Property is used to remove parentheses from variable expressions that contain both multiplication and addition.

INSTRUCTOR NOTE
Students frequently apply the Distributive Property incorrectly. Stress that *each* term within the parentheses must be multiplied by the term outside the parentheses.

> **Distributive Property**
> If a, b, and c are three numbers, then $a(b + c) = ab + ac$.

➡ Simplify: $4(z + 5)$

$4(z + 5) = 4z + 4(5)$ • The Distributive Property is used to rewrite the variable expression without parentheses.
$= 4z + 20$

➡ Simplify: $-3(2x + 7)$

$-3(2x + 7) = -3(2x) + (-3)(7)$ • Use the Distributive Property.
$= -6x + (-21)$
$= -6x - 21$ • Recall that $a + (-b) = a - b$.

The Distributive Property can also be stated in terms of subtraction.

$$a(b - c) = ab - ac$$

➡ Simplify: $8(2r - 3s)$

$$8(2r - 3s) = 8(2r) - 8(3s) = 16r - 24s$$

• Use the Distributive Property.

➡ Simplify: $-5(2x - 4y)$

$$-5(2x - 4y) = (-5) \cdot (2x) - (-5) \cdot (4y)$$
$$= -10x - (-20y)$$
$$= -10x + 20y$$

• Use the Distributive Property.

• Recall that $a - (-b) = a + b$.

INSTRUCTOR NOTE

A typical student error for the example at the right is to subtract 5 from 12 and then apply the Distributive Property. Remind students that the Order of Operations Agreement requires multiplication before addition or subtraction.

➡ Simplify: $12 - 5(m + 2) + 2m$

$$12 - 5(m + 2) + 2m = 12 - 5m + (-5)(2) + 2m$$
$$= 12 - 5m + (-10) + 2m$$

• Use the Distributive Property to simplify the expression $-5(m + 2)$.

$$= -5m + 2m + 12 + (-10)$$

• Use the Commutative and Associative Properties to group like terms.

$$= -3m + 2$$

• Combine like terms by adding their numerical coefficients. Add constant terms.

The answer $-3m + 2$ can also be written as $2 - 3m$. In this text, we will write answers with variable terms first, followed by the constant term.

Example 8
Simplify: $4(x - 3)$

Solution
$$4(x - 3) = 4x - 4(3)$$
$$= 4x - 12$$

You Try It 8
Simplify: $5(a - 2)$

Your solution $5a - 10$

Example 9
Simplify: $5n - 3(2n - 4)$

Solution
$$5n - 3(2n - 4) = 5n - 3(2n) - (-3)(4)$$
$$= 5n - 6n - (-12)$$
$$= 5n - 6n + 12$$
$$= -n + 12$$

You Try It 9
Simplify: $8s - 2(3s - 5)$

Your solution $2s + 10$

Example 10
Simplify:
$3(c - 2) + 2(c + 6)$

Solution
$$3(c - 2) + 2(c + 6) = 3c - 3(2) + 2c + 2(6)$$
$$= 3c - 6 + 2c + 12$$
$$= 3c + 2c - 6 + 12$$
$$= 5c + 6$$

You Try It 10
Simplify:
$4(x - 3) - 2(x + 1)$

Your solution $2x - 14$

Solutions on p. A47

11.1 Exercises

Objective A

Evaluate the variable expression when $a = -3$, $b = 6$, and $c = -2$.

1. $5a - 3b$
   -33
2. $4c - 2b$
   -20
3. $2a + 3c$
   -12
4. $2c + 4a$
   -16
5. $-c^2$
   -4
6. $-a^2$
   -9
7. $b - a^2$
   -3
8. $b - c^2$
   2
9. $ab - c^2$
   -22
10. $bc - a^2$
    -21
11. $2ab - c^2$
    -40
12. $3bc - a^2$
    -45
13. $a - (b \div a)$
    -1
14. $c - (b \div c)$
    1
15. $2ac - (b \div a)$
    14
16. $4ac \div (b \div a)$
    -12
17. $b^2 - c^2$
    32
18. $b^2 - a^2$
    27
19. $b^2 \div (ac)$
    6
20. $3c^2 \div (ab)$
    $-\frac{2}{3}$
21. $c^2 - (b \div c)$
    7
22. $a^2 - (b \div a)$
    11
23. $a^2 + b^2 + c^2$
    49
24. $a^2 - b^2 - c^2$
    -31
25. $ac + bc + ab$
    -24
26. $ac - bc - cb$
    30
27. $a^2 + b^2 - ab$
    63
28. $b^2 + c^2 - bc$
    52
29. $2b - (3c + a^2)$
    9
30. $\frac{2}{3}b + \left(\frac{1}{2}c - a\right)$
    6
31. $\frac{1}{3}a + \left(\frac{1}{2}b - \frac{2}{3}a\right)$
    4
32. $-\frac{2}{3}b - \left(\frac{1}{2}c + a\right)$
    0
33. $\frac{1}{6}b + \frac{1}{3}(c + a)$
    $-\frac{2}{3}$
34. $\frac{1}{2}c + \left(\frac{1}{3}b - a\right)$
    4

Evaluate the variable expression when $a = -\frac{1}{2}$, $b = \frac{3}{4}$, and $c = \frac{1}{4}$.

35. $a + (b - c)$
    0
36. $c + (b - 2c)$
    $\frac{1}{2}$
37. $4a + (3b - c)$
    0
38. $2b + (c - 3a)$
    $3\frac{1}{4}$
39. $2a - b^2 \div c$
    $-3\frac{1}{4}$
40. $b \div (-c) + 2a$
    -4

Evaluate the variable expression when $a = 3.72$, $b = -2.31$, and $c = -1.74$.

41. $a^2 - b^2$
8.5023

42. $a^2 - b \cdot c$
9.819

43. $3ac - (c \div a)$
-18.950658

44. $3b - (3b - a^2)$
13.8384

45. $2c + (b^2 - c)$
3.5961

46. $abc - 2(b \div c^2)$
16.478129

Objective B

Simplify.

47. $7z + 9z$
$16z$

48. $6x + 5x$
$11x$

49. $12m - 3m$
$9m$

50. $5y - 12y$
$-7y$

51. $5at + 7at$
$12at$

52. $12mn + 11mn$
$23mn$

53. $-4yt + 7yt$
$3yt$

54. $-12yt + 5yt$
$-7yt$

55. $-3x - 12y$
Unlike terms

56. $-12y - 7y$
$-19y$

57. $3t^2 - 5t^2$
$-2t^2$

58. $7t^2 + 8t^2$
$15t^2$

59. $6c - 5 + 7c$
$13c - 5$

60. $7x - 5 + 3x$
$10x - 5$

61. $2t + 3t - 7t$
$-2t$

62. $9x^2 - 5 - 3x^2$
$6x^2 - 5$

63. $7y^2 - 2 - 4y^2$
$3y^2 - 2$

64. $3w - 7u + 4w$
$7w - 7u$

65. $6w - 8u + 8w$
$14w - 8u$

66. $4 - 6xy - 7xy$
$-13xy + 4$

67. $10 - 11xy - 12xy$
$-23xy + 10$

68. $7t^2 - 5t^2 - 4t^2$
$-2t^2$

69. $3v^2 - 6v^2 - 8v^2$
$-11v^2$

70. $5ab - 7a - 10ab$
$-5ab - 7a$

71. $-10ab - 3a + 2ab$
$-8ab - 3a$

72. $-4x^2 - x + 2x^2$
$-2x^2 - x$

73. $-3y^2 - y + 7y^2$
$4y^2 - y$

74. $4x^2 - 8y - x^2 + y$
$3x^2 - 7y$

75. $2a - 3b^2 - 5a + b^2$
$-3a - 2b^2$

76. $8y - 4z - y + 2z$
$7y - 2z$

77. $3x^2 - 7x + 4x^2 - x$
$7x^2 - 8x$

78. $5y^2 - y + 6y^2 - 5y$
$11y^2 - 6y$

79. $6s - t - 9s + 7t$
$-3s + 6t$

80. $5w - 2v - 9w + 5v$
$-4w + 3v$

81. $4m + 8n - 7m + 2n$
$-3m + 10n$

82. $z + 9y - 4z + 3y$
$-3z + 12y$

83. $-5ab + 7ac + 10ab - 3ac$
$5ab + 4ac$

84. $-2x^2 - 3x - 11x^2 + 14x$
$-13x^2 + 11x$

Simplify.

85. $\frac{4}{9}a^2 - \frac{1}{5}b^2 + \frac{2}{9}a^2 + \frac{4}{5}b^2$

$\frac{2}{3}a^2 + \frac{3}{5}b^2$

86. $\frac{6}{7}x^2 + \frac{2}{5}x - \frac{3}{7}x^2 - \frac{4}{5}x$

$\frac{3}{7}x^2 - \frac{2}{5}x$

87. $4.235x - 0.297x + 3.056x$

$6.994x$

88. $8.092y - 3.0793y + 0.063y$

$5.0757y$

89. $7.81m + 3.42n - 6.25m - 7.19n$

$1.56m - 3.77n$

90. $8.34y^2 - 4.21y - 6.07y^2 - 5.39y$

$2.27y^2 - 9.60y$

Objective C

Simplify.

91. $5(x + 4)$
$5x + 20$

92. $3(m + 6)$
$3m + 18$

93. $(y - 3)4$
$4y - 12$

94. $(z - 3)7$
$7z - 21$

95. $-2(a + 4)$
$-2a - 8$

96. $-5(b + 3)$
$-5b - 15$

97. $3(5x + 10)$
$15x + 30$

98. $2(4m - 7)$
$8m - 14$

99. $5(3c - 5)$
$15c - 25$

100. $-4(w - 3)$
$-4w + 12$

101. $-3(y - 6)$
$-3y + 18$

102. $3m + 4(m + z)$
$7m + 4z$

103. $5x + 2(x + 7)$
$7x + 14$

104. $6z - 3(z + 4)$
$3z - 12$

105. $8y - 4(y + 2)$
$4y - 8$

106. $7w - 2(w - 3)$
$5w + 6$

107. $9x - 4(x - 6)$
$5x + 24$

108. $-5m + 3(m + 4)$
$-2m + 12$

109. $-2y + 3(y - 2)$
$y - 6$

110. $5m + 3(m + 4) - 6$
$8m + 6$

111. $4n + 2(n + 1) - 5$
$6n - 3$

112. $8z - 2(z - 3) + 8$
$6z + 14$

113. $9y - 3(y - 4) + 8$
$6y + 20$

114. $6 - 4(a + 4) + 6a$
$2a - 10$

115. $3x + 2(x + 2) + 5x$
$10x + 4$

116. $7x + 4(x + 1) + 3x$
$14x + 4$

117. $-7t + 2(t - 3) - t$
$-6t - 6$

118. $-3y + 2(y - 4) - y$
$-2y - 8$

119. $z - 2(1 - z) - 2z$
$z - 2$

120. $2y - 3(2 - y) + 4y$
$9y - 6$

121. $3(y - 2) - 2(y - 6)$
$y + 6$

122. $7(x + 2) + 3(x - 4)$
$10x + 2$

123. $2(t - 3) + 7(t + 3)$
$9t + 15$

124. $3(y - 4) - 2(y - 3)$
$y - 6$

125. $3t - 6(t - 4) + 8t$
$5t + 24$

126. $5x + 3(x - 7) - 9t$
$8x - 9t - 21$

APPLYING THE CONCEPTS

127. Tyrone has an after-school job that pays \$5 an hour. The table at the right shows the amount received for the number of hours worked. Complete the table. How much does he receive for working n hours? \$5n

Hours Worked	*Amount Received*
1	\$ 5
2	10
3	15
4	20
.	.
n	5n

128. The square and the rectangle at the right can be used to illustrate algebraic expressions. The illustration below represents the expression $2x + 1$.

x | x | 1

Using similar squares and rectangles, draw figures that represent the expressions $3 + 2x$, $4x + 6$, $3x + 2$, and $2x + 4$.

1 | 1 | 1 | x | x x | x | x | x | 1 | 1 | 1 | 1 | 1 | 1 x | x | x | 1 | 1 x | x | 1 | 1 | 1 | 1

129. Here the square and the rectangle introduced in Exercise 128 are used to form the expression $3(x + 1)$.

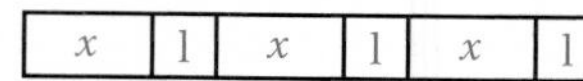

Rearrange these rectangles so that the x's are together and the 1's are together. Write a mathematical expression for the rearranged figure.

x | x | x | 1 | 1 | 1 $3x + 3$

130. [W] Using similar squares and rectangles from Exercise 128, draw figures that represent the expressions $2 + 3x$, $5x$, $2(2x + 3)$, $4x + 3$, and $4x + 6$.

a. Does the figure $2(2x + 3)$ equal the figure $4x + 6$? Explain how this is related to the Distributive Property. Yes

b. Does the figure $2 + 3x$ equal the figure $5x$? How is this related to combining like terms? No

131. [W] Simplifying variable expressions requires combining like terms. Give some examples of how this applies to everyday experience.

132. [W] Write a brief paragraph on the origin of the word *algebra*.

133. [W] It was stated in this section that the variable terms y^2 and y are not like terms. Use measurements of area and distance to show that these terms would not be combined as measurements.

134. [W] Explain why the simplification of the expression $2 + 3(2x + 4)$ shown at the right is incorrect. What is the correct simplification? $6x + 14$

Why is this incorrect?

$$2 + 3(2x + 4) = 5(2x + 4)$$
$$= 10x + 20$$

11.2 Introduction to Equations

Objective A *To determine whether a given value is a solution of an equation*

POINT OF INTEREST

Finding solutions of equations has been a principle aim of Mathematics for thousands of years. However, the equal sign did not occur in any text until 1557.

An **equation** expresses the equality of two mathematical expressions. The expressions can be either numerical or variable expressions.

$5 + 4 = 9$
$3x + 13 = x - 8$
$y^2 + 4 = 6y + 1$
$x = -3$
} Equations

In the equation at the right, if the variable is replaced by 4, the equation is true.

$x + 3 = 7$
$4 + 3 = 7$ A true equation

If the variable is replaced by 6, the equation is false.

$6 + 3 = 7$ A false equation

A **solution** of an equation is a value of the variable that results in a true equation. 4 is a solution of the equation $x + 3 = 7$. 6 is not a solution of the equation $x + 3 = 7$.

➡ Is -2 a solution of the equation $x^2 + 1 = 2x + 9$?

$x^2 + 1$	$= 2x + 9$
$(-2)^2 + 1$	$2(-2) + 9$
$4 + 1$	$-4 + 9$

$5 = 5$

- Replace the variable by the given value.
- Evaluate the numerical expressions.
- Compare the results. If the results are equal, the given value is a solution. If the results are not equal, the given value is not a solution.

Yes, -2 is a solution of the equation $x^2 + 1 = 2x + 9$.

Example 1 Is $\frac{1}{2}$ a solution of $2x(x + 2) = 3x + 1$?

Solution

$2x(x + 2)$	$= 3x + 1$
$2\left(\frac{1}{2}\right)\left(\frac{1}{2} + 2\right)$	$3\left(\frac{1}{2}\right) + 1$
$2\left(\frac{1}{2}\right)\left(\frac{5}{2}\right)$	$3\left(\frac{1}{2}\right) + 1$

$\frac{5}{2} = \frac{5}{2}$

Yes, $\frac{1}{2}$ is a solution.

You Try It 1 Is -2 a solution of $x(x + 3) = 4x + 6$?

Your solution Yes

Solution on p. A47

Example 2 Is 5 a solution of $(x - 2)^2 = x^2 - 4x + 2$?

Solution

$$
\begin{array}{r|l}
(x-2)^2 & x^2 - 4x + 2 \\
\hline
(5-2)^2 & 5^2 - 4(5) + 2 \\
3^2 & 25 - 4(5) + 2 \\
9 & 25 - 20 + 2 \\
 & 25 + (-20) + 2
\end{array}
$$

$9 \neq 7$ ($\neq$ means is not equal to)

No. 5 is not a solution.

You Try It 2 Is -3 a solution of $x^2 - x = 3x + 7$?

Your solution No

Solution on p. A47

Objective B To solve an equation of the form $x + a = b$

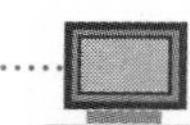

A solution of an equation is a value of the variable that, when substituted in the equation, results in a true equation. To **solve** an equation means to find a solution of the equation.

The simplest equation to solve is an equation of the form *variable* = *constant*. The constant is the solution of the equation.

If $x = 7$, then 7 is the solution of the equation because $7 = 7$ is a true equation.

In solving an equation of the form $x + a = b$, the goal is to simplify the given equation to one of the form *variable* = *constant*. The Addition Properties that follow are used to simplify equations to this form.

Addition Property of Zero

The sum of a term and zero is the term.

$a + 0 = a \qquad 0 + a = a$

Addition Property of Equations

If a, b, and c are algebraic expressions, then the equations $a = b$ and $a + c = b + c$ have the same solutions.

The Addition Property of Equations states that the same quantity can be added to each side of an equation without changing the solution of the equation.

In solving an equation, the goal is to rewrite the given equation in the form *variable* = *constant*. The Addition Property of Equations is used to remove a term from one side of the equation by adding the opposite of that term to each side of the equation.

➡ Solve: $x - 7 = -2$

$$x - 7 = -2$$

- The goal is to simplify the equation to one of the form *variable* = *constant*.

$$x - 7 + 7 = -2 + 7$$
$$x + 0 = 5$$
$$x = 5$$

- Add the opposite of the constant term −7 to each side of the equation. After simplifying and using the addition Property of Zero, the equation will be in the form *variable* = *constant*.

Check: $x - 7 = -2$

$5 - 7 \mid -2$

$-2 = -2$ A true equation

The solution is 5.

Because subtraction is defined in terms of addition, the Addition Property of Equations allows the same number to be subtracted from each side of an equation.

➡ Solve: $x + 8 = 5$

$$x + 8 = 5$$

- The goal is to simplify the equation to one of the form *variable* = *constant*.

$$x + 8 - 8 = 5 - 8$$
$$x + 0 = -3$$
$$x = -3$$

- Add the opposite of the constant term 8 to each side of the equation. This procedure is equivalent to subtracting 8 from each side of the equation.

The solution is −3. You should check this solution.

Example 3 Solve: $4 + m = -2$

Solution

$$4 + m = -2$$
$$4 - 4 + m = -2 - 4$$
$$0 + m = -6$$
$$m = -6$$

The solution is −6.

You Try It 3 Solve: $-2 + y = -5$

Your solution −3

Example 4 Solve: $3 = y - 2$

Solution

$$3 = y - 2$$
$$3 + 2 = y - 2 + 2$$
$$5 = y + 0$$
$$5 = y$$

The solution is 5.

You Try It 4 Solve: $7 = y + 8$

Your solution −1

Example 5 Solve: $\frac{2}{7} = \frac{5}{7} + t$

Solution

$$\frac{2}{7} = \frac{5}{7} + t$$
$$\frac{2}{7} - \frac{5}{7} = \frac{5}{7} - \frac{5}{7} + t$$
$$-\frac{3}{7} = 0 + t$$
$$-\frac{3}{7} = t$$

The solution is $-\frac{3}{7}$.

You Try It 5 Solve: $\frac{1}{5} = z + \frac{4}{5}$

Your solution $-\frac{3}{5}$

Solutions on p. A48

Objective C *To solve an equation of the form $ax = b$*

In solving an equation of the form $ax = b$, the goal is to simplify the given equation to one of the form *variable* = *constant*. The Multiplication Properties that follow are used to simplify equations to this form.

Multiplication Property of Reciprocals

The product of a non-zero term and its reciprocal equals 1.

$$a\left(\frac{1}{a}\right) = 1 \qquad \frac{1}{a}(a) = 1$$

or

$$\left(\frac{a}{b}\right)\left(\frac{b}{a}\right) = 1 \qquad \left(\frac{b}{a}\right)\left(\frac{a}{b}\right) = 1$$

Multiplication Property of One

The product of a term and 1 is the term.

$$a \cdot 1 = a \qquad 1 \cdot a = a$$

Multiplication Property of Equations

If a, b, and c are algebraic expressions and $c \neq 0$, then the equation $a = b$ has the same solutions as the equation $ac = bc$.

The Multiplication Property of Equations states that each side of an equation can be multiplied by the same non-zero number without changing the solutions of the equation.

Recall that the goal of solving an equation is to rewrite the equation in the form *variable* = *constant*. The Multiplication Property of Equations is used to rewrite an equation in this form by multiplying each side of the equation by the reciprocal of the coefficient.

➡ Solve: $\frac{2}{3}x = 8$

$$\frac{2}{3}x = 8$$

$$\left(\frac{3}{2}\right)\left(\frac{2}{3}\right)x = \left(\frac{3}{2}\right)8$$

$$1 \cdot x = 12$$

$$x = 12$$

• Multiply each side of the equation by $\frac{3}{2}$, the reciprocal of $\frac{2}{3}$. After simplifying, the equation will be in the form *variable* = *constant*.

Check: $\frac{2}{3}x = 8$

$\left(\frac{2}{3}\right)12$	8

$$8 = 8$$

The solution is 12.

Because division is defined in terms of multiplication, the Multiplication Property of Equations allows each side of an equation to be divided by the same non-zero quantity.

➡ Solve: $-4x = 24$

$-4x = 24$ • The goal is to rewrite the equation in the form *variable = constant.*

$\frac{-4x}{-4} = \frac{24}{-4}$ • Multiply each side of the equation by the reciprocal of -4. This is equivalent to dividing each side of the equation by -4. Then simplify.

$1x = -6$

$x = -6$

The solution is -6. You should check this solution.

When using the Multiplication Property of Equations, it is usually easier to multiply each side of the equation by the reciprocal of the coefficient when the coefficient is a fraction. Divide each side of the equation by the coefficient when the coefficient is an integer or a decimal.

Example 6 Solve: $-2x = 6$

Solution

$-2x = 6$

$\frac{-2x}{-2} = \frac{6}{-2}$

$1x = -3$

$x = -3$

The solution is -3.

You Try It 6 Solve: $4z = -20$

Your solution -5

Example 7 Solve: $-9 = \frac{3}{4}y$

Solution

$-9 = \frac{3}{4}y$

$\left(\frac{4}{3}\right)(-9) = \left(\frac{4}{3}\right)\left(\frac{3}{4}y\right)$

$-12 = 1y$

$-12 = y$

The solution is -12.

You Try It 7 Solve: $8 = \frac{2}{5}n$

Your solution 20

Example 8 Solve: $6z - 8z = -5$

Solution

$6z - 8z = -5$ Combine like terms.

$-2z = -5$

$\frac{-2z}{-2} = \frac{-5}{-2}$

$1z = \frac{5}{2}$

$z = \frac{5}{2}$

The solution is $\frac{5}{2}$ or $2\frac{1}{2}$.

You Try It 8 Solve: $\frac{2}{3}t - \frac{1}{3}t = -2$

Your solution -6

Solutions on p. A48

Objective D To solve application problems

Example 9
An accountant for an auto parts store found that the weekly profit for the store was \$850 and that the total amount spent during the week was \$1200. Use the formula $P = R - C$, where P is the profit, R is the revenue, and C is the amount spent, to find the revenue for the week.

Strategy
To find the revenue for the week, replace the variables P and C in the formula by the given values, and solve for R.

Solution

$$\begin{aligned} P &= R - C \\ 850 &= R - 1200 \\ 850 + 1200 &= R - 1200 + 1200 \\ 2050 &= R + 0 \\ 2050 &= R \end{aligned}$$

The revenue for the week was \$2050.

You Try It 9
A clothing store's regular price for a pair of slacks is \$30. During a storewide sale, the slacks were priced at \$22. Use the formula $S = R - D$, where S is the sale price, R is the regular price, and D is the discount, to find the discount.

Your strategy

Your solution
\$8

Example 10
A store manager uses the formula $S = R - D \cdot R$, where S is the sale price, R is the regular price, and D is the discount rate. During a clearance sale, all items are discounted 20%. Find the regular price of a jacket that is on sale for \$40.

Strategy
To find the regular price of the jacket, replace the variables S and D in the formula by the given values, and solve for R.

Solution

$$S = R - D \cdot R$$
$$40 = R - 0.20R$$
$$40 = 0.80R \qquad \bullet\ R - 0.20R = 1 \cdot R - 0.20R$$
$$\frac{40}{0.8} = \frac{0.80R}{0.80}$$
$$50 = R$$

The regular price of the jacket is \$50.

You Try It 10
An investor uses the formula $A = P + I \cdot P$, where A is the value of the investment in 1 year, P is the original investment, and I is the interest rate for the investment. Find the interest rate for an original investment of \$1000 that had a value of \$1120 after 1 year.

Your strategy

Your solution
12%

Solutions on p. A48

11.2 Exercises

Objective A

1. Is 2 a solution of $3x = 15$?
No

2. Is 4 a solution of $4x = 18$?
No

3. Is 2 a solution of $3x + 5 = 11$?
Yes

4. Is -3 a solution of $2x + 9 = 3$?
Yes

5. Is -2 a solution of $5x + 7 = 12$?
No

6. Is 2 a solution of $4 - 2x = 8$?
No

7. Is -2 a solution of $4x + 1 = 3x$?
No

8. Is 4 a solution of $5 - 2x = 4x$?
No

9. Is 3 a solution of $3x - 2 = x + 4$?
Yes

10. Is 2 a solution of $4x + 8 = 4 - 2x$?
No

11. Is 3 a solution of $3x - 7 = 5 - x$?
Yes

12. Is 3 a solution of $4 - 2x = 1 - x$?
Yes

13. Is 3 a solution of $x^2 - 5x + 1 = 10 - 5x$?
Yes

14. Is -5 a solution of $x^2 - 3x - 1 = 9 - 6x$?
Yes

15. Is -1 a solution of $x^2 - 4x + 4 = (x - 2)^2$?
Yes

16. Is 7 a solution of $x^2 + 2x + 1 = (x + 1)^2$?
Yes

17. Is -1 a solution of $2x(x - 1) = 3 - x$?
Yes

18. Is 2 a solution of $3x(x - 3) = x - 8$?
Yes

19. Is 2 a solution of $x(x - 2) = x^2 - 4$?
Yes

20. Is -4 a solution of $x(x + 4) = x^2 + 16$?
No

21. Is $-\frac{2}{3}$ a solution of $3x + 6 = 4$?
Yes

22. Is $\frac{1}{2}$ a solution of $2x - 7 = -3$?
No

23. Is $\frac{1}{4}$ a solution of $2x - 3 = 1 - 14x$?
Yes

24. Is $-\frac{1}{3}$ a solution of $5x - 2 = 1 - 2x$?
No

25. Is $-\frac{1}{4}$ a solution of $2 - 3x = 1 - 7x$?
Yes

26. Is $\frac{2}{3}$ a solution of $4 - 2x = 6 - 5x$?
Yes

27. Is $\frac{3}{4}$ a solution of $3x(x - 2) = x - 4$?
No

28. Is $\frac{2}{5}$ a solution of $5x(x + 1) = x + 3$?
No

29. Is $\frac{1}{2}$ a solution of $2x(x - 2) = x + 3$?
No

30. Is $-\frac{1}{3}$ a solution of $3x(2x + 4) = 3x - 7$?
No

31. Is 1.32 a solution of $x^2 - 3x = -0.8776 - x$?
No

32. Is -1.9 a solution of $x^2 - 3x = x + 3.8$?
No

33. Is 1.05 a solution of $x^2 + 3x = x(x + 3)$?
Yes

Objective B

Solve.

34. $x + 3 = 9$
6

35. $x + 7 = 5$
-2

36. $y - 6 = 16$
22

37. $z - 4 = 10$
14

38. $3 + n = 4$
1

39. $6 + x = 8$
2

40. $z + 7 = 2$
-5

41. $w + 9 = 5$
-4

42. $x - 3 = -7$
-4

43. $m - 4 = -9$
-5

44. $y + 6 = 6$
0

45. $t - 3 = -3$
0

46. $v - 7 = -4$
3

47. $x - 3 = -1$
2

48. $1 + x = 0$
-1

49. $3 + y = 0$
-3

50. $x - 10 = 5$
15

51. $y - 7 = 3$
10

52. $x + 4 = -7$
-11

53. $t - 3 = -8$
-5

54. $w + 5 = -5$
-10

55. $z + 6 = -6$
-12

56. $x + 7 = -8$
-15

57. $x + 2 = -5$
-7

58. $x + \frac{1}{2} = -\frac{1}{2}$
-1

59. $x - \frac{5}{6} = -\frac{1}{6}$
$\frac{2}{3}$

60. $y + \frac{7}{11} = -\frac{3}{11}$
$-\frac{10}{11}$

61. $\frac{2}{5} + x = -\frac{3}{5}$
-1

62. $\frac{7}{8} + y = -\frac{1}{8}$
-1

63. $\frac{1}{3} + x = \frac{2}{3}$
$\frac{1}{3}$

64. $x + \frac{1}{2} = -\frac{1}{3}$
$-\frac{5}{6}$

65. $y + \frac{3}{8} = \frac{1}{4}$
$-\frac{1}{8}$

66. $y + \frac{2}{3} = -\frac{3}{8}$
$-1\frac{1}{24}$

67. $t + \frac{1}{4} = -\frac{1}{2}$
$-\frac{3}{4}$

68. $x + \frac{1}{3} = \frac{5}{12}$
$\frac{1}{12}$

69. $y + \frac{2}{3} = -\frac{5}{12}$
$-1\frac{1}{12}$

Objective C

Solve.

70. $3y = 12$
4

71. $5x = 30$
6

72. $5z = -20$
-4

73. $3z = -27$
-9

74. $-2x = 6$
-3

75. $-4t = 20$
-5

76. $-5x = -40$
8

77. $-2y = -28$
14

78. $40 = 8x$ 5

79. $24 = 3y$ 8

80. $-24 = 4x$ −6

81. $-21 = 7y$ −3

82. $\frac{x}{3} = 5$ 15

83. $\frac{y}{2} = 10$ 20

84. $\frac{n}{4} = -2$ −8

85. $\frac{y}{7} = -3$ −21

86. $-\frac{x}{4} = 1$ −4

87. $\frac{-y}{3} = 5$ −15

88. $\frac{2}{3}w = 4$ 6

89. $\frac{5}{8}x = 10$ 16

90. $\frac{3}{4}v = -3$ −4

91. $\frac{2}{7}x = -12$ −42

92. $-\frac{1}{3}x = -2$ 6

93. $-\frac{1}{5}y = -3$ 15

94. $\frac{3}{8}x = -24$ −64

95. $\frac{5}{12}y = -16$ $-38\frac{2}{5}$

96. $-4 = -\frac{2}{3}z$ 6

97. $-8 = -\frac{5}{6}x$ $9\frac{3}{5}$

98. $-12 = -\frac{3}{8}y$ 32

99. $-9 = \frac{5}{6}t$ $-10\frac{4}{5}$

100. $\frac{2}{3}x = -\frac{2}{7}$ $-\frac{3}{7}$

101. $\frac{3}{7}y = \frac{5}{6}$ $1\frac{17}{18}$

102. $4x - 2x = 7$ $3\frac{1}{2}$

103. $3a - 6a = 8$ $-2\frac{2}{3}$

104. $\frac{4}{5}m - \frac{1}{5}m = 9$ 15

105. $\frac{1}{3}b - \frac{2}{3}b = -1$ 3

Objective D *Application Problems*

In Exercises 106 and 107, use the formula $A = P + I$, where A is the value of the investment after 1 year, P is the original investment, and I is the increase in value of the investment.

106. The value of an investment in a high-tech company after 1 year was \$17,700. The increase in value during the year was \$2700. Find the amount of the original investment. \$15,000

107. The original investment in a mutual fund was \$8000. The value of the mutual fund after 1 year was \$11,420. Find the increase in the value of the investment. \$3420

In Exercises 108 and 109, use the formula $V = N - D$, where V is the value of the car now, N is the original value of the car, and D is the depreciation.

108. The depreciation of a car that originally cost \$16,500 is \$8000. Find the value of the car now. \$8500

109. The value of a camper that originally cost \$22,400 is now \$16,000. Find the depreciation. \$6400

In Exercises 110 and 111, use the formula $D = M \cdot G$, where D is distance, M is miles per gallon, and G is the number of gallons.

110. Julio, a sales executive, averaged 27 mi/gal on a 621-mi trip. Find the number of gallons of gasoline used. 23 gal

111. The manufacturer of a subcompact car estimates that the car can travel 570 mi on a 15-gal tank of gas. Find the miles per gallon. 38 mi/gal

In Exercises 112 and 113, use the formula $S = \frac{D}{t}$, where S is the speed in miles per hour, D is the distance, and t is the time.

112. Allison drove 288 mi in 6 h. Find the average speed for her trip. 48 mi/h

113. A truck driver drove 12 h at an average speed of 54 mi/h. Find the distance the truck driver drove in the 12 h. 648 mi

In Exercises 114 and 115, use the formula $S = C + M$, where S is the selling price, C is the cost, and M is the markup.

114. A computer store sells a computer for \$2240. The computer has a markup of \$420. Find the cost of the computer. \$1820

115. A department store buys shirts for \$13.50 and sells the shirts for \$19.80. Find the markup on each shirt. \$6.30

In Exercises 116 and 117, use the formula $S = C + R \cdot C$, where S is the selling price, C is cost, and R is the markup rate.

116. A store manager uses a markup of 24% on all appliances. Find the cost of a blender that sells for \$52.70. \$42.50

117. A record store uses a markup rate of 30%. Find the cost of a compact disk that sells for \$13.39. \$10.30

APPLYING THE CONCEPTS

118. [W] Is 2 a solution of $x = x + 4$? Try -2, 0, 3, 6, and 10. Do you think there is a solution of this equation? Why or why not? No. The equation has no solution.

119. [W] Is 0 a solution of $\frac{4}{x} = 4$. Explain your answer. No. You cannot divide by zero.

120. [W] Make up an equation of the form $x + a = b$ that has 3 as its solution.

121. [W] Make up an equation of the form $a - x = b$ that has -2 as its solution.

122. [W] In your own words, state the Addition Property of Equations and the Multiplication Property of Equations.

123. Write out the steps for solving the equation $x - 3 = -5$. Identify each property of real numbers and each property of equations as you use it. The complete solution is in the Solutions Manual.

124. Write out the steps for solving the equation $\frac{3}{4}x = 6$. Identify each property of real numbers and each property of equations as you use it. The complete solution is in the Solutions Manual.

11.3 General Equations: Part I

Objective A To solve an equation of the form $ax + b = c$

POINT OF INTEREST

Evariste Galois, despite being killed in a duel at the age of 21, made significant contributions to solving equations. In fact, there is a branch of Mathematics called Galois Theory, showing what kind of equations can and cannot be solved.

To solve an equation of the form $ax + b = c$, it is necessary to use both the Addition and the Multiplication Properties to simplify the equation to one of the form *variable* = *constant*.

➡ Solve: $\frac{x}{4} - 1 = 3$

$\frac{x}{4} - 1 = 3$ • The goal is to simplify the equation to one of the form *variable* = *constant*.

$\frac{x}{4} - 1 + 1 = 3 + 1$ • Add the opposite of the constant term −1 to each side of the equation. Then simplify (Addition Properties).

$\frac{x}{4} + 0 = 4$

$\frac{x}{4} = 4$

$4 \cdot \frac{x}{4} = 4 \cdot 4$ • Multiply each side of the equation by the reciprocal of the numerical coefficient of the variable term. Then simplify (Multiplication Properties).

$1x = 16$

$x = 16$

The solution is 16. • Write the solution.

Example 1 Solve: $3x + 7 = 2$

Solution

$$3x + 7 = 2$$
$$3x + 7 - 7 = 2 - 7$$
$$3x = -5$$
$$\frac{3x}{3} = \frac{-5}{3}$$
$$x = -\frac{5}{3}$$

The solution is $-1\frac{2}{3}$.

You Try It 1 Solve: $5x + 8 = 6$

Your solution $-\frac{2}{5}$

Example 2 Solve: $5 - x = 6$

Solution

$$5 - x = 6$$
$$5 - 5 - x = 6 - 5$$
$$-x = 1$$
$$(-1)(-x) = (-1) \cdot 1$$
$$x = -1$$

The solution is −1.

You Try It 2 Solve: $7 - x = 3$

Your solution 4

Solutions on p. A49

Objective B To solve application problems

Example 3

Find the Celsius temperature when the Fahrenheit temperature is 212°. Use the formula $F = \frac{9}{5}C + 32$, where F is the Fahrenheit temperature and C is the Celsius temperature.

Strategy

To find the Celsius temperature, replace the variable F in the formula by the given value and solve for C.

Solution

$$F = \frac{9}{5}C + 32$$
$$212 = \frac{9}{5}C + 32$$
$$212 - 32 = \frac{9}{5}C + 32 - 32$$
$$180 = \frac{9}{5}C$$
$$\frac{5}{9} \cdot 180 = \frac{5}{9} \cdot \frac{9}{5}C$$
$$100 = C$$

The Celsius temperature is 100°.

You Try It 3

Find the Celsius temperature when the Fahrenheit temperature is $-22°$. Use the formula $F = \frac{9}{5}C + 32$, where F is the Fahrenheit temperature and C is the Celsius temperature.

Your strategy

Your solution

-30°C

Example 4

To find the total cost of production, an economist uses the formula $T = U \cdot N + F$, where T is the total cost, U is the cost per unit, N is the number of units made, and F is the fixed cost. Find the number of units made during a week when the total cost was $8000, the cost per unit was $16, and the fixed costs were $2000.

Strategy

To find the number of units made, replace the variables T, U, and F in the formula by the given values and solve for N.

Solution

$$T = U \cdot N + F$$
$$8000 = 16 \cdot N + 2000$$
$$8000 - 2000 = 16 \cdot N + 2000 - 2000$$
$$6000 = 16 \cdot N$$
$$\frac{6000}{16} = \frac{16 \cdot N}{16}$$
$$375 = N$$

The number of units made was 375.

You Try It 4

Find the cost per unit during a week when the total cost was $4500, the number of units produced was 250, and the fixed costs were $1500. Use the formula $T = U \cdot N + F$, where T is the total cost, U is the cost per unit, N is the number of units made, and F is the fixed cost.

Your strategy

Your solution

$12

Solutions on p. A49

11.3 Exercises

Objective A

Solve.

1. $3x + 5 = 14$
3

2. $5z + 6 = 31$
5

3. $2n - 3 = 7$
5

4. $4y - 4 = 20$
6

5. $5w + 8 = 3$
-1

6. $3x + 10 = 1$
-3

7. $3z - 4 = -16$
-4

8. $6x - 1 = -13$
-2

9. $5 + 2x = 7$
1

10. $12 + 7x = 33$
3

11. $6 - x = 3$
3

12. $4 - x = -2$
6

13. $3 - 4x = 11$
-2

14. $2 - 3x = 11$
-3

15. $5 - 4x = 17$
-3

16. $8 - 6x = 14$
-1

17. $3x + 6 = 0$
-2

18. $5x - 20 = 0$
4

19. $-3x - 4 = -1$
-1

20. $-7x - 22 = -1$
-3

21. $12x - 30 = 6$
3

22. $9x - 7 = 2$
1

23. $3x + 7 = 4$
-1

24. $8x + 13 = 5$
-1

25. $-2x + 11 = -3$
7

26. $-4x + 15 = -1$
4

27. $14 - 5x = 4$
2

28. $7 - 3x = 4$
1

29. $-8x + 7 = -9$
2

30. $-7x + 13 = -8$
3

31. $9x + 13 = 13$
0

32. $-2x + 7 = 7$
0

33. $7x - 14 = 0$
2

34. $5x + 10 = 0$
-2

35. $4x - 4 = -4$
0

36. $-13x - 1 = -1$
0

37. $3x + 5 = 7$
$\frac{2}{3}$

38. $4x + 6 = 9$
$\frac{3}{4}$

39. $2x - 3 = 15$
9

40. $6x - 1 = 16$
$2\frac{5}{6}$

Solve.

41. $3x - 4 = 17$
7

42. $12x - 3 = 7$
$\frac{5}{6}$

43. $2x - 3 = -8$
$-2\frac{1}{2}$

44. $5x - 3 = -12$
$-1\frac{4}{5}$

45. $-6x + 2 = -7$
$1\frac{1}{2}$

46. $-3x + 9 = -1$
$3\frac{1}{3}$

47. $-2x - 3 = -7$
2

48. $-5x - 7 = -4$
$-\frac{3}{5}$

49. $3x + 8 = 2$
-2

50. $2x - 9 = 8$
$8\frac{1}{2}$

51. $3x - 7 = 0$
$2\frac{1}{3}$

52. $7x - 2 = 0$
$\frac{2}{7}$

53. $-5x + 2 = 7$
-1

54. $-2x + 9 = 12$
$-1\frac{1}{2}$

55. $-7x + 3 = 1$
$\frac{2}{7}$

56. $-2x + 9 = 6$
$1\frac{1}{2}$

57. $12x - 12 = 1$
$1\frac{1}{12}$

58. $7x - 7 = 1$
$1\frac{1}{7}$

59. $4x - 7 = 13$
5

60. $7x - 9 = 21$
$4\frac{2}{7}$

61. $\frac{1}{2}x - 2 = 3$
10

62. $\frac{1}{3}x + 1 = 4$
9

63. $\frac{3}{5}w - 1 = 2$
5

64. $\frac{2}{5}w + 5 = 6$
$2\frac{1}{2}$

65. $\frac{2}{9}t - 3 = 5$
36

66. $\frac{5}{9}t - 3 = 2$
9

67. $\frac{y}{3} - 6 = -8$
-6

68. $\frac{y}{2} - 2 = 3$
10

69. $\frac{x}{3} - 2 = -5$
-9

70. $\frac{x}{4} - 3 = 5$
32

71. $\frac{5}{8}v + 6 = 3$
$-4\frac{4}{5}$

72. $\frac{2}{3}v - 4 = 3$
$10\frac{1}{2}$

73. $\frac{4}{7}z + 10 = 5$
$-8\frac{3}{4}$

74. $\frac{3}{8}v - 3 = 4$
$18\frac{2}{3}$

75. $\frac{2}{9}x - 3 = 5$
36

76. $\frac{1}{2}x + 3 = -8$
-22

77. $\frac{3}{4}x - 5 = -4$
$1\frac{1}{3}$

78. $\frac{2}{3}x - 5 = -8$
$-4\frac{1}{2}$

79. $1.5x - 0.5 = 2.5$
2

80. $2.5w - 1.3 = 3.7$
2

81. $0.8t + 1.1 = 4.3$
4

82. $0.3v + 2.4 = 1.5$
-3

83. $0.4x - 2.3 = 1.3$
9

84. $1.2t + 6.5 = 2.9$
-3

85. $3.5y - 3.5 = 10.5$
4

Solve.

86. $1.9x - 1.9 = -1.9$
0

87. $0.32x + 4.2 = 3.2$
-3.125

88. $2.2y - 12 = 3.4$
7

89. $1.4t - 8.4 = -14$
-4

90. $0.25x + 3 = -4.5$
-30

91. $5x - 3x + 2 = 8$
3

92. $6m + 2m - 3 = 5$
1

93. $4a - 7a - 8 = 4$
-4

94. $3y - 8y - 9 = 6$
-3

95. $x - 4x + 5 = 11$
-2

96. $-2y + y - 3 = 6$
-9

97. $-4y - y - 8 = 12$
-4

98. $1.94x - 3.925 = 16.94$
10.755154

99. $0.032x - 0.0194 = 0.139$
4.95

100. $423.8x - 49.6 = 182.75$
0.5482538

101. $-3.256x + 42.38 = -16.9$
18.206388

102. $6.09x + 17.33 = 16.805$
-0.0862068

103. $1.925x + 32.87 = -16.994$
-25.903376

Objective B Application Problems

In Exercises 104 and 105, use the relationship between Fahrenheit temperature and Celsius temperature, which is given by the formula $F = \frac{9}{5}C + 32$, where F is the Fahrenheit temperature and C is the Celsius temperature.

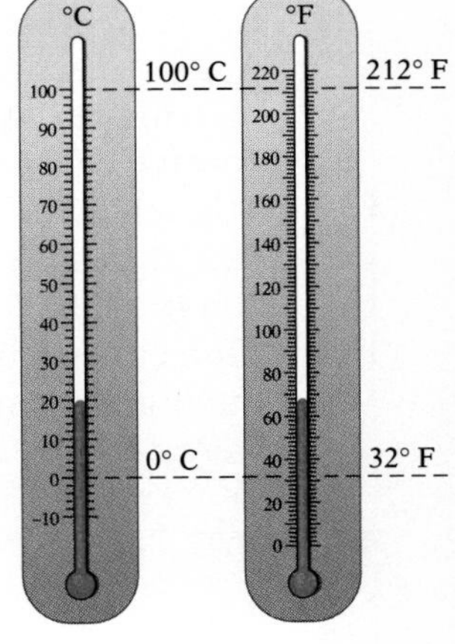

104. Find the Celsius temperature when the Fahrenheit temperature is $-40°$. $-40°C$

105. Find the Celsius temperature when the Fahrenheit temperature is 72°. Round to the nearest tenth of a degree. 22.2°C

In Exercises 106 and 107, use the formula $V = V_0 + 32t$, where V is the final velocity of a falling object, V_0 is the starting velocity of a falling object, and t is the time for the object to fall.

106. Find the time required for an object to reach a velocity of 480 ft/s when dropped from the top of a large building. (The starting velocity is 0 ft/s.) 15 s

107. Find the time required for a falling object to increase in velocity from 16 ft/s to 128 ft/s. 3.5 s

In Exercises 108 and 109, use the formula $T = U \cdot N + F$, where T is the total cost, U is the cost per unit, N is the number of units made, and F is the fixed cost.

108. Find the cost per unit during a week when the total cost was \$80,000, the total number of units produced was 500, and the fixed costs were \$15,000. \$130

109. Find the number of units made during a week when the total cost was \$25,000, the cost per unit was \$8, and the fixed costs were \$5000. 2500 units

In Exercises 110 and 111, use the formula $T = I \cdot R + B$, where T is the monthly tax, I is the monthly income, R is the income tax rate, and B is the base monthly tax.

110. The monthly tax that a mechanic pays is \$476. The mechanic's monthly tax rate is 22%, and the base monthly tax is \$80. Find the mechanic's monthly salary. \$1800

111. The monthly tax that Marcy, a teacher, pays is \$770. Her monthly income is \$3100, and the base monthly tax is \$150. Find Marcy's income tax rate. 20%

In Exercises 112–115, use the formula $M = S \cdot R + B$, where M is the monthly earnings, S is total sales, R is the commission rate, and B is the base monthly salary.

112. A book representative earns a base monthly salary of \$600 plus a 9% commission on total sales. Find the total sales during a month the representative earned \$3480. \$32,000

113. A sales executive earns a base monthly salary of \$1000 plus a 5% commission on total sales. Find the total sales during a month the executive earned \$2800. \$36,000

114. Miguel earns a base monthly salary of \$750. Find his commission rate during a month when total sales were \$42,000 and he earned \$2640. 4.5%

115. Tina earns a base monthly salary of \$500. Find her commission rate during a month when total sales were \$42,500 and her earnings were \$3560. 7.2%

APPLYING THE CONCEPTS

116. [W] Does the sentence "Solve $3x + 4(x - 3)$" make sense? Why or why not? No. You cannot *solve* an expression.

117. [W] Explain the difference between the word *equation* and the word *expression*. An equation has an equal sign, whereas an expression does not.

118. Make up an equation of the form $ax + b = c$ that has -3 as its solution. Answers will vary

119. Explain in your own words the steps you would take to solve the equation $\frac{2}{3}x - 4 = 10$. State the property of real numbers or the property of equations that is used at each step.

120. Do all equations have a solution? If so, explain why. If not, give an example of an equation with no solution.
No. $x = x + 1$ (Other answers are possible)

11.4 General Equations: Part II

Objective A To solve an equation of the form ax + b = cx + d

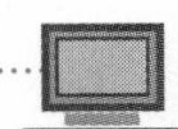

When a variable occurs on each side of an equation, the Addition Properties are used to rewrite the equation so that variable terms are on one side of the equation and constant terms are on the other side of the equation. Then the Multiplication Properties are used to simplify the equation to one of the form *variable = constant.*

➡ Solve: $4x - 6 = 8 - 3x$

$4x - 6 = 8 - 3x$

- **The goal is to write the equation in the form *variable = constant.***

$4x + 3x - 6 = 8 - 3x + 3x$
$7x - 6 = 8 + 0$
$7x - 6 = 8$

- **Add $3x$ to each side of the equation. Then simplify (Addition Properties). Now only one variable term occurs in the equation.**

$7x - 6 + 6 = 8 + 6$
$7x + 0 = 14$
$7x = 14$

- **Add 6 to each side of the equation. Then simplify (Addition Properties). Now only one constant term occurs in the equation.**

$\frac{7x}{7} = \frac{14}{7}$
$1x = 2$
$x = 2$

- **Divide each side of the equation by the numerical coefficient of the variable term. Then simplify (Multiplication Properties).**

The solution is 2.

- **Write the solution.**

Example 1

Solve: $\frac{2}{9}x - 3 = \frac{7}{9}x + 2$

Solution

$\frac{2}{9}x - 3 = \frac{7}{9}x + 2$

$\frac{2}{9}x - \frac{7}{9}x - 3 = \frac{7}{9}x - \frac{7}{9}x + 2$

$-\frac{5}{9}x - 3 = 2$

$-\frac{5}{9}x - 3 + 3 = 2 + 3$

$-\frac{5}{9}x = 5$

$\left(-\frac{9}{5}\right)\left(-\frac{5}{9}\right)x = \left(-\frac{9}{5}\right)5$

$x = -9$

The solution is -9.

You Try It 1

Solve: $\frac{1}{5}x - 2 = \frac{2}{5}x + 4$

Your solution

-30

Solution on p. A49

Objective B To solve an equation containing parentheses

When an equation contains parentheses, one of the steps in solving the equation requires use of the Distributive Property.

$$a(b + c) = ab + ac$$

The Distributive Property is used to rewrite a variable expression without parentheses.

➡ Solve: $4(3 + x) - 2 = 2(x - 4)$

$4(3 + x) - 2 = 2(x - 4)$
• The goal is to write the equation in the form *variable* = *constant.*

$12 + 4x - 2 = 2x - 8$
$10 + 4x = 2x - 8$
• Use the Distributive Property to simplify terms containing parentheses.

$10 + 4x - 2x = 2x - 2x - 8$
$10 + 2x = -8$
• Subtract $2x$ from each side of the equation.

$10 - 10 + 2x = -8 - 10$
$2x = -18$
• Subtract 10 from each side of the equation.

$\frac{2x}{2} = -\frac{18}{2}$
$x = -9$
• Divide each side of the equation by the numerical coefficient of the variable term.

The solution is -9.
• Write the solution.

Example 2 Solve: $3(x + 2) - x = 11$

Solution
$3(x + 2) - x = 11$
$3x + 6 - x = 11$
$2x + 6 = 11$
$2x + 6 - 6 = 11 - 6$
$2x = 5$
$\frac{2x}{2} = \frac{5}{2}$
$x = \frac{5}{2}$
$x = 2\frac{1}{2}$

The solution is $2\frac{1}{2}$.

You Try It 2 Solve: $4(x - 1) - x = 5$

Your solution 3

Solution on p. A49

11.4 Exercises

Objective A

Solve.

1. $6x + 3 = 2x + 5$
$\frac{1}{2}$

2. $7x + 1 = x + 19$
3

3. $3x + 3 = 2x + 2$
-1

4. $6x + 3 = 3x + 6$
1

5. $5x + 4 = x - 12$
-4

6. $3x - 12 = x - 8$
2

7. $7x - 2 = 3x - 6$
-1

8. $2x - 9 = x - 8$
1

9. $9x - 4 = 5x - 20$
-4

10. $8x - 7 = 5x + 8$
5

11. $2x + 1 = 16 - 3x$
3

12. $3x + 2 = -23 - 2x$
-5

13. $5x - 2 = -10 - 3x$
-1

14. $4x - 3 = 7 - x$
2

15. $2x + 7 = 4x + 3$
2

16. $7x - 6 = 10x - 15$
3

17. $x + 4 = 6x - 11$
3

18. $x - 6 = 4x - 21$
5

19. $3x - 7 = x - 7$
0

20. $2x + 6 = 7x + 6$
0

21. $3 - 4x = 5 - 3x$
-2

22. $6 - 2x = 9 - x$
-3

23. $7 + 3x = 9 + 5x$
-1

24. $12 + 5x = 9 - 3x$
$-\frac{3}{8}$

25. $5 + 2x = 7 + 5x$
$-\frac{2}{3}$

26. $9 + x = 2 + 3x$
$3\frac{1}{2}$

27. $8 - 5x = 4 - 6x$
-4

28. $9 - 4x = 11 - 5x$
2

29. $6x + 1 = 3x + 2$
$\frac{1}{3}$

30. $7x + 5 = 4x + 7$
$\frac{2}{3}$

31. $5x + 8 = x + 5$
$-\frac{3}{4}$

32. $9x + 1 = 3x - 4$
$-\frac{5}{6}$

33. $2x - 3 = 6x - 4$
$\frac{1}{4}$

Solve.

34. $2x - 3 = 4x + 3$
-3

35. $6x - 7 = 4x + 7$
7

36. $4 - x = 7 - 2x$
3

37. $7 - 2x = 5 - x$
2

38. $6 - 4x = -7 + 2x$
$2\frac{1}{6}$

39. $12 - 3x = -4 + 3x$
$2\frac{2}{3}$

40. $4 - 3x = 4 - 5x$
0

41. $6 - 3x = 6 - 5x$
0

42. $2x + 7 = 4x - 3$
5

43. $6x - 2 = 2x - 9$
$-1\frac{3}{4}$

44. $4x - 7 = -3x + 2$
$1\frac{2}{7}$

45. $6x - 3 = -5x + 8$
1

46. $7x - 5 = 3x + 9$
$3\frac{1}{2}$

47. $-6x - 2 = -8x - 4$
-1

48. $-7x + 2 = 3x - 8$
1

49. $-3 - 4x = 7 - 2x$
-5

50. $-8 + 5x = 8 + 6x$
-16

51. $3 - 7x = -2 + 5x$
$\frac{5}{12}$

52. $3x - 2 = 7 - 5x$
$1\frac{1}{8}$

53. $5x + 8 = 4 - 2x$
$-\frac{4}{7}$

54. $4 - 3x = 6x - 8$
$1\frac{1}{3}$

55. $12x - 9 = 3x + 12$
$2\frac{1}{3}$

56. $4x + 13 = -6x + 9$
$-\frac{2}{5}$

57. $\frac{5}{7}x - 3 = \frac{2}{7}x + 6$
21

58. $\frac{4}{5}x - 1 = \frac{1}{5}x + 5$
10

59. $\frac{3}{7}x + 5 = \frac{5}{7}x - 1$
21

60. $\frac{3}{4}x + 2 = \frac{1}{4}x - 9$
-22

Objective B

Solve.

61. $6x + 2(x - 1) = 14$
2

62. $3x + 2(x + 4) = 13$
1

63. $-3 + 4(x + 3) = 5$
-1

64. $8x - 3(x - 5) = 30$
3

65. $6 - 2(x + 4) = 6$
-4

66. $5 - 3(x + 2) = 8$
-3

Solve.

67. $5 + 7(x + 3) = 20$
$-\frac{6}{7}$

68. $6 - 3(x - 4) = 12$
2

69. $2x + 3(x - 5) = 10$
5

70. $3x - 4(x + 3) = 9$
-21

71. $3(x - 4) + 2x = 3$
3

72. $4 + 3(x - 9) = -12$
$3\frac{2}{3}$

73. $2x - 3(x - 4) = 12$
0

74. $4x - 2(x - 5) = 10$
0

75. $2x + 3(x + 4) = 7$
-1

76. $3(x + 2) + 7 = 12$
$-\frac{1}{3}$

77. $3(x - 2) + 5 = 5$
2

78. $4(x - 5) + 7 = 7$
5

79. $3x + 7(x - 2) = 5$
$1\frac{9}{10}$

80. $-3x - 3(x - 3) = 3$
1

81. $4x - 2(x + 9) = 8$
13

82. $3x - 6(x - 3) = 9$
3

83. $3x + 5(x - 2) = 10$
$2\frac{1}{2}$

84. $3x - 5(x - 1) = -5$
5

85. $3x + 4(x + 2) = 2(x + 9)$
2

86. $5x + 3(x + 4) = 4(x + 2)$
-1

87. $2x - 3(x - 4) = 2(x + 6)$
0

88. $3x - 4(x - 1) = 3(x - 2)$
$2\frac{1}{2}$

89. $7 - 2(x - 3) = 3(x - 1)$
$3\frac{1}{5}$

90. $4 - 3(x + 2) = 2(x - 4)$
$1\frac{1}{5}$

91. $6x - 2(x - 3) = 11(x - 2)$
4

92. $9x - 5(x - 3) = 5(x + 4)$
-5

Solve.

93. $6x - 3(x + 1) = 5(x + 2)$
$-6\frac{1}{2}$

94. $2x - 7(x - 2) = 3(x - 4)$
$3\frac{1}{4}$

95. $7 - (x + 1) = 3(x + 3)$
$-\frac{3}{4}$

96. $12 + 2(x - 9) = 3(x - 12)$
30

97. $2x - 3(x + 4) = 2(x - 5)$
$-\frac{2}{3}$

98. $3x + 2(x - 7) = 7(x - 1)$
$-3\frac{1}{2}$

99. $x + 5(x - 4) = 3(x - 8) - 5$
-3

100. $2x - 2(x - 1) = 3(x - 2) + 7$
$\frac{1}{3}$

101. $9x - 3(x - 4) = 13 + 2(x - 3)$
$-1\frac{1}{4}$

102. $3x - 4(x - 2) = 15 - 3(x - 2)$
$6\frac{1}{2}$

103. $3(x - 4) + 3x = 7 - 2(x - 1)$
$2\frac{5}{8}$

104. $2(x - 6) + 7x = 5 - 3(x - 2)$
$1\frac{11}{12}$

105. $3.67x - 5.3(x - 1.932) = 6.99$
1.9936196

106. $4.06x + 4.7(x + 3.22) = 1.774$
-1.5251141

107. $8.45(x - 10) = 3(x - 3.854)$
13.383119

108. $4(x - 1.99) - 3.92 = 3(x - 1.77)$
6.57

APPLYING THE CONCEPTS

109. If $2x - 2 = 4x + 6$, what is the value of $3x^2$? 48

110. If $3 + 2(4a - 3) = 5$ and $4 - 3(2 - 3b) = 11$, which is larger, a or b? b

111. [W] Explain what is wrong with the following demonstration, which suggests that $4 = 5$.

$$5x + 7 = 4x + 7$$
$$5x + 7 - 7 = 4x + 7 - 7 \quad \text{Subtract 7 from each side of the equation.}$$
$$5x = 4x$$
$$\frac{5x}{x} = \frac{4x}{x} \quad \text{Divide each side of the equation by } x.$$
$$5 = 4$$

112. [W] The equation $x = x + 1$ has no solution, whereas the solution of the equation $2x + 3 = 3$ is zero. Is there a difference between no solution and a solution of zero? Explain. Answers will vary

11.5 Translating Verbal Expressions into Mathematical Expressions

Objective A To translate a verbal expression into a mathematical expression given the variable

One of the major skills required in applied mathematics is to translate a verbal expression into a mathematical expression. Doing so requires recognizing the verbal phrases that translate into mathematical operations. Following is a partial list of the verbal phrases used to indicate the different mathematical operations:

Addition	more than	5 more than x	$x + 5$
	the sum of	the sum of w and 3	$w + 3$
	the total of	the total of 6 and z	$6 + z$
	increased by	x increased by 7	$x + 7$
Subtraction	less than	5 less than y	$y - 5$
	the difference between	the difference between w and 3	$w - 3$
	decreased by	8 decreased by x	$8 - x$
Multiplication	times	3 times x	$3x$
	the product of	the product of 4 and t	$4t$
	of	two-thirds of v	$\frac{2}{3}v$
Division	divided by	n divided by 3	$\frac{n}{3}$
	the quotient of	the quotient of z and 4	$\frac{z}{4}$
	the ratio of	the ratio of s to 6	$\frac{s}{6}$

Translating phrases that contain the words *sum, difference, product,* and *quotient* can sometimes cause a problem. In the examples at the right, note where the operation symbol is placed.

Note the placement of the fraction bar when translating the word *ratio.*

the *sum* of x *and* y	$x + y$
the *difference* of x *and* y	$x - y$
the *product* of x *and* y	$x \cdot y$
the *quotient* of x *and* y	$\frac{x}{y}$
the *ratio* of x *to* y	$\frac{x}{y}$

➡ Translate "the quotient of n and the sum of n and 6" into a mathematical expression.

the *quotient* of n *and* the *sum* of n *and* 6 $\qquad \frac{n}{n+6}$

Example 1 Translate "the sum of 5 and the product of 4 and n" into a mathematical expression.

Solution $5 + 4n$

You Try It 1 Translate "the difference between 8 and twice t" into a mathematical expression.

Your solution $8 - 2t$

Solution on p. A50

Example 2 Translate "the product of 3 and the difference between z and 4" into a mathematical expression.

Solution $3(z - 4)$

You Try It 2 Translate "the quotient of 5 and the product of 7 and x" into a mathematical expression.

Your solution $\frac{5}{7x}$

Solution on p. A50

Objective B To translate a verbal expression into a mathematical expression by assigning the variable

In most applications that involve translating phrases into mathematical expressions, the variable to be used is not given. To translate these phrases, we must assign a variable to the unknown quantity before writing the mathematical expression.

➡ Translate "the difference between seven and twice a number" into a mathematical expression.

The difference between seven and twice a number
- **Identify the phrases that indicate the mathematical operations.**

The unknown number: n
- **Assign a variable to one of the unknown quantities.**

Twice the number: $2n$
- **Use the assigned variable to write an expression for any other unknown quantity.**

$7 - 2n$
- **Use the identified operations to write the mathematical expression.**

Example 3
Translate "four less than some number" into a mathematical expression.

Solution
The unknown number: x

$x - 4$

You Try It 3
Translate "twelve decreased by some number" into a mathematical expression.

Your solution
$12 - x$

Example 4
Translate "the total of a number and the square of the number" into a mathematical expression.

Solution
The unknown number: c
The square of the number: c^2

$c + c^2$

You Try It 4
Translate "the product of a number and one-half of the number" into a mathematical expression.

Your solution
$(n)\left(\frac{n}{2}\right)$

Solutions on p. A50

11.5 Exercises

Objective A

Translate into a mathematical expression.

1. 9 less than y
$y - 9$

2. w divided by 7
$\frac{w}{7}$

3. z increased by 3
$z + 3$

4. the product of -2 and x
$-2x$

5. the sum of two-thirds of n and n
$\frac{2}{3}n + n$

6. the difference between the square of r and r
$r^2 - r$

7. the quotient of m and the difference between m and 3
$\frac{m}{m - 3}$

8. v increased by twice v
$v + 2v$

9. the product of 9 and the sum of 4 more than x
$9(x + 4)$

10. the total of a and the quotient of a and 7
$a + \frac{a}{7}$

11. the difference between n and the product of -5 and n
$n - (-5)n$

12. x decreased by the quotient of x and 2
$x - \frac{x}{2}$

13. the product of c and one-fourth of c
$c\left(\frac{1}{4}c\right)$

14. the quotient of 3 less than z and z
$\frac{z - 3}{z}$

15. the total of the square of m and twice the square of m
$m^2 + 2m^2$

16. the product of y and the sum of y and 4
$y(y + 4)$

17. 2 times the sum of t and 6
$2(t + 6)$

18. the quotient of r and the difference between 8 and r
$\frac{r}{8 - r}$

19. x divided by the total of 9 and x
$\frac{x}{9 + x}$

20. the sum of z and the product of 6 and z
$z + 6z$

21. three times the sum of b and 6
$3(b + 6)$

22. the ratio of w to the sum of w and 8
$\frac{w}{w + 8}$

Objective B

Translate into a mathematical expression.

23. the square of a number
x^2

24. five less than some number
$x - 5$

25. a number divided by 20
$\frac{x}{20}$

26. the difference between a number and twelve
$x - 12$

27. four times some number
$4x$

28. the quotient of five and a number
$\frac{5}{x}$

Translate into a mathematical expression.

29. three-fourths of a number
$\frac{3}{4}x$

30. the sum of a number and seven
$x + 7$

31. four increased by some number
$4 + x$

32. the ratio of a number and 9
$\frac{x}{9}$

33. the difference between five times a number and the number
$5x - x$

34. six less than the total of three and a number
$(3 + x) - 6$

35. the product of a number and two more than the number
$x(2 + x)$

36. the quotient of six and the sum of nine and a number
$\frac{6}{x + 9}$

37. seven times the total of a number and eight
$7(x + 8)$

38. the difference between ten and the quotient of a number and two
$10 - \frac{x}{2}$

39. the square of a number plus the product of three and the number
$x^2 + 3x$

40. a number decreased by the product of five and the number
$x - 5x$

41. the sum of three more than a number and one-half of the number
$(x + 3) + \frac{1}{2}x$

42. eight more than twice the sum of a number and seven
$2(x + 7) + 8$

43. the quotient of three times a number and the number
$\frac{3x}{x}$

44. the square of a number divided by the sum of a number and twelve
$\frac{x^2}{x + 12}$

APPLYING THE CONCEPTS

45. A wire whose length is given as x inches is bent into a square. Express the length of a side of the square in terms of x. $\frac{x}{4}$

46. If the length of the line AC at the right is 10 inches and the length of AB is x inches, find the length of BC in terms of x. $10 - x$

47. Translate the expressions $2x + 3$ and $2(x + 3)$ into phrases.
3 more than twice x; Twice the sum of x and 3

48. Translate the expressions $\frac{2x}{7}$ and $\frac{2 + x}{7}$ into phrases.
The quotient of twice x and 7; The quotient of the sum of 2 and x and 7

49. [W] In your own words, explain how variables are used. Answers will vary

50. The chemical formula for water is H_2O. This formula means that there are two hydrogen atoms and one oxygen atom in each molecule of water. If x represents the number of atoms of oxygen in a glass of pure water, express the number of hydrogen atoms in the glass of water. $2x$

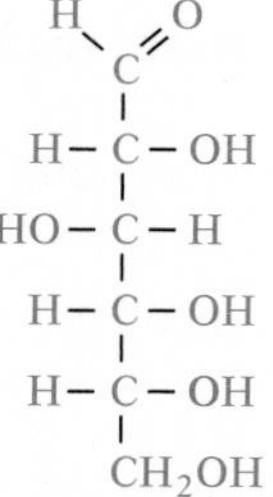

51. The chemical formula for one molecule of glucose (sugar) is $C_6H_{12}O_6$, where C is carbon, H is hydrogen, and O is oxygen. If x represents the number of atoms of hydrogen in a sample of pure sugar, express the number of carbon atoms and the number of oxygen atoms in terms of x.
$\frac{1}{2}x$; $\frac{1}{2}x$

11.6 Translating Sentences into Equations and Solving

Objective A To translate a sentence into an equation and solve

POINT OF INTEREST

Number puzzle problems similar to the one on this page have appeared in textbooks for hundreds of years. Here is one from a 1st century Chinese textbook: "When a number is divided by 3, the remainder is 2; when it is divided by 5, the remainder is 3; when it is divided by 7, the remainder is 2. Find the number." There are actually an infinite number of solutions to this problem. See if you can find one of them.

An equation states that two mathematical expressions are equal. Therefore, to translate a sentence into an equation requires recognition of the words or phrases that mean "equals." Some of these phrases are

equals
is
is equal to
amounts to
represents

} translate to =

Once the sentence is translated into an equation, the equation can be simplified to one of the form *variable* = *constant* and the solution found.

➡ Translate "three more than twice a number is seventeen" into an equation and solve.

The unknown number: n

- **Assign a variable to the unknown quantity.**

Three more than twice a number	is	seventeen

- **Find two verbal expressions for the same value.**

$$2n + 3 = 17$$
$$2n + 3 - 3 = 17 - 3$$
$$2n = 14$$
$$\frac{2n}{2} = \frac{14}{2}$$
$$n = 7$$

- **Write a mathematical expression for each verbal expression. Write the equals sign.**
- **Solve the resulting equation.**

The number is 7.

Example 1

Translate "a number decreased by six equals fifteen" into an equation and solve.

Solution

The unknown number: x

A number decreased by six	equals	fifteen

$$x - 6 = 15$$
$$x - 6 + 6 = 15 + 6$$
$$x = 21$$

The number is 21.

You Try It 1

Translate "a number increased by four equals twelve" into an equation and solve.

Your solution

8

Solution on p. A50

Example 2
The quotient of a number and six is five. Find the number.

Solution
The unknown number: z

The quotient of a number and six	is	five

$$\frac{z}{6} = 5$$
$$6 \cdot \frac{z}{6} = 6 \cdot 5$$
$$z = 30$$

The number is 30.

You Try It 2
The product of two and a number is ten. Find the number.

Your solution
5

Example 3
Eight decreased by twice a number is four. Find the number.

Solution
The unknown number: t

Eight decreased by twice a number	is	four

$$8 - 2t = 4$$
$$8 - 8 - 2t = 4 - 8$$
$$-2t = -4$$
$$\frac{-2t}{-2} = \frac{-4}{-2}$$
$$t = 2$$

The number is 2.

You Try It 3
The sum of three times a number and six equals four. Find the number.

Your solution
$-\frac{2}{3}$

Example 4
Three less than a number divided by seven is one. Find the number.

Solution
The unknown number: x

Three less than a number divided by seven	is	one

$$\frac{x}{7} - 3 = 1$$
$$\frac{x}{7} - 3 + 3 = 1 + 3$$
$$\frac{x}{7} = 4$$
$$7 \cdot \frac{x}{7} = 7 \cdot 4$$
$$x = 28$$

The number is 28.

You Try It 4
Three more than four times a number is ten. Find the number.

Your solution
$1\frac{3}{4}$

Solutions on p. A50

Objective B To solve application problems

Example 5
The cost of a television with remote control is \$649. This amount is \$125 more than the cost without remote control. Find the cost of the television without remote control.

Strategy
To find the cost of the television without remote control, write and solve an equation using C to represent the cost of the television without remote control.

Solution

\$649	is	\$125 more than the television without remote control

$$649 = C + 125$$
$$649 - 125 = C + 125 - 125$$
$$524 = C$$

The cost of the television without remote control is \$524.

You Try It 5
The sale price of a pair of slacks is \$18.95. This amount is \$6 less than the regular price. Find the regular price.

Your strategy

Your solution
\$24.95

Example 6
By purchasing a fleet of cars, a company receives a discount of \$972 on each car purchased. This amount is 18% off the regular price. Find the regular price.

Strategy
To find the regular price, write and solve an equation using P to represent the regular price of the car.

Solution

\$972	is	18% off the regular price

$$972 = 0.18 \cdot P$$
$$\frac{972}{0.18} = \frac{0.18P}{0.18}$$
$$5400 = P$$

The regular price is \$5400.

You Try It 6
At a certain speed, the engine rpm (revolutions per minute) of a car in fourth gear is 2500. This is two-thirds of the rpm of the engine in third gear. Find the rpm of the engine when it is in third gear.

Your strategy

Your solution
3750 rpm

Solutions on p. A50

Example 7
Ron Sierra charged \$815 for plumbing repairs in an office building. This charge included \$90 for parts and \$25 per hour for labor. Find the number of hours he worked in the office building.

Strategy
To find the number of hours worked, write and solve an equation using N to represent the number of hours worked.

Solution

\$815	included	\$90 for parts and \$25 per hour for labor

$$\begin{aligned} 815 &= 90 + 25N \\ 815 - 90 &= 90 - 90 + 25N \\ 725 &= 25N \\ \frac{725}{25} &= \frac{25N}{25} \\ 29 &= N \end{aligned}$$

Ron worked 29 h.

You Try It 7
Natalie Adams earned \$2500 last month for temporary accounting work. This amount was the sum of a base monthly salary of \$800 and an 8% commission on total sales. Find the total sales for the month.

Your strategy

Your solution
\$21,250

Example 8
The state income tax for Tim Fong last month was \$128. This amount is \$5 more than 8% of his monthly salary. Find Tim's monthly salary.

Strategy
To find Tim's monthly salary, write and solve an equation using S to represent his monthly salary.

Solution

\$128	is	\$5 more than 8% of the monthly salary

$$\begin{aligned} 128 &= 0.08 \cdot S + 5 \\ 128 - 5 &= 0.08 \cdot S + 5 - 5 \\ 123 &= 0.08 \cdot S \\ \frac{123}{0.08} &= \frac{0.08 \cdot S}{0.08} \\ 1537.50 &= S \end{aligned}$$

Tim's monthly salary is \$1537.50.

You Try It 8
The total cost to make a model ZY television is \$300. The cost includes \$100 for materials plus \$12.50 per hour for labor. How many hours of labor are required to make a model ZY television?

Your strategy

Your solution
16 h

Solutions on p. A51

11.6 Exercises

Objective A

1. The sum of a number and seven is twelve. Find the number.
$x + 7 = 12$; 5

2. A number decreased by seven is five. Find the number.
$x - 7 = 5$; 12

3. The product of three and a number is eighteen. Find the number.
$3x = 18$; 6

4. The quotient of a number and three is one. Find the number.
$\frac{x}{3} = 1$; 3

5. Five more than a number is three. Find the number.
$x + 5 = 3$; -2

6. A number divided by four is six. Find the number.
$\frac{x}{4} = 6$; 24

7. Six times a number is fourteen. Find the number.
$6x = 14$; $2\frac{1}{3}$

8. Seven less than a number is three. Find the number.
$x - 7 = 3$; 10

9. Five-sixths of a number is fifteen. Find the number.
$\frac{5}{6}x = 15$; 18

10. The total of twenty and a number is five. Find the number.
$20 + x = 5$; -15

11. The sum of three times a number and four is eight. Find the number.
$3x + 4 = 8$; $1\frac{1}{3}$

12. The sum of one-third of a number and seven is twelve. Find the number.
$\frac{1}{3}x + 7 = 12$; 15

13. Seven less than four times a number is nine. Find the number.
$4x - 7 = 9$; 4

14. The total of a number divided by four and nine is two. Find the number.
$\frac{x}{4} + 9 = 2$; -28

15. The ratio of a number to nine is fourteen. Find the number.
$\frac{x}{9} = 14$; 126

16. Five increased by five times a number is equal to 30. Find the number.
$5 + 5x = 30$; 5

17. Six less than the quotient of a number and four is equal to negative two. Find the number.
$\frac{x}{4} - 6 = -2$; 16

18. The product of a number plus three and two is eight. Find the number.
$(x + 3)2 = 8$; 1

19. The difference between seven and twice a number is thirteen. Find the number.
$7 - 2x = 13$; -3

20. Five more than the product of three and a number is eight. Find the number.
$3x + 5 = 8$; 1

21. Nine decreased by the quotient of a number and two is five. Find the number.
$9 - \frac{x}{2} = 5$; 8

22. The total of ten times a number and seven is twenty-seven. Find the number.
$10x + 7 = 27$; 2

23. The sum of three-fifths of a number and eight is two. Find the number.
$\frac{3}{5}x + 8 = 2$; -10

24. Five less than two thirds of a number is three. Find the number.
$\frac{2}{3}x - 5 = 3$; 12

25. The difference between a number divided by 4.186 and 7.92 is 12.529. Find the number.
$\frac{x}{4.186} - 7.92 = 12.529$; 85.599514

26. The total of 5.68 times a number and 132.7 is the number minus 29.265. Find the number.
$5.68x + 132.7 = x - 29.265$; -34.607905

Objective B Application Problems

Write an equation and solve.

27. Sears has a shoe on sale for $72.50. This amount is $4.25 less than the shoe sells for at Target. Find the price at Target. $76.75

28. As a restaurant manager, Uechi Kim is paid a salary of $832 a week. This is $58 more a week than the salary paid last year. Find the weekly salary paid to Uechi last year. $774

29. The value of a camper this year is $15,000, which is four-fifths of what its value was last year. Find the value of the camper last year. $18,750

30. The value of a house this year is $125,000. This amount is twice the value of the house 4 years ago. Find its value 4 years ago. $62,500

31. Suppose Kmart uses a markup rate of 40%. Find the selling price of a camcorder that cost $750. $1050

32. Sonia Parker works the night shift at Palomar Hospital and receives a salary of $3200 per month. This is four times the salary she was making 10 years ago. Find her salary 10 years ago. $800

33. The Manzanares family spends $680 on the house payment and utilities, which amounts to one-fourth of the family's monthly income. Find their monthly income. $2720

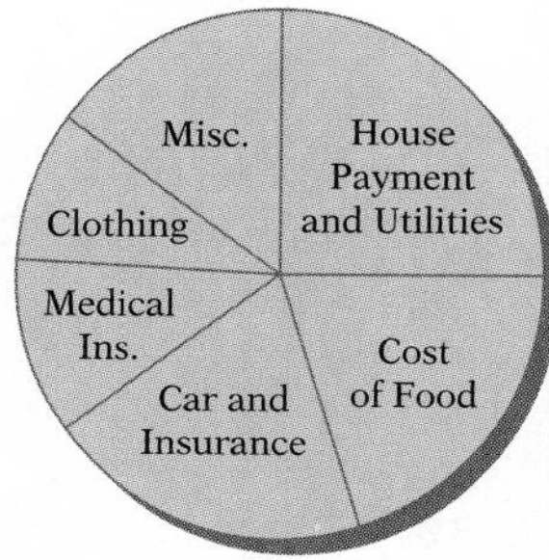

34. The cost of a scientific calculator is now one-third of what it cost 5 years ago. The cost of a scientific calculator is now $24. Find the cost of the calculator 5 years ago. $72

Write an equation and solve.

35. Assume that the Dell Computer Company has increased its output of computers by 400 computers per month. This amount represents an 8% increase over last year's production. Find the monthly output last year. 5000 computers

36. The population of a midwestern city has decreased by 12,000 in the last 5 years. This amount represents a 3% decrease. Find the city's population 5 years ago. 400,000 people

37. During a sale at Mervyn's, a camera is discounted $40, which is 20% off the regular price. Find the regular price. $200

38. The price of a pair of skis at the Solitude Ski Shop is $240. This price includes the store's cost for the skis plus a markup rate of 25%. Find Solitude's cost for the skis. $192

39. Budget Plumbing charged $385 for a water softener and installation. The charge included $310 for the water softener and $25 per hour for labor. How many hours of labor were charged? 3 h

40. Sandy's monthly salary as a sales representative was $2580. This amount included her base monthly salary of $600 plus a 3% commission on total sales. Find the total sales for the month. $66,000

41. Alpine Sporting Goods sold a pair of skates that cost $108 for $151.20. Find the markup rate on the pair of skates. 40%

42. A car was purchased for $5392. A down payment of $1000 was made. The remainder is to be paid in 36 equal monthly installments. Find the monthly payment. $122

43. There is a total of 600 ft^2 of window space in a house. This amount is 300 ft^2 less than 10% of the house's total wall space. Find the house's total wall space. 9000 ft^2

44. McPherson Cement charges $75 plus $24 for each yard of cement. How many yards of cement can be purchased for $363? 12 yd

45. A water flow restrictor has reduced the flow of water to 2 gal/min. This amount is 1 gal/min less than three-fifths the original flow rate. Find the original rate. 5 gal/min

Write an equation and solve.

46. The Fahrenheit temperature equals the sum of 32 and nine-fifths of the Celsius temperature. Find the Celsius temperature when the Fahrenheit temperature is 104°. 40°C

47. Assume that Paine Webber sales executives receive a base monthly salary of $600 plus an 8.25% commission on total sales per month. Find the total sales of a sales executive who receives a total of $4109.55 for the month. $42,540

48. A farmer harvested 28,336 bushels of corn. This amount represents a 12% increase over last year's crop. How many bushels of corn did the farmer harvest last year? 25,300 bushels

The graph at the right shows the cost of lunches at the executive dining room and the employee lunch room. Use data from this graph for Exercises 49 and 50.

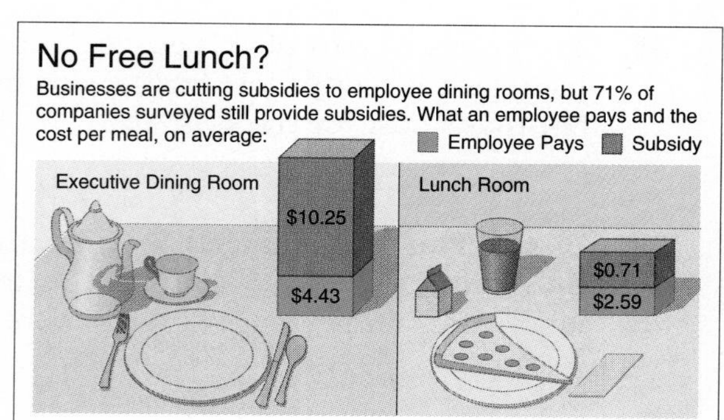

49. Find the amount subsidized in the executive dining room. What percent of the cost of the meal is subsidized? $10.25, 69.8%

50. Find the percent of the lunch room meal that is subsidized. Compare this percent with the percent in Exercise 49. 21.5%

APPLYING THE CONCEPTS

51. A man's boyhood lasted $\frac{1}{6}$ of his life, he played football for the next $\frac{1}{8}$ of his life, and he married 5 years after quitting football. A daughter was born after he had been married $\frac{1}{12}$ of his life. The daughter lived $\frac{1}{2}$ as many years as her father. The man died 6 years after his daughter. How old was the man when he died? Use a number line to illustrate the time. Then write an equation and solve it. 88 years old

52. [W] It is always important to check the answer to an application problem to be sure the answer makes sense. Consider the following problem. "A 4-quart mixture of fruit juices is made from apple juice and cranberry juice. There are 6 more quarts of apple juice than of cranberry juice. Write and solve an equation for the number of quarts of each juice used." Does the answer to this question make sense? Explain. $x + (x + 6) = 4$ No. There are more quarts of apple juice in the mixture than the total number allowed for the mixture.

53. [W] A formula is an equation that relates variables in a known way. Find two examples of formulas that are used in your college major. Explain what each of the variables represents. Answers will vary

Project in Mathematics

The y^x Key and Compound Interest

The y^x key is used to calculate powers of a number. For example, to evaluate 9^5, use the following keystrokes.

9 y^x 5 =

The display should be 59049.

The y^x key is useful in calculating compound interest.

Compound Interest Formula $P = A(1 + i)^n$

In this formula, A is the amount invested, i is the interest rate per period, n is the number of periods the amount is invested, and P is the principal or value of the investment at the end of n periods.

Example 1 An investment of $350 is placed in an account that earns 10% annual interest compounded quarterly. Find the principal after 5 years.

Because the interest is compounded quarterly (4 times a year), $i = \frac{0.10}{4} = 0.025$. In 5 years there are 20 quarters (4×5). $A = 350$. Enter the following on your calculator:

350 × (1 + .025) y^x 20 =

The display should be 573.5157.
The principal is $573.52, rounded to the nearest cent.

Example 2 An investment of $175 is placed in an account that earns 7% annual interest compounded daily. Find the principal after 3 years.

$i = \frac{0.07}{365} = 0.000192$; $n = 3 \times 365 = 1095$; $A = 175$

Enter the following on your calculator:

175 × (1 + 0.000192) y^x 1095 =

The display should be 215.9411.
The principal is $215.94, rounded to the nearest cent.

Chapter Summary

Key Words A *variable* is a letter that is used to stand for a quantity that is unknown.

A *variable expression* is an expression that contains one or more variables.

The *terms* of a variable expression are the addends of the expression.

A *variable term* is composed of a numerical coefficient and a variable part.

Like terms of a variable expression are the terms with the same variable part.

An *equation* expresses the equality of two mathematical expressions.

A *solution* of an equation is a value of the variable that results in a true equation.

Essential Rules ***The Commutative Property of Addition***

If a and b are two numbers, then $\boldsymbol{a + b = b + a}$.

The Associative Property of Addition

If a, b, and c are three numbers, then $\boldsymbol{a + (b + c) = (a + b) + c}$.

The Commutative Property of Multiplication

If a and b are two numbers, then $\boldsymbol{a \cdot b = b \cdot a}$.

The Associative Property of Multiplication

If a, b, and c are three numbers, then $\boldsymbol{a(b \cdot c) = (a \cdot b)c}$.

The Distributive Property

If a, b, and c are three numbers, then $\boldsymbol{a(b + c) = ab + bc}$.

Addition Property of Zero

The sum of a term and zero is the term. $\boldsymbol{a + 0 = a, 0 + a = a}$

Addition Property of Equations

If a, b, and c are algebraic expressions, then the equation $a = b$ has the same solution as the equation $a + c = b + c$.

Multiplication Property of Equations

If a, b, and c are algebraic expressions and $c \neq 0$, then the equation $a = b$ has the same solution as the equation $ac = bc$.

Multiplication Property of One

The product of a term and 1 is the term. $\boldsymbol{a \cdot 1 = a, 1 \cdot a = a}$

Chapter Review

SECTION 11.1

1. Evaluate $a^2 - 3b$ when $a = 2$ and $b = -3$.
13

2. Evaluate $a^2 - (b \div c)$ when $a = -2$, $b = 8$, and $c = -4$.
6

3. Simplify:
$6bc - 7bc + 2bc - 5bc$
$-4bc$

4. Simplify:
$\frac{1}{2}x^2 - \frac{1}{3}x^2 + \frac{1}{5}x^2 + 2x^2$
$\frac{71}{30}x^2$

5. Simplify: $-2(a - b)$
$-2a + 2b$

6. Simplify: $3x - 2(3x - 2)$
$-3x + 4$

SECTION 11.2

7. Is -2 a solution of $3x - 2 = -8$?
Yes

8. Is 5 a solution of $3x - 5 = -10$?
No

9. Solve: $x - 3 = -7$
-4

10. Solve: $x + 3 = -2$
-5

11. Solve: $-3x = 27$
-9

12. Solve: $-\frac{3}{8}x = -\frac{15}{32}$
$1\frac{1}{4}$

13. A tourist drove a rental car 621 mi on 27 gal of gas. Find the number of miles per gallon of gas. Use the formula $D = M \cdot G$, where D is distance, M is miles per gallon, and G is the number of gallons. 23 mi/gal

SECTION 11.3

14. Solve: $-2x + 5 = -9$
7

15. Solve: $35 - 3x = 5$
10

16. Solve: $\frac{2}{3}x + 3 = -9$

-18

17. Solve: $\frac{5}{6}x - 4 = 5$

$10\frac{4}{5}$

18. Find the Celsius temperature when the Fahrenheit temperature is 100°. Use the formula $F = \frac{9}{5}C + 32$, where F is the Fahrenheit temperature and C is the Celsius temperature. Round to the nearest tenth. 37.8°C

SECTION 11.4

19. Solve:
$7 - 3x = 2 - 5x$

$-2\frac{1}{2}$

20. Solve:
$6x - 9 = -3x + 36$

5

21. Solve:
$3(x - 2) + 2 = 11$

5

22. Solve:
$5x - 3(1 - 2x) = 4(2x - 1)$

$-\frac{1}{3}$

SECTION 11.5

23. Translate "the total of n and the quotient of n and 5" into a mathematical expression. $n + \frac{n}{5}$

24. Translate "the sum of five more than a number and one-third of the number" into a mathematical expression. $(n + 5) + \frac{1}{3}n$

SECTION 11.6

25. The difference between nine and twice a number is five. Find the number.
2

26. The product of five and a number is fifty. Find the number. 10

27. A compact disc player is now on sale for 40% off the regular price of \$380. Find the sale price. \$228

Chapter Test

1. Evaluate $c^2 - (2a + b^2)$ when $a = 3$, $b = -6$, and $c = -2$.
-38 [11.1A]

2. Evaluate $\frac{x^2}{y} - \frac{y^2}{x}$ for $x = 3$ and $y = -2$.
$-5\frac{5}{6}$ [11.1A]

3. Simplify: $3y - 2x - 7y - 9x$
$-4y - 11x$ [11.1B]

4. Simplify: $3y + 5(y - 3) + 8$
$8y - 7$ [11.1C]

5. Is 3 a solution of $x^2 + 3x - 7 = 3x - 2$?
No [11.2A]

6. Solve: $x - 12 = 14$
26 [11.2B]

7. Solve: $-5x = 14$
$-2\frac{4}{5}$ [11.2C]

8. Solve: $\frac{5}{8}x = -10$
-16 [11.2C]

9. A loan of \$6600 is to be paid in 48 equal monthly installments. Find the monthly payment. Use the formula $L = P \cdot N$, where L is the loan amount, P is the monthly payment, and N is the number of months. \$137.50 [11.2D]

10. Solve: $3x - 12 = -18$
-2 [11.3A]

11. Solve: $\frac{x}{5} - 12 = 7$
95 [11.3A]

12. Solve: $5 = 3 - 4x$

$-\frac{1}{2}$ [11.3A]

13. A clock manufacturer's fixed costs per month are \$5000. The unit cost for each clock is \$15. Find the number of clocks made during a month in which the total cost was \$65,000. Use the formula $T = U \cdot N + F$, where T is the total cost, U is the cost per unit, N is the number of units made, and F is the fixed cost. 4000 clocks [11.3B]

14. Solve: $8 - 3x = 2x - 8$

$3\frac{1}{5}$ [11.4A]

15. Solve: $3x - 4(x - 2) = 8$

0 [11.4B]

16. Translate "the sum of x and one-third of x" into a mathematical expression.

$x + \frac{1}{3}x$ [11.5A]

17. Translate "five times the sum of a number and three" into a mathematical expression.

$5(x + 3)$ [11.5B]

18. Translate "three less than two times a number is seven" into an equation and solve.

$2x - 3 = 7$; 5 [11.6A]

19. The total of five and three times a number is the number minus two. Find the number.

$-2\frac{1}{3}$ [11.6A]

20. Eduardo Santos earned \$3600 last month. This salary is the sum of the base monthly salary of \$1200 and a 6% commission on total sales. Find his total sales for the month. \$40,000 [11.6B]

Cumulative Review

1. Simplify:
$6^2 - (18 - 6) \div 4 + 8$
41 [1.6B]

2. Subtract: $3\frac{1}{6} - 1\frac{7}{15}$
$1\frac{7}{10}$ [2.5C]

3. Simplify:
$\left(\frac{3}{8} - \frac{1}{4}\right) \div \frac{3}{4} + \frac{4}{9}$
$\frac{11}{18}$ [2.8C]

4. Multiply: 9.67×0.0049
0.047383 [3.4A]

5. Write "$84 earned in 20 hours" as a unit rate.
$4.20/h [4.2B]

6. Solve the proportion $\frac{2}{3} = \frac{n}{40}$.
Round to the nearest hundredth.
$n = 26.67$ [4.3B]

7. Write $5\frac{1}{3}\%$ as a fraction.
$\frac{4}{75}$ [5.1A]

8. What percent of 30 is 42?
140% [5.3A]

9. 8 is 125% of what number?
6.4 [5.4A]

10. Multiply: 3 ft 9 in. × 5
18 ft 9 in. [8.1B]

11. Convert $1\frac{3}{8}$ lb to ounces.
22 oz [8.2A]

12. Convert 2 g 82 mg to grams.
2.082 g [9.2A]

13. Add: $-2 + 5 + (-8) + 4$
-1 [10.2A]

14. Find -6 less than 13.
19 [10.2B]

15. Simplify: $(-2)^2 - (-8) \div (3 - 5)^2$
6 [10.5A]

16. Evaluate $3ab - 2ac$ when $a = -2$, $b = 6$, and $c = -3$.
-48 [11.1A]

17. Simplify: $3z - 2x + 5z - 8x$
$-10x + 8z$ [11.1B]

18. Simplify: $6y - 3(y - 5) + 8$
$3y + 23$ [11.1C]

19. Solve: $2x - 5 = -7$
$x = -1$ [11.3A]

20. Solve: $7x - 3(x - 5) = -10$
$-6\frac{1}{4}$ [11.4B]

21. Solve: $-\frac{2}{3}x = 5$

$-7\frac{1}{2}$ [11.2C]

22. Solve: $\frac{x}{3} - 5 = -12$

$x = -21$ [11.3A]

23. In a Mathematics class of 34 students, 6 received an A grade. Find the percent of the students in the Mathematics class who received an A grade. Round to the nearest tenth of a percent. 17.6% [5.3B]

24. The manager of a pottery store used a markup rate of 40%. Find the price of a piece of pottery that cost the store $28.50. $39.90 [6.2B]

25. A department store has a suit regularly priced at $450 on sale for $369.
a. What is the discount? $81
b. What is the discount rate? 18% [6.2D]

26. A toy store borrowed $80,000 at a simple interest rate of 11% for 4 months. What is the simple interest due on the loan? Round to the nearest cent. $2933.33 [6.3A]

27. Translate "The sum of three times a number and 4" into a mathematical expression. $3n + 4$ [11.5B]

28. A car travels 318 mi in 6 h. Find the average speed in miles per hour. Use the formula $S = \frac{D}{t}$, where S is the average speed, D is the distance traveled, and t is the time of travel. 53 mi/h [11.6B]

29. Sunah Yee, a sales executive, receives a base salary of $800 plus an 8% commission on total sales. Find the total sales during a month in which Sunah earned $3400. Use the formula $M = S \cdot R + B$, where M is the monthly earnings, S is the total sales, R is the commission rate, and B is the base monthly salary. $32,500 [11.6B]

30. Three less than eight times a number is three more than five times the number. Find the number. 2 [11.6A]

Chapter

12

Geometry

Objectives

Section 12.1

To define and describe lines and angles

To define and describe geometric figures

To solve problems involving angles formed by intersecting lines

Section 12.2

To find the perimeter of plane geometric figures

To find the perimeter of composite geometric figures

To solve application problems

Section 12.3

To find the area of geometric figures

To find the area of composite geometric figures

To solve application problems

Section 12.4

To find the volume of geometric solids

To find the volume of composite geometric solids

To solve application problems

Section 12.5

To use a table to find the square root of a number

To find the unknown side of a right triangle using the Pythagorean Theorem

To solve application problems

Section 12.6

To solve similar and congruent triangles

To solve application problems

Mobius Strips and Klein Bottles

Some geometric shapes have very unusual characteristics. Among these figures are Mobius strips and Klein bottles.

A Mobius strip is formed by taking a long strip of paper and twisting it one-half turn. The resulting figure is called a "one-sided" surface. It is one-sided in the sense that if you tried to paint the strip in one continuous motion beginning at one spot, the entire surface would be painted the same color, unlike a strip that has not been twisted.

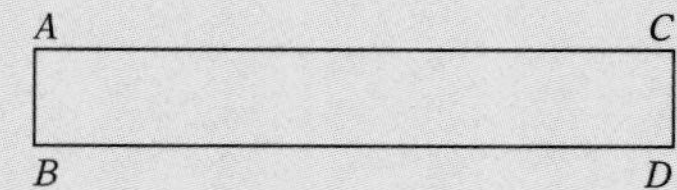

Another remarkable fact of being one-sided can be demonstrated by making a Mobius strip and sealing the junction with tape. Now cut the strip by cutting along the center of the strip. Try this; you will be amazed at the result.

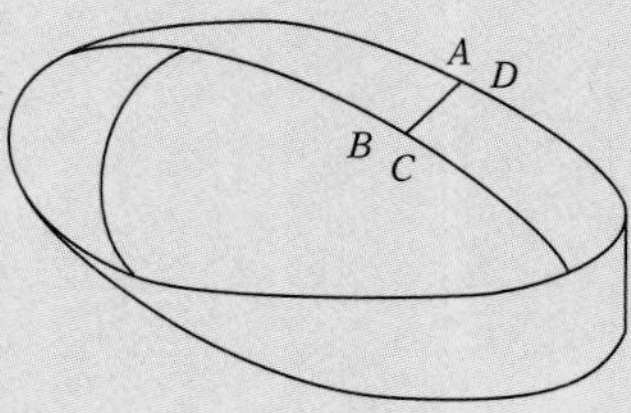

A second interesting surface is called a Klein bottle, which is a one-sided surface with no edges and no "inside" or "outside." A Klein bottle is formed by pulling the small open end of a tapering tube through the side of the tube and joining the ends of the small open end to the ends of the larger open end.

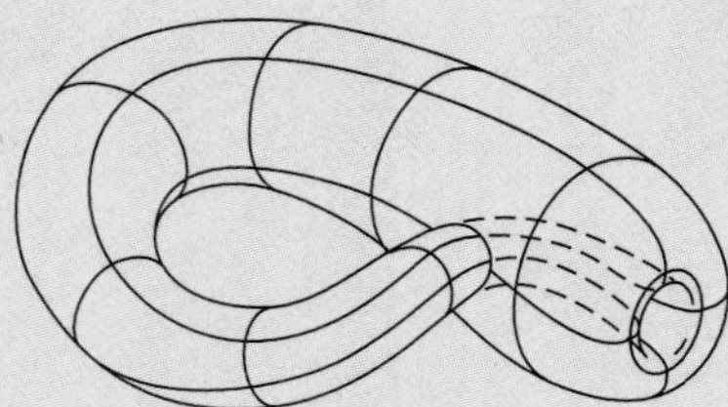

12.1 Angles, Lines, and Geometric Figures

Objective A To define and describe lines and angles

POINT OF INTEREST
Geometry is one of the oldest branches of Mathematics. Around 350 B.C., the Greek mathematician, Euclid, wrote the *Elements* which contained all of the known concepts of Geometry. Euclid's contribution was to unify various concepts into a single deductive system that was based on a set of axioms.

The word "geometry" comes from the Greek words for "earth" (*geo*) and "measure." The original purpose of geometry was to measure land. Today geometry is used in many sciences, such as physics, chemistry, and geology, and in applied fields such as mechanical drawing and astronomy. Geometric form is used in art and design.

Two basic geometric concepts are plane and space.

A **plane** is a flat surface, such as a tabletop or a blackboard. Figures that can lie totally in a plane are called **plane figures.**

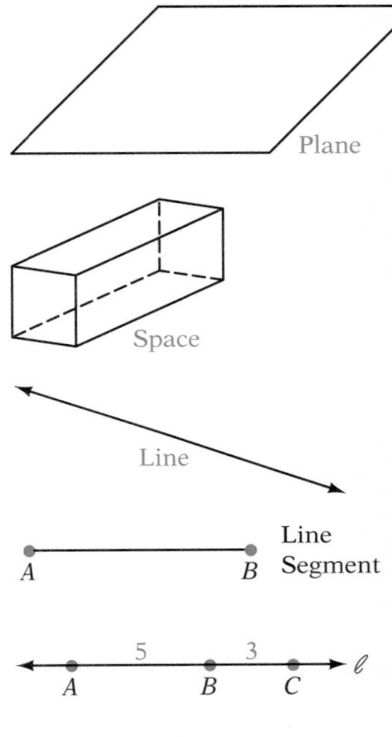

Space extends in all directions. Objects in space, such as trees, ice cubes, and doors, are called **solids.**

A **line** extends indefinitely in two directions in a plane. A line has no width.

A **line segment** is part of a line and has two endpoints. The line segment *AB* is shown in the figure.

The length of a line segment is the distance between the endpoints of the line segment. The length of a line segment may be expressed as the sum of two or more shorter line segments, as shown. For this example, $AB = 5$, $BC = 3$, and $AC = AB + BC = 5 + 3 = 8$. When no units, such as feet or meters, are given for the length, the distances are assumed to be in the same units of length.

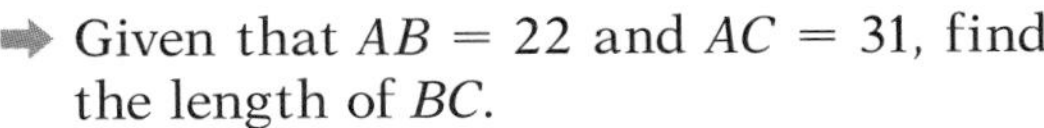

➡ Given that $AB = 22$ and $AC = 31$, find the length of BC.

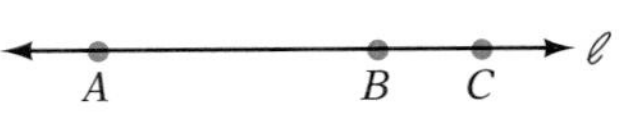

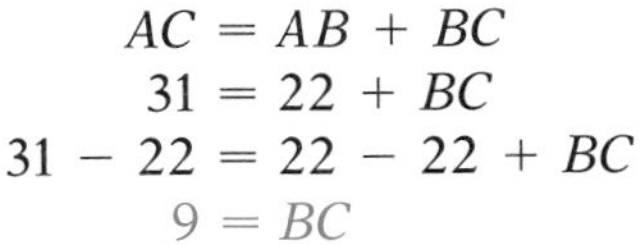

$$AC = AB + BC$$
$$31 = 22 + BC$$
$$31 - 22 = 22 - 22 + BC$$
$$9 = BC$$

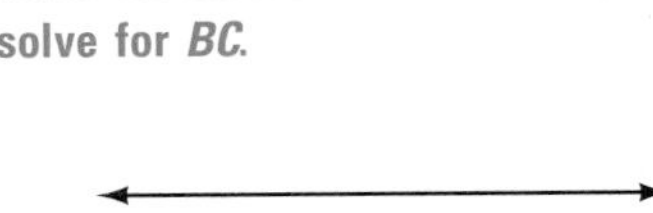

• **Substitute 22 for *AB* and 31 for *AC*, and solve for *BC*.**

Lines in a plane can be parallel or intersecting. **Parallel lines** never meet; the distance between them is always the same. **Intersecting lines** cross at a point in the plane.

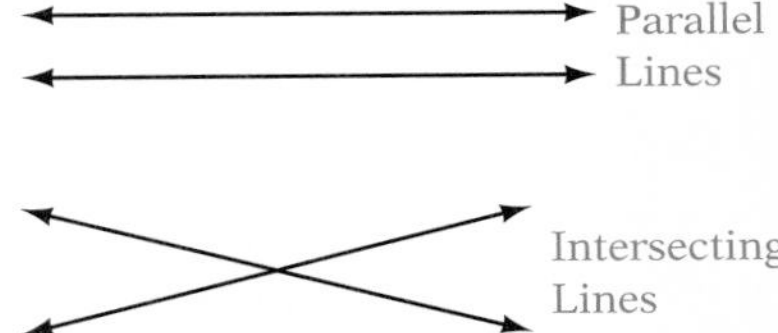

The symbol ∥ means "is parallel to." In the accompanying figure, $AB \parallel CD$ and $p \parallel q$. Note that line p contains line segment AB and that line q contains line segment CD. Parallel lines contain parallel line segments.

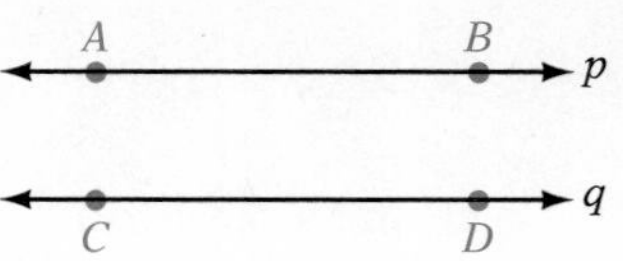

A **ray** starts at a point and extends indefinitely in one direction.

Ray

An **angle** is formed when two rays start from the same point. Rays r_1 and r_2 start from point B. The common endpoint is called the **vertex** of the angle.

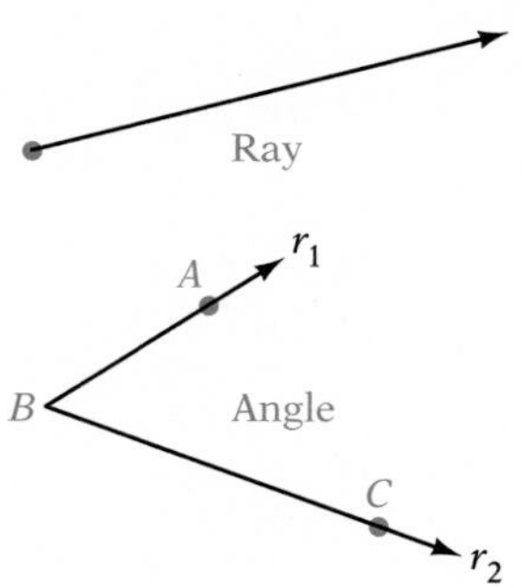

INSTRUCTOR NOTE

Show students that two naming conventions are used so that there is no confusion for figures such as the figure below.

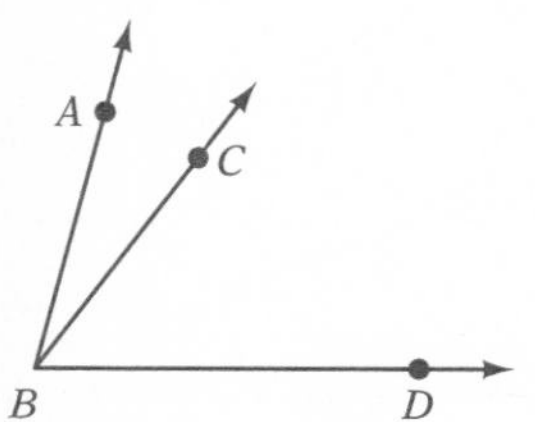

Just writing $\angle B$ does not identify a specific angle.

If A and C are points on rays r_1 and r_2 above, respectively, then the angle is called $\angle ABC$ or $\angle B$, where $\angle$ is the symbol for angle. Note that an angle is named by giving three points, with the vertex as the second point listed or, by the point at the vertex.

An angle can also be named by a variable written between the rays close to the vertex. In the figure, $\angle x = \angle QRS = \angle SRQ$ and $\angle y = \angle SRT = \angle TRS$. Note that in this figure, more than two rays meet at the vertex. In this case, the vertex cannot be used to name the angle.

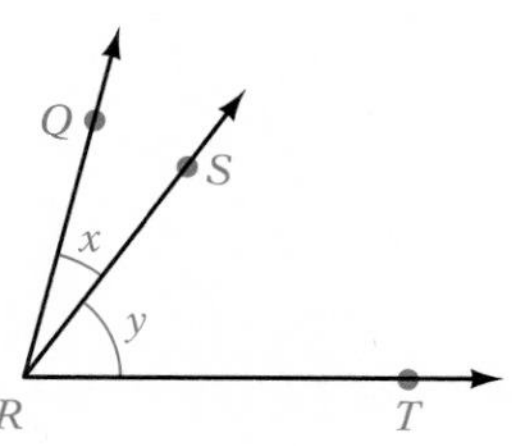

An angle is measured in **degrees.** The symbol for degree is °. One complete revolution is 360° (360 degrees).

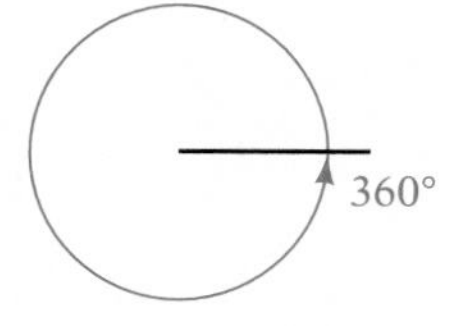

POINT OF INTEREST

The Babylonians knew that the earth was in approximately the same position in the sky every 365 days. Historians suggest that the reason one complete revolution of a circle is 360° is that 360 is the closest number to 365 that is divisible by many numbers.

$\frac{1}{4}$ of a revolution is 90°. A 90° angle is called a **right angle.** The symbol ∟ represents a right angle.

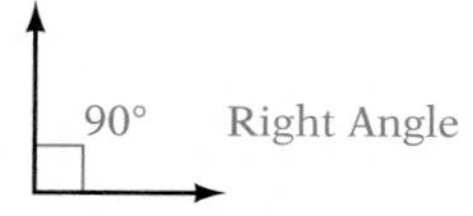

Perpendicular lines are intersecting lines that form right angles.

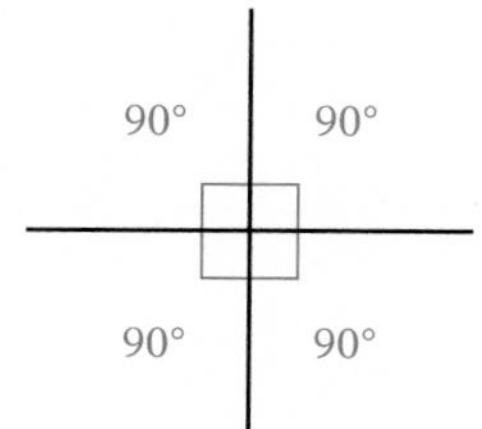

The symbol ⊥ means "is perpendicular to." In the accompanying figure, $AB \perp CD$ and $p \perp q$. Note that line p contains line segment AB and line q contains line segment CD. Perpendicular lines contain perpendicular line segments.

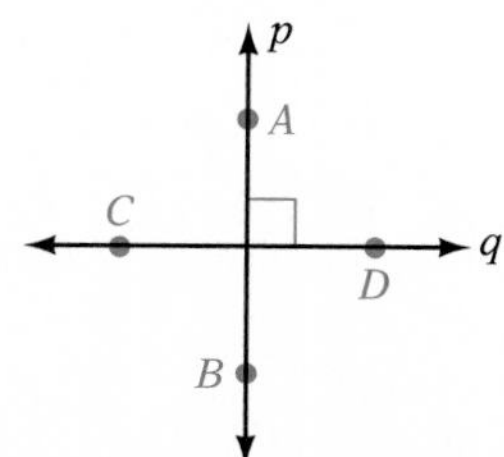

INSTRUCTOR NOTE
The corner of a page of this book serves as a good model for a 90° angle.

Complementary angles are two angles whose sum is 90°.

$\angle A + \angle B = 70° + 20° = 90°$
$\angle A$ and $\angle B$ are complementary angles.

$\frac{1}{2}$ of a revolution is 180°. A 180° angle is called a **straight angle.** $\angle AOB$ in the figure is a straight angle.

Supplementary angles are two angles whose sum is 180°.

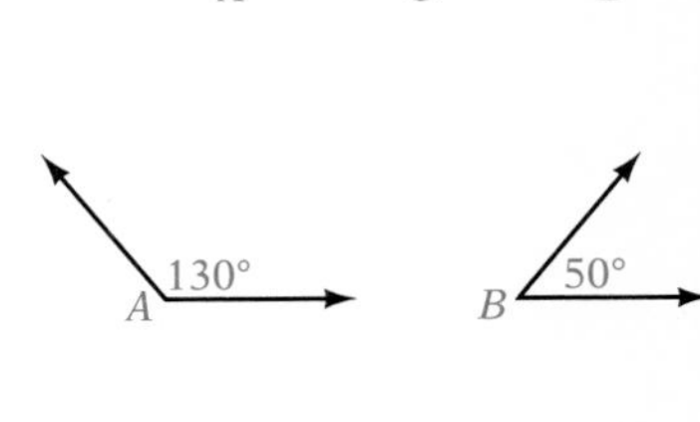

$\angle A + \angle B = 130° + 50° = 180°$
$\angle A$ and $\angle B$ are supplementary angles.

An **acute angle** is an angle whose measure is between 0° and 90°. $\angle B$ in the figure above is an acute angle. An **obtuse angle** is an angle whose measure is between 90° and 180°. $\angle A$ in the figure above is an obtuse angle.

In the accompanying figure, $\angle DAC = 45°$ and $\angle CAB = 55°$.

$\angle DAB = \angle DAC + \angle CAB = 45° + 55° = 100°$

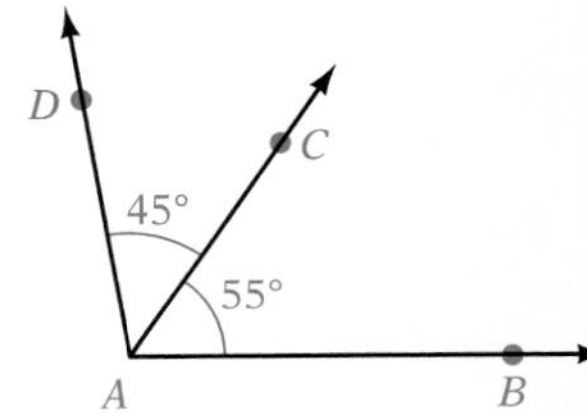

Example 1
Given that $MN = 15$, $NO = 18$, and $MP = 48$, find the length of OP.

Solution

$$\begin{aligned} MP &= MN + NO + OP \\ 48 &= 15 + 18 + OP \\ 48 &= 33 + OP \\ 48 - 33 &= 33 - 33 + OP \\ 15 &= OP \end{aligned}$$

You Try It 1
Given that $QR = 24$, $ST = 17$, and $QT = 62$, find the length of RS.

Your solution
21

Example 2
Find the complement of a 32° angle.

Solution
Let x represent the complement of 32°.

$$\begin{aligned} x + 32° &= 90° \\ x + 32° - 32° &= 90° - 32° \\ x &= 58° \end{aligned}$$

• The sum of complementary angles is 90°.

58° is the complement of 32°.

You Try It 2
Find the supplement of a 32° angle.

Your solution
148°

Solutions on p. A51

Example 3
Find the measure of $\angle x$ in the figure.

Solution
$\angle x = 90° - 47°$
$\angle x = 43°$

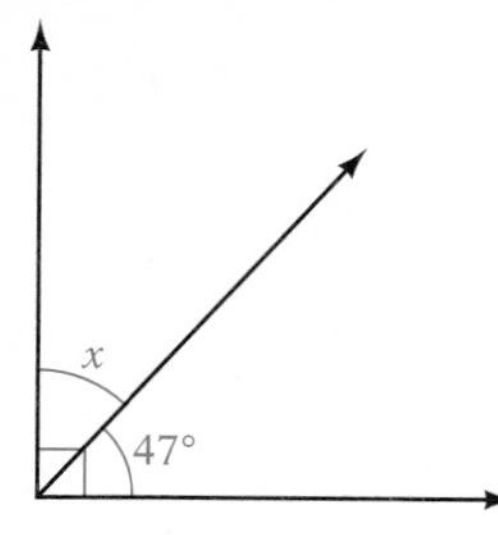

You Try It 3
Find the measure of $\angle a$ in the figure.

Your solution
50°

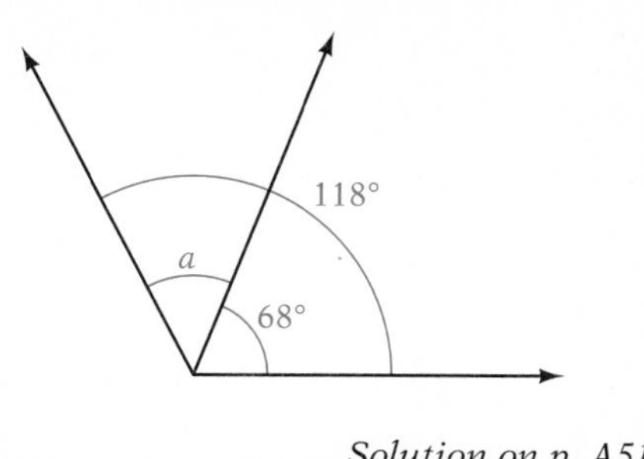

Solution on p. A51

Objective B *To define and describe geometric figures*

INSTRUCTOR NOTE
Much of this objective is the description of various geometric figures. After completing a discussion of these figures, ask questions such as "Is a cube a plane figure?" or "What geometric figure approximates the shape of the earth?"

A **triangle** is a closed three-sided plane figure. Figure *ABC* is a triangle, *AB* is called the **base.** The line *CD,* perpendicular to the base, is called the **height.**

The sum of the three angles in a triangle is 180°.

Angle A + Angle B + Angle C = 180°

➡ In the figure, $\angle A = 32°$ and $\angle B = 88°$. Find the measure of angle C.

$\angle A + \angle B + \angle C = 180°$ • The sum of the measures of the angles of a triangle is 180°.

$32° + 88° + \angle C = 180°$ • $\angle A = 32°$, $\angle B = 88°$.

$120° + \angle C = 180°$ • Solve for $\angle C$.

$120° - 120° + \angle C = 180° - 120°$

$\angle C = 60°$

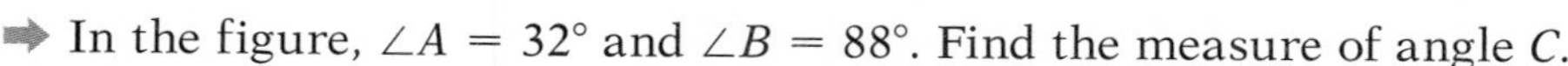

A **right triangle** contains one right angle. The side opposite the right angle is called the **hypotenuse.** The other two sides are called **legs.** In a right triangle, the two acute angles are complementary.

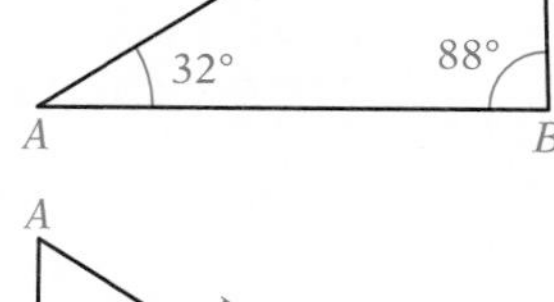

$\angle A + \angle B = 90°$

➡ In right triangle *ACB,* angle $A = 30°$. Find the measure of angle B.

$\angle A + \angle B = 90°$ • The two acute angles are complementary.

$30° + \angle B = 90°$ • $\angle A = 30°$.

$30° - 30° + \angle B = 90° - 30°$ • Solve for $\angle B$.

$\angle B = 60°$

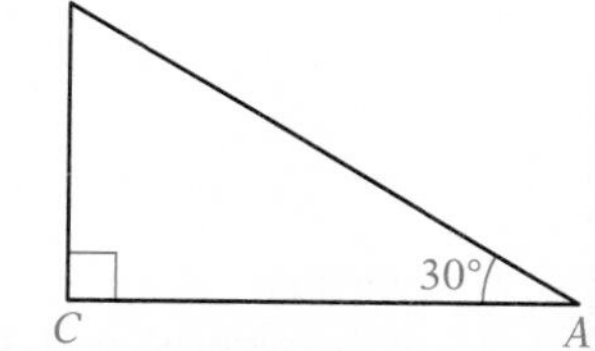

A **quadrilateral** is a closed four-sided plane figure. Three quadrilaterals with special characteristics are described here.

A **parallelogram** has opposite sides parallel and equal. The distance *AE* between the parallel sides is called the **height.**

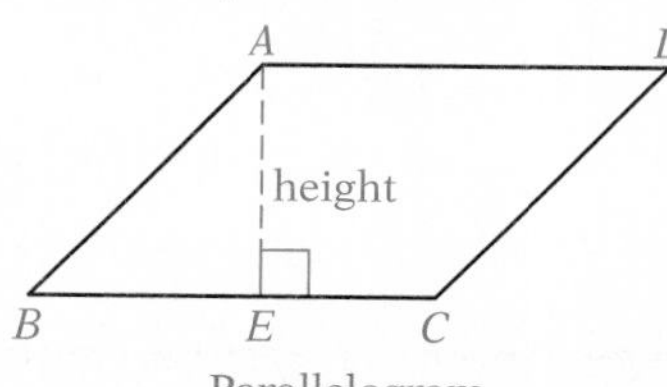

Parallelogram

A **rectangle** is a parallelogram that has four right angles.

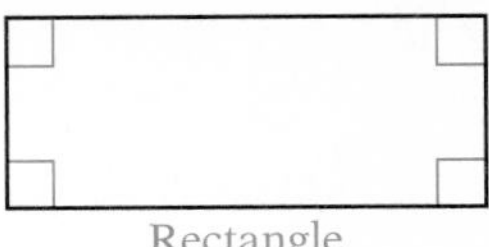
Rectangle

A **square** is a rectangle that has four equal sides.

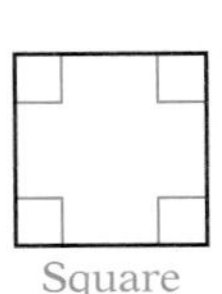
Square

A **circle** is a plane figure in which all points are the same distance from point O, which is called the **center** of the circle.

The **diameter** (d) is a line segment with endpoints on the circle. AB is a diameter of the circle.

The **radius** is a line segment from the center to a point on the circle. OC is a radius of the circle.

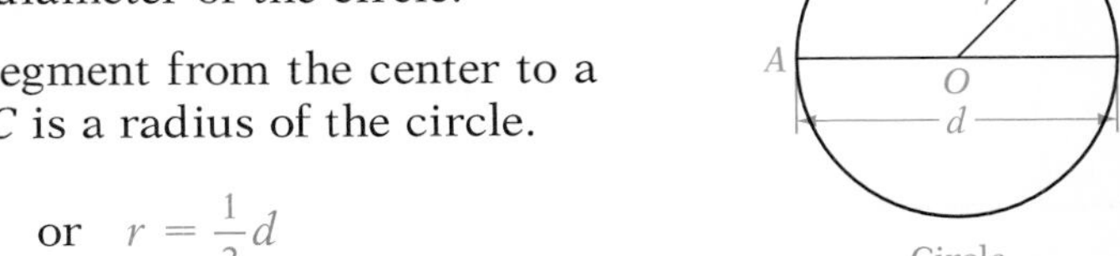

$$d = 2r \quad \text{or} \quad r = \frac{1}{2}d$$

Circle

➡ The line segment AB is a diameter of the circle shown. Find the radius of the circle.

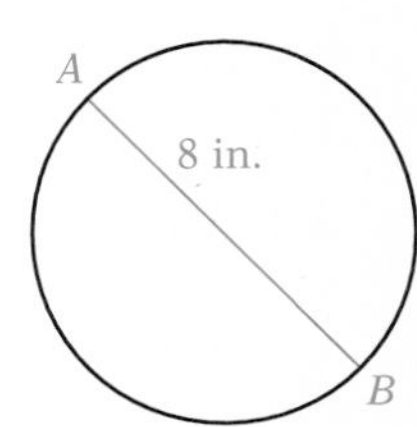

The radius is one-half the diameter. Therefore,

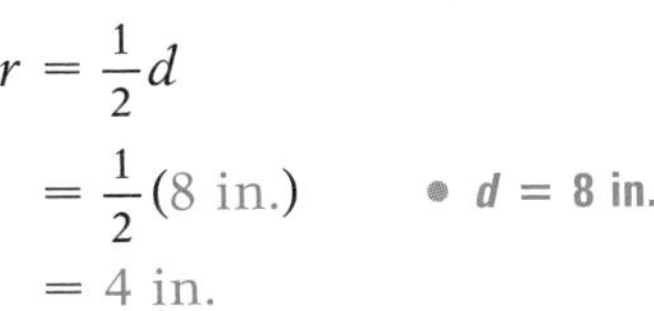

$$\begin{aligned} r &= \frac{1}{2}d \\ &= \frac{1}{2}(8 \text{ in.}) \qquad \bullet\ d = 8 \text{ in.} \\ &= 4 \text{ in.} \end{aligned}$$

Geometric solids are figures in space, or space figures. Four common space figures are the rectangular solid, cube, sphere, and cylinder.

A **rectangular solid** is a solid in which all six faces are rectangles.

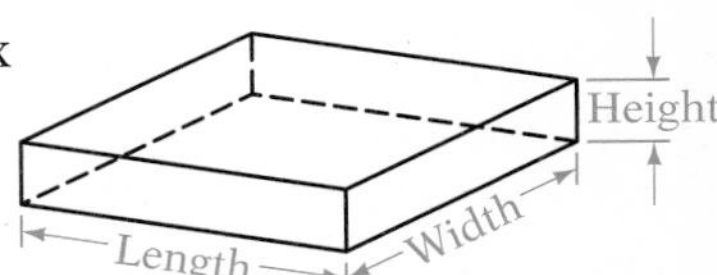

Rectangular Solid

A **cube** is a rectangular solid in which all six faces are squares.

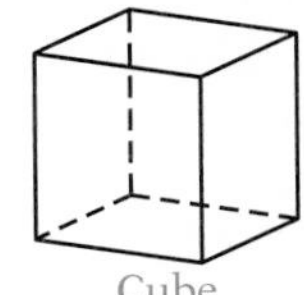
Cube

A **sphere** is a solid in which all points are the same distance from point O, which is called the **center** of the sphere.

The **diameter** of the sphere is a line segment going through the center with endpoints on the sphere. AB is a diameter of the sphere.

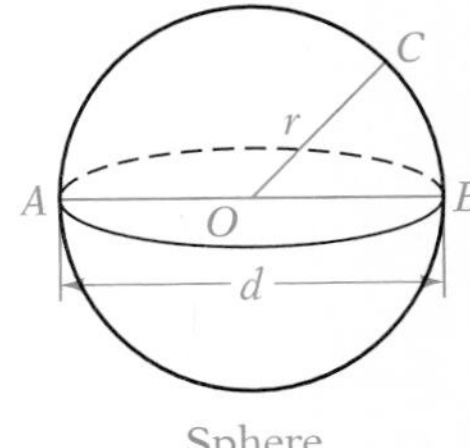

Sphere

The **radius** of the sphere is a line segment from the center to a point on the sphere. OC is a radius of the sphere.

$$r = \frac{1}{2}d \quad \text{or} \quad d = 2r$$

➡ The radius of the sphere shown is 5 cm. Find the diameter of the sphere.

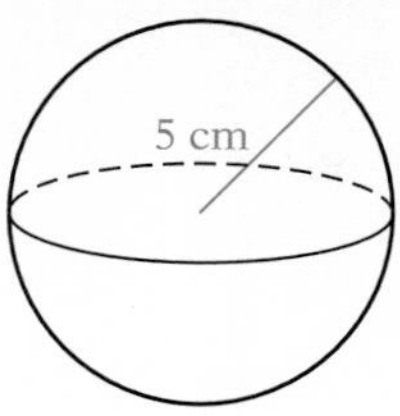

The diameter is equal to twice the radius.

$$d = 2r$$
$$= 2(5 \text{ cm}) \quad \bullet\ r = 5 \text{ cm}$$
$$= 10 \text{ cm}$$

The diameter is 10 cm.

The most common **cylinder** is one in which the bases are circles and are perpendicular to the height.

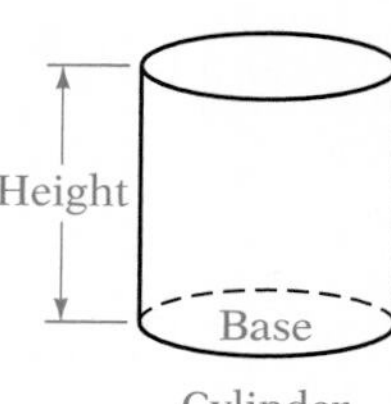

Cylinder

Example 4
One angle in a right triangle is equal to 30°. Find the measure of the other angles.

Solution
In a right triangle, the two acute angles are complementary.

$$\angle A + \angle B = 90°$$
$$\angle A + 30° = 90°$$
$$\angle A + 30° - 30° = 90° - 30°$$
$$\angle A = 60°$$

The other angles of the triangle are 90° and 60°.

You Try It 4
A right triangle has one angle equal to 7°. Find the measure of the other angles.

Your solution
90° and 83°

Example 5
Two angles of a triangle are 42° and 103°. Find the measure of the third angle of the triangle.

Solution
The sum of the three angles of a triangle is 180°.

$$\angle A + \angle B + \angle C = 180°$$
$$\angle A + 42° + 103° = 180°$$
$$\angle A + 145° = 180°$$
$$\angle A + 145° - 145° = 180° - 145°$$
$$\angle A = 35°$$

The third angle is 35°.

You Try It 5
Two angles of a triangle are 62° and 45°. Find the measure of the other angle.

Your solution
73°

Example 6
A circle has a radius of 8 cm. Find the diameter.

Solution
$$d = 2r$$
$$= 2 \cdot 8 \text{ cm} = 16 \text{ cm}$$

The diameter is 16 cm.

You Try It 6
A circle has a diameter of 8 in. Find the radius.

Your solution
4 in.

Solutions on p. A51

Objective C To solve problems involving angles formed by intersecting lines

Four angles are formed by the intersection of two lines. If the two lines are perpendicular, each of the four angles is a right angle. If the two lines are not perpendicular, then two of the angles formed are acute angles and two of the angles are obtuse angles. The two acute angles are always opposite each other, and the two obtuse angles are always opposite each other.

In the figure, $\angle w$ and $\angle y$ are acute angles. $\angle x$ and $\angle z$ are obtuse angles. Two angles that are on opposite sides of the intersection of two lines are called **vertical angles.** Vertical angles have the same measure. $\angle w$ and $\angle y$ are vertical angles. $\angle x$ and $\angle z$ are vertical angles.

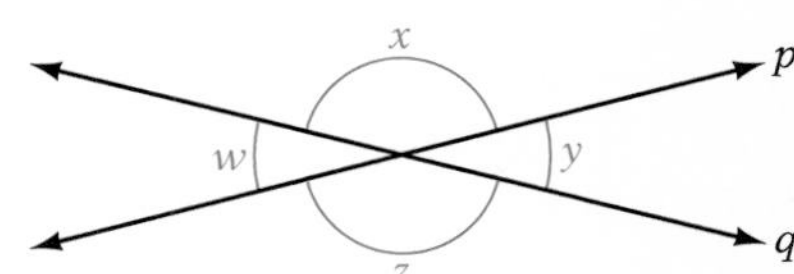

$\angle w = \angle y$
$\angle x = \angle z$

Two angles that share a common side are called **adjacent angles.** In the previous figure, $\angle x$ and $\angle y$ are adjacent angles, as are $\angle y$ and $\angle z$, $\angle w$ and $\angle x$, and $\angle w$ and $\angle z$. Adjacent angles of intersecting lines are supplementary angles.

$\angle x + \angle y = 180°$
$\angle y + \angle z = 180°$
$\angle z + \angle w = 180°$
$\angle w + \angle x = 180°$

➡ In the accompanying figure, given that $\angle c = 65°$, find the measure of angles a, b, and d.

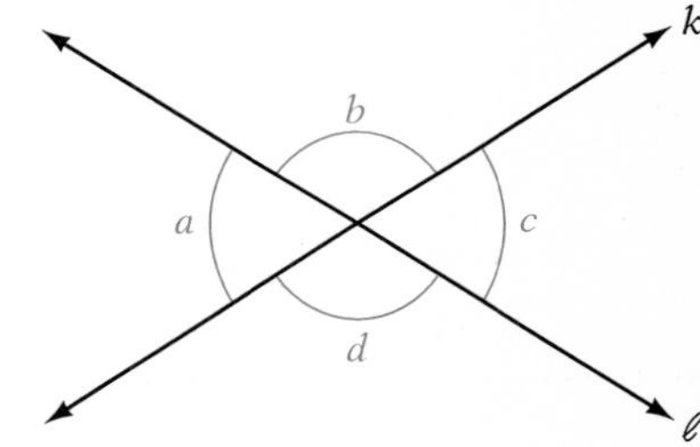

$\angle a = 65°$ • $\angle c = \angle a$ because $\angle c$ and $\angle a$ are vertical angles.

$\angle b + \angle c = 180°$ • $\angle c$ is supplementary to $\angle b$ because $\angle c$ and $\angle b$ are adjacent angles.

$\angle b + 65° = 180°$ • $\angle c = 65°$.

$\angle b + 65° - 65° = 180° - 65°$

$\angle b = 115°$

$\angle d = 115°$ • $\angle b = \angle d$ because $\angle b$ and $\angle d$ are vertical angles.

A line intersecting two other lines at two different points is called a **transversal.**

If the lines cut by a transversal are parallel lines and the transversal is perpendicular to the parallel lines, all eight angles formed are right angles.

If the lines cut by a transversal are parallel lines and the transversal is not perpendicular to the parallel lines, all four acute angles have the same measure and all four obtuse angles have the same measure. For the accompanying figure:

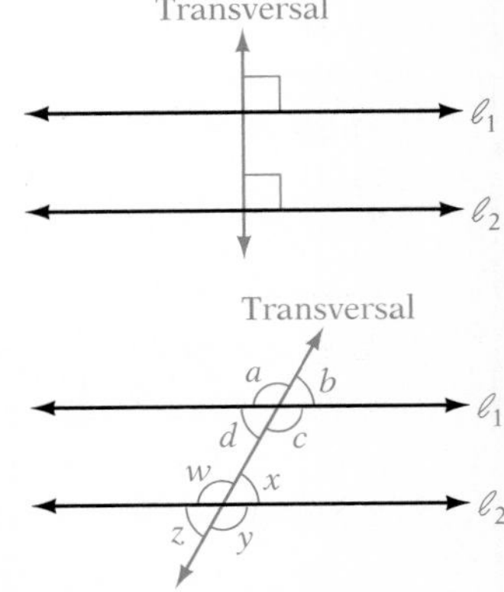

$\angle a = \angle c = \angle w = \angle y$ and $\angle b = \angle d = \angle x = \angle z$

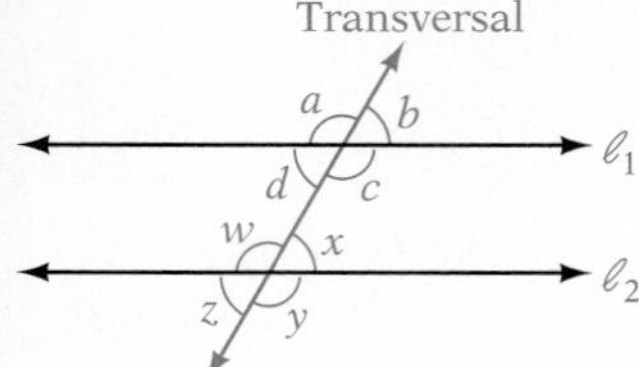

Alternate interior angles are two angles that are on opposite sides of the transversal and between the parallel lines. For the figure at the left, $\angle c$ and $\angle w$ are alternate interior angles. $\angle d$ and $\angle x$ are alternate interior angles. Alternate interior angles have the same measure.

Alternate exterior angles are two angles that are on opposite sides of the transversal and outside the parallel lines. For the figure at the left, $\angle a$ and $\angle y$ are alternate exterior angles. $\angle b$ and $\angle z$ are alternate exterior angles. Alternate exterior angles have the same measure.

Corresponding angles are two angles that are on the same side of the transversal and are both acute angles or are both obtuse angles. For the figure at the left, the following pairs of angles are corresponding angles: $\angle a$ and $\angle w$, $\angle d$ and $\angle z$, $\angle b$ and $\angle x$, $\angle c$ and $\angle y$. Corresponding angles have the same measure.

➡ In the accompanying figure, given that $\ell_1 \parallel \ell_2$ and $\angle c = 58°$, find the measures of $\angle f$, $\angle h$, and $\angle g$.

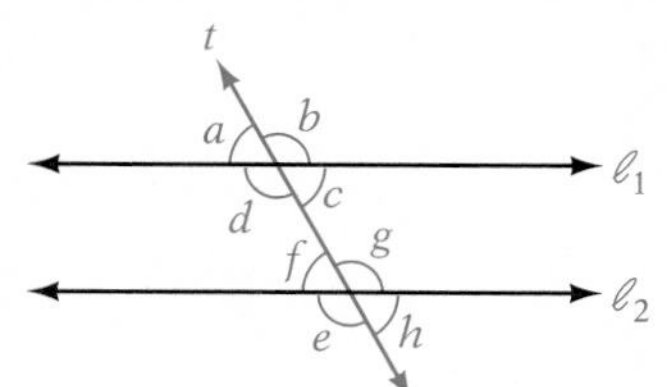

$\angle c = 58°$ • **$\angle c = \angle f$ because $\angle f$ and $\angle c$ are alternate interior angles.**

$\angle h = 58°$ • **$\angle c = \angle h$ because $\angle c$ and $\angle h$ are corresponding angles.**

$\angle g + \angle h = 180°$ • **$\angle g$ is supplementary to $\angle h$.**

$\angle g + 58° = 180°$ • **$\angle h = 58°$.**

$\angle g + 58° - 58° = 180° - 58°$

$\angle g = 122°$

Example 7
In the figure, angle $a = 75°$. Find angle b.

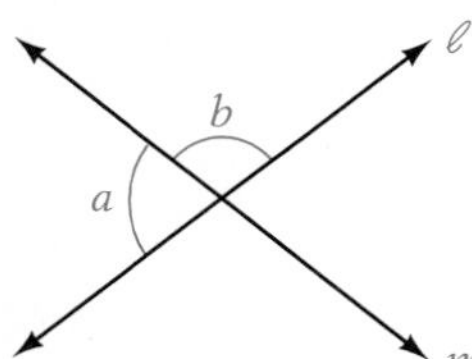

Solution
Angles a and b are supplementary angles.

$$\angle a + \angle b = 180°$$
$$75° + \angle b = 180°$$
$$75° - 75° + \angle b = 180° - 75°$$
$$\angle b = 105°$$

You Try It 7
In the figure, angle $a = 125°$. Find angle b.

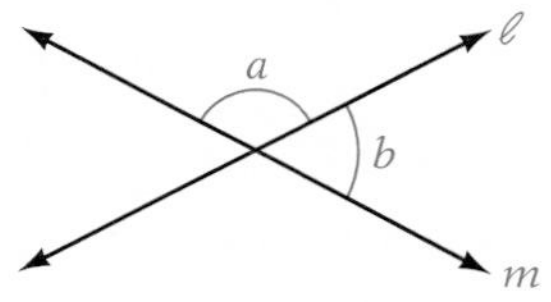

Your solution
55°

Example 8
In the figure, $\ell_1 \parallel \ell_2$ and angle $a = 70°$. Find angle b.

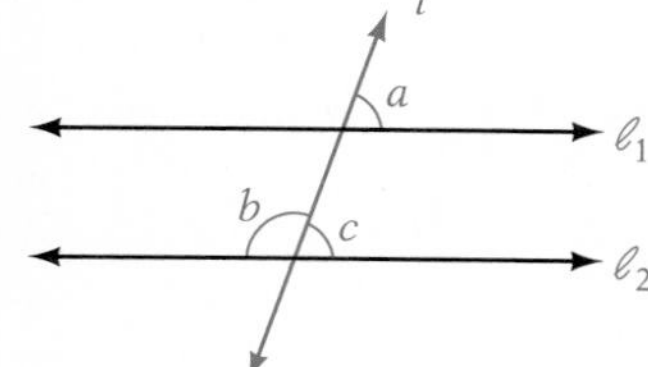

Solution
Angles b and c are supplementary angles.

$$\angle b + \angle c = 180°$$
$$\angle b + 70° = 180°$$ • **$\angle c = \angle a = 70°$**
$$\angle b + 70° - 70° = 180° - 70°$$
$$\angle b = 110°$$

You Try It 8
In the figure, $\ell_1 \parallel \ell_2$ and angle $a = 120°$. Find angle b.

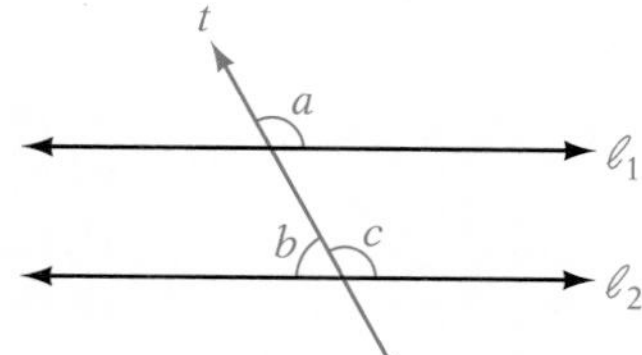

Your solution
60°

Solutions on p. A51

12.1 Exercises

Objective A

1. The measure of an acute angle is between _____ and _____.
   0°, 90°

2. The measure of an obtuse angle is between _____ and _____.
   90°, 180°

3. How many degrees are in a straight angle?
   180°

4. Two lines that intersect at right angles are _____ lines.
   Perpendicular

5. In the figure, $EF = 20$ and $FG = 10$. Find the length of EG. 30

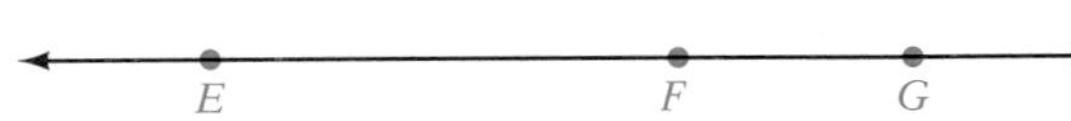

6. In the figure, $EF = 18$ and $FG = 6$. Find the length of EG. 24

7. In the figure, it is given that $QR = 7$ and $QS = 28$. Find the length of RS. 21

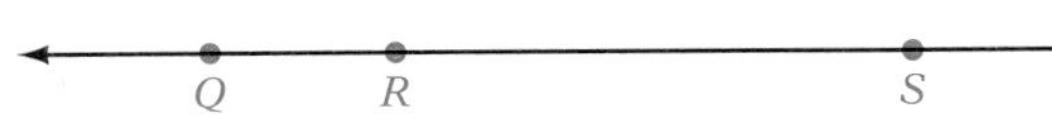

8. In the figure, it is given that $QR = 15$ and $QS = 45$. Find the length of RS. 30

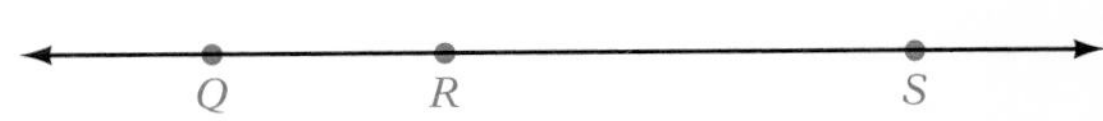

9. In the figure, it is given that $AB = 12$, $CD = 9$, and $AD = 35$. Find the length of BC. 14

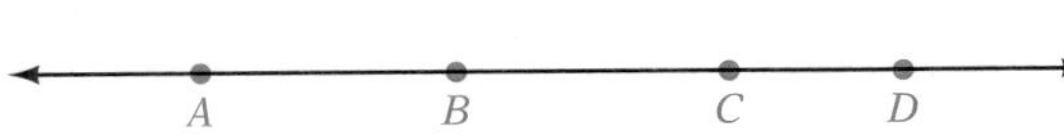

10. In the figure, it is given that $AB = 21$, $BC = 14$, and $AD = 54$. Find the length of CD. 19

11. Find the complement of a 31° angle.
    59°

12. Find the complement of a 62° angle.
    28°

13. Find the supplement of a 72° angle.
    108°

14. Find the supplement of a 162° angle.
    18°

15. Find the complement of a 13° angle.
    77°

16. Find the complement of an 88° angle.
    2°

17. Find the supplement of a 127° angle.
    53°

18. Find the supplement of a 7° angle.
    173°

19. In the figure, find the measure of angle AOB. 77°

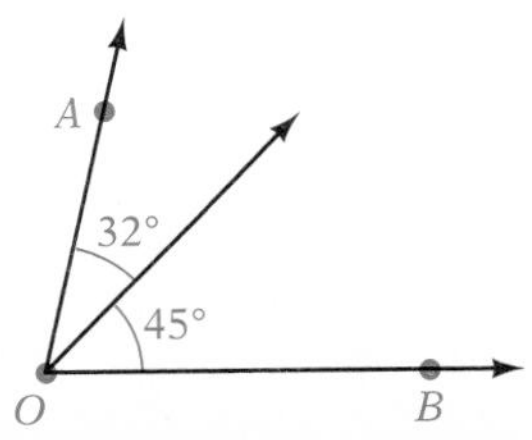

20. In the figure, find the measure of angle AOB. 136°

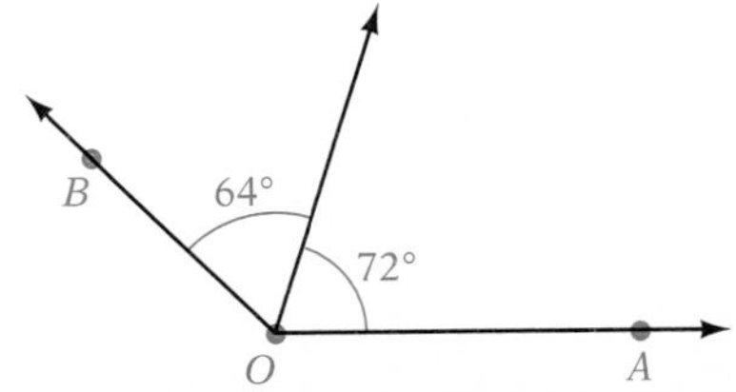

21. Find the measure of angle a in the figure.
118°

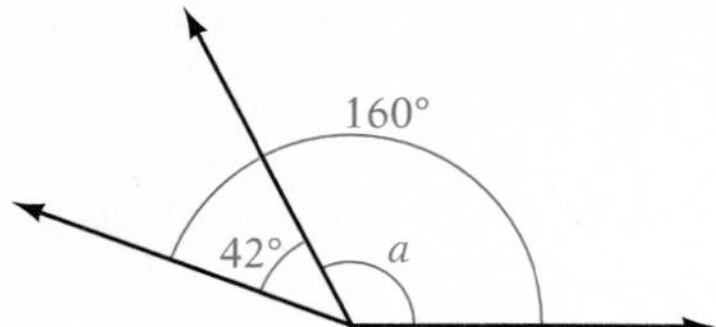

22. Find the measure of angle a in the figure.
33°

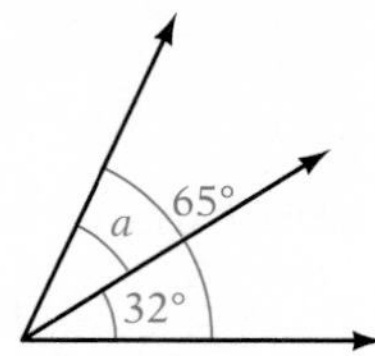

23. Find the measure of angle a in the figure.
133°

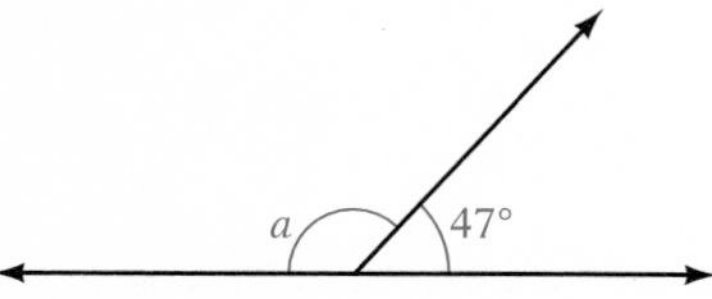

24. Find the measure of angle a in the figure.
77°

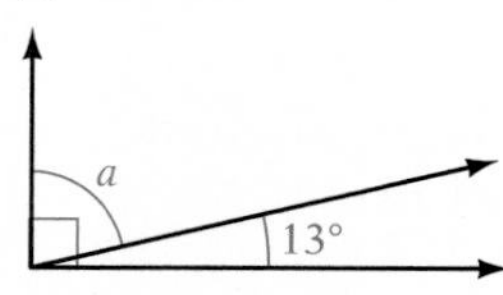

25. In the figure, it is given that $\angle LOM = 53°$ and $\angle LON = 139°$. Find the measure of $\angle MON$. 86°

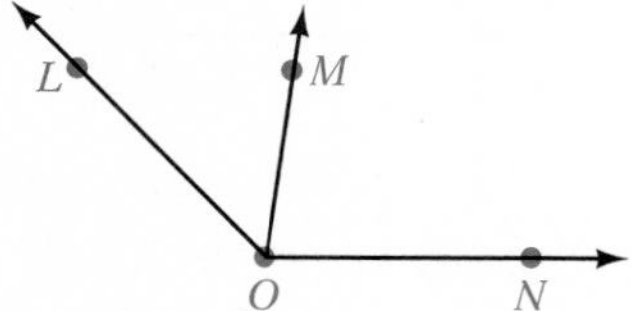

26. In the figure, it is given that $\angle MON = 38°$ and $\angle LON = 85°$. Find the measure of $\angle LOM$. 47°

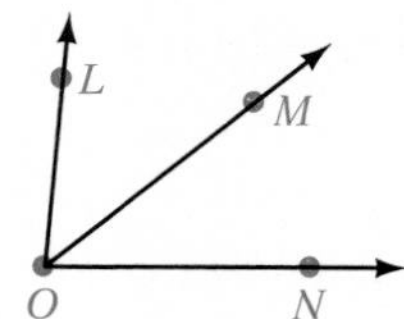

Objective B

27. What is the sum of the three angles of a triangle?
180°

28. Name the side opposite the right angle in a right triangle.
Hypotenuse

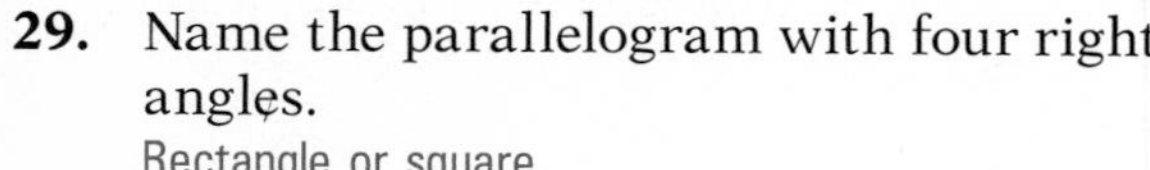

29. Name the parallelogram with four right angles.
Rectangle or square

30. Name the rectangle with four equal sides.
Square

31. Name the rectangular solid in which all six faces are squares.
Cube

32. Name the solid in which all points are the same distance from the center.
Sphere

33. Name the quadrilateral in which opposite sides are parallel and equal.
Parallelogram, rectangle, or square

34. Name the plane figure in which all points are the same distance from the center.
Circle

35. Name the solid in which the bases are circular and perpendicular to the height.
Cylinder

36. Name the solid in which all the faces are rectangles.
Rectangular solid or cube

37. A triangle has a 13° angle and a 65° angle. Find the measure of the other angle. 102°

38. A triangle has a 105° angle and a 32° angle. Find the measure of the other angle. 43°

39. A right triangle has a 45° angle. Find the measure of the other two angles.
90° and 45°

40. A right triangle has a 62° angle. Find the measure of the other two angles.
90° and 28°

41. A triangle has a 62° angle and a 104° angle. Find the measure of the other angle.
14°

42. A triangle has a 30° angle and a 45° angle. Find the measure of the other angle.
105°

43. A right triangle has a 25° angle. Find the measure of the other two angles.
90° and 65°

44. Two angles of a triangle are 42° and 105°. Find the measure of the other angle.
33°

45. Find the radius of a circle with a diameter of 16 in.
8 in.

46. Find the radius of a circle with a diameter of 9 ft.
$4\frac{1}{2}$ ft

47. Find the diameter of a circle with a radius of $2\frac{1}{3}$ ft.
$4\frac{2}{3}$ ft

48. Find the diameter of a circle with a radius of 24 cm.
48 cm

49. The radius of a sphere is 3.5 cm. Find the diameter.
7 cm

50. The radius of a sphere is $1\frac{1}{2}$ ft. Find the diameter.
3 ft

51. The diameter of a sphere is 4 ft 8 in. Find the radius.
2 ft 4 in.

52. The diameter of a sphere is 1.2 m. Find the radius.
0.6 m

Objective C

53. Find the measures of angles a and b in the figure.

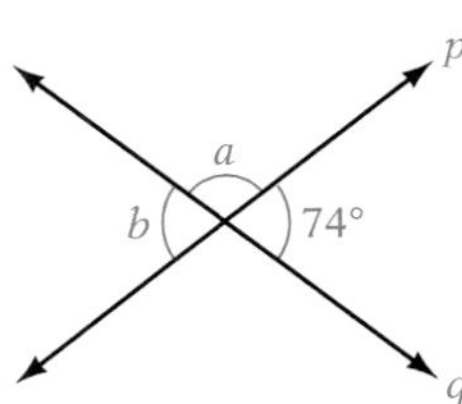

$a = 106°$
$b = 74°$

54. Find the measures of angles a and b in the figure.

$a = 131°$
$b = 49°$

55. Find the measures of angles a and b in the figure.

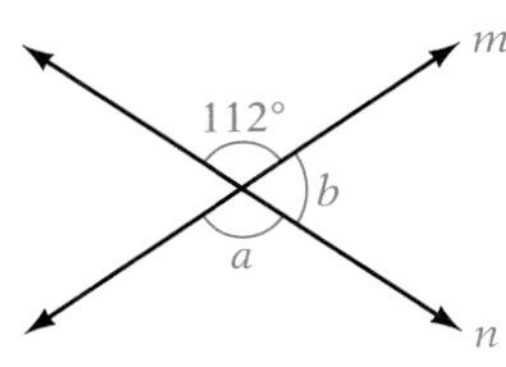

$a = 112°$
$b = 68°$

56. Find the measures of angles a and b in the figure.

$a = 131°$
$b = 49°$

57. In the figure, it is given that $\ell_1 \parallel \ell_2$. Find the measures of angles a and b.

$a = 38°$
$b = 142°$

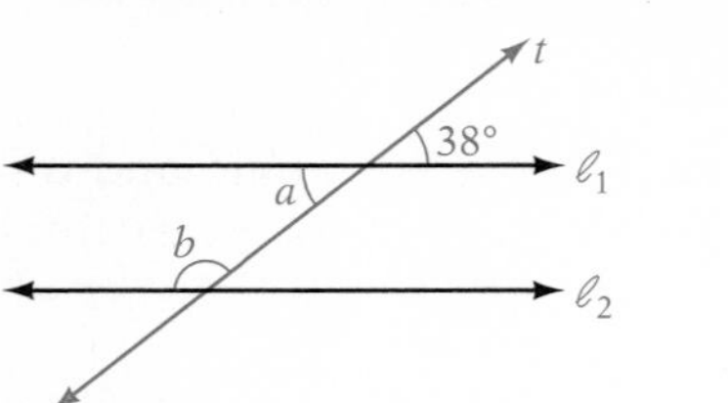

58. In the figure, it is given that $\ell_1 \parallel \ell_2$. Find the measures of angles a and b.

$a = 44°$
$b = 44°$

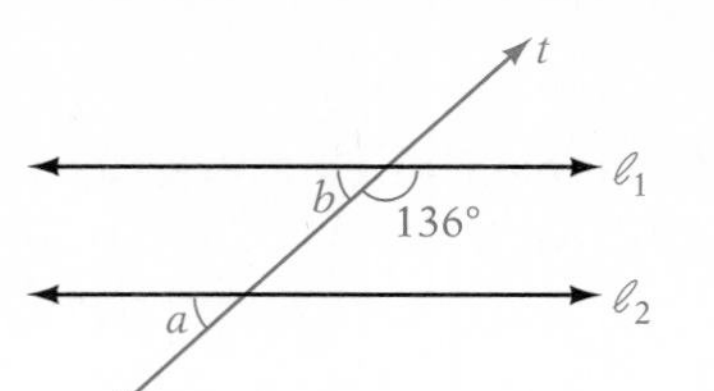

59. In the figure, it is given that $\ell_1 \parallel \ell_2$. Find the measures of angles a and b.

$a = 58°$
$b = 58°$

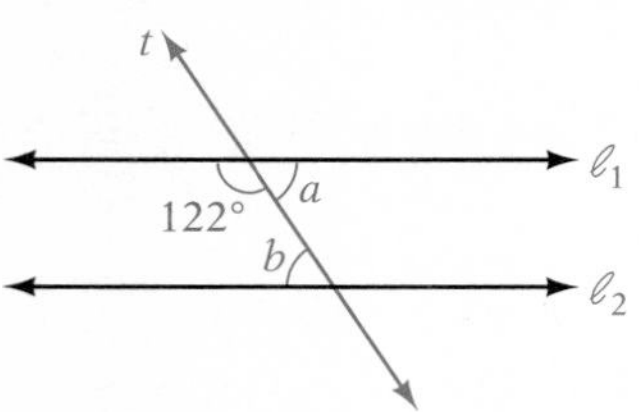

60. In the figure, it is given that $\ell_1 \parallel \ell_2$. Find the measures of angles a and b.

$a = 55°$
$b = 125°$

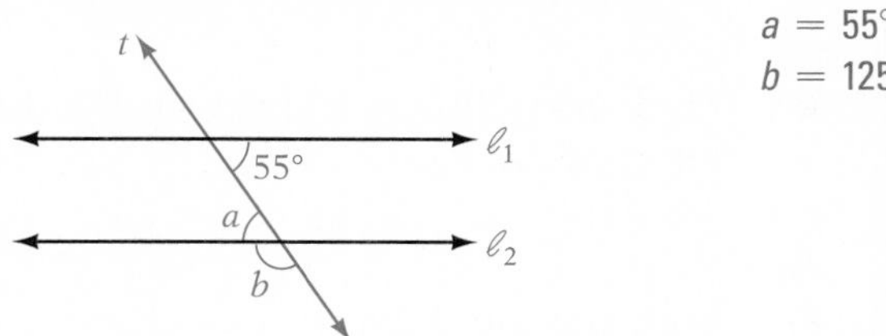

61. In the figure, it is given that $\ell_1 \parallel \ell_2$. Find the measures of angles a and b.

$a = 152°$
$b = 152°$

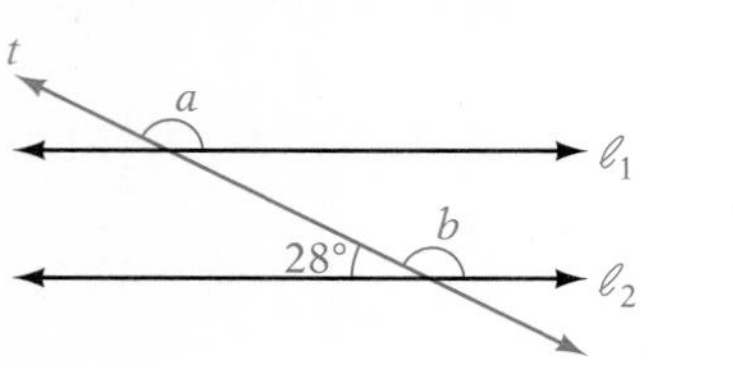

62. In the figure, it is given that $\ell_1 \parallel \ell_2$. Find the measures of angles a and b.

$a = 105°$
$b = 75°$

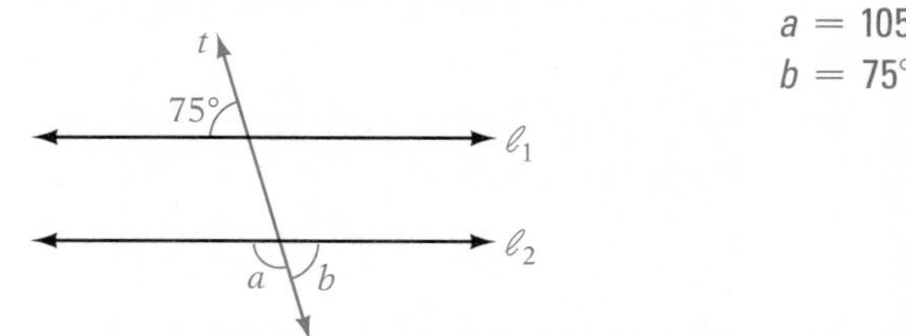

63. In the figure, it is given that $\ell_1 \parallel \ell_2$. Find the measures of angles a and b.

$a = 130°$
$b = 50°$

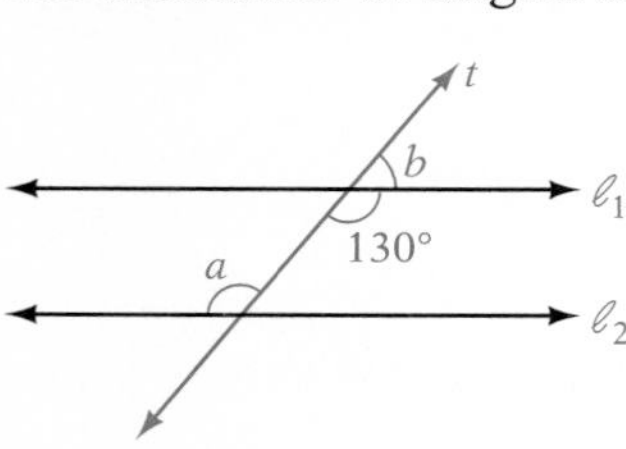
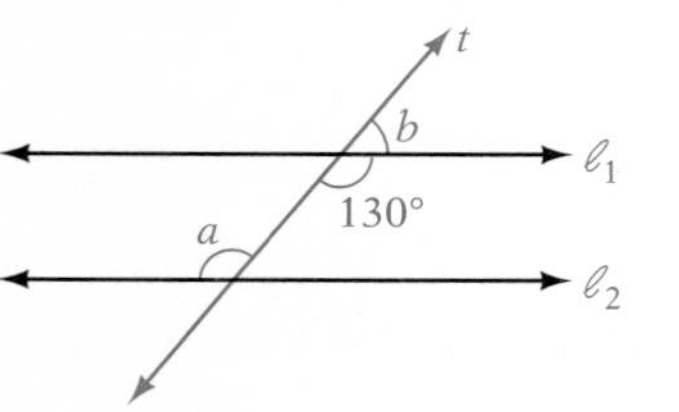

64. In the figure, it is given that $\ell_1 \parallel \ell_2$. Find the measures of angles a and b.

$a = 62°$
$b = 118°$

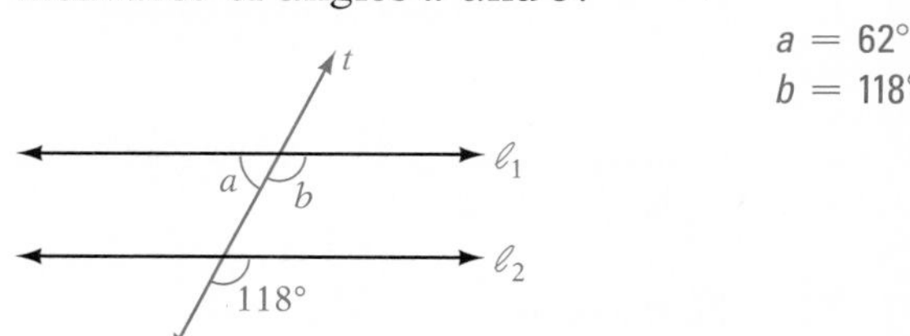

APPLYING THE CONCEPTS

65. **a.** What is the smallest possible whole number of degrees in an angle of a triangle? 1°

b. What is the largest possible whole number of degrees in an angle of a right triangle? 90°

66. Determine whether the statement is always true, sometimes true, or never true.

a. Two lines that are both parallel to a third line are parallel to each other. Always true

b. A triangle contains at least two acute angles. Always true

c. Vertical angles are complementary angles. Sometimes true

67. [W] If AB and CD intersect at point O, and $\angle AOC = \angle BOC$, explain why AB is perpendicular to CD. $\angle AOC$ and $\angle BOC$ are supplementary. Since $\angle AOC = \angle BOC$ is given, each angle is 90° and the lines are perpendicular.

68. [W] What are the meanings of the words *acute* and *obtuse* in describing a person? Answers will vary

12.2 Plane Geometric Figures

Objective A To find the perimeter of plane geometric figures

A **polygon** is a closed figure determined by three or more line segments that lie in a plane. The line segments that form the polygon are called its **sides.** The figures below are examples of polygons.

A

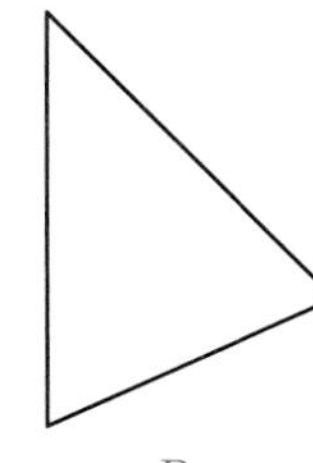

B

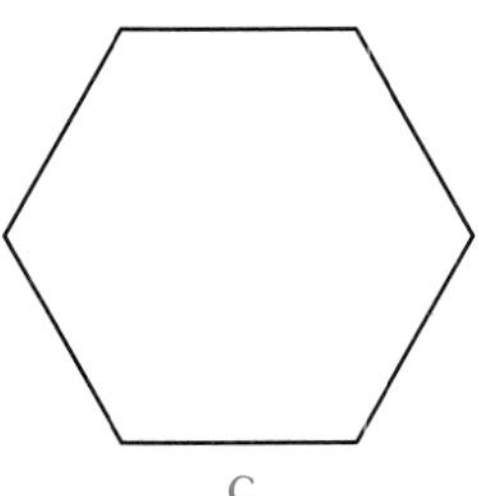

C

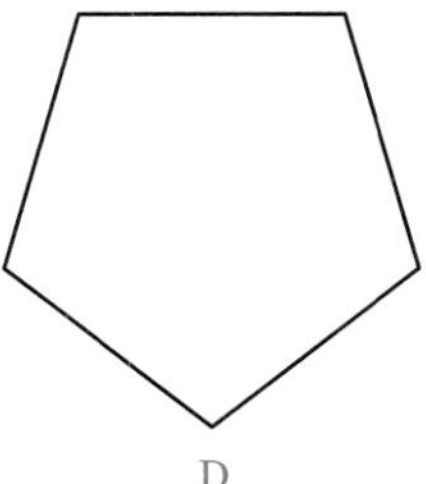

D

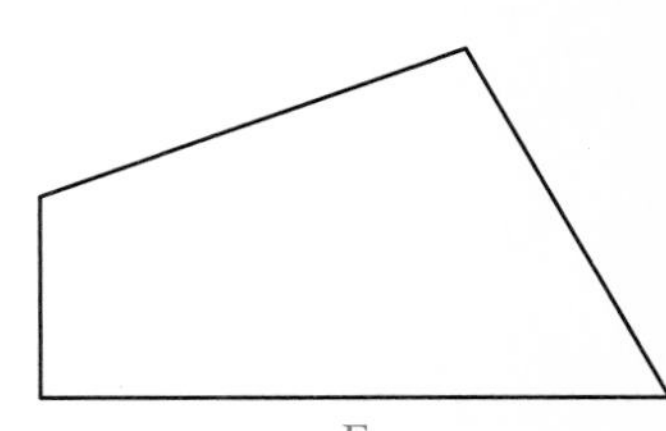

E

POINT OF INTEREST

Although a polygon is defined in terms of its *sides* (see the definition above), the word actually comes from the Latin word *polygonum* which means having many *angles*. This is certainly the case for a polygon.

A **regular polygon** is one in which each side has the same length and each angle has the same measure. The polygons in Figures A, C, and D above are regular polygons.

The name of a polygon is based on the number of its sides. The table below lists the names of polygons that have from 3 to 10 sides.

Number of Sides	*Name of the Polygon*
3	Triangle
4	Quadrilateral
5	Pentagon
6	Hexagon
7	Heptagon
8	Octagon
9	Nonagon
10	Decagon

Triangles and quadrilaterals are two of the most common types of polygons. Triangles are distinguished by the number of equal sides and also by the measures of their angles.

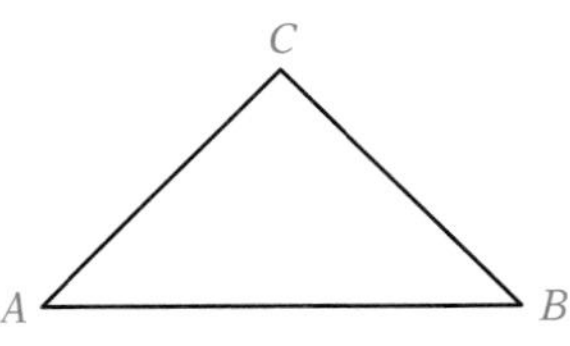

An **isosceles triangle** has two sides of equal length. The angles opposite the equal sides are of equal measure.
$AC = BC$
$\angle A = \angle B$

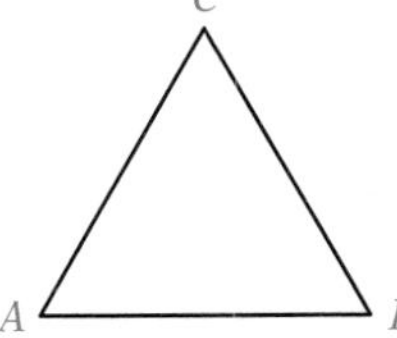

The three sides of an **equilateral triangle** are of equal length. The three angles are of equal measure.
$AB = BC = AC$
$\angle A = \angle B = \angle C$

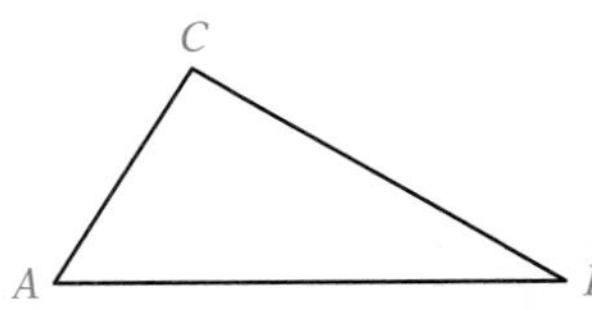

A **scalene triangle** has no two sides of equal length. No two angles are of equal measure.

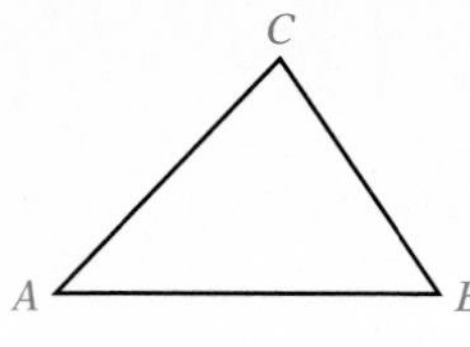

An acute triangle has three acute angles.

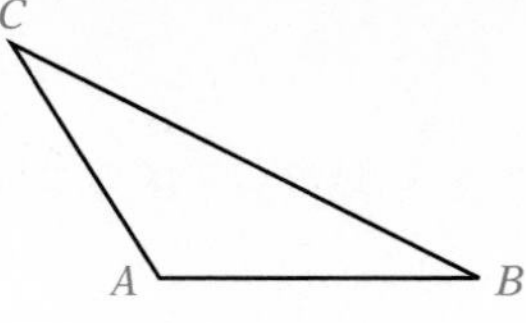

An obtuse triangle has one obtuse angle.

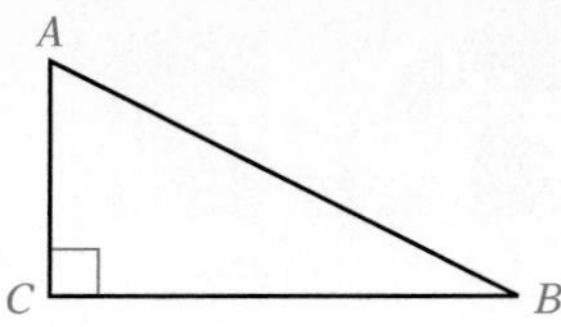

A right triangle has a right angle.

INSTRUCTOR NOTE

The diagram below shows the relationships between all quadrilaterals. The description of each quadrilateral is within an example of that quadrilateral.

Quadrilaterals are also distinguished by their sides and angles, as shown below. Note that a rectangle, a square, and a rhombus are different forms of a parallelogram.

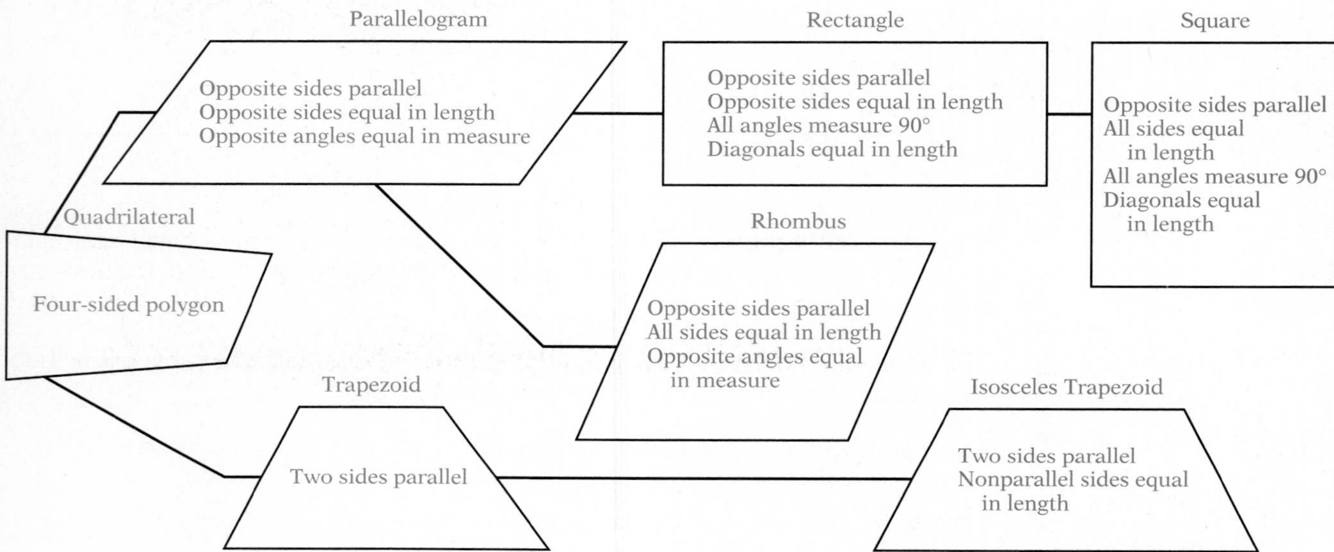

The **perimeter** of a plane geometric figure is a measure of the distance around the figure. Perimeter is used in buying fencing for a lawn or determining how much baseboard is needed for a room.

The perimeter of a triangle is the sum of the lengths of the three sides.

Perimeter of a Triangle

$$P = a + b + c$$

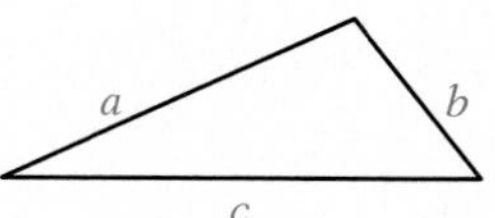

➡ Find the perimeter of the triangle shown at the right.

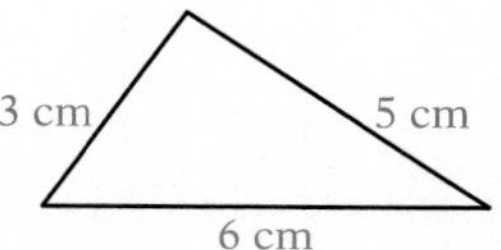

$$\begin{aligned} P &= a + b + c \\ &= 3\text{ cm} + 5\text{ cm} + 6\text{ cm} \\ &= 14\text{ cm} \end{aligned}$$

The perimeter of the triangle is 14 cm.

The perimeter of a square is the sum of the four equal sides.

Perimeter of a Square

$$P = 4s$$

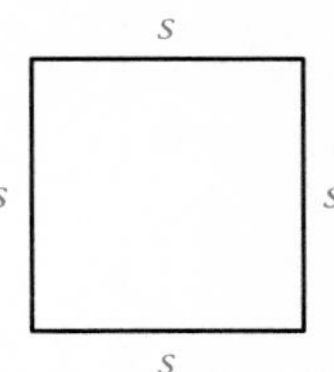

➡ Find the perimeter of the square shown at the right.

$$\begin{aligned} P &= 4s \\ &= 4(3 \text{ ft}) \qquad \bullet\ s = 3 \text{ ft.} \\ &= 12 \text{ ft} \end{aligned}$$

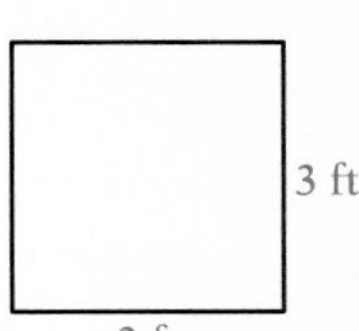

The perimeter of the square is 12 ft.

The perimeter of a quadrilateral is the sum of the lengths of the four sides.

A rectangle is a quadrilateral with opposite sides of equal length. The length of a rectangle refers to the longer side, and the width refers to the length of the shorter side.

Perimeter of a Rectangle

$$P = 2L + 2W$$

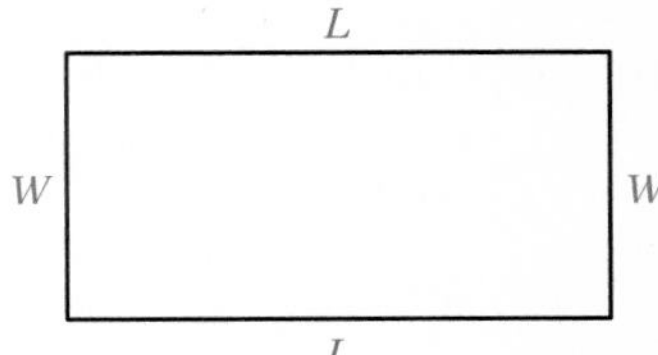

INSTRUCTOR NOTE

Explain to students that computations with π produce approximate solutions. If students are using a calculator, the answers they obtain may be different from our answer. For consistency, we always approximate π as 3.14.

➡ Find the perimeter of the rectangle shown at the right.

$$\begin{aligned} P &= 2L + 2W \\ &= 2(6 \text{ m}) + 2(3 \text{ m}) \qquad \bullet\ L = 6 \text{ m},\ W = 3 \text{ m.} \\ &= 12 \text{ m} + 6 \text{ m} \\ &= 18 \text{ m} \end{aligned}$$

3 m

6 m

The perimeter of the rectangle is 18 m.

The distance around a circle is called the **circumference.** The circumference of a circle is equal to the product of π (pi) and the diameter.

POINT OF INTEREST

Archimedes (c. 287–212 B.C.) was the mathematician who gave us the approximate value of π as $\frac{22}{7} = 3\frac{1}{7}$. He actually showed that π was between $3\frac{10}{71}$ and $3\frac{1}{7}$. The approximation $3\frac{10}{71}$ is actually closer to the exact value of π, but it is more difficult to use.

Circumference of a Circle

$$C = \pi d$$

or

$$C = 2\pi r \qquad \text{Because diameter} = 2r$$

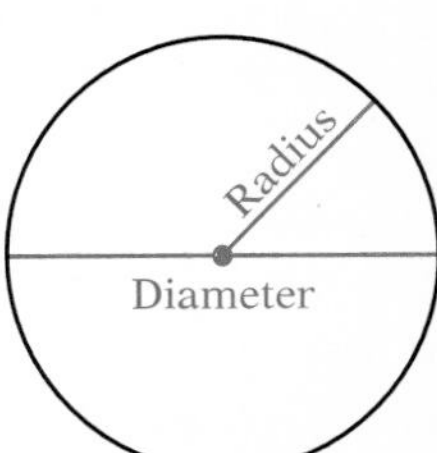

The formula for circumference uses the number π (pi). The value of π can be approximated by a fraction or a decimal.

$$\pi \approx 3.14 \qquad \pi \approx \frac{22}{7}$$

The π key on a calculator gives a closer approximation of π than 3.14.

➡ Find the circumference of the circle shown at the right.

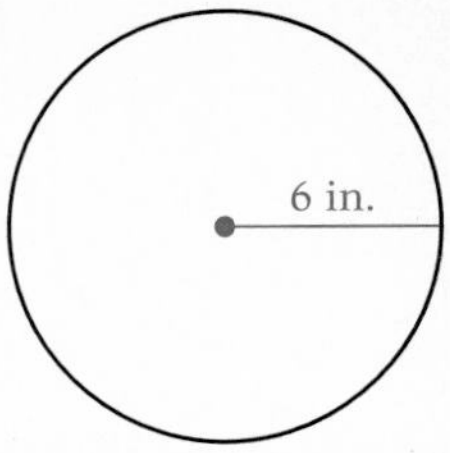

$$\begin{aligned} C &= 2\pi r \\ &\approx 2 \cdot 3.14 \cdot 6 \text{ in.} \\ &= 37.68 \text{ in.} \end{aligned}$$

• $r = 6$ in.

The circumference of the circle is approximately 37.68 in.

Example 1
Find the perimeter of a rectangle with a width of $\frac{2}{3}$ ft and a length of 2 ft.

Solution

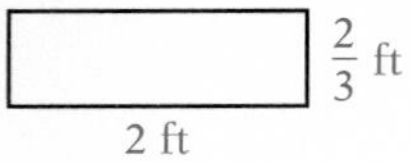

$$\begin{aligned} P &= 2L + 2W \\ &= (2 \cdot 2 \text{ ft}) + \left(2 \cdot \frac{2}{3} \text{ ft}\right) \\ &= 4 \text{ ft} + \frac{4}{3} \text{ ft} \\ &= 5\frac{1}{3} \text{ ft} \end{aligned}$$

• $L = 2$ ft, $W = \frac{2}{3}$ ft

The perimeter of the rectangle is $5\frac{1}{3}$ ft.

You Try It 1
Find the perimeter of a rectangle with a length of 2 m and a width of 0.85 m.

Your solution
5.7 m

Example 2
Find the perimeter of a triangle with sides 5 in., 7 in., and 8 in.

Solution

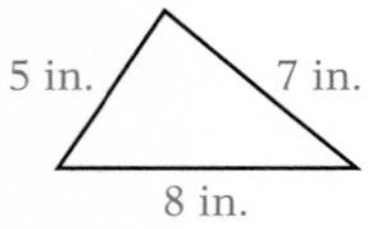

$$\begin{aligned} P &= a + b + c \\ &= 5 \text{ in.} + 7 \text{ in.} + 8 \text{ in.} \\ &= 20 \text{ in.} \end{aligned}$$

The perimeter of the triangle is 20 in.

You Try It 2
Find the perimeter of a triangle with sides 12 cm, 15 cm, and 18 cm.

Your solution
45 cm

Example 3
Find the circumference of a circle with a radius of 18 cm. Use $\pi \approx 3.14$.

Solution

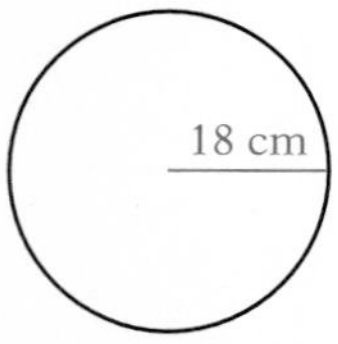

$$\begin{aligned} C &= 2\pi r \\ &\approx 2 \cdot 3.14 \cdot 18 \text{ cm} \\ &= 113.04 \text{ cm} \end{aligned}$$

The circumference is approximately 113.04 cm.

You Try It 3
Find the circumference of a circle with a diameter of 6 in. Use $\pi \approx 3.14$.

Your solution
18.84 in.

Solutions on p. A52

Objective B To find the perimeter of composite geometric figures

Composite geometric figures are figures made from two or more geometric figures. The following composite is made from part of a rectangle and part of a circle:

Perimeter of the composite figure $= 3$ sides of a rectangle $+ \frac{1}{2}$ the circumference of a circle

Perimeter of the composite figure $= 2L + W + \frac{1}{2}\pi d$

The perimeter of the composite figure below is found by adding the measures of twice the length plus the width plus one-half the circumference of the circle.

$$P = 2L + 2W + \frac{1}{2}\pi d$$

$$\approx 2(12 \text{ m}) + 4 \text{ m} + \frac{1}{2}(3.14)(4 \text{ m}) \qquad \bullet\ L = 12 \text{ m},\ W = 4 \text{ m},\ d = 4\text{m}.$$

$$= 34.28 \text{ m}$$

The perimeter is approximately 34.28 m.

Example 4
Find the perimeter of the composite figure.
Use $\pi \approx \frac{22}{7}$.

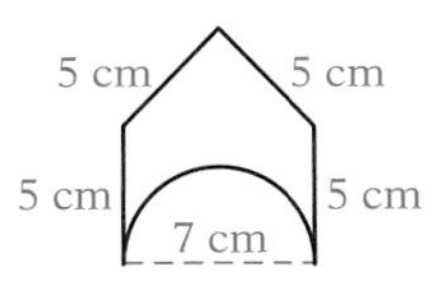

Solution

Perimeter of composite figure $=$ sum of lengths of the 4 sides $+ \frac{1}{2}$ the circumference of the circle

$$P = 4s + \frac{1}{2}\pi d$$

$$\approx 4 \cdot 5 \text{ cm} + \frac{1}{2}\left(\frac{22}{7}\right)(7 \text{ cm})$$

$$= 20 \text{ cm} + 11 \text{ cm} = 31 \text{ cm}$$

The perimeter is approximately 31 cm.

You Try It 4
Find the perimeter of the composite figure.
Use $\pi \approx 3.14$.

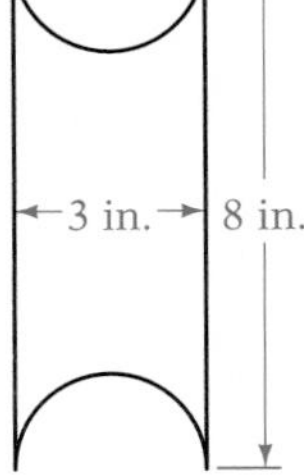

Your solution
25.42 in.

Solution on p. A52

Objective C To solve application problems

Example 5
The dimensions of a triangular sail are 18 ft, 11 ft, and 15 ft. What is the perimeter of the sail?

Strategy

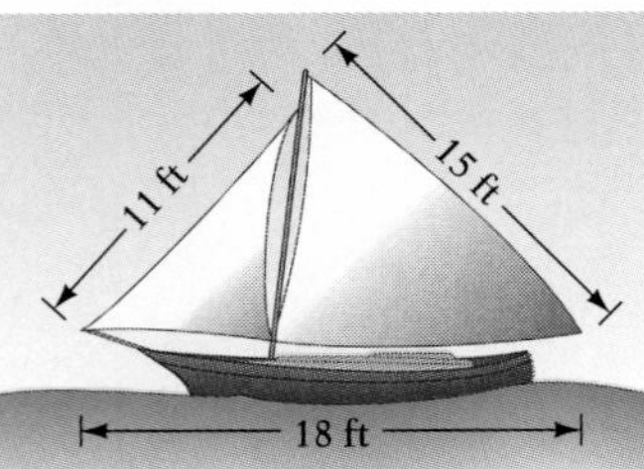

To find the perimeter, use the formula for the perimeter of a triangle.

Solution
$P = a + b + c$
$= 18 \text{ ft} + 11 \text{ ft} + 15 \text{ ft}$
$= 44 \text{ ft}$

The perimeter of the sail is 44 ft.

You Try It 5
What is the perimeter of a standard piece of typing paper that measures $8\frac{1}{2}$ in. by 11 in.?

Your strategy

Your solution
39 in.

Example 6
If fencing costs $2.75 per foot, how much will it cost to fence a rectangular lot that is 108 ft wide and 240 ft long?

Strategy

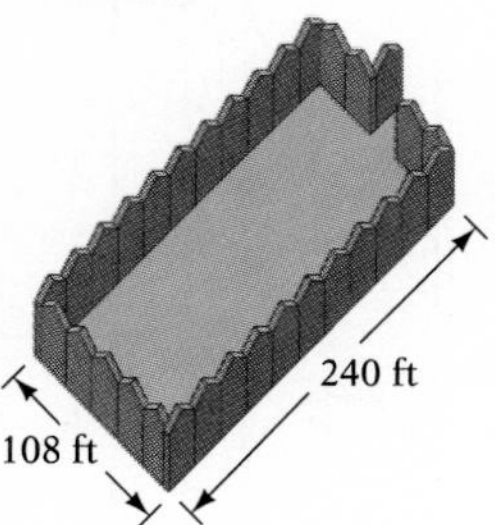

To find the cost of the fence:

- Find the perimeter of the lot.
- Multiply the perimeter by the per-foot cost of fencing.

Solution
$P = 2L + 2W$
$= 2 \cdot 240 \text{ ft} + 2 \cdot 108 \text{ ft}$
$= 480 \text{ ft} + 216 \text{ ft}$
$= 696 \text{ ft}$

Cost $= 696 \times 2.75 = 1914$

The cost is $1914.

You Try It 6
A metal strip is being installed around a workbench that is 0.74 m wide and 3 m long. At $1.76 per meter, find the cost of the metal stripping. Round to the nearest cent.

Your strategy

Your solution
$13.16

Solutions on p. A52

12.2 Exercises

Objective A

In Exercises 1 to 8, find the perimeter or circumference of the given figures.

1.

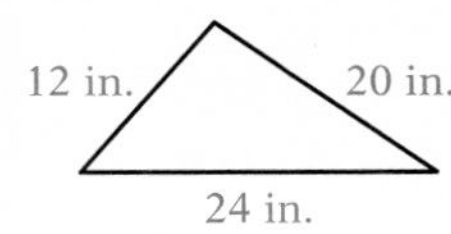

56 in.

2.

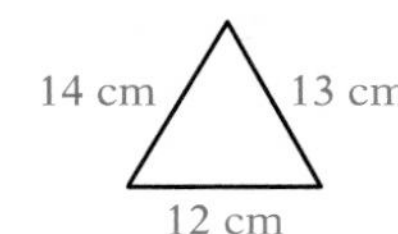

39 cm

3.

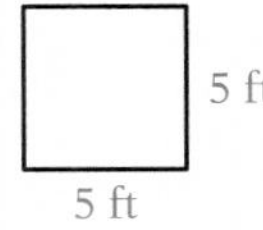

20 ft

4.

8 m

5.

92 cm

6.

5 ft

18 ft

46 ft

7.

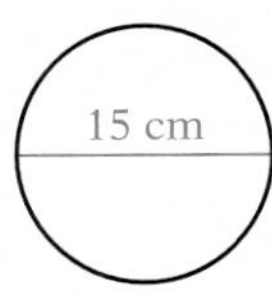

47.1 cm

8.

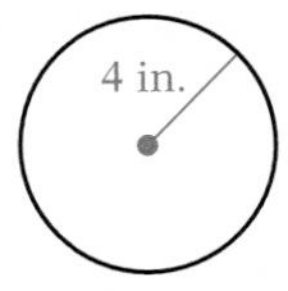

25.12 in.

9. Find the perimeter of a triangle with sides 2 ft 4 in., 3 ft, and 4 ft 6 in.

9 ft 10 in.

10. Find the perimeter of a rectangle with a length of 2 m and a width of 0.8 m.

5.6 m

11. Find the circumference of a circle with a radius of 8 cm. Use $\pi \approx 3.14$.

50.24 cm

12. Find the circumference of a circle with a diameter of 14 in. Use $\pi \approx \frac{22}{7}$.

44 in.

13. Find the perimeter of a square in which each side is equal to 60 m.
240 m

14. Find the perimeter of a triangle in which each side is $1\frac{2}{3}$ ft.
5 ft

15. Find the perimeter of a five-sided figure with sides of 22 cm, 47 cm, 29 cm, 42 cm, and 17 cm.
157 cm

16. Find the perimeter of a rectangular farm that is $\frac{1}{2}$ mi wide and $\frac{3}{4}$ mi long.
$2\frac{1}{2}$ mi

Objective B

In the figures accompanying Exercises 17 to 24, find the perimeter. Use $\pi \approx 3.14$.

17.

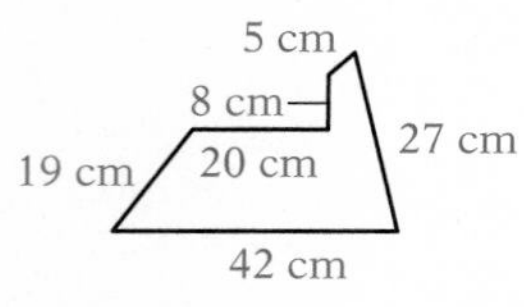

121 cm

18.

$6\frac{3}{4}$ ft
$1\frac{1}{2}$ ft
3 ft
$2\frac{3}{4}$ ft
$2\frac{2}{3}$ ft
$3\frac{1}{2}$ ft

$20\frac{1}{6}$ ft

19.

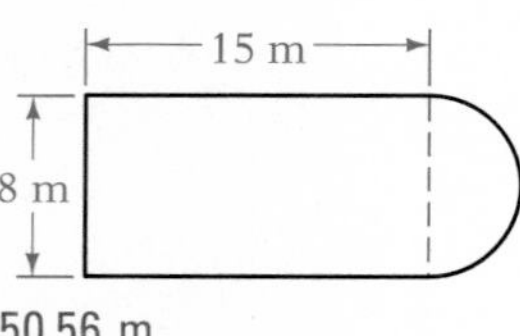

50.56 m

20.

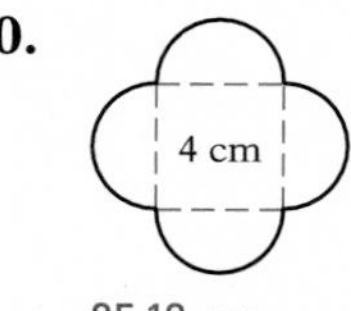

25.12 cm

21.

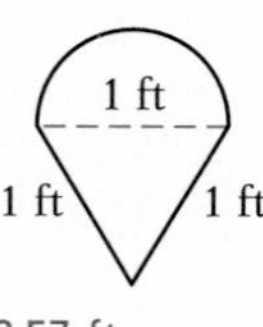

3.57 ft

22.

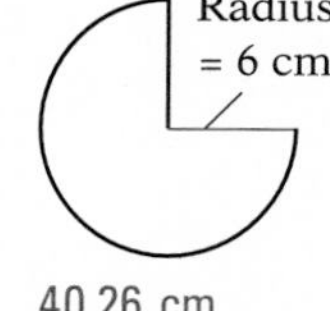

40.26 cm

23.

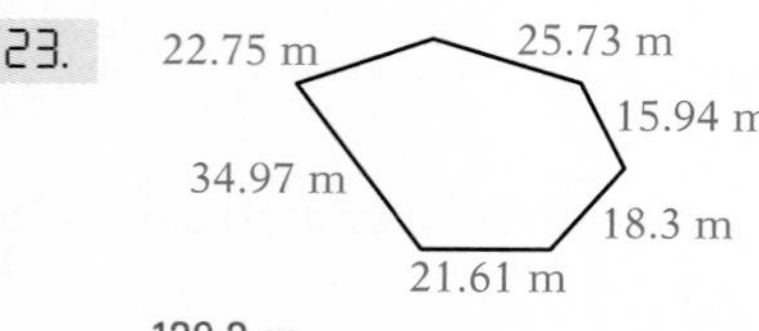

139.3 m

24.

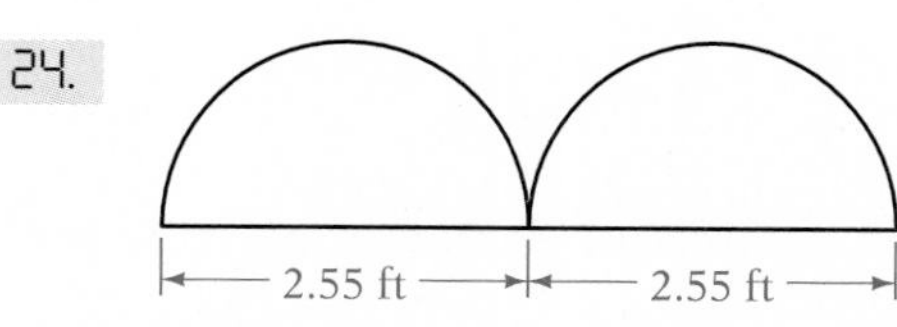

13.107 ft

Objective C *Application Problems*

25. Find the amount of fencing needed to fence a farm that is $1\frac{1}{2}$ mi long and $\frac{3}{4}$ mi wide. $4\frac{1}{2}$ mi

26. Find the number of feet of framing needed to frame a picture that is $2\frac{1}{2}$ ft by $1\frac{2}{3}$ ft. $8\frac{1}{3}$ ft

27. How much binding is needed to bind the outside of a circular rug that is 5 m in diameter? Use $\pi \approx 3.14$. 15.7 m

28. Find the length of molding needed to put around a circular table that is 3.8 ft in diameter. Use $\pi \approx 3.14$. (Round to hundredths.) 11.93 ft

29. A bicycle tire has a diameter of 24 in. How many feet does the bicycle travel if the wheel makes 5 revolutions? Leave the answer in terms of π. 10π ft

30. A tricycle tire has a diameter of 12 in. How many feet does the tricycle travel if the wheel makes 8 revolutions? Leave the answer in terms of π. 8π ft

31. Find the length of weather stripping installed around the door shown in the figure at the right. Use $\pi \approx 3.14$. 20.71 ft

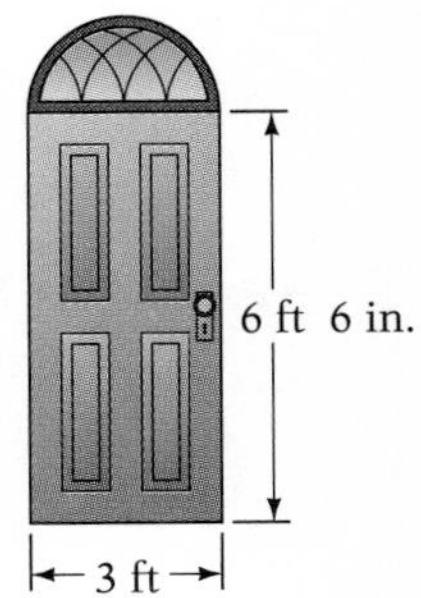

32. Find the perimeter of a roller skating rink with the dimensions shown in the figure. Use $\pi \approx 3.14$. 81.4 m

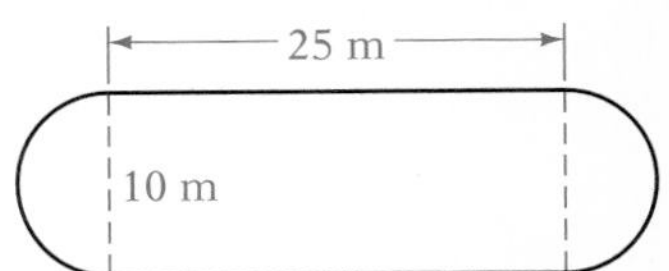

33. The rectangular lot shown in the figure at the right is being fenced. The fencing along the road is to cost \$2.20 per foot. The rest of the fencing costs \$1.85 per foot. Find the total cost to fence the lot. \$8022.50

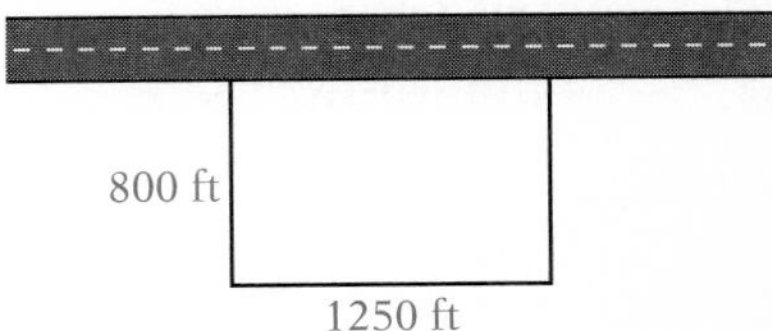

34. A rain gutter is being installed on a home that has the dimensions shown at the right. At \$8.29 per meter, how much will it cost to install the rain gutter? \$364.76

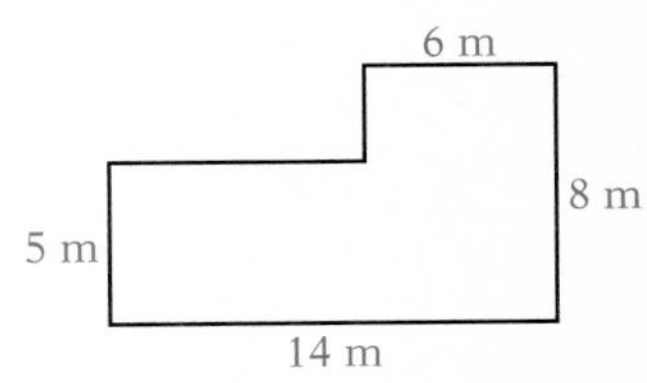

35. The distance from the earth to the sun is 93,000,000 mi. Approximate the distance the earth travels in making 1 revolution about the sun. Use $\pi \approx 3.14$. 584,040,000 mi

36. The distance from the surface to the center of the earth is 6356 km. Find the circumference of the earth. Use $\pi \approx 3.14$. 39,915.68 km

APPLYING THE CONCEPTS

37. If the diameter of a circle is doubled, how many times larger is the resulting circumference? 2 times

38. If the radius of a circle is doubled, how many times larger is the resulting circumference? 2 times

39. In the following pattern, the length of one side of a square is 1 unit. Find the perimeter of the eighth figure in the pattern. 22

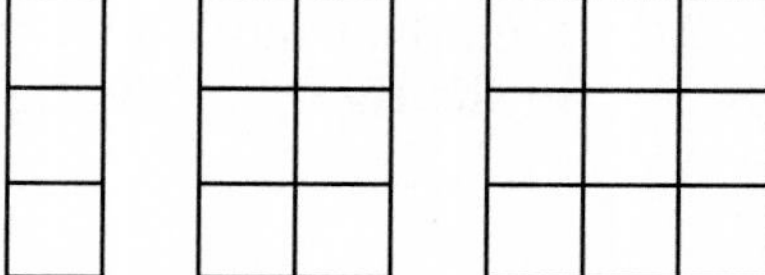

40. Remove six toothpicks from the figure at the right in such a way as to leave two squares.

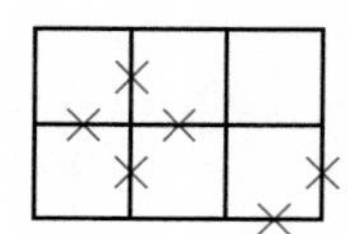

41. An equilateral triangle is placed inside an equilateral triangle as shown below. Now three more equilateral triangles are placed inside the unshaded equilateral triangles. The process is repeated again. Determine the perimeter of all the shaded triangles in Figure C.

$20\frac{1}{4}$ cm

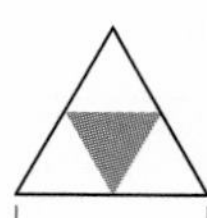
2 cm
Figure A

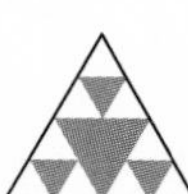
Figure B

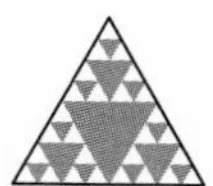
Figure C

42. [W] A forest ranger must determine the diameter of a redwood tree. Explain how the ranger could do this without cutting down the tree.
Answers will vary

12.3 Area

Objective A To find the area of geometric figures

INSTRUCTOR NOTE

The concepts of square units and area are difficult for students. After introducing these ideas, ask students questions such as "Would I measure the distance from Chicago to Boston in miles or square miles?" or "Is the amount of land cultivated by a gardener measured in feet or square feet?"

Area is a measure of the amount of surface in a region. Area can be used to describe the size of a rug, a parking lot, a farm, or a national park. Area is measured in square units.

A square that measures 1 in. on each side has an area of 1 square inch, written 1 in^2.

A square that measures 1 cm on each side has an area of 1 square centimeter, written 1 cm^2.

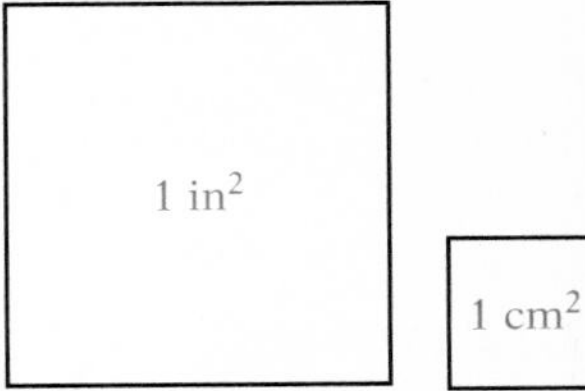

Larger areas can be measured in square feet (ft^2), square meters (m^2), square miles (mi^2), acres (43,560 ft^2), or any other square unit.

The area of a geometric figure is the number of squares that are necessary to cover the figure. In the figures below, two rectangles have been drawn and covered with squares. In the figure on the left, 12 squares, each of area 1 cm^2, were used to cover the rectangle. The area of the rectangle is 12 cm^2. In the figure on the right, 6 squares, each of area 1 in^2, were used to cover the rectangle. The area of the rectangle is 6 in^2.

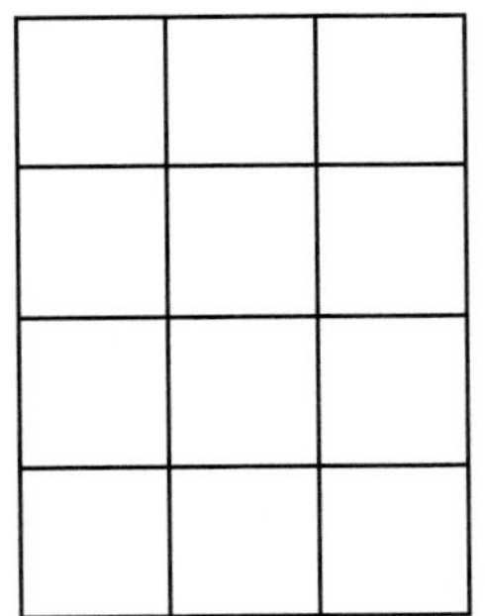

The area of the rectangle is 12 cm^2.

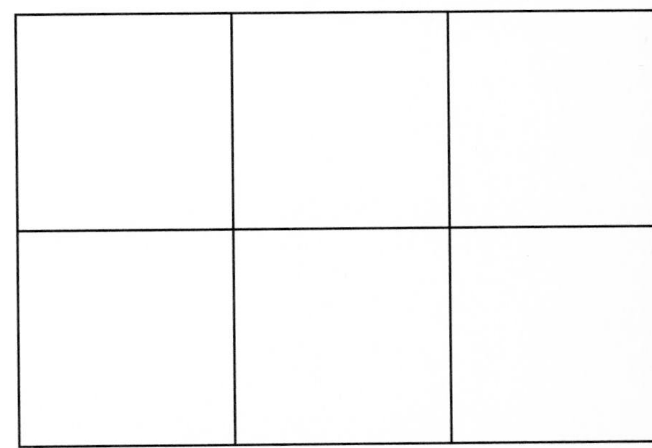

The area of the rectangle is 6 in^2.

Note from the above figures that the area of a rectangle can be found by multiplying the length of the rectangle by its width.

Area of a Rectangle

$A = LW$

➡ Find the area of the rectangle shown at the right.

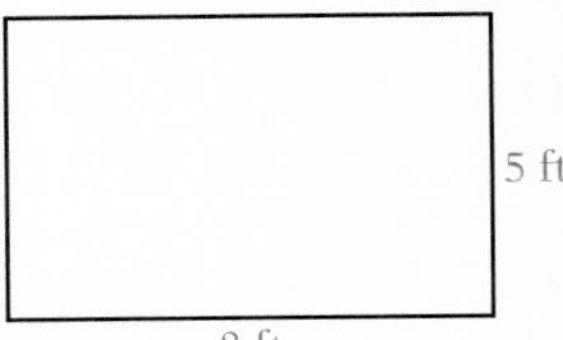

$$\begin{aligned} A &= LW \\ &= 8 \text{ ft} \cdot 5 \text{ ft} \\ &= 40 \text{ ft}^2 \end{aligned}$$

• $L = 8$ ft, $W = 5$ ft.

The area of the rectangle is 40 ft^2.

A square is a rectangle in which all sides are the same length. Therefore, both the length and width can be represented by a side.

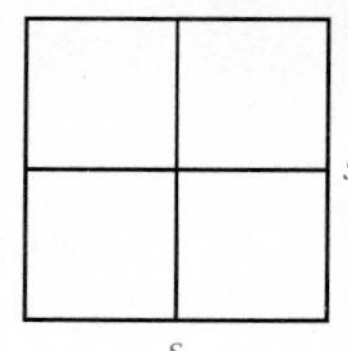

Area of a Square

$A = s \cdot s = s^2$

➡ Find the area of the square shown at the right.

$A = s^2$
$= (14 \text{ cm})^2$ • $s = 14$ cm.
$= 196 \text{ cm}^2$

The area of the square is 196 cm^2.

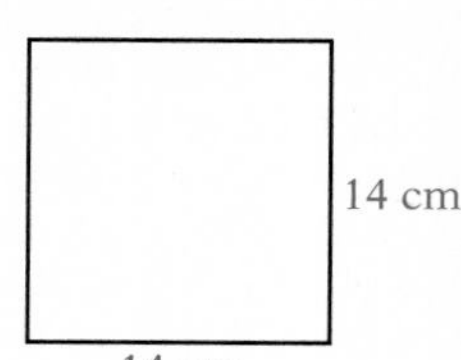

INSTRUCTOR NOTE
Ancient Egyptians gave the formula for the area of a circle as $\left(\frac{8}{9}d\right)^2$. As an enrichment problem, ask students if this formula gives an area that is less than or greater than the correct formula.

The area of a circle is equal to the product of π and the square of the radius.

Area of a Circle

$A = \pi r^2$

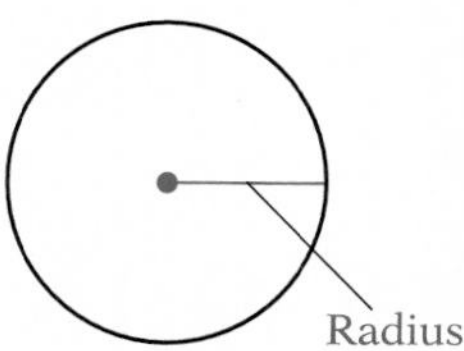

➡ Find the area of the circle shown at the right.

$A = \pi r^2$
$= \pi(8 \text{ in.})^2 = 64\pi \text{ in}^2$
$\approx 64 \cdot 3.14 \text{ in}^2 = 200.96 \text{ in}^2$

The area is exactly 64π in^2.
The area is approximately 200.96 in^2.

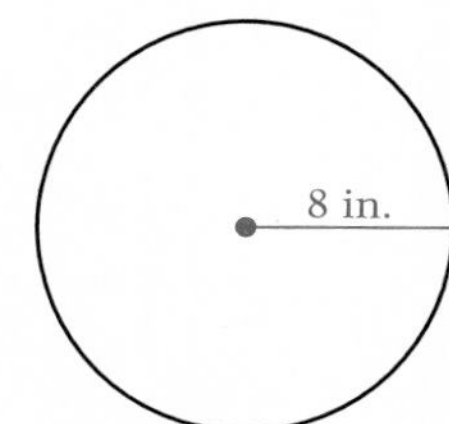

In the figure below, AB is the base of the triangle, and CD, which is perpendicular to the base, is the height.

The area of a triangle is one-half the product of the base and the height.

Area of a Triangle

$A = \frac{1}{2}bh$

C
h
A b D B

➡ Find the area of the triangle shown at the right.

$A = \frac{1}{2}bh$
$= \frac{1}{2}(20 \text{ m})(5 \text{ m})$ • $b = 20$ m, $h = 5$ m.
$= 50 \text{ m}^2$

The area of the triangle is 50 m^2.

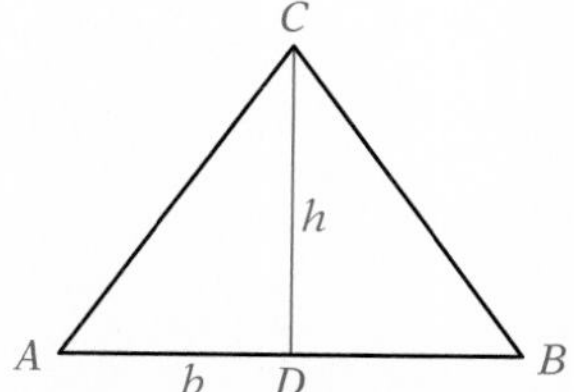

Example 1 Find the area of a circle with a diameter of 9 cm. Use $\pi \approx 3.14$.

Solution

$r = \frac{1}{2}d$

$r = \frac{1}{2} \cdot 9 \text{ cm} = 4.5 \text{ cm}$

$A = \pi r^2$
$\approx 3.14(4.5 \text{ cm})^2$
$= 63.585 \text{ cm}^2$

You Try It 1 Find the area of a triangle with a base of 24 in. and a height of 14 in.

Your solution 168 in^2

Solution on p. A52

Objective B To find the area of composite geometric figures

The area of the composite figure shown below is found by calculating the area of the rectangle and then subtracting the area of the triangle.

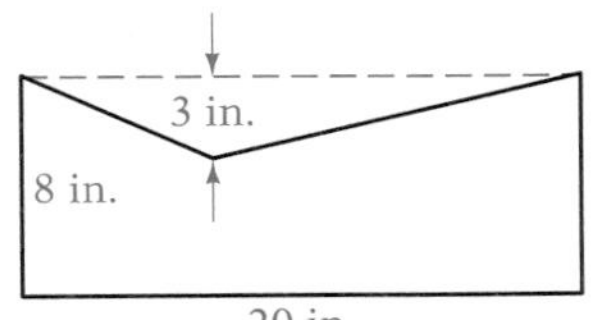

= 8 in. 20 in. −

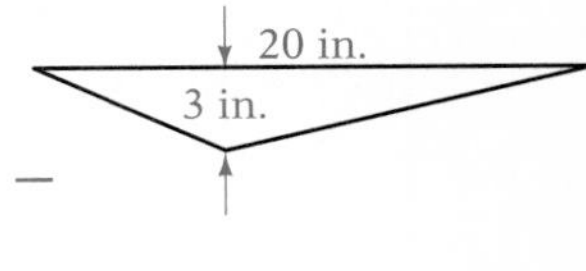

$A = LW - \frac{1}{2}bh$

$A = 20 \text{ in.} \cdot 8 \text{ in.} - \frac{1}{2} \cdot 20 \text{ in.} \cdot 3 \text{ in.} = 160 \text{ in}^2 - 30 \text{ in}^2 = 130 \text{ in}^2$

Example 2
Find the area of the shaded portion of the figure. Use $\pi \approx 3.14$.

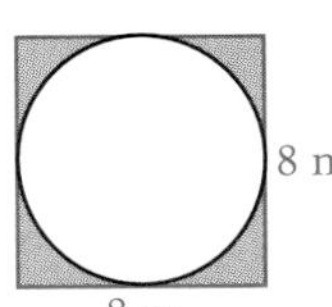

Solution

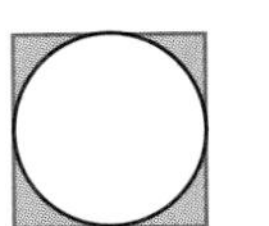 =  −

Area of shaded portion = area of square − area of circle

$A = s^2 - \pi r^2$
$A = (8 \text{ m})^2 - \pi(4 \text{ m})^2$
$\approx 64 \text{ m}^2 - 3.14(16 \text{ m}^2)$
$= 64 \text{ m}^2 - 50.24 \text{ m}^2$
$\approx 13.76 \text{ m}^2$

You Try It 2
Find the area of the composite figure.

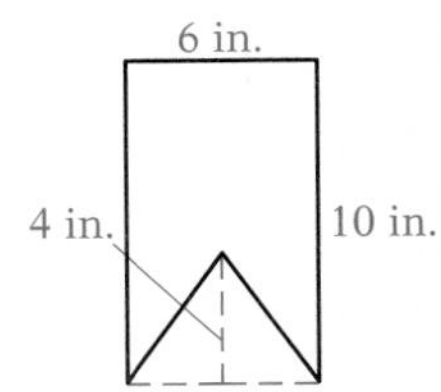
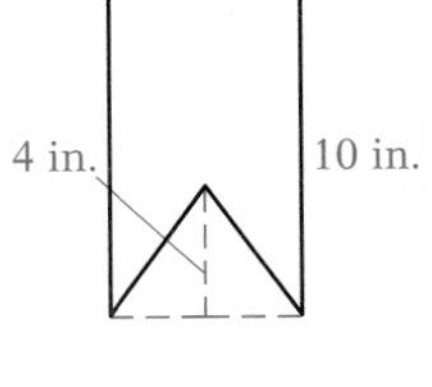

Your solution
48 in^2

Solution on p. A52

Objective C To solve application problems

Example 3

A walkway 2 m wide is built along the front and along both sides of a building, as shown in the figure. Find the area of the walkway.

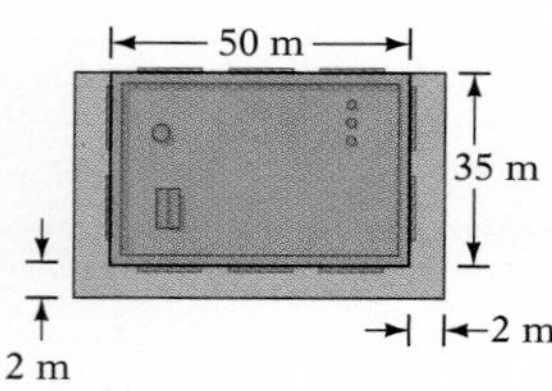

Strategy

To find the area of walkway, add the area of the front section (54 m · 2 m) and the area of the two side sections (each 35 m · 2 m).

Solution

$$\text{Area of walkway} = \underbrace{\text{area of front section}} + \underbrace{2(\text{area of one side section})}$$

$$\begin{aligned} A &= 54 \text{ m} \cdot 2 \text{ m} + 2(35 \text{ m} \cdot 2 \text{ m}) \\ &= 108 \text{ m}^2 + 140 \text{ m}^2 \\ &= 248 \text{ m}^2 \end{aligned}$$

The area of the walkway is 248 m^2.

You Try It 3

New carpet is installed in a room measuring 9 ft by 12 ft. Find the area of the room in square yards. (9 ft^2 = 1 yd^2)

Your strategy

Your solution

12 yd^2

Solution on p. A53

12.3 Exercises

Objective A

In Exercises 1 to 8, find the area of the given figures.

1. 6 ft; 24 ft
144 ft^2

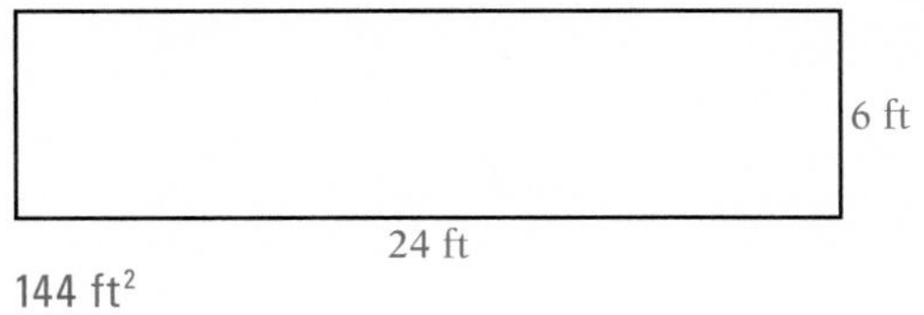

2. 8 in.; 18 in.
144 in^2

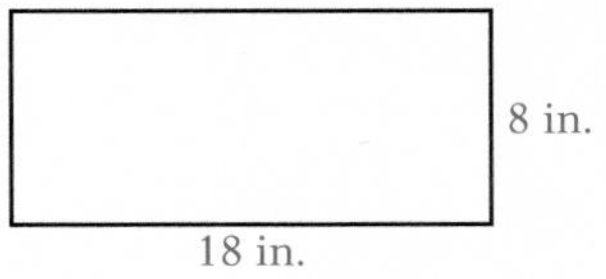

3. 9 in.; 9 in.
81 in^2

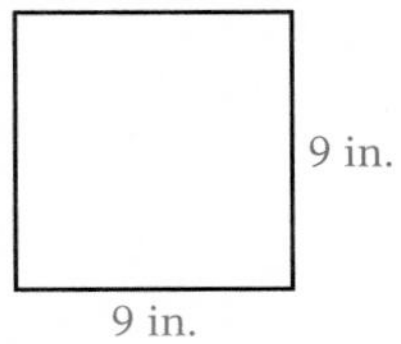

4. 4 in.; 4 in.
16 in^2

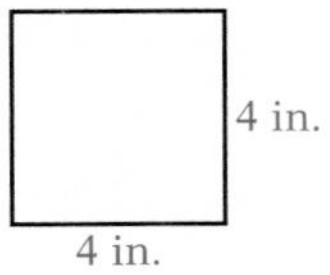

5. 4 ft
50.24 ft^2

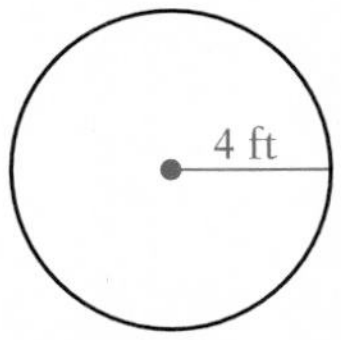

6. 3 cm
28.26 cm^2

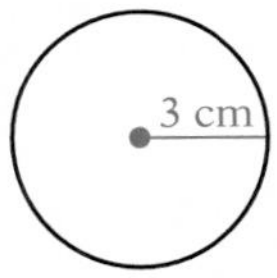

7. 4 in.; 10 in.
20 in^2

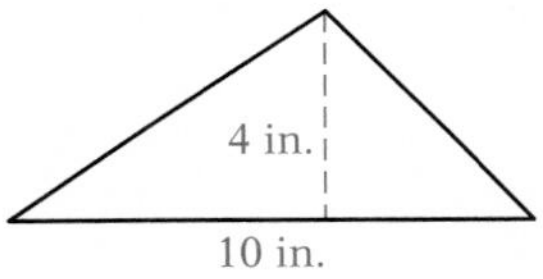

8. 6 m; 7 m
21 m^2

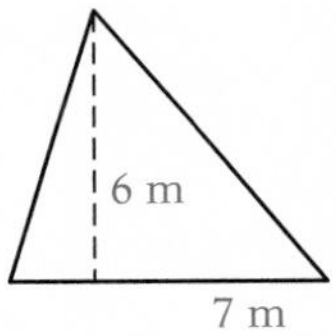

9. Find the area of a right triangle with a base of 3 cm and a height of 1.42 cm.
2.13 cm^2

10. Find the area of a triangle with a base of 3 ft and a height of $\frac{2}{3}$ ft.
1 ft^2

11. Find the area of a square with a side of 4 ft. 16 ft^2

12. Find the area of a square with a side of 10 cm. 100 cm^2

13. Find the area of a rectangle with a length of 43 in. and a width of 19 in.
817 in^2

14. Find the area of a rectangle with a length of 82 cm and a width of 20 cm.
1640 cm^2

15. Find the area of a circle with a radius of 7 in. Use $\pi \approx \frac{22}{7}$.
154 in^2

16. Find the area of a circle with a diameter of 40 cm. Use $\pi \approx 3.14$.
1256 cm^2

Objective B

In the figures accompanying Exercises 17 to 24, find the area. Use $\pi \approx 3.14$.

17.

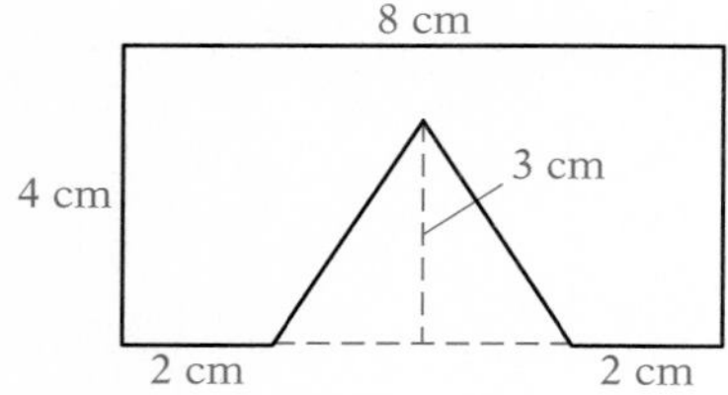

26 cm^2

18.

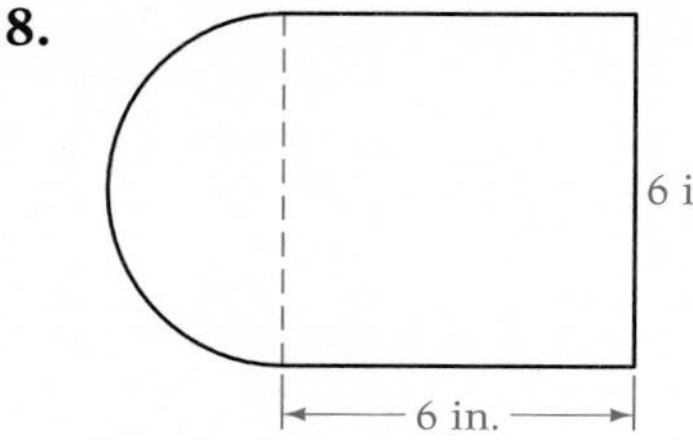

50.13 in^2

19.

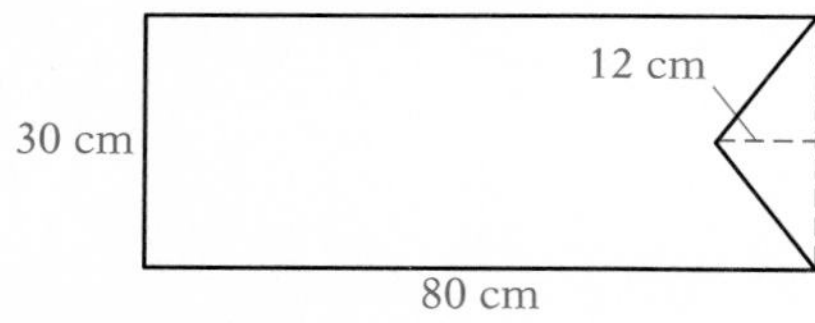

2220 cm^2

20.

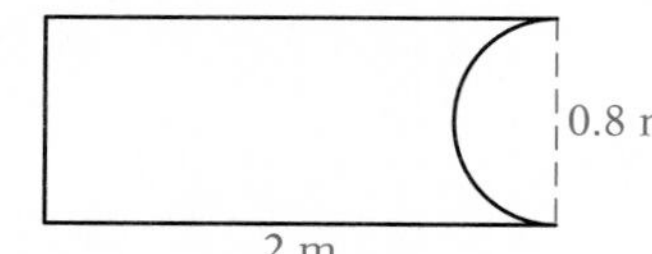

1.3488 m^2

21.

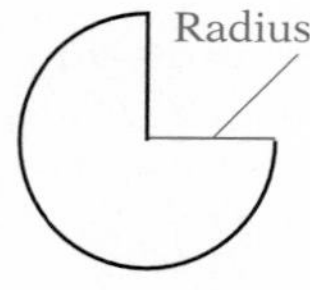

150.72 in^2

22.

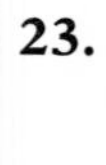
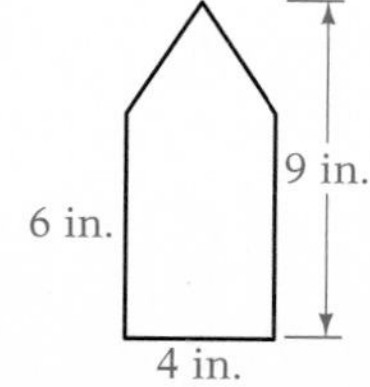

30 in^2

23.

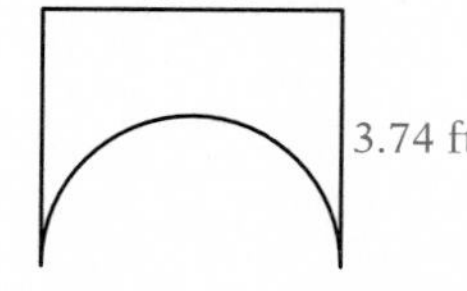

8.851323 ft^2

24.

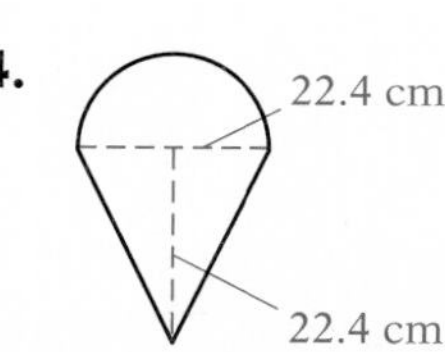

447.8208 cm^2

Objective C *Application Problems*

25. Find the area of a rectangular park that has a length of 54 yd and a width of 32 yd. 1728 yd^2

26. A rectangular garden has a length of 17 ft and a width of 12 ft. Find the area of the garden. 204 ft^2

27. The telescope lens located on Mt. Palomar has a diameter of 200 in. Find the area of the lens. Leave the answer in terms of π. 10,000π in^2

28. An irrigation system waters a circular field that has a 50-ft radius. Find the area watered by the irrigation system. Use $\pi \approx 3.14$. 7850 ft^2

29. A carpet is to be placed in one room and a hallway, as shown in the diagram at the right. At \$16.50 per square meter, how much will it cost to carpet the area? \$570.90

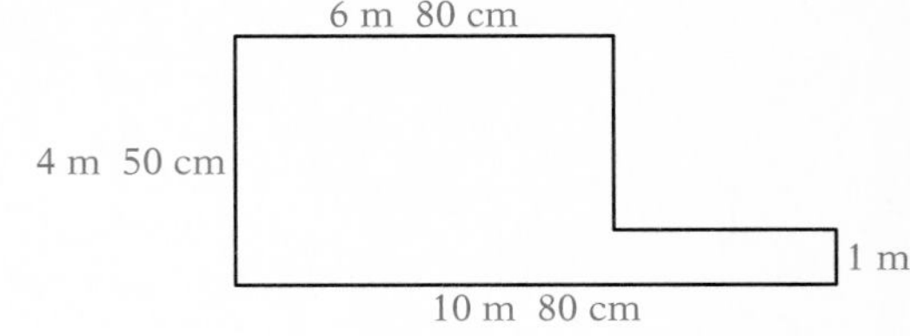

30. Find the area of the concrete driveway with the measurements shown in the figure. 1250 ft^2

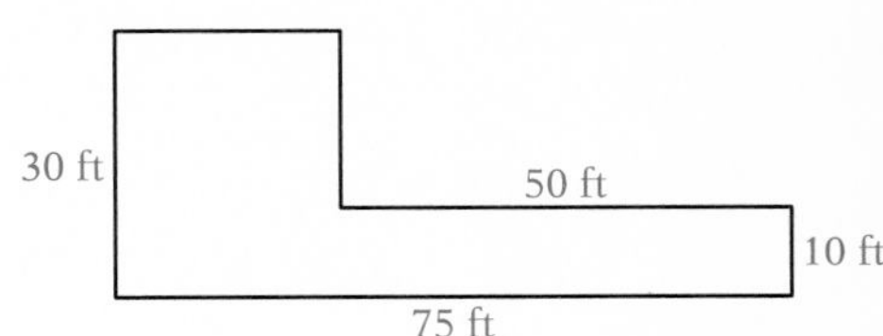

31. Find the area of the 2-m boundary around the swimming pool shown in the figure. 68 m^2

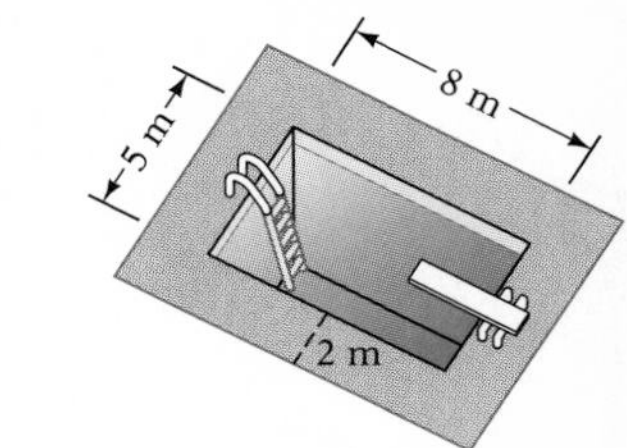

32. How much hardwood floor is needed to cover the roller rink shown in the figure? Use $\pi \approx 3.14$. 19,024 ft^2

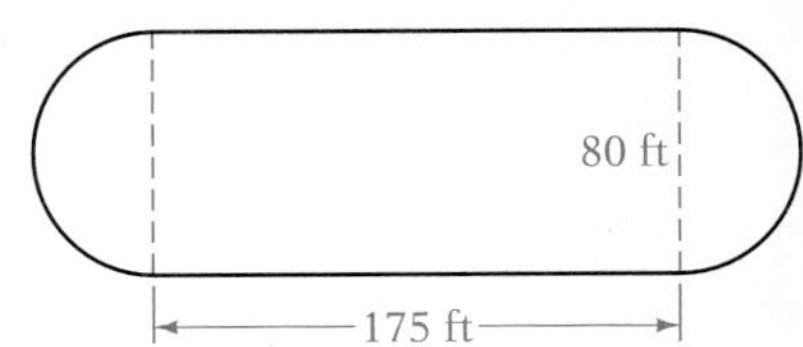

33. Find the total area of the national park with dimensions shown in the figure. Use $\pi \approx 3.14$. 125.1492 mi^2

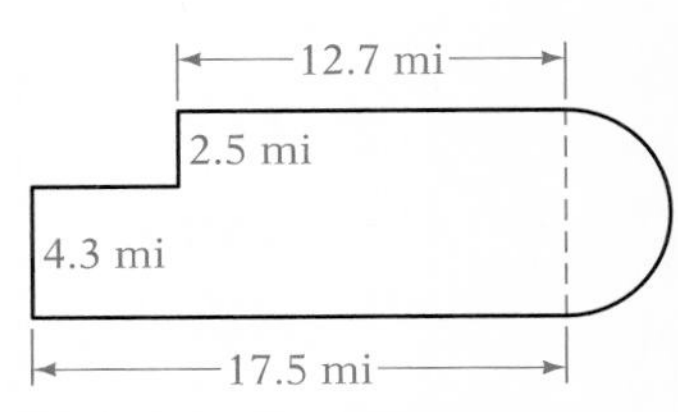

34. A circle has a radius of 8 in. Find the increase in area when the radius is increased by 2 in. Use $\pi \approx 3.14$. 113.04 in^2

35. A circle has a radius of 5 cm. Find the increase in area when the radius is doubled. Use $\pi \approx 3.14$. 235.5 cm^2

36. Find the cost of plastering the walls of a room 22 ft long, 25 ft 6 in. wide, and 8 ft high. Subtract 120 ft^2 for windows and doors. The cost is $.75 per square foot. $480.00

APPLYING THE CONCEPTS

37. If both the length and the width of a rectangle are doubled, how many times larger is the area of the resulting rectangle? 4 times

38. If the radius of a circle is doubled, what happens to the area? What happens to the area when the diameter is doubled?
quadrupled quadrupled

39. Determine whether the statement is always true, sometimes true, or never true.
a. If two triangles have the same perimeter, then they have the same area. Sometimes true
b. If two rectangles have the same area, then they have the same perimeter. Sometimes true
c. If two squares have the same area, then the sides of the squares have the same length. Always true

40. What fractional part of the area of the larger of the two squares is the shaded area? Write your answer as a fraction in simplest form. This problem appeared in *Math Teacher,* vol. 86, No. 3 (September 1993).
$\frac{1}{64}$

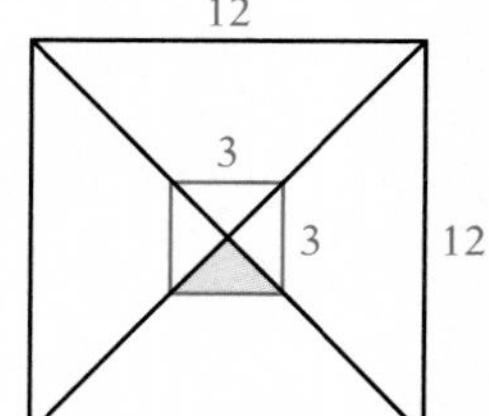

41. [W] The circles at the right are identical. Is the area in the circles to the left of the line equal to, less than, or greater than the area in the circles to the right of the line? Explain your answer. Equal

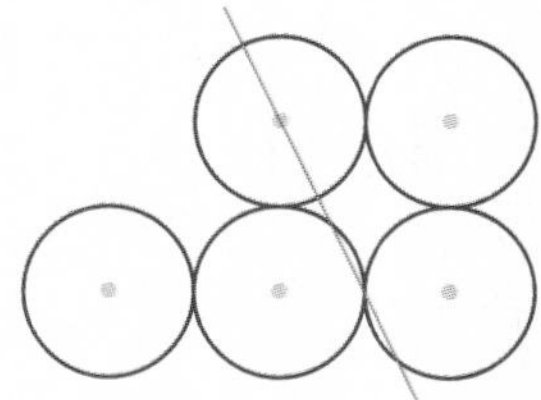

42. All of these dots are equally spaced, horizontally and vertically, 1 inch apart. What is the area of the triangle? $2\frac{1}{2}$ in^2

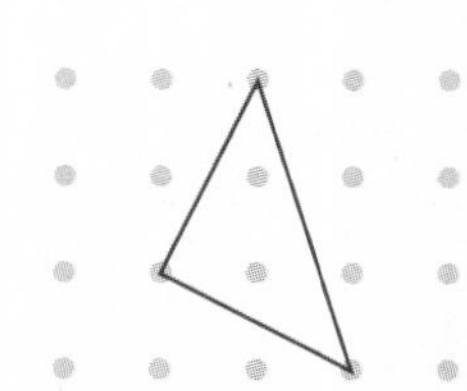

12.4 Volume

Objective A To find the volume of geometric solids

INSTRUCTOR NOTE

The difficulties students have distinguishing linear measure from square measure is compounded with volume measure. Ask students to give examples of things that would be measured in, for instance, ft, ft^2, and ft^3. For instance, the length of a classroom, the area of the floor, the volume of air in the room.

Volume is a measure of the amount of space inside a figure in space. Volume can be used to describe the amount of heating gas used for cooking, the amount of concrete delivered for the foundation of a house, or the amount of water in storage for a city's water supply.

A cube that is 1 ft on each side has a volume of 1 cubic foot, which is written 1 ft^3. A cube that measures 1 cm on each side has a volume of 1 cubic centimeter, written 1 cm^3.

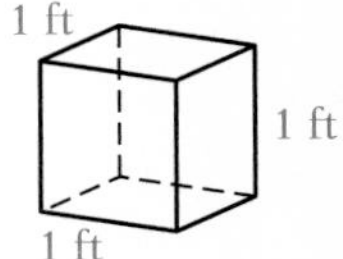

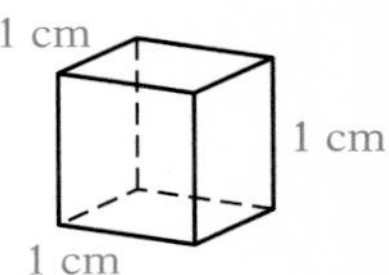

The volume of a solid is the number of cubes that are necessary to fill the solid exactly. The volume of the rectangular solid at the right is 24 cm^3 because it will hold exactly 24 cubes, each 1 cm on a side. Note that the volume can be found by multiplying the length times the width times the height.

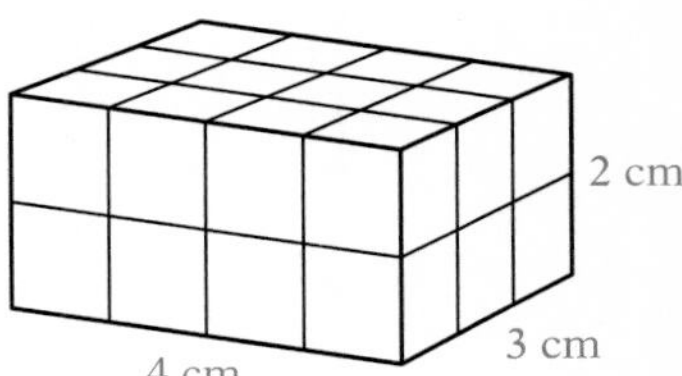

Volume of a Rectangular Solid

$V = LWH$

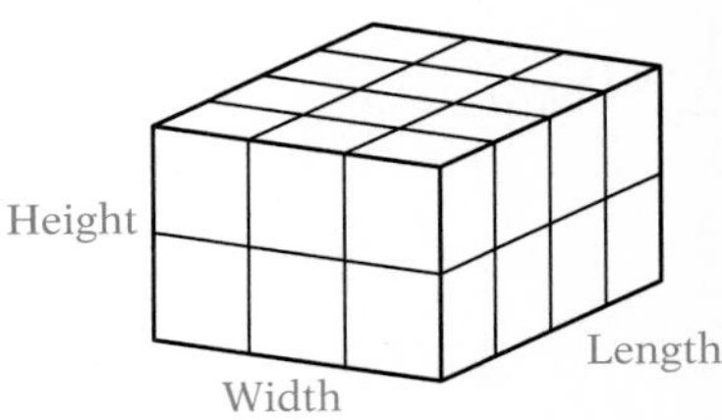

➡ Find the volume of a rectangular solid with a length of 9 in., a width of 3 in., and a height of 4 in.

$$\begin{aligned} V &= LWH \\ &= (9\text{ in.})(3\text{ in.})(4\text{ in.}) \\ &= 108\text{ in}^3 \end{aligned}$$

• *L* = 9 in., *W* = 3 in., *H* = 4 in.

4 in.

9 in.

3 in.

The volume of the rectangular solid is 108 in^3.

The length, width, and height of a cube have the same measure. The volume of a cube is found by multiplying the side of the cube by itself three times (side cubed).

Volume of a Cube

$V = s^3$

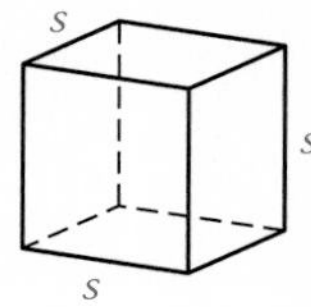

➡ Find the volume of the cube shown at the right.

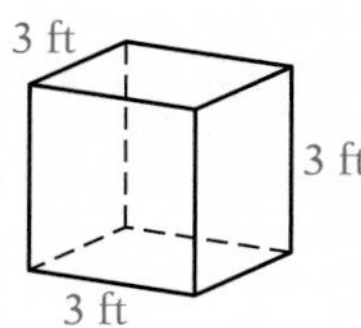

$V = s^3$
$= (3 \text{ ft})^3$ • $s = 3$ ft.
$= 27 \text{ ft}^3$

The volume of the cube is 27 ft^3.

The volume of a sphere is found by multiplying four-thirds times pi (π) times the radius cubed.

Volume of a Sphere

$V = \frac{4}{3}\pi r^3$

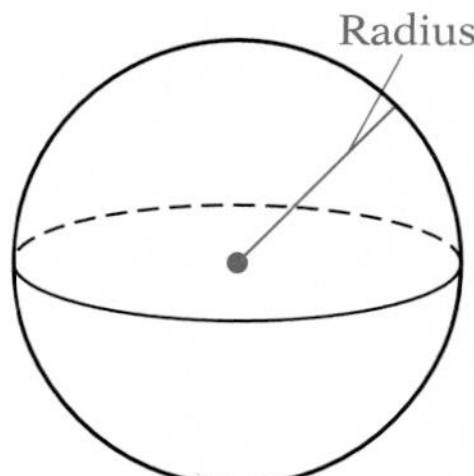

➡ Find the volume of the sphere shown at the right. Use $\pi \approx 3.14$.

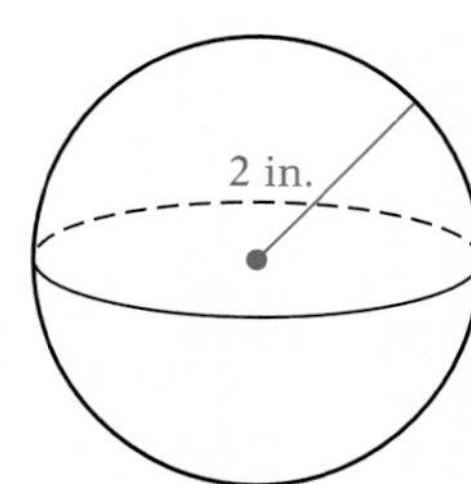

$V = \frac{4}{3}\pi r^3$
$\approx \frac{4}{3} \cdot 3.14 \cdot (2 \text{ in.})^3$ • $r = 2$ in.
$= \frac{4}{3} \cdot 3.14 \cdot 8 \text{ in}^3$
$\approx 33.49 \text{ in}^3$

The volume is approximately 33.49 in^3.

The volume of a cylinder is found by multiplying the base (a circle) of the cylinder times the height.

Volume of a Cylinder

$V = \pi r^2 h$

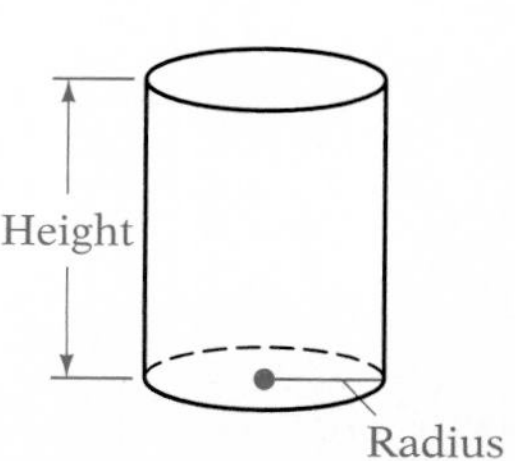

➡ Find the volume of the cylinder shown at the right. Use $\pi \approx 3.14$.

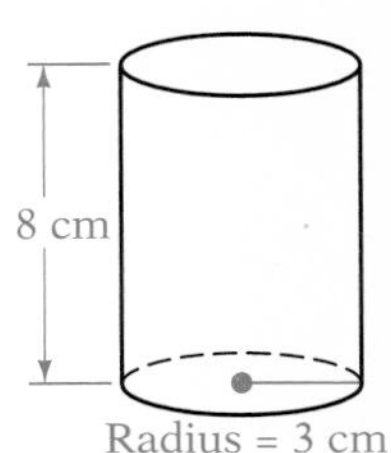

$$\begin{aligned} V &= \pi r^2 h \\ &\approx 3.14(3 \text{ cm})^2(8 \text{ cm}) \\ &= 3.14 \cdot 9 \text{ cm}^2 \cdot 8 \text{ cm} \\ &= 226.08 \text{ cm}^3 \end{aligned}$$

• $r = 3$ cm; $h = 8$ cm.

The volume of the cylinder is 226.08 cm^3.

Example 1
Find the volume of a rectangular solid with a length of 3 ft, a width of 1.5 ft, and a height of 2 ft.

Solution

$$\begin{aligned} V &= LWH \\ &= 3 \text{ ft} \cdot 1.5 \text{ ft} \cdot 2 \text{ ft} \\ &= 9 \text{ ft}^3 \end{aligned}$$

You Try It 1
Find the volume of a cube with a side of 5 cm.

Your solution
125 cm^3

Example 2
Find the volume of a cylinder with a radius of 12 cm and a height of 65 cm. Use $\pi \approx 3.14$.

Solution

$$\begin{aligned} V &= \pi r^2 h \\ &\approx 3.14 \cdot (12 \text{ cm})^2 \cdot 65 \text{ cm} \\ &= 29{,}390.4 \text{ cm}^3 \end{aligned}$$

You Try It 2
Find the volume of a cylinder with a diameter of 14 in. and a height of 15 in. Use $\pi \approx \frac{22}{7}$.

Your solution
2310 in^3

Example 3
Find the volume of a sphere with a diameter of 12 in. Use $\pi \approx 3.14$.

Solution
$r = 6$ in.

$$\begin{aligned} V &= \frac{4}{3}\pi r^3 \\ &\approx \frac{4}{3} \cdot 3.14 \cdot (6 \text{ in.})^3 \\ &= 904.32 \text{ in}^3 \end{aligned}$$

You Try It 3
Find the volume of a sphere with a radius of 3 m. Use $\pi \approx 3.14$.

Your solution
113.04 m^3

Solutions on p. A53

Objective B To find the volume of composite geometric solids

Composite geometric solids are solids made from two or more geometric solids. The solid shown is made from a cylinder and one-half of a sphere.

$$= \quad + $$

Volume of the composite solid = volume of the cylinder + one-half the volume of the sphere

➡ Find the volume of the composite solid shown above if the radius of the base of the cylinder is 3 in. and the height of the cylinder is 10 in. Use $\pi \approx 3.14$.

The volume equals the volume of a cylinder plus one-half the volume of a sphere. The radius of the sphere equals the radius of the base of the cylinder.

$$V = \pi r^2 h + \frac{1}{2}\left(\frac{4}{3}\pi r^3\right)$$

$$V \approx 3.14 \cdot (3 \text{ in.})^2 \cdot 10 \text{ in.} + \frac{2}{3} \cdot 3.14 \cdot (3 \text{ in.})^3$$

$$\approx 282.6 \text{ in}^3 + 56.25 \text{ in}^3$$

$$= 339.12 \text{ in}^3$$

Example 4
Find the volume of the solid in the figure. Use $\pi \approx 3.14$.

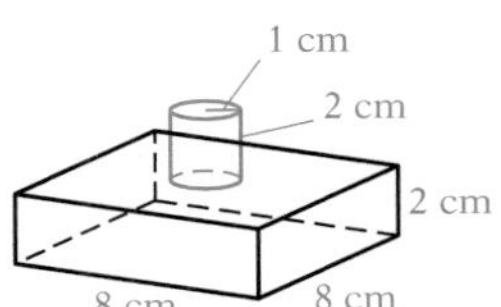

Solution

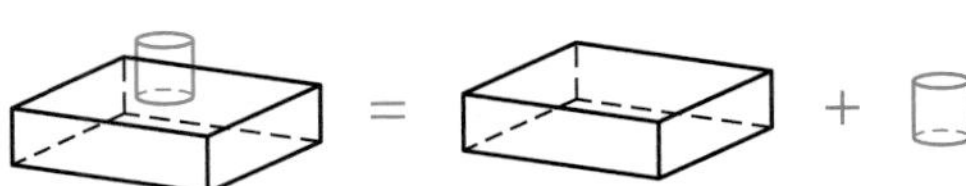

Volume of the solid	=	Volume of rectangular solid	+	Volume of cylinder

$$
\begin{aligned}
V &= LWH + \pi r^2 h \\
&\approx 8 \text{ cm} \cdot 8 \text{ cm} \cdot 2 \text{ cm} + 3.14 \cdot (1 \text{ cm})^2 \cdot 2 \text{ cm} \\
&= 128 \text{ cm}^3 + 6.28 \text{ cm}^3 \\
&= 134.28 \text{ cm}^3
\end{aligned}
$$

Volume $\approx 134.28 \text{ cm}^3$

You Try It 4
Find the volume of the solid in the figure. Use $\pi \approx 3.14$.

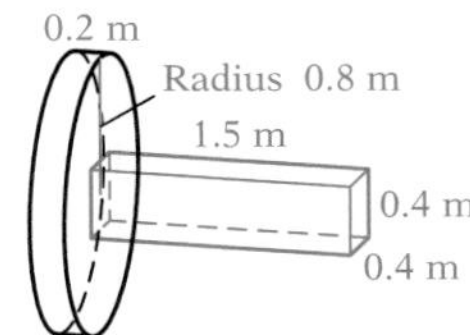

Your solution

0.64192 m^3

Example 5
Find the volume of the solid in the figure. Use $\pi \approx 3.14$.

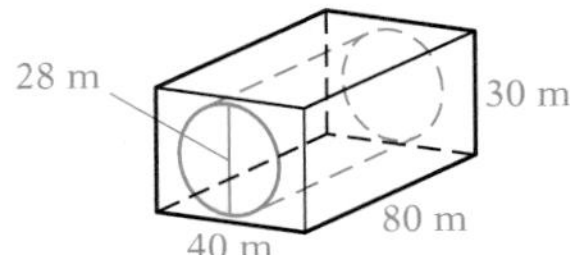

Solution

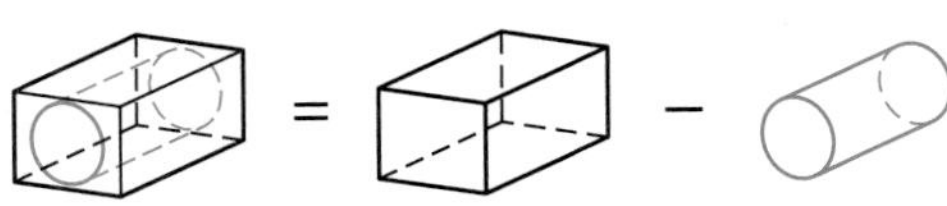

Volume of the solid	=	Volume of rectangular solid	−	Volume of cylinder

$$
\begin{aligned}
V &= LWH - \pi r^2 h \\
&\approx 80 \text{ m} \cdot 40 \text{ m} \cdot 30 \text{ m} - 3.14 \cdot (14 \text{ m})^2 \cdot 80 \text{ m} \\
&= 96{,}000 \text{ m}^3 - 49{,}235.2 \text{ m}^3 \\
&= 46{,}764.8 \text{ m}^3
\end{aligned}
$$

Volume $\approx 46{,}764.8 \text{ m}^3$

You Try It 5
Find the volume of the solid in the figure. Use $\pi \approx 3.14$.

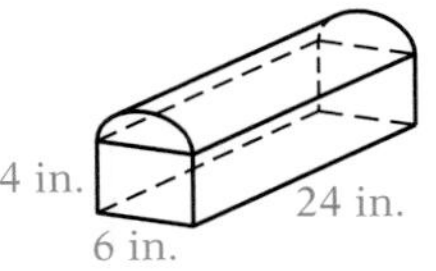

Your solution

915.12 in^3

Solutions on p. A53

Objective C To solve application problems

Example 6
An aquarium is 28 in. long, 14 in. wide, and 16 in. high. Find the volume of the aquarium.

Strategy
To find the volume of the aquarium, use the formula for the volume of a rectangular solid.

Solution

$V = LWH$
$= 28 \text{ in.} \cdot 14 \text{ in.} \cdot 16 \text{ in.}$
$= 6272 \text{ in}^3$

The volume of the aquarium is 6272 in^3.

You Try It 6
Find the volume of a freezer that is 7 ft long, 3 ft high, and 2.5 ft wide.

Your strategy

Your solution
52.5 ft^3

Example 7
Find the volume of the bushing shown in the figure below. Use $\pi \approx 3.14$.

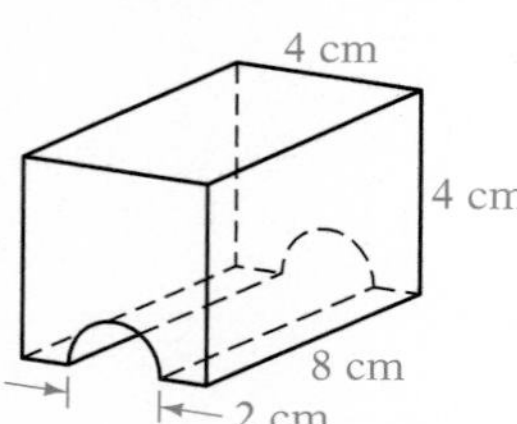

Strategy
To find the volume of the bushing, subtract the volume of the half-cylinder from the volume of the rectangular solid.

Solution

Volume of the bushing = volume of rectangular solid − $\frac{1}{2}$ of the volume of cylinder

$V = LWH - \frac{1}{2}\pi r^2 h$
$\approx 4 \text{ cm} \cdot 4 \text{ cm} \cdot 8 \text{ cm} - \frac{1}{2} \cdot 3.14 \cdot (1 \text{ cm})^2 \cdot 8 \text{ cm}$
$\approx 128 \text{ cm}^3 - 12.56 \text{ cm}^3$
$\approx 115.44 \text{ cm}^3$

The volume of the bushing is approximately 115.44 cm^3.

You Try It 7
Find the volume of the channel iron shown in the figure below.

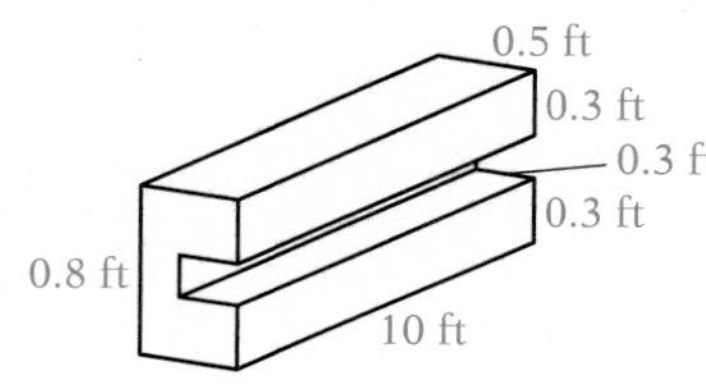

Your strategy

Your solution
3.4 ft^3

Solutions on pp. A53 and A54

12.4 Exercises

Objective A

In Exercises 1 to 8, find the volume of the solids in the accompanying figures. Round to the nearest hundredth. Use $\pi \approx 3.14$.

1.

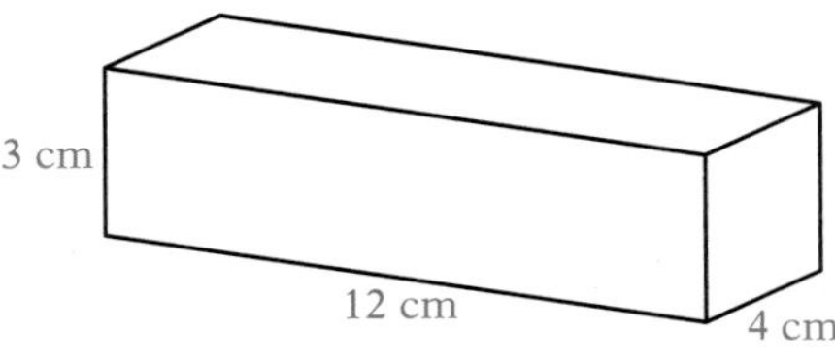

144 cm^3

2.

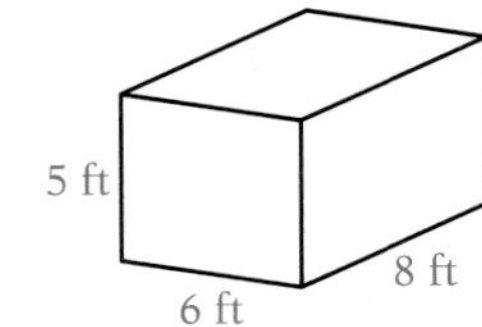

240 ft^3

3.

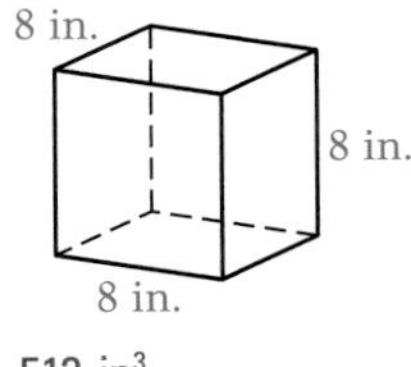

512 in^3

4.

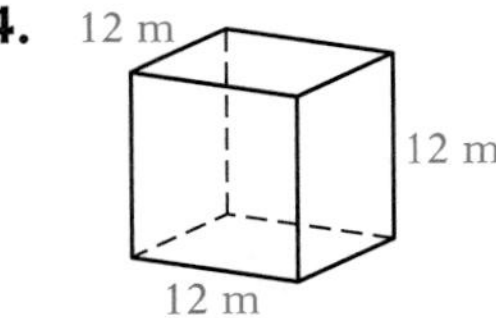

1728 m^3

5.

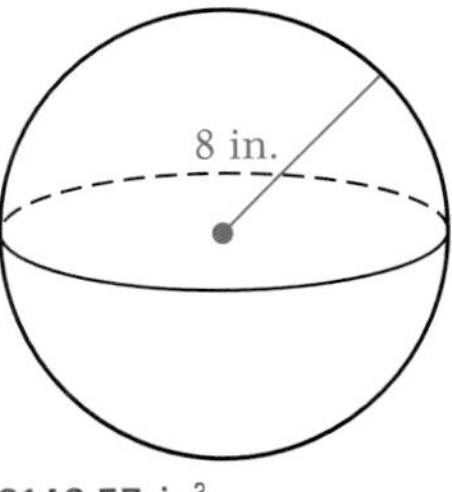

2143.57 in^3

6.

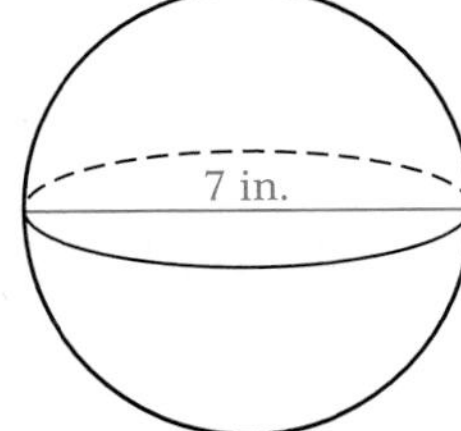

179.50 in^3

7.

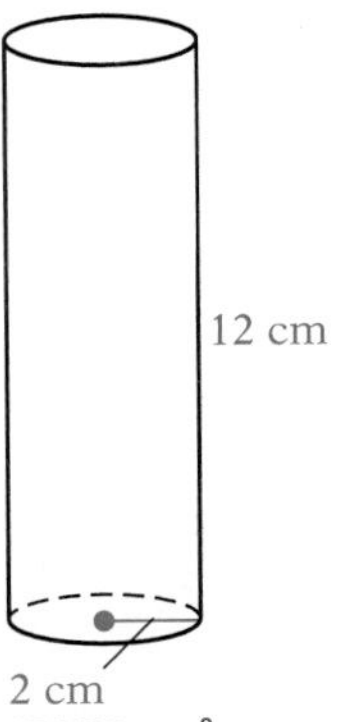

150.72 cm^3

8.

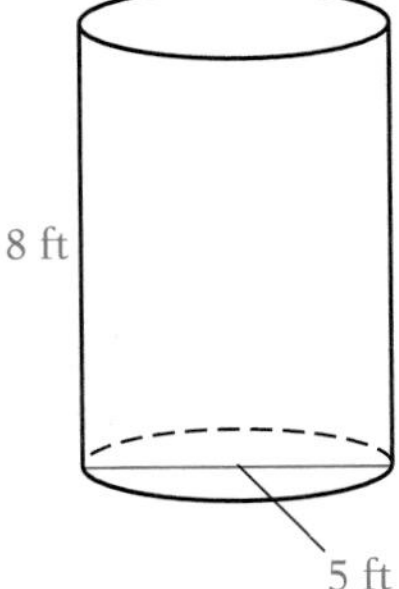

157 ft^3

In Exercises 9 to 16, find the volume. Write measurements involving two units in terms of the larger unit before working the problem.

9. Find the volume of a rectangular solid with a length of 2 m, a width of 80 cm, and a height of 4 m.
$6.4\ m^3$

10. Find the volume of a cylinder with a radius of 7 cm and a height of 14 cm. Use $\pi \approx \frac{22}{7}$.
$2156\ cm^3$

11. Find the volume of a sphere with a 11-mm radius. Use $\pi \approx 3.14$. Round to the nearest hundredth.
$5572.45\ mm^3$

12. Find the volume of a cube with a side of 2 m 14 cm. Round to the nearest tenth.
$9.8\ m^3$

13. Find the volume of a cylinder with a diameter of 12 ft and a height of 30 ft. Use $\pi \approx 3.14$.
$3391.2\ ft^3$

14. Find the volume of a sphere with a 6-ft diameter. Use $\pi \approx 3.14$.
$113.04\ ft^3$

15. Find the volume of a cube with a side of $3\frac{1}{2}$ ft.
$42\frac{7}{8}\ ft^3$

16. Find the volume of a rectangular solid with a length of 1 m 15 cm, a width of 60 cm, and a height of 25 cm.
$0.1725\ m^3$

Objective B

In Exercises 17 to 22, find the volume of the accompanying figures. Write measurements involving two units in terms of the larger unit before working the problem. Use $\pi \approx 3.14$.

17.

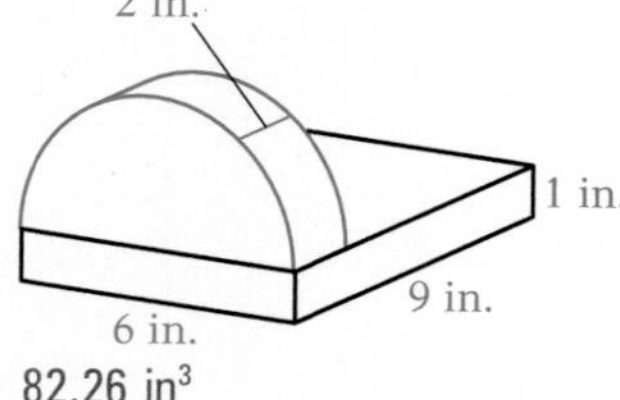

$82.26\ in^3$

18.

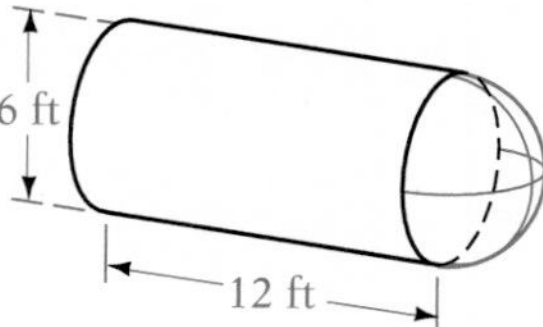

$395.64\ ft^3$

19.

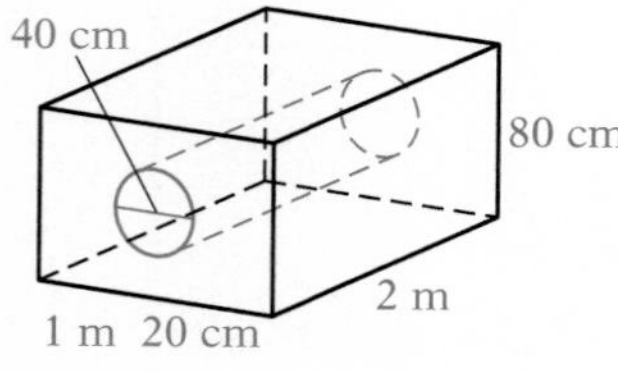

$1.6688\ m^3$

20.

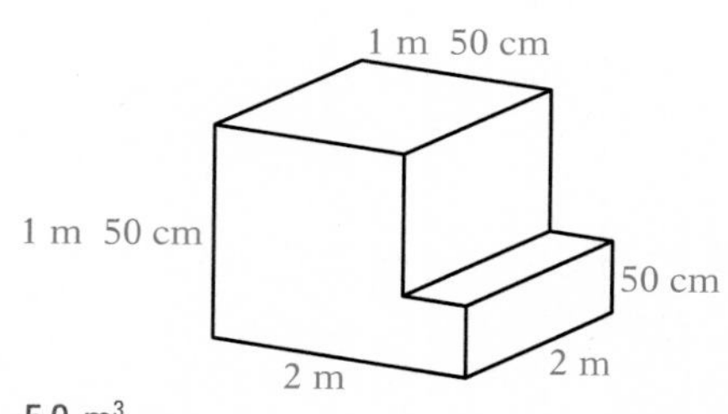

$5.0\ m^3$

21.

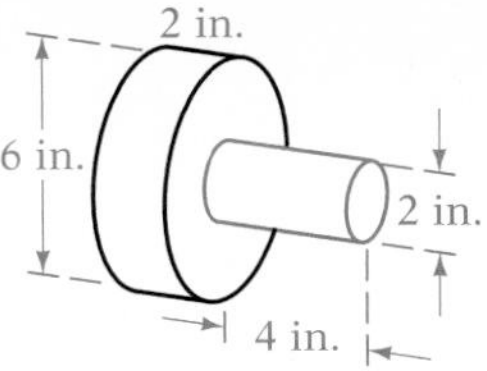

69.08 in^3

22.

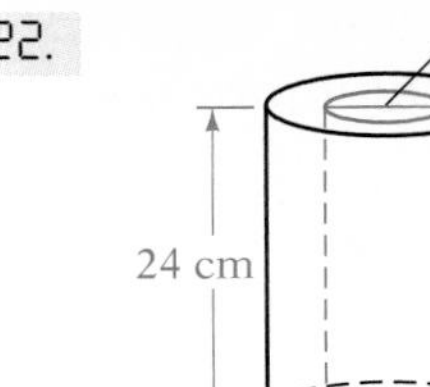

4578.12 cm^3

Objective C Application Problems

23. A rectangular tank at the fish hatchery is 9 m long, 3 m wide, and 1.5 m deep. Find the volume of the water in the tank when the tank is full. 40.5 m^3

24. A fuel tank in a booster rocket is a cylinder 10 ft in diameter and 52 ft high. Find the volume of the fuel tank. Use $\pi \approx 3.14$. 4082 ft^3

25. A hot air balloon is in the shape of a sphere. Find the volume of a hot air balloon that is 32 ft in diameter. Use $\pi \approx 3.14$. (Round to the nearest hundredth.) 17,148.59 ft^3

26. A storage tank for propane is in the shape of a sphere. Find the volume of a storage tank that is 9 m in diameter. Leave the answer in terms of π. 121.5π m^3

27. A silo, which is in the shape of a cylinder, is 16 ft in diameter and has a height of 36 ft. Find the volume of the silo. Leave the answer in terms of π. 2304π ft^3

28. An oil storage tank, which is in the shape of a cylinder, is 4 m high and has a 6-m diameter. Find the volume of the oil storage tank. Use $\pi \approx 3.14$. 113.04 m^3

29. How many gallons of water will fill an aquarium that is 12 in. wide, 18 in. long, and 16 in. high? Round to the nearest tenth. (1 gal = 231 in^3) 15.0 gal

30. How many gallons of water will fill a fish tank that is 12 in. long, 8 in. wide, and 9 in. high? Round to the nearest tenth. (1 gal = 231 in^3) 3.7 gal

31. A swimming pool contains 65,000 ft^3 of water. Find the total weight of the water in the swimming pool. (1 ft^3 of water weighs 62.4 lb.)
4,056,000 lb

32. An architect is designing the heating for an auditorium and needs to know the volume of the structure. Find the volume of the auditorium with the measurements shown in the figure. Use $\pi \approx 3.14$. 809,516.25 ft^3

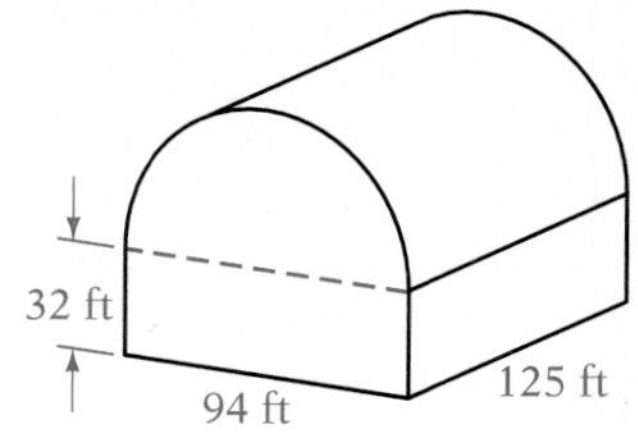

33. Find the volume of the bushing shown at the right. Use $\pi \approx 3.14$. 212.64 in^3

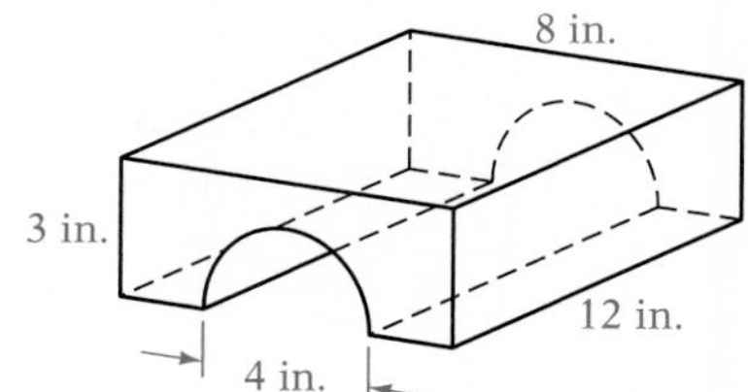

34. The floor of a building is shown at the right. The concrete floor is 6 in. thick. Find the cost of the floor at $2.75 per cubic foot. Use $\pi \approx 3.14$. Round to the nearest cent.
$3067.97

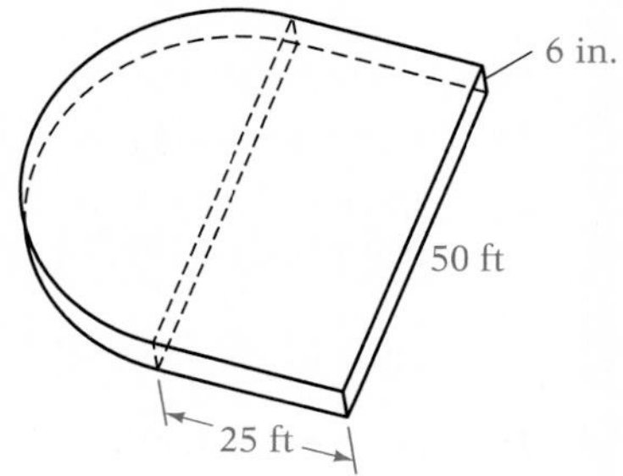

APPLYING THE CONCEPTS

35. Half a sphere is called a hemisphere. Derive a formula for the volume of a hemisphere. $\frac{2}{3}\pi r^3$

36. The side of a cube is doubled. How many times larger is the volume?
8 times larger

37. If the length, width, and height of a rectangular solid are all doubled, how many times larger is the resulting rectangular solid?
8 times larger

38. [W] Suppose a cylinder is cut into 16 equal pieces, which are then arranged as shown at the right. The figure resembles a rectangular solid. What variable expressions could be used to represent the length, width, and height of the rectangular solid? Explain how the formula for the volume of a cylinder is derived from this approach.

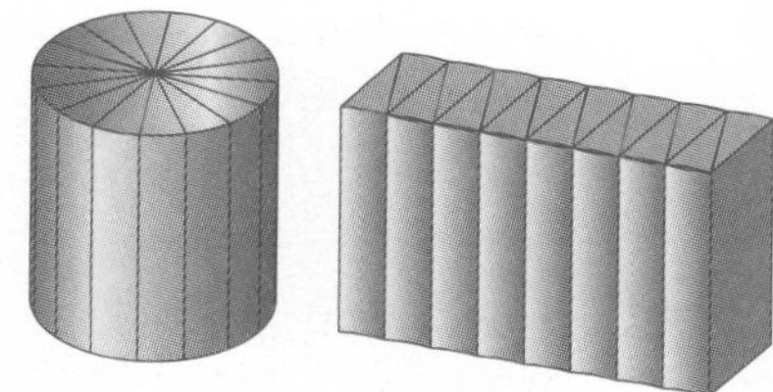

12.5 The Pythagorean Theorem

Objective A To use a table to find the square root of a number

The area of the square is 36 in^2. What is the length of one side?

Area =
36 in^2

Area of the square = $(\text{side})^2$
$36 = \text{side} \cdot \text{side}$

What number multiplied times itself equals 36?

$$36 = 6 \cdot 6$$

The side of the square is 6 in.

The **square root** of a number is one of two identical factors of that number. The square root symbol is $\sqrt{\ }$.

The square root of 36 is 6.

$$\sqrt{36} = 6$$

POINT OF INTEREST
There is evidence that the ancient Greeks knew of irrational numbers by 500 B.C. These numbers were not very well understood, and they were given the name *numerus surdus.* This phrase comes from the Latin word, *surdus,* which means deaf or mute. Thus irrational numbers were "inaudible numbers."

A **perfect square** is the product of a whole number times itself.

$1 \cdot 1 = 1$ | $\sqrt{1} = 1$
$2 \cdot 2 = 4$ | $\sqrt{4} = 2$
$3 \cdot 3 = 9$ | $\sqrt{9} = 3$
$4 \cdot 4 = 16$ | $\sqrt{16} = 4$
$5 \cdot 5 = 25$ | $\sqrt{25} = 5$
$6 \cdot 6 = 36$ | $\sqrt{36} = 6$

1, 4, 9, 16, 25, and 36 are perfect squares.

The square root of a perfect square is a whole number.

If a number is not a perfect square, its square root can only be approximated. The approximate square roots of the whole numbers up to 200 are found in the appendix on page A3. The square roots have been rounded to the nearest thousandth.

Number	*Square Root*	
33	5.745	$\sqrt{33} \approx 5.745$
34	5.831	$\sqrt{34} \approx 5.831$
35	5.916	$\sqrt{35} \approx 5.916$

Example 1 Find the square roots of the perfect squares 49 and 81.

Solution $\sqrt{49} = 7$ $\quad \sqrt{81} = 9$

You Try It 1 Find the square roots of the perfect squares 16 and 169.

Your solution $\sqrt{16} = 4$ $\quad \sqrt{169} = 13$

Solution on p. A54

Example 2	Find the square roots of 27 and 108. Use the table on page A3.	**You Try It 2**	Find the square roots of 32 and 162. Use the table on page A3.
Solution	$\sqrt{27} \approx 5.196$ $\sqrt{108} \approx 10.392$	**Your solution**	$\sqrt{32} \approx 5.657$ $\sqrt{162} \approx 12.728$

Solution on p. A54

Objective B *To find the unknown side of a right triangle using the Pythagorean Theorem*

POINT OF INTEREST

The first known proof of the Pythagorean Theorem is in a Chinese textbook which dates from 150 B.C. The book is called the *Nine Chapters on the Mathematical Art.* The diagram below is from that book and was used in the proof of the theorem.

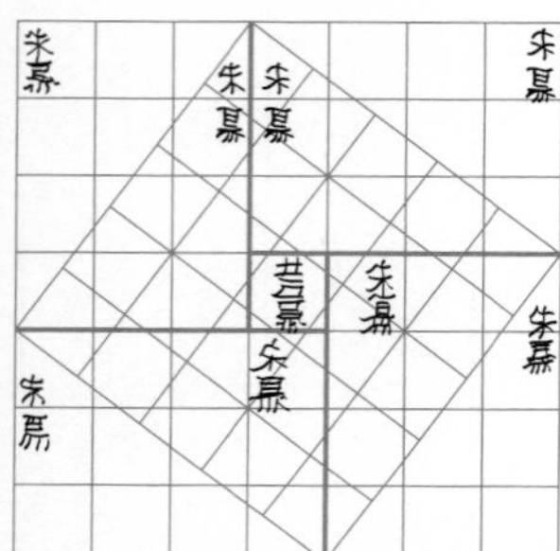

The Greek mathematician, Pythagoras, is generally credited with the discovery that the square of the hypotenuse of a right triangle is equal to the sum of the squares of the two legs. This is called the **Pythagorean Theorem.** However, the Babylonians used this theorem more than 1000 years before Pythagoras lived.

Square of the hypotenuse	equals	sum of the squares of the two legs
5^2	$=$	$3^2 + 4^2$
25	$=$	$9 + 16$
25	$=$	25

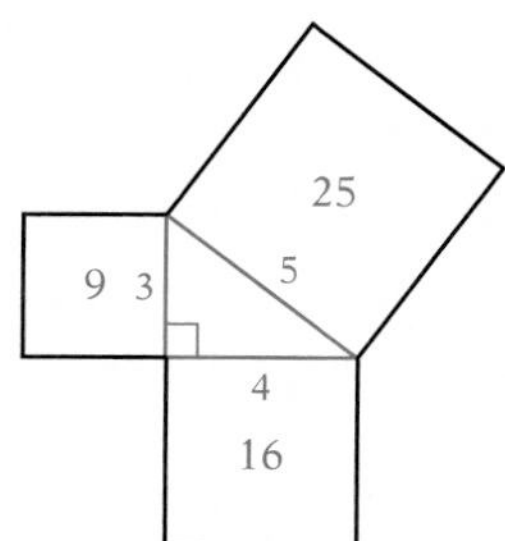

If the length of one side of a right triangle is unknown, one of the following formulas can be used to find its length.

If the hypotenuse is unknown, use

$$\begin{aligned}\text{Hypotenuse} &= \sqrt{(\text{leg})^2 + (\text{leg})^2}\\ &= \sqrt{(3)^2 + (4)^2}\\ &= \sqrt{9 + 16}\\ &= \sqrt{25}\\ &= 5\end{aligned}$$

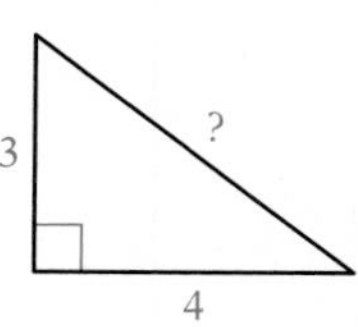

If the length of a leg is unknown, use

$$\begin{aligned}\text{Leg} &= \sqrt{(\text{hypotenuse})^2 - (\text{leg})^2}\\ &= \sqrt{(5)^2 - (4)^2}\\ &= \sqrt{25 - 16}\\ &= \sqrt{9}\\ &= 3\end{aligned}$$

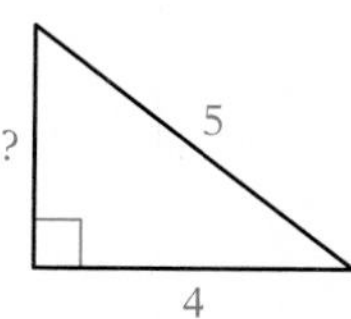

By using the Pythagorean Theorem and several facts from geometry, a relationship among the sides of two special right triangles can be found.

The first special right triangle has two 45° angles and is called a **45°-45°-90° triangle.** The sides opposite the 45° angles are equal.

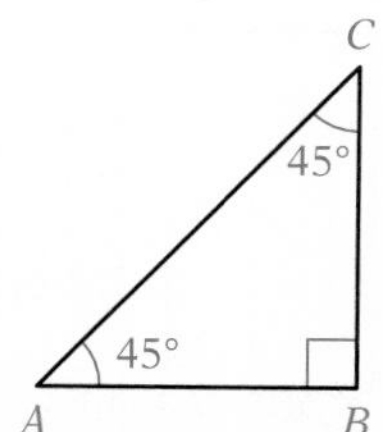

In a 45°-45°-90° triangle, the hypotenuse is equal to $\sqrt{2} \times$ (length of the leg).

➡ Find the length of the hypotenuse for a 45°-45°-90° triangle in which the length of one leg is 26 m.

$$\begin{aligned}\text{Hypotenuse} &= \sqrt{2} \times (\text{leg}) \\ &= \sqrt{2} \times 26 \text{ m} \\ &\approx 1.414 \times 26 \text{ m} \\ \text{Hypotenuse} &\approx 36.764 \text{ m}\end{aligned}$$

The second special right triangle is the **30°-60°-90° triangle.**

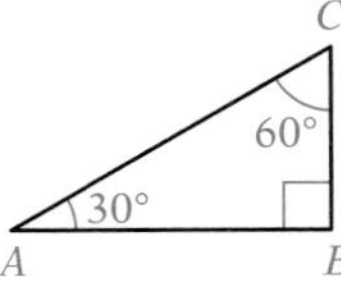

For a 30°-60°-90° triangle, the length of the leg opposite the 30° angle is one-half the length of the hypotenuse.

$$BC = \frac{1}{2} \times AC$$

➡ Find the length of the two legs of a 30°-60°-90° triangle that has a hypotenuse 16 cm in length.

$$\begin{aligned}\text{Leg} &= \frac{1}{2} \times 16 \text{ cm} \\ \text{Leg} &= 8 \text{ cm}\end{aligned}$$

- Because the right triangle is a 30°-60°-90° triangle, the leg opposite the 30° angle is one-half the hypotenuse.

$$\begin{aligned}\text{Leg} &= \sqrt{(\text{hypotenuse})^2 - (\text{leg})^2} \\ &= \sqrt{16^2 - 8^2} = \sqrt{256 - 64} \\ &= \sqrt{192} \\ &= 13.856 \\ \text{Leg} &\approx 13.856 \text{ cm}\end{aligned}$$

- Use the Pythagorean Theorem to find the other leg.

The lengths of the legs are 8 cm and 13.856 cm.

Example 3
Find the hypotenuse of the triangle in the figure.

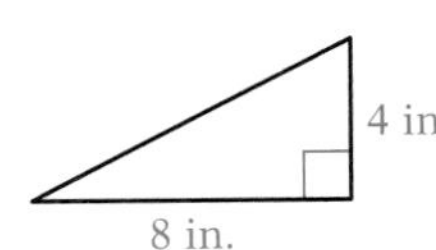

Solution

$$\begin{aligned}\text{Hypotenuse} &= \sqrt{(\text{leg})^2 + (\text{leg})^2} \\ &= \sqrt{(8 \text{ in.})^2 + (4 \text{ in.})^2} \\ &= \sqrt{64 \text{ in}^2 + 16 \text{ in}^2} \\ &= \sqrt{80 \text{ in}^2}\end{aligned}$$

Hypotenuse ≈ 8.944 in.

You Try It 3
Find the hypotenuse of the triangle in the figure.

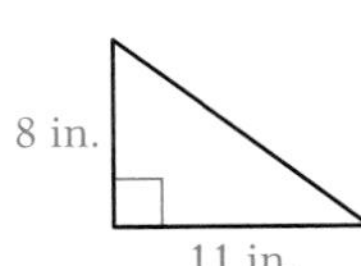

Your solution
13.601 in.

Solution on p. A54

Example 4
Find the length of the leg of the triangle in the figure.

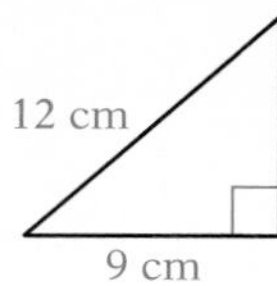

Solution
Leg $= \sqrt{(\text{hypotenuse})^2 - (\text{leg})^2}$
$= \sqrt{(12 \text{ cm})^2 - (9 \text{ cm})^2}$
$= \sqrt{144 \text{ cm}^2 - 81 \text{ cm}^2}$
$= \sqrt{63 \text{ cm}^2}$

Leg $\approx$ 7.937 cm

You Try It 4
Find the length of the leg of the triangle in the figure.

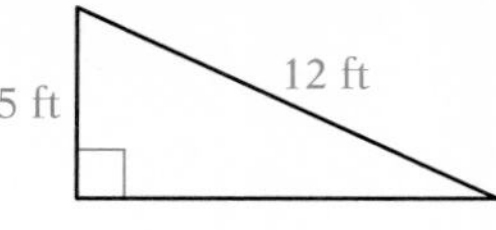

Your solution
10.909 ft

Solution on p. A54

Objective C To solve application problems

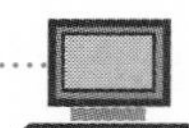

Example 5
A 25-ft ladder is placed against a building at a point 21 ft from the ground as shown in the figure. Find the distance from the base of the building to the base of the ladder.

Strategy
To find the distance from the base of the building to the base of the ladder, use the Pythagorean Theorem. The hypotenuse is the length of the ladder (25 ft). One leg is the distance along the building from the ground to the top of the ladder. The distance from the base of the building to the base of the ladder is the unknown leg.

Solution
Leg $= \sqrt{(\text{hypotenuse})^2 - (\text{leg})^2}$
$= \sqrt{(25 \text{ ft})^2 - (21 \text{ ft})^2}$
$= \sqrt{625 \text{ ft}^2 - 441 \text{ ft}^2}$
$= \sqrt{184 \text{ ft}^2}$
Leg $\approx$ 13.565 ft

The distance is 13.565 ft.

You Try It 5
Find the distance between the centers of the holes in the metal plate in the figure.

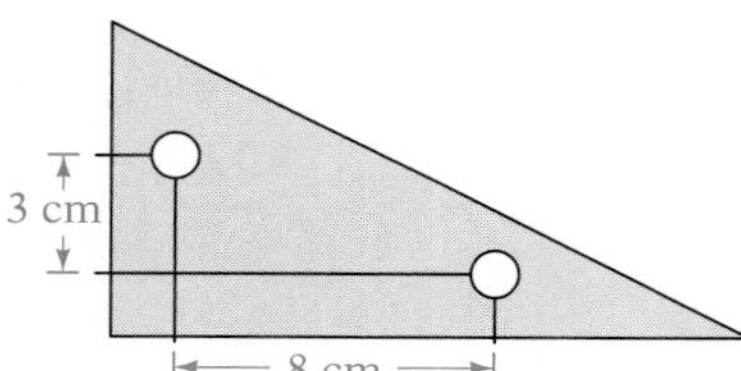

Your strategy

Your solution
8.544 cm

Solution on p. A54

12.5 Exercises

Objective A

Find the square root. Use the table on page A3.

1. 7 2.646	**2.** 34 5.831	**3.** 42 6.481	**4.** 64 8
5. 165 12.845	**6.** 144 12	**7.** 189 13.748	**8.** 130 11.402

Objective B

Find the unknown side of the triangle in the figures accompanying Exercises 9 to 17.

9.

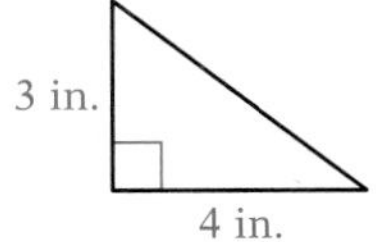

5 in.

10.

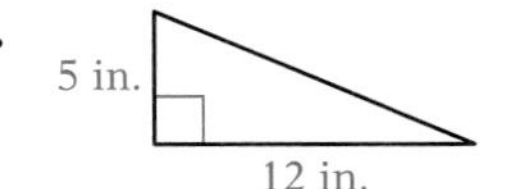

13 in.

11.

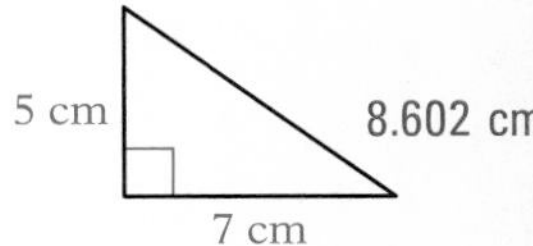

8.602 cm

12.

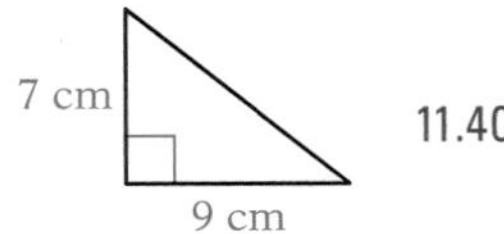

11.402 cm

13.

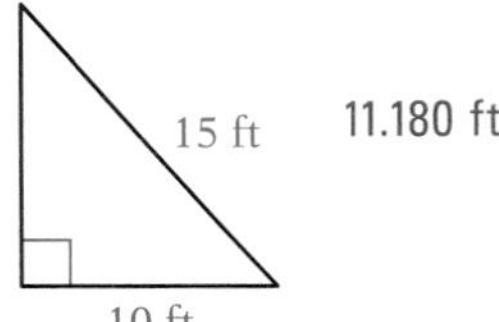

11.180 ft

14.

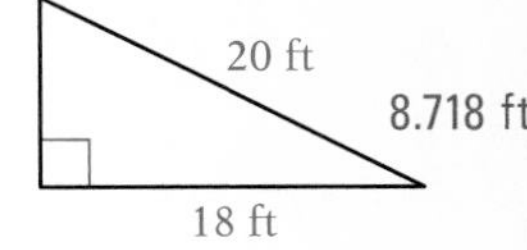

8.718 ft

15.

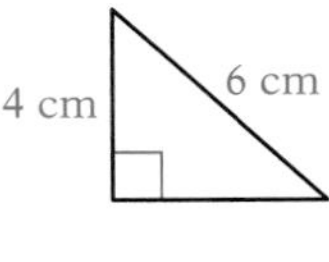

4.472 cm

16.

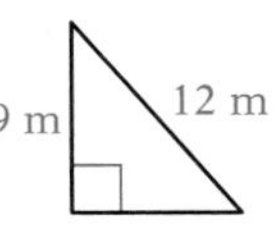

7.937 m

17.

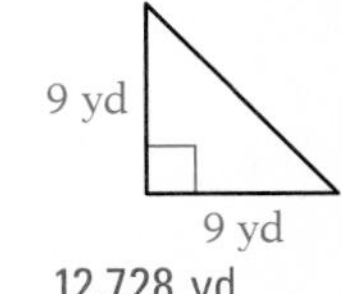

12.728 yd

Find the lengths of the two legs in the figures accompanying Exercises 18 to 20.

18.

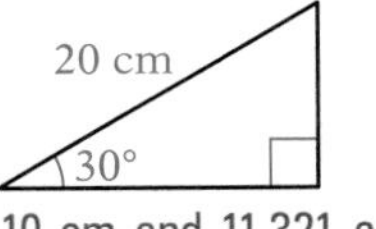

10 cm and 11.321 cm

19.

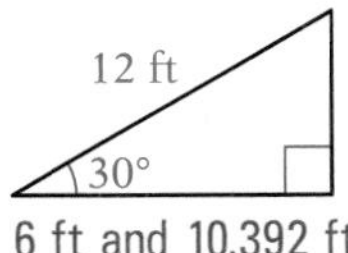

6 ft and 10.392 ft

20.

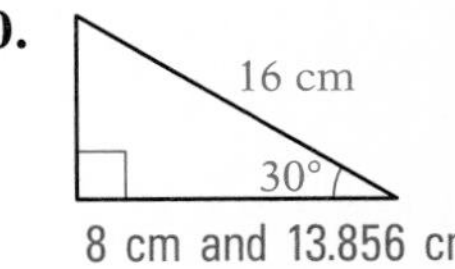

8 cm and 13.856 cm

Find the hypotenuse of the triangle in the figures accompanying Exercises 21 to 26.

21.

21.21 cm

22.

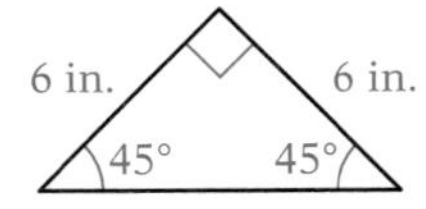

8.484 in.

23.

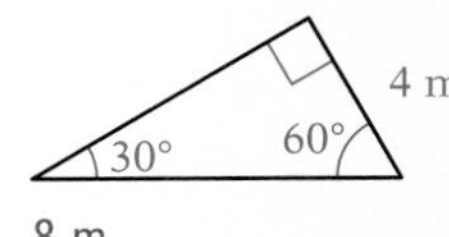

8 m

24.

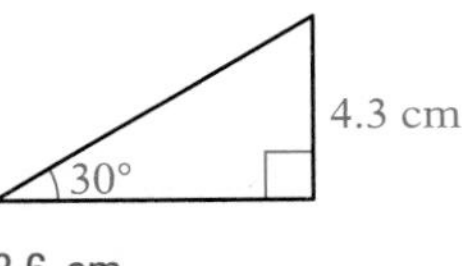

8.6 cm

25.

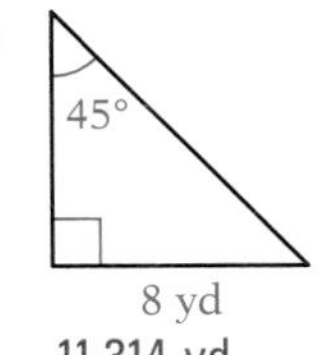

11.314 yd

26.

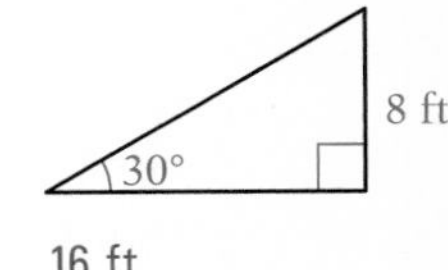

16 ft

Objective C *Application Problems*

27. Find the distance between the holes in the metal plate in the figure shown. Round to the nearest hundredth. 6.32 in.

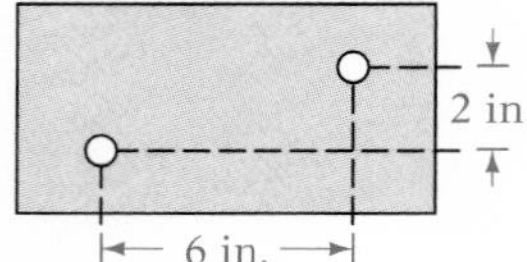

28. Find the length of the ramp used to roll barrels up to the 3.5-ft-high loading ramp shown in the figure. Round to the nearest hundredth. 9.66 ft

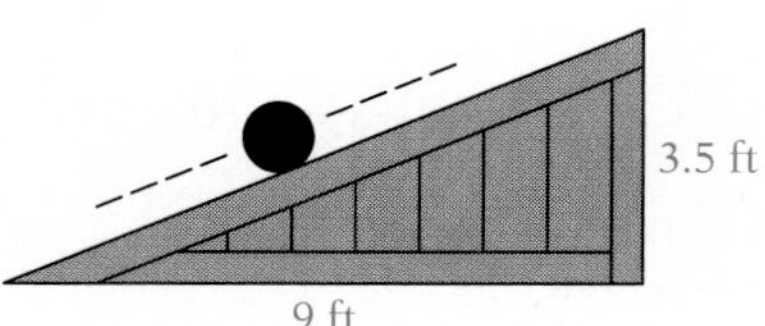

29. A fence is built around the plot shown in the figure. At $4.20 per meter, how much did it cost to fence the plot? Use the Pythagorean Theorem to find the unknown length. $109.20

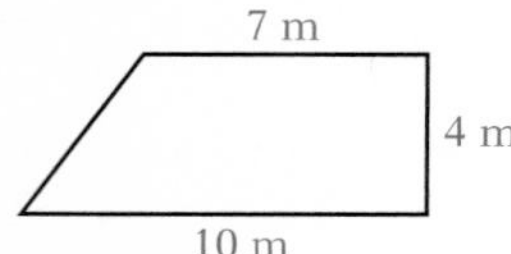

30. Four holes are drilled in the circular plate in the figure shown. The centers of the holes are 3 in. from the center. Find the distance between the centers of adjacent holes. 4.243 in.

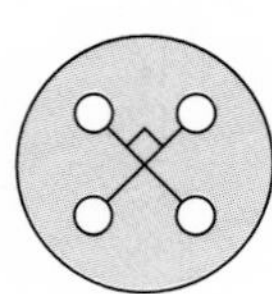

31. A conduit is run in a building as shown in the figure. Find the total length of the conduit. 35 ft

32. Find the offset distance of the length of pipe shown in the diagram below. The total length of the pipe is 62 in. $3\frac{3}{4}$ in.

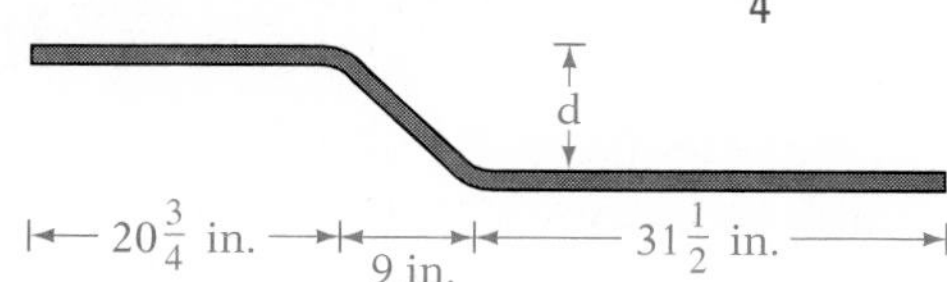

APPLYING THE CONCEPTS

33. Determine whether the statement is always true, sometimes true, or never true.

a. A positive number is a perfect square. Sometimes true

b. The sum of the lengths of two sides of a triangle is greater than the length of the third side of the triangle. Always true

c. The hypotenuse is the largest side of a right triangle. Always true

34. [W] Graph $\sqrt{2}$ on the number line. *Hint:* What is the length of the hypotenuse of an isosceles right triangle with a leg of 1 unit?

35. [W] What is a Pythagorean triple? Provide at least three examples of Pythagorean triples.

36. A construction company must lay a pipeline between the buildings labeled *A* and *B* at the right. Because the construction costs of laying the pipe under the river are high, the construction plan is to connect the buildings as shown. What is the total length of the pipe necessary to connect the buildings? 8 mi

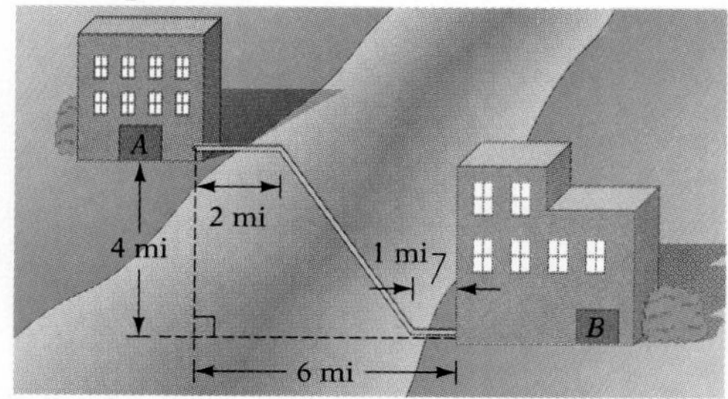

12.6 Similar and Congruent Triangles

Objective A To solve similar and congruent triangles

INSTRUCTOR NOTE

The concepts of similar triangles and the fact that ratios of corresponding sides are equal is not obvious to students. Nonetheless, these concepts are essential in many practical applications.

Similar objects have the same shape but not necessarily the same size. A baseball is similar to a basketball. A model airplane is similar to an actual airplane.

Similar objects have corresponding parts; for example, the propellers on the model airplane correspond to the propellers on the actual airplane. The relationship between the sizes of each of the corresponding parts can be written as a ratio, and all such ratios will be the same. If the propellers on the model plane are $\frac{1}{50}$ the size of the propellers on the actual plane, then the model wing is $\frac{1}{50}$ the size of the actual wing, the model fuselage is $\frac{1}{50}$ the size of the actual fuselage, and so on.

The two triangles ABC and DEF shown are similar. The ratios of corresponding sides are equal.

$$\frac{AB}{DE} = \frac{2}{6} = \frac{1}{3}, \frac{BC}{EF} = \frac{3}{9} = \frac{1}{3}, \text{ and } \frac{AC}{DF} = \frac{4}{12} = \frac{1}{3}$$

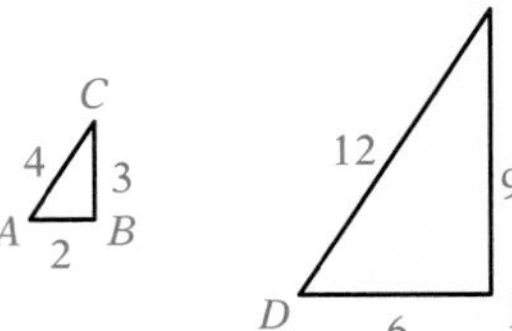

The ratio of corresponding sides $= \frac{1}{3}$.

Because the ratios of corresponding sides are equal, three proportions can be formed:

$$\frac{AB}{DE} = \frac{BC}{EF}, \frac{AB}{DE} = \frac{AC}{DF}, \text{ and } \frac{BC}{EF} = \frac{AC}{DF}$$

The ratio of corresponding heights equals the ratio of corresponding sides, as shown in the figure.

Ratio of corresponding sides $= \frac{1.5}{6} = \frac{1}{4}$

Ratio of heights $= \frac{2}{8} = \frac{1}{4}$

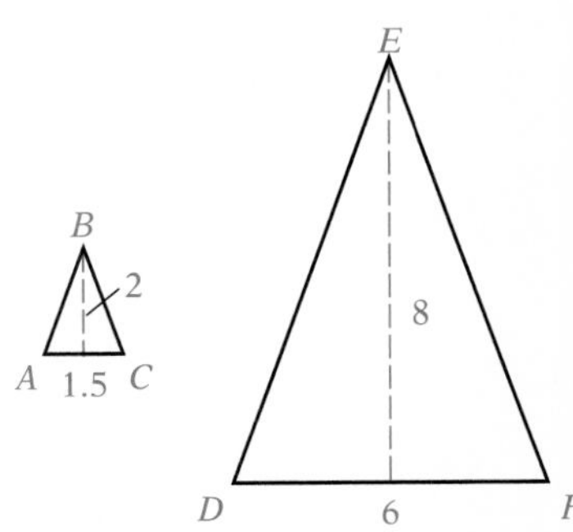

Congruent objects have the same shape *and* the same size.

The two triangles shown are congruent. They have exactly the same size.

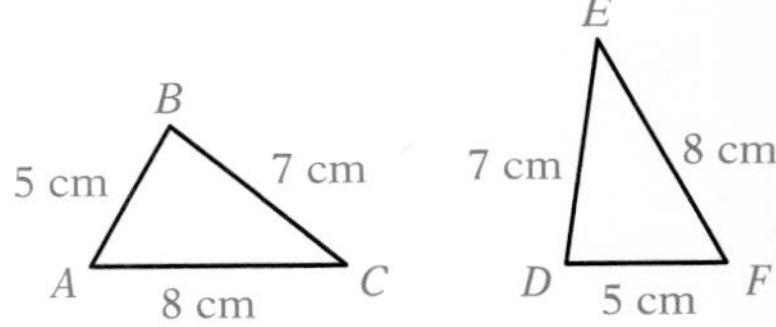

For triangles, congruent means that the corresponding sides *and* angles of the triangle are equal, unlike similar triangles, which have corresponding angles equal but corresponding sides are not necessarily equal.

Here are two major rules that can be used to determine whether two triangles are congruent.

Side-Side-Side Rule (SSS)

Two triangles are congruent if three sides of one triangle equal the corresponding sides of the second triangle.

In the two triangles at the right, $AB = DE$, $AC = DF$, and $BC = EF$. The corresponding sides of triangles ABC and DEF are equal. The triangles are congruent by the SSS rule.

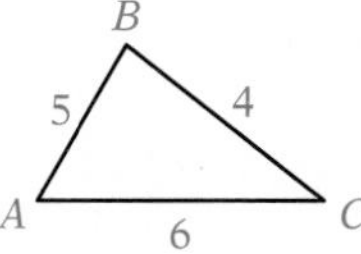

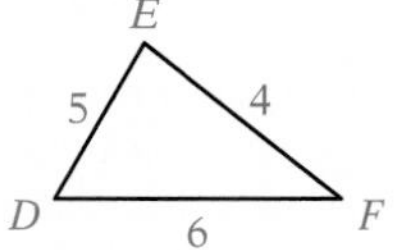

Side-Angle-Side Rule (SAS)

Two triangles are congruent if two sides and the included angle of one triangle equal the corresponding sides and included angle of the second triangle.

In the two triangles at the right, $AB = EF$, $AC = DE$, and angle BAC = angle DEF. The triangles are congruent by the SAS rule.

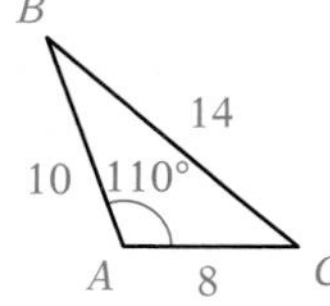

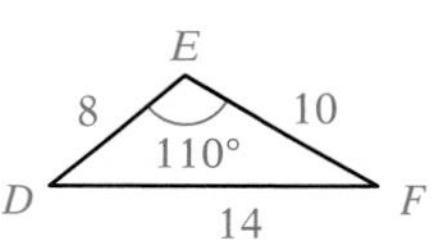

➡ Determine whether the two triangles in the adjacent figure are congruent.

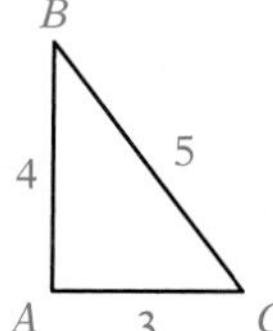

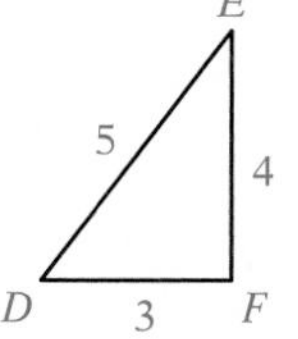

Because $AC = DF$, $AB = FE$, and $BC = DE$ all three sides of one triangle equal the corresponding sides of the second triangle, the triangles are congruent by the SSS rule.

Example 1

Find the ratio of corresponding sides for the similar triangles ABC and DEF in the figure.

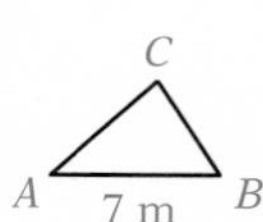

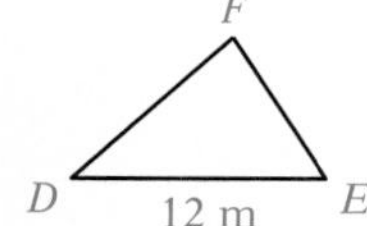

Solution $\frac{7\ \cancel{m}}{12\ \cancel{m}} = \frac{7}{12}$

You Try It 1

Find the ratio of corresponding sides for the similar triangles ABC and DEF in the figure.

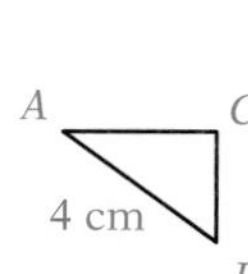

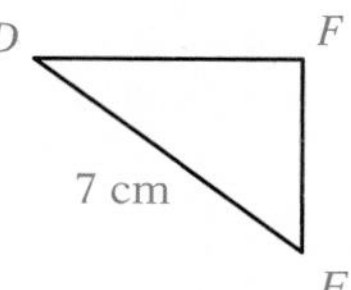

Your solution $\frac{4}{7}$

Solution on p. A55

Example 2
Triangles *ABC* and *DEF* in the figure are similar. Find side *EF*.

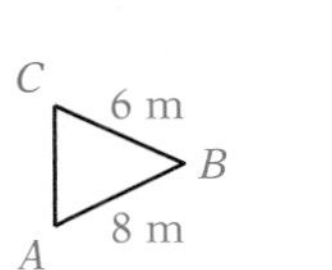

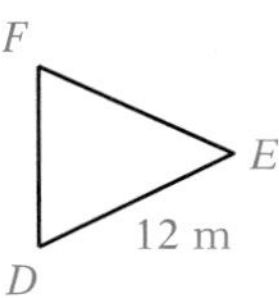

Solution
Let x represent the length of the unknown side *EF*.

$\frac{AB}{DE} = \frac{BC}{x}$ • The ratios of corresponding sides of similar triangles are equal.

$$\frac{8 \text{ m}}{12 \text{ m}} = \frac{6 \text{ m}}{x}$$
$$8x = 12 \cdot 6 \text{ m}$$
$$8x = 72 \text{ m}$$
$$\frac{8x}{8} = \frac{72 \text{ m}}{8}$$
$$x = 9 \text{ m}$$

Side *EF* is 9 m.

You Try It 2
Triangles *ABC* and *DEF* in the figure are similar. Find side *DF*.

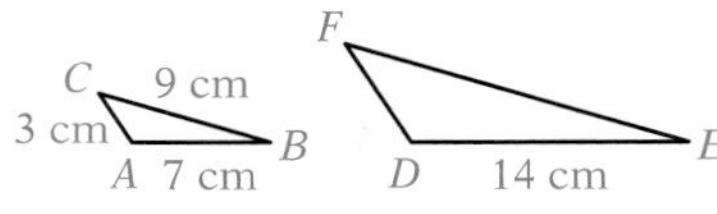

Your solution 6 cm

Example 3
Determine whether triangle *ABC* in the figure is congruent to triangle *DEF*.

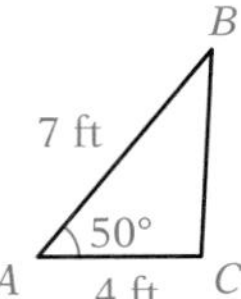

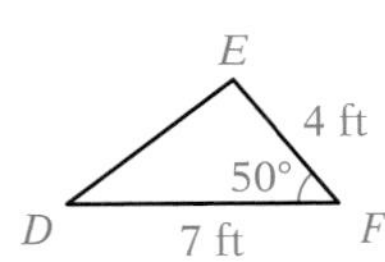

Solution
Because $AB = DF$, $AC = EF$, and angle BAC = angle DFE, the triangles are congruent by the SAS rule.

You Try It 3
Determine whether triangle *ABC* in the figure is congruent to triangle *DEF*.

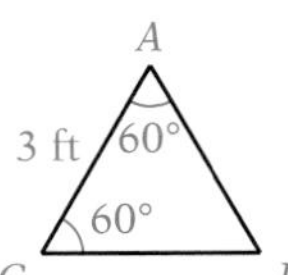

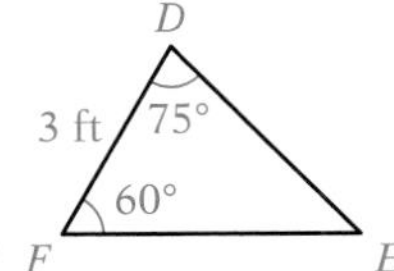

Your solution Not congruent

Example 4
Triangles *ABC* and *DEF* in the figure are similar. Find the height of *FG*.

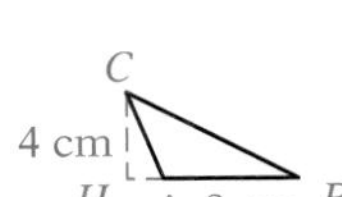

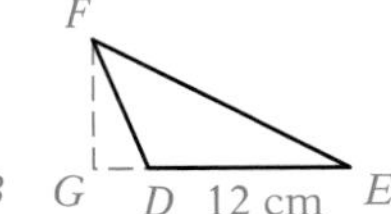

Solution
Let x represent the height *FG*.

$$\frac{8 \text{ cm}}{12 \text{ cm}} = \frac{4 \text{ cm}}{x}$$
$$8x = 12 \cdot 4 \text{ cm}$$
$$8x = 48 \text{ cm}$$
$$\frac{8x}{8} = \frac{48 \text{ cm}}{8}$$
$$x = 6 \text{ cm}$$

• The ratios of corresponding sides of similar triangles equal the ratio of corresponding heights.

The height of *FG* is 6 cm.

You Try It 4
Triangles *ABC* and *DEF* in the figure are similar. Find the height of *FG*.

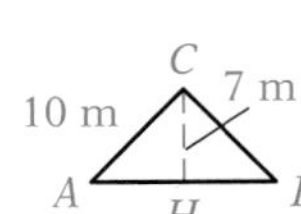

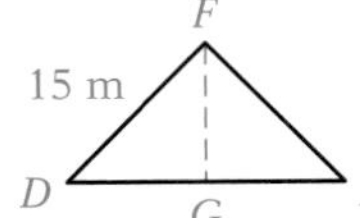

Your solution 10.5 m

Solutions on p. A55

Objective B To solve application problems

Example 5

Triangles ABC and DEF in the figure are similar. Find the area of triangle DEF.

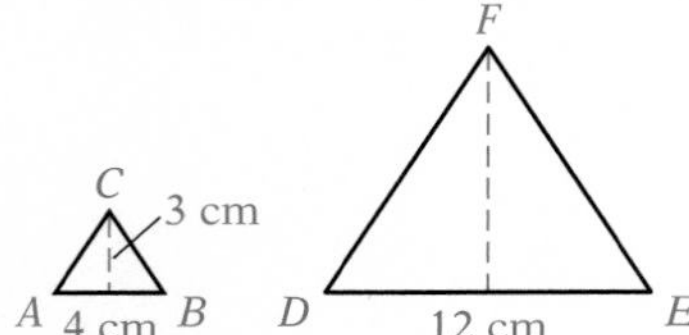

Strategy

To find the area of triangle DEF:

- Solve a proportion to find the height of triangle DEF. Let h = the height.
- Use the formula $A = \frac{1}{2}bh$.

Solution

$$\frac{AB}{DE} = \frac{\text{height of triangle } ABC}{\text{height of triangle } DEF}$$

$$\frac{4 \text{ cm}}{12 \text{ cm}} = \frac{3 \text{ cm}}{h}$$

$$4h = 12 \cdot 3 \text{ cm}$$

$$4h = 36 \text{ cm}$$

$$\frac{4h}{4} = \frac{36 \text{ cm}}{4}$$

$$h = 9 \text{ cm}$$

$$A = \frac{1}{2}bh$$

$$= \frac{1}{2} \cdot 12 \text{ cm} \cdot 9 \text{ cm}$$

$$= 54 \text{ cm}^2$$

The area is 54 cm^2.

You Try It 5

Triangles ABC and DEF in the figure are similar. Find the perimeter of triangle ABC.

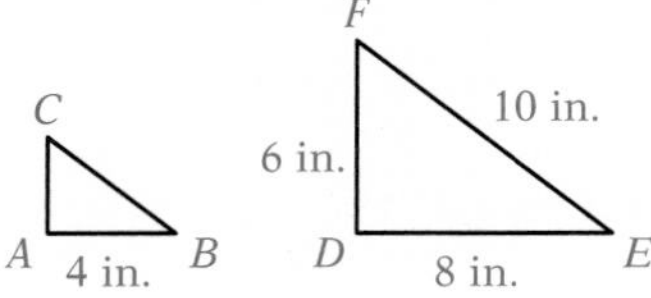

Your strategy

Your solution 12 in.

Solution on p. A55

12.6 Exercises

Objective A

Find the ratio of corresponding sides for the similar triangles in the figures accompanying Exercises 1 to 4.

1.

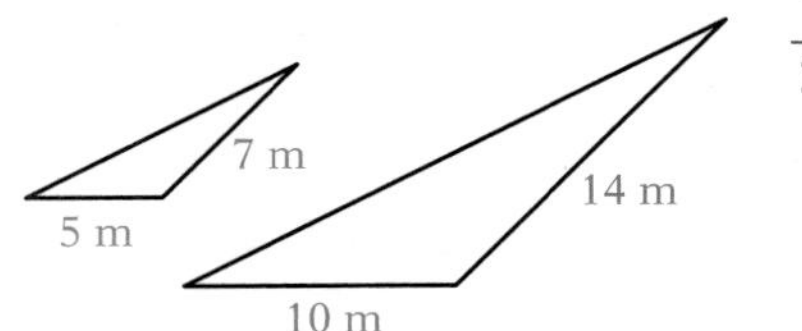

$\frac{1}{2}$

2.

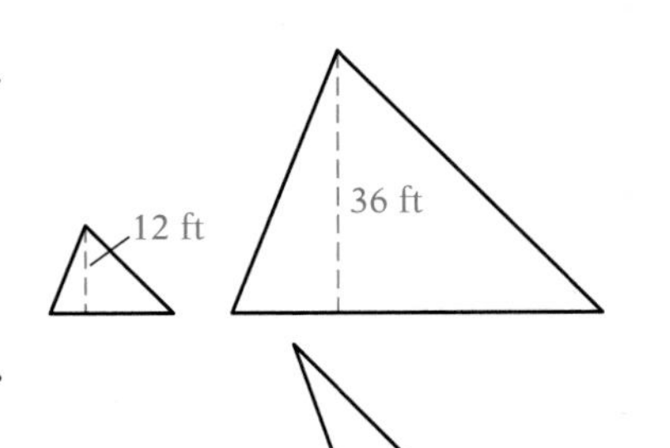

$\frac{1}{3}$

3.

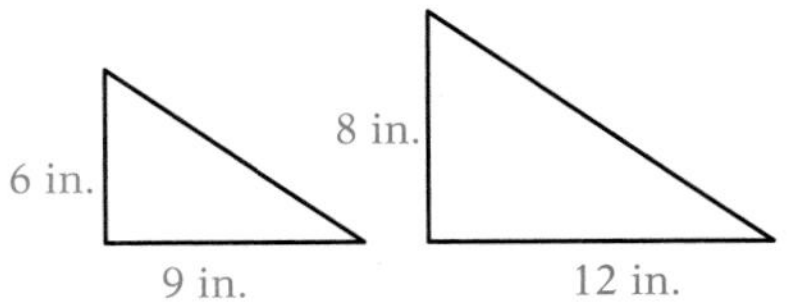

$\frac{3}{4}$

4.

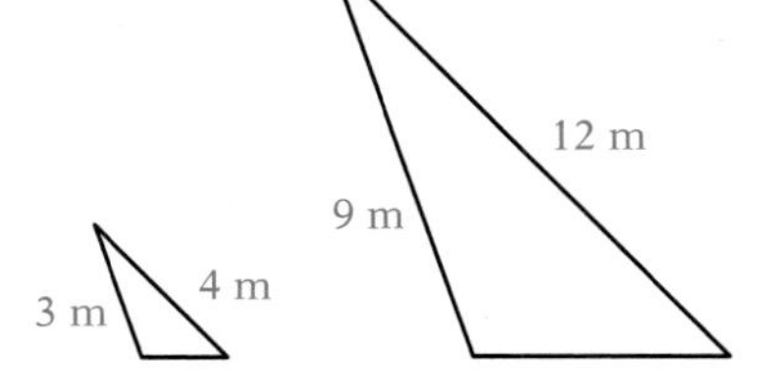

$\frac{1}{3}$

Determine whether the two triangles in Exercises 5 to 8 are congruent.

5.

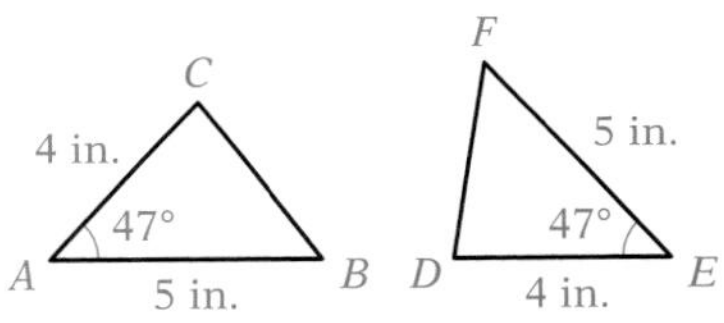

Yes

6.

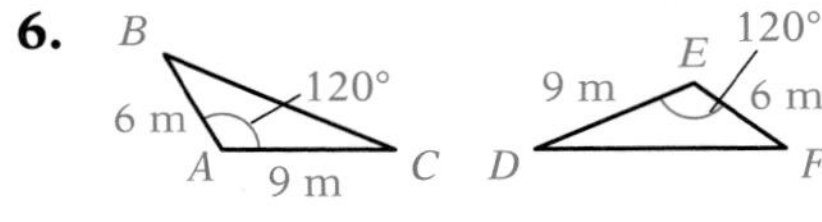

Yes

7.

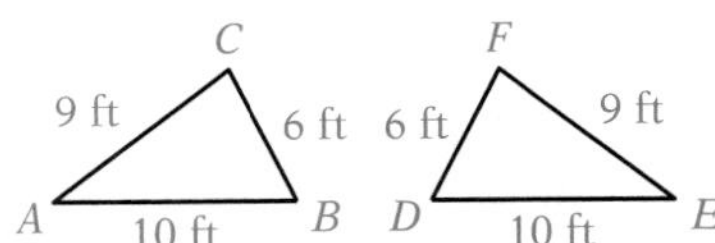

Yes

8.

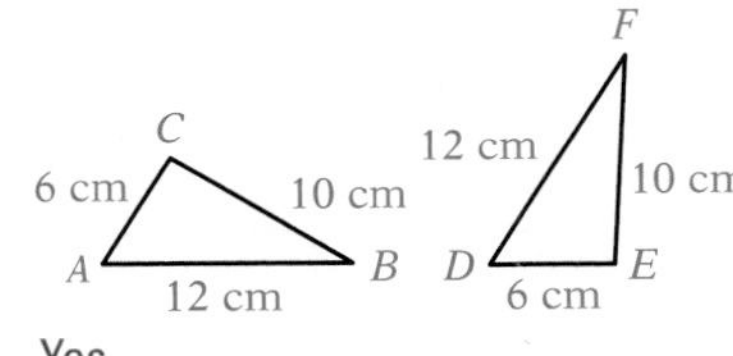

Yes

Triangles *ABC* and *DEF* in Exercises 9 to 12 are similar. Find the indicated distance. Round to the nearest tenth.

9. Find side *DE*.

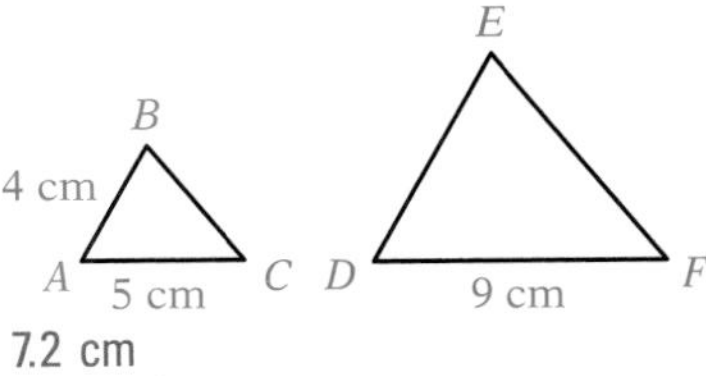

7.2 cm

10. Find side *DE*.

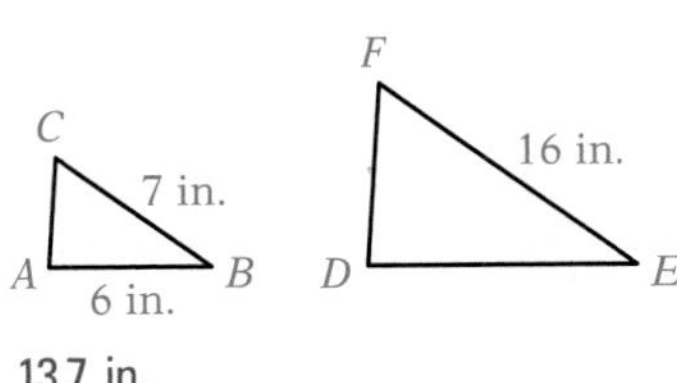

13.7 in.

11. Find the height of triangle *DEF*.

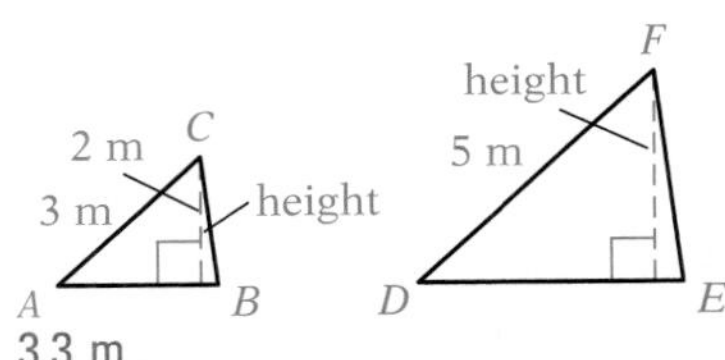

3.3 m

12. Find the height of triangle *ABC*.

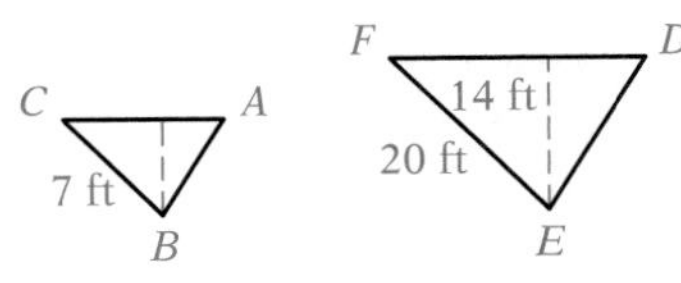

4.9 ft

Objective B *Application Problems*

The sun's rays, objects on earth, and the shadows cast by them form similar triangles.

13. Find the height of the building shown.

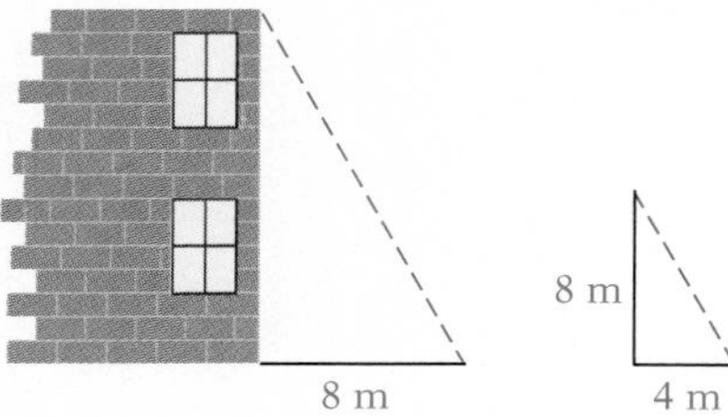

16 m

14. Find the height of the building shown.

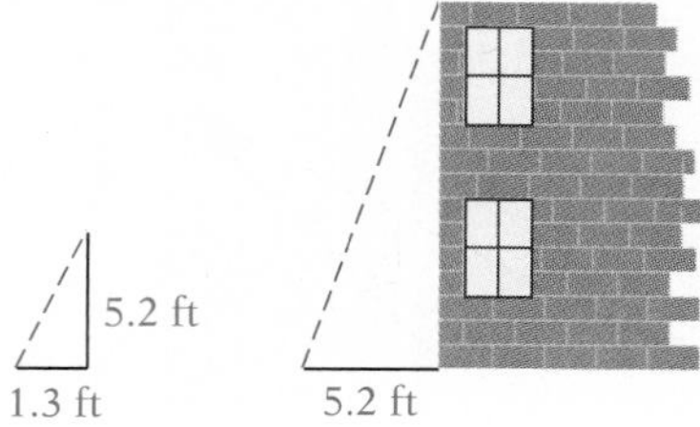

20.8 ft

In Exercises 15 to 18, triangles *ABC* and *DEF* are similar.

15. Find the perimeter of triangle *ABC*.

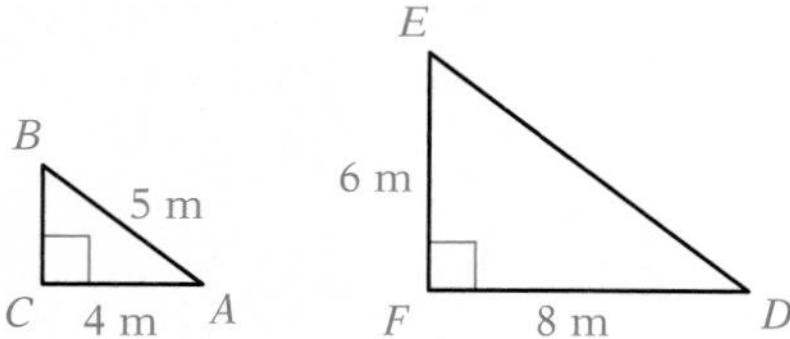

12 m

16. Find the perimeter of triangle *DEF*.

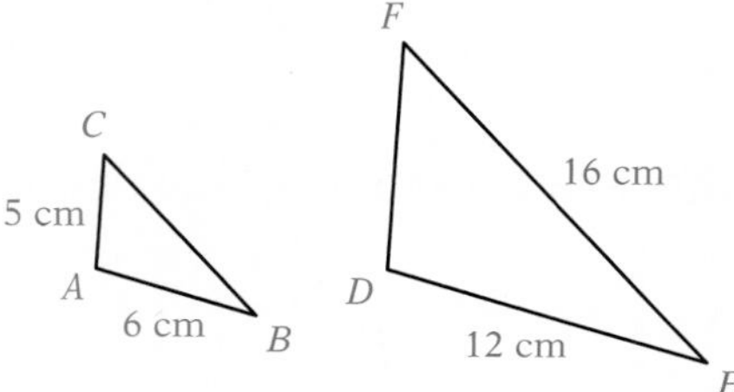

38 cm

17. Find the area of triangle *ABC*.

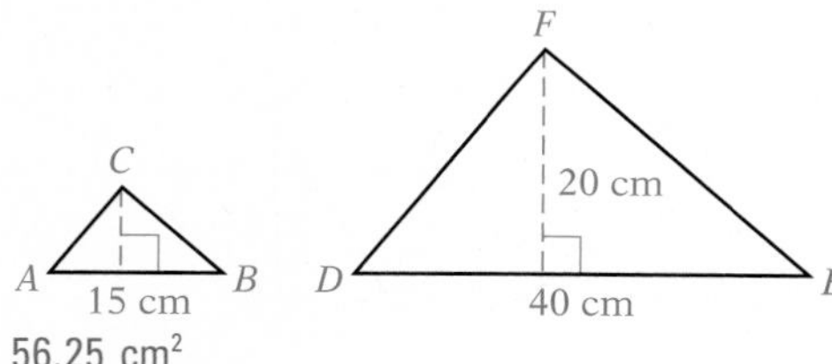

56.25 cm^2

18. Find the area of triangle *DEF*.

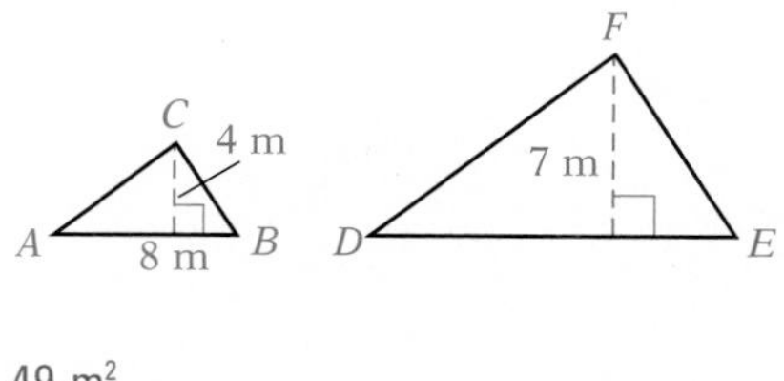

49 m^2

APPLYING THE CONCEPTS

19. Determine whether the statement is always true, sometimes true, or never true.

a. If two angles of one triangle are equal to two angles of a second triangle, then the triangles are similar triangles. Always true

b. Two isosceles triangles are similar triangles. Sometimes true

c. Two equilateral triangles are similar triangles. Always true

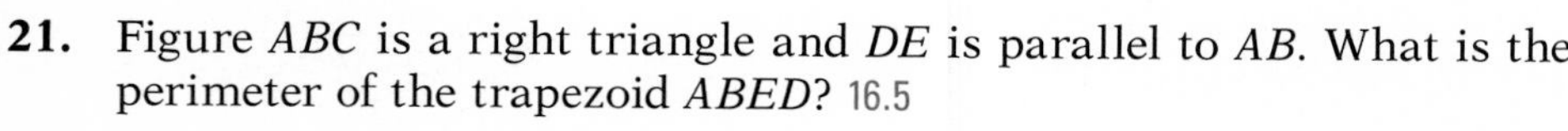

20. Are all squares similar? Are all rectangles similar? Explain. Use a
[W] drawing in your explanation. Yes. No. Answers will vary

21. Figure *ABC* is a right triangle and *DE* is parallel to *AB*. What is the perimeter of the trapezoid *ABED*? 16.5

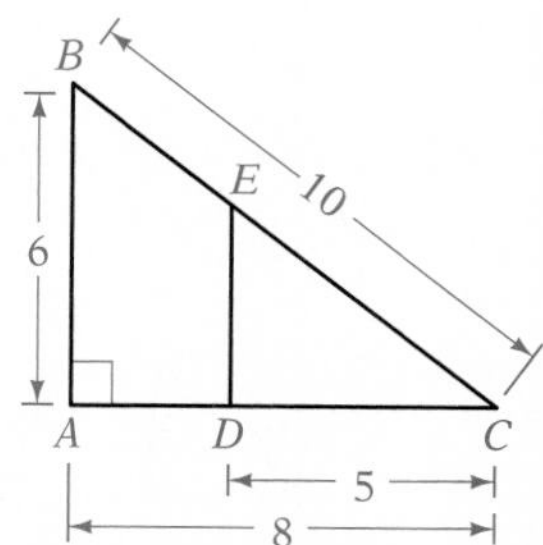

22. Explain how, by using only a yardstick, you could determine the ap-
[W] proximate height of a tree without climbing it. Answers will vary

Project in Mathematics

Investigating Perimeter

The perimeter of the square at the right is 4 units.

If two squares are joined along one of the sides, the perimeter is 6 cm. Note that it does not matter which sides are joined; the perimeter is still 6 units.

If three squares are joined, the perimeter of the resulting figure is 8 units for each possible placement of the squares.

Four squares can be joined in five different ways as shown. There are two possible perimeters, 10 units for A, B, C, and D, and 8 units for E.

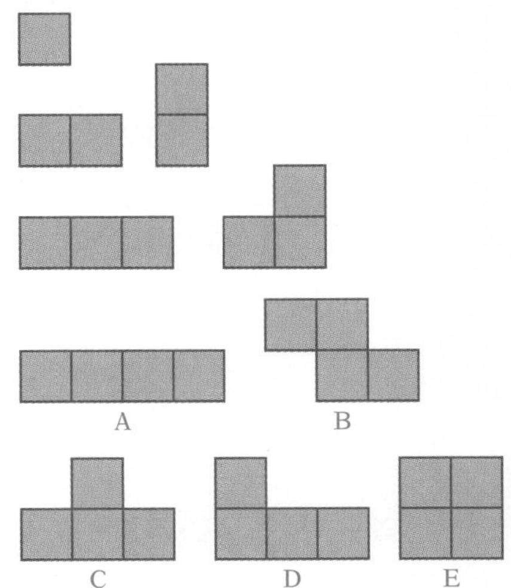

Exercises:

Answer the following questions about joining more than four squares.

1. If five squares are joined, what is the maximum perimeter possible?
2. If five squares are joined, what is the minimum perimeter possible?
3. If six squares are joined, what is the maximum perimeter possible?
4. If six squares are joined, what is the minimum perimeter possible?

Chapter Summary

Key Words

An *angle* is formed when two rays start from the same point. An angle is measured in *degrees*.

A 90° angle is called a *right angle*.

Perpendicular lines are intersecting lines that form right angles.

Complementary angles are two angles whose sum is 90°.

Supplementary angles are two angles whose sum is 180°.

A *right triangle* contains one right angle. The side opposite the right angle in a right triangle is called the *hypotenuse*. The other two sides are called *legs*.

A *45°-45°-90° triangle* is a special right triangle in which the sides opposite the 45° angles are equal.

A *30°-60°-90° triangle* is a special right triangle in which the length of the leg opposite the 30° angle is one-half the length of the hypotenuse.

A *quadrilateral* is a four-sided plane figure.

A *rectangle* is a parallelogram that has four right angles.

A *square* is a rectangle that has four equal sides.

A *circle* is a plane figure in which all points are the same distance from the center of the circle. The *diameter* is a line segment across a circle going through the center. The *radius* is equal to one-half the diameter.

A *rectangular solid* is a solid in which all six faces are rectangles.

A *cube* is a rectangular solid in which all six faces are squares.

A *sphere* is a solid in which all points are the same distance from the center of the sphere.

Perimeter is the distance around a plane figure.

Area is a measure of the amount of surface in a region.

Volume is a measure of the amount of space inside a closed surface.

Composite geometric solids are solids made from two or more geometric solids.

The *square root* of a number is one of two identical factors of that number.

Similar objects have the same shape but not necessarily the same size.

Congruent objects have the same shape and the same size.

Essential Rules

Perimeter Equations

Triangle: $P = a + b + c$
Square: $P = 4s$
Rectangle: $P = 2L + 2W$
Circle: $C = 2\pi r$ or $C = \pi d$

Area Equations

Triangle: $A = \frac{1}{2}bh$
Square: $A = s^2$
Rectangle: $A = LW$
Circle: $A = \pi r^2$

Volume Equations

Rectangular solid: $V = LWH$
Cube: $V = s^3$
Sphere: $V = \frac{4}{3}\pi r^3$
Cylinder: $V = \pi r^2 h$

Pythagorean Theorem

Hypotenuse $= \sqrt{\text{leg}^2 + \text{leg}^2}$

Side-Side-Side (SSS)

Two triangles are congruent when three sides of one triangle equal the corresponding sides of the second triangle.

Side-Angle-Side (SAS)

Two triangles are congruent when two sides and the included angle of one triangle equal the corresponding sides and included angle of the second triangle.

Chapter Review

SECTION 12.1

1. Given $AB = 15$, $CD = 6$, and $AD = 24$, find the length of BC. 3

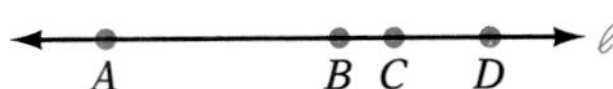

2. Find the supplement of a 105° angle. 75°

3. The diameter of a sphere is 1.5 m. Find the radius of the sphere. 0.75 m

4. A right triangle has a 32° angle. Find the measures of the other two angles. 58° and 90°

5. Given that $\ell_1 \parallel \ell_2$, find the measures of angles a and b.

$a = 135°$
$b = 45°$

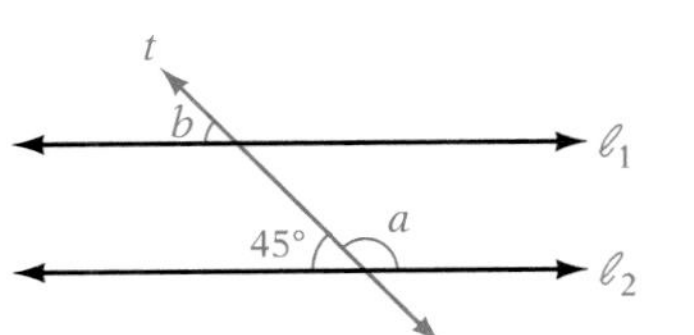

6. Given that $\ell_1 \parallel \ell_2$, find the measures of angles a and b.

$a = 100°$
$b = 80°$

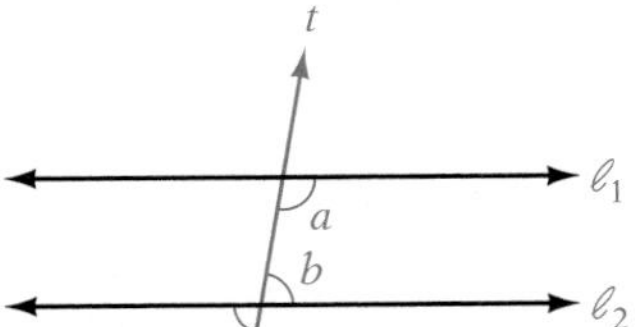

SECTION 12.2

7. Find the perimeter of the rectangle in the figure below. 26 ft

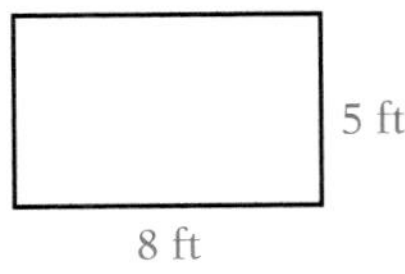

8. Find the circumference of the circle in the figure below. Use $\pi = 3.14$. 31.4 cm

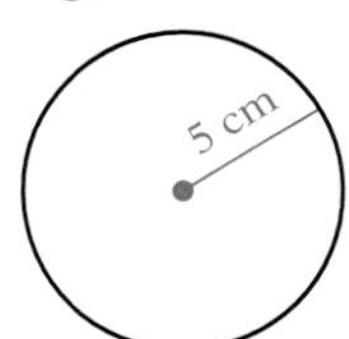

9. Find the perimeter of the composite figure shown. Use $\pi \approx 3.14$. 47.7 in.

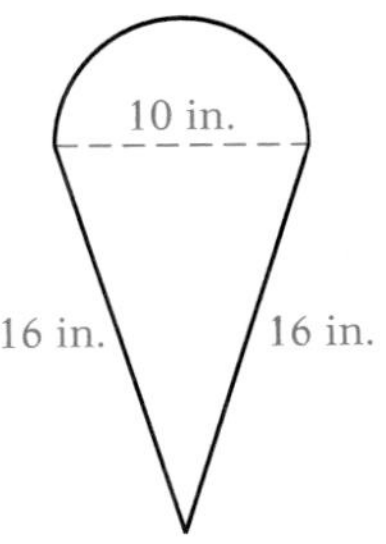

10. A bicycle tire has a diameter of 28 in. How many feet does the bicycle travel if the wheel makes 10 revolutions? Use $\pi \approx 3.14$. Round to the nearest tenth of a foot. 73.3 ft

SECTION 10.3

11. Find the area of the rectangle shown in the figure. 55 m^2

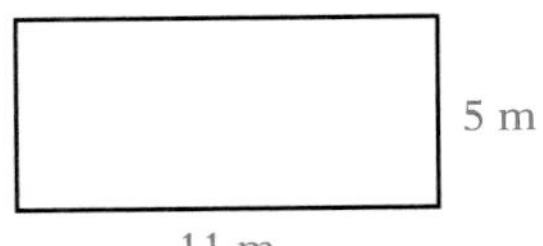

12. Find the area of the circle shown in the figure. Use $\pi \approx 3.14$. 63.585 cm^2

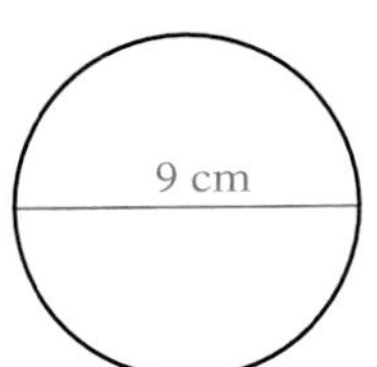

13. Find the area of the composite figure shown. Use $\pi \approx 3.14$. 57.12 in²

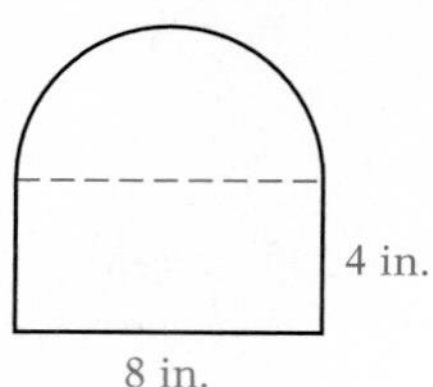

14. New carpet is installed in a room measuring 18 ft by 14 ft. Find the area of the room in square yards. ($9\text{ ft}^2 = 1\text{ yd}^2$) 28 yd²

SECTION 12.4

15. Find the volume of the rectangular solid shown in the figure. 200 ft³

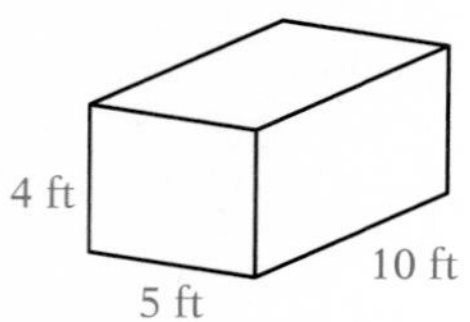

16. Find the volume of a sphere with a diameter of 8 ft. Use $\pi \approx 3.14$. Round to the nearest tenth. 267.9 ft³

17. Find the volume of the composite figure shown below. 240 in³

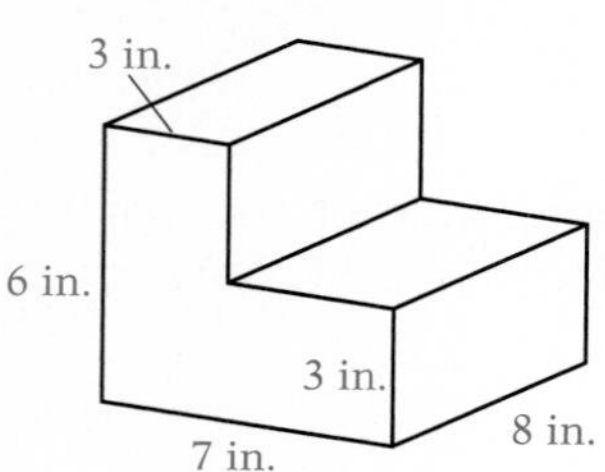

18. A silo, which is in the shape of a cylinder, is 9 ft in diameter and has a height of 18 ft. Find the volume of the silo. Use $\pi \approx 3.14$. 1144.53 ft³

SECTION 12.5

19. Find the square root of 15. Use the table on page A3. 3.873

20. Find the unknown side of the triangle in the figure below. 26 cm

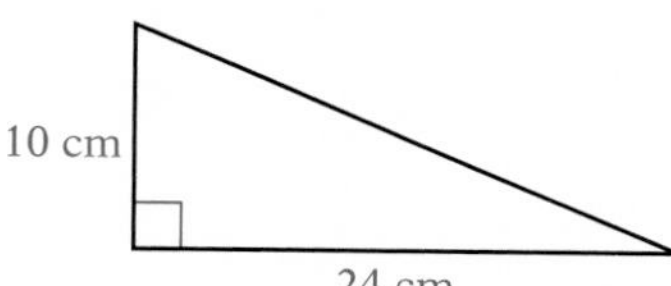

21. How high on a building will a 17-ft ladder reach when the bottom of the ladder is 8 ft from the building? 15 ft

SECTION 12.6

22. Triangles *ABC* and *DEF* are similar. Find the height of triangle *DEF*. 16 cm

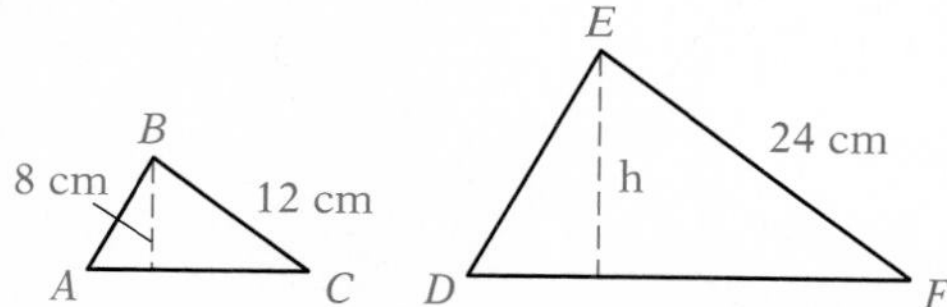

23. Triangles *ABC* and *DEF* are similar. Find the area of triangle *DEF*. 64.8 m²

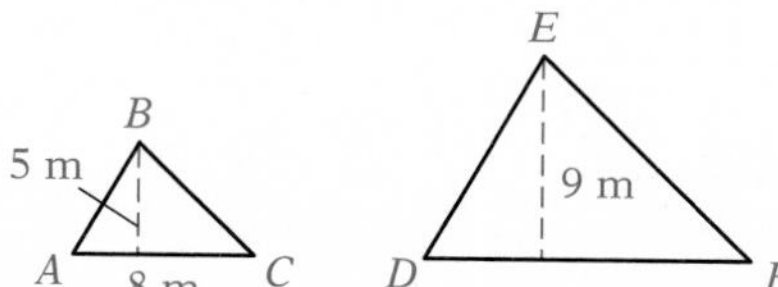

Chapter Test

1. Find the complement of a 32° angle.
58° [12.1A]

2. A right triangle has a 40° angle. Find the measure of the other two angles.
90° and 50° [12.1B]

3. In the accompanying figure, lines ℓ_1 and ℓ_2 are parallel. $\angle x = 30°$. Find $\angle y$.
150° [12.1C]

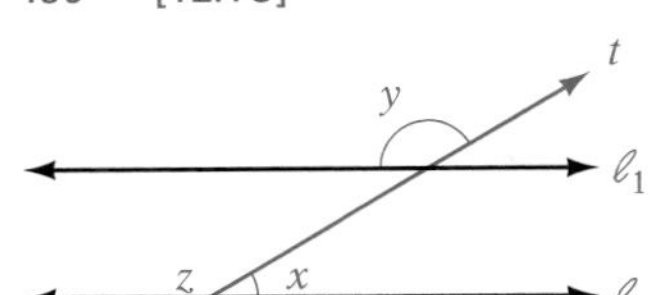

4. In the accompanying figure, lines ℓ_1 and ℓ_2 are parallel. $\angle x = 45°$. Find $\angle a + \angle b$.
180° [12.1C]

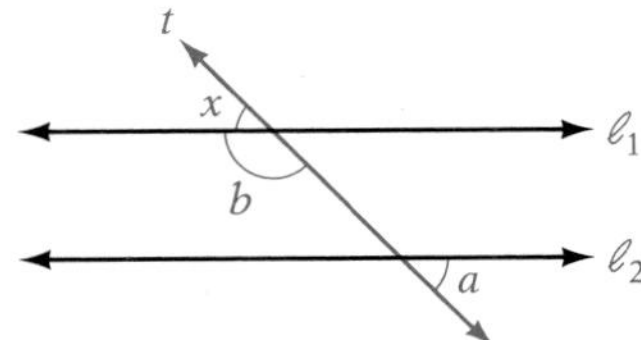

5. Find the perimeter of a rectangle with a length of 2 m and a width of 1.4 m.
6.8 m [12.2A]

6. Find the perimeter of the composite figure. Use $\pi \approx 3.14$. 15.85 ft [12.2B]

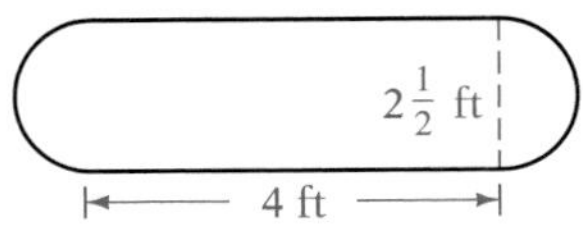

7. A carpet is to be placed as shown in the diagram below. At $13.40 per square yard, how much will it cost to carpet the area? Round to the nearest cent. Hint: $9\text{ ft}^2 = 1\text{ yd}^2$.
$556.84 [12.2C]

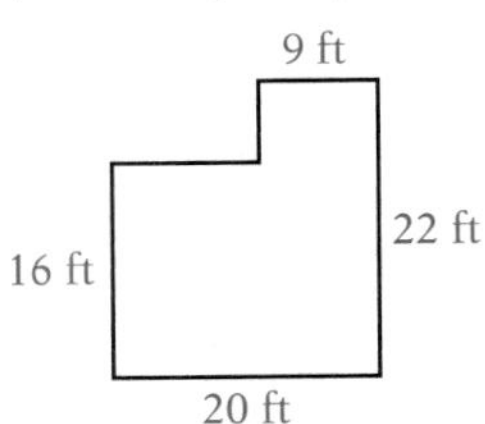

8. Find the area of a circle with a diameter of 2 m. Use $\frac{22}{7}$ for π. $3\frac{1}{7}\text{ m}^2$ [12.3A]

9. Find the area of the composite figure.
10.125 ft^2 [12.3B]

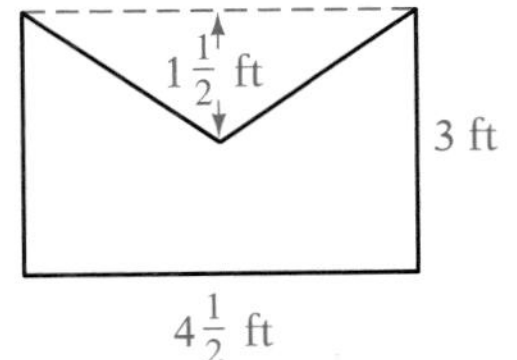

10. Find the cross-sectional area of a redwood tree that is 11 ft 6 in. in diameter. Use 3.14 for π. (Round to hundredths.) 103.82 ft^2 [12.3C]

11. How much more pizza is contained in a pizza with radius 10 in. than in one with radius 8 in.? Use $\pi \approx 3.14$. 113.04 in^2 [12.3C]

12. Find the volume of a cylinder with a height of 6 m and a radius of 3 m. Use 3.14 for π. 169.56 m^3 [12.4A]

13. Find the volume of the composite figure. 1406.72 cm^3 [12.4B]

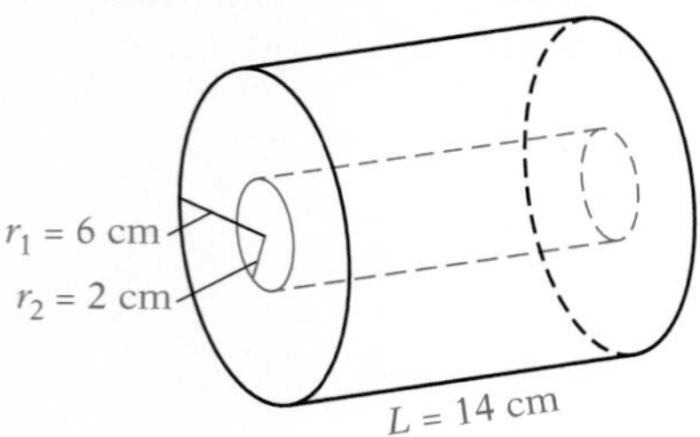

14. A toolbox is 1 ft 2 in. long, 9 in. wide, and 8 in. high. The sides and bottom of the toolbox are $\frac{1}{2}$ in. thick. Find the volume of the interior of the toolbox in cubic inches. 780 in^3 [12.4C]

15. Find the square root of 189. Use the table on page A3. 13.748 [12.5A]

16. Find the unknown side of the triangle shown in the figure. 9.747 ft [12.5B]

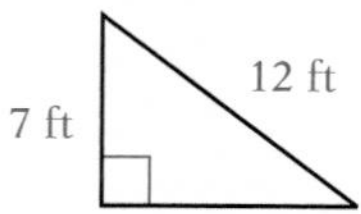

17. Find the length of the rafter for the roof shown in the figure. 15 ft [12.5C]

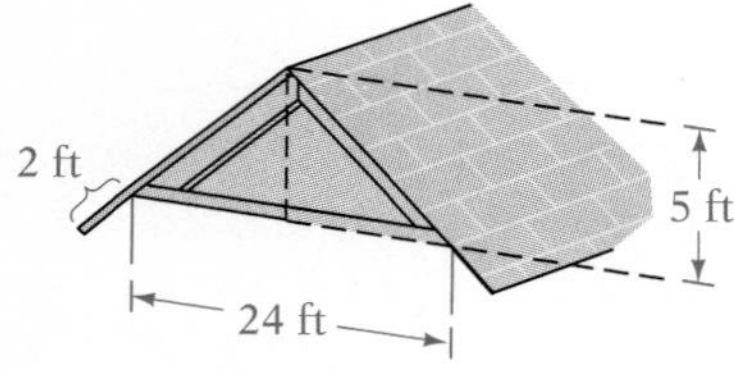

18. Triangles ABC and DEF below are similar. Find side BC. $1\frac{1}{5}$ ft [12.6A]

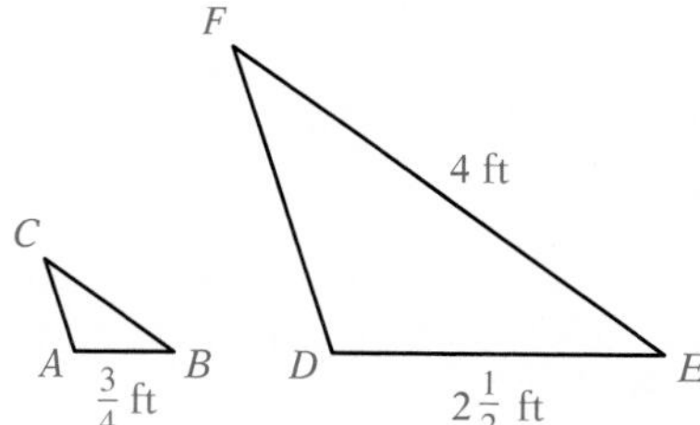

19. Find the width of the canal shown in the figure below. 25 ft [12.6B]

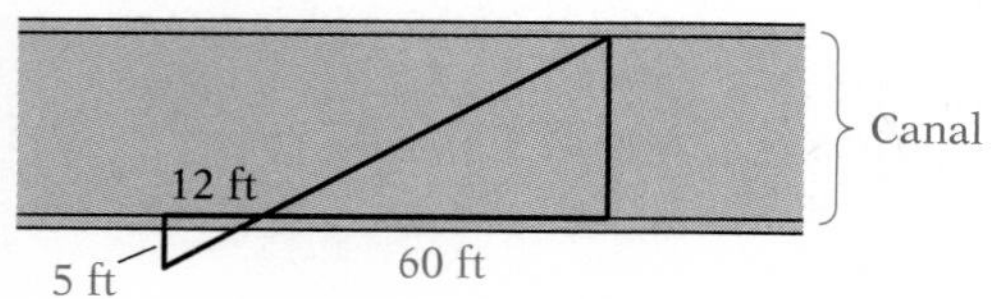

20. In the accompanying figure, triangles ABC and DEF are congruent right triangles. Find the length of FE. 10 m [12.6A]

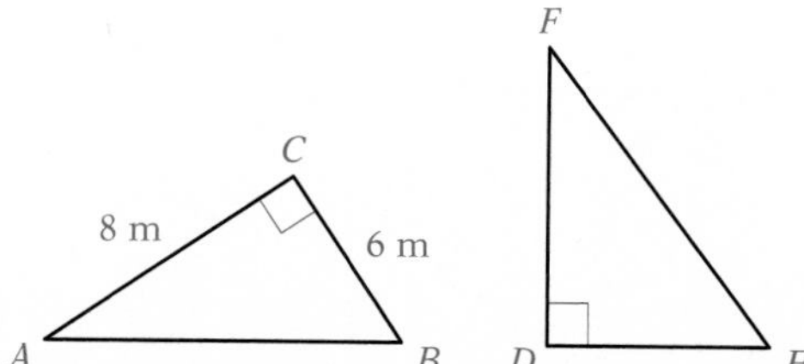

Cumulative Review

1. Find the GCF of 96 and 144.
48 [2.1B]

2. Add: $3\frac{5}{12} + 2\frac{9}{16} + 1\frac{7}{8}$
$7\frac{41}{48}$ [2.4C]

3. Find the quotient of $4\frac{1}{3}$ and $6\frac{2}{9}$.
$\frac{39}{56}$ [2.7B]

4. Simplify: $\left(\frac{2}{3}\right)^2 \div \left(\frac{1}{3} + \frac{1}{2}\right) - \frac{2}{5}$
$\frac{2}{15}$ [2.8C]

5. Simplify: $-\frac{2}{3} - \left(-\frac{5}{8}\right)$
$-\frac{1}{24}$ [10.4A]

6. Write "$348.80 earned in 40 hours" as a unit rate.
$8.72/h [4.2B]

7. Solve the proportion $\frac{3}{8} = \frac{n}{100}$.
$n = 37.5$ [4.3B]

8. Write $37\frac{1}{2}\%$ as a fraction.
$\frac{3}{8}$ [5.1A]

9. Evaluate $a^2 - (b^2 - c)$ when $a = 2$, $b = -2$, and $c = -4$.
-4 [11.1A]

10. 30.94 is 36.4% of what number?
85 [5.4A]

11. Solve: $\frac{x}{3} + 3 = 1$
-6 [11.3A]

12. Solve: $2(x - 3) + 2 = 5x - 8$
$\frac{4}{3}$ [11.4B]

13. Convert 32.5 km to meters.
32,500 m [9.1A]

14. Subtract: 32 m − 42 cm
31.58 m [9.1B]

15. Solve: $\frac{2}{3}x = -10$
-15 [11.2C]

16. Solve: $2x - 4(x - 3) = 8$
2 [11.4B]

17. You bought a car for $7516 and made a down payment of $1000. You paid the balance in 36 equal monthly installments. Find the monthly payment.
$181 [1.5D]

18. The sales tax on a suit costing $175 is $6.75. At the same rate, find the sales tax on a stereo system costing $1220. $47.06 [4.3C]

19. A heavy equipment operator receives an hourly wage of $16.06 an hour after receiving a 10% wage increase. Find the operator's hourly wage before the increase. $14.60 [5.4B]

20. An after-Christmas sale has a markdown rate of 55%. Find the sale price of a dress that had a regular price of $120. $54 [6.2D]

21. An IRA pays 10% annual interest compounded daily. What would the value on an investment of $25,000 after 20 years be? Use the table on page A4. $184,675.75 [6.3B]

22. A 4 × 4-in. tile weighs 6 oz. Find the weight of a package of 144 such tiles in pounds. 54 lb [8.2C]

23. Twenty-five rivets are used to fasten two steel plates together. The plates are 5 m 40 cm long, and the rivets are equally spaced with a rivet at each end. Find the distance in centimeters between the rivets. 22.5 cm [9.1C]

24. The total of two and four times a number is −6. Find the number. −2 [11.6A]

25. The lines ℓ_1 and ℓ_2 in the figure below are parallel. Find the angles a and b. $a = 74°$ $b = 106°$ [12.1A]

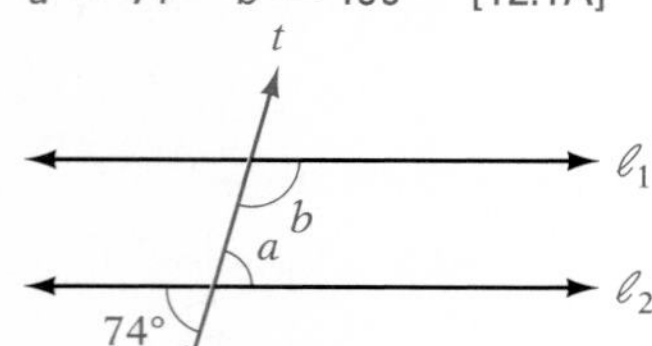

26. Find the perimeter of the composite figure. Use 3.14 for π. 29.42 cm [12.2B]

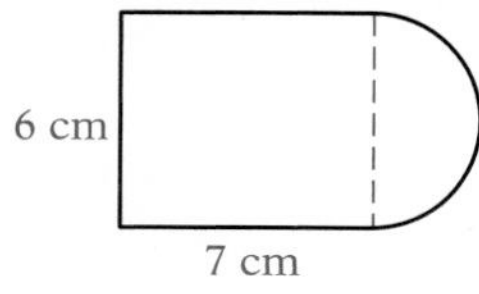

27. Find the area of the composite figure. 50 in² [12.3B]

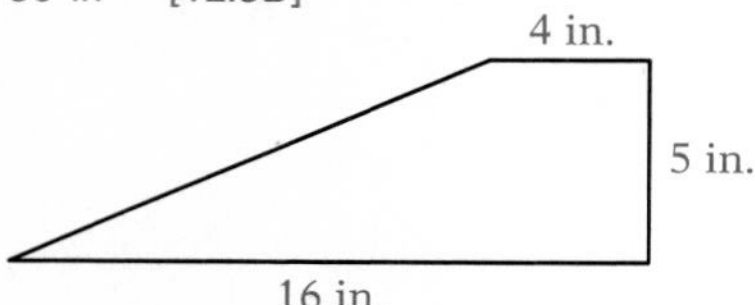

28. Find the volume of the composite figure. Use 3.14 for π. 92.86 in³ [12.4B]

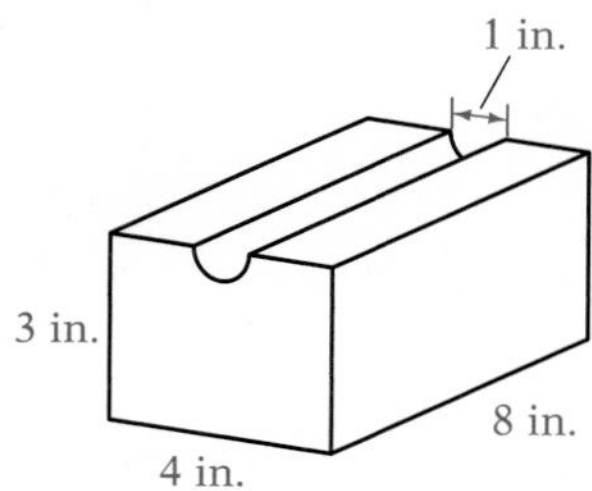

29. Find the unknown side of the triangle shown in the figure below. Round to the nearest hundredth. 10.63 ft [12.5B]

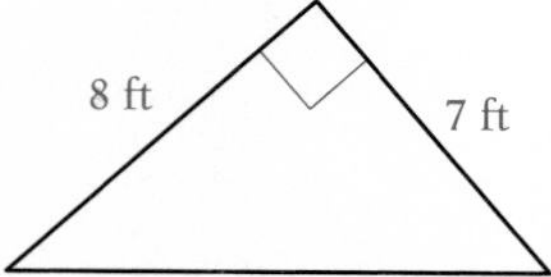

30. Triangles ABC and DEF below are similar. Find the perimeter of DEF. 36 cm [12.6B]

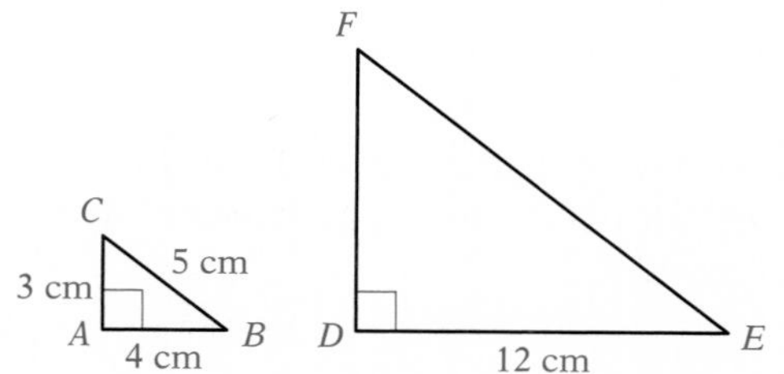

Final Examination

1. Subtract:
$100{,}914 - 97{,}655$
3259 [1.3B]

2. Find 34,821 divided by 657.
53 [1.5C]

3. Simplify:
$3^2 \cdot (5 - 3)^2 \div 3 + 4$
16 [1.6B]

4. Find the LCM of 9, 12, and 16.
144 [2.1A]

5. Add: $\frac{3}{8} + \frac{5}{6} + \frac{1}{5}$
$1\frac{49}{120}$ [2.4B]

6. Subtract: $7\frac{5}{12} - 3\frac{13}{16}$
$3\frac{29}{48}$ [2.5C]

7. Find the product of $3\frac{5}{8}$ and $1\frac{5}{7}$.
$6\frac{3}{14}$ [2.6B]

8. Divide: $1\frac{2}{3} \div 3\frac{3}{4}$
$\frac{4}{9}$ [2.7B]

9. Simplify: $\left(\frac{2}{3}\right)^3 \cdot \left(\frac{3}{4}\right)^2$
$\frac{1}{6}$ [2.8B]

10. Simplify: $\left(\frac{2}{3}\right)^2 \div \left(\frac{3}{4} + \frac{1}{3}\right) - \frac{1}{3}$
$\frac{1}{13}$ [2.8C]

11. Add:
$$\begin{array}{r} 4.972 \\ 28.6 \\ 1.88 \\ +\ 128.725 \\ \hline \end{array}$$
164.177 [3.2A]

12. Find 90,001 decreased by 29,796.
60,205 [3.3A]

13. Multiply:
$$\begin{array}{r} 2.97 \\ \times\ 0.0094 \\ \hline \end{array}$$
0.027918 [3.4A]

14. Divide: $0.062\overline{)0.0426}$
Round to the nearest hundredth.
0.69 [3.5A]

15. Convert 0.45 to a fraction in simplest form.
$\frac{9}{20}$ [3.6B]

16. Write "323.4 miles on 13.2 gallons of gas" as a unit rate.
24.5 mi/gal [4.2B]

17. Solve the proportion $\frac{12}{35} = \frac{n}{160}$.
Round to the nearest tenth.
$n = 54.9$ [4.3B]

18. Write $22\frac{1}{2}\%$ as a fraction.
$\frac{9}{40}$ [5.1A]

19. Write 1.35 as a percent.
135% [5.1B]

20. Write $\frac{5}{4}$ as a percent.
125% [5.1B]

21. Find 120% of 30.
36 [5.2A]

22. 12 is what percent of 9?
$133\frac{1}{3}\%$ [5.3A]

23. 42 is 60% of what number?
70 [5.4A]

24. Convert $1\frac{2}{3}$ ft to inches.
20 in. [8.1A]

25. Subtract:
3 ft 2 in. − 1 ft 10 in.
1 ft 4 in. [8.1B]

26. Convert 40 oz to pounds.
2.5 lb [8.2A]

27. Find the sum of 3 lb 12 oz and 2 lb 10 oz.
6 lb 6 oz [8.2B]

28. Convert 18 pt to gallons.
2.25 gal [8.3A]

29. Divide: $3\overline{)5 \text{ gal } 1 \text{ qt}}$
1 gal 3 qt [8.3B]

30. Convert 2.48 m to centimeters.
248 cm [9.1A]

31. Multiply: $\begin{array}{r} 4 \text{ m } 62 \text{ cm} \\ \times \quad 5 \\ \hline \end{array}$
23.10 m [9.1B]

32. Convert 1 kg 614 g to kilograms.
1.614 kg [9.2A]

33. Find 2 kg decreased by 742 g.
1.258 kg [9.2B]

34. Convert 2 L 67 ml to milliliters.
2067 ml [9.3A]

35. Divide: 3L 642 ml $\div$ 4
0.9105 L [9.3B]

36. Convert 55 mi to kilometers. Round to the nearest hundredth. (1.61 km = 1 mi)
88.55 km [9.5A]

37. Find the perimeter of a rectangle with a length of 1.2 m and a width of 0.75 m.
3.9 m [12.2A]

38. Find the area of a rectangle with a length of 9 in. and a width of 5 in.
45 in^2 [12.3A]

39. Find the volume of a box with a length of 20 cm, a width of 12 cm, and a height of 5 cm.
1200 cm^3 [12.4A]

40. Add: $-2 + 8 + (-10)$
-4 [10.2A]

41. Subtract: $-30 - (-15)$
-15 [10.2B]

42. Multiply: $2\frac{1}{2} \times \left(-\frac{1}{5}\right)$
$-\frac{1}{2}$ [10.4B]

43. Find the quotient of $-1\frac{3}{8}$ and $5\frac{1}{2}$.
$-\frac{1}{4}$ [10.4B]

44. Simplify:
$(-4)^2 \div (1 - 3)^2 - (-2)$
6 [10.5A]

45. Simplify:
$2x - 3(x - 4) + 5$
$-x + 17$ [11.1C]

46. Solve: $\frac{2}{3}x = -12$
$x = -18$ [11.2C]

47. Solve: $3x - 5 = 10$
$x = 5$ [11.3A]

48. Solve: $8 - 3x = x + 4$
$x = 1$ [11.4A]

49. You have \$872.48 in your checking account. You write checks of \$321.88 and \$34.23 and then make a deposit of \$443.56. Find your new checking account balance. \$959.93 [6.7A]

50. In a pre-election survey, it is estimated that 5 out of 8 eligible voters will vote in an election. How many people will vote in an election with 102,000 eligible voters? 63,750 people [4.3C]

51. A new computer has a sales price of $1800. The price of the computer is 40% of what the price was 4 years ago. What was the price of the computer 4 years ago? $4500 [5.4B]

52. A sales executive received commissions of $4320, $3572, $2864, and $4420 during a 4-month period. Find the average income for the 4 months. $3794 [7.4A]

53. A contractor borrows $120,000 for 9 months at an annual interest rate of 10%. What is the simple interest due on the loan? $9000 [6.3A]

54. The circle graph shows the population of the five most populous countries. Find what percent the population of China is of the total population of the top five countries. Round to the nearest tenth of a percent. 41.4% [7.1B]

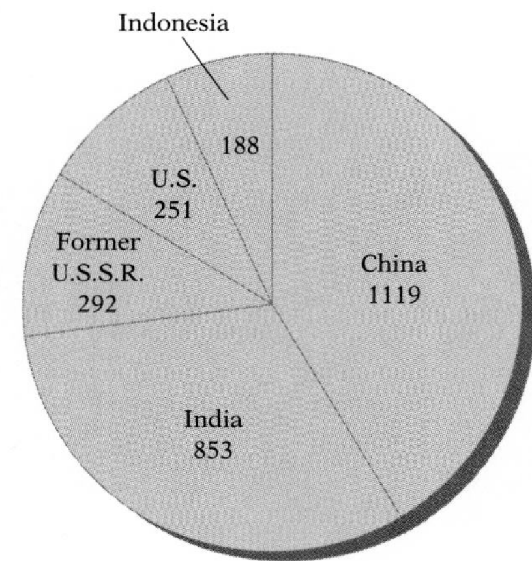

Population in Millions of People.

55. A compact disc player that regularly sells for $314.00 is on sale for $226.08. What is the discount rate? 28% [6.2D]

56. An 8 × 8-in. tile weighs 9 oz. Find the weight in pounds of a box containing 144 tiles. 81 lb [8.2C]

57. Find the perimeter of the composite figure. Use $\pi \approx 3.14$. 28.56 in. [12.2B]

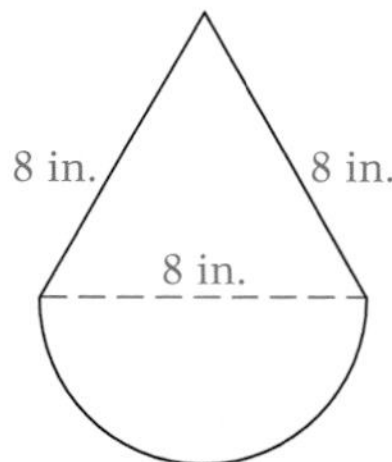

58. Find the area of the composite figure. Use $\pi \approx 3.14$. 16.86 cm^2 [12.3B]

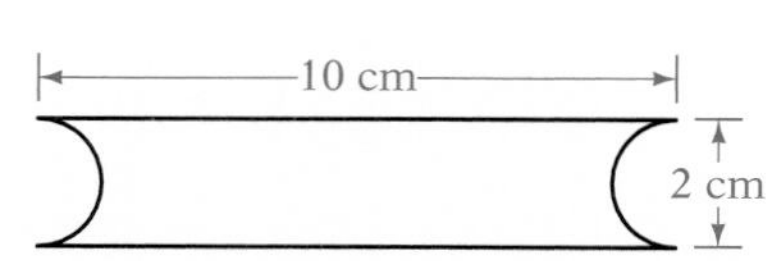

59. Five less than the quotient of a number and two is equal to three. Find the number. 16 [11.6A]

Appendix

Addition Table

+	0	1	2	3	4	5	6	7	8	9	10	11	12
0	0	1	2	3	4	5	6	7	8	9	10	11	12
1	1	2	3	4	5	6	7	8	9	10	11	12	13
2	2	3	4	5	6	7	8	9	10	11	12	13	14
3	3	4	5	6	7	8	9	10	11	12	13	14	15
4	4	5	6	7	8	9	10	11	12	13	14	15	16
5	5	6	7	8	9	10	11	12	13	14	15	16	17
6	6	7	8	9	10	11	12	13	14	15	16	17	18
7	7	8	9	10	11	12	13	14	15	16	17	18	19
8	8	9	10	11	12	13	14	15	16	17	18	19	20
9	9	10	11	12	13	14	15	16	17	18	19	20	21
10	10	11	12	13	14	15	16	17	18	19	20	21	22
11	11	12	13	14	15	16	17	18	19	20	21	22	23
12	12	13	14	15	16	17	18	19	20	21	22	23	24

Multiplication Table

+	0	1	2	3	4	5	6	7	8	9	10	11	12
0	0	0	0	0	0	0	0	0	0	0	0	0	0
1	0	1	2	3	4	5	6	7	8	9	10	11	12
2	0	2	4	6	8	10	12	14	16	18	20	22	24
3	0	3	6	9	12	15	18	21	24	27	30	33	36
4	0	4	8	12	16	20	24	28	32	36	40	44	48
5	0	5	10	15	20	25	30	35	40	45	50	55	60
6	0	6	12	18	24	30	36	42	48	54	60	66	72
7	0	7	14	21	28	35	42	49	56	63	70	77	84
8	0	8	16	24	32	40	48	56	64	72	80	88	96
9	0	9	18	27	36	45	54	63	72	81	90	99	108
10	0	10	20	30	40	50	60	70	80	90	100	110	120
11	0	11	22	33	44	55	66	77	88	99	110	121	132
12	0	12	24	36	48	60	72	84	96	108	120	132	144

Table of Square Roots

Decimal approximations have been rounded to the nearest thousandth.

Number	Square Root	Number	Square Root	Number	Square Root	Number	Square Root
1	1	51	7.141	101	10.050	151	12.288
2	1.414	52	7.211	102	10.100	152	12.329
3	1.732	53	7.280	103	10.149	153	12.369
4	2	54	7.348	104	10.198	154	12.410
5	2.236	55	7.416	105	10.247	155	12.450
6	2.449	56	7.483	106	10.296	156	12.490
7	2.646	57	7.550	107	10.344	157	12.530
8	2.828	58	7.616	108	10.392	158	12.570
9	3	59	7.681	109	10.440	159	12.610
10	3.162	60	7.746	110	10.488	160	12.649
11	3.317	61	7.810	111	10.536	161	12.689
12	3.464	62	7.874	112	10.583	162	12.728
13	3.606	63	7.937	113	10.630	163	12.767
14	3.742	64	8	114	10.677	164	12.806
15	3.873	65	8.062	115	10.724	165	12.845
16	4	66	8.124	116	10.770	166	12.884
17	4.123	67	8.185	117	10.817	167	12.923
18	4.243	68	8.246	118	10.863	168	12.961
19	4.359	69	8.307	119	10.909	169	13
20	4.472	70	8.367	120	10.954	170	13.038
21	4.583	71	8.426	121	11	171	13.077
22	4.690	72	8.485	122	11.045	172	13.115
23	4.796	73	8.544	123	11.091	173	13.153
24	4.899	74	8.602	124	11.136	174	13.191
25	5	75	8.660	125	11.180	175	13.229
26	5.099	76	8.718	126	11.225	176	13.267
27	5.196	77	8.775	127	11.269	177	13.304
28	5.292	78	8.832	128	11.314	178	13.342
29	5.385	79	8.888	129	11.358	179	13.379
30	5.477	80	8.944	130	11.402	180	13.416
31	5.568	81	9	131	11.446	181	13.454
32	5.657	82	9.055	132	11.489	182	13.491
33	5.745	83	9.110	133	11.533	183	13.528
34	5.831	84	9.165	134	11.576	184	13.565
35	5.916	85	9.220	135	11.619	185	13.601
36	6	86	9.274	136	11.662	186	13.638
37	6.083	87	9.327	137	11.705	187	13.675
38	6.164	88	9.381	138	11.747	188	13.711
39	6.245	89	9.434	139	11.790	189	13.748
40	6.325	90	9.487	140	11.832	190	13.784
41	6.403	91	9.539	141	11.874	191	13.820
42	6.481	92	9.592	142	11.916	192	13.856
43	6.557	93	9.644	143	11.958	193	13.892
44	6.633	94	9.695	144	12	194	13.928
45	6.708	95	9.747	145	12.042	195	13.964
46	6.782	96	9.798	146	12.083	196	14
47	6.856	97	9.849	147	12.124	197	14.036
48	6.928	98	9.899	148	12.166	198	14.071
49	7	99	9.950	149	12.207	199	14.107
50	7.071	100	10	150	12.247	200	14.142

Compound Interest Table

Compounded Annually

	4%	5%	6%	7%	8%	9%	10%
1 year	1.04000	1.05000	1.06000	1.07000	1.08000	1.09000	1.10000
5 years	1.21665	1.27628	1.33823	1.40255	1.46933	1.53862	1.61051
10 years	1.48024	1.62890	1.79085	1.96715	2.15893	2.36736	2.59374
15 years	1.80094	2.07893	2.39656	2.75903	3.17217	3.64248	4.17725
20 years	2.19112	2.65330	3.20714	3.86968	4.66095	5.60441	6.72750

Compounded Semiannually

	4%	5%	6%	7%	8%	9%	10%
1 year	1.04040	1.05062	1.06090	1.07123	1.08160	1.09203	1.10250
5 years	1.21899	1.28008	1.34392	1.41060	1.48024	1.55297	1.62890
10 years	1.48595	1.63862	1.80611	1.98979	2.19112	2.41171	2.65330
15 years	1.81136	2.09757	2.42726	2.80679	3.24340	3.74531	4.32194
20 years	2.20804	2.68506	3.26204	3.95926	4.80102	5.81634	7.03999

Compounded Quarterly

	4%	5%	6%	7%	8%	9%	10%
1 year	1.04060	1.05094	1.06136	1.07186	1.08243	1.09308	1.10381
5 years	1.22019	1.28204	1.34686	1.41478	1.48595	1.56051	1.63862
10 years	1.48886	1.64362	1.81402	2.00160	2.20804	2.43519	2.68506
15 years	1.81670	2.10718	2.44322	2.83182	3.28103	3.80013	4.39979
20 years	2.21672	2.70148	3.29066	4.00639	4.87544	5.93015	7.20957

Compounded Daily

	4%	5%	6%	7%	8%	9%	10%
1 year	1.04080	1.05127	1.06183	1.07250	1.08328	1.09416	1.10516
5 years	1.22139	1.28400	1.34983	1.41902	1.49176	1.56823	1.64861
10 years	1.49179	1.64866	1.82203	2.01362	2.22535	2.45933	2.71791
15 years	1.82206	2.11689	2.45942	2.85736	3.31968	3.85678	4.48077
20 years	2.22544	2.71810	3.31979	4.05466	4.95217	6.04830	7.38703

To use this table:

1. Locate the section which gives the desired compounding period.
2. Locate the interest rate in the top row of that section.
3. Locate the number of years in the left-hand column of that section.
4. Locate the number where the interest-rate column and the number-of-years row meet. This is the compound interest factor.

Example An investment yields an annual interest rate of 10% compounded quarterly for 5 years.
The compounding period is "compounded quarterly."
The interest rate is 10%.
The number of years is 5.
The number where the row and column meet is 1.63862. This is the compound interest factor.

Compound Interest Table

Compounded Annually

	11%	12%	13%	14%	15%	16%	17%
1 year	1.11000	1.12000	1.13000	1.14000	1.15000	1.16000	1.17000
5 years	1.68506	1.76234	1.84244	1.92542	2.01136	2.10034	2.19245
10 years	2.83942	3.10585	3.39457	3.70722	4.04556	4.41144	4.80683
15 years	4.78459	5.47357	6.25427	7.13794	8.13706	9.26552	10.53872
20 years	8.06239	9.64629	11.52309	13.74349	16.36654	19.46076	23.10560

Compounded Semiannually

	11%	12%	13%	14%	15%	16%	17%
1 year	1.11303	1.12360	1.13423	1.14490	1.15563	1.16640	1.17723
5 years	1.70814	1.79085	1.87714	1.96715	2.06103	2.15893	2.26098
10 years	2.91776	3.20714	3.52365	3.86968	4.24785	4.66096	5.11205
15 years	4.98395	5.74349	6.61437	7.61226	8.75496	10.06266	11.55825
20 years	8.51331	10.28572	12.41607	14.97446	18.04424	21.72452	26.13302

Compounded Quarterly

	11%	12%	13%	14%	15%	16%	17%
1 year	1.11462	1.12551	1.13648	1.14752	1.15865	1.16986	1.18115
5 years	1.72043	1.80611	1.89584	1.98979	2.08815	2.19112	2.29891
10 years	2.95987	3.26204	3.59420	3.95926	4.36038	4.80102	5.28497
15 years	5.09225	5.89160	6.81402	7.87809	9.10513	10.51963	12.14965
20 years	8.76085	10.64089	12.91828	15.67574	19.01290	23.04980	27.93091

Compounded Daily

	11%	12%	13%	14%	15%	16%	17%
1 year	1.11626	1.12747	1.13880	1.15024	1.16180	1.17347	1.18526
5 years	1.73311	1.82194	1.91532	2.01348	2.11667	2.22515	2.33918
10 years	3.00367	3.31946	3.66845	4.05411	4.48031	4.95130	5.47178
15 years	5.20569	6.04786	7.02625	8.16288	9.48335	11.01738	12.79950
20 years	9.02203	11.01883	13.45751	16.43582	20.07316	24.51534	29.94039

Monthly Payment Table

	4%	5%	6%	7%	8%	9%
1 year	0.0851499	0.0856075	0.0860664	0.0865267	0.0869884	0.0874515
2 years	0.0434249	0.0438714	0.0443206	0.0447726	0.0452273	0.0456847
3 years	0.0295240	0.0299709	0.0304219	0.0308771	0.0313364	0.0317997
4 years	0.0225791	0.0230293	0.0234850	0.0239462	0.0244129	0.0248850
5 years	0.0184165	0.0188712	0.0193328	0.0198012	0.0202764	0.0207584
20 years	0.0060598	0.0065996	0.0071643	0.0077530	0.0083644	0.0089973
25 years	0.0052784	0.0058459	0.0064430	0.0070678	0.0077182	0.0083920
30 years	0.0047742	0.0053682	0.0059955	0.0066530	0.0073376	0.0080462

	10%	11%	12%	13%
1 year	0.0879159	0.0883817	0.0888488	0.0893173
2 years	0.0461449	0.0466078	0.0470735	0.0475418
3 years	0.0322672	0.0327387	0.0332143	0.0336940
4 years	0.0253626	0.0258455	0.0263338	0.0268275
5 years	0.0212470	0.0217424	0.0222445	0.0227531
20 years	0.0096502	0.0103219	0.0110109	0.0117158
25 years	0.0090870	0.0098011	0.0105322	0.0112784
30 years	0.0087757	0.0095232	0.0102861	0.0110620

To use this table:

1. Locate the desired interest rate in the top row.
2. Locate the number of years in the left-hand column.
3. Locate the number where the interest-rate column and the number-of-years row meet. This is the monthly payment factor.

Example: A home has a 30-year mortgage at an annual interest rate of 12%.
The interest rate is 12%.
The number of years is 30.
The number where the row and column meet is 0.0102861. This is the monthly payment factor.

SOLUTIONS to Chapter 1 "You Try It"

SECTION 1.1 *pages 3–6*

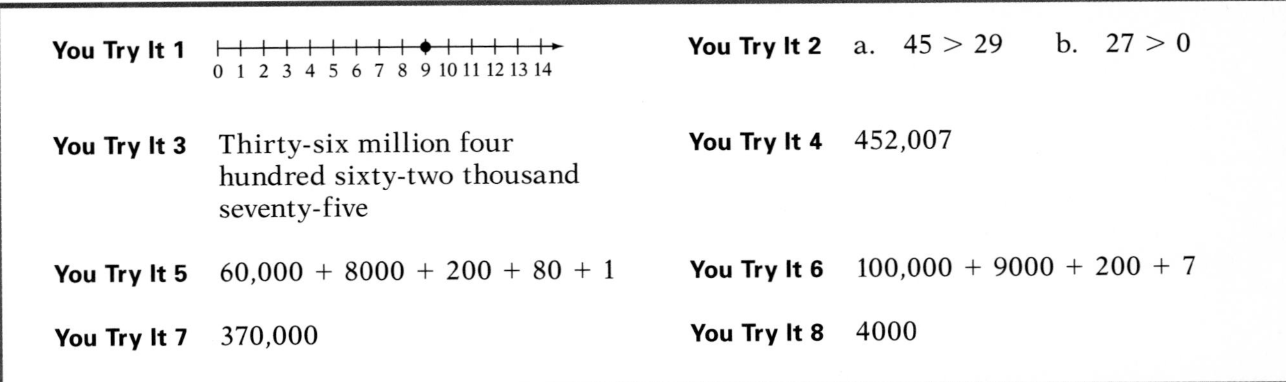

You Try It 1 (number line 0–14 with a dot at 9)

You Try It 2 a. $45 > 29$ b. $27 > 0$

You Try It 3 Thirty-six million four hundred sixty-two thousand seventy-five

You Try It 4 452,007

You Try It 5 $60{,}000 + 8000 + 200 + 80 + 1$

You Try It 6 $100{,}000 + 9000 + 200 + 7$

You Try It 7 370,000

You Try It 8 4000

SECTION 1.2 *pages 9–12*

You Try It 1

$$\begin{array}{r} 347 \\ +\ 12453 \\ \hline 12{,}800 \end{array}$$

347 increased by 12,453 is 12,800.

You Try It 2

$$\begin{array}{r} {}^{2} \\ 95 \\ 88 \\ +\ 67 \\ \hline 250 \end{array}$$

You Try It 3

$$\begin{array}{r} {}^{1\,1\ \,2\,1} \\ 392 \\ 4{,}079 \\ 89{,}035 \\ +\ \ 4{,}992 \\ \hline 98{,}498 \end{array}$$

You Try It 4

Strategy To find the total amount budgeted for the three items each month, add the three amounts (\$475, \$275, and \$120).

Solution

$$\begin{array}{r} \$475 \\ 275 \\ +\ 120 \\ \hline \$870 \end{array}$$

The total amount budgeted for the three items is \$870.

SECTION 1.3 *pages 17–20*

You Try It 1

$$\begin{array}{r} 8925 \\ -\ 6413 \\ \hline 2512 \end{array}$$

You Try It 2

$$\begin{array}{r} 17{,}504 \\ -\ \ 9{,}302 \\ \hline 8{,}202 \end{array}$$

You Try It 3

$$\begin{array}{r} \scriptstyle 2\ 14\ 7\ 11 \\ \not{3}\ \not{4}\ \not{8}\ \not{1} \\ -\quad 8\ 6\ 5 \\ \hline 2\ 6\ 1\ 6 \end{array}$$

You Try It 4

$$\begin{array}{r} \scriptstyle 15 \\ \scriptstyle 4\ \not{5}\ 12 \\ 5\ 4{,}\not{5}\ \not{6}\ \not{2} \\ -\ 1\ 4{,}4\ 8\ 5 \\ \hline 4\ 0{,}0\ 7\ 7 \end{array}$$

You Try It 5

$$\begin{array}{r} \scriptstyle 13\ 9\ 9 \\ \scriptstyle 5\ \not{3}\ \not{10}\ \not{10}\ 13 \\ \not{6}\ \not{4}{,}\not{0}\ \not{0}\ \not{3} \\ -\ 5\ 4{,}9\ 3\ 6 \\ \hline 9\ 0\ 6\ 7 \end{array}$$

You Try It 6

Strategy To find the difference in value, subtract the value of the takeovers in 1992 ($82 billion) from the value of the takeovers in 1993 ($135 billion).

Solution

$$\begin{array}{r} 135 \\ -\ \ 82 \\ \hline 53 \end{array}$$

The difference in value is $53 billion.

You Try It 7

Strategy To find your take-home pay:

- Add to find the total of the deductions ($127 + $18 + $35).
- Subtract the total of the deductions from your total salary ($638).

Solution

$$\begin{array}{r} 127 \\ 18 \\ +\ 35 \\ \hline 180 \end{array} \text{ deductions} \qquad \begin{array}{r} 638 \\ -\ 180 \\ \hline 458 \end{array}$$

Your take-home pay is $458.

SECTION 1.4 *pages 25–28*

You Try It 1

$$\begin{array}{r} \scriptstyle 3\,5\ \ \\ 648 \\ \times\quad 7 \\ \hline 4536 \end{array}$$

You Try It 2

$$\begin{array}{r} 756 \\ \times\ 305 \\ \hline 3780 \\ 22680\ \ \\ \hline 230{,}580 \end{array}$$

You Try It 3

Strategy
To find the number of cars the dealer will receive in 12 months, multiply the number of months (12) by the number of cars received each month (37).

Solution

$$\begin{array}{r} 37 \\ \times\ 12 \\ \hline 74 \\ 37\ \ \\ \hline 444 \end{array}$$

The dealer will receive 444 cars in 12 months.

You Try It 4

Strategy
To find the total cost of the order:

- Find the cost of the sports jackets by multiplying the number of jackets (25) by the cost for each jacket ($23).
- Add the product to the cost for the suits ($4800).

Solution

$$\begin{array}{r} \$23 \\ \times\ 25 \\ \hline 115 \\ 46\ \ \\ \hline \$575 \end{array} \text{ cost for jackets} \qquad \begin{array}{r} \$4800 \\ +\ \ 575 \\ \hline \$5375 \end{array}$$

The total cost of the order is $5375.

SECTION 1.5 *pages 33–40*

You Try It 1

$$9\overline{)63}\ \text{quotient } 7$$

Check: $7 \times 9 = 63$

You Try It 2

$$\begin{array}{r} 453 \\ 9\overline{)\ 4077} \\ -36\ \ \\ \hline 47\ \\ -45\ \\ \hline 27 \\ -27 \\ \hline 0 \end{array}$$

Check: $453 \times 9 = 4077$

You Try It 3

$$\begin{array}{r} 705 \\ 9\overline{)\ 6345} \\ -63\ \ \\ \hline 04\ \\ -\ 0\ \\ \hline 45 \\ -45 \\ \hline 0 \end{array}$$

Check: $705 \times 9 = 6345$

You Try It 4

$$\begin{array}{r} 870 \text{ r5} \\ 6\overline{)\ 5225}\ \ \ \\ -48\ \ \ \ \ \ \\ \hline 42\ \ \ \ \ \\ -42\ \ \ \ \ \\ \hline 05\ \ \ \\ -\ 0\ \ \ \\ \hline 5\ \ \ \end{array}$$

Check: $(870 \times 6) + 5 =$
$5220 + 5 = 5225$

You Try It 5

$$\begin{array}{r}
3{,}058 \text{ r3} \\
7\overline{)\ 21{,}409} \\
-21 \\
\hline
0\,4 \\
-\ \ 0 \\
\hline
40 \\
-35 \\
\hline
59 \\
-56 \\
\hline
3
\end{array}$$

Check: $(3058 \times 7) + 3 =$
$21{,}406 + 3 = 21{,}409$

You Try It 6

$$\begin{array}{r}
109 \\
42\overline{)\ 4578} \\
-42 \\
\hline
37 \\
-\ 0 \\
\hline
378 \\
-378 \\
\hline
0
\end{array}$$

Check: $109 \times 42 = 4578$

You Try It 7

$$\begin{array}{r}
470 \text{ r29} \\
39\overline{)\ 18{,}359} \\
-15\,6 \\
\hline
2\,75 \\
-2\,73 \\
\hline
29 \\
-\ \ 0 \\
\hline
29
\end{array}$$

Check: $(470 \times 39) + 29 =$
$18{,}330 + 29 = 18{,}359$

You Try It 8

$$\begin{array}{r}
62 \text{ r111} \\
534\overline{)\ 33{,}219} \\
-32\,04 \\
\hline
1\,179 \\
-1\,068 \\
\hline
111
\end{array}$$

Check: $(62 \times 534) + 111 =$
$33{,}108 + 111 = 33{,}219$

You Try It 9

$$\begin{array}{r}
421 \text{ r33} \\
515\overline{)\ 216{,}848} \\
-206\,0 \\
\hline
10\,84 \\
-10\,30 \\
\hline
548 \\
-515 \\
\hline
33
\end{array}$$

Check: $(421 \times 515) + 33 =$
$216{,}815 + 33 = 216{,}848$

You Try It 10

Strategy To find the number of tires to be stored on each shelf, divide the number of tires (270) by the number of shelves (15).

Solution

$$\begin{array}{r} 18 \\ 15\overline{)\ 270} \\ -15 \\ \hline 120 \\ -120 \\ \hline 0 \end{array}$$

Each shelf can store 18 tires.

You Try It 11

Strategy To find the number of cases produced in 8 hours:

- Find the number of cases produced in one hour by dividing the number of cans produced (12,600) by the number of cans to a case (24).
- Multiply the number of cases produced in one hour by 8.

Solution

$$\begin{array}{r} 525 \\ 24\overline{)\ 12{,}600} \\ -12\ 0 \\ \hline 60 \\ -48 \\ \hline 120 \\ -120 \\ \hline 0 \end{array}$$

cases produced in one hour

$$\begin{array}{r} 525 \\ \times\quad 8 \\ \hline 4200 \end{array}$$

In 8 hours, 4200 cases are produced.

SECTION 1.6 *pages 45–46*

You Try It 1 $2^4 \cdot 3^3$

You Try It 2 10^7

You Try It 3 $(2 \cdot 2 \cdot 2) \cdot (5 \cdot 5) = 8 \cdot 25$
$= 200$

You Try It 4
$5 \cdot (8 - 4) \div 4 - 2$
$5 \cdot 4 \div 4 - 2$
$20 \div 4 - 2$
$5 - 2$
3

SECTION 1.7 *pages 49–50*

You Try It 1 1, 2, 4, 5, 8, 10, 20 and 40 are factors of 40.

You Try It 2 $44 = 2 \cdot 2 \cdot 11$

You Try It 3 $177 = 3 \cdot 59$

SOLUTIONS to Chapter 2 "You Try It"

SECTION 2.1 *pages 61–62*

You Try It 1

	2	3	5	7
50 =	2		(5 · 5)	
84 =	(2 · 2)	3		(7)
135 =		(3 · 3 · 3)	5	

The LCM = 2 · 2 · 3 · 3 · 3 · 5 · 5 · 7
= 18,900

You Try It 2

	2	3	5
36 =	(2 · 2)	3 · 3	
60 =	2 · 2	(3)	5
72 =	2 · 2 · 2	3 · 3	

The GCF = 2 · 2 · 3 = 12.

You Try It 3

	2	3	5	11
11 =				11
24 =	2 · 2 · 2	3		
30 =	2	3	5	

Since no numbers are circled, the GCF = 1.

SECTION 2.2 *pages 65–66*

You Try It 1 $4\frac{1}{4}$

You Try It 2 $\frac{17}{4}$

You Try It 3 $4\frac{2}{5}$

You Try It 4 4

You Try It 5 $\frac{117}{8}$

SECTION 2.3 *pages 69–70*

You Try It 1 $45 \div 5 = 9$ $\quad \frac{3}{5} = \frac{3 \cdot 9}{5 \cdot 9} = \frac{27}{45}$

$\frac{27}{45}$ is equivalent to $\frac{3}{5}$.

You Try It 2 Write 6 as $\frac{6}{1}$.

$18 \div 1 = 18 \quad 6 = \frac{6 \cdot 18}{1 \cdot 18} = \frac{108}{18}$

$\frac{108}{18}$ is equivalent to 6.

You Try It 3 $\frac{16}{24} = \frac{\overset{1}{\cancel{2}} \cdot \overset{1}{\cancel{2}} \cdot \overset{1}{\cancel{2}} \cdot 2}{\underset{1}{\cancel{2}} \cdot \underset{1}{\cancel{2}} \cdot \underset{1}{\cancel{2}} \cdot 3} = \frac{2}{3}$

You Try It 4 $\frac{8}{56} = \frac{\overset{1}{\cancel{2}} \cdot \overset{1}{\cancel{2}} \cdot \overset{1}{\cancel{2}}}{\underset{1}{\cancel{2}} \cdot \underset{1}{\cancel{2}} \cdot \underset{1}{\cancel{2}} \cdot 7} = \frac{1}{7}$

You Try It 5 $\frac{15}{32} = \frac{3 \cdot 5}{2 \cdot 2 \cdot 2 \cdot 2 \cdot 2} = \frac{15}{32}$

You Try It 6 $\frac{48}{36} = \frac{\overset{1}{\cancel{2}} \cdot \overset{1}{\cancel{2}} \cdot 2 \cdot 2 \cdot \overset{1}{\cancel{3}}}{\underset{1}{\cancel{2}} \cdot \underset{1}{\cancel{2}} \cdot \underset{1}{\cancel{3}} \cdot 3} = \frac{4}{3} = 1\frac{1}{3}$

SECTION 2.4 *pages 73–76*

You Try It 1

$$\begin{array}{r} \frac{3}{8} \\ + \frac{7}{8} \\ \hline \frac{10}{8} = \frac{5}{4} = 1\frac{1}{4} \end{array}$$

You Try It 2

$$\begin{array}{rcl} \frac{5}{12} &=& \frac{20}{48} \\ + \frac{9}{16} &=& \frac{27}{48} \\ \hline && \frac{47}{48} \end{array}$$

You Try It 3

$$\begin{array}{rcl} \frac{7}{8} &=& \frac{105}{120} \\ + \frac{11}{15} &=& \frac{88}{120} \\ \hline && \frac{193}{120} = 1\frac{73}{120} \end{array}$$

You Try It 4

$$\begin{array}{rcl} \frac{3}{4} &=& \frac{30}{40} \\ \frac{4}{5} &=& \frac{32}{40} \\ + \frac{5}{8} &=& \frac{25}{40} \\ \hline && \frac{87}{40} = 2\frac{7}{40} \end{array}$$

You Try It 5 $7 + \frac{6}{11} = 7\frac{6}{11}$

You Try It 6

$$\begin{array}{r} 29 \\ + 17\frac{5}{12} \\ \hline 46\frac{5}{12} \end{array}$$

You Try It 7

$$\begin{array}{rcl} 7\frac{4}{5} &=& 7\frac{24}{30} \\ 6\frac{7}{10} &=& 6\frac{21}{30} \\ + 13\frac{11}{15} &=& 13\frac{22}{30} \\ \hline && 26\frac{67}{30} = 28\frac{7}{30} \end{array}$$

You Try It 8

$$\begin{array}{rcl} 9\frac{3}{8} &=& 9\frac{45}{120} \\ 17\frac{7}{12} &=& 17\frac{70}{120} \\ + 10\frac{14}{15} &=& 10\frac{112}{120} \\ \hline && 36\frac{227}{120} = 37\frac{107}{120} \end{array}$$

You Try It 9

Strategy
To find the total time spent on the activities, add the three times $\left(4\frac{1}{2}, 3\frac{3}{4}, 1\frac{1}{3}\right)$.

Solution

$$\begin{array}{r} 4\frac{1}{2} = 4\frac{6}{12} \\ 3\frac{3}{4} = 3\frac{9}{12} \\ +\ 1\frac{1}{3} = 1\frac{4}{12} \\ \hline 8\frac{19}{12} = 9\frac{7}{12} \end{array}$$

The total time spent on the three activities was $9\frac{7}{12}$ hours.

You Try It 10

Strategy
To find the overtime pay:

- Find the total number of overtime hours $\left(1\frac{2}{3} + 3\frac{1}{3} + 2\right)$.
- Multiply the total number of hours by the overtime hourly wage ($24).

Solution

$$\begin{array}{r} 1\frac{2}{3} \\ 3\frac{1}{3} \\ +\ 2 \\ \hline 6\frac{3}{3} = 7 \text{ hours} \end{array} \qquad \begin{array}{r} \$24 \\ \times\ 7 \\ \hline \$168 \end{array}$$

Jeff earned $168 in overtime pay.

SECTION 2.5 *pages 81–84*

You Try It 1

$$\begin{array}{r} \frac{16}{27} \\ -\frac{7}{27} \\ \hline \frac{9}{27} = \frac{1}{3} \end{array}$$

You Try It 2

$$\begin{array}{r} \frac{13}{18} = \frac{52}{72} \\ -\frac{7}{24} = \frac{21}{72} \\ \hline \frac{31}{72} \end{array}$$

You Try It 3

$$\begin{array}{r} 17\frac{5}{9} = 17\frac{20}{36} \\ -\ 11\frac{5}{12} = 11\frac{15}{36} \\ \hline 6\frac{5}{36} \end{array}$$

You Try It 4

$$\begin{array}{r} 8 = 7\frac{13}{13} \\ -\ 2\frac{4}{13} = 2\frac{4}{13} \\ \hline 5\frac{9}{13} \end{array}$$

You Try It 5

$$\begin{array}{r} 21\frac{7}{9} = 21\frac{28}{36} = 20\frac{64}{36} \\ -\ 7\frac{11}{12} = 7\frac{33}{36} = 7\frac{33}{36} \\ \hline 13\frac{31}{36} \end{array}$$

You Try It 6

Strategy
To find the time remaining before the plane lands, subtract the number of hours already in the air $\left(2\frac{3}{4}\right)$ from the total time of the trip $\left(5\frac{1}{2}\right)$.

Solution

$$\begin{array}{r} 5\frac{1}{2} = 5\frac{2}{4} = 4\frac{6}{4} \\ -\,2\frac{3}{4} = 2\frac{3}{4} = 2\frac{3}{4} \\ \hline 2\frac{3}{4}\text{ hours} \end{array}$$

The plane will land in $2\frac{3}{4}$ hours.

You Try It 7

Strategy
To find the amount of weight to be lost during the third month:

- Find the total weight loss during the first two months $\left(7\frac{1}{2} + 5\frac{3}{4}\right)$.
- Subtract the total weight loss from the goal (24 pounds).

Solution

$$\begin{array}{r} 7\frac{1}{2} = 7\frac{2}{4} \\ +\,5\frac{3}{4} = 5\frac{3}{4} \\ \hline 12\frac{5}{4} = 13\frac{1}{4}\text{ pounds lost} \end{array}$$

$$\begin{array}{r} 24 = 23\frac{4}{4} \\ -\,13\frac{1}{4} = 13\frac{1}{4} \\ \hline 10\frac{3}{4}\text{ pounds} \end{array}$$

The patient must lose $10\frac{3}{4}$ pounds to achieve the goal.

SECTION 2.6 *pages 89–92*

You Try It 1

$$\frac{4}{21} \times \frac{7}{44} = \frac{4 \cdot 7}{21 \cdot 44} = \frac{\overset{1}{\cancel{2}} \cdot \overset{1}{\cancel{2}} \cdot \overset{1}{\cancel{7}}}{3 \cdot \underset{1}{\cancel{7}} \cdot \underset{1}{\cancel{2}} \cdot \underset{1}{\cancel{2}} \cdot 11} = \frac{1}{33}$$

You Try It 2

$$\frac{2}{21} \times \frac{10}{33} = \frac{2 \cdot 10}{21 \cdot 33} = \frac{2 \cdot 2 \cdot 5}{3 \cdot 7 \cdot 3 \cdot 11} = \frac{20}{693}$$

You Try It 3

$$\frac{16}{5} \times \frac{15}{24} = \frac{16 \cdot 15}{5 \cdot 24} = \frac{\overset{1}{\cancel{2}} \cdot \overset{1}{\cancel{2}} \cdot \overset{1}{\cancel{2}} \cdot 2 \cdot \overset{1}{\cancel{3}} \cdot \overset{1}{\cancel{5}}}{\underset{1}{\cancel{5}} \cdot \underset{1}{\cancel{2}} \cdot \underset{1}{\cancel{2}} \cdot \underset{1}{\cancel{2}} \cdot \underset{1}{\cancel{3}}} = 2$$

You Try It 4

$$5\frac{2}{5} \times \frac{5}{9} = \frac{27}{5} \times \frac{5}{9} = \frac{27 \cdot 5}{5 \cdot 9} = \frac{\overset{1}{\cancel{3}} \cdot \overset{1}{\cancel{3}} \cdot 3 \cdot \overset{1}{\cancel{5}}}{\underset{1}{\cancel{5}} \cdot \underset{1}{\cancel{3}} \cdot \underset{1}{\cancel{3}}} = 3$$

You Try It 5

$$3\frac{2}{5} \times 6\frac{1}{4} = \frac{17}{5} \times \frac{25}{4} = \frac{17 \cdot 25}{5 \cdot 4}$$

$$= \frac{17 \cdot \overset{1}{\cancel{5}} \cdot 5}{\underset{1}{\cancel{5}} \cdot 2 \cdot 2} = \frac{85}{4} = 21\frac{1}{4}$$

You Try It 6

$$3\frac{2}{7} \times 6 = \frac{23}{7} \times \frac{6}{1} = \frac{23 \cdot 6}{7 \cdot 1}$$

$$= \frac{23 \cdot 3 \cdot 2}{7 \cdot 1} = \frac{138}{7} = 19\frac{5}{7}$$

You Try It 7

Strategy
To find the value of the house today, multiply the old value of the house (\$30,000) by $3\frac{1}{2}$.

Solution

$$\frac{30{,}000}{1} \times \frac{7}{2} = \frac{30{,}000 \cdot 7}{1 \cdot 2}$$

$$= 105{,}000$$

The value of the house today is \$105,000.

You Try It 8

Strategy
To find the cost of the air compressor:

- Multiply to find the value of the drying chamber $\left(\frac{4}{5} \times \$60{,}000\right)$.
- Subtract the value of the drying chamber from the total value of the two items (\$60,000).

Solution

$$\frac{4}{5} \times \frac{\$60{,}000}{1} = \frac{\$240{,}000}{5}$$

$$= \$48{,}000$$

value of the drying chamber

$$\begin{array}{r} \$60{,}000 \\ -\ 48{,}000 \\ \hline \$12{,}000 \end{array}$$

The cost of the air compressor was \$12,000.

SECTION 2.7 *pages 97–100*

You Try It 1

$$\frac{3}{7} \div \frac{2}{3} = \frac{3}{7} \times \frac{3}{2} = \frac{3 \cdot 3}{7 \cdot 2} = \frac{9}{14}$$

You Try It 2

$$\frac{3}{4} \div \frac{9}{10} = \frac{3}{4} \times \frac{10}{9}$$

$$= \frac{3 \cdot 10}{4 \cdot 9} = \frac{\overset{1}{\cancel{3}} \cdot \overset{1}{\cancel{2}} \cdot 5}{\underset{1}{\cancel{2}} \cdot 2 \cdot \underset{1}{\cancel{3}} \cdot 3} = \frac{5}{6}$$

You Try It 3

$$\frac{5}{7} \div 6 = \frac{5}{7} \div \frac{6}{1}$$

$$= \frac{5}{7} \times \frac{1}{6} = \frac{5}{7 \cdot 2 \cdot 3} = \frac{5}{42}$$

You Try It 4

$$12\frac{3}{5} \div 7 = \frac{63}{5} \div \frac{7}{1} = \frac{63}{5} \times \frac{1}{7}$$

$$= \frac{63 \cdot 1}{5 \cdot 7} = \frac{3 \cdot 3 \cdot \overset{1}{\cancel{7}}}{5 \cdot \underset{1}{\cancel{7}}} = \frac{9}{5} = 1\frac{4}{5}$$

You Try It 5

$$3\frac{2}{3} \div 2\frac{2}{5} = \frac{11}{3} \div \frac{12}{5}$$
$$= \frac{11}{3} \times \frac{5}{12} = \frac{11 \cdot 5}{3 \cdot 12}$$
$$= \frac{11 \cdot 5}{3 \cdot 2 \cdot 2 \cdot 3} = \frac{55}{36} = 1\frac{19}{36}$$

You Try It 6

$$2\frac{5}{6} \div 8\frac{1}{2} = \frac{17}{6} \div \frac{17}{2}$$
$$= \frac{17}{6} \times \frac{2}{17} = \frac{17 \cdot 2}{6 \cdot 17}$$
$$= \frac{\overset{1}{\cancel{17}} \cdot \overset{1}{\cancel{2}}}{\underset{1}{\cancel{2}} \cdot 3 \cdot \underset{1}{\cancel{17}}} = \frac{1}{3}$$

You Try It 7

$$6\frac{2}{5} \div 4 = \frac{32}{5} \div \frac{4}{1}$$
$$= \frac{32}{5} \times \frac{1}{4} = \frac{32 \cdot 1}{5 \cdot 4}$$
$$= \frac{2 \cdot 2 \cdot 2 \cdot \overset{1}{\cancel{2}} \cdot \overset{1}{\cancel{2}}}{5 \cdot \underset{1}{\cancel{2}} \cdot \underset{1}{\cancel{2}}} = \frac{8}{5} = 1\frac{3}{5}$$

You Try It 8

Strategy
To find the price of one ounce of gold, divide the total price of the coin ($195) by the number of ounces $\left(\frac{1}{2}\right)$.

Solution

$$195 \div \frac{1}{2} = \frac{195}{1} \div \frac{1}{2}$$
$$= \frac{195}{1} \times \frac{2}{1} = \frac{195 \cdot 2}{1 \cdot 1} = 390$$

The price of one ounce of gold is $390.

You Try It 9

Strategy
To find the length of the remaining piece:

- Divide the total length of the board (16 feet) by the length of each shelf $\left(3\frac{1}{3}\text{ feet}\right)$.
- Multiply the fractional part of the result by the length of one shelf to determine the length of the remaining piece.

Solution

$$16 \div 3\frac{1}{3} = 16 \div \frac{10}{3}$$
$$= \frac{16}{1} \times \frac{3}{10} = \frac{16 \cdot 3}{1 \cdot 10}$$
$$= \frac{\cancel{2} \cdot 2 \cdot 2 \cdot 2 \cdot 3}{\cancel{2} \cdot 5} = \frac{24}{5} = 4\frac{4}{5}$$

$$\frac{4}{5} \times \frac{10}{3} = \frac{40}{15} = 2\frac{2}{3}$$

The length of the piece remaining is $2\frac{2}{3}$ feet.

SECTION 2.8 *pages 105–106*

You Try It 1 $\frac{9}{14} > \frac{13}{21}$

You Try It 2

$$\left(\frac{7}{11}\right)^2 \cdot \left(\frac{2}{7}\right) = \left(\frac{7}{11} \cdot \frac{7}{11}\right) \cdot \left(\frac{2}{7}\right) = \frac{\cancel{7} \cdot 7 \cdot 2}{11 \cdot 11 \cdot \cancel{7}} = \frac{14}{121}$$

You Try It 3

$$\left(\frac{1}{13}\right)^2 \cdot \left(\frac{1}{4} + \frac{1}{6}\right) \div \frac{5}{13}$$

$$\left(\frac{1}{13}\right)^2 \cdot \left(\frac{5}{12}\right) \div \frac{5}{13}$$

$$\left(\frac{1}{169}\right) \cdot \left(\frac{5}{12}\right) \div \frac{5}{13}$$

$$\left(\frac{1 \cdot 5}{13 \cdot 13 \cdot 12}\right) \div \frac{5}{13}$$

$$\frac{1 \cdot \overset{1}{\cancel{5}} \cdot \overset{1}{\cancel{13}}}{\underset{1}{\cancel{13}} \cdot 13 \cdot 12 \cdot \underset{1}{\cancel{5}}}$$

$$\frac{1}{156}$$

SOLUTIONS to Chapter 3 "You Try It"

SECTION 3.1 *pages 119–120*

You Try It 1 Two hundred nine and five thousand eight hundred thirty-eight hundred-thousandths

You Try It 2 42,000.000207

You Try It 3 4.35

You Try It 4 3.29053

SECTION 3.2 *pages 123–124*

You Try It 1

```
  1 2
   4.62
  27.9
+  0.62054
----------
  33.14054
```

You Try It 2

```
   1
   6.05
  12.
+  0.374
--------
  18.424
```

You Try It 3

Strategy
To find the total income, add the four commissions ($485.60, $599.46, $326.75, and $725.42) to the salary ($425.00).

Solution

```
   $485.60
    599.46
    326.75
    725.42
+   425.00
  --------
  $2562.23
```

Anita's total income was $2562.23.

SECTION 3.3 *pages 127–128*

You Try It 1

```
   11 9
  6 1 10 13                     1 1 1
  7 2.0 3 9     Check:           8.47
−     8.4 7                  + 63.569
 -----------                 --------
  6 3.5 6 9                    72.039
```

You Try It 2

```
   14 9
  2 4 10 10                    1 1 1
  3 5.0 0       Check:          9.67
−   9.6 7                    + 25.33
 ---------                   -------
  2 5.3 3                      35.00
```

You Try It 3

```
   16 9 9
  2 6 10 10 10                  1 111
  3.7 0 0 0     Check:         1.9715
− 1.9 7 1 5                  + 1.7285
 -----------                 --------
  1.7 2 8 5                    3.7000
```

You Try It 4

Strategy
To find the amount of change, subtract the amount paid ($3.85) from $5.00.

Solution

```
  $5.00
− 3.85
 ------
  $1.15
```

Your change was $1.15.

You Try It 5

Strategy
To find the new balance:

- Add to find the total of the three checks ($1025.60 + $79.85 + $162.47).
- Subtract the total from the previous balance ($2472.69).

Solution

```
  $1025.60              $2472.69
     79.85            − 1267.92
+   162.47            ---------
 ---------              $1204.77
  $1267.92
```

The new balance is $1204.77.

SECTION 3.4 *pages 131–134*

You Try It 1

$$\begin{array}{r} 870 \\ \times\ 4.6 \\ \hline 5220 \\ 3480 \\ \hline 4002.0 \end{array}$$

You Try It 2

$$\begin{array}{r} 0.000086 \\ \times\ \ 0.057 \\ \hline 602 \\ 430 \\ \hline 0.000004902 \end{array}$$

You Try It 3

$$\begin{array}{r} 4.68 \\ \times\ 6.03 \\ \hline 1404 \\ 28\ 080 \\ \hline 28.2204 \end{array}$$

You Try It 4 $6.9 \times 1000 = 6900$

You Try It 5 $4.0273 \times 10^2 = 402.73$

You Try It 6

Strategy
To find the cost of water:

- Find the number of gallons of water used by multiplying the number of gallons used per day (5000) by the number of days (62).
- Then multiply the cost per 1000 gallons (1.39) by the number of gallons used.
- Add the cost of the water to the meter fee ($133.70).

Solution
Number of gallons $= 5000(62) = 310{,}000$

Cost of water $= \frac{310{,}000}{1000} \times 1.39 = 430.90$

Total cost $= 133.70 + 430.90 = 564.60$

The total cost is $564.60.

You Try It 7

Strategy
To find the cost of running the freezer for 210 hours, multiply the hourly cost ($.035) by the number of hours (210).

Solution

$$\begin{array}{r} \$.035 \\ \times \ \ 210 \\ \hline \$7.35 \end{array}$$

The cost of running the freezer for 210 hours is $7.35.

You Try It 8

Strategy
To find the total cost of the stereo:

- Multiply the monthly payment ($37.18) by the number of months (18).
- Add the total to the down payment ($175.00).

Solution

$$\begin{array}{r} \$37.18 \\ \times \ \ \ 18 \\ \hline \$669.24 \end{array} \qquad \begin{array}{r} \$699.24 \\ +\ 175.00 \\ \hline \$844.24 \end{array}$$

The total cost of the stereo is $844.24.

SECTION 3.5 *pages 141–144*

You Try It 1

$$\begin{array}{r} 2.7 \\ 0.052.\overline{)0.140.4} \\ -104 \\ \hline 36\ 4 \\ -36\ 4 \\ \hline 0 \end{array}$$

You Try It 2

$$\begin{array}{r} 0.4873 \approx 0.487 \\ 76\overline{)37.0420} \\ -30\ 4 \\ \hline 6\ 64 \\ -6\ 08 \\ \hline 562 \\ -532 \\ \hline 300 \\ -228 \\ \hline \end{array}$$

You Try It 3

$$\begin{array}{r} 72.73 \approx 72.7 \\ 5.09.\overline{)370.20.00} \\ -356\ 3 \\ \hline 13\ 90 \\ 10\ 18 \\ \hline 3\ 720 \\ -3\ 563 \\ \hline 1570 \\ -1527 \\ \hline \end{array}$$

You Try It 4 $309.21 \div 10{,}000 = 0.030921$

You Try It 5 $42.93 \div 10^4 = 0.004293$

You Try It 6

Strategy
To find the amount he paid in gasoline taxes.

- Find the total number of gallons of gas he used by dividing his total miles (13,400) by the number of miles traveled per gallon of gasoline (26).
- Multiply the total tax ($.4013) by the total number of gallons of gasoline used.

Solution
$13{,}400 \div 26 \approx 515.385$

$0.4013 \times 515.385 \approx 206.842$

Jefferson paid approximately $207 for gasoline taxes.

You Try It 7

Strategy
To find the average number of people watching TV:

- Add the number of people watching each day of the week.
- Divide the total number of people watching by 7.

Solution
$91.9 + 89.8 + 90.6 + 93.9 + 78.0 + 77.1 + 87.7 = 609$

$\frac{609}{7} = 87$

An average of 87 million people watch television per day.

SECTION 3.6 *pages 149–150*

You Try It 1 $16\overline{)9.00}$ gives $0.56 \approx 0.6$

You Try It 2 $4\frac{1}{6} = \frac{25}{6}$ $6\overline{)25.00}$ gives $4.166 \approx 4.17$

You Try It 3 $0.56 = \frac{56}{100} = \frac{14}{25}$

$5.35 = 5\frac{35}{100} = 5\frac{7}{20}$

You Try It 4 $0.12\frac{7}{8} = \frac{12\frac{7}{8}}{100} = 12\frac{7}{8} \div 100$

$= \frac{103}{8} \times \frac{1}{100} = \frac{103}{800}$

You Try It 5 $0.63 > \frac{5}{8}$

SOLUTIONS to Chapter 4 "You Try It"

SECTION 4.1 *pages 163–164*

You Try It 1

$\frac{20 \text{ pounds}}{24 \text{ pounds}} = \frac{20}{24} = \frac{5}{6}$

20 pounds:24 pounds = 20:24 = 5:6

20 pounds to 24 pounds = 20 to 24
= 5 to 6

You Try It 2

$\frac{64 \text{ miles}}{8 \text{ miles}} = \frac{64}{8} = \frac{8}{1}$

64 miles:8 miles = 64:8 = 8:1

64 miles to 8 miles = 64 to 8 = 8 to 1

You Try It 3

Strategy
To find the ratio, write the ratio of board feet of cedar (12,000) to board feet of ash (18,000) in simplest form.

Solution
$\frac{12{,}000}{18{,}000} = \frac{2}{3}$

The ratio is $\frac{2}{3}$.

You Try It 4

Strategy
To find the ratio, write the ratio of the amount spent on radio advertising ($15,000) to the amount spent on radio and television advertising ($15,000 + $20,000).

Solution
$\frac{\$15{,}000}{\$15{,}000 + \$20{,}000} = \frac{\$15{,}000}{\$35{,}000} = \frac{3}{7}$

The ratio is $\frac{3}{7}$.

SECTION 4.2 *pages 167–168*

You Try It 1 $\frac{15 \text{ pounds}}{12 \text{ trees}} = \frac{5 \text{ pounds}}{4 \text{ trees}}$

You Try It 2 $\frac{260 \text{ miles}}{8 \text{ hours}}$

$8\overline{)260.0}$ = 32.5

32.5 miles/hour

You Try It 3

Strategy
To find Erik's profit per ounce:

- Find the total profit by subtracting the cost ($1625) from the selling price ($1720).
- Find the profit per ounce by dividing the total profit by the number of ounces (5).

Solution

$$\begin{array}{r} 1720 \\ -\ 1625 \\ \hline 95 \end{array}$$ total profit

$5\overline{)95}$ = 19

The profit per ounce is $19.

SECTION 4.3 *pages 171–174*

You Try It 1 $\frac{6}{10} \times \frac{9}{15}$ → $10 \times 9 = 90$; → $6 \times 15 = 90$

The proportion is true.

You Try It 2 $\frac{32}{6} \times \frac{90}{8}$ → $6 \times 90 = 540$; → $32 \times 8 = 256$

The proportion is not true.

You Try It 3

$n \times 7 = 14 \times 3$
$n \times 7 = 42$
$n = 42 \div 7$
$n = 6$

Check: $\frac{6}{14} \times \frac{3}{7}$ → $14 \times 3 = 42$; → $6 \times 7 = 42$

You Try It 4

$5 \times 20 = 8 \times n$
$100 = 8 \times n$
$100 \div 8 = n$
$12.5 = n$

You Try It 5

$15 \times n = 20 \times 12$
$15 \times n = 240$
$n = 240 \div 15$
$n = 16$

Check: $\frac{15}{20} \times \frac{12}{16}$ → $20 \times 12 = 240$; → $15 \times 16 = 240$

You Try It 6

$12 \times 4 = 7 \times n$
$48 = 7 \times n$
$48 \div 7 = n$
$6.86 \approx n$

You Try It 7

$n \times 1 = 12 \times 4$
$n \times 1 = 48$
$n = 48 \div 1$
$n = 48$

Check: $\frac{48}{12} \times \frac{4}{1}$ → $12 \times 4 = 48$; → $48 \times 1 = 48$

You Try It 8

$3 \times n = 12 \times 8$
$3 \times n = 96$
$n = 96 \div 3$
$n = 32$

Check: $\frac{3}{8} \times \frac{12}{32}$ → $8 \times 12 = 96$; → $3 \times 32 = 96$

You Try It 9

Strategy
To find the number of jars that can be packed in 15 boxes, write and solve a proportion using n to represent the number of jars.

Solution

$$\frac{24 \text{ jars}}{6 \text{ boxes}} = \frac{n \text{ jars}}{15 \text{ boxes}}$$

$$24 \times 15 = 6 \times n$$
$$360 = 6 \times n$$
$$360 \div 6 = n$$
$$60 = n$$

60 jars can be packed in 15 boxes.

You Try It 10

Strategy
To find the number of tablespoons of fertilizer needed, write and solve a proportion using n to represent the number of tablespoons of fertilizer.

Solution

$$\frac{3 \text{ tablespoons}}{4 \text{ gallons}} = \frac{n \text{ tablespoons}}{10 \text{ gallons}}$$

$$3 \times 10 = 4 \times n$$
$$30 = 4 \times n$$
$$30 \div 4 = n$$
$$7\frac{1}{2} = n$$

For 10 gallons of water, $7\frac{1}{2}$ tablesppons of fertilizer are required.

SOLUTIONS to Chapter 5 "You Try It"

SECTION 5.1 *pages 189–190*

You Try It 1 $125\% = 125 \times \frac{1}{100} = \frac{125}{100} = 1\frac{1}{4}$

$125\% = 125 \times 0.01 = 1.25$

You Try It 2 $33\frac{1}{3}\% = 33\frac{1}{3} \times \frac{1}{100}$

$= \frac{100}{3} \times \frac{1}{100}$

$= \frac{100}{300} = \frac{1}{3}$

You Try It 3 $0.25\% = 0.25 \times 0.01 = 0.0025$

You Try It 4 $0.048 = 0.048 \times 100\% = 4.8\%$

You Try It 5 $3.67 = 3.67 \times 100\% = 367\%$

You Try It 6 $0.62\frac{1}{2} = 0.62\frac{1}{2} \times 100\%$

$= 62\frac{1}{2}\%$

You Try It 7 $\frac{5}{6} = \frac{5}{6} \times 100\% = \frac{500\%}{6} = 83\frac{1}{3}\%$

You Try It 8 $1\frac{4}{9} = \frac{13}{9} = \frac{13}{9} \times 100\%$

$= \frac{1300\%}{9} \approx 144.4\%$

SECTION 5.2 *pages 193–194*

You Try It 1 $n = 0.063 \times 150$

$n = 9.45$

You Try It 2 $n = \frac{1}{6} \times 66$

$n = 11$

You Try It 3

Strategy

To find the new hourly wage:

- Find the amount of the raise. Write and solve a basic percent equation using n to represent the amount of the raise (amount). The percent is 8%. The base is \$13.50.
- Add the amount of the raise to the old wage.

Solution

$8\% \times \$13.50 = n$

$0.08 \times \$13.50 = n$

$\$1.08 = n$

$$\begin{array}{r} \$13.50 \\ +\ \ 1.08 \\ \hline \$14.58 \end{array}$$

The new hourly wage is \$14.58.

SECTION 5.3 *pages 197–198*

You Try It 1 $n \times 32 = 16$

$n = 16 \div 32$

$n = 0.50$

$n = 50\%$

You Try It 2 $n \times 15 = 48$

$n = 48 \div 15$

$n = 3.20$

$n = 320\%$

You Try It 3 $30 = n \times 45$

$30 \div 45 = n$

$0.66\frac{2}{3} = n$

$66\frac{2}{3}\% = n$

You Try It 4

Strategy
To find what percent of the income the income tax is, write and solve a basic percent equation using n to represent the percent. The base is \$20,000 and the amount is \$3000.

Solution

$$n \times \$20{,}000 = \$3000$$
$$n = \$3000 \div \$20{,}000$$
$$n = 0.15 = 15\%$$

The income tax is 15% of the income.

You Try It 5

Strategy
To find the percent that did not favor Minna Oliva:

- Subtract to find the number of people that did not favor her (1000 − 667).
- Write and solve a basic percent equation using n to represent the percent. The base is 1000 and the amount is the number of people who did not favor her.

Solution

$1000 - 667 = 333$ number who did not favor the candidate

$$n \times 1000 = 333$$
$$n = 333 \div 1000$$
$$n = 0.333 = 33.3\%$$

33.3% of the people did not favor Minna Oliva.

SECTION 5.4 *pages 201–202*

You Try It 1

$$0.86 \times n = 215$$
$$n = 215 \div 0.86$$
$$n = 250$$

You Try It 2

$$0.025 \times n = 15$$
$$n = 15 \div 0.025$$
$$n = 600$$

You Try It 3

$$\frac{1}{6} \times n = 5$$
$$n = 5 \div \frac{1}{6}$$
$$n = 30$$

You Try It 4

Strategy
To find the original value of the car, write and solve a basic percent equation using n to represent the original value (base). The percent is 51%. The amount is \$3876.

Solution

$$
\begin{aligned}
51\% \times n &= \$3876 \\
0.51 \times n &= \$3876 \\
n &= \$3876 \div 0.51 \\
n &= \$7600
\end{aligned}
$$

The original value of the car was \$7600.

You Try It 5

Strategy
To find the difference between the original price and the selling price:

- Find the original price. Write and solve a basic percent equation using n to represent the original price (base). The percent is 80%. The amount is \$44.80.
- Subtract the sale price (\$44.80) from the original price.

Solution

$$
\begin{aligned}
80\% \times n &= \$44.80 \\
0.80 \times n &= \$44.80 \\
n &= \$44.80 \div 0.80 \\
n &= \$56.00 \text{ original price}
\end{aligned}
$$

$\$56.00 - \$44.80 = \$11.20$

The difference between the original price and the sale price is \$11.20.

SECTION 5.5 *pages 205–206*

You Try It 1

$$
\begin{aligned}
\frac{26}{100} &= \frac{22}{n} \\
26 \times n &= 100 \times 22 \\
26 \times n &= 2200 \\
n &= 2200 \div 26 \\
n &\approx 84.62
\end{aligned}
$$

You Try It 2

$$
\begin{aligned}
\frac{16}{100} &= \frac{n}{132} \\
16 \times 132 &= 100 \times n \\
2112 &= 100 \times n \\
2112 \div 100 &= n \\
21.12 &= n
\end{aligned}
$$

You Try It 3

Strategy
To find the number of days it snowed, write and solve a proportion using n to represent the number of days (amount). The percent is 64%. The base is 150.

Solution

$$\frac{64}{100} = \frac{n}{150}$$

$64 \times 150 = 100 \times n$
$9600 = 100 \times n$
$9600 \div 100 = n$
$96 = n$

It snowed 96 days.

You Try It 4

Strategy
To find the percent of pens which were not defective:

- Subtract to find the number of pens which were not defective (200 − 5).
- Write and solve a proportion using n to represent the percent of pens which were not defective. The base is 200 and the amount is the number of pens not defective.

Solution
$200 - 5 = 195$ number of pens not defective

$$\frac{n}{100} = \frac{195}{200}$$

$200 \times n = 195 \times 100$
$200 \times n = 19500$
$n = 19500 \div 200$
$n = 97.5\%$

97.5% of the pens were not defective.

SOLUTIONS to Chapter 6 "You Try It"

SECTION 6.1 *pages 219–220*

You Try It 1

a. $6.25 \div 5 = 1.25$
Unit cost: \$1.25 per quart

b. $85 \div 4 = 21.25$
Unit cost: 21.3¢ per ear

You Try It 2

Strategy
To find the more economical purchase, compare unit costs.

Solution
$2.52 \div 6 = 0.42$
$1.66 \div 4 = 0.415$
$\$.415 < \$.42$

The more economical purchase is 4 cans for \$1.66.

You Try It 3

Strategy
To find the total cost, multiply the unit cost (\$4.96) by the number of units (7).

Solution
$4.96 \times 7 = 34.72$

The total cost is \$34.72.

SECTION 6.2 *pages 223–228*

You Try It 1

Strategy
To find the percent increase:
- Find the amount of increase.
- Solve the basic percent equation for *percent*.

Solution

$$\begin{array}{r} 2250 \\ -\ 2000 \\ \hline 250 \end{array} \qquad \begin{aligned} n \times 2000 &= 250 \\ n &= 250 \div 2000 \\ n &= 0.125 = 12.5\% \end{aligned}$$

The percent increase was 12.5%.

You Try It 2

Strategy
To find the new hourly wage:
- Solve the basic percent equation for *amount*.
- Add the amount of increase to the original wage.

Solution

$0.14 \times 6.50 = n$
$0.91 = n$
$6.50 + 0.91 = 7.41$

The new hourly wage is \$7.41.

You Try It 3

Strategy
To find the markup, solve the basic percent equation for *amount*.

Solution

$0.20 \times 8 = n$
$1.60 = n$

The markup is \$1.60.

You Try It 4

Strategy
To find the selling price:
- Find the markup by solving the basic percent equation for *amount*.
- Add the markup to the cost.

Solution

$0.55 \times 72 = n$
$39.60 = n$
$72 + 39.60 = 111.60$

The selling price is \$111.60.

You Try It 5

Strategy
To find the percent decrease:
- Find the amount of decrease.
- Solve the basic percent equation for *percent*.

Solution

$150 - 120 = 30$
$n \times 150 = 30$
$n = 30 \div 150$
$n = 0.2 = 20\%$

The percent decrease is 20%.

You Try It 6

Strategy
To find the visibility:
- Find the amount of decrease by solving the basic percent equation for *amount*.
- Subtract the amount of decrease from the original visibility.

Solution

$0.40 \times 5 = n$
$2 = n$
$5 - 2 = 3$

The visibility was 3 miles.

You Try It 7

Strategy
To find the discount rate, solve the basic percent equation for *percent.*

Solution
$n \times 125 = 25$
$n = 0.20 = 20\%$

The discount rate is 20%.

You Try It 8

Strategy
To find the sale price:

- Find the discount by solving the basic percent equation for *amount.*
- Subtract to find the sale price.

Solution
$0.20 \times 10.25 = n$
$2.05 = n$
$10.25 - 2.05 = 8.20$

The sale price is $8.20.

SECTION 6.3 *pages 233–234*

You Try It 1

Strategy
To find the simple interest, multiply the principal by the annual interest rate by the time (in years).

Solution
$15{,}000 \times 0.08 \times 1.5 = 1800$

The interest due is $1800.

You Try It 2

Strategy
To find the interest due, multiply the principal by the monthly interest rate by the time (in months).

Solution
$400 \times 0.012 \times 2 = 9.60$

The interest charge is $9.60.

You Try It 3

Strategy
To find the interest earned:

- Find the new principal by multiplying the original principal by the factor found in the compound interest table.
- Subtract the original principal from the new principal.

Solution
$1000 \times 3.29066 = 3290.66$

The new principal is $3290.66.

$3290.66 - 1000 = 2290.66$

The interest earned is $2290.66.

SECTION 6.4 *pages 237–240*

You Try It 1

Strategy
To find the mortgage:

- Find the down payment by solving the basic percent equation for *amount*.
- Subtract the down payment from the purchase price.

Solution
$0.25 \times 216{,}000 = n$
$54{,}000 = n$

The down payment is $54,000.

$216{,}000 - 54{,}000 = 162{,}000$

The mortgage is $162,000.

You Try It 2

Strategy
To find the loan origination fee, solve the basic percent equation for *amount*.

Solution
$0.045 \times 80{,}000 = n$
$3600 = n$

The loan origination fee was $3600.

You Try It 3

Strategy
To find the monthly mortgage payment:

- Subtract the down payment from the purchase price to find the mortgage.
- Multiply the mortgage by the factor found in the monthly payment table.

Solution
$75{,}000 - 15{,}000 = 60{,}000$

The mortgage is $60,000.

$60{,}000 \times 0.0089973 = 539.838$

The monthly mortgage payment is $539.84.

You Try It 4

Strategy
To find the interest:

- Multipy the mortgage by the factor found in the monthly payment table to find the monthly mortgage payment.
- Subtract the principal from the monthly mortgage payment.

Solution
$125{,}000 \times 0.0083920 = 1049$

The monthly mortgage payment is $1049.

$1049 - 492.65 = 556.35$

The interest on the mortgage is $556.35.

You Try It 5

Strategy
To find the monthly payment:

- Divide the annual property tax by 12 to find the monthly property tax.
- Add the monthly property tax to the monthly mortgage payment.

Solution
$744 \div 12 = 62$

The monthly property tax is $62.

$415.20 + 62 = 477.20$

The total monthly payment is $477.20.

SECTION 6.5 *pages 243–244*

You Try It 1

Strategy
To find the amount financed:

- Find the down payment by solving the basic percent equation for *amount*.
- Subtract the down payment from the purchase price.

Solution

$$0.20 \times 9200 = n$$
$$1840 = n$$

The down payment is \$1840.

$9200 - 1840 = 7360$

The amount financed is \$7360.

You Try It 2

Strategy
To find the license fee, solve the basic percent equation for *amount*.

Solution

$$0.015 \times 7350 = n$$
$$110.25 = n$$

The license fee is \$110.25.

You Try It 3

Strategy
To find the cost of operating the car, multiply the cost per mile by the number of miles driven.

Solution
$23{,}000 \times 0.22 = 5060$

The cost of operating the car is \$5060.

You Try It 4

Strategy
To find the cost per mile for the car insurance, divide the cost for insurance by the number of miles driven.

Solution
$360 \div 15{,}000 = 0.024$

The cost per mile for insurance is \$0.024.

You Try It 5

Strategy
To find the monthly payment:

- Subtract the down payment from the purchase price to find the amount financed.
- Multiply the amount financed by the factor found in the monthly payment table.

Solution
$15{,}900 - 3975 = 11{,}925$

The amount financed is \$11,925.

$11{,}925 \times 0.0244129 = 291.123$

The monthly payment is \$291.12.

SECTION 6.6 *pages 247–248*

You Try It 1

Strategy
To find the worker's earnings:

- Find the worker's overtime wage by multiplying the hourly wage by 2.
- Multiply the number of hours worked by the overtime hours.

Solution
$8.50 \times 2 = 17$

The hourly wage for overtime is $17.

$17 \times 8 = 136$

The construction worker earns $136.

You Try It 2

Strategy
To find the salary per month, divide the annual salary by the number of months in a year (12).

Solution
$28{,}224 \div 12 = 2352$

The contractor's monthly salary is $2352.

You Try It 3

Strategy
To find the total earnings:

- Find the commission by multiplying the commission rate by the sales over $50,000.
- Add the commission to the annual salary.

Solution
$175{,}000 - 50{,}000 = 125{,}000$

Sales over $50,000 totaled $125,000.

$125{,}000 \times 0.095 = 11{,}875$

Earnings from commissions totaled $11,875.

$12{,}000 + 11{,}875 = 23{,}875$

The insurance agent earned $23,875.

SECTION 6.7 *pages 251–256*

You Try It 1

Strategy
To find the current balance:

- Subtract the amount of the check from the old balance.
- Add the amount of each deposit.

Solution

302.46	
− 20.59	check
281.87	
176.86	first deposit
+ 94.73	second deposit
553.46	

The current checking account balance is \$553.46.

You Try It 2

Current checkbook balance:	653.41
Check: 237	+ 48.73
	702.14
Interest:	+ 2.11
	704.25
Deposit:	− 523.84
	180.41

Closing bank balance from bank statement: \$180.41.

Checkbook balance: \$180.41.

The bank statement and checkbook balance.

SOLUTIONS to Chapter 7 "You Try It"

SECTION 7.1 *pages 271–274*

You Try It 1

Strategy
To find what percent (N) of the total number of housing starts (30) the number of July housing starts (12) is, use the basic percent equation to solve for N.

Solution

$$\textit{Percent} \times \textit{base} = \textit{amount}$$
$$n \times 30 = 12$$
$$n = \frac{12}{30} = \frac{4}{10}$$
$$n = 40\%$$

The number of July housing starts is 40% of the total number of housing starts.

You Try It 2

Strategy
To find the ratio:

- Locate the annual cost of insurance and the annual cost of maintenance in the circle graph.
- Write the ratio of insurance cost to maintenance cost in simplest form.

Solution
Annual insurance cost: \$400
Annual maintenance cost: \$500

$$\frac{\$400}{\$500} = \frac{4}{5}$$

The ratio is $\frac{4}{5}$.

You Try It 3

Strategy
To find the Federal income tax:

- Locate the percent of the distribution which is Federal income tax in the circle graph.
- Solve the basic percent for amount.

Solution
Federal income tax: 15%

$$\text{Percent} \times \text{base} = \text{amount}$$
$$0.15 \times 2000 = 300$$

The Federal income tax is \$300.

SECTION 7.2 *pages 279–280*

You Try It 1

Strategy
To find the increase in sales:

- Subtract the sales for 1993 from the sales in 1994.
- Compare the increases for the four companies.

Solution
GM: $350{,}000 - 300{,}000 = 50{,}000$
Ford: $260{,}000 - 230{,}000 = 30{,}000$
Chrysler: $160{,}000 - 140{,}000 = 20{,}000$
Toyota: $70{,}000 - 60{,}000 = 10{,}000$

GM had the largest increase in sales from January 1993 to January 1994.

You Try It 2

Strategy
To find which years had the largest net income increase:

- Read the line graphs to find the net income for each year.
- Subtract to find the difference between each year.

Solution

1990	\$1 million
1991	\$3 million
1992	\$5 million
1993	\$8 million
1994	\$11 million

between 1990 and 1991: $3 - 1 = 2$
between 1991 and 1992: $5 - 3 = 2$
between 1992 and 1993: $8 - 4.5 = 3.5$
between 1993 and 1994: $11 - 8 = 3$

The net income of Math Associates increased the most between the years 1992 and 1993.

SECTION 7.3 *pages 283–284*

You Try It 1

Strategy
To find the number of employees:

- Read the histogram to find the number of employees whose hourly wage is between \$8 and \$10 and the number whose hourly wage is between \$10 and \$12.
- Add to find the number of employees whose wage is between \$8 and \$12.

Solution
Number whose wage is between
\$8 and \$10: 15
\$10 and \$12: 17

151 + 7 = 32

32 employees earn between \$8 and \$12.

You Try It 2

Strategy
To find the number of people who scored between 50 and 70 on the exam:

- Read the frequency polygon to find the number of people who scored between 50 and 60 and the number who scored between 60 and 70.
- Add to find the number of people who scored between 50 and 70.

Solution
The number who scored between
50 and 60: 5
60 and 70: 6

$5 + 6 = 11$

11 people scored between 50 and 70.

SECTION 7.4 *pages 287–288*

You Try It 1

Strategy
To find the monthly mean cost for gasoline:

- Find the sum of the costs.
- Divide the sum by the number of months (6).

Solution

$$\begin{array}{r} \$118 \\ 130 \\ 109 \\ 141 \\ 134 \\ +\ 136 \\ \hline \$768 \end{array} \qquad \begin{array}{r} \$128 \\ 6\overline{)\$768} \end{array}$$

The monthly mean cost for gasoline is \$128.

You Try It 2

Strategy
To find the median number of books borrowed, arrange the number of books from smallest to largest. The median is the average of the two middle numbers.

Solution

$$\left.\begin{array}{r} 375 \\ 420 \end{array}\right\} \text{2 numbers}$$
$$\left.\begin{array}{r} 450 \\ 480 \end{array}\right\} \text{middle numbers}$$
$$\left.\begin{array}{r} 490 \\ 500 \end{array}\right\} \text{2 numbers}$$

$$\frac{450 + 480}{2} = 465$$

The median number of books borrowed was 465.

SOLUTIONS to Chapter 8 "You Try It"

SECTION 8.1 *pages 301–304*

You Try It 1 $60 \text{ in.} = 60\,\cancel{\text{in.}} \times \dfrac{1 \text{ ft}}{12\,\cancel{\text{in.}}} = 5 \text{ ft}$

You Try It 2 $14 \text{ ft} = 14\,\cancel{\text{ft}} \times \dfrac{1 \text{ yd}}{3\,\cancel{\text{ft}}} = 4\frac{2}{3} \text{ yd}$

You Try It 3 $9800 \text{ ft} = 9800\,\cancel{\text{ft}} \times \dfrac{1 \text{ mi}}{5280\,\cancel{\text{ft}}} = 1\frac{113}{132} \text{ mi}$

You Try It 4

$$\begin{array}{r} 3 \text{ ft } 6 \text{ in.} \\ 12\overline{)\ 42} \\ \underline{-36} \\ 6 \end{array}$$

42 in. = 3 ft 6 in.

You Try It 5

$$\begin{array}{r} 4 \text{ yd } 2 \text{ ft} \\ 3\overline{)\ 14} \\ \underline{-12} \\ 2 \end{array}$$

14 ft = 4 yd 2 ft

You Try It 6

$$\begin{array}{r} 3 \text{ ft} \quad 5 \text{ in.} \\ \underline{+\ 4 \text{ ft} \quad 9 \text{ in.}} \\ 7 \text{ ft} \ 14 \text{ in.} \end{array} = 8 \text{ ft } 2 \text{ in.}$$

You Try It 7

$$\begin{array}{r} \overset{3 \text{ ft}}{\cancel{4 \text{ ft}}} \quad \overset{14 \text{ in.}}{\cancel{2 \text{ in.}}} \\ \underline{-\ 1 \text{ ft} \quad 8 \text{ in.}} \\ 2 \text{ ft} \quad 6 \text{ in.} \end{array}$$

You Try It 8

$$\begin{array}{r} 4 \text{ yd } 1 \text{ ft} \\ \underline{\times \qquad 8} \\ 32 \text{ yd } 8 \text{ ft} \end{array} = 34 \text{ yd } 2 \text{ ft}$$

You Try It 9

$$\begin{array}{r} 3 \text{ yd} \quad 2 \text{ ft} \\ 2\overline{)\ 7 \text{ yd} \quad 1 \text{ ft}} \\ \underline{-6 \text{ yd}} \qquad\quad \\ 1 \text{ yd} = \underline{3 \text{ ft}} \\ 4 \text{ ft} \\ \underline{-4 \text{ ft}} \\ 0 \end{array}$$

You Try It 10

$$\begin{array}{r} 6\frac{1}{4} \text{ ft} = 6\frac{3}{12} \text{ ft} = 5\frac{15}{12} \text{ ft} \\ \underline{-3\frac{2}{3} \text{ ft} = 3\frac{8}{12} \text{ ft} = 3\frac{8}{12} \text{ ft}} \\ 2\frac{7}{12} \text{ ft} \end{array}$$

You Try It 11

Strategy
To find the width of the storage room:

- Multiply the number of tiles (8) by the width of each tile (9 in.)
- Divide the result by the number of inches in one foot (12) to find the width in feet.

Solution
9 in. × 8 = 72 in.

72 ÷ 12 = 6

The width is 6 ft.

You Try It 12

Strategy
To find the length of each piece, divide the total length (9 ft 8 in.) by the number of pieces (4).

Solution

$$\begin{array}{r} 2 \text{ ft} \quad 5 \text{ in.} \\ 4\overline{)\ 9 \text{ ft} \quad 8 \text{ in.}} \\ \underline{-8 \text{ ft}} \qquad\quad \\ 1 \text{ ft} = \underline{12 \text{ in.}} \\ 20 \text{ in.} \\ \underline{-20 \text{ in.}} \\ 0 \end{array}$$

Each piece is 2 ft 5 in. long.

SECTION 8.2 *pages 307–308*

You Try It 1 $3\text{ lb} = 3\not{\text{lb}} \times \frac{16\text{ oz}}{1\not{\text{lb}}} = 48\text{ oz}$

You Try It 2 $4200\text{ lb} = 4200\not{\text{lb}} \times \frac{1\text{ ton}}{2000\not{\text{lb}}}$

$= 2\frac{1}{10}\text{ tons}$

You Try It 3

$$\begin{array}{rr} & \overset{6\text{ lb}}{\cancel{7\text{ lb}}} \quad \overset{17\text{ oz}}{\cancel{1\text{ oz}}} \\ - & 3\text{ lb} \quad 4\text{ oz} \\ \hline & 3\text{ lb} \quad 13\text{ oz} \end{array}$$

You Try It 4

$$\begin{array}{lr} & 3\text{ lb } 6\text{ oz} \\ \times & 4 \\ \hline \end{array}$$

$12\text{ lb } 24\text{ oz} = 13\text{ lb } 8\text{ oz}$

You Try It 5

Strategy

To find the weight of 12 bars of soap:

- Multiply the number of bars (12) by the weight of each bar (9 oz).
- Convert the number of ounces to pounds.

Solution

$$\begin{array}{rr} & 12 \\ \times & 9\text{ oz} \\ \hline & 108\text{ oz} \end{array}$$

$108\not{\text{oz}} \times \frac{1\text{ lb}}{16\not{\text{oz}}} = 6\frac{3}{4}\text{ lb}$

The 12 bars of soap weigh $6\frac{3}{4}$ lb.

SECTION 8.3 *pages 311–312*

You Try It 1 $18\text{ pt} = 18\not{\text{pt}} \times \frac{1\not{\text{qt}}}{2\not{\text{pt}}} \times \frac{1\text{ gal}}{4\not{\text{qt}}}$

$= \frac{9\text{ gal}}{4} = 2\frac{1}{4}\text{ gal}$

You Try It 2

$$\begin{array}{r} 1\text{ gal} \quad 2\text{ qt} \\ 3\overline{)\,4\text{ gal} \quad 2\text{ qt}} \\ \underline{-3\text{ gal}} \qquad\quad \\ 1\text{ gal} = \underline{4\text{ qt}} \\ 6\text{ qt} \\ \underline{-6\text{ qt}} \\ 0 \end{array}$$

You Try It 3

Strategy
To find the number of gallons of water needed:

- Find the number of quarts required by multiplying the number of quarts one students needs (1) by the number of students (5) by the number of days (3).
- Convert the number of quarts to gallons.

Solution

$$\begin{array}{r} 5 \\ \times\ 1\ \text{qt} \\ \hline 5\ \text{qt} \end{array} \qquad \begin{array}{r} 5\ \text{qt} \\ \times\ 3 \\ \hline 15\ \text{qt} \end{array}$$

$$15\ \cancel{\text{qt}} \cdot \frac{1\ \text{gal}}{4\ \cancel{\text{qt}}} = 3\frac{3}{4}\ \text{gal}$$

The students should take $3\frac{3}{4}$ gal of water.

SECTION 8.4 *pages 315–317*

You Try It 1 $4.5\ \text{BTU} = 4.5\ \cancel{\text{BTU}} \times \frac{778\ \text{ft} \cdot \text{lb}}{1\ \cancel{\text{BTU}}}$
$= 3501\ \text{ft} \cdot \text{lb}$

You Try It 2 $800\ \text{lb} \times 16\ \text{ft} = 12{,}800\ \text{ft} \cdot \text{lb}$

You Try It 3 $56{,}000\ \text{BTU} =$
$56{,}000\ \cancel{\text{BTU}} \times \frac{778\ \text{ft} \cdot \text{lb}}{1\ \cancel{\text{BTU}}} =$
$43{,}568{,}000\ \text{ft} \cdot \text{lb}$

You Try It 4 $\text{Power} = \frac{90\ \text{ft} \times 1200\ \text{lb}}{24\ \text{s}}$
$= 4500\ \frac{\text{ft} \cdot \text{lb}}{\text{s}}$

You Try It 5 $\frac{3300}{550} = 6\ \text{hp}$

SOLUTIONS to Chapter 9 "You Try It"

SECTION 9.1 *pages 229–230*

You Try It 1 $3.07\ \text{m} = 307\ \text{cm}$

You Try It 2 $7\ \text{cm} = 0.07\ \text{m}$
$3\ \text{m}\ 7\ \text{cm} = 3\ \text{m} + 0.07\ \text{m}$
$= 3.07\ \text{m}$

You Try It 3

$$\begin{array}{r} 3\ \text{m} = 3.00\ \text{m} \\ -\ 42\ \text{cm} = 0.42\ \text{m} \\ \hline 2.58\ \text{m} \end{array}$$

You Try It 4

Strategy
To find the cost of the shelves:

- Multiply the length of the bookcase (1 m 75 cm) by the number of shelves (4).
- Multiply the product by the cost per meter.

Solution

$$\begin{array}{r} 1\ \text{m}\ 75\ \text{cm} = 1.75\ \text{m} \\ \times\quad 4 \quad = \quad 4 \\ \hline 7.00\ \text{m} \end{array}$$

$$\begin{array}{r} \$11.75 \\ \times\quad 7 \\ \hline \$82.25 \end{array}$$

The cost is $82.25.

SECTION 9.2 *pages 333–334*

You Try It 1 42.3 mg = 0.0423 g

You Try It 2

$$\begin{aligned} 3\text{g} &= 3000\ \text{mg} \\ 3\ \text{g}\ 54\ \text{mg} &= 3000\ \text{mg} + 54\ \text{mg} \\ &= 3054\ \text{mg} \end{aligned}$$

You Try It 3

$$\begin{array}{r} 4\ \text{g}\ 620\ \text{mg} = 4.620\ \text{g} \\ \times\quad 8 \quad = \quad 8 \\ \hline 36.960\ \text{g} \end{array}$$

You Try It 4

Strategy
To find how much fertilizer is required:

- Convert 300 g to kilograms.
- Multiply the number of kilograms by the number of trees (400).

Solution
300 g = 0.3 kg

$$\begin{array}{r} 400 \\ \times\ 0.3\ \text{kg} \\ \hline 120.0\ \text{kg} \end{array}$$

To fertilize the trees, 120 kg of fertilizer are required.

SECTION 9.3 *pages 337–338*

You Try It 1

$$2 \text{ kl} = 2000 \text{ L}$$
$$2 \text{ kl} + 167 \text{ L} = 2000 \text{ L} + 167 \text{ L} = 2167 \text{ L}$$

You Try It 2 $325 \text{ cm}^3 = 325 \text{ ml} = 0.325 \text{ L}$

You Try It 3

$$12\overline{)22.992 \text{ kl}} = 1.916 \text{ kl}$$

You Try It 4

Strategy
To find the profit:
- Find the number of containers by dividing the total amount of lotion (5 L) by the amount per container (125 ml).
- Multiply the number of containers by the price per container ($2.79).
- Subtract the cost ($42.50) from the total income.

Solution

$$\frac{5 \text{ L}}{125 \text{ ml}} = \frac{5 \text{ L}}{0.125 \text{ L}} = 40$$

$40 \times \$2.79 = \$111.60$

$\$111.60 - \$42.50 = \$69.10$

The profit is $69.10.

SECTION 9.4 *pages 341–342*

You Try It 1

Strategy
To find the number of calories:
- Multiply to find the number of hours worked in 5 days.
- Multiply the product by the number of calories burned off per hour.

Solution

$1\frac{1}{2} \times 5 = \frac{3}{2} \times 5 = \frac{15}{2} = 7\frac{1}{2}$

$7\frac{1}{2} \times 240 = 1800 \text{ cal}$

1800 cal are used.

You Try It 2

Strategy
To find the number of kilowatt-hours:
- Multiply to find the number of watthours used.
- Convert to kilowatt-hours.

Solution

$150 \text{ W} \times 200 \text{ h} = 30{,}000 \text{ Wh}$

$30{,}000 \text{ Wh} = 30 \text{ kWh}$

30 kWh of energy are used.

You Try It 3

Strategy
To find the cost:

- Convert 20 min to hours.
- Multiply to find the total number of hours the oven is used.
- Multiply the number of hours used by the number of watts to find the watt-hours.
- Convert to kilowatt-hours.
- Multiply the number of kilowatt-hours by the cost per kilowatt-hour.

Solution

$$20 \text{ min} = 20 \cancel{\text{min}} \times \frac{1 \text{ h}}{60 \cancel{\text{min}}} = \frac{20}{60} \text{ h} = \frac{1}{3} \text{ h}$$

$\frac{1}{3} \text{ h} \times 30 = 10 \text{ h}$

$10 \text{ h} \times 500 \text{ W} = 5000 \text{ Wh}$

$5000 \text{ Wh} = 5 \text{ kWh}$

$5 \times 8.7¢ = 43.5¢$

The cost is 43.5¢.

SECTION 9.5 *pages 345–346*

You Try It 1

$$10 \text{ c} \approx 10 \cancel{\text{c}} \times \frac{1 \cancel{\text{qt}}}{4 \cancel{\text{c}}} \times \frac{1 \text{ L}}{1.06 \cancel{\text{qt}}} = \frac{10 \text{ L}}{4.24} = 2.36 \text{ L}$$

$10 \text{ c} \approx 2.36 \text{ L}$

You Try It 2

$$\frac{60 \text{ ft}}{\text{s}} \approx \frac{60 \cancel{\text{ft}}}{\text{s}} \times \frac{1 \text{ m}}{3.28 \cancel{\text{ft}}} = \frac{60 \text{ m}}{3.28 \text{ s}} = 18.29 \text{ m/s}$$

$60 \text{ ft/s} \approx 18.29 \text{ m/s}$

You Try It 3

$$\frac{\$1.86}{\text{gal}} \approx \frac{\$1.86}{\cancel{\text{gal}}} \times \frac{\cancel{\text{gal}}}{4 \cancel{\text{qt}}} \times \frac{1.06 \cancel{\text{qt}}}{1 \text{ L}} = \frac{\$1.9716}{4 \text{ L}} \approx \frac{\$0.49}{\text{L}}$$

$\$1.86/\text{gal} \approx \$0.49/\text{L}$

You Try It 4

$$45 \text{ cm} \approx 45 \cancel{\text{cm}} \times \frac{0.39 \text{ in.}}{1 \cancel{\text{cm}}} = 17.55 \text{ in.}$$

$45 \text{ cm} \approx 17.55 \text{ in.}$

You Try It 5

$$\frac{75 \text{ km}}{\text{h}} \approx \frac{75 \cancel{\text{km}}}{\text{h}} \times \frac{1 \text{ mi}}{1.61 \cancel{\text{km}}} = 46.58 \text{ mi/h}$$

$75 \text{ km/h} \approx 46.58 \text{ mi/h}$

You Try It 6

$$\frac{\$1.50}{\text{L}} \approx \frac{\$1.50}{\cancel{\text{L}}} \times \frac{1 \cancel{\text{L}}}{1.06 \cancel{\text{qt}}} \times \frac{4 \cancel{\text{qt}}}{1 \text{ gal}} = \frac{\$6}{1.06 \text{ gal}} = \$5.66/\text{gal}$$

$\$1.50/\text{L} \approx \$5.66/\text{gal}$

SOLUTIONS to Chapter 10 "You Try It"

SECTION 10.1 *pages 359–360*

You Try It 1 -232 ft

You Try It 2

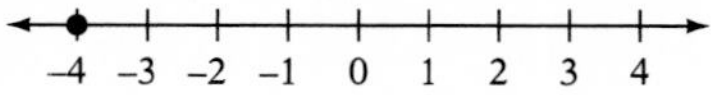

You Try It 3
a. $-12 < -8$
b. $-5 < 0$

You Try It 4
$|-7| = 7$
$|21| = 21$

You Try It 5
$|2| = 2$
$|-9| = 9$

You Try It 6 $-|-12| = -12$

SECTION 10.2 *pages 363–366*

You Try It 1
$-154 + (-37)$
-191

You Try It 2
$-5 + (-2) + 9 + (-3)$
$-7 + 9 + (-3)$
$2 + (-3)$
-1

You Try It 3
$-8 - 14$
$-8 + (-14)$
-22

You Try It 4
$3 - (-4) - 15$
$3 + 4 + (-15)$
$7 + (-15)$
-8

You Try It 5
$4 - (-3) - 12 - (-7) - 20$
$4 + 3 + (-12) + 7 + (-20)$
$7 + (-12) + 7 + (-20)$
$-5 + 7 + (-20)$
$2 + (-20)$
-18

You Try It 6

Strategy
To find the temperature, add the increase (12) to the previous temperature (-10).

Solution
$-10 + 12 = 2$

After an increase of 12°C, the temperature is 2°C.

SECTION 10.3 *pages 371–374*

You Try It 1
$(-3) \cdot 4 \cdot (-5)$
$(-12) \cdot (-5)$
60

You Try It 2
$-38 \cdot 51$
-1938

You Try It 3
$-6 \cdot 8 \cdot (-11) \cdot 3$
$-48 \cdot (-11) \cdot 3$
$528 \cdot 3$
1584

You Try It 4
$-7(-8)(9)(-2)$
$56(9)(-2)$
$504(-2)$
-1008

You Try It 5 $(-135) \div (-9)$
15

You Try It 6 $-72 \div 4$
-18

You Try It 7 $84 \div (-6)$
-14

You Try It 8

Strategy
To find the melting point of argon, multiply the melting point of mercury $(-38°)$ by 5.

Solution
$5(-38) = -190$

The melting point of argon is $-190°$C.

You Try It 9

Strategy
To find the average daily low temperature:

- Add the seven temperature readings.
- Divide by 7.

Solution
$-6 + (-7) + 1 + 0 + (-5) + (-10) + (-1)$
$-13 + 1 + 0 + (-5) + (-10) + (-1)$
$-12 + 0 + (-5) + (-10) + (-1)$
$-12 + (-5) + (-10) + (-1)$
$-17 + (-10) + (-1)$
$-27 + (-1)$
-28

$-28 \div 7 = -4$

The average daily low temperature was $-4°$.

SECTION 10.4 *pages 379–382*

You Try It 1 The LCM of 9 and 12 is 36.

$$\frac{5}{9} - \frac{11}{12} = \frac{20}{36} - \frac{33}{36} = \frac{20}{36} + \frac{-33}{36}$$
$$= \frac{20 - 33}{36} = \frac{-13}{36}$$
$$= -\frac{13}{36}$$

You Try It 2 The LCM of 8, 6 and 3 is 24.

$$-\frac{7}{8} - \frac{5}{6} + \frac{2}{3}$$
$$= \frac{-21}{24} - \frac{20}{24} + \frac{16}{24}$$
$$= \frac{-21}{24} + \frac{-20}{24} + \frac{16}{24}$$
$$= \frac{-21 - 20 + 16}{24}$$
$$= -\frac{25}{24} = -1\frac{1}{24}$$

You Try It 3

$$\begin{array}{r} 67.910 \\ -\ 16.127 \\ \hline 51.783 \end{array}$$

$16.127 - 67.91 = -51.783$

You Try It 4 $2.7 + (-9.44) + 6.2$
$-6.74 + 6.2$
-0.54

You Try It 5 The product is positive.

$$\left[-\frac{2}{3}\right]\left[-\frac{9}{10}\right] = \frac{2 \cdot 9}{3 \cdot 10}$$
$$= \frac{18}{30} = \frac{3}{5}$$

You Try It 6 The quotient is negative.

$$-\frac{5}{8} \div \frac{5}{40} = -\frac{5}{8} \cdot \frac{40}{5}$$
$$= -\frac{5 \cdot 40}{8 \cdot 5}$$
$$= -\frac{200}{40} = -5$$

You Try It 7

$$\begin{array}{r} 5.44 \\ \times\ 3.8 \\ \hline 4352 \\ 1632 \\ \hline 20.672 \end{array}$$

$-5.44 \times 3.8 = -20.672$

You Try It 8 $3.44 \times (-1.7) \times 0.6$
$(-5.848) \times 0.6$
-3.5088

You Try It 9

$$\begin{array}{r} 0.231 \\ 1.7.\overline{)0.3.940} \\ -3\,4 \\ \hline 54 \\ -51 \\ \hline 30 \\ -17 \\ \hline 13 \end{array}$$

$-0.394 \div 1.7 \approx -0.23$

SECTION 10.5 *pages 387–388*

You Try It 1 $9 - 9 \div (-3)$
$9 - (-3)$
$9 + 3$
12

You Try It 2 $8 \div 4 \cdot 4 - (-2)^2$
$8 \div 4 \cdot 4 - 4$
$2 \cdot 4 - 4$
$8 - 4$
$8 + (-4)$
4

You Try It 3 $8 - (-15) \div (2 - 7)$
$8 - (-15) \div (-5)$
$8 - 3$
$8 + (-3)$
5

You Try It 4 $(-2)^2 \times (3 - 7)^2 - (-16) \div (-4)$
$(-2)^2 \times (-4)^2 - (-16) \div (-4)$
$4 \times 16 - (-16) \div (-4)$
$64 - 4$
$64 + (-4)$
60

You Try It 5 $7 \div \left(\frac{1}{7} - \frac{3}{14}\right) - 9$
$7 \div \left(-\frac{1}{14}\right) - 9$
$-98 - 9$
$-98 + (-9)$
-107

SOLUTIONS to Chapter 11 "You Try It"

SECTION 11.1 *pages 401–406*

You Try It 1

$6a - 5b$

$$\begin{aligned} 6(-3) - 5(4) &= -18 - 20 \\ &= -18 + (-20) \\ &= -38 \end{aligned}$$

You Try It 2

$-3s^2 - 12 \div t$

$$\begin{aligned} -3(-2)^2 - 12 \div 4 &= -3(4) - 12 \div 4 \\ &= -12 - 12 \div 4 \\ &= -12 - 3 \\ &= -12 + (-3) \\ &= -15 \end{aligned}$$

You Try It 3

$-\frac{2}{3}m + \frac{3}{4}n^3$

$$\begin{aligned} -\frac{2}{3}(6) + \frac{3}{4}(2)^3 &= -\frac{2}{3}(6) + \frac{3}{4}(8) \\ &= -4 + 6 \\ &= 2 \end{aligned}$$

You Try It 4

$-3yz - z^2 + y^2$

$$\begin{aligned} &-3\left(-\frac{2}{3}\right)\left(\frac{1}{3}\right) - \left(\frac{1}{3}\right)^2 + \left(-\frac{2}{3}\right)^2 \\ &= -3\left(-\frac{2}{3}\right)\left(\frac{1}{3}\right) - \frac{1}{9} + \frac{4}{9} \\ &= \frac{2}{3} - \frac{1}{9} + \frac{4}{9} \\ &= 1 \end{aligned}$$

You Try It 5

$$\begin{aligned} 5a^2 - 6b^2 + 7a^2 - 9b^2 \\ &= 5a^2 + (-6)b^2 + 7a^2 + (-9)b^2 \\ &= 5a^2 + 7a^2 + (-6)b^2 + (-9)b^2 \\ &= 12a^2 + (-15)b^2 \\ &= 12a^2 - 15b^2 \end{aligned}$$

You Try It 6

$$\begin{aligned} -6x + 7 + 9x - 10 \\ &= (-6)x + 7 + 9x + (-10) \\ &= (-6)x + 9x + 7 + (-10) \\ &= 3x + (-3) \\ &= 3x - 3 \end{aligned}$$

You Try It 7

$$\begin{aligned} \frac{3}{8}w + \frac{1}{2} - \frac{1}{4}w - \frac{2}{3} &= \frac{3}{8}w - \frac{1}{4}w + \frac{1}{2} - \frac{2}{3} \\ &= \frac{3}{8}w - \frac{2}{8}w + \frac{3}{6} - \frac{4}{6} \\ &= \frac{1}{8}w - \frac{1}{6} \end{aligned}$$

You Try It 8

$$\begin{aligned} 5(a - 2) &= 5a - 5(2) \\ &= 5a - 10 \end{aligned}$$

You Try It 9

$$\begin{aligned} 8s - 2(3s - 5) &= 8s + (-2)(3s) - (-2)(5) \\ &= 8s + (-6s) + 10 \\ &= 2s + 10 \end{aligned}$$

You Try It 10

$$\begin{aligned} 4(x - 3) - 2(x + 1) &= 4x - 4(3) - 2x - 2(1) \\ &= 4x - 12 - 2x - 2 \\ &= 4x - 2x - 12 - 2 \\ &= 2x - 14 \end{aligned}$$

SECTION 11.2 *pages 411–416*

You Try It 1

$x(x + 3) = 4x + 6$	
$(-2)(-2 + 3)$	$4(-2) + 6$
$(-2)(1)$	$(-8) + 6$
$-2 = -2$	

Yes, -2 is a solution.

You Try It 2

$x^2 - x = 3x + 7$	
$(-3)^2 - (-3)$	$3(-3) + 7$
$9 + 3$	$-9 + 7$
$12 \neq -2$	

No, -3 is not a solution.

You Try It 3

$$\begin{aligned} -2 + y &= -5 \\ -2 + 2 + y &= -5 + 2 \\ 0 + y &= -3 \\ y &= -3 \end{aligned}$$

The solution is −3.

You Try It 4

$$\begin{aligned} 7 &= y + 8 \\ 7 - 8 &= y + 8 - 8 \\ -1 &= y + 0 \\ -1 &= y \end{aligned}$$

The solution is −1.

You Try It 5

$$\begin{aligned} \frac{1}{5} &= z + \frac{4}{5} \\ \frac{1}{5} - \frac{4}{5} &= z + \frac{4}{5} - \frac{4}{5} \\ -\frac{3}{5} &= z + 0 \\ -\frac{3}{5} &= z \end{aligned}$$

The solution is $-\frac{3}{5}$.

You Try It 6

$$\begin{aligned} 4z &= -20 \\ \frac{4z}{4} &= \frac{-20}{4} \\ 1z &= -5 \\ z &= -5 \end{aligned}$$

The solution is −5.

You Try It 7

$$\begin{aligned} 8 &= \frac{2}{5}n \\ \left(\frac{5}{2}\right)(8) &= \left(\frac{5}{2}\right)\frac{2}{5}n \\ 20 &= 1n \\ 20 &= n \end{aligned}$$

The solution is 20.

You Try It 8

$$\begin{aligned} \frac{2}{3}t - \frac{1}{3}t &= -2 \\ \frac{1}{3}t &= -2 \\ \left(\frac{3}{1}\right)\frac{1}{3}t &= \left(\frac{3}{1}\right)(-2) \\ 1t &= -6 \\ t &= -6 \end{aligned}$$

The solution is −6.

You Try It 9

Strategy
To find the discount, replace the variables S and R in the formula by the given values and solve for D.

Solution

$$\begin{aligned} S &= R - D \\ 22 &= 30 - D \\ 22 - 30 &= 30 - 30 - D \\ -8 &= -D \\ -8(-1) &= -D(-1) \\ 8 &= D \end{aligned}$$

The discount is $8.

You Try It 10

Strategy
To find the interest, replace the variables P and A in the formula by the given values and solve for I.

Solution

$$\begin{aligned} A &= P + I \cdot P \\ 1120 &= 1000 + 1000I \\ 1120 - 1000 &= 1000 - 1000 + 1000I \\ 120 &= 1000I \\ 120\left(\frac{1}{1000}\right) &= \left(\frac{1}{1000}\right)1000I \\ 0.12 &= I \end{aligned}$$

The interest rate is 12%.

SECTION 11.3 *pages 421–422*

You Try It 1

$$5x + 8 = 6$$
$$5x + 8 - 8 = 6 - 8$$
$$5x = -2$$
$$\frac{5x}{5} = \frac{-2}{5}$$
$$x = -\frac{2}{5}$$

The solution is $-\frac{2}{5}$.

You Try It 2

$$7 - x = 3$$
$$7 - 7 - x = 3 - 7$$
$$-x = -4$$
$$(-1)(-x) = (-1)(-4)$$
$$x = 4$$

The solution is 4.

You Try It 3

Strategy
To find the Celsius temperature, replace the variable F in the formula by the given value and solve for C.

Solution

$$F = \frac{9}{5}C + 32$$
$$-22 = \frac{9}{5}C + 32$$
$$-22 - 32 = \frac{9}{5}C + 32 - 32$$
$$-54 = \frac{9}{5}C$$
$$\left(\frac{5}{9}\right)(-54) = \left(\frac{5}{9}\right)\frac{9}{5}C$$
$$-30 = C$$

The Celsius temperature is $-30°$.

You Try It 4

Strategy
To find the cost per unit, replace the variables T, N, and F in the formula by the given values and solve for U.

Solution

$$T = U \cdot N + F$$
$$4500 = 250U + 1500$$
$$4500 - 1500 = 250U + 1500 - 1500$$
$$3000 = 250U$$
$$\frac{3000}{250} = \frac{250U}{250}$$
$$12 = U$$

The cost per unit is \$12.

SECTION 11.4 *pages 427–428*

You Try It 1

$$\frac{1}{5}x - 2 = \frac{2}{5}x + 4$$
$$\frac{1}{5}x - \frac{2}{5}x - 2 = \frac{2}{5}x - \frac{2}{5}x + 4$$
$$-\frac{1}{5}x - 2 = 4$$
$$-\frac{1}{5}x - 2 + 2 = 4 + 2$$
$$-\frac{1}{5}x = 6$$
$$(-5)\left(-\frac{1}{5}x\right) = (-5)6$$
$$x = -30$$

The solution is -30.

You Try It 2

$$4(x - 1) - x = 5$$
$$4x - 4 - x = 5$$
$$3x - 4 = 5$$
$$3x - 4 + 4 = 5 + 4$$
$$3x = 9$$
$$\frac{3x}{3} = \frac{9}{3}$$
$$x = 3$$

The solution is 3.

SECTION 11.5 *pages 433–434*

You Try It 1
$8 - 2t$

You Try It 2
$\frac{5}{7x}$

You Try It 3
Twelve *decreased* by some number
The unknown number: x
$12 - x$

You Try It 4
The *product* of a number and one-half of the number.
The unknown number: n
One-half the number: $\frac{n}{2}$
$(n)\left(\frac{n}{2}\right)$

SECTION 11.6 *pages 437–440*

You Try It 1 The unknown number: x

$$\begin{aligned} x + 4 &= 12 \\ x + 4 - 4 &= 12 - 4 \\ x &= 8 \end{aligned}$$

The number is 8.

You Try It 2 The unknown number: x

$$\begin{aligned} 2x &= 10 \\ \frac{2x}{2} &= \frac{10}{2} \\ x &= 5 \end{aligned}$$

The number is 5.

You Try It 3 The unknown number: x

$$\begin{aligned} 3x + 6 &= 4 \\ 3x + 6 - 6 &= 4 - 6 \\ 3x &= -2 \\ \frac{3x}{3} &= \frac{-2}{3} \\ x &= -\frac{2}{3} \end{aligned}$$

The number is $-\frac{2}{3}$.

You Try It 4 The unknown number: x

$$\begin{aligned} 4x + 3 &= 10 \\ 4x + 3 - 3 &= 10 - 3 \\ 4x &= 7 \\ \frac{4x}{4} &= \frac{7}{4} \\ x &= \frac{7}{4} \end{aligned}$$

The number is $1\frac{3}{4}$.

You Try It 5

Strategy
To find the regular price, write and solve an equation using R to replace the regular price of the slacks.

Solution

$$\begin{aligned} 18.95 &= R - 6 \\ 18.95 + 6 &= R - 6 + 6 \\ 24.95 &= R \end{aligned}$$

The regular price of the slacks is \$24.95.

You Try It 6

Strategy
To find the rpm, write and solve an equation using R to represent the rpm of the engine when in third gear.

Solution

$$\begin{aligned} 2500 &= \frac{2}{3}R \\ \frac{3}{2}(2500) &= \left(\frac{3}{2}\right)\frac{2}{3}R \\ 3750 &= R \end{aligned}$$

The rpm of the engine when in third gear is 3750.

You Try It 7

Strategy
To find the total sales, write and solve an equation using S to represent the total sales.

Solution

$$2500 = 800 + 0.08S$$
$$2500 - 800 = 800 - 800 + 0.08S$$
$$1700 = 0.08S$$
$$\frac{1700}{0.08} = \frac{0.08S}{0.08}$$
$$21{,}250 = S$$

The total sales are $21,250.

You Try It 8

Strategy
To find the number of hours, write and solve an equation using H to represent the number of hours of labor required.

Solution

$$300 = 100 + 12.50H$$
$$300 - 100 = 100 - 100 + 12.50H$$
$$200 = 12.50H$$
$$\frac{200}{12.50} = \frac{12.50H}{12.50}$$
$$16 = H$$

The number of hours of labor required is 16.

SOLUTIONS to Chapter 12 "You Try It"

SECTION 12.1 *pages 455–462*

You Try It 1

$$RS = QT - QR - ST$$
$$RS = 62 - 24 - 17$$
$$RS = 38 - 17$$
$$RS = 21$$

You Try It 2 $180° - 32° = 148°$

You Try It 3

$$\angle a = 118° - 68°$$
$$\angle a = 50°$$

You Try It 4 $90° - 7° = 83°$
The other angles are 90° and 83°.

You Try It 5 $62° + 45° = 107°$
$180° - 107° = 73°$
The third angle is 73°.

You Try It 6

$$\text{Radius} = \frac{1}{2} \times \text{diameter}$$
$$= \frac{1}{2} \times 8 \text{ in.}$$
$$= 4 \text{ in.}$$

You Try It 7 Angles a and b are supplementary angles.

Angle $b = 180° - 125° = 55°$

You Try It 8 Angles a and c are corresponding angles. Therefore angle $c = 120°$.

Angles b and c are supplementary angles. Therefore $\angle b = 180° - 120° = 60°$.

SECTION 12.2 *pages 467–472*

You Try It 1 $P = 2L + 2W$
$= (2 \cdot 2 \text{ m}) + (2 \cdot 0.85 \text{ m})$
$= 4 \text{ m} + 1.7 \text{ m}$
$= 5.7 \text{ m}$

The perimeter of the rectangle is 5.7 m.

You Try It 2 $P = a + b + c$
$= 12 \text{ cm} + 15 \text{ cm} + 18 \text{ cm}$
$= 45 \text{ cm}$

The perimeter of the triangle is 45 cm.

You Try It 3 $C = \pi d$
$\approx 3.14 \cdot 6 \text{ in.}$
$\approx 18.84 \text{ in.}$

The circumference is approximately 18.84 in.

You Try It 4 $P = 2L + \pi d$
$\approx (2 \cdot 8 \text{ in.}) + (3.14 \cdot 3 \text{ in.})$
$= 16 \text{ in.} + 9.42 \text{ in.}$
$\approx 25.42 \text{ in.}$

The perimeter is approximately 25.42 in.

You Try It 5

Strategy
To find the perimeter use the formula for the perimeter of a rectangle.

Solution
$P = 2L + 2W$
$= 2(11 \text{ in.}) + 2\left(8\frac{1}{2} \text{ in.}\right)$
$= 22 \text{ in.} + 17 \text{ in.}$
$= 39 \text{ in.}$

The perimeter of the typing paper is 39 in.

You Try It 6

Strategy
To find the cost:

- Find the perimeter of the table.
- Multiply the perimeter by the per-meter cost of the stripping.

Solution
$P = 2L + 2W$
$= (2 \cdot 3 \text{ m}) + (2 \cdot 0.74 \text{ m})$
$= 6 \text{ m} + 1.48 \text{ m}$
$= 7.48 \text{ m}$

$\$1.76 \times 7.48 = \$13.1648$

The cost is $13.16.

SECTION 12.3 *pages 477–480*

You Try It 1

$A = \frac{1}{2}bh = \frac{1}{2} \cdot 24 \text{ in.} \cdot 14 \text{ in.} = 168 \text{ in}^2$

You Try It 2
A = area of rectangle − area of triangle

$A = LW - \frac{1}{2}bh$
$= (10 \text{ in.} \times 6 \text{ in.}) - \left(\frac{1}{2} \cdot 6 \text{ in.} \times 4 \text{ in.}\right)$
$= 60 \text{ in}^2 - 12 \text{ in}^2$
$= 48 \text{ in}^2$

You Try It 3

Strategy
To find the area of the rug:

- Find the area in square feet.
- Convert to square yards.

Solution

$$\begin{aligned} A &= LW \\ &= 12 \text{ ft} \cdot 9 \text{ ft} \\ &= 108 \text{ ft}^2 \end{aligned}$$

$$108 \cancel{\text{ft}^2} \times \frac{1 \text{ yd}^2}{9 \cancel{\text{ft}^2}} = \frac{108}{9} \text{ yd}^2 = 12 \text{ yd}^2$$

The area of the room is 12 yd^2.

SECTION 12.4 *pages 485–490*

You Try It 1

$$\begin{aligned} V = s^3 &= (5 \text{ cm})^3 \\ &= 125 \text{ cm}^3 \end{aligned}$$

You Try It 2

$$\begin{aligned} V = \pi r^2 h &\approx \frac{22}{7}(7 \text{ in.})^2 \cdot (15 \text{ in.}) \\ &\approx 2310 \text{ in}^3. \end{aligned}$$

You Try It 3

$$V = \frac{4}{3}r^3 \approx \frac{4}{3} \cdot 3.14(3 \text{ m})^3$$

Volume $\approx 113.04 \text{ m}^3$

You Try It 4

V = volume of rectangular solid + volume of cylinder

$$\begin{aligned} V &= LWH + \pi r^2 h \\ &\approx 1.5 \text{ m} \cdot 0.4 \text{ m} \cdot 0.4 \text{ m} + 3.14 \cdot (0.8 \text{ m})^2 \cdot 0.2 \text{ m} \\ &= 0.24 \text{ m}^3 + 0.40192 \text{ m}^3 \end{aligned}$$

Volume $\approx 0.64192 \text{ m}^3$

You Try It 5

V = volume of rectangular solid + $\frac{1}{2}$ of the volume of cylinder

$$\begin{aligned} V &= LWH + \frac{1}{2}r^2 h \\ &\approx (24 \text{ in.} \cdot 6 \text{ in.} \cdot 4 \text{ in.}) + \frac{1}{2} \cdot 3.14 \cdot (3 \text{ in.})^2 \cdot 24 \text{ in.} \\ &= 576 \text{ in}^3 + 339.12 \text{ in}^3 \end{aligned}$$

Volume $\approx 915.12 \text{ in}^3$

You Try It 6

Strategy
To find the volume of the freezer, use the formula for the volume of a rectangular solid.

Solution

$$\begin{aligned} V &= LWH \\ &= 7 \text{ ft} \cdot 2.5 \text{ ft} \cdot 3 \text{ ft} \\ &= 52.5 \text{ ft}^3 \end{aligned}$$

The volume of the freezer is 52.5 ft^3.

You Try It 7

Strategy
To find the volume of the channel iron, add the volumes of the three rectangular solids.

Solution

$$\begin{aligned} V &= 2LWH + LWH \\ &= 2(10 \text{ ft} \cdot 0.5 \text{ ft} \cdot 0.3 \text{ ft}) + (10 \text{ ft} \cdot 0.2 \text{ ft} \cdot 0.2 \text{ ft}) \\ &= 3 \text{ ft}^3 + 0.4 \text{ ft}^3 \\ &= 3.4 \text{ ft}^3 \end{aligned}$$

The volume of the channel iron is 3.4 ft^3.

SECTION 12.5 *pages 495–498*

You Try It 1 $\sqrt{16} = 4$ $\sqrt{169} = 13$

You Try It 2 $\sqrt{32} \approx 5.657$
$\sqrt{162} \approx 12.728$

You Try It 3

$$\begin{aligned} \text{Hypotenuse} &= \sqrt{(\text{leg})^2 + (\text{leg})^2} \\ &= \sqrt{(8 \text{ in.})^2 + (11 \text{ in.})^2} \\ &= \sqrt{64 \text{ in}^2 + 121 \text{ in}^2} \\ &= \sqrt{185 \text{ in}^2} \\ \text{Hypotenuse} &\approx 13.601 \text{ in.} \end{aligned}$$

You Try It 4

$$\begin{aligned} \text{Leg} &= \sqrt{(\text{hypotenuse})^2 - (\text{leg})^2} \\ &= \sqrt{(12 \text{ ft})^2 - (5 \text{ ft})^2} \\ &= \sqrt{144 \text{ ft}^2 - 25 \text{ ft}^2} \\ &= \sqrt{119 \text{ ft}^2} \\ \text{Leg} &\approx 10.909 \text{ ft} \end{aligned}$$

You Try It 5

Strategy
To find the distance between the holes, use the Pythagorean Theorem. The hypotenuse is the distance between the holes. The length of each leg is given (3 cm and 8 cm).

Solution

$$\begin{aligned} \text{Hypotenuse} &= \sqrt{(\text{leg})^2 + (\text{leg})^2} \\ &= \sqrt{(3 \text{ cm})^2 + (8 \text{ cm})^2} \\ &= \sqrt{9 \text{ cm}^2 + 64 \text{ cm}^2} \\ &= \sqrt{73 \text{ cm}^2} \\ \text{Hypotenuse} &\approx 8.544 \text{ cm} \end{aligned}$$

The distance is 8.544 cm.

SECTION 12.6 *pages 501–504*

You Try It 1 $\frac{4\text{ cm}}{7\text{ cm}} = \frac{4}{7}$

You Try It 2 Let x represent the side DF.

$$\frac{AB}{DE} = \frac{AC}{x}$$

$$\frac{7\text{ cm}}{14\text{ cm}} = \frac{3\text{ cm}}{x}$$

$$7x = 14 \cdot 3\text{ cm}$$

$$7x = 42\text{ cm}$$

$$\frac{7x}{7} = \frac{42\text{ cm}}{7}$$

$$x = 6\text{ cm}$$

Side DF is 6 cm.

You Try It 3 $AC = DF$, Angle $ACB =$ Angle DFE, but $CB \neq EF$, therefore the triangles are not congruent.

You Try It 4

Strategy
To find the height FG:

- Solve a proportion to find the height.

Solution
Let x represent the height FG.

$$\frac{AC}{DF} = \frac{\text{height } CH}{\text{height } FG}$$

$$\frac{10\text{ cm}}{15\text{ cm}} = \frac{7\text{ cm}}{x}$$

$$10x = 15 \cdot 7\text{ cm}$$

$$10x = 105\text{ cm}$$

$$\frac{10x}{10} = \frac{105\text{ cm}}{10}$$

$$x = 10.5\text{ cm}$$

The height FG is 10.5 cm.

You Try It 5

Strategy
To find the perimeter of triangle ABC:

- Solve a proportion to find the lengths of sides BC and AC.
- Use the formula: Perimeter = side AB + side BC + side AC.

Solution

$$\frac{BC}{EF} = \frac{AB}{DE}$$

$$\frac{BC}{10\text{ in.}} = \frac{4\text{ in.}}{8\text{ in.}}$$

$$8BC = 10\text{ in.} \cdot 4$$

$$8BC = 40\text{ in.}$$

$$\frac{8BC}{8} = \frac{40\text{ in.}}{8}$$

$$BC = 5\text{ in.}$$

$$\frac{AC}{DF} = \frac{AB}{DE}$$

$$\frac{AC}{6\text{ in.}} = \frac{4\text{ in.}}{8\text{ in.}}$$

$$8AC = 6\text{ in.} \cdot 4$$

$$8AC = 24\text{ in.}$$

$$\frac{8AC}{8} = \frac{24\text{ in.}}{8}$$

$$AC = 3\text{ in.}$$

$$\text{Perimeter} = 4\text{ in.} + 5\text{ in.} + 3\text{ in.} = 12\text{ in.}$$

The perimeter of triangle ABC is 12 in.

ANSWERS to Chapter 1 Odd-Numbered Exercises

SECTION 1.1 *pages 7–8*

1. 0 1 2 3 4 5 6 7 8 9 10 11 12 **3.** 0 1 2 3 4 5 6 7 8 9 10 11 12

5. $37 < 49$ **7.** $101 > 87$ **9.** $245 > 158$ **11.** $0 < 45$ **13.** $815 < 928$
15. Three thousand seven hundred ninety **17.** Fifty-eight thousand four hundred seventy-three
19. Four hundred ninety-eight thousand five hundred twelve
21. Six million eight hundred forty-two thousand seven hundred fifteen **23.** 357 **25.** 63,780
27. 7,024,709 **29.** 6000 + 200 + 90 + 5 **31.** 400,000 + 50,000 + 3000 + 900 + 20 + 1
33. 300,000 + 1000 + 800 + 9 **35.** 3,000,000 + 600 + 40 + 2 **37.** 850 **39.** 4000
41. 53,000 **43.** 250,000 **45.** False. 8270 rounded to the nearest hundred is 8300.
47. Yes. IX = 9 XI = 11

SECTION 1.2 *pages 13–16*

1. 28 **3.** 125 **5.** 102 **7.** 154 **9.** 1489 **11.** 828 **13.** 1584 **15.** 1219
17. 102,317 **19.** 79,326 **21.** 1804 **23.** 1579 **25.** 19,740 **27.** 7420
29. 120,570 **31.** 207,453 **33.** 24,218 **35.** 11,974 **37.** 9323 **39.** 77,139
41. 14,383 **43.** 9473 **45.** 33,247 **47.** 5058 **49.** 1992 **51.** 68,263
53. Est.: 17,700 Cal.: 17,754 **55.** Est.: 2900 Cal.: 2872 **57.** Est.: 101,000 Cal.: 101,712 **59.** Est.: 158,000 Cal.: 158,763 **61.** Est.: 260,000 Cal.: 261,595
63. Est.: 940,000 Cal.: 946,718 **65.** Est.: 33,000,000 Cal.: 32,691,621 **67.** Est.: 34,000,000 Cal.: 34,420,922
69. The total attendance was 5792 people. **71.** The total number of yards gained by passing was 307. **73.** The total commission Ken received was $6973. **75a.** The total paid attendance for the fifth and sixth games was 104,031 people. **75b.** The total paid attendance for the entire series was 311,460 people. **77a.** The total amount deposited was $1664. **77b.** The new checking account balance is $3799. **79.** 90,900 **81.** No. 0 + 2 = 2 **83.** Answers will vary **85.** The total average amount for all Americans is $4527. **87.** The total average amount for Americans ages 16 to 34 is $7838.

SECTION 1.3 *pages 21–24*

1. 4 **3.** 4 **5.** 10 **7.** 7 **9.** 11 **11.** 9 **13.** 22 **15.** 60 **17.** 66 **19.** 33
21. 501 **23.** 962 **25.** 5002 **27.** 1513 **29.** 9 **31.** 7 **33.** 31 **35.** 47
37. 925 **39.** 4561 **41.** 3205 **43.** 1222 **45.** 3021 **47.** 3022 **49.** 3040
51. 212 **53.** 60,245 **55.** 65 **57.** 17 **59.** 8 **61.** 37 **63.** 353 **65.** 57
67. 160 **69.** 337 **71.** 1423 **73.** 754 **75.** 2179 **77.** 5726 **79.** 3347 **81.** 6489
83. 889 **85.** 71,129 **87.** 698 **89.** 29,405 **91.** 49,624 **93.** 628 **95.** 6532
97. 4286 **99.** 4042 **101.** 5209 **103.** 10,378 **105.** 1024 **107.** 556 **109.** 126,504
111. 38,208 **113.** 17,438 **115.** Est.: 60,000 Cal.: 60,427 **117.** Est.: 20,000 Cal.: 21,613 **119.** Est.: 360,000 Cal.: 358,979
121. There is $165 left in your checking account. **123.** The amount that remains to be paid is $899. **125.** The length of the trip was 3162 miles. **127.** Florida has lost 9,286,713 acres of wetlands over the last 200 years. **129.** It took 91 months longer to withdraw principal and interest in 1992. **131.** **a.** True **b.** False **c.** False **133.** Answers will vary **135.** The difference in cost between the two cars is $210. **137.** The SAT scores increased the most between 1992 and 1993. The amount of the increase was 4 points.

SECTION 1.4 *pages 29–32*

1. 12 **3.** 35 **5.** 25 **7.** 0 **9.** 72 **11.** 198 **13.** 335 **15.** 2492 **17.** 5463
19. 4200 **21.** 6327 **23.** 1896 **25.** 5056 **27.** 3450 **29.** 2674 **31.** 2096
33. 5046 **35.** 18,040 **37.** 21,980 **39.** 18,450 **41.** 140 **43.** 47,736
45. 22,456 **47.** 18,630 **49.** 336 **51.** 910 **53.** 1794 **55.** 1541 **57.** 63,063

59. 33,520 **61.** 380,834 **63.** 541,164 **65.** 400,995 **67.** 105,315 **69.** 428,770 **71.** 260,000 **73.** 344,463 **75.** 41,808 **77.** 189,500 **79.** 401,880 **81.** 1,052,763 **83.** 4,198,388 **85.** 259,335 **87.** 1292 **89.** 151,152 **91.** 701,634,445 **93.** Est.: 540,000 Cal.: 550,935 **95.** Est.: 1,200,000 Cal.: 1,138,134 **97.** Est.: 600,000 Cal.: 605,022 **99.** Est.: 3,000,000 Cal.: 3,174,425 **101.** Est.: 18,000,000 Cal.: 19,212,228 **103.** Est.: 72,000,000 Cal.: 73,427,310 **105.** The car could travel 516 miles on 12 gallons of gas. **107.** In 40 hours, the machine can fill 168,000 bottles of coke. **109.** There are 307,200 pixels on the screen. **111.** The total cost of the car is $11,280. **113.** The lighting designer can save $84. **115.** Carlos paid the 4 plumbers $1380. **117.** The total cost of the 4 components is $383. **119.** 12 accidental deaths occurred each hour. 288 accidental deaths occurred each day. 105,120 accidental deaths occurred each year. **121.** Answers will vary

SECTION 1.5 *pages 41–44*

1. 2 **3.** 6 **5.** 5 **7.** 16 **9.** 80 **11.** 210 **13.** 44 **15.** 206 **17.** 530 **19.** 902 **21.** 21,560 **23.** 3580 **25.** 482 **27.** 1075 **29.** 2 r1 **31.** 5 r2 **33.** 13 r1 **35.** 10 r3 **37.** 90 r2 **39.** 230 r1 **41.** 204 r3 **43.** 1347 r3 **45.** 778 r2 **47.** 391 r4 **49.** 1160 r4 **51.** 708 r2 **53.** 3825 **55.** 9044 r2 **57.** 388 r3 **59.** 11,430 **61.** 510 **63.** 3 r15 **65.** 2 r3 **67.** 21 r36 **69.** 34 r2 **71.** 8 r8 **73.** 4 r49 **75.** 200 r25 **77.** 203 r2 **79.** 35 r47 **81.** 271 **83.** 4484 r6 **85.** 608 **87.** 15 r7 **89.** 1 r563 **91.** 50 r92 **93.** 40 r7 **95.** 258 r14 **97.** 517 r70 **99.** 500 **101.** Est.: 5000 Cal.: 5129 **103.** Est.: 20,000 Cal.: 21,968 **105.** Est.: 22,500 Cal.: 24,596 **107.** Est.: 3000 Cal.: 2836 **109.** Est.: 3000 Cal.: 3024 **111.** Est.: 30,000 Cal.: 32,036 **113.** Each organization received $137,000. **115.** The monthly contribution was $298. **117.** The annual contribution in 1993 was 79 times larger. **119.** Each disk can store 368,640 bytes of information. **121.** The amount of each payment is $119. **123.** Oprah Winfrey's average hourly wage is $49,000 per hour. **125.** Steven Spielberg's annual income is 2 times larger than Madonna's. **127.** The smallest three-digit palindromic number is 212. **129.** The largest possible number divisible by 4 is 91,180. **131.** The answer to a division problem can be checked by adding the remainder to the product of the quotient and divisor.

SECTION 1.6 *pages 47–48*

1. 2^3 **3.** $6^3 \cdot 7^4$ **5.** $6^3 \cdot 11^5$ **7.** $3 \cdot 10^4$ **9.** $2^2 \cdot 3^3 \cdot 5^4$ **11.** $2 \cdot 3^2 \cdot 7^2 \cdot 11$ **13.** $2^2 \cdot 7^4 \cdot 11^4$ **15.** 8 **17.** 400 **19.** 900 **21.** 972 **23.** 120 **25.** 360 **27.** 0 **29.** 90,000 **31.** 540 **33.** 4050 **35.** 11,025 **37.** 25,920 **39.** 4,320,000 **41.** 92,160 **43.** 5 **45.** 10 **47.** 47 **49.** 8 **51.** 5 **53.** 8 **55.** 6 **57.** 53 **59.** 44 **61.** 19 **63.** 37 **65.** 168 **67.** 27 **69.** 14 **71.** 10 **73.** 9 **75.** 1024 **77.** Yes. Answers will vary

SECTION 1.7 *pages 51–52*

1. 1, 2, 4 **3.** 1, 2, 5, 10 **5.** 1, 5 **7.** 1, 2, 3, 4, 6, 12 **9.** 1, 2, 4, 8 **11.** 1, 13 **13.** 1, 2, 3, 6, 9, 18 **15.** 1, 5, 25 **17.** 1, 2, 3, 4, 6, 9, 12, 18, 36 **19.** 1, 2, 4, 7, 14, 28 **21.** 1, 29 **23.** 1, 2, 11, 22 **25.** 1, 2, 4, 11, 22, 44 **27.** 1, 7, 49 **29.** 1, 37 **31.** 1, 3, 19, 57 **33.** 1, 2, 3, 4, 6, 8, 12, 16, 24, 48 **35.** 1, 3, 29, 87 **37.** 1, 2, 23, 46 **39.** 1, 2, 5, 10, 25, 50 **41.** 1, 2, 3, 6, 11, 22, 33, 66 **43.** 1, 2, 4, 5, 8, 10, 16, 20, 40, 80 **45.** 1, 2, 3, 4, 6, 7, 12, 14, 21, 28, 42, 84 **47.** 1, 5, 17, 85 **49.** 1, 101 **51.** $2 \cdot 3$ **53.** Prime **55.** $2 \cdot 2 \cdot 2 \cdot 2$ **57.** $2 \cdot 2 \cdot 3$ **59.** $3 \cdot 3$ **61.** $2 \cdot 2 \cdot 3 \cdot 3$ **63.** Prime **65.** $2 \cdot 5 \cdot 5$ **67.** $5 \cdot 13$ **69.** $2 \cdot 2 \cdot 2 \cdot 2 \cdot 5$ **71.** $2 \cdot 3 \cdot 3$ **73.** $2 \cdot 2 \cdot 7$ **75.** Prime **77.** $2 \cdot 3 \cdot 7$ **79.** $3 \cdot 3 \cdot 3 \cdot 3$ **81.** $2 \cdot 11$ **83.** Prime **85.** $2 \cdot 5 \cdot 7$ **87.** $2 \cdot 43$ **89.** $5 \cdot 19$ **91.** Prime **93.** $5 \cdot 11$ **95.** $2 \cdot 61$ **97.** $2 \cdot 59$ **99.** $2 \cdot 3 \cdot 5 \cdot 5$ **101.** $2 \cdot 2 \cdot 2 \cdot 3 \cdot 5$ **103.** $2 \cdot 2 \cdot 2 \cdot 2 \cdot 2 \cdot 5$ **105.** $7 \cdot 7 \cdot 7$ **107.** $2 \cdot 2 \cdot 2 \cdot 2 \cdot 5 \cdot 5$ **109.** $3 \cdot 3 \cdot 5 \cdot 5$ **111.** 3 and 5, 5 and 7, 11 and 13. Other answers are possible. **113.** Answers will vary

CHAPTER REVIEW *pages 55–56*

1. 101 > 87 **2.** Two hundred seventy-six thousand fifty-seven **3.** 2,011,044 **4.** 10,000 + 300 + 20 + 7 **5.** 1081 **6.** 12,493 **7.** The total income from commissions was \$2567. **8.** The total amount was \$301. The new checking account balance is \$817. **9.** 1749 **10.** 5409 **11.** A U.S. company spends \$1985 per person more than an Italian company. **12.** A U.S. company spends \$563 more for business travel and entertainment. **13.** No **14.** 22,761 **15.** 619,833 **16.** The total of the car payments is \$1476. **17.** Vincent's total pay for the last week was \$384. **18.** 2135 **19.** 1306 r59 **20.** The car drove 27 miles per gallon of gasoline. **21.** The monthly car payment is \$155. **22.** $2^4 \cdot 5^3$ **23.** $5^2 \cdot 7^5$ **24.** 600 **25.** 2 **26.** 17 **27.** 8 **28.** 1, 2, 3, 6, 9, 18 **29.** 1, 2, 3, 5, 6, 10, 15, 30 **30.** $2 \cdot 3 \cdot 7$ **31.** $2 \cdot 2 \cdot 2 \cdot 3 \cdot 3$

CHAPTER TEST *pages 57–58*

1. 21 > 19 (1.1A) **2.** Two hundred seven thousand sixty-eight (1.1B) **3.** 1,204,006 (1.1B) **4.** 900,000 + 6000 + 300 + 70 + 8 (1.1C) **5.** 75,000 (1.1D) **6.** 96,798 (1.2A) **7.** 135,915 (1.2A) **8a.** They drove 855 miles (1.2C) **8b.** The odometer reading was 48,481 miles. (1.2B) **9.** 9333 (1.3A) **10.** 19,922 (1.3B) **11.** The difference is \$25. (1.3C) **12.** The difference is \$91. (1.3C) **13.** 726,104 (1.4A) **14.** 6,854,144 (1.4B) **15.** The investor will receive \$2844 over 12 months. (1.4C) **16.** 703 (1.5A) **17.** 8710 r2 (1.5B) **18.** 1,121 r27 (1.5C) **19.** The farmer needed 3000 boxes. (1.5D) **20.** $3^3 \cdot 7^2$ (1.6A) **21.** 432 (1.6A) **22.** 9 (1.6B) **23.** 7 (1.6B) **24.** 1, 2, 4, 5, 10, 20 (1.7A) **25.** $2 \cdot 2 \cdot 3 \cdot 7$ (1.7B)

ANSWERS to Chapter 2 Odd-Numbered Exercises

SECTION 2.1 *pages 63–64*

1. 40 **3.** 24 **5.** 30 **7.** 12 **9.** 24 **11.** 60 **13.** 56 **15.** 9 **17.** 32 **19.** 36 **21.** 660 **23.** 9384 **25.** 24 **27.** 30 **29.** 24 **31.** 576 **33.** 1680 **35.** 1 **37.** 3 **39.** 5 **41.** 25 **43.** 1 **45.** 4 **47.** 4 **49.** 2 **51.** 48 **53.** 2 **55.** 1 **57.** 7 **59.** 4 **61.** 2 **63.** 1 **65.** 14 **67.** 4 **69.** 1 **71.** 12 **73.** In 12 days, Joe and his friend will have another day off together. **75.** The GCF of 3 and 5 is 1, of 7 and 11 is 1, of 29 and 43 is 1. The GCF of 3 prime numbers is 1. **77.** 4

SECTION 2.2 *pages 67–68*

1. $\frac{3}{4}$ **3.** $\frac{7}{8}$ **5.** $1\frac{1}{2}$ **7.** $2\frac{5}{8}$ **9.** $3\frac{3}{5}$ **11.** $\frac{5}{4}$ **13.** $\frac{8}{3}$ **15.** $\frac{28}{8}$

17.  **19.** 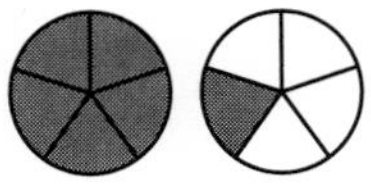 **21.** $5\frac{1}{3}$ **23.** 2 **25.** $3\frac{1}{4}$ **27.** $14\frac{1}{2}$ **29.** 17

31. $1\frac{7}{9}$ **33.** $1\frac{4}{5}$ **35.** 23 **37.** $1\frac{15}{16}$ **39.** $6\frac{1}{3}$ **41.** 5 **43.** 1 **45.** $\frac{14}{3}$ **47.** $\frac{26}{3}$ **49.** $\frac{59}{8}$ **51.** $\frac{25}{4}$ **53.** $\frac{121}{8}$ **55.** $\frac{41}{12}$ **57.** $\frac{34}{9}$ **59.** $\frac{38}{3}$ **61.** $\frac{38}{7}$ **63.** $\frac{63}{5}$ **65.** $\frac{41}{9}$ **67.** $\frac{117}{14}$ **69.** Answers will vary

SECTION 2.3 *pages 71–72*

1. $\frac{5}{10}$ **3.** $\frac{9}{48}$ **5.** $\frac{12}{32}$ **7.** $\frac{9}{51}$ **9.** $\frac{12}{16}$ **11.** $\frac{27}{9}$ **13.** $\frac{20}{60}$ **15.** $\frac{44}{60}$ **17.** $\frac{12}{18}$ **19.** $\frac{35}{49}$ **21.** $\frac{10}{18}$ **23.** $\frac{21}{3}$ **25.** $\frac{35}{45}$ **27.** $\frac{60}{64}$ **29.** $\frac{21}{98}$ **31.** $\frac{30}{48}$ **33.** $\frac{15}{42}$ **35.** $\frac{102}{144}$ **37.** $\frac{153}{408}$

39. $\frac{340}{800}$ **41.** $\frac{1}{3}$ **43.** $\frac{1}{2}$ **45.** $\frac{1}{6}$ **47.** $\frac{2}{3}$ **49.** $1\frac{1}{2}$ **51.** 0 **53.** $\frac{9}{22}$ **55.** 3
57. $\frac{4}{21}$ **59.** $\frac{5}{11}$ **61.** $\frac{2}{9}$ **63.** $\frac{3}{4}$ **65.** $1\frac{1}{3}$ **67.** $\frac{3}{5}$ **69.** $\frac{1}{11}$ **71.** 4 **73.** $\frac{1}{3}$ **75.** $\frac{3}{5}$
77. $2\frac{1}{4}$ **79.** $\frac{1}{5}$ **81.**

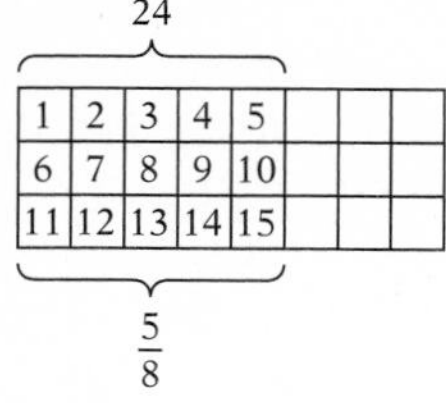

83. Answers will vary

SECTION 2.4 *pages 77–80*

1. $\frac{3}{7}$ **3.** 1 **5.** $1\frac{4}{11}$ **7.** $3\frac{2}{5}$ **9.** $2\frac{4}{5}$ **11.** $2\frac{1}{4}$ **13.** $1\frac{3}{8}$ **15.** $1\frac{7}{15}$ **17.** $\frac{15}{16}$
19. $1\frac{4}{11}$ **21.** 1 **23.** $1\frac{7}{8}$ **25.** $1\frac{1}{6}$ **27.** $\frac{13}{14}$ **29.** $1\frac{1}{45}$ **31.** $\frac{17}{18}$ **33.** $\frac{35}{48}$
35. $\frac{23}{60}$ **37.** $\frac{56}{57}$ **39.** $\frac{13}{21}$ **41.** $2\frac{1}{12}$ **43.** $\frac{277}{315}$ **45.** $2\frac{2}{15}$ **47.** $2\frac{17}{120}$ **49.** $1\frac{89}{150}$
51. $1\frac{31}{72}$ **53.** $1\frac{5}{36}$ **55.** $1\frac{5}{36}$ **57.** $1\frac{29}{40}$ **59.** $1\frac{1}{42}$ **61.** $5\frac{7}{10}$ **63.** $5\frac{11}{16}$ **65.** $8\frac{23}{60}$
67. $18\frac{8}{9}$ **69.** $11\frac{1}{36}$ **71.** $24\frac{11}{12}$ **73.** $12\frac{17}{22}$ **75.** $15\frac{29}{160}$ **77.** $20\frac{37}{48}$ **79.** $44\frac{17}{84}$
81. $11\frac{7}{18}$ **83.** $8\frac{1}{12}$ **85.** $12\frac{71}{105}$ **87.** $15\frac{1}{4}$ **89.** $12\frac{7}{24}$ **91.** $29\frac{227}{240}$ **93.** $12\frac{17}{48}$
95. $9\frac{5}{24}$ **97.** $14\frac{1}{18}$ **99.** $12\frac{3}{4}$ **101.** The veneer is $1\frac{5}{16}$ inches thick. **103.** The length of the shaft is $1\frac{5}{16}$ inches. **105.** The price of the stock at the end of the month is $\$187\frac{1}{8}$.
107a. Fred worked 12 hours of overtime. **107b.** Fred received \$264 in overtime pay.
109. The price of the stock at the end of the three months was $\$30\frac{5}{8}$. **111.** Answers will vary.
113. No. These activities account for only 22 hours.

SECTION 2.5 *pages 85–88*

1. $\frac{2}{17}$ **3.** $\frac{1}{3}$ **5.** $\frac{1}{10}$ **7.** $\frac{7}{11}$ **9.** $\frac{1}{4}$ **11.** $\frac{2}{7}$ **13.** $\frac{4}{7}$ **15.** $\frac{1}{4}$ **17.** $\frac{9}{23}$ **19.** $\frac{1}{4}$
21. $\frac{1}{2}$ **23.** $\frac{19}{56}$ **25.** $\frac{1}{2}$ **27.** $\frac{4}{45}$ **29.** $\frac{11}{18}$ **31.** $\frac{5}{48}$ **33.** $\frac{37}{51}$ **35.** $\frac{17}{70}$ **37.** $\frac{49}{120}$
39. $\frac{8}{45}$ **41.** $\frac{11}{21}$ **43.** $\frac{23}{60}$ **45.** $\frac{1}{18}$ **47.** $5\frac{1}{5}$ **49.** $10\frac{9}{17}$ **51.** $4\frac{7}{8}$ **53.** $\frac{16}{21}$ **55.** $5\frac{1}{2}$
57. $5\frac{4}{7}$ **59.** $7\frac{5}{24}$ **61.** $48\frac{31}{70}$ **63.** $85\frac{2}{9}$ **65.** $9\frac{5}{13}$ **67.** $15\frac{11}{20}$ **69.** $4\frac{23}{24}$ **71.** $4\frac{37}{45}$
73. $1\frac{19}{24}$ **75.** $9\frac{1}{2}$ inches **77.** Ritter jumped $13\frac{7}{8}$ inches further than Coachman. Ritter jumped $10\frac{3}{4}$ inches further than McDaniel. **79a.** It is $7\frac{17}{24}$ miles to the second checkpoint.
79b. It is $4\frac{7}{24}$ miles from the second checkpoint to the finish line. **81.** The wrestler must lose $3\frac{1}{4}$ pounds the third week. **83.** $\$27\frac{5}{8}$ **85.** $\$46\frac{1}{2}$ **87.** $6\frac{1}{8}$ **89.** $\frac{11}{15}$

SECTION 2.6 *pages 93–96*

1. $\frac{7}{12}$ **3.** $\frac{7}{48}$ **5.** $\frac{5}{12}$ **7.** $\frac{1}{3}$ **9.** $\frac{11}{20}$ **11.** $\frac{1}{7}$ **13.** $\frac{1}{8}$ **15.** $\frac{5}{12}$ **17.** $\frac{9}{50}$ **19.** $\frac{5}{32}$
21. 6 **23.** $\frac{2}{3}$ **25.** $\frac{15}{44}$ **27.** $\frac{2}{45}$ **29.** $\frac{5}{14}$ **31.** 10 **33.** $\frac{1}{15}$ **35.** $\frac{10}{51}$ **37.** $\frac{45}{328}$
39. $\frac{10}{27}$ **41.** 4 **43.** $\frac{100}{357}$ **45.** $\frac{27}{100}$ **47.** $1\frac{1}{3}$ **49.** $2\frac{1}{2}$ **51.** $\frac{9}{34}$ **53.** 10 **55.** $16\frac{2}{3}$
57. 1 **59.** $\frac{1}{2}$ **61.** $17\frac{1}{2}$ **63.** 30 **65.** 42 **67.** $12\frac{2}{3}$ **69.** $1\frac{4}{5}$ **71.** $1\frac{2}{3}$
73. $12\frac{2}{9}$ **75.** $1\frac{2}{3}$ **77.** 0 **79.** $\frac{63}{184}$ **81.** $27\frac{2}{3}$ **83.** $17\frac{85}{128}$ **85.** $\frac{2}{5}$ **87.** $8\frac{1}{16}$
89. 18 **91.** 8 **93.** $15\frac{3}{4}$ **95.** 40 **97.** $8\frac{1}{28}$ **99.** Maria can walk $1\frac{1}{6}$ miles. **101.** $1\frac{11}{12}$ feet of the board is cut off. **103.** 535,000 gallons of propellant is used before burnout. **105a.** Ann read 288 pages. **105b.** 144 pages remain to be read. **107.** \$267,000 of the monthly income remained. **109.** The weight is $54\frac{19}{36}$ pounds. **111.** $\frac{1}{2}$ **113.** Yes
115. $\frac{4}{9}$ **117.** $\frac{3}{16}$ **119.**

$\frac{2}{3}$	$\frac{3}{4}$	$\frac{5}{9}$
$\frac{10}{9}$	$\frac{1}{6}$	$\frac{3}{2}$
$\frac{9}{4}$	$\frac{5}{18}$	$\frac{4}{9}$

SECTION 2.7 *pages 101–104*

1. $\frac{5}{6}$ **3.** $1\frac{5}{88}$ **5.** 0 **7.** $1\frac{1}{3}$ **9.** $8\frac{4}{5}$ **11.** $2\frac{73}{256}$ **13.** $1\frac{1}{9}$ **15.** $2\frac{1}{13}$ **17.** $\frac{2}{3}$
19. $\frac{9}{10}$ **21.** 2 **23.** 2 **25.** $\frac{1}{6}$ **27.** 6 **29.** $\frac{1}{15}$ **31.** 2 **33.** $2\frac{1}{2}$ **35.** 3 **37.** 6
39. $\frac{1}{2}$ **41.** $1\frac{1}{6}$ **43.** $3\frac{1}{3}$ **45.** $\frac{1}{30}$ **47.** $1\frac{4}{5}$ **49.** $8\frac{8}{9}$ **51.** $\frac{1}{6}$ **53.** $1\frac{1}{11}$ **55.** 3
57. $\frac{1}{5}$ **59.** 8 **61.** $\frac{7}{26}$ **63.** $\frac{9}{44}$ **65.** 120 **67.** $\frac{11}{40}$ **69.** $\frac{2}{35}$ **71.** $4\frac{4}{5}$ **73.** 24
75. $\frac{13}{32}$ **77.** $10\frac{2}{3}$ **79.** $6\frac{9}{10}$ **81.** $9\frac{7}{13}$ **83.** Undefined **85.** $3\frac{1}{25}$ **87.** $1\frac{139}{394}$ **89.** $8\frac{2}{7}$
91. $1\frac{1}{2}$ **93.** 6 **95.** $1\frac{10}{11}$ **97.** $1\frac{3}{5}$ **99.** $\frac{9}{34}$ **101.** The box contains 16 servings. **103.** The cost is \$24,000 per acre. **105.** The capacity of the theater is 420. **107a.** The total weight of the fat and bone is $1\frac{5}{12}$ pounds. **107b.** There will be 28 servings. **109.** There will be 1 foot of the board remaining. **111.** There is 740 miles between the two cities.
113. a. $\frac{2}{3}$ **b.** $\frac{21}{8}$ **115.** The quotient is greater. **117.** Yes. Answers will vary

SECTION 2.8 *pages 107–108*

1. $\frac{11}{40} < \frac{19}{40}$ **3.** $\frac{2}{3} < \frac{5}{7}$ **5.** $\frac{5}{8} > \frac{7}{12}$ **7.** $\frac{7}{9} < \frac{11}{12}$ **9.** $\frac{13}{14} > \frac{19}{21}$ **11.** $\frac{7}{24} < \frac{11}{30}$ **13.** $\frac{9}{64}$
15. $\frac{8}{729}$ **17.** $\frac{2}{9}$ **19.** $\frac{3}{125}$ **21.** $\frac{8}{245}$ **23.** $\frac{1}{121}$ **25.** $\frac{16}{1225}$ **27.** $\frac{4}{49}$ **29.** $\frac{9}{125}$ **31.** $\frac{27}{88}$
33. $\frac{4}{121}$ **35.** $\frac{1}{30}$ **37.** $1\frac{1}{2}$ **39.** $\frac{20}{21}$ **41.** $\frac{12}{125}$ **43.** $\frac{11}{32}$ **45.** $\frac{17}{24}$ **47.** $\frac{14}{15}$ **49.** $2\frac{7}{10}$
51. $\frac{21}{44}$ **53.** $\frac{25}{39}$ **55.** Between $\frac{2}{3}$ and $\frac{3}{4}$.

CHAPTER REVIEW *pages 111–112*

1. 36 **2.** 54 **3.** 4 **4.** 5 **5.** $\frac{13}{4}$ **6.** $1\frac{7}{8}$ **7.** $3\frac{2}{5}$ **8.** $\frac{19}{7}$ **9.** $\frac{24}{36}$ **10.** $\frac{32}{44}$ **11.** $\frac{2}{3}$ **12.** $\frac{4}{11}$ **13.** $1\frac{1}{8}$ **14.** $1\frac{13}{18}$ **15.** $5\frac{7}{8}$ **16.** $18\frac{13}{54}$ **17.** The total rainfall was $21\frac{7}{24}$ inches. **18.** $\frac{1}{3}$ **19.** $\frac{25}{48}$ **20.** $14\frac{19}{42}$ **21.** $10\frac{1}{8}$ **22.** The second checkpoint is $4\frac{3}{4}$ miles from the finish line. **23.** $\frac{1}{15}$ **24.** $\frac{1}{8}$ **25.** $9\frac{1}{24}$ **26.** $16\frac{1}{2}$ **27.** The car can travel 243 miles. **28.** 2 **29.** $\frac{3}{4}$ **30.** 2 **31.** $3\frac{1}{3}$ **32.** Each acre cost $36,000. **33.** $\frac{11}{18} < \frac{17}{24}$ **34.** $\frac{5}{16}$ **35.** $\frac{5}{36}$ **36.** $\frac{1}{15}$

CHAPTER TEST *pages 113–114*

1. 120 (2.1A) **2.** 8 (2.1B) **3.** $\frac{11}{4}$ (2.2A) **4.** $3\frac{3}{5}$ (2.2B) **5.** $\frac{49}{5}$ (2.2B) **6.** $\frac{5}{8} = \frac{45}{72}$ (2.3A) **7.** $\frac{5}{8}$ (2.3B) **8.** $1\frac{11}{12}$ (2.4A) **9.** $1\frac{61}{90}$ (2.4B) **10.** $22\frac{4}{15}$ (2.4C) **11.** The total rainfall was $21\frac{11}{24}$ inches. (2.4D) **12.** $\frac{1}{4}$ (2.5A) **13.** $\frac{7}{48}$ (2.5B) **14.** $13\frac{81}{88}$ (2.5C) **15.** The value of one share is $\$27\frac{7}{8}$. (2.5D) **16.** $\frac{4}{9}$ (2.6A) **17.** 8 (2.6B) **18.** The electrician's total earnings are $420. (2.6C) **19.** $1\frac{3}{7}$ (2.7A) **20.** $2\frac{2}{19}$ (2.7B) **21.** There were 11 lots available for sale. (2.7C) **22.** $\frac{3}{8} < \frac{5}{12}$ (2.8A) **23.** $\frac{1}{6}$ (2.8B) **24.** $\frac{7}{24}$ (2.8C) **25.** $\frac{5}{6}$ (2.8C)

CUMULATIVE REVIEW *pages 115–116*

1. 290,000 (1.1D) **2.** 291,278 (1.3B) **3.** 73,154 (1.4B) **4.** 540 R12 (1.5C) **5.** 1 (1.6B) **6.** $2 \cdot 2 \cdot 11$ (1.7B) **7.** 210 (2.1A) **8.** 20 (2.1B) **9.** $\frac{23}{3}$ (2.2B) **10.** $6\frac{1}{4}$ (2.2B) **11.** $\frac{15}{48}$ (2.3A) **12.** $\frac{2}{5}$ (2.3B) **13.** $1\frac{7}{48}$ (2.4B) **14.** $14\frac{11}{48}$ (2.4C) **15.** $\frac{13}{24}$ (2.5B) **16.** $1\frac{7}{9}$ (2.5C) **17.** $\frac{7}{20}$ (2.6A) **18.** $7\frac{1}{2}$ (2.6B) **19.** $1\frac{1}{20}$ (2.7A) **20.** $2\frac{5}{8}$ (2.7B) **21.** $\frac{1}{9}$ (2.8B) **22.** $5\frac{5}{24}$ (2.8C) **23.** There is $862 in the checking account at the end of the week. (1.3C) **24.** The total income is $705. (1.4C) **25.** The total weight is $12\frac{1}{24}$ pounds. (2.4D) **26.** The length of the remaining piece is $4\frac{17}{24}$ feet. (2.5D) **27.** The car travels 225 miles. (2.6C) **28.** 25 parcels can be sold from the remaining land. (2.7C)

ANSWERS to Chapter 3 Odd-Numbered Exercises

SECTION 3.1 *pages 121–122*

1. Twenty-seven hundredths **3.** One and five thousandths
5. Thirty-six and four tenths **7.** Thirty-five hundred-thousandths
9. Ten and seven thousandths **11.** Fifty-two and ninety-five hundred-thousandths
13. Two hundred ninety-three ten-thousandths
15. Six and three hundred twenty-four thousandths

17. Two hundred seventy-six and three thousand two hundred ninety-seven ten-thousandths
19. Two hundred sixteen and seven hundred twenty-nine ten-thousandths
21. Four thousand six hundred twenty-five and three hundred seventy-nine ten thousandths
23. One and one hundred-thousandths **25.** 0.762 **27.** 0.000062 **29.** 8.0304
31. 304.07 **33.** 362.048 **35.** 3048.2002 **37.** 7.4 **39.** 23.0 **41.** 22.68
43. 480.33 **45.** 1.039 **47.** 1946.375 **49.** 0.0299 **51.** 2 **53.** 0.010290
55. Hundredths of a second. Answers will vary.
57. **a.** Last two zeros **b.** First zero **c.** All zeros needed **d.** First and last zero
59. 10.15–10.24

SECTION 3.2 *pages 125–126*

1. 150.1065 **3.** 95.8446 **5.** 69.644 **7.** 92.883 **9.** 113.205 **11.** 0.69 **13.** 16.305
15. 110.7666 **17.** 104.4959 **19.** Est.: 234 Cal.: 234.192 **21.** Est.: 782 Cal.: 781.943
23. The length of the shaft is 5.65 inches. **25** The total amount of gas used is 47.8 gallons. **27.** The amount in the checking account is $3664.20. **29.** The total number of television viewers is 366.2 million. **31.** No, the customer does not have enough money. **33.** No, the rope cannot be wrapped around the box.

SECTION 3.3 *pages 129–130*

1. 5.627 **3.** 113.6427 **5.** 6.7098 **7.** 215.697 **9.** 53.8776 **11.** 72.7091 **13.** 4.685
15. 10.0365 **17.** 0.7727 **19.** 3.273 **21.** 791.247 **23.** 547.951 **25.** 403.8557
27. 22.479 **29.** Est.: 600 Cal.: 590.25 **31.** Est.: 30 Cal.: 35.194 **33.** Est.: 3 Cal.: 2.74506 **35.** Est.: 7 Cal.: 7.14925
37. The amount of sales was $469.61. **39.** The missing dimension is 2.59 feet. **41.** The price of gasoline was $1.03 per gallon. **43.** The wage earner would need an additional $20,085.58. **45.** 0.05, 0.005, 0.0005 **47.** The difference in average speeds was 41.983 mph.

SECTION 3.4 *pages 135–140*

1. 0.36 **3.** 0.30 **5.** 0.25 **7.** 0.45 **9.** 6.93 **11.** 1.84 **13.** 4.32 **15.** 0.74
17. 39.5 **19.** 2.72 **21.** 0.603 **23.** 0.096 **25.** 13.50 **27.** 79.80 **29.** 4.316
31. 1.794 **33.** 0.06 **35.** 0.072 **37.** 0.1323 **39.** 0.03568 **41.** 0.0784 **43.** 0.076
45. 34.48 **47.** 580.5 **49.** 20.148 **51.** 0.04255 **53.** 0.17686 **55.** 0.19803
57. 14.8657 **59.** 0.0006608 **61.** 53.9961 **63.** 0.536335 **65.** 0.429 **67.** 2.116
69. 0.476 **71.** 1.022 **73.** 37.96 **75.** 2.318 **77.** 3.2 **79.** 6.5 **81.** 6285.6
83. 3200 **85.** 35,700 **87.** 6.3 **89.** 3.9 **91.** 49,000 **93.** 6.7 **95.** 0.012075
97. 0.117796 **99.** 0.31004 **101.** 0.082845 **103.** 5.175
105. Est.: 90 Cal.: 91.2 **107.** Est.: 0.8 Cal.: 1.0472 **109.** Est.: 4.5 Cal.: 3.897 **111.** Est.: 12 Cal.: 11.2406
113. Est.: 0.32 Cal.: 0.371096 **115.** Est.: 30 Cal.: 31.8528 **117.** Est.: 2000 Cal.: 1941.0695 **119.** Est.: 0.00005 Cal.: 0.0000435
121. The cost to rent the car is $20.52. **123.** The amount received from recycling is $14.06. **125.** The broker's fee is $173.25. **127a.** The amount of the payments is $4590. **127b.** The total cost of the car is $6590. **129.** The total cost to rent the car is $64.20. **131.** The value of the shares is $2047.39. **133.** Grade 1 of steel costs $21.12, Grade 2 costs $29.84, and Grade 3 costs $201.66. **135a.** The cost will be $52.90. **135b.** The cost will be $79.60. **135c.** The cost will be $61.45. **137.** The total cost of the parts is $260.60. **139.** The total cost for the parts and labor is $1290.48. **141.** Answers will vary

SECTION 3.5 *pages 145–148*

1. 0.82 **3.** 4.8 **5.** 89 **7.** 60 **9.** 84.3 **11.** 32.3 **13.** 5.06 **15.** 1.3
17. 0.11 **19.** 3.8 **21.** 6.3 **23.** 0.6 **25.** 2.5 **27.** 1.1 **29.** 130.6 **31.** 0.81
33. 42.40 **35.** 40.70 **37.** 0.46 **39.** 0.019 **41.** 0.087 **43.** 0.360 **45.** 0.103
47. 0.009 **49.** 1 **51.** 3 **53.** 1 **55.** 57 **57.** 0.407 **59.** 4.267 **61.** 0.01037

63. 0.008295 **65.** 0.82537 **67.** 0.032 **69.** 0.23627 **71.** 0.000053 **73.** 0.0018932
75. 18.42 **77.** 16.07 **79.** 0.0135 **81.** 0.023678 **83.** 0.112 **85.** Est.: 10 Cal.: 11.1632
87. Est.: 1000 Cal.: 884.0909 **89.** Est.: 1.5 Cal.: 1.8269 **91.** Est.: 50 Cal.: 58.8095 **93.** Est.: 100 Cal.: 72.3053 **95.** Est.: 0.0025 Cal.: 0.0023
97. Ramon averaged 6.23 yards per carry. **99.** Each car insurance payment was \$236.90.
101. Three shelves can be cut from the board. **103.** The amount of the dividend is \$1.72.
105a. The amount to be paid in monthly payments is \$842.58. **105b.** The monthly payment is \$46.81. **107.** The possible answers are 1¢, 3¢, 5¢, 9¢, 15¢, or 45¢. **109.** Answers will vary
111. The average yogurt consumption is 5.16 pounds per person. **113.** × **115.** ×
117. ÷ **119.** 2.53

SECTION 3.6 *pages 151–152*

1. 0.625 **3.** 0.667 **5.** 0.167 **7.** 0.417 **9.** 1.750 **11.** 1.500 **13.** 4.000
15. 0.003 **17.** 7.080 **19.** 37.500 **21.** 0.375 **23.** 0.208 **25.** 3.333
27. 5.444 **29.** 0.313 **31.** $\frac{4}{5}$ **33.** $\frac{8}{25}$ **35.** $\frac{1}{8}$ **37.** $1\frac{1}{4}$ **39.** $16\frac{9}{10}$ **41.** $8\frac{2}{5}$
43. $8\frac{437}{1000}$ **45.** $2\frac{1}{4}$ **47.** $\frac{23}{150}$ **49.** $\frac{703}{800}$ **51.** $1\frac{17}{25}$ **53.** $\frac{9}{200}$ **55.** $16\frac{18}{25}$
57. $\frac{33}{100}$ **59.** $\frac{1}{3}$ **61.** $0.15 < 0.5$ **63.** $6.65 > 6.56$ **65.** $2.504 > 2.054$
67. $\frac{3}{8} > 0.365$ **69.** $\frac{2}{3} > 0.65$ **71.** $\frac{5}{9} > 0.55$ **73.** $0.62 > \frac{7}{15}$ **75.** $0.161 > \frac{1}{7}$
77. $0.86 > 0.855$ **79.** $1.005 > 0.5$ **81.** **a.** False **b.** False **c.** True
83. No. 0.0402 rounded to hundredths is 0.04, to thousandths is 0.040. **85.** Answers will vary

CHAPTER REVIEW *pages 155–156*

1. Twenty-two and ninety-two ten-thousandths **2.** Three hundred forty-two and thirty-seven hundredths **3.** 34.025 **4.** 3.06753 **5.** 7.94 **6.** 0.05678 **7.** 36.714 **8.** 833.958
9. The checking account balance is \$478.02. **10.** 22.8635 **11.** 4.8785 **12.** The new checking account balance is \$661.51. **13.** 8.932 **14.** 25.7446 **15.** The amount of income tax paid was \$5600. **16.** 54.5 **17.** 6.594 **18.** The amount of each monthly payment is \$123.45. **19.** 0.778 **20.** 2.33 **21.** $\frac{3}{8}$ **22.** $\frac{2}{3}$ **23.** $0.055 < 0.1$ **24.** $\frac{5}{8} > 0.62$

CHAPTER TEST *pages 157–158*

1. Forty-five and three hundred two ten-thousandths (3.1A) **2.** 209.07086 (3.1A)
3. 7.095 (3.1B) **4.** 0.0740 (3.1B) **5.** 255.957 (3.2A) **6.** 458.581 (3.2A)
7. The total income was \$1543.57. (3.2B) **8.** 4.087 (3.3A) **9.** 27.76626 (3.3A)
10. 1.37 inch (3.3B) **11.** 0.00548 (3.4A) **12.** Los Angeles had the highest rate for a gallon of gasoline. (3.2B) **13.** Boston had the lowest tax for a gallon of gasoline. (3.2B) **14.** The cost of the long distance call is \$4.63. (3.4B) **15.** 232 (3.5A) **16.** 1.538 (3.5A)
17. The amount of each monthly payment is \$142.85. (3.5B) **18.** 0.692 (3.6A)
19. $\frac{33}{40}$ (3.6B) **20.** $0.66 < 0.666$ (3.6C)

CUMULATIVE REVIEW *pages 159–160*

1. 235 r17 (1.5C) **2.** 128 (1.6A) **3.** 3 (1.6B) **4.** 72 (2.1A) **5.** $4\frac{2}{5}$ (2.2B)
6. $\frac{37}{8}$ (2.2B) **7.** $\frac{25}{60}$ (2.3A) **8.** $1\frac{17}{48}$ (2.4B) **9.** $8\frac{35}{36}$ (2.4C) **10.** $5\frac{23}{36}$ (2.5C)
11. $\frac{1}{12}$ (2.6A) **12.** $9\frac{1}{8}$ (2.6B) **13.** $1\frac{2}{9}$ (2.7A) **14.** $\frac{19}{20}$ (2.7B) **15.** $\frac{3}{16}$ (2.8B)
16. $2\frac{5}{18}$ (2.8C) **17.** Sixty-five and three hundred nine ten-thousandths (3.1A)

18. 504.6991 (3.2A) **19.** 21.0764 (3.3A) **20.** 55.26066 (3.4A) **21.** 2.154 (3.5A) **22.** 0.733 (3.6A) **23.** $\frac{1}{6}$ (3.6B) **24.** $\frac{8}{9} < 0.98$ (3.6C) **25.** There were 234 passengers on the continuing flight. (1.3C) **26.** The value of each share is $\$32\frac{7}{8}$. (2.5D) **27.** The checking account balance was $617.38. (3.3B) **28.** The resulting thickness of the bushing was 1.395 inches. (3.3B) **29.** The amount of income tax paid last year was $6008.80. (3.4B) **30.** The monthly payment is $23.87. (3.5B)

ANSWERS to Chapter 4 Odd-Numbered Exercises

SECTION 4.1 *pages 165–166*

1. $\frac{1}{5}$ 1:5 1 TO 5 **3.** $\frac{2}{1}$ 2:1 2 TO 1 **5.** $\frac{3}{8}$ 3:8 3 TO 8 **7.** $\frac{37}{24}$ 37:24 37 TO 24 **9.** $\frac{1}{1}$ 1:1 1 TO 1 **11.** $\frac{7}{10}$ 7:10 7 TO 10 **13.** $\frac{1}{2}$ 1:2 1 TO 2 **15.** $\frac{2}{1}$ 2:1 2 TO 1 **17.** $\frac{3}{4}$ 3:4 3 TO 4 **19.** $\frac{5}{3}$ 5:3 5 TO 3 **21.** $\frac{2}{3}$ 2:3 2 TO 3 **23.** $\frac{2}{1}$ 2:1 2 TO 1 **25.** The ratio of housing costs to total income is $\frac{1}{3}$. **27.** The ratio of utilities cost to food cost is $\frac{3}{8}$. **29.** The ratio of College Senior football to High School Senior football players is $\frac{1}{30}$. **31.** The ratio of turns in the primary coil to the turns in the secondary coil is $\frac{1}{12}$. **33a.** The amount of increase is $20,000. **33b.** The ratio of the increase to the original value is $\frac{2}{9}$. **35.** The ratio of the increase in price to the original price is $\frac{5}{16}$. **37.** No. $\frac{265}{778}$ is greater than $\frac{1}{3}$ **39.** No. Answers will vary

SECTION 4.2 *pages 169–170*

1. $\frac{\text{3 pounds}}{\text{4 people}}$ **3.** $\frac{\$20}{\text{3 boards}}$ **5.** $\frac{\text{20 miles}}{\text{1 gallon}}$ **7.** $\frac{\text{5 children}}{\text{2 families}}$ **9.** $\frac{\text{8 gallons}}{\text{1 hour}}$ **11.** 2.5 feet/second **13.** $325/week **15.** 110 trees/acre **17.** $4.71/hour **19.** 52.4 miles/hour **21.** 28 miles/gallon **23.** $1.65/pound **25.** The student drives 28.4 miles per gallon of gas. **27.** The rocket uses 213,600 gallons of fuel per minute. **29.** The dividend is $1.60 per share. **31a.** There were 4878 compact discs meeting company standards. **31b.** The cost was $5.44 per disc. **33.** The profit was $0.60 per box of strawberries. **35.** The cost was $8.33 per thousand viewers. **37.** 60 minutes had the lowest cost per thousand viewers.

SECTION 4.3 *pages 175–178*

1. True **3.** Not true **5.** Not true **7.** True **9.** True **11.** Not true **13.** True **15.** True **17.** True **19.** Not true **21.** True **23.** Not true **25.** $n = 3$ **27.** $n = 6$ **29.** $n = 20$ **31.** $n = 2$ **33.** $n = 8$ **35.** $n = 4$ **37.** $n = 4.38$ **39.** $n = 9.6$ **41.** $n = 10.67$ **43.** $n = 6.67$ **45.** $n = 26.25$ **47.** $n = 96$ **49.** $n = 9.78$ **51.** $n = 3.43$ **53.** $n = 1.34$ **55.** $n = 50.4$ **57.** There are 50 calories in the serving of cereal. **59.** There were 50 pounds of fertilizer used. **61.** The property tax is $2500. **63.** It would take 2496 bricks. **65.** The distance between the two cities is 16 miles. **67.** 1.25 ounces of medication are required. **69.** There would be 160,000 people voting in the election. **71.** There would be 960 braking defects. **73.** You own 400 shares of stock. **75.** Students used the computer 300 hours. **77.** The national debt would have been $6.76 trillion in 1992. The difference is $2.76 trillion. **79.** The bowling ball would weigh $2\frac{2}{3}$ pounds on the moon. **81.** Answers will vary

CHAPTER REVIEW *pages 181–182*

1. $\frac{2}{7}$, 2:7, 2 TO 7 **2.** $\frac{2}{5}$, 2:5, 2 TO 5 **3.** $\frac{1}{1}$, 1:1, 1 TO 1 **4.** $\frac{2}{5}$, 2:5, 2 TO 5 **5.** The ratio of the high temperature to the low temperature is 2:1. **6.** The ratio of increase to original value is $\frac{1}{2}$. **7.** The ratio of decrease in price to original price is $\frac{2}{5}$. **8.** The ratio of T.V. advertising to newpaper advertising is $\frac{5}{2}$. **9.** $\frac{100 \text{ miles}}{3 \text{ hours}}$ **10.** $\frac{\$15}{4 \text{ hours}}$ **11.** 62.5 miles/hour **12.** $7.50/hour **13.** 27.2 miles/gallon **14.** $1.75/pound **15.** The cost is $44.75 per share. **16.** Mahesh drove an average of 56.8 miles/hour. **17.** The cost is $.68/pound. **18.** The cost per radio is $37.50. **19.** True **20.** True **21.** True **22.** Not true **23.** $n = 36$ **24.** $n = 68$ **25.** $n \approx 65.45$ **26.** $n \approx 19.44$ **27.** There were 22.5 pounds of fertilizer used. **28.** The cost of the insurance is $193.50. **29.** The property tax is $2400. **30.** It would take 1344 blocks to build the wall.

CHAPTER TEST *pages 183–184*

1. $\frac{3}{2}$, 3:2, 3 TO 2 (4.1A) **2.** $\frac{3}{5}$, 3:5, 3 TO 5 (4.1A) **3.** $\frac{1}{3}$, 1:3, 1 TO 3 (4.1A) **4.** $\frac{1}{6}$, 1:6, 1 TO 6 (4.1A) **5.** The ratio of the city temperature to the desert temperature is $\frac{43}{56}$. (4.1B) **6.** The ratio of the cost of radio advertising to the total cost of advertising is $\frac{8}{13}$. (4.1B) **7.** $\frac{9 \text{ supports}}{4 \text{ feet}}$ (4.2A) **8.** $\frac{\$27}{4 \text{ boards}}$ (4.2A) **9.** $1836.40/month (4.2B) **10.** 30.5 miles/gallon (4.2B) **11.** The cost of the lumber is $1.73/foot. (4.2C) **12.** The plane's speed is 538 miles/hour. (4.2C) **13.** True (4.3A) **14.** Not true (4.3A) **15.** $n = 40.5$ (4.3B) **16.** $n = 144$ (4.3B) **17.** The student's body contains 132 pounds of water. (4.3C) **18.** The person requires 0.875 ounce of medication. (4.3C) **19.** The property tax is $2800. (4.3C) **20.** The dividend would be $625. (4.3C)

CUMULATIVE REVIEW *pages 185–186*

1. 9158 (1.3B) **2.** $2^4 \cdot 3^3$ (1.6A) **3.** 3 (1.6B) **4.** $2 \cdot 2 \cdot 2 \cdot 2 \cdot 2 \cdot 5$ (1.7B) **5.** 36 (2.1A) **6.** 14 (2.1B) **7.** $\frac{5}{8}$ (2.3B) **8.** $8\frac{3}{10}$ (2.4C) **9.** $5\frac{11}{18}$ (2.5C) **10.** $2\frac{5}{6}$ (2.6B) **11.** $4\frac{2}{3}$ (2.7B) **12.** $\frac{23}{30}$ (2.8C) **13.** Four and seven hundred nine ten-thousandths (3.1A) **14.** 2.10 (3.1B) **15.** 1.990 (3.5A) **16.** $\frac{1}{15}$ (3.6B) **17.** $\frac{1}{8}$ (4.1A) **18.** $\frac{29¢}{2 \text{ bars}}$ (4.2A) **19.** 33.4 miles/gallon (4.2B) **20.** $n = 4.25$ (4.3B) **21.** 57.2 miles/hour (4.2C) **22.** $n = 36$ (4.3B) **23.** The new balance is $744. (1.3C) **24.** The monthly payment is $370. (1.5D) **25.** There are 105 pages remaining to be read. (2.6C) **26.** The cost for each acre is $36,000. (2.7C) **27.** The amount of change is $19.62. (3.3B) **28.** The player's batting average is 0.271. (3.5B) **29.** There will be 25 inches eroded. (4.3C) **30.** The person needs 1.6 ounces of medication. (4.3C)

ANSWERS to Chapter 5 Odd-Numbered Exercises

SECTION 5.1 *pages 191–192*

1. $\frac{1}{4}$, 0.25 **3.** $1\frac{3}{10}$, 1.30 **5.** 1, 1.00 **7.** $\frac{73}{100}$, 0.73 **9.** $3\frac{83}{100}$, 3.83 **11.** $\frac{7}{10}$, 0.70 **13.** $\frac{22}{25}$, 0.88 **15.** $\frac{8}{25}$, 0.32 **17.** $\frac{2}{3}$ **19.** $\frac{5}{6}$ **21.** $\frac{1}{9}$ **23.** $\frac{5}{11}$ **25.** $\frac{3}{70}$ **27.** $\frac{1}{15}$

29. 0.065 **31.** 0.0055 **33.** 0.0825 **35.** 0.0675 **37.** 0.0045 **39.** 0.804 **41.** 16% **43.** 5% **45.** 1% **47.** 70% **49.** 124% **51.** 0.4% **53.** 0.6% **55.** 310.6% **57.** 54% **59.** 33.3% **61.** 62.5% **63.** 16.7% **65.** 17.5%

67. 177.8% **69.** 30% **71.** $23\frac{1}{3}\%$ **73.** $237\frac{1}{2}\%$ **75.** $216\frac{2}{3}\%$

77. **a.** False; 4(200%) = 8 **b.** False; $\frac{4}{200\%} = 2$ **c.** True **d.** False; 125% = 1.25

79. Answers will vary **81.** $\frac{1}{2}$ **83.** No; 0.495 **85.** The fraction is less than $\frac{1}{100}$.

SECTION 5.2 *pages 195–196*

1. 8 **3.** 10.8 **5.** 0.075 **7.** 80 **9.** 51.895 **11.** 7.5 **13.** 13 **15.** 3.75 **17.** 20 **19.** 210 **21.** 5% of 95 **23.** 82% of 16 **25.** 84% of 32 **27.** 25,805.0324 **29.** The amount deducted for income tax is $403.20. **31.** The profit is $4000. **33.** The rebate is $1680. **35a.** The sales tax is $570. **35b.** The total cost including sales tax is $10,070. **37.** The total number of employees needed is 671. **39.** There were 23 permanent injuries obtained playing offense. **41.** The difference between the number of viewers is 752,000. **43.** The increase in price is $204.26.

SECTION 5.3 *pages 199–200*

1. 32% **3.** $16\frac{2}{3}\%$ **5.** 200% **7.** 37.5% **9.** 18% **11.** 0.25% **13.** 20%

15. 400% **17.** 2.5% **19.** 37.5% **21.** 0.25% **23.** 2.4% **25.** 9.6% **27.** The company spent 5% of its budget for advertising. **29.** The dividend is 3.2%. **31.** 62.5% of the families liked the TV show. **33a.** The fat is 20% of the roast. **33b.** The side of beef would weigh 110 pounds. **35.** 98.5% of the concrete slabs met safetly requirements. **37.** Answers will vary **39.** The government spent 20.9% of its income on public debt. **41.** Home heating consisted of 56.7% of the utility bill.

SECTION 5.4 *pages 203–204*

1. 75 **3.** 50 **5.** 100 **7.** 85 **9.** 1200 **11.** 19.2 **13.** 7.5 **15.** 32 **17.** 200 **19.** 80 **21.** 9 **23.** 504 **25.** 108 **27.** 7122.1548 **29.** The estimated life of the brakes is 50,000 miles. **31.** The city's population was 56,000 people. **33.** The price at the competitor's store is $46.50. **35a.** The manufacturer tested 3000 computer boards. **35b.** There were 2976 computer boards that were not defective. **37.** 6.6 grams **39.** No

SECTION 5.5 *pages 207–208*

1. 65 **3.** 25% **5.** 75 **7.** 12.5% **9.** 400 **11.** 19.5 **13.** 14.8% **15.** 62.62 **17.** 5 **19.** 45 **21.** 15 **23.** $24,500 **25.** $71.25 **27.** Enrico typed 98% of the words correctly. **29.** The winner's share increased by 5.6%. **31.** The harvest decreased by 19.1%. **33.** The current yield is 7.6%. **35.** The projected increase is 13.7%. **37.** The potency of the solution is reduced by 87.5%.

CHAPTER REVIEW *pages 211–212*

1. $\frac{3}{25}$ **2.** $\frac{1}{6}$ **3.** 0.42 **4.** 0.076 **5.** 38% **6.** 150% **7.** $n = 60$ **8.** $n = 5.4$

9. $n = 19.36$ **10.** $n = 77.5$ **11.** The company spent $4500 for T.V. advertising. **12.** The total cost is $1041.25. **13.** $n = 150\%$. **14.** $n = 20\%$ **15.** $n = 7.3\%$ **16.** $n = 613.3\%$ **17.** The dividend is 4.6% of the stock price. **18.** The percent increase in value was 64%. **19.** $n = 75$ **20.** $n = 10.9$ **21.** $n = 157.5$ **22.** $n = 504$ **23.** The city's population was 70,000 people. **24.** The previous year's batting average was 0.252. **25.** $n = 198.4$ **26.** $n = 160\%$ **27.** The computer cost $3000. **28.** Trent answered 85% of the questions correctly.

CHAPTER TEST *pages 213–214*

1. 0.973 (5.1A) **2.** $\frac{1}{6}$ (5.1A) **3.** 30% (5.1B) **4.** 163% (5.1B) **5.** 150% (5.1B)

6. $66\frac{2}{3}$% (5.1B) **7.** $n = 50.05$ (5.2A) **8.** $n = 61.36$ (5.2A) **9.** 76% of 13 (5.2A) **10.** 212% of 12 (5.2A) **11.** The company spends $4500 for advertising. (5.2B) **12.** There were 1170 pounds of vegetables unspoiled. (5.2B) **13.** One serving provides 14.7% of the daily recommended amount of potassium. (5.3B) **14.** One serving provides 9.1% of the daily recommended number of calories. (5.3B) **15.** The store's temporary employees were 16% of the total number of permanent employees. (5.3B) **16.** Conchita answered 91.3% of the questions correctly. (5.3B) **17.** $n = 80$ (5.4A) **18.** $n = 28.3$ (5.4A) **19.** There were 32,000 transistors tested. (5.4B) **20.** The increase is 60% of the original price. (5.4B) **21.** $n = 143.0$ (5.5A) **22.** 1000% (5.5A) **23.** The increase in the hourly wage is $1.02. (5.5B) **24.** The population now is 220% of the population 10 years ago. (5.5B) **25.** The value of the car is $6500. (5.5B)

CUMULATIVE REVIEW *pages 215–216*

1. 4 (1.6B) **2.** 240 (2.1A) **3.** $10\frac{11}{24}$ (2.4C) **4.** $12\frac{41}{48}$ (2.5C) **5.** $12\frac{4}{7}$ (2.6B)

6. $\frac{7}{24}$ (2.7B) **7.** $\frac{1}{3}$ (2.8B) **8.** $\frac{13}{36}$ (2.8C) **9.** 3.08 (3.1B) **10.** 1.1196 (3.3A)

11. 34.2813 (3.5A) **12.** 3.625 (3.6A) **13.** $1\frac{3}{4}$ (3.6B) **14.** $\frac{3}{8} < 0.87$ (3.6C)

15. $n = 53.3$ (4.3B) **16.** $9.60/hour (4.2B) **17.** $\frac{11}{60}$ (5.1A) **18.** $83\frac{1}{3}$% (5.1B)

19. 19.56 (5.2A) **20.** $133\frac{1}{3}$% (5.3A) **21.** 9.92 (5.4A) **22.** 342.9% (5.5A)

23. Sergio's take-home pay is $592. (2.6C) **24.** The monthly payment is $92.25. (3.5B) **25.** There were 420 gallons of gasoline used during the month. (3.5B) **26.** The real estate tax is $3000. (4.3C) **27.** The sales tax is 6% of the purchase price. (5.3B) **28.** 45% of the people surveyed did not favor the candidate. (5.3B) **29.** The 1990 value is 250% of the 1985 value. (5.5B) **30.** 18% of the children tested had levels of lead that exceeded federal standards. (5.5B)

ANSWERS to Chapter 6 Odd-Numbered Exercises

SECTION 6.1 *pages 221–222*

1. The unit cost is $0.056 per ounce. **3.** The unit cost is $0.149 per ounce. **5.** The unit cost is $0.033 per screw. **7.** The unit cost is $0.195 per foot. **9.** The unit cost is $0.037 per ounce. **11.** The unit cost is $2.995 per clamp. **13.** The more economical purchase is 12 ounces for $1.79. **15.** The more economical purchase is 12 ounces for $2.64. **17.** The more economical purchase is 46 ounces for $0.69. **19.** The more economical purchase is 2 quarts for $4.79. **21.** The more economical purchase is 200 tablets for $9.98. **23.** The more economical purchase is 16 ounces for $1.32. **25.** The total cost is $8.37. **27.** The total cost is $2.94. **29.** The total cost is $3.51. **31.** The total cost is $0.40. **33.** The total cost is $2.01. **35.** Answers will vary

SECTION 6.2 *pages 229–232*

1. The percent increase is 8%. **3.** The percent increase is $33\frac{1}{3}$%. **5a.** The amount of increase is $2925. **5b.** The new salary is $35,425. **7.** The dividend per share of stock is $1.60. **9.** The home is 240 square feet larger. **11.** The markup is $71.25. **13.** The markup is $12.60. **15.** The markup rate is 30%. **17a.** The markup is $77.76. **17b.** The selling price is $239.76. **19a.** The markup is $21.70. **19b.** The selling price is $83.70. **21.** The selling price is $227.20. **23.** The percent decrease is 40%. **25.** The percent decrease is 40%.

27. The car loses $3360 in value. **29a.** The amount of the decrease is 12 cameras. **29b.** The percent decrease is 60%. **31a.** The amount of decrease was $15.20. **31b.** The average monthly bill is $60.80. **33.** The percent decrease is 20%. **35.** The discount rate is $33\frac{1}{3}$%. **37.** The discount is $67.50. **39.** The discount rate is 25%. **41a.** The discount is $0.17 per pound. **41b.** The sale price is $0.68 per pound. **43a.** The discount is $4. **43b.** The discount rate is 25%. **45.** The percent increase was 21.1%. **47.** WstwOn **49a.** The total cost is $172. **49b.** The total selling price is $258. **49c.** The florist expects to sell 186 roses. **49d.** The selling price is $1.39 per rose. **51.** Answers will vary

SECTION 6.3 *pages 235–236*

1. The simple interest due is $6750. **3.** The simple interest due is $16. **5a.** The interest on the loan is $72,000. **5b.** The monthly payment is $6187.50. **7a.** The simple interest is $1080. **7b.** The monthly payment is $545. **9.** The monthly payment is $1332.50. **11.** The value will be $1414.78. **13.** The value will be $12,380.43. **15.** The value will be $28,212. **17a.** The value of the investment will be $6040.86. **17b.** The interest earned on the investment is $3040.86. **19.** Answers will vary **21a.** The value after the deposit will be $200.50. **21b.** The value on March 1 will be $301.50.

SECTION 6.4 *pages 241–242*

1. The mortgage is $82,450. **3.** The down payment is $7500. **5.** The down payment is $212,500. **7.** The loan origination fee is $3750. **9a.** The down payment is $4750. **9b.** The mortgage is $90,250. **11.** The mortgage is $86,400. **13.** The monthly mortgage payment is $1157.73. **15.** No, the couple cannot afford to buy the home. **17.** The monthly property tax is $112.35. **19a.** The monthly mortgage payment is $1678.40. **19b.** The interest is $736.68 per month. **21.** The total monthly payment (mortgage and property tax) is $982.70. **23.** The monthly mortgage payment is $1430.83. **25.** $88,917.60 of interest can be saved.

SECTION 6.5 *pages 245–246*

1. Amanda does not have enough money for the down payment. **3.** The sales tax is $742.50. **5.** The license fee is $250. **7a.** The sales tax is $420. **7b.** The total cost of the sales tax and the license fee is $595. **9a.** The down payment is $550. **9b.** The amount financed is $1650. **11.** The amount financed is $28,000. **13.** The monthly truck payment is $348.39. **15.** The cost to operate the car is $5120. **17.** The cost per mile is $0.11. **19.** The interest is $74.75. **21a.** The amount financed is $76,600. **21b.** The monthly truck payment is $1590.09. **23.** The monthly car payment is $709.88. **25.** The 10% loan with the application fee has a lower loan cost.

SECTION 6.6 *pages 249–250*

1. Lewis earns $220. **3.** The commission earned is $2955. **5.** The commission received was $84. **7.** The teacher received $3244 per month. **9.** The electrician's overtime hourly wage was $31.60. **11.** The commission earned is $112.50. **13.** The typist earned $393.75. **15.** The consultant's hourly wage is $85. **17a.** Mark's hourly wage is $12.75. **17b.** Mark earned $102 on Saturday. **19a.** The increase in pay is $0.85 per hour. **19b.** The nurse's hourly pay is $9.35. **21.** The commission earned is $375. **23.** The difference in monthly pay is $7408.33. **25.** The projected increase is 500,500 analysts.

SECTION 6.7 *pages 257–260*

1. The new balance is $486.32. **3.** The balance in the account is $1222.47. **5.** The current balance is $825.27. **7.** The current balance is $3000.82. **9.** Yes, there is enough money to purchase a refrigerator. **11.** Yes, there is enough money to make both purchases. **13.** The bank statement and checkbook balance. **15.** The bank statement and checkbook balance. **17.** Added **19.** Subtract

CHAPTER REVIEW *pages 263–264*

1. 14.5¢/ounce **2.** 60 ounces for $8.40 **3.** The percent increase was 30.4%. **4.** The yearly percent increase was 200%. **5.** The selling price is $2079. **6.** The percent increase in earnings is 15%. **7.** The markup is $72. **8.** The sale price is $141. **9.** The simple interest due is $6750. **10.** The simple interest due is $1200. **11.** The value of the investment is $45,550.75. **12.** The value of the investment will be $53,593. **13.** The down payment is $18,750. **14.** The loan origination fee is $1875. **15.** The total monthly payment is $578.53. **16.** The monthly mortgage payment is $934.08. **17.** The total cost is $1158.75. **18.** The cost for the 4 items is 12.1¢/mile. **19.** The amount of interest is $97.33. **20.** The monthly payment is $330.82. **21.** The commission received was $3240. **22.** Richard's total income was $655.20. **23.** The current balance is $943.68. **24.** The current balance is $8866.58.

CHAPTER TEST *pages 265–266*

1. The cost per foot is $6.92. (6.1A) **2.** The more economical purchase is 5 pounds for $1.65. (6.1B) **3.** The cost is $14.53. (6.1C) **4.** The percent increase is 20%. (6.2A) **5.** The percent increase was 150%. (6.2A) **6.** The selling price is $301. (6.2B) **7.** The selling price is $6.25. (6.2B) **8.** The percent decrease was 57.6%. (6.2C) **9.** The percent decrease was 20%. (6.2C) **10.** The sale price is $209.30. (6.2D) **11.** The discount rate is 40%. (6.2D) **12.** The simple interest due is $2000. (6.3A) **13.** The interest earned was $24,420.60. (6.3B) **14.** The loan origination fee is $3350. (6.4A) **15.** The monthly mortgage payment is $1713.44. (6.4B) **16.** The amount financed is $14,350. (6.5A) **17.** The monthly car payment is $301.82. (6.5B) **18.** The nurse earns $703.50. (6.6A) **19.** The current balance is $6612.25. (6.7A) **20.** The bank statement and checkbook balance. (6.7B)

CUMULATIVE REVIEW *pages 267–268*

1. 13 (1.6B) **2.** $8\frac{13}{24}$ (2.4C) **3.** $2\frac{37}{48}$ (2.5C) **4.** 9 (2.6B) **5.** 2 (2.7B) **6.** 5 (2.8C) **7.** 52.2 (3.5A) **8.** 1.417 (3.6A) **9.** $51.25/hour (4.2B) **10.** $n = 10.94$ (4.3B) **11.** 62.5% (5.1B) **12.** 27.3 (5.2A) **13.** 0.182 (5.1A) **14.** 42% (5.3A) **15.** 250 (5.4A) **16.** 154.76 (5.5A) **17.** The total rainfall was $13\frac{11}{12}$ inches. (2.4D) **18.** The amount the family paid in taxes was $570. (2.6C) **19.** The ratio of decrease in price to original price is $\frac{3}{5}$. (4.1B) **20.** The car drove 33.4 miles per gallon. (4.2C) **21.** The unit cost is $0.93. (4.2C) **22.** The dividend is $280. (4.3C) **23.** The sale price is $720. (5.2B) **24.** The selling price is $119. (6.2B) **25.** The percent increase is 8%. (6.2A) **26.** The simple interest due is $6000. (6.3A) **27.** The monthly payment is $381.60. (6.5B) **28.** The new balance is $2243.77. (6.7A) **29.** The cost per mile is $0.20. (6.5B) **30.** The monthly mortgage payment is $743.18. (6.4B)

ANSWERS to Chapter 7 Odd-Numbered Exercises

SECTION 7.1 *pages 275–278*

1. The service station sold 19,000 gallons of gasoline. **3.** 21.1% of the gasoline sold was sold during the first week. **5.** 15% of the budget comes from the federal government. **7.** The number of fish caught was 4000. **9.** 27.5% of the fish caught were caught in 1994. **11.** The ratio of the units is $\frac{1}{3}$. **13.** 9.4% of the required units were in mathematics. **15.** The ratio of the population of Asia to Africa is $\frac{60}{13}$. **17.** The ratio of the population of North America to South America is $\frac{11}{18}$. **19.** The total value of bonds rated A/A or higher was $94,026,000. **21.** The total value of bonds rated A/A or AA/Aa was $72,343,000. **23.** The amount spent for student services was 1,244,452. **25.** The total amount spent for institutional support and the physical plant was \$2,074,086. **27.** The ratio of the land area of North America to South America is $\frac{314}{229}$. **29.** The ratio of the land area of Australia to the total land area of the seven continents is $\frac{11}{212}$. **31.** \$5112 of the family's income was spent for food. **33.** \$3408 of the family's income was spent on clothing and entertainment. **35.** Answers will vary

SECTION 7.2 *pages 281–282*

1. 40 cars were sold in May. **3.** The ratio of the number of cars sold in February to the number of cars sold in June is $\frac{3}{10}$. **5.** 200 burglaries per 100,000 of population was committed. **7.** The ratio of auto thefts in 1990 to the number of auto thefts in 1994 is $\frac{10}{7}$. **9.** There was a 44% increase in the rate of inflation for prescription drugs. **11.** There was a 14% decrease in the rate of inflation for physicians services. **13.** 60 inches of snow fell during January. **15.** A total of 85 inches of snow fell during November and December. **17.** The difference between the number of business and residential calls was 25,000 calls. **19.** 195,000 residential calls were made between 9 A.M. and noon. **21.** Coal: 19%, Wood: 3%, Natural Gas: 20%, Petroleum: 50%, Hydroelectric: 5%, Nuclear, 4%; your estimate should be approximately 100%. **23.** Answers will vary

SECTION 7.3 *pages 285–286*

1. 13 students scored between 60 and 80. **3.** 11 students scored above 80. **5.** 36 customers made purchases between \$20 and \$30. **7.** 66 customers made purchases of more than \$30. **9.** 20 cars sold for more than \$13,000. **11.** The ratio of the number of cars is $\frac{4}{7}$. **13.** 4 runners ran the 100 yard dash in less than 11 seconds. **15.** 19 runners ran the race between 11 and 12 seconds. **17.** 5 families spent more than 8% of their income on vacations. **19.** The ratio is $\frac{3}{1}$. **21.** The ratio of the number of students scoring between 500 and 600 to the number tested is $\frac{11}{24}$. **23.** 525 students scored above 400. **25.** Answers will vary

SECTION 7.4 *pages 289–290*

1. The mean grade is 83. **3.** The mean height is 77.4 inches. **5.** The mean low temperature was 17.4°F. **7.** The mean monthly pay check was \$1240. **9.** The mean braking distance was 218 feet. **11.** The mean annual rainfall was 460 inches. **13.** The median temperature was 95°F. **15.** The median hourly wage was \$7.16. **17.** $7\frac{1}{2}$ hours was the median number of hours worked. **19.** 33.5 requests was the median number for a 6-day period. **21.** The median utility bill was \$86.48. **23a.** False **23b.** True **25.** Median

CHAPTER REVIEW *pages 293–294*

1. 44 students received grades. **2.** The ratio of the number of students receiving B grades to D grades is $\frac{7}{3}$. **3.** 29.5% of the students received C grades. **4.** The baseball team's total income was \$965,000,000. **5.** The ratio of income from tickets sold to income from local broadcasting is $\frac{20}{9}$. **6.** 22.8% of the income was received from national broadcasting. **7.** The difference in temperature was 30°. **8.** The ratio of the maximum to minimum temperature was $\frac{7}{4}$. **9.** The lowest temperature was on Wednesday, and it was 15°. **10.** The difference in profit was \$20 million. **11.** The total profit in the third and fourth quarters was \$85 million. **12.** The ratio of profit in the third quarters of 1993 and 1994 was $\frac{5}{8}$. **13.** 9 trees were over 72 inches tall. **14.** The ratio is $\frac{9}{10}$.

15. 31.7% of the trees had a height between 66 to 69 inches. **16.** In 28 games, less than 100 points was scored. **17.** The ratio was $\frac{5}{8}$. **18.** In 7.3% of the games, over 120 points was scored. **19.** The mean score was 82.4. **20.** The median number of students was 26 students.

CHAPTER TEST *pages 295–296*

1. 30 students were in the geology class. (7.1A) **2.** The ratio was $\frac{1}{2}$. (7.1A) **3.** 30% of the students received a B grade. (7.1A) **4.** The ratio is $\frac{1}{16}$. (7.1B) **5.** 53.75% of the budget comes from state funds. (7.1B) **6.** The ratio is $\frac{43}{5}$. (7.1B) **7.** 27,000 cars were sold in November. (7.2A) **8.** The difference in the number of cars sold is 5000 cars. (7.2A) **9.** The difference between the first three month's sales was 19,000 cars. (7.2A) **10.** The third quarter income was \$5,000,000. (7.2B) **11.** The ratio was $\frac{3}{4}$. (7.2B) **12.** The difference in fourth-quarter incomes was \$1,000,000. (7.2B) **13.** 21 employees receive a salary over \$25,000. (7.3A) **14.** The ratio is $\frac{10}{51}$. (7.3A) **15.** 53% of the employees receive salaries between \$20,000 and \$30,000. (7.3A) **16.** 55 families watched between 15 and 25 hours of television. (7.3B) **17.** The ratio was $\frac{3}{5}$. (7.3B) **18.** 60% of the families watched over 15 hours of television per week. (7.3B) **19.** The mean number of miles was 189 miles. (7.4A) **20.** The median score was 77. (7.4B)

CUMULATIVE REVIEW *pages 297–298*

1. 540 (1.6A) **2.** 14 (1.6B) **3.** 120 (2.1A) **4.** $\frac{5}{12}$ (2.3B) **5.** $12\frac{3}{40}$ (2.4C) **6.** $4\frac{17}{24}$ (2.5C) **7.** 2 (2.6B) **8.** $\frac{64}{85}$ (2.7B) **9.** $8\frac{1}{4}$ (2.8C) **10.** 209.305 (3.1A) **11.** 2.82348 (3.4A) **12.** 16.67 (3.6A) **13.** 26.4 miles/gallon (4.2B) **14.** $n = 3.2$ (4.3B) **15.** 80% (5.1B) **16.** 80 (5.4A) **17.** 16.34 (5.2A) **18.** 40% (5.3A) **19.** Tanim had \$27,500 in sales. (6.6A) **20.** The life insurance costs \$207.50. (4.3C) **21.** The interest due is \$6875. (6.3A) **22.** The markup rate is 55%. (6.2B) **23.** \$570 is budgeted for food. (7.1B) **24.** The difference between the number of problems answered correctly was 12 problems. (7.2B) **25.** The mean high temperature was 69.6°. (7.4A) **26.** The median salary was \$21,650. (7.4B)

ANSWERS to Chapter 8 Odd-Numbered Exercises

SECTION 8.1 *pages 305–306*

1. 72 in. **3.** $2\frac{1}{2}$ ft **5.** 39 ft **7.** $5\frac{1}{3}$ yd **9.** 84 in. **11.** $3\frac{1}{3}$ yd **13.** 10,560 ft **15.** $\frac{5}{8}$ ft **17.** 1 mi 1120 ft **19.** 9 ft 11 in. **21.** 2 ft 9 in. **23.** 14 ft 6 in. **25.** 2 ft 8 in. **27.** $11\frac{1}{6}$ ft **29.** 4 mi 2520 ft **31.** 14 tiles can be placed along one row. **33.** The missing dimension is $1\frac{5}{6}$ ft. **35.** The distance between the holes is $7\frac{1}{2}$ in. **37.** Each piece is $1\frac{2}{3}$ ft long. **39.** 6 ft 6 in. of framing is needed. **41.** The wall is $33\frac{3}{4}$ ft in length. **43.** The stack would reach 12,375,000,000 ft into space.

SECTION 8.2 *pages 309–310*

1. 4 lb **3.** 64 oz **5.** $1\frac{3}{5}$ tons **7.** 12,000 lb **9.** $4\frac{1}{8}$ lb **11.** 24 oz **13.** 2600 lb **15.** $\frac{1}{4}$ tons **17.** $11\frac{1}{4}$ lb **19.** 4 tons 1000 lb **21.** 2 lb 8 oz **23.** 5 tons 400 lb **25.** 1 ton 1700 lb **27.** 33 lb **29.** 14 lb **31.** 9 oz **33.** 1 lb 7 oz **35.** The load will weigh 2000 lb. **37.** The weight of 144 tiles is 63 lb. **39.** A case of soft drinks weighs 9 lb. **41.** 1 lb 5 oz of shampoo is in each container. **43.** The ham costs $13.40. **45.** The cost of mailing the manuscript is $8.75. **47.** Answers will vary

SECTION 8.3 *pages 313–314*

1. $7\frac{1}{2}$ c **3.** 24 fl oz **5.** 4 pt **7.** 7 c **9.** $5\frac{1}{2}$ gal **11.** 9 qt **13.** $3\frac{3}{4}$ qt **15.** $1\frac{1}{4}$ pt **17.** $4\frac{1}{4}$ qt **19.** 3 gal 2 qt **21.** 2 qt 1 pt **23.** 7 qt **25.** 6 fl oz **27.** $17\frac{1}{2}$ pt **29.** $\frac{7}{8}$ gal **31.** 5 gal 1 qt **33.** 1 gal 2 qt **35.** $2\frac{3}{4}$ gal **37.** $7\frac{1}{2}$ gallons of coffee should be prepared. **39.** The solution contains $4\frac{1}{4}$ quarts. **41.** The farmer used $8\frac{3}{4}$ gallons of oil. **43.** 1 quart for $1.20 is the more economical purchase. **45.** $43.50 of profit was made on the 5-quart package. **47.** grain = 0.002285 ounce; dram = 0.0625 ounce; furlong = 0.125 mile; rod = 16.5 feet

SECTION 8.4 *pages 317–318*

1. 19,450 ft · lb **3.** 19,450,000 ft · lb **5.** 1500 ft · lb **7.** 29,700 ft · lb **9.** 30,000 ft · lb **11.** 25,500 ft · lb **13.** 35,010,000 ft · lb **15.** 9,336,000 ft · lb **17.** 2 hp **19.** 8 hp **21.** $2750 \frac{\text{ft} \cdot \text{lb}}{\text{s}}$ **23.** $3850 \frac{\text{ft} \cdot \text{lb}}{\text{s}}$ **25.** $500 \frac{\text{ft} \cdot \text{lb}}{\text{s}}$ **27.** $4800 \frac{\text{ft} \cdot \text{lb}}{\text{s}}$ **29.** $1440 \frac{\text{ft} \cdot \text{lb}}{\text{s}}$ **31.** 3 hp **33.** 12 hp

CHAPTER REVIEW *pages 321–322*

1. 48 in. **2.** $4\frac{2}{3}$ yd **3.** 9 ft 3 in. **4.** 1 yd 2 ft **5.** 13 ft 4 in. **6.** 2 ft 6 in. **7.** A 3 ft 6 in. piece of board remains. **8.** 54 oz **9.** $1\frac{1}{5}$ tons **10.** 9 lb 3 oz **11.** 1 ton 1000 lb **12.** 43 lb **13.** 2 lb 7 oz **14.** The cost of mailing the book is $6.30. **15.** 3 qt **16.** 40 fl oz **17.** $13\frac{1}{2}$ qt **18.** 16 gallons of milk were sold. **19.** 38,900 ft · lb **20.** 1600 ft · lb **21.** 27,230,000 ft · lb **22.** 7 hp **23.** $1375 \frac{\text{ft} \cdot \text{lb}}{\text{s}}$ **24.** $480 \frac{\text{ft} \cdot \text{lb}}{\text{s}}$

CHAPTER TEST *pages 323–324*

1. 30 in. (8.1A) **2.** 2 ft 5 in. (8.1B) **3.** Each piece is $1\frac{1}{3}$ ft. (8.1C) **4.** The wall is 48 ft in length. (8.1C) **5.** 46 oz (8.2A) **6.** 2 lb 8 oz (8.2A) **7.** 17 lb 1 oz (8.2B) **8.** 1 lb 11 oz (8.2B) **9.** The books weigh 750 lb. (8.2C) **10.** The class earned $28.13. (8.2C) **11.** $3\frac{1}{4}$ gal (8.3A) **12.** 28 pt (8.3A) **13.** $12\frac{1}{4}$ gal (8.3B) **14.** 8 gal 1 qt (8.3B) **15.** There are 60 cups of juice in 24 cans. (8.3C) **16.** Nick made a profit of $126. (8.3C) **17.** 3750 ft · lb (8.4A) **18.** 31,120,000 ft · lb (8.4A) **19.** $160 \frac{\text{ft} \cdot \text{lb}}{\text{s}}$ (8.4B) **20.** 4 hp (8.4B)

CUMULATIVE REVIEW *pages 325–326*

1. 180 (2.1A) **2.** $5\frac{3}{8}$ (2.2B) **3.** $3\frac{7}{24}$ (2.5C) **4.** 2 (2.7B) **5.** $4\frac{3}{8}$ (2.8C) **6.** 2.10 (3.1B) **7.** 0.038808 (3.4A) **8.** $n = 8.8$ (4.3B) **9.** 1.25 (5.2A) **10.** 42.86 (5.4A) **11.** \$2.15/lb (6.1A) **12.** $8\frac{11}{15}$ in. (8.1B) **13.** 1 lb 8 oz (8.2B) **14.** 31 lb 8 oz (8.2B) **15.** $2\frac{1}{2}$ qt (8.3B) **16.** 1 lb 12 oz (8.2B) **17.** The dividend would be \$280. (4.3C) **18.** The new checkbook balance is \$642.79. (6.7A) **19.** The monthly income is \$3100. (6.6A) **20.** 2425 pounds of the shipment were sold. (5.2B) **21.** 18% of the class received scores between 80% and 90%. (7.3A) **22.** The selling price is \$308. (6.2B) **23.** The interest paid was \$14,666.67. (6.3A) **24.** Each student received \$1267. (8.2C) **25.** The cost to mail the books was \$10.80. (8.2C) **26.** 36 oz for \$2.70 is the better buy. (6.1B) **27.** The contractor saves \$1080. (8.3C) **28.** 3200 ft · lb (8.4A) **29.** $400 \frac{\text{ft} \cdot \text{lb}}{\text{s}}$ (8.4B)

ANSWERS to Chapter 9 Odd-Numbered Exercises

SECTION 9.1 *pages 331–332*

1. 420 mm **3.** 8.1 cm **5.** 6.804 km **7.** 2109 m **9.** 4.32 m **11.** 88 cm **13.** 6.42 m 642 cm **15.** 42.6 cm 426 mm **17.** 8.42 m **19.** 78.85 m **21.** 29.6 cm **23.** 0.414 km **25.** 18.126 km **27.** 901,161.28 km **29.** The carpenter needs 69 m of ceiling joist. **31.** The missing dimension is 7.8 cm. **33.** The race was 30.5 km. **35.** The distance between the rivets is 17.9 cm. **37.** 15.6 m of fencing is left on the roll. **39.** Light travels 9.4608×10^{15} m in one year. **41.** You would use kilometers to measure the distance from Los Angeles to New York; centimeters to measure your waist; millimeters to measure a hair; centimeters to measure your height; and kilometers the distance to the grocery store. **43.** Answers will vary

SECTION 9.2 *pages 335–336*

1. 0.420 kg **3.** 0.127 g **5.** 4200 g **7.** 450 mg **9.** 1.856 kg **11.** 4.057 g **13.** 3.922 kg 3922 g **15.** 7.891 g 7891 mg **17.** 3.308 g **19.** 4.692 kg **21.** 736 g **23.** 2236.017 kg **25.** 1.947 kg **27.** The patient needs to lose another 6.2 kg. **29.** The box of tiles weighs 15.84 kg. **31.** The daily dosage is 500 mg. **33.** 1600 g of grass seed is needed. **35.** The profit is \$33.50. **37.** 45.59% of the fiber production is cotton. **39.** Answers will vary

SECTION 9.3 *pages 339–340*

1. 4.2 L **3.** 3420 ml **5.** 423 cm^3 **7.** 642 ml **9.** 0.042 L **11.** 435 cm^3 **13.** 1267 cm^3 **15.** 3023 cm^3 **17.** 35 L **19.** 3.042 L 3042 ml **21.** 3.004 kl 3004 L **23.** 12.438 L **25.** 37.944 L **27.** 6.254 L **29.** 1131.372 L **31.** 0.5545 L **33.** The auxiliary needs 6.6 L of punch. **35.** 4000 patients can be immunized. **37.** The students will use 1.08 L of acid. **39.** The amount of oxygen in 50 L of air is 9.5 L. **41.** There are 7 bottles of cough syrup still in stock. **43.** The profit made was \$8715. **45.** Answers will vary

SECTION 9.4 *pages 343–344*

1. You can eliminate 3300 cal. **3.** The person would need 3000 cal per day. **5.** The percent of the total daily intake of calories left is 44.4%. **7.** You burn up 10,125 cal in 30 days. **9.** Ruben lost 410 cal. **11.** Avi would have to hike 2.6 h. **13.** The oven used 1250 Wh of energy. **15.** The TV set uses 2.205 kWh of energy. **17.** The cost of listening to the stereo is \$.32. **19.** The hair dryer uses 6 kWh of energy each week. **21.** The percent decrease is 17.0%. **23.** The cost of using the welder is \$109.98. **25.** Answers will vary

SECTION 9.5 *pages 347–348*

1. 91 m **3.** 1.93 m **5.** 5.45 kg **7.** 235.85 ml **9.** 104.65 km/h **11.** $3.72/kg **13.** $0.39/L **15.** The weight loss will be 1 lb. **17.** 40,068.07 km **19.** 328 ft **21.** 1.59 gal **23.** 4920 ft **25.** 12.72 gal **27.** 1.86 pt **29.** 49.69 mi/h **31.** $1.45/gal **33.** 1000 kg **35.** **a.** False **b.** False **c.** True **d.** True **e.** False

CHAPTER REVIEW *pages 351–352*

1. 3.7 mm **2.** 1250 m **3.** 9.55 m **4.** 20.5 cm **5.** 36.4 m **6.** 3.5 km **7.** There are 37.2 m of wire fence left on the roll. **8.** 450 mg **9.** 4050 g **10.** 14.243 g **11.** 12.46 kg **12.** 37.95 kg **13.** 2.519 g **14.** The total cost of the turkey is $20.37. **15.** 5.6 ml **16.** 1200 cm^3 **17.** 10.613 L **18.** 6.597 L **19.** 102.44 L **20.** 1.81 L **21.** There should be 50 L of coffee prepared. **22.** You can eliminate 2700 cal. **23.** It costs $3.42 to run the T.V. set. **24.** 1098.9 yd **25.** 4.19 lb **26.** The ham costs $7.48/kg. **27.** It is necessary to cycle 8.75 h.

CHAPTER TEST *pages 353–354*

1. 4.26 cm (9.1A) **2.** 5038 m (9.1A) **3.** 4.65 m (9.1B) **4.** 22.16 m (9.1B) **5.** The total length of rafters needed is 114 m. (9.1C) **6.** The distance between the rivets is 17.5 cm. (9.1C) **7.** 3290 g (9.2A) **8.** 3.089 g (9.2A) **9.** 24.36 kg (9.2B) **10.** 7.297 g (9.2B) **11.** The weight of the box of tiles is 36 kg. (9.2C) **12.** It costs $540 to fertilize the apple orchard. (9.2C) **13.** 3250 ml (9.3A) **14.** 1600 cm^3 (9.3A) **15.** 0.75 L (9.3B) **16.** 3.931 L (9.3B) **17.** The clinic needs 5.2 L of vaccine. (9.3C) **18.** It costs $16.32 to operate the TV set. (9.4A) **19.** The record ski jump for men is 193.90 m. (9.5A) **20.** The record ski jump for women is 360.80 ft. (9.5B)

CUMULATIVE REVIEW *pages 355–356*

1. 6 (1.6B) **2.** $12\frac{13}{36}$ (2.4C) **3.** $\frac{29}{36}$ (2.5C) **4.** $3\frac{1}{14}$ (2.7B) **5.** 1 (2.8B) **6.** 2.0702 (3.3A) **7.** $n = 31.3$ (4.3B) **8.** 175% (5.1B) **9.** 145 (5.4A) **10.** 2.25 gal (8.3A) **11.** 18.75 m (9.1A) **12.** 2.528 km (9.1B) **13.** 5.05 kg (9.2A) **14.** 3.672 g (9.2B) **15.** 0.83 kg (9.2B) **16.** 5.548 L (9.3B) **17.** There is $1683 left after the rent is paid. (2.6C) **18.** The amount of income tax paid is $7207.20. (3.4B) **19.** The property tax is $1500. (4.3C) **20.** The new car buyer would receive a $1620 rebate. (5.2B) **21.** The dividend is 6.5% of the investment. (5.3B) **22.** Your average grade is 76. (7.4A) **23.** The salary next year will be $25,200. (5.2B) **24.** The discount rate is 22%. (6.2D) **25.** The length of the wall is 36 feet. (8.1C) **26.** The number of quarts of apple juice is 12. (8.3C) **27.** The profit is $59.20. (8.3C) **28.** There are 24 L of chlorine used. (9.3C) **29.** It costs $1.89 to operate the hair dryer. (9.4A) **30.** The conversion is 96.6 km/h. (9.5A)

ANSWERS to Chapter 10 Odd-Numbered Exercises

SECTION 10.1 *pages 361–362*

1. $3 < 5$ **3.** $-2 > -5$ **5.** $-16 < 1$ **7.** $3 > -7$ **9.** $-11 < -8$ **11.** $-1 > -6$ **13.** $0 > -3$ **15.** $6 > -8$ **17.** $-14 < 16$ **19.** $35 > 28$ **21.** $-42 < 27$ **23.** $21 > -34$ **25.** $-27 > -39$ **27.** $-87 < 63$ **29.** $68 > -79$ **31.** $-62 > -84$ **33.** $94 > 83$ **35.** $59 > -67$ **37.** $-93 < -55$ **39.** $-88 < 57$ **41.** −4 **43.** 2 **45.** −22 **47.** 31 **49.** 2 **51.** 6 **53.** 8 **55.** 9 **57.** −1 **59.** −5 **61.** 16 **63.** 12 **65.** −29 **67.** −14 **69.** 15 **71.** −33 **73.** −36 **75.** 32 **77.** 38 **79.** −37 **81.** −42 **83.** 44 **85.** 74 **87.** −88 **89.** −2, 8; −4, 2 **91.** Abdul

SECTION 10.2 *pages 367–370*

1. −2 **3.** 20 **5.** −11 **7.** −9 **9.** −3 **11.** 1 **13.** −5 **15.** −30 **17.** 9 **19.** 1 **21.** −10 **23.** −28 **25.** −41 **27.** −392 **29.** −20 **31.** −23 **33.** −2 **35.** −6 **37.** −6 **39.** 8 **41.** −7 **43.** −9 **45.** 9 **47.** −3 **49.** 18 **51.** −9 **53.** 11 **55.** 0 **57.** 11 **59.** 2 **61.** −138 **63.** −8 **65.** −337 **67.** 4 **69.** 6 **71.** −12 **73.** 4 **75.** 3 **77.** The temperature rose to −1°C. **79.** Nick's score was −15 points. **81.** The difference in temperatures is 465°F. **83.** The difference between the high and low temperature is 17°C. **85.** The difference in elevation is 20,602 ft. **87.** The difference in elevation is 30,314 ft. **89.** The difference in elevation is 9688 ft. **91.** 22; 2 **93.** Answers will vary **95.**

−3	2	1
4	0	−4
−1	−2	3

97. Concord, NH experienced the smallest decrease in low temperature. **99.** The difference was $190.

SECTION 10.3 *pages 375–378*

1. 42 **3.** −24 **5.** 6 **7.** 18 **9.** −20 **11.** −16 **13.** 25 **15.** 0 **17.** −72 **19.** −102 **21.** 140 **23.** −228 **25.** −320 **27.** −156 **29.** −70 **31.** 162 **33.** 120 **35.** 36 **37.** 192 **39.** −108 **41.** −2100 **43.** 20 **45.** −48 **47.** 140 **49.** −2 **51.** 8 **53.** 0 **55.** −9 **57.** −9 **59.** 9 **61.** −24 **63.** −12 **65.** 31 **67.** 17 **69.** 15 **71.** −13 **73.** −18 **75.** 19 **77.** 13 **79.** −19 **81.** 17 **83.** 26 **85.** 23 **87.** 25 **89.** −34 **91.** 11 **93.** −13 **95.** 13 **97.** 12 **99.** −14 **101.** −14 **103.** −6 **105.** −4 **107.** The average high temperature was −3°. **109.** The boiling point of neon is −245°C. **111.** The golfer's average score was −4.5. **113.** **a.** True **b.** Never true **115.** −17 **117.** Answers will vary **119.** 132

SECTION 10.4 *pages 383–386*

1. $1\frac{1}{12}$ **3.** $-\frac{5}{24}$ **5.** $-\frac{19}{24}$ **7.** $\frac{5}{26}$ **9.** $\frac{7}{24}$ **11.** $\frac{7}{24}$ **13.** $-\frac{8}{15}$ **15.** $-\frac{1}{12}$ **17.** $\frac{3}{4}$ **19.** $\frac{17}{18}$ **21.** $-\frac{47}{48}$ **23.** $\frac{3}{8}$ **25.** $-\frac{7}{60}$ **27.** $-\frac{1}{16}$ **29.** $-\frac{7}{24}$ **31.** $\frac{13}{24}$ **33.** $-1\frac{1}{10}$ **35.** −1.63 **37.** 3.5 **39.** 1.91 **41.** −1.22 **43.** −181.51 **45.** −22.804 **47.** −0.6 **49.** −10.03 **51.** −37.19 **53.** −17.5 **55.** −1.867 **57.** −12.932 **59.** −2.0867 **61.** $-\frac{3}{8}$ **63.** $\frac{1}{10}$ **65.** $-\frac{4}{9}$ **67.** $-\frac{7}{30}$ **69.** $\frac{2}{27}$ **71.** $-\frac{45}{328}$ **73.** 10 **75.** $\frac{45}{328}$ **77.** $-\frac{3}{7}$ **79.** $\frac{3}{2} = 1\frac{1}{2}$ **81.** $-\frac{8}{9}$ **83.** $\frac{7}{4} = 1\frac{3}{4}$ **85.** $\frac{5}{6}$ **87.** $-\frac{16}{7} = -2\frac{2}{7}$ **89.** $-\frac{27}{2} = -13\frac{1}{2}$ **91.** 3.78 **93.** 31.15 **95.** 74.88 **97.** 0.117 **99.** 0.0363 **101.** 97 **103.** 2.2 **105.** −4.14 **107.** −1.7 **109.** −42.40 **111.** 4.07 **113.** −0.46 **115.** Always true; Always true **117.**

	Add	*Subtract*	*Multiply*	*Divide*
Whole numbers	Y	N	Y	N
Integers	Y	Y	Y	N
Rational numbers	Y	Y	Y	Y

119. Answers will vary

SECTION 10.5 *pages 389–390*

1. 4 **3.** 0 **5.** 12 **7.** −6 **9.** −5 **11.** 2 **13.** −3 **15.** 1 **17.** 2 **19.** −3 **21.** 14 **23.** −4 **25.** 33 **27.** −13 **29.** −12 **31.** 17 **33.** 0 **35.** 30 **37.** 94 **39.** −8 **41.** 39 **43.** 0.21 **45.** −0.96 **47.** −0.29 **49.** −1.76 **51.** 2.1 **53.** $\frac{3}{16}$ **55.** $\frac{7}{16}$ **57.** $-\frac{5}{16}$ **59.** $-\frac{5}{8}$ **61.** **a.** 100 **b.** −100 **c.** 225 **d.** −225

CHAPTER REVIEW *pages 393–394*

1. $-2 > -40$ **2.** $0 > -3$ **3.** 4 **4.** -22 **5.** 5 **6.** -6 **7.** -26 **8.** 1 **9.** -26 **10.** 6 **11.** $-4°$ **12.** $-\frac{5}{24}$ **13.** $-\frac{1}{4}$ **14.** $\frac{17}{24}$ **15.** $\frac{1}{36}$ **16.** 0.21 **17.** -7.4 **18.** $-\frac{1}{4}$ **19.** $\frac{3}{40}$ **20.** $\frac{8}{25}$ **21.** $\frac{15}{32}$ **22.** $1\frac{5}{8}$ **23.** -0.042 **24.** -1.28 **25.** -0.08 **26.** 10 **27.** 8 **28.** 18 **29.** 2 **30.** 24 **31.** 1.33 **32.** $-\frac{7}{18}$ **33.** $\frac{1}{3}$

CHAPTER TEST *pages 395–396*

1. $-8 > -10$ (10.1A) **2.** $0 > -4$ (10.1A) **3.** -2 (10.1B) **4.** -7 (10.2A) **5.** -14 (10.2A) **6.** 3 (10.2B) **7.** 10 (10.2B) **8.** -48 (10.3A) **9.** 90 (10.3A) **10.** -9 (10.3B) **11.** The temperature rose to 7°C. (10.2C) **12.** The average low temperature was $-2°$. (10.3C) **13.** $\frac{1}{15}$ (10.4A) **14.** $\frac{1}{12}$ (10.4A) **15.** $\frac{7}{24}$ (10.4A) **16.** $\frac{3}{10}$ (10.4A) **17.** $\frac{1}{12}$ (10.4B) **18.** $-\frac{4}{5}$ (10.4B) **19.** -1.88 (10.4A) **20.** -4.014 (10.4A) **21.** 3.213 (10.4A) **22.** -0.0608 (10.4B) **23.** 3.4 (10.4B) **24.** -26 (10.5A) **25.** -4 (10.5A)

CUMULATIVE REVIEW *pages 397–398*

1. 0 (1.6B) **2.** $4\frac{13}{14}$ (2.5C) **3.** $2\frac{7}{12}$ (2.7B) **4.** $1\frac{2}{7}$ (2.8C) **5.** 1.80939 (3.3A) **6.** $n = 18.67$ (4.3B) **7.** 13.75 (5.4A) **8.** 1 gal 3 qt (8.3A) **9.** 6.692 L (9.3A) **10.** 1.28 m (9.5A) **11.** 57.6 (5.2A) **12.** 340% (5.1B) **13.** -3 (10.2A) **14.** $-3\frac{3}{8}$ (10.4A) **15.** $-10\frac{13}{24}$ (10.4A) **16.** 19 (10.5A) **17.** 3.488 (10.4B) **18.** $31\frac{1}{2}$ (10.4B) **19.** -7 (10.3B) **20.** $\frac{25}{42}$ (10.4B) **21.** -4 (10.5A) **22.** -4 (10.5A) **23.** The length of board remaining is $2\frac{1}{3}$ ft. (2.5D) **24.** Nimisha's new checkbook balance is \$803.31. (6.7A) **25.** The percent decrease in price is 27.3%. (6.2C) **26.** There should be 10 gal of coffee prepared. (8.3C) **27.** The dividend per share is \$1.68. (6.2A) **28.** The median hourly pay is \$9.40. (7.4B) **29.** There would be 600,000 people voting in the city election. (4.3C) **30.** The average high temperature was $-4°$. (10.3C)

ANSWERS to Chapter 11 Odd-Numbered Exercises

SECTION 11.1 *pages 407–410*

1. -33 **3.** -12 **5.** -4 **7.** -3 **9.** -22 **11.** -40 **13.** -1 **15.** 14 **17.** 32 **19.** 6 **21.** 7 **23.** 49 **25.** -24 **27.** 63 **29.** 9 **31.** 4 **33.** $-\frac{2}{3}$ **35.** 0 **37.** 0 **39.** $-3\frac{1}{4}$ **41.** 8.5023 **43.** -18.950658 **45.** 3.5961 **47.** $16z$ **49.** $9m$ **51.** $12at$ **53.** $3yt$ **55.** Unlike terms **57.** $-2t^2$ **59.** $13c - 5$ **61.** $-2t$ **63.** $3y^2 - 2$ **65.** $14w - 8u$ **67.** $-23xy + 10$ **69.** $-11v^2$ **71.** $-8ab - 3a$ **73.** $4y^2 - y$ **75.** $-3a - 2b^2$ **77.** $7x^2 - 8x$ **79.** $-3s + 6t$ **81.** $-3m + 10n$ **83.** $5ab + 4ac$ **85.** $\frac{2}{3}a^2 + \frac{3}{5}b^2$ **87.** $6.994x$ **89.** $1.56m - 3.77n$ **91.** $5x + 20$ **93.** $4y - 12$ **95.** $-2a - 8$ **97.** $15x + 30$ **99.** $15c - 25$ **101.** $-3y + 18$ **103.** $7x + 14$ **105.** $4y - 8$ **107.** $5x + 24$ **109.** $y - 6$ **111.** $6n - 3$ **113.** $6y + 20$ **115.** $10x + 4$ **117.** $-6t - 6$ **119.** $z - 2$ **121.** $y + 6$ **123.** $9t + 15$ **125.** $5t + 24$ **127.** Tyrone will receive \$5n for working n hours. **129.** | x | x | x | 1 | 1 | 1 | $3x + 3$ **131.** Answers will vary **133.** Answers will vary

SECTION 11.2 *pages 417–420*

1. No **3.** Yes **5.** No **7.** No **9.** Yes **11.** Yes **13.** Yes **15.** Yes **17.** Yes **19.** Yes **21.** Yes **23.** Yes **25.** Yes **27.** No **29.** No **31.** No **33.** Yes **35.** -2 **37.** 14 **39.** 2 **41.** -4 **43.** -5 **45.** 0 **47.** 2 **49.** -3 **51.** 10 **53.** -5 **55.** -12 **57.** -7 **59.** $\frac{2}{3}$ **61.** -1 **63.** $\frac{1}{3}$ **65.** $-\frac{1}{8}$ **67.** $-\frac{3}{4}$ **69.** $-1\frac{1}{12}$ **71.** 6 **73.** -9 **75.** -5 **77.** 14 **79.** 8 **81.** -3 **83.** 20 **85.** -21 **87.** -15 **89.** 16 **91.** -42 **93.** 15 **95.** $-38\frac{2}{5}$ **97.** $9\frac{3}{5}$ **99.** $-10\frac{4}{5}$ **101.** $1\frac{17}{18}$ **103.** $-2\frac{2}{3}$ **105.** 3 **107.** The increase in value is \$3420. **109.** The depreciation is \$6400. **111.** The car travels 38 mi/gal. **113.** The distance traveled was 648 miles. **115.** The markup on each shirt is \$6.30. **117.** The cost of the compact disk is \$10.30. **119.** No, you cannot divide by zero. **121.** Answers will vary **123.** Answers will vary

SECTION 11.3 *pages 423–426*

1. 3 **3.** 5 **5.** -1 **7.** -4 **9.** 1 **11.** 3 **13.** -2 **15.** -3 **17.** -2 **19.** -1 **21.** 3 **23.** -1 **25.** 7 **27.** 2 **29.** 2 **31.** 0 **33.** 2 **35.** 0 **37.** $\frac{2}{3}$ **39.** 9 **41.** 7 **43.** $-2\frac{1}{2}$ **45.** $1\frac{1}{2}$ **47.** 2 **49.** -2 **51.** $2\frac{1}{3}$ **53.** -1 **55.** $\frac{2}{7}$ **57.** $1\frac{1}{12}$ **59.** 5 **61.** 10 **63.** 5 **65.** 36 **67.** -6 **69.** -9 **71.** $-4\frac{4}{5}$ **73.** $-8\frac{3}{4}$ **75.** 36 **77.** $1\frac{1}{3}$ **79.** 2 **81.** 4 **83.** 9 **85.** 4 **87.** -3.125 **89.** -4 **91.** 3 **93.** -4 **95.** -2 **97.** -4 **99.** 4.95 **101.** 18.206388 **103.** -25.903376 **105.** The Celsius temperature is 22.2°C. **107.** The time required is 3.5 seconds. **109.** There were 2500 units produced. **111.** Marcy's income tax rate is 20%. **113.** The total sales for the month were \$36,000. **115.** Tina's commission rate is 7.2%. **117.** An equation has an equal sign, whereas an expression does not. **119.** Answers will vary

SECTION 11.4 *pages 429–432*

1. $\frac{1}{2}$ **3.** -1 **5.** -4 **7.** -1 **9.** -4 **11.** 3 **13.** -1 **15.** 2 **17.** 3 **19.** 0 **21.** -2 **23.** -1 **25.** $-\frac{2}{3}$ **27.** -4 **29.** $\frac{1}{3}$ **31.** $-\frac{3}{4}$ **33.** $\frac{1}{4}$ **35.** 7 **37.** 2 **39.** $2\frac{2}{3}$ **41.** 0 **43.** $-1\frac{3}{4}$ **45.** 1 **47.** -1 **49.** -5 **51.** $\frac{5}{12}$ **53.** $-\frac{4}{7}$ **55.** $2\frac{1}{3}$ **57.** 21 **59.** 21 **61.** 2 **63.** -1 **65.** -4 **67.** $-\frac{6}{7}$ **69.** 5 **71.** 3 **73.** 0 **75.** -1 **77.** 2 **79.** $1\frac{9}{10}$ **81.** 13 **83.** $2\frac{1}{2}$ **85.** 2 **87.** 0 **89.** $3\frac{1}{5}$ **91.** 4 **93.** $-6\frac{1}{2}$ **95.** $-\frac{3}{4}$ **97.** $-\frac{2}{3}$ **99.** -3 **101.** $-1\frac{1}{4}$ **103.** $2\frac{5}{8}$ **105.** 1.9936196 **107.** 13.383119 **109.** 48 **111.** Answers will vary

SECTION 11.5 *pages 435–436*

1. $y - 9$ **3.** $z + 3$ **5.** $\frac{2}{3}n + n$ **7.** $\frac{m}{m - 3}$ **9.** $9(x + 4)$ **11.** $n - (-5)n$ **13.** $c\left(\frac{1}{4}c\right)$ **15.** $m^2 + 2m^2$ **17.** $2(t + 6)$ **19.** $\frac{x}{9 + x}$ **21.** $3(b + 6)$ **23.** x^2 **25.** $\frac{x}{20}$ **27.** $4x$ **29.** $\frac{3}{4}x$ **31.** $4 + x$ **33.** $5x - x$ **35.** $x(2 + x)$ **37.** $7(x + 8)$ **39.** $x^2 + 3x$

41. $(x + 3) + \frac{1}{2}x$ **43.** $\frac{3x}{x}$ **45.** $\frac{x}{4}$ **47.** 3 more than twice x; twice the sum of x and 3

49. Answers will vary **51.** carbon $\frac{1}{2}x$; oxygen $\frac{1}{2}x$

SECTION 11.6 *pages 441–444*

1. $x + 7 = 12; 5$ **3.** $3x = 18; 6$ **5.** $5 + x = 3; -2$ **7.** $6x = 14; 2\frac{1}{3}$ **9.** $\frac{5}{6}x = 15; 18$

11. $3x + 4 = 8; 1\frac{1}{3}$ **13.** $4x - 7 = 9; 4$ **15.** $\frac{x}{9} = 14; 126$ **17.** $\frac{x}{4} - 6 = -2; 16$

19. $7 - 2x = 13; -3$ **21.** $9 - \frac{x}{2} = 5; 8$ **23.** $\frac{3}{5}x + 8 = 2; -10$

25. $\frac{x}{4.186} - 7.92 = 12.529; 85.599514$ **27.** Target's price is $76.75. **29.** The value of the camper last year was $18,750. **31.** The selling price is $1050. **33.** The Manzanares' monthly income is $2720. **35.** 5000 computers **37.** The regular price is $200. **39.** It took the plumber 3 h. **41.** The markup rate is 40%. **43.** The total wall space is 9000 ft^2. **45.** The original rate was 5 gal/min. **47.** Total monthly sales were $42,540. **49.** The amount subsidized was $10.25; 69.8% of the cost was subsidized. **51.** The man died at the age of 88 years. **53.** Answers will vary

CHAPTER REVIEW *pages 447–448*

1. 13 **2.** 6 **3.** $-4bc$ **4.** $\frac{71}{30}x^2$ **5.** $-2a + 2b$ **6.** $-3x + 4$ **7.** Yes **8.** No

9. -4 **10.** -5 **11.** -9 **12.** $1\frac{1}{4}$ **13.** The car traveled 23 mi/gal. **14.** 7 **15.** 10

16. -18 **17.** $10\frac{4}{5}$ **18.** The Celsius temperature is 37.8°C. **19.** $-2\frac{1}{2}$ **20.** 5 **21.** 5

22. $-\frac{1}{3}$ **23.** $n + \frac{n}{5}$ **24.** $(n + 5) + \frac{1}{3}n$ **25.** 2 **26.** 10 **27.** The sale price is $228.

CHAPTER TEST *pages 449–450*

1. -38 (11.1A) **2.** $-5\frac{5}{6}$ (11.1A) **3.** $-4y - 11x$ (11.1B) **4.** $8y - 7$ (11.1C)

5. No (11.2A) **6.** 26 (11.2B) **7.** $-2\frac{4}{5}$ (11.2C) **8.** -16 (11.2C) **9.** The monthly payment is $137.50. (11.2D) **10.** -2 (11.3A) **11.** 95 (11.3A) **12.** $-\frac{1}{2}$ (11.3A)

13. There were 4000 clocks made. (11.3B) **14.** $3\frac{1}{5}$ (11.4A) **15.** 0 (11.4B)

16. $x + \frac{1}{3}x$ (11.5A) **17.** $5(x + 3)$ (11.5B) **18.** $2x - 3 = 7; 5$ (11.6A) **19.** $-2\frac{1}{3}$ (11.6A)

20. Eduardo's total sales were $40,000. (11.6B)

CUMULATIVE REVIEW *pages 451–452*

1. 41 (1.6B) **2.** $1\frac{7}{10}$ (2.5C) **3.** $\frac{11}{18}$ (2.8C) **4.** 0.047383 (3.4A) **5.** $4.20/h (4.2B)

6. $n = 26.67$ (4.3B) **7.** $\frac{4}{75}$ (5.1A) **8.** 140% (5.3A) **9.** 6.4 (5.4A)

10. 18 ft 9 in. (8.1B) **11.** 22 oz (8.2A) **12.** 2.082 g (9.2A) **13.** -1 (10.2A)
14. 19 (10.2B) **15.** 6 (10.5A) **16.** -48 (11.1A) **17.** $-10x + 8z$ (11.1B)

18. $3y + 23$ (11.1C) **19.** $x = -1$ (11.3A) **20.** $-6\frac{1}{4}$ (11.4B) **21.** $-7\frac{1}{2}$ (11.2C)

22. $x = -21$ (11.3A) **23.** The percent was 17.6%. (5.3B) **24.** The price of the pottery was \$39.90. (6.2B) **25a.** The discount is \$81. (6.2D) **25b.** The discount rate is 18%. (6.2D) **26.** The simple interest due is \$2933.33. (6.3A) **27.** The mathematical expression is $3n + 4$. (11.5B) **28.** The average speed is 53 mi/h (11.6B) **29.** The total monthly sales is \$32,500. (11.6B) **30.** The number is 2. (11.6A)

ANSWERS to Chapter 12 Odd-Numbered Exercises

SECTION 12.1 *pages 463–466*

1. 0°, 90° **3.** 180° **5.** 30 **7.** 21 **9.** 14 **11.** 59° **13.** 108° **15.** 77° **17.** 53° **19.** 77° **21.** 118° **23.** 133° **25.** 86° **27.** 180° **29.** Rectangle or square **31.** Cube **33.** Square **35.** Cylinder **37.** 102° **39.** 90° and 45° **41.** 14° **43.** 90° and 65° **45.** 8 in. **47.** $4\frac{2}{3}$ ft **49.** 7 cm **51.** 2 ft 4 in. **53.** $a = 106°$ $b = 74°$ **55.** $a = 112°$ $b = 68°$ **57.** $a = 38°$ $b = 142°$ **59.** $a = 58°$ $b = 58°$ **61.** $a = 152°$ $b = 152°$ **63.** $a = 130°$ $b = 50°$ **65a.** 1° **65b.** 90° **67.** $\angle AOC$ and $\angle BOC$ are supplementary. Since $\angle AOC = \angle BOC$ is given, each angle is 90° and the lines are perpendicular.

SECTION 12.2 *pages 473–476*

1. 56 in. **3.** 20 ft **5.** 92 cm **7.** 47.1 cm **9.** 9 ft 10 in. **11.** 50.24 cm **13.** 240 m **15.** 157 cm **17.** 121 cm **19.** 50.56 m **21.** 3.57 ft **23.** 139.3 m **25.** They need $4\frac{1}{2}$ mi of fencing. **27.** They need 15.7 m of binding. **29.** The bicycle travels 10π ft. **31.** The length of weather stripping is 20.71 ft. **33.** The total cost to fence the lot is \$8022.50. **35.** The distance for one revolution is 584,040,000 mi. **37.** The resulting circumference is doubled. **39.** The perimeter of the eighth figure is 22. **41.** The perimeter of the shaded triangles is $20\frac{1}{4}$ cm.

SECTION 12.3 *pages 481–484*

1. 144 ft^2 **3.** 81 $in.^2$ **5.** 50.24 ft^2 **7.** 20 in^2 **9.** 2.13 cm^2 **11.** 16 ft^2 **13.** 817 $in.^2$ **15.** 154 in^2 **17.** 26 cm^2 **19.** 2220 cm^2 **21.** 150.72 in^2 **23.** 8.851323 ft^2 **25.** The area is 1728 yd^2. **27.** The area of the telescope is $10{,}000\pi$ in^2. **29.** It will cost \$570.90 to carpet the area. **31.** The area of the boundary is 68 m^2. **33.** The total area of the park is 125.1492 m^2. **35.** The increase in area is 235.5 cm^2. **37.** The area of the resulting rectangle is quadrupled. **39.** **a.** Sometimes true; **b.** Sometimes true; **c.** Always true **41.** Equal

SECTION 12.4 *pages 491–494*

1. 144 cm^3 **3.** 512 $in.^3$ **5.** 2143.57 $in.^3$ **7.** 150.72 cm^3 **9.** 6.4 m^3 **11.** 5572.45 mm^3 **13.** 3391.2 ft^3 **15.** $42\frac{7}{8}$ ft^3 **17.** 82.26 in^3 **19.** 1.6688 m^3 **21.** 69.08 in^3 **23.** The volume is 40.5 m^3. **25.** The volume is 17,148.59 ft^3. **27.** The volume is 2304π ft^3. **29.** The aquarium will be filled by 15.0 gal of water. **31.** The total weight of the water is 4,056,000 lb. **33.** The volume of the bushing is 212.64 in^3. **35.** $\frac{2}{3}\pi r^3$ **37.** 8 times larger

SECTION 12.5 *pages 499–500*

1. 2.646 **3.** 6.481 **5.** 12.845 **7.** 13.748 **9.** 5 in. **11.** 8.602 cm **13.** 11.180 ft **15.** 4.472 cm **17.** 12.728 yd **19.** 6 ft and 10.392 ft **21.** 21.21 cm **23.** 8 m **25.** 11.314 yd **27.** The distance between the holes is 6.32 in. **29.** It cost \$109.20 to fence the plot. **31.** The total length of the conduit is 35 ft. **33.** **a.** Sometimes true; **b.** Always true; **c.** Always true **35.** Answers will vary

SECTION 12.6 *pages 505–506*

1. $\frac{1}{2}$ **3.** $\frac{3}{4}$ **5.** Yes **7.** Yes **9.** 7.2 cm **11.** 3.3 m **13.** The height of the building is 16 m. **15.** The perimeter is 12 m. **17.** The area is 56.25 cm^2. **19.** Always true; Sometimes true; Always true **21.** 16.5

CHAPTER REVIEW *pages 509–510*

1. 3 **2.** 75° **3.** 0.75 m **4.** 58° and 90° **5.** $a = 135°$ $b = 45°$ **6.** $a = 100°$ $b = 80°$ **7.** 26 ft **8.** 31.4 cm **9.** 47.7 in. **10.** The bicycle travels 73.3 ft. **11.** 55 m^2 **12.** 63.585 cm^2 **13.** 57.12 in^2 **14.** The area is 28 yd^2. **15.** 200 ft^3 **16.** 267.9 ft^3 **17.** 240 in^3 **18.** The volume is 1144.53 ft^3. **19.** 3.873 **20.** 26 cm **21.** The ladder will reach up to 15 ft. **22.** 16 cm **23.** 64.8 m^2

CHAPTER TEST *pages 511–512*

1. 58° (12.1A) **2.** 90° and 50° (12.1B) **3.** 150° (12.1C) **4.** 180° (12.1C) **5.** 6.8 m (12.2A) **6.** 15.85 ft (12.2B) **7.** It will cost $556.84 to carpet the area. (12.2C) **8.** $3\frac{1}{7}$ m^2 (12.3A) **9.** 10.125 ft^2 (12.3B) **10.** The cross-sectional area is 103.82 ft^2. (12.3C) **11.** There are 113.04 in^2 more of the pizza. (12.3C) **12.** 169.56 m^3 (12.4A) **13.** 1406.72 cm^3 (12.4B) **14.** The volume is 780 in^3. (12.4C) **15.** 13.748 (12.5A) **16.** 9.747 ft (12.5B) **17.** The length of the rafter is 15 ft. (12.5C) **18.** $1\frac{1}{5}$ ft (12.6A) **19.** The width of the canal is 25 ft. (12.6B) **20.** 10 m (12.6A)

CUMULATIVE REVIEW *pages 513–514*

1. 48 (2.1B) **2.** $7\frac{41}{48}$ (2.4C) **3.** $\frac{39}{56}$ (2.7B) **4.** $\frac{2}{15}$ (2.8C) **5.** $-\frac{1}{24}$ (10.4A) **6.** $8.72/h (4.2B) **7.** $n = 37.5$ (4.3B) **8.** $\frac{3}{8}$ (5.1A) **9.** −4 (11.1A) **10.** 85 (5.4A) **11.** −6 (11.3A) **12.** $\frac{4}{3}$ (11.4B) **13.** 32,500 m (9.1A) **14.** 31.58 m (9.1B) **15.** −15 (11.2C) **16.** 2 (11.4B) **17.** The monthly payment is $181. (1.5D) **18.** The sales tax is $47.06. (4.3C) **19.** The operator's hourly wage was $14.60. (5.4B) **20.** The sale price is $54. (6.2D) **21.** The value after 20 years would be $184,675.75. (6.3B) **22.** The weight of the package is 54 lb. (8.2C) **23.** The distance between the rivets is 22.5 cm. (9.1C) **24.** −2 (11.6A) **25.** $a = 74°$ $b = 106°$ (12.1A) **26.** 29.42 cm (12.2B) **27.** 50 in^2 (12.3B) **28.** 92.86 in^3 (12.4B) **29.** 10.63 ft (12.5B) **30.** 36 cm (12.6B)

FINAL EXAMINATION *pages 515–518*

1. 3259 (1.3B) **2.** 53 (1.5C) **3.** 16 (1.6B) **4.** 144 (2.1A) **5.** $1\frac{49}{120}$ (2.4B) **6.** $3\frac{29}{48}$ (2.5C) **7.** $6\frac{3}{14}$ (2.6B) **8.** $\frac{4}{9}$ (2.7B) **9.** $\frac{1}{6}$ (2.8B) **10.** $\frac{1}{13}$ (2.8C) **11.** 164.177 (3.2A) **12.** 60,205 (3.3A) **13.** 0.027918 (3.4A) **14.** 0.69 (3.5A) **15.** $\frac{9}{20}$ (3.6B) **16.** 24.5 mi/gal (4.2B) **17.** $n \approx 54.9$ (4.3B) **18.** $\frac{9}{40}$ (5.1A) **19.** 135% (5.1B) **20.** 125% (5.1B) **21.** 36 (5.2A) **22.** $133\frac{1}{3}\%$ (5.3A) **23.** 70 (5.4A) **24.** 20 in. (8.1A) **25.** 1 ft 4 in. (8.1B) **26.** 2.5 lb (8.2A) **27.** 6 lb 6 oz (8.2B) **28.** 2.25 gal (8.3A) **29.** 1 gal 3 qt (8.3B) **30.** 248 cm (9.1A) **31.** 23.10 m (9.1B) **32.** 1.614 kg (9.2A) **33.** 1.258 kg (9.2B) **34.** 2067 ml (9.3A) **35.** 0.9105 L (9.3B) **36.** 88.55 km (9.5A) **37.** The perimeter is 3.9 m. (12.2A) **38.** The area is 45 in^2. (12.3A) **39.** The volume is 1200 cm^3. (12.4A) **40.** −4 (10.2A)

41. -15 (10.2B) **42.** $-\frac{1}{2}$ (10.4B) **43.** $-\frac{1}{4}$ (10.4B) **44.** 6 (10.5A)
45. $-x + 17$ (11.1C) **46.** $x = -18$ (11.2C) **47.** $x = 5$ (11.3A) **48.** $x = 1$ (11.4A)
49. Your new balance is $959.93. (6.7A) **50.** There will be 63,750 people voting in the election. (4.3C) **51.** The price was $4500. (5.4B) **52.** The average income was $3794. (7.4A) **53.** The simple interest due is $9000. (6.3A) **54.** The percent is 41.4%. (7.1B) **55.** The discount rate is 28%. (6.2D) **56.** The weight of the tiles is 81 lb. (8.2C) **57.** The perimeter is 28.56 in. (12.2B) **58.** The area is 16.86 cm^2. (12.3B) **59.** The number is 16. (11.6A)

Glossary

absolute value of a number The distance of the number from zero on the number line. (Sec. 10.1)

acute angle An angle whose measure is between 0° and 90°. (Sec. 12.1)

acute triangle A triangle that has three acute angles. (Sec. 12.2)

addend In addition, one of the numbers added. (Sec. 1.2)

addition The process of finding the total of two numbers. (Sec. 1.2)

addition property of zero Zero added to a number does not change the number. (Sec. 1.2)

adjacent angles Two angles that share a common side. (Sec. 12.1)

alternate exterior angles Two angles that are on opposite sides of the transversal and outside the parallel lines. (Sec. 12.1)

alternate interior angles Two angles that are on opposite sides of the transversal and between the parallel lines. (Sec. 12.1)

angle An angle is formed when two rays start at the same point; it is measured in degrees. (Sec. 12.1)

area A measure of the amount of surface in a region. (Sec. 12.3)

associative property of addition Numbers to be added can be grouped (with parentheses, for example) in any order; the sum will be the same. (Sec. 1.2)

associative property of multiplication Numbers to be multiplied can be grouped (with parentheses, for example) in any order; the product will be the same. (Sec. 1.4)

average value The sum of all values divided by the number of those values; also known as the mean value. (Sec. 7.4)

balancing a checkbook Determining if the checking account balance is accurate. (Sec. 6.7)

bar graph A graph that represents data by the height of the bars. (Sec. 7.1)

basic percent equation Percent times base equals amount. (Sec. 5.2)

British Thermal Unit A unit of energy. 1 British Thermal Unit = 778 foot-pounds. (Sec. 8.4)

broken-line graph A graph that represents data by the position of the lines and shows trends and comparisons. (Sec. 7.2)

calorie The basic unit of energy in the metric system. (Sec. 9.4)

capacity A measure of liquid substances. (Sec. 8.3)

circle A plane figure in which all points are the same distance from point O, which is called the center of the circle. (Sec. 12.1)

circle graph A graph that represents data by the size of the sectors. (Sec. 7.1)

circumference The distance around a circle. (Sec. 12.2)

class frequency The number of occurrences of data in each class interval on a histogram; represented by the height of each bar. (Sec. 7.3)

class interval Range of numbers represented by the width of a bar on a histogram. (Sec. 7.3)

class midpoint The center of a class interval in a frequency polygon. (Sec. 7.3)

commission That part of the pay earned by a salesperson that is calculated as a percent of the salesperson's sales. (Sec. 6.6)

common factor A number that is a factor of two or more numbers is a common factor of those numbers. (Sec. 2.1)

common multiple A number that is a multiple of two or more numbers is a common multiple of those numbers. (Sec. 2.1)

commutative property of addition Two numbers can be added in either order; the sum will be the same. (Sec. 1.2)

commutative property of multiplication Two numbers can be multiplied in either order; the product will be the same. (Sec. 1.4)

complementary angles Two angles whose sum is 90°. (Sec. 12.1)

composite geometric figure A figure made from two or more geometric figures. (Sec. 12.2)

composite geometric solid A solid made from two or more geometric solids. (Sec. 12.4)

composite number A number that has whole-number factors besides 1 and itself. For instance, 18 is a composite number. (Sec. 1.7)

compound interest Interest computed not only on the original amount but also on interest already earned. (Sec. 6.3)

congruent objects Objects that have the same shape and the same size. (Sec. 12.6)

constant term A term that has no variables. (Sec. 11.1)

corresponding angles Two angles that are on the same side of the transversal and are both acute angles or are both obtuse angles. (Sec. 12.1)

cost The price that a business pays for a product. (Sec. 6.2)

cube A rectangular solid in which all six faces are squares. (Sec. 12.1)

cubic centimeter A unit of capacity equal to 1 milliliter. (Sec. 9.3)

cup A U.S. Customary measure of capacity. 2 cups = 1 pint. (Sec. 8.3)

data Numerical information. (Sec. 7.1)

decimal A number written in decimal notation. (Sec. 3.1)

decimal notation Notation in which a number consists of a whole-number part, a decimal point, and a decimal part. (Sec. 3.1)

decimal part In decimal notation, that part of the number that follows the decimal point. (Sec. 3.1)

decimal point In decimal notation, the point that separates the whole-number part from the decimal part. (Sec. 3.1)

degree Unit used to measure angles; one complete revolution is 360°. (Sec. 12.1)

denominator The part of a fraction that appears below the fraction bar. (Sec. 2.2)

diameter of a circle A line segment with endpoints on the circle and going through the center. (Sec. 12.1)

diameter of a sphere A line segment with endpoints on the sphere and going through the center. (Sec. 12.1)

difference In subtraction, the result of subtracting two numbers. (Sec. 1.3)

discount The difference between the regular price and the sale price. (Sec. 6.2)

discount rate The percent of a product's regular price that is represented by the discount. (Sec. 6.2)

dividend In division, the number into which the divisor is divided to yield the quotient. (Sec. 1.5)

division The process of finding the quotient of two numbers. (Sec. 1.5)

divisor In division, the number that is divided into the dividend to yield the quotient. (Sec. 1.5)

energy The ability to do work. (Sec. 8.4)

equation A statement of the equality of two mathematical expressions. (Sec. 11.2)

equilateral triangle A triangle that has three sides of equal length; the three angles are also equal. (Sec. 12.2)

equivalent fractions Equal fractions with different denominators. (Sec. 2.3)

evaluating the variable expression Replacing the variable or variables and then simplifying the resulting numerical expression. (Sec. 11.1)

expanded form The number 46,208 can be written in expanded form as 40,000 + 6000 + 0 + 8. (Sec. 1.1)

exponent In exponential notation, the raised number that indicates how many times the number to which it is attached is taken as a factor. (Sec. 1.6)

exponential notation The expression of a number to some power, indicated by an exponent. (Sec. 1.6)

factors In multiplication, the numbers that are multiplied. (Sec. 1.4)

fluid ounce A U.S. Customary measure of capacity. 8 fluid ounces = 1 cup. (Sec. 8.3)

foot A U.S. Customary unit of length. 3 feet = 1 yard. (Sec. 8.1)

foot-pound A unit of energy. One foot-pound of energy is required to lift 1 pound a distance of 1 foot. (Sec. 8.4)

foot-pounds per second A unit of power. (Sec. 8.4)

fraction Notation used to represent the number of equal parts of a whole. (Sec. 2.2)

fraction bar Bar that separates the numerator of a fraction from the denominator. (Sec. 2.2)

frequency polygon A graph that displays information similarly to a histogram. (Sec. 7.3)

gallon A U.S. Customary measure of capacity. 1 gallon = 4 quarts. (Sec. 8.3)

geometric solid A figure in space. (Sec. 12.1)

gram The basic unit of mass in the metric system. (Sec. 9.2)

graph A display that provides a pictorial representation of data. (Sec. 7.1)

graph of a whole number A heavy dot placed directly above that number on the number line. (Sec. 1.1)

greater than A number that appears to the right of a given number on the number line is greater than that given number. (Sec. 1.1)

greatest common factor The largest common factor of two or more numbers. (Sec. 2.1)

histogram A bar graph in which the width of each bar corresponds to a range of numbers called a class interval. (Sec. 7.3)

horsepower The U.S. Customary unit of power. 1 horsepower = 550 foot-pounds per second. (Sec. 8.4)

hourly wage Pay calculated on the basis of a certain amount for each hour worked. (Sec. 6.6)

hypotenuse The side opposite the right angle in a right triangle. (Sec. 12.1)

improper fraction A fraction greater than or equal to 1. (Sec. 2.2)

inch A U.S. Customary unit of length. 12 inches = 1 foot. (Sec. 8.1)

integers The integers are ... −3, −2, −1, 0, 1, 2, 3, ... (Sec. 10.1)

interest The amount of money paid for the privilege of using someone else's money. (Sec. 6.3)

interest rate The percent used to determine the amount of interest. (Sec. 6.3)

intersecting lines Lines that cross at a point in the plane. (Sec. 12.1)

inverting a fraction Interchanging the numerator and denominator. (Sec. 2.7)

isosceles triangle A triangle that has two sides of equal length; the angles opposite the equal sides are of equal measure. (Sec. 12.2)

least common denominator The least common multiple of denominators. (Sec. 2.4)

least common multiple The smallest common multiple of two or more numbers. (Sec. 2.1)

legs The two sides of a right triangle that are not opposite the right angle. (Sec. 12.1)

less than A number that appears to the left of a given number on the number line is less than that given number. (Sec. 1.1)

like terms Terms of a variable expression that have the same variable part. (Sec. 11.1)

line A line extends indefinitely in two directions in a plane; it has no width. (Sec. 12.1)

line segment Part of a line; it has two endpoints. (Sec. 12.1)

liter The basic unit of capacity in the metric system. (Sec. 9.3)

markup The difference between selling price and cost. (Sec. 6.2)

markup rate The percent of a product's cost that is represented by the markup. (Sec. 6.2)

mass The amount of material in an object. On the surface of the earth, mass is the same as weight. (Sec. 9.2)

mean value The sum of all values divided by the number of those values; also known as the average value. (Sec. 7.4)

measurement A measurement has both a number and a unit. Examples include 7 feet, 4 ounces, and 0.5 gallon. (Sec. 8.1)

median value The value that separates a list of values in such a way that there are the same number of values below the median as above it. (Sec. 7.4)

meter The basic unit of length in the metric system. (Sec. 9.1)

metric system A system of measurement based on the decimal system. (Sec. 9.1)

mile A U.S. Customary unit of length. 5280 feet = 1 mile. (Sec. 8.1)

minuend In subtraction, the number from which another number (the subtrahend) is subtracted. (Sec. 1.3)

mixed number A number greater than 1 that has a whole-number part and a fractional part. (Sec. 2.2)

mortgage The amount borrowed to buy real estate. (Sec. 6.4)

multiples of a number The products of that number and the numbers 1, 2, 3, ... (Sec. 2.1)

multiplication The process of finding the product of two numbers. (Sec. 1.4)

multiplication property of one The product of a number and one is the number. (Sec. 1.4)

multiplication property of zero The product of a number and zero is zero. (Sec. 1.4)

negative integers The integers to the left of zero on the number line. (Sec. 10.1)

negative numbers Numbers less than zero. (Sec. 10.1)

number line A line on which a number can be graphed. (Sec. 1.1)

numerator The part of a fraction that appears above the fraction bar. (Sec. 2.2)

numerical coefficient The number part of a variable term. When the numerical coefficient is 1 or −1, the 1 is usually not written. (Sec. 11.1)

obtuse angle An angle whose measure is between 90° and 180°. (Sec. 12.1)

obtuse triangle A triangle that has one obtuse angle. (Sec. 12.2)

opposites Two numbers that are the same distance from zero on the number line, but on opposite sides. (Sec. 10.1)

Order of Operations Agreement A set of rules that tell us in what order to perform the operations that occur in a numerical expression. (Sec. 1.6)

ounce A U.S. Customary unit of weight. 16 ounces = 1 pound. (Sec. 8.2)

parallel lines Lines that never meet; the distance between them is always the same. (Sec. 12.1)

parallelogram A quadrilateral that has opposite sides equal and parallel. (Sec. 12.1)

percent Parts per hundred. (Sec. 5.1)

percent decrease A decrease of a quantity expressed as a portion of its original value. (Sec. 6.2)

percent increase An increase of a quantity expressed as a portion of its original value. (Sec. 6.2)

perfect square The product of a whole number and itself. (Sec. 12.5)

perimeter The distance around a plane figure. (Sec. 12.2)

period In a number written in standard form, each group of digits separated by a comma. (Sec. 1.1)

perpendicular lines Intersecting lines that form right angles. (Sec. 12.1)

pictograph A graph that uses symbols to represent information. (Sec. 7.1)

pint A U.S. Customary measure of capacity. 2 pints = 1 quart. (Sec. 8.3)

place value The position of each digit in a number in standard form determines that digit's place value. (Sec. 1.1)

plane A flat surface. (Sec. 12.1)

plane figure A figure that lies totally in a plane. (Sec. 12.1)

points A term banks use to mean percent of a mortgage; used to express the loan origination fee. (Sec. 6.4)

polygon A closed figure determined by three or more line segments that lie in a plane. (Sec. 12.2)

positive integers The integers to the right of zero on the number line; also called natural numbers. (Sec. 10.1)

positive numbers Numbers greater than zero. (Sec. 10.1)

pound A U.S. Customary unit of weight. 1 pound = 16 ounces. (Sec. 8.2)

power The rate at which work is done or energy is released. (Sec. 8.4)

prime factorization The expression of a number as the product of its prime factors. (Sec. 1.7)

prime number A number whose only whole-number factors are 1 and itself. For instance, 13 is a prime number. (Sec. 1.7)

principal The amount of money originally deposited or borrowed. (Sec. 6.3)

product In multiplication, the result of multiplying two numbers. (Sec. 1.4)

proper fraction A fraction less than 1. (Sec. 2.2)

proportion An expression of the equality of two ratios or rates. (Sec. 4.3)

Pythagorean Theorem The square of the hypotenuse of a right triangle is equal to the sum of the squares of the two legs. (Sec. 12.5)

quadrilateral A four-sided closed figure. (Sec. 12.1)

quart A U.S. Customary measure of capacity. 4 quarts = 1 gallon. (Sec. 8.3)

quotient In division, the result of dividing the divisor into the dividend. (Sec. 1.5)

radius of a circle A line segment going from the center to a point on the circle. (Sec. 12.1)

radius of a sphere A line segment going from the center to a point on the sphere. (Sec. 12.1)

rate A comparison of two quantities that have different units. (Sec. 4.2)

rate in simplest form A rate in which the numbers that form the rate have no common factors. (Sec. 4.2)

ratio A comparison of two quantities that have the same units. (Sec. 4.1)

ratio in simplest form A ratio in which the two numbers that form the ratio do not have a common factor. (Sec. 4.1)

rational number A number that can be written as the ratio of two integers, where the denominator is not zero. (Sec. 10.4)

ray A ray starts at a point and extends indefinitely in one direction. (Sec. 12.1)

reciprocal of a fraction The fraction with the numerator and denominator interchanged. (Sec. 2.7)

rectangle A parallelogram that has four right angles. (Sec. 12.1)

rectangular solid A solid in which all six faces are rectangles. (Sec. 12.1)

regular polygon A polygon in which each side has the same length and each angle has the same measure. (Sec. 12.2)

remainder In division, the quantity left over when it is not possible to separate objects or numbers into a whole number of even groups. (Sec. 1.5)

right angle A 90° angle. (Sec. 12.1)

right triangle A triangle that contains one right angle. (Sec. 12.1)

right triangle 30°-60°-90° A special right triangle in which the length of the leg opposite the 30° angle is one-half the length of the hypotenuse. (Sec. 12.5)

right triangle 45°-45°-90° A special right triangle in which the sides opposite the 45° angles are equal. (Sec. 12.5)

rounding Giving an approximate value of an exact number. (Sec. 1.1)

salary Pay based on a weekly, biweekly, monthly, or annual time schedule. (Sec. 6.6)

sale price The reduced price. (Sec. 6.2)

scalene triangle A triangle that has no sides of equal length; no two of its angles are of equal measure. (Sec. 12.2)

selling price The price for which a business sells a product to a customer. (Sec. 6.2)

similar objects Objects that have the same shape but not necessarily the same size. (Sec. 12.6)

simple interest Interest computed on the original amount. (Sec. 6.3)

simplest form A fraction is in simplest form when there are no common factors in the numerator and the denominator. (Sec. 2.3)

simplifying a variable expression Combining like terms by adding their numerical coefficients. (Sec. 11.1)

solid An object in space. (Sec. 12.1)

solution of an equation A number that, when substituted for the variable, results in a true equation. (Sec. 11.2)

solving a proportion Finding a number to replace the unknown so that the proportion is true. (Sec. 4.3)

solving an equation Finding a solution of the equation. (Sec. 11.2)

space Space extends in all directions. (Sec. 12.1)

sphere A solid in which all points are the same distance from point *O*, which is called the center of the sphere. (Sec. 12.1)

square A rectangle that has four equal sides. (Sec. 12.1)

square root A square root of a number is one of two identical factors of that number. (Sec. 12.5)

standard form A whole number is in standard form when it is written using the digits 0, 1, 2, . . . , 9. An example is 46,208. (Sec. 1.1)

statistics The branch of mathematics concerned with data, or numerical information. (Sec. 7.1)

straight angle A 180° angle. (Sec. 12.1)

subtraction The process of finding the difference between two numbers. (Sec. 1.3)

subtrahend In subtraction, the number that is subtracted from another number (the minuend). (Sec. 1.3)

sum In addition, the total of the numbers added. (Sec. 1.2)

supplementary angles Two angles whose sum is 180°. (Sec. 12.1)

terms of a variable expression The addends of the expression. (Sec. 11.1)

ton A U.S. Customary unit of weight. 1 ton = 2000 pounds. (Sec. 8.2)

transversal A line intersecting two other lines at two different points. (Sec. 12.1)

triangle A three-sided closed figure. (Sec. 12.1)

true proportion A proportion in which the fractions are equal when written in lowest terms. (Sec. 4.3)

unit cost The cost of one item. (Sec. 6.1)

unit rate A rate in which the number in the denominator is 1. (Sec. 4.2)

units In the quantity 3 feet, feet are the units in which the measurement is made. (Sec. 4.1)

variable A letter used to stand for a quantity that is unknown or that can change. (Sec. 11.1)

variable expression An expression that contains one or more variables. (Sec. 11.1)

variable part In a variable term, the variable or variables and their exponents. (Sec. 11.1)

variable term A term composed of a numerical coefficient and a variable part. (Sec. 11.1)

vertex The common endpoint of two rays that form an angle. (Sec. 12.1)

vertical angles Two angles that are on opposite sides of the intersection of two lines. (Sec. 12.1)

volume A measure of the amount of space inside a closed surface. (Sec. 12.4)

watt-hour A unit of electrical energy in the metric system. (Sec. 9.4)

weight A measure of how strongly the earth is pulling on an object. (Sec. 8.2)

whole numbers The whole numbers are 0, 1, 2, 3, ... (Sec. 1.1)

whole-number part In decimal notation, that part of the number that appears before the decimal point. (Sec. 3.1)

yard A U.S. Customary unit of length. 36 inches = 1 yard. (Sec. 8.1)

Index